雷达目标检测与恒虚警处理

（第3版）

何友 关键 黄勇 简涛◎著

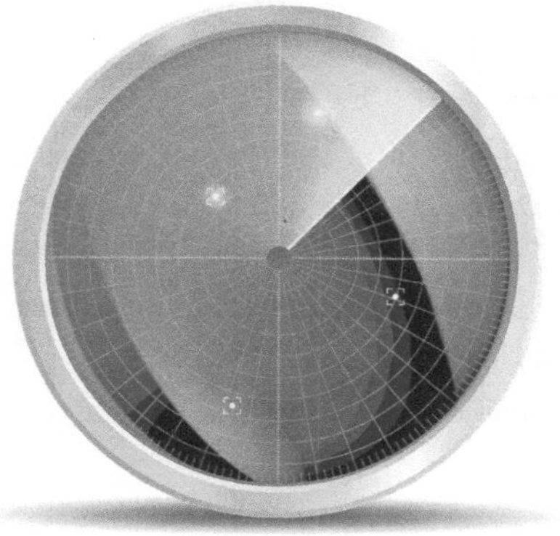

清华大学出版社
北京

内 容 简 介

本书是关于雷达目标检测与恒虚警(CFAR)处理理论及方法的一部专著。书中总结了数十年来,在这一领域国际上的研究进展及大量研究成果。全书由16章组成,主要内容有:经典的固定门限检测,均值类CFAR处理方法,有序统计类CFAR处理方法,采用自动筛选技术的GOS类CFAR检测器,自适应CFAR检测器,经典非高斯杂波背景中的CFAR检测器,复合高斯杂波中的CFAR处理,非参量CFAR处理,杂波图CFAR处理,变换域CFAR处理,高分辨率雷达目标检测,多传感器分布式CFAR处理,多维CFAR处理,以及基于特征的CFAR处理,最后是本书的回顾、建议与展望。

本书可供从事雷达工程、声呐、电子工程、信号与信息处理等专业的科技人员阅读和参考,也可作为上述专业的研究生教材。

图书在版编目(CIP)数据

雷达目标检测与恒虚警处理/何友等著. —3版. —北京:清华大学出版社,2023.1(2023.7重印)
ISBN 978-7-302-61830-0

Ⅰ.①雷… Ⅱ.①何… Ⅲ.①雷达目标—目标检测—研究②恒虚警检测器—研究 Ⅳ.①TN951②TN957.51

中国版本图书馆CIP数据核字(2022)第169291号

责任编辑:崔 彤
封面设计:李召霞
责任校对:申晓焕
责任印制:杨 艳

出版发行:清华大学出版社
网 址:http://www.tup.com.cn,http://www.wqbook.com
地 址:北京清华大学学研大厦A座 **邮 编**:100084
社 总 机:010-83470000 **邮 购**:010-62786544
投稿与读者服务:010-62776969,c-service@tup.tsinghua.edu.cn
质量反馈:010-62772015,zhiliang@tup.tsinghua.edu.cn
课件下载:http://www.tup.com.cn,010-83470236
印 装 者:三河市君旺印务有限公司
经 销:全国新华书店
开 本:203mm×260mm **印 张**:33.75 **字 数**:927千字
版 次:1999年3月第1版 2023年2月第3版 **印 次**:2023年7月第4次印刷
印 数:3101~4100
定 价:128.00元

产品编号:092969-01

Abstract

This book is a monograph about radar target detection and constant false alarm rate (CFAR) processing theory and technique. In this book, the international advances in this research field for decades and many fruits of authors' research are summarized. The book is composed of sixteen chapters. The main contents are classical detection with fixed threshold, mean-level type CFAR processing methods, order-statistics type CFAR processing methods, generalized order statistics CFAR detectors with automatic censoring technique, adaptive CFAR detectors, classical CFAR processing in non-Gaussian clutter, CFAR processing in compound Gaussian clutter, nonparametric CFAR processing, clutter map CFAR processing, CFAR processing in transform domain, detection of high resolution radar target, distributed CFAR processing with multisensor, multi-dimension CFAR processing and feature-based CFAR processing, the last is a brief review, suggestion and prospect.

This book could be used by the scientific and technical staffs engaged in radar engineering, sonar technique, electronic engineering, signal and information processing etc. to read and consult, and can also serve as textbook of graduate students of the above professions.

第3版前言

雷达信号处理是雷达软件化的重要支撑。雷达目标检测则是雷达信号处理的重要环节，恒虚警(CFAR)处理在目标检测中起着不可或缺的作用。数十年来，雷达目标检测与 CFAR 处理技术逐渐发展成为国际雷达信号处理界的热点研究领域。

《雷达目标检测与恒虚警处理》于 1999 年出版第 1 版、2011 年出版第 2 版，受到了广大读者的厚爱与关怀，作者再次表示衷心感谢。近十年来，雷达目标检测与 CFAR 处理理论又有了很大的发展，使我们认识到应当对本书继续进行扩展和完善，以适应雷达目标检测领域快速发展的需要。本书第 3 版结合本领域近十多年的研究成果对第 2 版做了修订，全书总删减、新增、扩展和调整内容约达 54.9%，第 3 版修订后全书共 16 章。第 3 版主要做了如下修订。

(1) 充实了有序统计类 CFAR 检测器的内容，如扩充了对 MX-CMLD、OSGO、OSSO、S-CFAR 检测器的性能分析，增加了对 CATM、SOSGO、MS-CFAR 检测器的讨论。

(2) 充实了 GOSCA、GOSGO、GOSSO、MOSCA、OSCAGO、OSCASO-CFAR 检测器的性能分析结果，使得这部分的内容更加系统和完善。

(3) 对自适应 CFAR 检测器一章进行了较大修改，新增了“基于回波形状信息的删除单元平均 CFAR 检测器”一节，拓展了参考距离单元自适应筛选的方法。

(4) 对韦布尔和对数正态杂波背景中 CFAR 检测器一章进行了较大修改，拓展为经典非高斯杂波背景中的 CFAR 检测器，补充了韦布尔杂波下多脉冲二进制积累的 OS 和 OSGO 检测性能分析，作为 α 稳定分布的两个特例增加了 Pearson 分布和 Cauchy 分布背景下的 CFAR 检测器。

(5) 充实了复合高斯杂波中的 CFAR 处理，增加了球不变随机杂波下相参 CFAR 检测技术，补充了近年来复合高斯杂波下 CFAR 检测的研究成果。

(6) 扩充了非参量 CFAR 处理章节，主要内容包括韦布尔杂波下非参量检测器的性能分析，以及利用逆正态得分函数修正秩的非参量检测器和相应的性能分析。

(7) 充实了杂波图 CFAR 处理章节，补充了 Nitzberg 杂波图在韦布尔分布中的性能分析，增加了双参数杂波图检测技术。

(8) 对变换域 CFAR 方法进行重新梳理，在介绍频域 CFAR、小波域 CFAR 方法基础上，新增了分数阶傅里叶变换域、Hilbert-Huang 变换域和稀疏表示域目标检测，重点讨论多种变换域的检测器设计及相应的检测方法。

(9) 对距离扩展目标检测一章进行了较大修改，拓展为高分辨率雷达目标检测，新增了 SAR 图像 CFAR 检测，重点讨论了 SAR 图像 CFAR 检测的杂波单元选取、基于广义 Gamma 杂波模型的 SAR 图像 CFAR 检测、基于语义知识辅助的 SAR 图像 CFAR 检测、基于密度特征的 SAR 图像 CFAR 检测快速实现等内容。

(10) 对多传感器分布式 CFAR 处理一章进行了较大修改，新增了“分布式 MIMO 雷达 CFAR 检测”一节内容,以分布式 MIMO 雷达为研究对象,讨论了 AMF CFAR 检测器的设计及性能。

(11) 新增了多维 CFAR 处理一章,重点从阵列雷达 CFAR 检测的角度讨论了“阵元-脉冲”二维联合 CFAR 检测,从基于自适应空时编码设计的 MIMO 阵列雷达目标检测的角度讨论了“阵元-波形”二维联合 CFAR 检测,并从空时距三维联合自适应处理的角度讨论了“阵元-脉冲-波形”三维联合 CFAR 检测。

(12) 新增了基于特征的 CFAR 处理一章,重点讨论了特征提取与 CFAR 处理相结合的检测器,包括利用频域分形特征的目标 CFAR 检测方法和基于深度循环神经网络的脉压、检测一体化处理方法。

(13) 删除了第 2 版的第 14 章,将 α 稳态分布编入第 7 章,将应用图像处理技术的方法与 SAR 图像 CFAR 处理结合归入第 12 章,将“阵列”和“极化”内容编入新增的第 14 章。

(14) 全面修改了“回顾、建议和展望”一章,结合雷达目标检测领域的新发展,系统地总结了全书的内容,并为今后继续开展该领域的工作给出了建议。

在本书第 3 版出版之际,本书作者再次衷心感谢德国汉堡工业大学 Hermann Rohling 教授、清华大学陆大绘教授、彭应宁教授,以及国内著名雷达专家王越院士、保铮院士、刘永坦院士、王小谟院士、郭桂蓉院士、毛二可院士、贲德院士、张光义院士、黄培康院士、吴一戎院士、吴曼青院士、王永良院士、黎湘院士、丁赤彪院士、龙腾院士、吴剑旗院士等,多年来对我们的研究工作所给予的指导、支持、关心和帮助。

在本书的撰写出版过程中,中国人民解放军海军航空大学刘宁波博士、陈小龙博士,清华大学王学谦博士等也参与了撰写工作,丁昊博士、王国庆博士、董云龙博士、薛永华博士、张建博士、周伟博士、张林博士、于恒力博士等参与了试验数据采集与处理分析等工作。电子科技大学蔡德强教授在文献检索和查询中给予了众多帮助,在此一并表示衷心的感谢。此外,还要感谢清华大学出版社,特别是盛东亮和崔彤编辑对本书按期高质量出版的大力支持。

我们希望本书的再版能够继续给从事雷达目标检测与 CFAR 处理的广大科技人员和研究生提供有益的参考及帮助。恳请广大学者一如既往地关心本书,并提出宝贵意见和建议。

何友 关键 黄勇 简涛

2022 年 1 月

于清华大学和海军航空大学

Foreword of the Third Edition

Radar signal processing is an important support of software defined radar. Radar target detection is an important part of radar signal processing and the constant false alarm rate (CFAR) processing plays an indispensable role in radar target detection. For decades, radar target detection and CFAR processing technology have gradually developed into a hot research field in the international radar signal processing community.

Since the first edition of *Radar Target Detection and CFAR Processing* was published in 1999 and the second edition in 2011, this book has received great love and care from readers. The authors would like to express our sincere gratitude again. In the past ten years, the theory of radar target detection and CFAR processing has made great progress, which makes us recognize the necessity to continue to expand and improve this book, in order to adapt to the rapid development of the research field of radar target detection. The second edition is revised based on the research results in this field for more than ten years. The total contents of the book are deleted, added, expanded and adjusted by about 54.9%. The revised third edition consists of 16 chapters. The main work is as follows.

(1) The contents of ordered statistics type CFAR detectors are enriched. For example, the performance analysis of MX-CMLD, OSGO, OSSO and S-CFAR detectors is expanded, and the discussion of CATM, SOSGO and MS-CFAR detectors is added.

(2) The performance analysis results of GOSCA, GOSGO, GOSSO, MOSCA, OSCAGO and OSCASO-CFAR detectors were enriched, which makes this part more systematic and perfect.

(3) The chapter on Adaptive CFAR Processing was greatly modified with adding a new section named "Envelope Shape-based Excision Cell Average CFAR Detector", which expands the adaptive censoring method of reference range cells.

(4) The chapter on CFAR Detector in the Background of Weibull and Lognormal Clutter is greatly modified and extended to the CFAR Detector in the Background of Classical Non-Gaussian Clutter, which supplements the detection performance analysis of OS and OSGO based on multi-pulse binary accumulation under Weibull clutter, and adds the CFAR detectors under Pearson distribution and Cauchy distribution background as two special cases under α stable distribution.

(5) The CFAR processing in compound Gaussian clutter is enriched, which adds the coherent CFAR detection technique under spherically invariant random clutter, and complements the research results of CFAR detection under compound Gaussian clutter in recent years.

(6) The chapter on nonparametric CFAR processing is extended, which mainly includes the performance analysis of nonparametric detectors under Weibull clutter, and nonparametric detectors

based on the rank modified by inverse normal score function and corresponding performance analysis.

(7) The chapter on Clutter Map CFAR Processing is enriched, which supplements the performance analysis of the Nitzberg's clutter map technique in Weibull distribution clutter, and the detection technique of two-parameter clutter map.

(8) The CFAR methods in the transform domain is reorganized. On the basis of elaborating CFAR methods in the frequency domain and wavelet domain, target detection in the fractional Fourier transform domain, Hilbert-Huang transform domain and sparse representation domain are added, which focuses on detector design and corresponding detection methods in a variety of transform domains.

(9) The chapter on Detection of Range-Distributed Target is greatly modified, which is extended to CFAR detection of high-resolution radar targets, with adding the CFAR detection in SAR images. The selection of clutter cells for CFAR detection in SAR images, CFAR detection in SAR images based on generalized Gamma clutter model, CFAR detection in SAR images based on semantic knowledge aids, and fast implementation of CFAR detection in SAR images based on density features are discussed.

(10) The chapter on Distributed CFAR Processing with Multisensor is greatly modified with adding a new section named "CFAR Detection of Distributed MIMO Radar", in which the design and performance of AMF CFAR detector is discussed for distributed MIMO radar.

(11) A new chapter on Multi-dimensional CFAR Processing is added, which focuses on discussing the "array element-pulse" two-dimensional joint CFAR detection from the perspective of array radar CFAR detection, the "array element-waveform" two-dimensional joint CFAR detection from the perspective of MIMO array radar target detection based on adaptive space-time coding design, and the "array element-pulse-waveform" three-dimensional joint CFAR detection from the perspective of space-time-range joint adaptive processing.

(12) A new chapter on CFAR processing based on feature is added, and the detector combining feature extraction with CFAR processing is highlighted, which includes target CFAR detection method using frequency domain fractal features and the integration of pulse compression and detection based on RNN.

(13) The Chapter 14 in the second edition is deleted, in which α stable distribution is incorporated into the Chapter 7, the image processing methods and SAR image CFAR processing are incorporated into the Chapter 12, and the contents about "array" and "polarization" are incorporated into the new Chapter 14.

(14) The chapter on "Review, Suggestion and Prospect" is overhauled. By combining with the new development of radar target detection, the chapter systematically summarizes the content of the book, and gives suggestions for the future work in this field.

At the moment of the publication of the third edition, the authors again would like to express sincere thanks to Professor Hermann Rohling in Technical University Hamburg-Harburg of Germany, Professor Lu Dajin, Peng Yingning in Tsinghua University, as well as other domestic well-known radar experts, Wang Yue academician, Bao Zheng academician, Liu Yongtan academician, Wang Xiaomo academician, Guo Guirong academician, Mao Erke academician, Ben De academician, Zhang Guangyi academician, Huang Peikang academician, Wu Yirong academician, Wu Manqing academician, Wang Yongliang academician, Li Xiang academician, Ding Chibiao academician, Long Teng academician and Wu Jianqi academician, et al. Thanks for their guidance, support, care and help to our scientific research work over the years.

In the process of writing and publishing of this book, Dr. Liu Ningbo and Chen Xiaolong of Naval Aeronautical University, as well as Dr. Wang Xueqian of Tsinghua University also participate in the writing work, Dr. Ding Hao, Dr. Wang Guoqing, Dr. Dong Yunlong, Dr. Xue Yonghua, Dr. Zhang Jian, Dr. Zhou Wei, Dr. Zhang Lin, and Dr. Yu Hengli, et al. participate in the test data acquisition and processing analysis, and Professor Cai Deqiang of University of Electronic Science and Technology of China gives a lot of help in literature retrieval and query. Here, we would like to express our sincere thanks to all of them. Also, we would like to give thanks to the Tsinghua University Press, especially the great support of Editor Sheng Dongliang and Cui Tong who guarantee the publication on schedule with high quality.

We hope that the republication of this book will continue to provide useful reference and help to all the technologist and graduate students engaged in radar target detection and CFAR processing. We sincerely hope that all scholars will continue to care about this book and put forward valuable opinions and suggestions.

He You　Guan Jian　Huang Yong　Jian Tao

January 2022

At Tsinghua University and Naval Aviation University

第2版前言

雷达自动检测和跟踪中的虚警问题是每个雷达系统和设计人员不可回避的重要问题之一。三十多年来，雷达自动检测与CFAR处理技术逐渐发展成为国际雷达信号处理界的一大热门研究领域和关键性课题。CFAR技术是雷达自动检测系统中控制虚警率的最重要手段，它在雷达自动检测过程中起着极其重要的作用。

《雷达自动检测与恒虚警处理》自从1999年出版第1版以来，受到广大读者的厚爱与关怀，作者对此表示衷心的感谢。雷达自动检测与恒虚警处理的理论、算法和应用的不断发展，使我们迫切感觉到要对本书进行全面扩展、完善、修改和补充新的内容，以适应当前雷达目标检测领域发展的需要。

本书是在1999年由我们所著《雷达自动检测与恒虚警处理》的基础上进行系统地扩展、完善并加以修改和增订而成的。这次结合近十多年的最新研究成果对原书做了较大幅度的修订。主要做了如下工作。

(1) 将书名修订为《雷达目标检测与恒虚警处理》，因为在以前的雷达目标自动检测技术中强调“自动”，而现在的雷达目标检测方法的“自动”功能是不言而喻的，因此还是强调目标检测，而不用“雷达自动检测”，以区别于雷达的自动测试。

(2) 综合近十几年雷达目标检测技术的发展，对雷达目标检测的研究现状给予了更全面的阐述。

(3) 充实了雷达目标自动检测所需的基础理论，对第1章雷达自动检测原理和第2章均值类CFAR检测器两章内容进行了完善和扩充，特别是补充了多脉冲和相干脉冲条件下雷达目标检测的原理和方法，并将“雷达自动检测原理”更名为“经典的固定门限检测”。

(4) 充实了有序统计类CFAR检测器的研究内容，增加了一些理论分析。

(5) 增加了采用自动筛选技术的MTM-CFAR、TMGO-CFAR和TMSO-CFAR检测器的研究内容，使得这部分的内容更加系统和完善。

(6) 对自适应CFAR处理一章进行了较大修改，大幅删减了E-CFAR检测器、VTM-CFAR检测器以及Himonas的一系列CFAR检测器，主要以文字和框图的形式对这些检测器进行了简要阐述，并补充了一些新的自适应CFAR检测器，归纳总结了“自适应类”CFAR检测器的基本方法。

(7) 将韦布尔和对数正态杂波背景中的CFAR技术以及两参数CFAR技术两章内容合并为一章，首先研究了在韦布尔和对数正态杂波中能提供CFAR检测的经典Log-t检测器，再按照非高斯背景中单参数和双参数检测方法的次序对原有内容进行了重新组织和改写。

(8) 将“K分布和莱斯分布杂波中的CFAR处理”更名为“复合高斯分布杂波中的CFAR处理”，从“复合高斯分布”出发，将K分布模型作为特例，重点介绍K分布杂波中的CFAR处理，并补充了近年来意大利Gini等关于复合高斯杂波中CFAR检测的研究成果。

(9) 基于非参量检测最新的理论研究成果和工程实际中的应用状况，对原书中的内容进行了大幅度的更替；扩展了时频域CFAR处理的最新研究成果，补充了其他变换域CFAR处理的内容。

(10) 增加了杂波图 CFAR 处理章节，主要内容包括 Nitzberg 杂波图处理、杂波图处理的点技术和面技术，以及空域和时域混合处理的 CM/L-CFAR 检测器和相应内容的性能分析。

(11) 新增了距离扩展目标检测技术，重点讨论了复合高斯杂波、复合高斯杂波加热噪声及 SαS 分布杂波背景中距离扩展目标的检测问题。

(12) 新增了多传感器分布式 CFAR 处理，重点讨论了两类多传感器分布式 CFAR 检测器，具体包括分布式 CA-CFAR 检测器、分布式 OS-CFAR 检测器以及基于 R 类、S 类和 P 类局部检测统计量的分布式 CFAR 检测器。

(13) 全面修改了“回顾、建议和展望”一章，结合雷达目标自动检测技术的新发展，系统地总结了全书的内容，并为今后继续开展该领域的工作给予了指导性意见。此外，根据最近十多年来国内外的研究成果，本书增加了必要的参考文献，并对一些文字叙述不正确之处进行了修正，旨在进一步提高本书的可读性。

在本书第 2 版出版之际，本书作者在此衷心感谢德国汉堡工业大学 Hermann Rohling 教授、清华大学陆大绘教授、彭应宁教授，以及国内著名雷达专家郭桂蓉院士、王越院士、保铮院士、毛二可院士、刘永坦院士、王小谟院士、张光义院士、贲德院士、黄培康院士、吴一戎院士、吴曼青院士等，多年来对我们的科学研究工作所给予的指导、关心和帮助。

在本书撰写出版过程中，海军航空工程学院黄勇博士、简涛博士、曲付勇博士等参与了本书的修订和文字录入工作，在此向他们表示谢意，还要感谢清华大学出版社，特别是王一玲编辑对本书按期高质量出版的大力支持。

我们希望本书的出版，不仅给广大从事雷达目标自动检测和 CFAR 处理的科技人员提供一本可读性较好的参考书，也给学习雷达目标自动检测和 CFAR 处理的研究生提供一本内容比较全面的教材。

恳请广大学者能一如既往地关心本书，并提出宝贵的意见和建议。

何友　关键　孟祥伟

2010 年 10 月

于海军航空工程学院

Foreword of the Second Edition

The false alarm problem of automatic radar detection and tracking is one of the important issues which can not be avoided by each radar system and designer. The automatic radar detection and CFAR processing technology have gradually developed into a popular research area and key issues in the international radar signal processing community over the past three decades. CFAR technology is the most important tool in the control of false alarm rate in automatic radar detection system, which plays an important role in the process of automatic radar detection.

Since 1999, the first edition of *Automatic Radar Detection and CFAR Processing* has been winning the readers' love and care. The authors would like to express our sincere gratitude. With the development of the theories, algorithms and applications of automatic radar detection and CFAR processing, we urgently feel it is necessary to expand, improve, modify and add new content to meet the current needs of developments in the field of radar target detection.

The content of this book have been expanded systematically, improved, revised and enlarged based on our book *Automatic Radar Detection and CFAR Processing* written in 1999. We have revised the first edition extensively with the latest research over the past ten years. The revisions are as follows.

(1) The book is renamed as *Radar Target Detection and CFAR Processing*. Because the past automatic radar target detection stresses the word "automatic", while the "automatic" function of present radar target detection methods is self-evident. Thus, we emphasize the radar detection instead of "automatic radar detection" in order to differ from automatic radar test.

(2) Combing with the development of radar target detection algorithms over the past ten years, more comprehensive elaboration has been made on the present study of automatic radar detection algorithms.

(3) The necessary basic theories of automatic radar target detection are enlarged. The principles of automatic radar detection in Chapter 1 and the mean-level type CFAR detectors in Chapter 2 are improved and enlarged. Especially, the theories and methods of radar target detection with multiple pulses and coherent pulses are supplemented and as well "principles of automatic radar detection" is renamed as "classic fixed threshold detection".

(4) The content of "order statistics type CFAR detectors" are extended with some theoretical analysis.

(5) The MTM-CFAR, TMGO-CFAR and TMSO-CFAR detectors of automatic censoring technique are added so as to make the content more systematical and substantial.

(6) The chapter named as adaptive CFAR detectors is revised greatly and the E-CFAR detector,

VTM-CFAR detector and a series of Himonas CFAR detectors are scaled back with brief descriptions in the form of text and diagram. Some novel adaptive CFAR detectors are added and the fundamental algorithms of adaptive CFAR detectors are summarized.

(7) The chapter named as CFAR technique in Weibull and log-normal clutter background and the chapter named as biparametric CFAR techniques are combined into one chapter. The classical Log-t CFAR detector in Weibull and log-normal clutter is researched and then the original content are reorganized and rewritten according to the order of single parametric detection and biparametric detection.

(8) "CFAR processing in K distribution and Rice distribution clutter" is renamed as "CFAR processing in compound Gaussian distribution clutter". Beginning with the "compound Gaussian distribution",CFAR detection in K distribution clutter is emphasized with regarding the K distribution model as the special case. Also the recent research results of the Italian professor Gini about the CFAR detection in compound Gaussian clutter are added.

(9) Based on the recent theories and research results of nonparametric CFAR processing and its engineering applications,the content of the original book are revised extensively,the recent research results about the CFAR processing in time-frequency domain and the CFAR processing technology in other transform domains are added.

(10) Clutter map CFAR processing technology is added, and main content includes Nitzberg's clutter map technique,clutter map point/plane-detection technique,spatial and temporal hybrid CM/L-CFAR clutter map detection technique as well as corresponding performance analysis.

(11) Detection technology of range-spread target is added,and the range-spread target detection issues in compound Gaussian clutter, compound Gaussian clutter plus thermal noise as well as SαS clutter are mainly discussed.

(12) Multisensor distributed CFAR processing is added and two kinds of detectors are mainly discussed,and the main content includes the distributed CA-CFAR detector,the distributed OS-CFAR detector and the distributed CFAR detectors based on kinds of R,S,and P local test statistic.

(13) The chapter named as "review,suggestion and prospect" is revised Comprehen sively. The overall content of this book is summarized systematically with the new development of automatic radar detection technologies,which provides guidance for the future work in this area. In addition,according to the domestic and foreign research results over the past ten years,necessary references are added in the book and some mistakes on word description are revised. All of these aim at increasing the readability of this book.

At the moment of the publication of the second edition,the authors would like to express sincere thanks to Professor Hermann Rohling in Technical University Hamburg-Harburg of Germany, Professor Lu Dajin, Peng Yingning in Tsinghua University, and other domestic well-known radar experts, Guo Guirong academician, Wang Yue academician, Bao Zheng academician, Mao Erke academician,Liu Yongtan academician,Wang Xiaomo academician,Zhang Guangyi academician,Ben De academician,Huang Peikang academician,Wu Yirong academician and Wu Manqing academician,et al. Thanks for their guidance,care and help to our scientific research over the years.

In the process of writing and publishing of this book, Dr. Huang Yong, Jian Tao, Qu Fuyong et al of Naval Aeronautical and Astronautical University in Yantai are thanked for their jobs of revising and word typing. Also, we would like to give thanks to the Tsinghua University Press, especially the great support of Editor Wang Yiling who guarantees the publication on schedule with high quality.

We hope that the publication of this book will not only provide a readable reference book for the scientific workers engaged in the automatic radar target detection and CFAR processing but also provide a comprehensive teaching material for the graduate students engaged in the automatic radar target detection and CFAR processing.

Sincerely hope the majority of scholars continue to care about the book and make valuable comments and suggestions.

He You Guan Jian Meng Xiangwei

October 2010

At Naval Aeronautical and Astronautical University

第1版前言

在自然界中，无论是在人造的机器还是在生物体中，检测都是广泛存在的行为方式，它根据一定的准则做出事件是否存在的判决。由于环境的不稳定和准则的不精确，在此过程中事件并没有发生而被错误判决为发生的概率，即虚警概率。

雷达是军事和民用领域中探测目标的主要工具。它的主要用途是在存在干扰的背景中检测出有用目标，这些干扰包括接收机内部热噪声、地物、雨雪、海浪等杂波干扰和电子对抗措施(ECM)、人工有源和无源干扰以及与有用目标混杂在一起的邻近干扰目标和它的旁瓣等。在雷达自动检测系统中，通常是将自动检测和恒虚警(Constant False Alarm Rate，CFAR)技术结合使用以保持在变化的杂波环境中获得可预测的检测性能和恒定虚警率。

雷达自动检测和跟踪中的虚警问题是每个雷达系统和设计人员不可回避的重要问题之一。二十多年来，雷达自动检测与 CFAR 处理技术逐渐发展成为国际雷达信号处理界的一大热门研究领域和关键性课题。CFAR 技术是雷达自动检测系统中控制虚警率的最重要手段，它在雷达自动检测过程中起着极其重要的作用。现在 CFAR 研究已经出现了多个研究方向，根据模拟杂波背景所使用的杂波分布模型分为瑞利分布、韦布尔分布、对数正态分布、K 分布和莱斯分布模型中的 CFAR 研究；按照数据处理方式分为参量和非参量 CFAR 技术；按处理所在的数域分为时域和频域 CFAR 研究方法；根据数据的形式分为标量和向量 CFAR 技术；根据信号的相关程度分为相关和不相关信号及部分相关信号的 CFAR 方法。此外，还可分为单参数和多参数 CFAR 技术，单传感器和多传感器分布式 CFAR 技术，以及其他的一些研究方法。

涉足该领域的雷达系统研究与设计人员，多年来一直渴望有一部全面、系统介绍自动检测重要准则和 CFAR 处理模型的专著。本书试图较全面、系统地向读者介绍当代雷达自动检测原理与 CFAR 处理技术的发展与最新研究成果，并重点介绍均值(ML)类 CFAR 检测器、有序统计量(OS)类 CFAR 检测器、具有自动筛选技术的广义有序统计量 CFAR 检测器、自适应 CFAR 检测器、韦布尔和对数正态杂波中的 CFAR 处理、两参数 CFAR 技术、莱斯和 K 分布杂波中的 CFAR 处理、非参量 CFAR 处理和频域中的 CFAR 处理方法等，同时研究它们在均匀杂波背景、多目标环境和杂波边缘环境这三种典型信号背景中的性能。在撰写过程中，我们尽可能地搜集了大量有关 CFAR 和自动检测理论的文献，并通过各章节的合理编排使读者对这些种类繁多的 CFAR 处理方法有一个比较清晰和系统的了解。

应当指出，雷达自动检测理论和 CFAR 处理技术当前仍处于迅速发展阶段，由于篇幅的限制，本书不可能对这些发展做出统览无余的介绍。为此，我们在每章的最后都进行了归纳和总结，指出一些重要的新发展，并罗列了相应的参考文献，供读者进一步阅览和研究参考。

本书第一作者在德国不伦瑞克工业大学进修期间，在 CFAR 领域得到了 Hermann Rohling 教授开拓性的指导,他对本书的出版给予了极大的鼓励和支持,在此表示最衷心的谢意。

著名雷达专家郦能敬教授和北京航空航天大学毛士艺教授对本书提出了一些宝贵意见。海军航空工程学院刘永硕士参与了本书的出版工作,并为本书绘制了插图。在此一并表示感谢。

何友　关键　彭应宁　陆大绘

1998年10月

于清华大学和海军航空工程学院

Foreword of the First Edition

In whatever man-made machines or organisms all over the world, detection is a widespread behavior pattern. It is a judgment for the presence of an event, based on a given criterion. Due to the instability of the environment and the impreciseness of the criteria, the probability of not-happening events, which are judged to happen, is the false alarm probability.

Radar is the main tool of detecting targets in the martial and civilian applications. Its main purpose is to detect the interesting targets in the disturbances including internal thermal noise of receiver; clutter from ground, rain, snow and sea waves; electronic counter measures (ECM); active and passive interferences; neighboring useless targets and its side lobes. In radar systems, the automatic detection is usually combined with constant false alarm rate (CFAR) techniques to obtain the predictable detection performance, and at the same time, to keep the false alarm probability constant in changing clutter environment.

The false alarm problem in radar automatic detection and track is one of the important issues, which can not be avoided in the radar system design. In the past two decades, radar automatic detection and CFAR processing techniques gradually developed into popular research fields, and now is the key issues in the international radar signal processing community. More precisely, CFAR technique is the most important means to control the false alarm rate in radar system, and it plays an important role in the radar automatic detection process. At present, several research directions have appeared in the CFAR processing. According to the clutter distribution models, CFAR technique involves the Rayleigh, Weibull, lognormal, Rician and K distributions. According to the data processing methods, it includes parameters CFAR techniques and Non-parametric ones. According to the processing domains, it includes time domain CFAR methods and frequency domain ones. According to the data forms, it contains scalar CFAR techniques and vector ones. According to the signal correlation, it involves relevant, irrelevant and partly relevant signals. In addition, it also includes single-parameter and multi-parameter CFAR techniques, single-sensor and multi-sensor distributed CFAR techniques, as well as other methods.

For those who are engaged in the fields of radar system research and design, a comprehensive and systematic monograph introducing the important criteria for automatic detection and CFAR processing model is desirable for them. In this book, we attempt to introduce the contemporary theories and the latest research results of radar CFAR processing techniques, comprehensively and systematically. It covers the mean lever (ML) CFAR detector, order statistics (OS) CFAR detector, CFAR detector based on generalized order statistics with automatic excision, adaptive CFAR detector, CFAR processing in Weibull and lognormal clutter, two-parameters CFAR technique, CFAR processing in Rician and K distributed clutter, non-parametric CFAR processing and frequency domain processing methods. Moreover, the performances for these CFAR detectors are investigated in homogeneous backgrounds, multiple target situations and clutter edge situations. In the writing process, we collected a wealth of literatures about automatic CFAR detection theory, and arranged each chapter appropriately so that the readers can understand various CFAR detectors clearly and systemically.

It should be noted that, both of radar automatic detection theory and CFAR processing techniques are still in rapid development. For the space limitations, we hardly give a fine-drawn description in this book. Hence, we provided the corresponding comparison and summary at the end of each chapter; and pointed out some new developments, while listing the corresponding references for readers' further reading and research.

The first author of this book was guided by Prof. H. Rohling, in the study period Technical University of Braunschweig in Germany. Herein we express our honestly appreciations to Prof. H. Rohling for his great encouragements and supports for the book publication.

The famous radar expert Prof. Li Nengjing, and Prof. Mao Shiyi of Beijing University of Aeronautics and Astronautics, gave lots of valuable suggestions for this book. Moreover, Liu Yong, master of Naval Aeronautical and Astronautical Engineering University, has worked for the book publication and has drawn figures for this book. Thank them together.

He You   Guan Jian   Peng Yingning   Lu Dajin

October 1998

At Tsinghua University and Naval Aeronautical and Astronautical Engineering University

目 录

CONTENTS

第1章 绪论

CHAPTER 1

早期雷达的功能主要是发现目标和测量目标的空间位置，其信号处理功能也很简单，一般将接收的所有信息都直接送至视频显示器。噪声、杂波、干扰和目标回波信号同时被显示出来，对目标的检测能力取决于雷达操作员。为了从背景噪声、杂波和干扰中发现目标，操作员要监视显示器回波信号的变化。尽管还有一些雷达系统仍然直接显示这些原始数据，但目标自动检测和跟踪功能已成为现代雷达系统的一个重要特点，并且已得到广泛应用[1]。

所谓"自动检测"是指雷达在感兴趣的分辨单元(如距离-角度-多普勒单元)中根据回波信号按照预定程序进行目标检测的过程，无须操作员参与就能自动地提供目标报告，使雷达系统可以同时处理多批目标。雷达目标自动检测过程中一个重要环节就是恒虚警率(Constant False Alarm Rate，CFAR)处理。CFAR 处理提供的检测阈值可以在一定程度上减少背景噪声、杂波和干扰变化的影响，使自动检测在均匀背景中具有恒定的虚警概率。雷达系统在所选择的分辨单元中，把匹配滤波器的输出与自动适应于背景噪声、杂波和干扰变化的阈值进行比较，获得具有恒定虚警率的目标检测能力。显示器显示的不再是原始数据，而是分辨单元中超过阈值的检测标志。本书将讨论雷达目标自动检测的经典方法和现代雷达技术中的 CFAR 处理技术。

雷达目标自动检测的最基本问题是确定检测准则，使它能在某种约束条件下提供对目标的自动检测，即在预定的最大范围内和系统参数条件下，对于特定目标、背景噪声、杂波或干扰的统计模型，使检测具有特定虚警概率 P_{fa} 和检测概率 P_d。自 20 世纪 40 年代初开始，North[2] 和 Rice[3] 等对雷达最优接收问题进行了严格理论分析。早期工作主要集中在确知恒定高斯白噪声(接收机噪声)中对非起伏或 Swerling 起伏目标的单脉冲或非相参脉冲串的最优检测。除单脉冲线性检测外，还有多脉冲线性、平方律检测和双门限二元检测策略。为了解决距离模糊和速度模糊，脉冲多普勒雷达系统需要在每个分辨角度发射具有不同脉冲重复频率的多个相参脉冲串。在接收机噪声背景中，多脉冲线性检测利用一组相参脉冲串能实现对非起伏或 Swerling 起伏目标的近似最优检测策略，但在一个分辨角度的所有相参处理间隔(Coherent Processing Interval，CPI)内必须存储所有的距离-多普勒频率样本，故脉冲多普勒雷达系统线性检测的数据存储要求可能很大，而次优的双门限检测(二元检测)策略通常在非瑞利包络干扰环境中提供较好的检测性能。关于在平均功率水平确知恒定的白高斯噪声或杂波环境中的多脉冲线性检测策略和二元检测策略将在第 2 章讨论，从第 3 章开始讨论平均功率水平未知的噪声或杂波背景中的 CFAR 检测。

对于瑞利包络噪声或杂波背景，第 3 章中讨论最经典的均值(Mean Level，ML)类 CFAR 检测。与固定门限检测相比，因引入 CFAR 而导致额外增加的信噪(杂)比需求被称作 CFAR 损失。不同的 CFAR 策略可通过 CFAR 损失来表征其性能。第 4～6 章讨论有序统计量(Ordered Statistics，OS)类、广义有序统计量(Generalized OS,GOS)类和自适应类 CFAR 检测器的 CFAR 损失。此外,第 4 章和第 5 章讨论的 CFAR 处理方法都是以有序统计处理为基础的。第 4 章详细讨论分析最经典的 OS-CFAR 检测器,然后分析了派生的 CMLD、TM、MX-CMLD、OSGO、OSSO-CFAR 检测器,它们都是以对参考单元采样进行排序处理为基础的,并结合了其他的删除干扰目标的方法。第 5 章在局部有序处理的基础上引入自动筛选,称为广义有序统计(GOS)类 CFAR 处理方法;并将 TM 推广到局部处理中,建立了具有自动筛选技术的广义有序统计类 CFAR 处理的统一模型[4]。第 3～5 章的 CFAR 处理参数很多是事先设定的,但 CFAR 处理面对的是复杂多变的杂波和干扰背景,有些参数应根据实际情况进行自适应估计,如应删除的干扰目标数和边缘杂波的位置等,将这类 CFAR 处理称为自适应类 CFAR,将在第 6 章进行讨论分析。

在许多实际检测问题中,杂波和干扰的包络概率密度函数可能不服从瑞利分布。许多实测数据表明,用对数正态(Log-Normal)分布、韦布尔(Weibull)分布、Pearson 分布和 Cauchy 分布等模型能更好地描述许多类型的干扰或杂波[5-7]。第 7 章将研究这些杂波环境中的 CFAR 检测方法。此外,复合高斯分布模型可在很宽范围内很好地描述高分辨率杂波包络特性[8],并且还能很好地反映脉冲间相关性。K 分布即复合高斯分布的一种特例。有关复合高斯分布杂波中的检测及 CFAR 处理问题也已引起国际雷达界的重视和兴趣[9-10]。第 8 章将讨论复合高斯分布杂波中的 CFAR 处理。

在噪声或杂波的统计模型不确定或变化时,非参量 CFAR 检测器仍可以提供 CFAR 能力,它对背景噪声或杂波分布不敏感。最常见的非参量 CFAR 处理技术是利用参考单元样本求取检测单元的秩。与参量检测策略相比,非参量检测策略在非瑞利包络噪声或杂波环境中通常能提供较高的检测性能和更稳定的虚警控制[11]。非参量检测策略的代价是,与为特定噪声或杂波环境设计的 CFAR 检测相比,在此噪声或杂波环境中检测时的性能有所降低。非参量 CFAR 处理方法将在第 9 章中讨论分析。

CFAR 处理的主要过程是形成杂波平均功率水平估计,相应的估计方法有两大类:一类是空域方法,即利用与检测单元在空间上邻近的参考单元样本值,这类方法适用于在空域上平稳的杂波;另一类是时域的方法,即杂波图,其利用检测单元以往多次扫描测量值形成杂波平均功率水平估计,并依靠新的测量值迭代更新[12],一般可分为单参数和双参数两类。第 10 章中将详细讨论杂波图 CFAR 处理技术。

雷达在进行实际 CFAR 处理时,还面临复杂背景回波的干扰,包括噪声、地杂波、海杂波、气象杂波等,这时首先需将这些噪声或杂波去除,或找到有效特征将背景干扰和目标区分开,改善信噪比或信杂比后,再进行 CFAR 检测。变换域处理是信噪比或信杂比提升的有效手段,针对雷达回波的不同特性,通过滤波、相参积累、特征提取、稀疏化等处理,可提高雷达在复杂背景下的目标检测能力。因此,除了在时域上进行 CFAR 处理,还可把数据变换到频域、时频域、小波域、分数阶傅里叶变换域、Hilbert-Huang 变换域、稀疏表示域等其他变换域上,再做 CFAR 处理[13-16]。第 11 章中将讨论这些变换域 CFAR 技术。

雷达作为一种探测装置,人们不仅希望它可以探测到感兴趣的目标,而且还希望它能够对目标的类别属性进行判别,获得目标更多的细节信息,因此提高其分辨率始终是雷达技术的一个重要发展方向。随着雷达成像和高分辨率雷达技术的快速发展,一般的目标在视频层上已不再是点目标,而是分布在多个分辨单元中的空间扩展目标。扩展目标的 CFAR 处理不同于点目标,又多了分辨单元这一维,如何在分辨单元间进行积累是主要问题之一[17-18]。将高距离分辨率与高方位分辨率相结合,还可获得高分辨的雷达二维图像,其中以合成孔径雷达(SAR)应用最为广泛[19]。由于增加方位信息,SAR 目标检测需在二维图像处理的基础上开展,为 CFAR 处理带来了新的挑战。第 12 章将研究高分辨率雷达目标检测。

现代雷达面临着电子干扰、反辐射武器、隐身目标和低空突防目标的威胁。多基地雷达分布在范围很广的空间中，可以增强抗干扰和抗摧毁能力，且空间大范围布站可在多个角度上对目标进行探测，对于对抗隐身目标和低空突防目标也大有好处。因此，多传感器分布式检测受到了重视[20-22]。第 13 章讨论多传感器分布式 CFAR 检测。

现代雷达体制向多维度探测发展，如数字阵列、MIMO、多极化等[23-25]雷达体制提供了比脉冲更多维度的信息，由此引发的多维 CFAR 处理将在第 14 章中进行讨论。

上述 CFAR 检测主要是以幅度作为特征进行处理的，此外，一些非线性特征，如分形等特征也可作为检测统计量。第 15 章将讨论这类特征的 CFAR 处理[26]。

为了总结过去，展望未来，第 16 章对本书讨论的内容作了简要回顾，提出了一些建议和有待进一步研究的方向。

参考文献

[1] Skolnik M I. Introduction to radar systems[M]. 3rd ed. New York: McGraw-Hill, 2002.

[2] North D O. An analysis of the factors which determine signal/noise distribution in pulsed carrier system[J]. Proceedings of the IEEE, 1963,51(7): 1015-1028.

[3] Rice S O. Mathematical analysis of random noise[J]. Bell System Technical Journal, 1994, 23: 282-332.

[4] 孟祥伟，何友. 最小选择恒虚警检测器的统一模型——TMSO[J]. 仪器仪表学报，1997，18(5)：481-485.

[5] 何友，Rohling H. 有序统计恒虚警(OS-CFAR)检测器在韦布尔干扰中的性能[J]. 电子学报，1995，23(1)：79-84.

[6] Gonzalez J G, Paredes J L, Arce G R. Zero-order statistics a mathematical framework for the processing and characterization of very impulsive signals[J]. IEEE Transactions on Signal Processing, 2006, 54(10): 3839-3851.

[7] Achim A, Kuruoglu E E, Zerubia J. SAR image filtering based on the heavy-tailed Rayleigh model[J]. IEEE Transactions on Image Processing, 2006, 15(9): 2686-2693.

[8] Ward K D. Comparison representation of high resolution sea clutter[J]. Electronics Letters, 1981, 16: 561-563.

[9] Gini F, Farina A. Vector subspace detection in compound-Gaussian clutter part I: Survey and new results[J]. IEEE Transactions on AES, 2002, 38(4): 1295-1311.

[10] De Maio A, Greco M S. Modern radar detection theory[M]. Hampshire: Scitech Publishing, 2015.

[11] Meng X W. Rank sum nonparametric CFAR detector in nonhomogeneous background[J]. IEEE Transactions on AES, 2021, 57(1): 397-403.

[12] 何友，刘永，孟祥伟. 杂波图 CFAR 平面技术在均匀背景中的性能[J]. 电子学报，1999，27(3)：119-120.

[13] 简涛，何友，苏峰，等. 一种基于小波变换的信号恒虚警率检测方法[J]. 信号处理，2006，22(3)：430-433.

[14] Chen X L , Guan J, Chen W S, et al. Sparse long-time coherent integration-based detection method for radar low-observable maneuvering target[J]. IET Radar Sonar and Navigation, 2019, 14(4): 538-546.

[15] 陈小龙，关键，黄勇，等. 雷达低可观测目标探测技术[J]. 科技导报，2017，35(11)：30-38.

[16] 陈小龙，刘宁波，黄勇，等. 雷达目标检测分数域理论及应用[M]. 北京：科学出版社，2022.

[17] He Y, Jian T, Su F, et al. Novel range-spread target detectors in non-Gaussian clutter[J]. IEEE Transactions on Aerospace and Electronic Systems, 2010, 46(3): 1312-1328.

[18] Guan J, Zhang Y F, Huang Y. Adaptive subspace detection of range-distributed target in compound-Gaussian clutter[J]. Digital Signal Processing, 2009, 19(1): 66-78.

[19] Wang X Q, Li G, Zhang X P, et al. A fast CFAR algorithm based on density-censoring operation for ship detection in SAR images[J]. IEEE Signal Process. Letters, 2021, 28: 1085-1089.

[20] 关键. 多传感器分布式恒虚警率检测(CFAR)算法研究[D]. 北京：清华大学，2000.
[21] Guan J，Peng Y N，He Y. Three types of distributed CFAR detection based on local test statistic[J]. IEEE Trans. on AES，2002，38(1)：278-288.
[22] 刘向阳，许稼，彭应宁. 多传感器分布式信号检测理论与方法[M]. 北京：国防工业出版社，2017.
[23] 张光义，赵玉洁. 相控阵雷达技术[M]. 北京：电子工业出版社，2006.
[24] 吴曼青. 数字阵列雷达的发展与构想[J]. 雷达科学与技术，2008，6(6)：401-405.
[25] 王雪松. 雷达极化技术研究现状与展望[J]. 雷达学报，2016，5(2)：119-131.
[26] 关键，刘宁波，黄勇，等. 雷达目标检测的分形理论及应用[M]. 北京：电子工业出版社，2011.

第 2 章 经典的固定门限检测

CHAPTER 2

在不用人工干预的情况下，把统计判决理论应用到目标检测问题中，便形成了雷达目标自动检测理论。本章讨论经典的固定门限检测，就是在统计特性确知的接收机噪声背景中对各种 Swerling 起伏目标进行自动检测的理论与方法。其中，“统计特性确知”是指以下两种含义之一：一是通过测量得到噪声的统计分布；二是针对具有某种统计分布的噪声背景来讨论雷达目标自动检测问题。Swerling 根据目标回波幅度的不同起伏特征（包括非起伏情况）定义了 5 种统计模型[1-2]。对于每种模型，根据贝叶斯估计准则，在确知恒定的接收机（白高斯）噪声环境中推导了单脉冲（或单个相参脉冲串）最优检测的检测概率 P_d 与虚警概率 P_{fa} 和信噪比之间的关系，并把经典的单脉冲（或单个相参脉冲串）最优检测原理扩展到了多脉冲检测。多脉冲检测中首先讨论了各种 Swerling 起伏模型条件下非相参脉冲串的二元检测及准最优线性检测策略，然后在 Swerling 0、Swerling Ⅰ和 Swerling Ⅲ型起伏模型条件下简要讨论了相参脉冲串的最优检测问题。

2.1 雷达目标自动检测的基本问题

为研究一个雷达目标自动检测系统，一般要预先指定目标雷达横截面积（Radar Cross Section，RCS）、背景噪声、杂波或干扰，以及虚警概率 P_{fa}、检测概率 P_d 和最大检测距离。理论上，检测策略或雷达设计应使接收机结构能够在给定的最大距离处对具有给定平均 RCS 的目标提供最优的（或准最优的）自动检测，并满足给定的 P_{fa}（当没有目标存在时）和 P_d（当一个具有给定平均 RCS 的目标在给定的最大距离时）。P_{fa} 和 P_d 与检测的观测区间有关。在目标 RCS 起伏模型和背景噪声、杂波或干扰的统计模型给定情况下，对于给定的 P_{fa}、P_d 和最大检测距离，雷达系统设计需要为最优（或准最优）检测策略确定参数，例如发射机峰值功率、波形、天线增益等，以便完成以下两个功能。

（1）当目标出现在指定的最大距离内时，在观测区间的最大距离上以预定的 P_d 检测具有给定平均 RCS 的目标。

（2）当没有目标存在时，根据指定的 P_{fa} 在观测区间上控制虚警率。

在给定的系统参数，如目标 RCS 起伏模型、背景噪声、杂波或干扰的统计模型，以及指定的 P_{fa} 和 P_d 条件下，要对预定最大距离处具有给定平均 RCS 的目标在观测区间上进行某种最优意义上的检测，需要确定检测策略，这是雷达目标自动检测的基本问题。检测策略是根据某些统计判决准则和给定的目标 RCS 起伏模型以及背景噪声、杂波或干扰的统计模型而确定的，其中还包含选择门限的原则。对来自每个分辨单元（距离-角度-多普勒频率）的回波信号幅度进行采样，并与相应的门限进行比较，其中门限是根据判决准则和背景噪声、杂波或干扰的统计模型确定的。在给定统计模型假设下，检测门限应使虚警概率小于或等于设计值。检测概率取决于目标、环境、最大检测距离和雷达系统参数。

检测距离、目标和雷达系统参数间的确切关系由雷达距离方程给出，该方程在不同环境下具有不同形式。对雷达目标检测问题来说，最重要的 5 种环境包括：确知恒定的内部(接收机)热噪声、未知或变化的内部热噪声、区域杂波、空间杂波和干扰。

在确知恒定的内部噪声中进行检测的情况是很少的。但是，在这种假设下接收机噪声中目标检测问题的经典分析(Swerling 曲线)可以很方便地用于研究接收机在未知或变化的接收机噪声环境中的检测性能。当一个目标所处的位置使检测必须在杂波背景(如大地、树木回波及宇宙辐射等)中进行时，并且杂波功率远大于噪声功率时，检测将在区域杂波中进行，这时应使用另一种雷达距离方程。类似地，当检测是在箔条、角反、雨或某些其他空间杂波存在的环境中进行，并且空间杂波功率远大于噪声功率时，也应该重新描述雷达距离方程。在严重的干扰机干扰情况下，即干扰功率远大于噪声功率时，应该使用适合于干扰机环境的雷达距离方程。

2.1.1 最大检测距离

各种环境下的检测一般都要求给出发射和接收一个单脉冲时可能获得的最大检测距离 R 的表示式。下面是几种环境下 R 表示式，各式中均假定了雷达接收机与接收的单脉冲波形是匹配的。

在接收机内部热噪声环境中[3]

$$R^4 = \frac{P_t G_t G_r \lambda^2 \sigma F_t^2 F_r^2}{(4\pi)^3 L k T_n B_n (\mathrm{SNR})} \tag{2.1}$$

其中，P_t 对于脉冲体制雷达是峰值发射功率，对于连续波体制雷达是平均发射功率；G_t 和 G_r 分别是发射和接收天线增益；λ 是发射载波波长；σ 是目标 RCS；F_t 是从发射天线到目标的方向图传播因子；F_r 是从目标到发射天线的方向图传播因子；L 是考虑雷达各部分的损耗而引入的累积损失系数；k 是玻耳兹曼常数；T_n 是系统噪声温度；B_n 是接收机噪声带宽；SNR 是接收信号的信噪比(目标信号平均功率与噪声平均功率之比)。对于一个检测方法和给定的目标类型及虚警概率设计值，要达到预定的检测概率所需信噪比的最小值 $\mathrm{SNR}_{\min}$ 对应的距离就是在该条件下的最大检测距离。或者反过来说，根据对最大检测距离的要求可以确定所需接收信号的最小信噪比 $\mathrm{SNR}_{\min}$，进而再根据目标类型、虚警概率设计值和检测概率的要求确定相应的检测方法。因此，$\mathrm{SNR}_{\min}$ 是与最大检测距离、检测方法、目标类型、虚警概率和检测概率相联系的。

在区域杂波和低入射角情况下[3]

$$R = \frac{\sigma_t F_t^2 F_r^2}{\sigma_0 \theta (c\tau/2)(\sec\psi) L (\mathrm{SCR})} \tag{2.2}$$

其中，σ_t 是目标 RCS；σ_0 是单位表面积的雷达杂波截面积；θ 是雷达的 3dB 水平波束宽度；τ 是脉冲雷达的脉宽，对于连续波雷达是多普勒带宽的倒数；c 是光速；ψ 是辐射入射角；SCR 是接收信号的信杂比(目标信号平均功率与杂波平均功率之比)，它与最大检测距离、检测方法、目标类型、虚警概率和检测概率的关系类似于前面在接收机内部热噪声条件下对信噪比的讨论。

在区域杂波和高入射角情况下[3]

$$R = \left[\frac{\sigma_t F_t^2 F_r^2 (\sin\psi)}{\sigma_0 \theta (\beta\pi/4) L (\mathrm{SCR})}\right]^{1/2} \tag{2.3}$$

其中，β 是 3dB 仰角波束宽度。σ_t、ψ、σ_0、θ、L 和 SCR 的定义同前。

在空间杂波环境中[3]

$$R^2=\frac{\sigma_t}{(c\tau/2)(\beta\theta\pi/4)\Sigma_\sigma L(\mathrm{SCR})_{\min}} \tag{2.4}$$

其中，Σ_σ 是散射体的后向散射系数；σ_t、c、τ、β、θ、L 和 SCR 的定义同前。

在干扰机环境中[3]

$$R^4=\frac{P_tG_tLR_j^2}{4\pi B_n(\mathrm{SIR})_{\min}}\frac{\sigma_t}{P_jG_j}\frac{G_r}{G_r'} \tag{2.5}$$

其中，G_j 是干扰机天线增益；R_j 是干扰源与雷达之间的距离；B_n 是接收机噪声带宽；P_j 是干扰机功率密度；G_r 是目标方向的接收天线增益，G_r'是雷达接收天线对着干扰机方向的增益；SIR 是接收信号的信干比(目标信号平均功率与干扰平均功率之比)，它与最大检测距离、检测方法、目标类型、虚警概率和检测概率的关系类似于前面在接收机内部热噪声条件下对信噪比的讨论。

在特定环境中，雷达系统探测目标的能力完全由检测中对雷达距离 R 有影响的参数和措施决定。为了在给定条件下确定最大检测距离，必须评价雷达检测策略，以确定最小可检测的信噪比、信杂比或信干比。在给定的雷达特性、目标 RCS 和环境条件下，最大检测距离可以通过把最小的信噪比、信杂比或信干比代入相应的雷达距离方程中得到。

目标 RCS、杂波、噪声和干扰机干扰都具有随机起伏性，因此从雷达方程确定的最大可探测距离不是确定量，它需要用统计方法描述。一般来说，雷达距离方程中的 RCS 值为目标 RCS 分布的平均值。对于指定的 P_{fa} 和 P_d，最小的信噪比、信杂比或信干比值是根据雷达检测策略和目标 RCS 幅度起伏、杂波、噪声或干扰机干扰的统计特征确定的。对于在某一观测区间上的检测来说，最小信噪比、信杂比或信干比值是与指定的 P_{fa}(当没有目标存在时)和 P_d(当目标在指定的最大距离时)相联系的。对于每种检测策略，一般将(P_d，P_{fa}，SNR、SCR 或 SIR)间的这些关系绘成曲线。给定检测策略及 P_{fa} 和 P_d 值，可以从有关该检测策略的曲线中获得信噪比、信杂比或信干比值。把给定的最大检测距离、在该距离处检测的平均 RCS 及与指定的 P_{fa} 和 P_d 对应的信噪比、信杂比或信干比值代入相应环境的雷达距离方程中，可以确定发射波形的参数及其他雷达系统参数。

2.1.2 虚警率

为了确定一个准则以获得令人满意的虚警率(False Alarm Rate，FAR)，通常要指定一个 P_{fa} 值，该值与检测的某一观测区间有关。对于单脉冲线性检测策略，观测区间是一个脉冲宽度(或者对相参系统来说是单个 CPI)。对于多脉冲线性检测策略，观测区间是非相参脉冲串的持续时间。对于指定的 P_{fa}，观测区间一般指与每个分辨单元的检测策略相对应的观测区间。

FAR 是分辨单元数、观测区间及在观测区间上指定的 P_{fa} 三者的函数，它可以表示为

$$\mathrm{FAR}=\frac{NP_{fa}}{\tau} \tag{2.6}$$

其中，N 是分辨单元数；τ 是观测区间；P_{fa} 是在观测区间上进行检测的虚警概率。故虚警率是每秒的虚警数。

2.1.3 目标雷达截面积的 Swerling 起伏模型

目标的 RCS 是随视角变化而随机起伏的。Swerling 模型是关于目标 RCS 起伏的统计和相关特性的 5 种标准统计假设，即 Swerling 0、Swerling Ⅰ、Swerling Ⅱ、Swerling Ⅲ和 Swerling Ⅳ型，它们已经被证明适用于多种雷达目标幅度的变化。在确知恒定的接收机噪声(白高斯)条件下，对于这 5 种

Swerling 模型,Marcum[4]和 Swerling[5]都建立了一组曲线,用以表示单脉冲最优检测和多脉冲准最优检测的 P_{fa}、P_d 与最优单脉冲检波输出的信噪比三者之间的关系。对每种 Swerling 模型和非相参脉冲串中脉冲数 N,建立不同的(P_{fa},P_d,SNR)曲线,这些曲线被称作 Swerling 曲线。

Swerling 0 型:有时也称作 Swerling Ⅴ型,假设目标 RCS 没有起伏,即非起伏目标。该模型最初由 Marcum[4]、Rice[6]和 North[7]用于评价接收机噪声中的最优检测策略。Swerling Ⅰ～Swerling Ⅳ都是起伏模型,假设目标 RCS 起伏服从瑞利分布或一主加瑞利分布,并且具有扫描间或脉冲间的统计独立性。

Swerling Ⅰ型:假设接收到的目标回波振幅在任意一次扫描期间都是恒定的(完全相关),但是从一次扫描到下一次扫描之间是独立的(不相关),是服从瑞利分布的随机变量。

Swerling Ⅱ型:假设脉冲串中的每个脉冲的振幅是相互统计独立的瑞利分布随机变量。Swerling Ⅱ型假设脉冲间是按瑞利分布起伏的非相参脉冲串。

Swerling Ⅲ型:假设接收到的目标回波振幅在任意一次扫描期间都是恒定的(完全相关),但是从一次扫描到下一次扫描之间是独立的(不相关)、服从一主加瑞利分布的随机变量。Swerling Ⅲ型仅是在回波幅度分布形式上与 Swerling Ⅰ型不同。

Swerling Ⅳ型:假设脉冲串中的每个脉冲的振幅是相互统计独立的一主加瑞利分布随机变量。Swerling Ⅳ型假设的是脉冲间按一主加瑞利分布起伏的非相参脉冲串。Swerling Ⅳ型仅是在分布形式上与 Swerling Ⅱ型不同。

上述定义中的“扫描”是指对监视区域的一次完整搜索。假设目标回波信号幅度在监视区域中的一个分辨角度的整个观测时间上是常量,但是在两次扫描间存在瑞利或一主加瑞利随机起伏,称这种随机变量为扫描间起伏。“脉冲间”起伏是在非相参脉冲串中的两个脉冲间发生的目标回波幅度起伏。

在现代数字脉冲多普勒雷达系统中,一般在每个分辨角度上发射一组相参脉冲串,并对多个 CPI 上获得的处理结果进行非相参或二元积累。在这两种情况下都像 Marcum 和 Swerling 模型假定的那样,接收机与每个相参脉冲串匹配,而不是简单地与单个脉冲匹配。因此,在关于目标 RCS 起伏和接收机噪声的一般假设下,由 Marcum 和 Swerling 在确知恒定接收机噪声条件下获得的最优或准最优检测结果(Swerling 曲线)适用于整个相参脉冲串的检测。

对于多个相参脉冲串之间的非相参积累,由于接收机匹配于每个相参脉冲串,所以可以认为每个相参脉冲串是具有脉冲间起伏、扫描间起伏的 Swerling 类型脉冲。Swerling 曲线提供了单个 CPI 上的(P_{fa},P_d,SNR)关系,其中 SNR 是最优单脉冲检波器输出的整个相参脉冲串的平均功率与噪声的平均功率之比。对于指定的 P_{fa} 和 P_d 值,从曲线中获得的信噪比值可以代入雷达距离方程(对于接收机噪声环境)中来确定在给定最大距离处在单个 CPI 上提供目标检测所需的相参脉冲串能量。当以上述方式应用 Swerling 曲线时,“脉冲间”变化不是指相参脉冲串中的相邻脉冲之间,而是指一个分辨角度上不同的相参脉冲串之间。

2.1.4 自动检测的经典问题——固定门限检测

从 1943 年 North[7]的研究开始,人们已经对不同的目标 RCS 起伏模型及背景噪声和杂波统计模型条件下的最优(或准最优)自动检测问题进行了充分的研究。这一领域的早期工作还包括 Rice[6]、Marcum[4]和 Swerling[5]的工作。他们都假设目标 RCS 起伏服从 5 种 Swerling 统计模型之一,并且检测是在确知恒定的接收机噪声(白高斯)环境中进行的。自动检测的经典问题就是指这种条件下的目标自动检测问题,均为固定门限检测,这是本章将要讨论的主要内容。固定门限检测与 CFAR 检测的区别在于,前者的门限是根据确知的背景统计特性及设定的虚警概率 P_{fa} 得出来的,而且该门限一旦设定就固定不变；而 CFAR 检测的门限则可以随着背景的变化而自适应地变化，以保持恒定的虚警率。因此，固定门限检测的最大缺点就是不能自动适应背景的变化，会带来大量虚警而导致过载，因而在实际中一般不采用固定门限检测。但对固定门限检测的研究与分析有助于理解雷达目标自动检测的基本原理，是深入学习 CFAR 检测理论的基础。

为了进行目标自动检测，假设在某一分辨单元中的某一观测区间 $0\leqslant t\leqslant\tau$ 上的接收信号为 $v(t)$，自动检测问题就是确定最优判决准则，以便由观测 $v(t)$ 判别该分辨单元中是否存在目标回波信号。本节假定观测区间 τ（即 $v(t)$ 的持续时间）是已知的。$v(t)$ 一般由两个可能的分量组成，一个来自目标的反射信号，表示为 $s(t)$，另一个来自噪声环境，表示为 $n(t)$。在有目标条件下

$$v(t)=s(t)+n(t)=v_1(t) \tag{2.7a}$$

在没有目标条件下

$$v(t)=n(t)=v_0(t) \tag{2.7b}$$

因此，$v(t)$ 可以被看作来自两个不同随机过程之一的样本函数。其一表示为 $v_0(t)$，是 $v(t)$ 在目标不存在情况下的样本；其二表示为 $v_1(t)$，是 $v(t)$ 在目标信号与噪声同时存在条件下的样本。

上述假设下的自动检测问题可以利用统计判决理论，由某一分辨单元（距离-角度-多普勒频率）中的接收信号 $v(t)$ 的采样形成一个检测统计量，然后通过判决确定该分辨单元中是否存在目标，并且满足某种意义上的最优。为了选择一个最优准则，必须评价各种准则的相对性能。“损失”和“代价”是评价不同判决准则性能的两个重要的统计概念。

“损失”是一个定义在信号空间和判决空间上与最优检测策略有关的函数。对于所选择的损失函数，在某种意义上使损失函数极小化的策略称为最优检测策略（或判决准则）。因此，为确定最优检测准则，必须定义一个损失函数，然后确定使该损失函数极小化的检测准则。在雷达应用中，贝叶斯最小风险和 Neyman-Pearson 准则是两个与损失函数有关的重要的最优准则。在一般的条件下，二者都可以简化成似然比检验，都是对于给定的 P_{fa} 使 P_{d} 极大化的检测策略。相对于其他检测准则，它们提供了较高的检测概率和较低的虚警概率。

典型的雷达目标检测问题是在每个分辨单元中检测目标反射信号 $s(t)$ 的存在与否。所以，信号空间包含了两部分：$s(t)=0$ 或 $s(t)\neq 0$。同样，判决空间也包括两个判决：$s(t)$ 不存在（用 H_0 假设表示）或 $s(t)$ 存在（用 H_1 假设表示）。正确判决的代价一般取作零，所以只考虑两个损失：当 $s(t)$ 不存在而判决为 $s(t)$ 存在时产生的损失，用 L_0 表示；当 $s(t)$ 存在而判决为 $s(t)$ 不存在时产生的损失，用 L_1 表示。可以证明[8]，平均损失函数可以用 L_0、L_1、P_{fa} 和 P_{d} 表示。虚警概率和检测概率可定义为

$$P_{\mathrm{fa}}=\Pr[\text{判决 } s(t) \text{ 存在} \mid s(t) \text{ 不存在}] \tag{2.8}$$

$$P_{\mathrm{d}}=\Pr[\text{判决 } s(t) \text{ 存在} \mid s(t) \text{ 存在}] \tag{2.9}$$

于是，平均损失为

$$L=qP_{\mathrm{fa}}L_0+p(1-P_{\mathrm{d}})L_1 \tag{2.10}$$

其中，p 是目标信号存在的先验概率，q 是目标信号不存在的先验概率。

由极小化平均损失产生的贝叶斯判决准则是一种似然比检验[8]，用 $l(v)$ 表示似然比

$$l(v)=\frac{f(v\mid \mathrm{H}_1)}{f(v\mid \mathrm{H}_0)} \tag{2.11}$$

其中，$f(v|\mathrm{H}_1)$ 是在 $s(t)$ 存在的条件下在观测区间 τ 上的接收波形 $v(t)$ 的概率密度函数，$f(v|\mathrm{H}_0)$ 是在 $s(t)$ 不存在条件下在观测区间 τ 上的接收波形 $v(t)$ 的概率密度函数。

贝叶斯判决准则可以表示为

$$\begin{cases} l(v) \geqslant l, & \text{判决目标存在} \\ l(v) < l, & \text{判决目标不存在} \end{cases} \tag{2.12}$$

其中

$$l = \frac{qL_0}{pL_1} \tag{2.13}$$

所以,若已知先验概率 p 和 q 及损失函数,则可以确定 l,这样目标的存在与否可通过将 $l(v)$ 与 l 进行比较而最优地确定。然而目标是否存在的先验概率 p 和 q 及错误判决的代价往往是未知的,一般还不能由式(2.13)确定 l,需要根据 $s(t)$ 和 $n(t)$ 的统计特征间接地确定 l。

由于 $l(v)$ 是观测时间 τ 内信号 $v(t)$ 的函数,并且 $v(t)$ 可以被认为是两个随机过程 $v_0(t)$ 和 $v_1(t)$ 之一的样本函数。那么,当 τ 固定时,$l(v)$ 可以被认为是两个随机变量 $l_0(t)$ 和 $l_1(t)$ 之一。其中,$l_0(t)$ 是在没有目标信号存在条件下 $l(v)$ 的样本;$l_1(t)$ 是在目标信号和噪声同时存在情况下 $l(v)$ 的样本。并且设 $l_0(t)$ 和 $l_1(t)$ 的概率密度函数(Probability Density Function,PDF)分别为 $f_{l_0}(x)$ 和 $f_{l_1}(x)$。由于

$$P_{\mathrm{fa}} = \Pr[l(v) \geqslant l \mid \mathrm{H}_0] = 1 - F_{l_0}(l) \tag{2.14}$$

通过求解方程

$$P_{\mathrm{fa}} = \int_l^{\infty} f_{l_0}(x)\mathrm{d}x \tag{2.15}$$

可以解得阈值 l。类似地,P_{d} 与 l 的关系为

$$P_{\mathrm{d}} = \Pr[l(v) \geqslant l \mid \mathrm{H}_1] = 1 - F_{l_1}(l) = \int_l^{\infty} f_{l_1}(x)\mathrm{d}x \tag{2.16}$$

在一般情况下,要得到 $f_{l_0}(x)$ 和 $f_{l_1}(x)$ 的表达式并不容易,甚至 $l(v)$ 的表达式也是如此。因此,需要找一个等价的方法,如假定表达式 $g(v)$ 和 l' 具有以下特性。

(1) $g(v) \geqslant l'$ 与 $l(v) \geqslant l$ 等价。

(2) $g(v)$ 是不复杂的表达式。

(3) $g(v)$ 的概率密度函数 $f_{g_0}(x)$ 和 $f_{g_1}(x)$ 的确定要比 $f_{l_0}(x)$ 和 $f_{l_1}(x)$ 更容易。

那么,贝叶斯判决准则简化成

$$\begin{cases} g(v) \geqslant l', & \text{判决目标存在} \\ g(v) < l', & \text{判决目标不存在} \end{cases} \tag{2.17}$$

其中,阈值 l' 通过求解式(2.18)确定

$$P_{\mathrm{fa}} = \int_{l'}^{\infty} f_{g_0}(x)\mathrm{d}x \tag{2.18}$$

并且

$$P_{\mathrm{d}} = \Pr[g(v) \geqslant l' \mid \mathrm{H}_1] = 1 - F_{g_1}(l') = \int_{l'}^{\infty} f_{g_1}(x)\mathrm{d}x \tag{2.19}$$

式(2.19)确立了 P_{d}、P_{fa} 和 SNR 之间的关系。

2.2 匹配滤波

目标回波信号总是混杂在背景噪声、杂波或干扰中,目标信号与背景的平均功率之比是检测中的重要指标,该比值越大,则检测概率越高。一般期望用滤波器使雷达接收机输出端的该比值增至最大。匹

配滤波器就是一种最佳线性滤波器，在输入为确知信号加白噪声的情况下，所得输出在某一时刻的信噪比能达到最大。本节介绍白噪声背景下的匹配滤波原理，以及匹配滤波与相关接收的关系和相参脉冲信号的匹配滤波。

2.2.1 白噪声背景下的匹配滤波

设线性非时变滤波器的输入端由信号和噪声混合输入

$$v(t)=s(t)+n(t) \tag{2.20}$$

其中，噪声为平稳白噪声，其双边功率谱密度为

$$P_{\mathrm{n}}(f)=\frac{N_0}{2} \tag{2.21}$$

而确知信号 $s(t)$ 的频谱为

$$S(f)=\int_{-\infty}^{\infty} s(t)\mathrm{e}^{-\mathrm{j}2\pi ft}\mathrm{d}t \tag{2.22}$$

当滤波器的频率响应为

$$H(f)=kS^{*}(f)\mathrm{e}^{-\mathrm{j}2\pi ft_0} \tag{2.23}$$

时，在滤波器输出端能够得到最大信噪比。这个滤波器是最大信噪比准则下的最佳滤波器，称为匹配滤波器。匹配滤波器的频率特性与输入信号的频谱呈复共轭。式(2.23)中，k 为常数，t_0 为使滤波器物理可实现所附加的延迟。

匹配滤波器输出端信噪比的最大值为

$$d_{\max}=\frac{2E}{N_0} \tag{2.24}$$

式中，E 为输入信号能量，匹配滤波器输出端的最大信噪比只取决于输入信号的能量 E 和输入噪声功率谱密度 $N_0/2$，而与输入信号形式无关。无论什么信号，只要它们所含能量相同，那么通过与之对应的匹配滤波器后输出的最大信噪比都是一样的。

由式(2.23)可见，滤波器幅频特性与有用信号 $s(t)$ 的幅频特性相同，各频率分量在频率轴上的分布是不均匀的，而噪声频谱是均匀的。因此，滤波器的幅频特性对不同频率分量进行加权，使有用信号分量强的地方增益大，弱的地方增益小，结果在输出端就相对地加强了信号而减弱了噪声的影响。此外，滤波器相频特性与有用信号 $s(t)$ 的频谱的相频特性相反，并有一个附加的延迟项。因此，通过此滤波器后，有用信号中各频率成分的相位一致，只保留一个线性相位项。这表示这些不同频率成分在特定的时间 t_0 全部同相相加，从而在输出端信号形成峰值。而噪声各频率分量间的相位是随机的，不会同相相加积累。

匹配滤波器的频率响应是输入信号频谱的复共轭，因此信号幅度大小不影响滤波器的形式。当两信号只有时间差别时，也可用同一匹配滤波器，只不过输出端有相应的时间差，即匹配滤波器对时延具有适应性。但对于频移后的信号，由于其信号频谱发生频移，则匹配滤波器的频率特性与该信号频谱的共轭不同，因此各频率分量没有得到合适的加权，且在输出端得不到信号各频率分量全部同相相加形成的峰值。所以，匹配滤波器对于信号的频移不具有适应性，因而当回波有多普勒频移时将产生失配。

2.2.2 匹配滤波与相关接收

相关器与匹配滤波器对信号的处理本质是一致的，即匹配滤波器等效为互相关器，它的输出是信号

的自相关函数和信号与噪声的互相关函数之和。

首先,匹配滤波器的冲激响应函数为

$$h(t)=Ks^*(t_0-t) \tag{2.25}$$

其中,信号持续时间为 $0\leqslant t\leqslant\tau$,那么匹配滤波器的输出为

$$\begin{aligned}y_0(t)&=\int_0^\tau s^*(t_0-t+u)v(u)\mathrm{d}u\\&=\int_0^\tau s^*(t_0-t+u)s(u)\mathrm{d}u+\int_0^\tau s^*(t_0-t+u)n(u)\mathrm{d}u\\&=R_{ss}(t-t_0)+R_{sn}(t-t_0)\end{aligned} \tag{2.26}$$

式(2.26)表明,匹配滤波器的输出是信号的自相关函数和信号与噪声的互相关函数之和。通常选 t_0 为信号结束之时,即 $t_0=\tau$,且在信号结束时,匹配滤波器的输出能达到最大信噪比,那么在 $t=\tau$ 时刻,匹配滤波器的输出为

$$y_0=\int_0^\tau s^*(t)v(t)\mathrm{d}t \tag{2.27}$$

此式在判决准则的推导过程中经常可以见到。

在具体应用中,二者考虑的出发点和实现方法有所差别。互相关器主要考虑输入信号在时域上的特性,而匹配滤波器则主要考虑输入信号频率域上的特性。在实际运用中,应该根据输入信号的时间函数和其频谱函数的特点来选用匹配滤波器或互相关器。此外,匹配滤波器的输出为连续且实时的,给出了互相关函数全景图。相关接收器仅对单一时延(对应距离)进行检测,对其他时延进行检测就需要多通道,这就增加了相关接收的复杂性。

2.2.3 相参脉冲串信号的匹配滤波

一组脉冲组成脉冲串,如果各脉冲信号间彼此有确定的相位关系,那么就称为相参脉冲串,其典型形式可表示为

$$s(t)=u(t)\cos(\omega_0 t) \tag{2.28}$$

其中,$u(t)$为调制函数。

$$u(t)=\mathrm{rect}\left(\frac{t}{\tau}\right)+\mathrm{rect}\left(\frac{t-T_\mathrm{r}}{\tau}\right)+\cdots+\mathrm{rect}\left(\frac{t-NT_\mathrm{r}}{\tau}\right)+\mathrm{rect}\left(\frac{t+T_\mathrm{r}}{\tau}\right)+\cdots+\mathrm{rect}\left(\frac{t+NT_\mathrm{r}}{\tau}\right) \tag{2.29}$$

式中,$\mathrm{rect}(x)$为矩形函数,T_r 为脉冲重复周期。由式(2.28)和式(2.29)可以求得该相参脉冲串的频谱,进而可以设计它的匹配滤波器。该匹配滤波器的频率响应为

$$H(f)=\frac{\sin(2N+1)\pi fT_\mathrm{r}}{\sin\pi fT_\mathrm{r}}\mathrm{e}^{-\mathrm{j}2\pi NfT_\mathrm{r}} \tag{2.30}$$

相参脉冲串的频谱可以看成单个中频脉冲的频谱和等幅梳齿频谱相乘得到,因而滤波器也可以用两个级联滤波器来实现:第一个实现对单个中频脉冲的匹配,第二个实现等幅梳齿形滤波。

中频相参积累滤波器要求 $T_\mathrm{r}=mT_0$,其中 T_0 是中频周期。在早期雷达中,很难将中频迟延线的迟延长度准确地做成中频周期的整数倍,因此中频上实现相参积累很不容易,若将其降至零中频进行处理也可以完全等效于中频处理。在动目标检测中用多普勒滤波器组实现相参积累,不同中心频率的滤波器分别适应于不同多普勒频移的信号,以实现对不同多普勒频率相参脉冲串的匹配滤波。

2.3 单脉冲检测

单脉冲检测是指对单个射频脉冲或者由一系列相参射频脉冲组成的一个相参脉冲串的检测。单脉冲检测问题就是判定单脉冲反射信号 $s(t)$是否存在，即在某一分辨单元的某一观测区间上观测得到接收信号 $v(t)$，并且从观测 $v(t)$中确定信号 $s(t)$在该分辨单元中是否存在。$s(t)$的持续时间一般是已知的，并设置等于观测区间 τ。除了回波脉冲到达时间(距离延迟)、多普勒频移、到达角度、初始射频相位和幅度，单脉冲检测是针对确知形式的射频回波信号进行的。

假设天线在搜索空间内指向某一任意方向，并假设只针对零多普勒信号进行检测，待检测的距离单元为

$$\mathrm{RB}=\frac{ct}{2}, \quad t_{\mathrm{d}} \leqslant t \leqslant t_{\mathrm{d}}+\Delta\tau \tag{2.31}$$

其中，t_{d} 是反射信号的延迟时间，假设为已知；$\Delta\tau$ 是单(相参)脉冲宽度(对于大时宽带宽积信号来说，是指脉冲压缩之后的脉冲宽度)；c 是光速。在距离、角度和多普勒频率已知的情况下，检测问题是指在确知恒定的白高斯噪声环境中建立对幅度和初始射频相位均未知的反射信号进行检测的贝叶斯判决准则，以确定在距离-角度-多普勒频率分辨单元中是否存在反射信号。在没有关于初始相位的先验信息的条件下，通常假设接收信号的初始相位是在[0,2π]上均匀分布的随机变量。

目标反射信号 $s(t)$是雷达发射信号 $s_{\mathrm{t}}(t)$的映像，$s(t)$在时间上比 $s_{\mathrm{t}}(t)$延迟了 t_{d}，并且在频率上有多普勒频移(这里取为零)，在幅度上也发生了改变。一个单脉冲或一个相参脉冲串发射信号可以表示为

$$s_{\mathrm{t}}(t)=A'a(t)\cos[\omega_0 t+\theta(t)+\beta'], \quad 0\leqslant t\leqslant\tau \tag{2.32}$$

其中，A'是发射信号的幅值；$a(t)$是对 A'的幅度调制；ω_0 是载波频率(常数)；$\theta(t)$是关于 $\omega_0 t$ 的相位调制；β'是在 $t=0$ 时刻的初始相位；τ 是波形持续时间；函数 $a(t)$和 $\theta(t)$与 $\cos(\omega_0 t)$相比都是慢变的。对于零多普勒频率，雷达反射信号可以写为

$$s(t)=Aa(t-t_{\mathrm{d}})\cos[\omega_0(t-t_{\mathrm{d}})+\theta(t-t_{\mathrm{d}})+\beta], \quad t_{\mathrm{d}}\leqslant t\leqslant t_{\mathrm{d}}+\tau \tag{2.33}$$

其中，A 是接收信号的幅值；t_{d} 是目标回波延迟；β 是在 $t=t_{\mathrm{d}}$ 时刻的初始相位；$\theta(t)$、$a(t)$、τ 和 ω_0 的定义同前。

为了确定最优贝叶斯检测准则，假设 t_{d} 是已知的。不失一般性，可以设置 t_{d} 等于零。这相当于在时间轴上的一个平移，使反射信号到达接收天线时从 $t=0$ 开始。于是，$s(t)$可以表示为

$$s(t)=Aa(t)\cos[\omega_0 t+\theta(t)+\beta], \quad 0\leqslant t\leqslant\tau \tag{2.34}$$

对于式(2.34)形式的 $s(t)$，由文献[9]可知，式(2.11)中的似然比为

$$l(v)=\exp\left[-\lambda+\frac{2}{N_0}\int_0^{\tau}v(t)s(t)\mathrm{d}t\right] \tag{2.35}$$

其中，$\lambda=E/N_0$ 是检波前的接收信号与噪声的能量之比，$\frac{2}{N_0}$是噪声的双边功率谱密度；信号 $s(t)$的能量 E 为

$$E=\int_0^{\tau}s^2(t)\mathrm{d}t \tag{2.36}$$

在下面的讨论中，对幅值 A 做下述几种假设：①A 是未知、非起伏的；②A 是瑞利分布的随机变量；③A 是一主加瑞利分布的随机变量。

2.3.1 对非起伏目标的单脉冲线性检测

设信号 $s(t)$ 的同相和正交分量分别为

$$u_{\mathrm{I}}(t)=a(t)\cos[\omega_0 t+\theta(t)],\quad 0\leqslant t\leqslant\tau \tag{2.37}$$

$$u_{\mathrm{Q}}(t)=a(t)\sin[\omega_0 t+\theta(t)],\quad 0\leqslant t\leqslant\tau \tag{2.38}$$

则

$$s(t)=Au_{\mathrm{I}}(t)\cos\beta-Au_{\mathrm{Q}}(t)\sin\beta \tag{2.39}$$

$s(t)$ 的总能量为

$$E=A^2\int_0^\tau u_{\mathrm{I}}^2(t)\mathrm{d}t=A^2\int_0^\tau u_{\mathrm{Q}}^2(t)\mathrm{d}t=A^2\varepsilon \tag{2.40}$$

对于给定的 β,由式(2.35)和式(2.39)可得似然比为

$$l(v\mid\beta)=\mathrm{e}^{-\lambda}\exp\left[\frac{2A\cos\beta}{N_0}\int_0^\tau v(t)u_{\mathrm{I}}(t)\mathrm{d}t-\frac{2A\sin\beta}{N_0}\int_0^\tau v(t)u_{\mathrm{Q}}(t)\mathrm{d}t\right] \tag{2.41}$$

对于 $k\neq0$,令

$$I(v)=k\int_0^\tau v(t)u_{\mathrm{I}}(t)\mathrm{d}t,\quad Q(v)=k\int_0^\tau v(t)u_{\mathrm{Q}}(t)\mathrm{d}t \tag{2.42}$$

由文献[8]可知,$I(v)$ 和 $Q(v)$ 分别看成增益为 k 并匹配于 $u_{\mathrm{I}}(t)$ 和 $u_{\mathrm{Q}}(t)$ 的滤波器的输出。设

$$D(v)=\sqrt{I^2(v)+Q^2(v)},\quad \Phi(v)=\arctan\left[\frac{Q(v)}{I(v)}\right] \tag{2.43}$$

则

$$l(v\mid\beta)=\mathrm{e}^{-\lambda}\exp\left\{\frac{2A}{kN_0}D(v)\cos[\beta+\Phi(v)]\right\} \tag{2.44}$$

因此,将式(2.44)对 β 取平均得

$$l(v)=\mathrm{e}^{-\lambda}I_0\left[\frac{2A}{kN_0}D(v)\right] \tag{2.45}$$

其中,$I_0(\cdot)$ 是第一类零阶修正的贝塞尔函数,且

$$I_0(x)=\frac{1}{2\pi}\int_0^{2\pi}\exp[x\cos(\alpha+t)]\mathrm{d}t$$

由式(2.12)和式(2.45)可得,在确知恒定的零均值白高斯噪声背景中,对具有未知非起伏幅值和均匀分布初始相位的目标反射信号进行最优单脉冲检测的贝叶斯判决准则为

$$\begin{cases}I_0\left[\dfrac{2A}{kN_0}D(v)\right]\geqslant\mathrm{e}^{\lambda}l, & \text{判决目标存在}\\[2ex] I_0\left[\dfrac{2A}{kN_0}D(v)\right]<\mathrm{e}^{\lambda}l, & \text{判决目标不存在}\end{cases} \tag{2.46}$$

其中,l 在式(2.13)中给出。

利用 $I_0(x)$ 是 x 的单调递增函数这一特性可以得到等价关系:$D(v)\geqslant S$ 与 $l(v)\geqslant l$ 等价,以及 $D(v)<S$ 与 $l(v)<l$ 等价。其中,S 是检测阈值。于是,贝叶斯判决准则可以简化为

$$\begin{cases}D(v)\geqslant S, & \text{判决目标存在}\\ D(v)<S, & \text{判决目标不存在}\end{cases} \tag{2.47}$$

图 2.1 中给出了利用式(2.47)作为判决准则的最优单脉冲线性检测器框图。此结构比采用修正贝

塞尔函数的结构简单得多,可以比较容易地用数字器件实现。

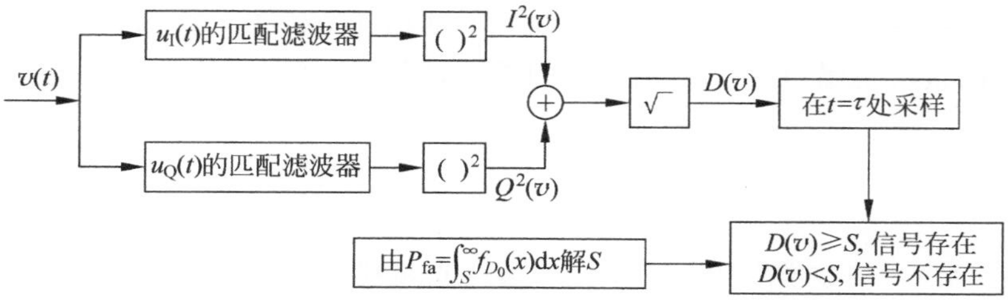

图 2.1 最优单脉冲线性检测器框图

由于 $D(v)$ 是 $v(t)$ 和信号持续时间 τ 的函数,并且可以认为 $v(t)$ 是两个随机过程 $v_0(t)$ 和 $v_1(t)$ 之一(见式(2.7))。当 τ 固定时,$D(v)$ 可以被认为是两个随机变量 $D_0(v)$ 和 $D_1(v)$ 之一的样本函数。其中,$D_0(v)$ 是在没有目标存在条件下的 $D(v)$,而 $D_1(v)$ 是目标和噪声同时存在情况下的 $D(v)$。用 $f_{D_0}(x)$ 和 $f_{D_1}(x)$ 分别表示 $D_0(v)$ 和 $D_1(v)$ 的概率密度函数,于是可以确定阈值 S 及(P_d,P_{fa},SNR)关系。对于贝叶斯判决,利用区间 τ 上的观测 $v(t)$ 对信号 $s(t)$ 进行检测的虚警概率和检测概率分别为

$$P_{fa}=\Pr[D(v)\geqslant S \mid H_0]=1-F_{D_0}(S) \tag{2.48a}$$

$$P_d=\Pr[D(v)\geqslant S \mid H_1]=1-F_{D_1}(S) \tag{2.48b}$$

给定检测波形 $s(t)$ 和 P_{fa},式(2.48a)中检测阈值 S 可以通过求解式(2.49)获得。

$$P_{fa}=\int_S^\infty f_{D_0}(x)\mathrm{d}x \tag{2.49}$$

在确定 S 以后,(P_d,P_{fa},SNR)间的关系可以由式(2.50)给出。

$$P_d=\int_S^\infty f_{D_1}(x)\mathrm{d}x \tag{2.50}$$

在观测区间 τ 内,当分辨单元中不存在 $s(t)$ 时,观测信号为

$$v(t)=n(t),\quad 0\leqslant t\leqslant\tau \tag{2.51}$$

则由式(2.42)得 $I(v)$ 和 $Q(v)$ 分别为

$$I(v)=k\int_0^\tau n(t)u_I(t)\mathrm{d}t,\quad Q(v)=k\int_0^\tau n(t)u_Q(t)\mathrm{d}t \tag{2.52}$$

对于增益 $k\neq 0$ 的滤波器,由式(2.43)得 $D_0(v)$ 为

$$D_0(v)=\sqrt{\left[k\int_0^\tau n(t)u_I(t)\mathrm{d}t\right]^2+\left[k\int_0^\tau n(t)u_Q(t)\mathrm{d}t\right]^2} \tag{2.53}$$

由于 $n(t)$ 是零均值白高斯过程,所以 $I(v)$ 和 $Q(v)$ 是具有零均值和方差为 $N_0k^2\varepsilon/2$ 的相互统计独立的高斯随机变量。其中,$N_0/2$ 是 $n(t)$ 的双边功率谱密度;ε 是由式(2.40)确定的已知参数。因此,$D_0(v)$ 是瑞利分布的随机变量,其概率密度函数为

$$f_{D_0}(x)=\frac{2x}{N_0k^2\varepsilon}\exp\left(-\frac{x^2}{N_0k^2\varepsilon}\right),\quad x\geqslant 0 \tag{2.54}$$

于是,由式(2.49)和式(2.54)得

$$P_{fa}=\int_S^\infty \frac{2x}{N_0k^2\varepsilon}\exp\left(-\frac{x^2}{N_0k^2\varepsilon}\right)\mathrm{d}x=\exp\left(-\frac{S^2}{N_0k^2\varepsilon}\right) \tag{2.55}$$

所以

$$S=\sqrt{N_0k^2\varepsilon\ln(P_{\text{fa}}^{-1})}=\sigma\sqrt{2\ln(P_{\text{fa}}^{-1})} \tag{2.56}$$

其中，$\sigma=k\sqrt{N_0\varepsilon/2}$是瑞利分布的参数。如果已知 N_0 和 ε，并给定 P_{fa}，检测阈值 S 可以由式(2.56)确定。对于确知恒定的接收机噪声，N_0 是可测量的常数，则由式(2.56)给出的检测阈值 S 也是常量，可以表示为

$$S=\mathrm{E}\{D_0(v)\}\sqrt{\frac{4}{\pi}\ln(P_{\text{fa}}^{-1})} \tag{2.57}$$

其中，$\mathrm{E}\{D_0(v)\}$是 $D_0(v)$的数学期望。在变化的噪声或瑞利包络杂波环境中，当不能测量或精确地估计 N_0 时，可用式(2.57)估计检测阈值 S。一般在这些情况下，通过由检测单元周围的分辨单元形成 $\mathrm{E}\{D_0(v)\}$的估计来产生 S。这就是第 3 章中将要讨论的单元平均恒虚警处理的基本原理。

在观测区间 τ 上，当信号 $s(t)$存在于分辨单元中时，观测信号为

$$v(t)=s(t)+n(t),\quad 0\leqslant t\leqslant\tau \tag{2.58}$$

因此

$$\begin{aligned}I(v)&=k\int_0^\tau[s(t)+n(t)]u_{\mathrm{I}}(t)\mathrm{d}t\\&=kA\varepsilon\cos\beta-kA\sin\beta\int_0^\tau u_{\mathrm{Q}}(t)u_{\mathrm{I}}(t)\mathrm{d}t+k\int_0^\tau n(t)u_{\mathrm{I}}(t)\mathrm{d}t\end{aligned}$$

可以证明，上式第二项近似等于零。因此，

$$I(v)\approx kA\varepsilon\cos\beta+k\int_0^\tau n(t)u_{\mathrm{I}}(t)\mathrm{d}t \tag{2.59}$$

类似可得

$$Q(v)\approx-kA\varepsilon\sin\beta+k\int_0^\tau n(t)u_{\mathrm{Q}}(t)\mathrm{d}t \tag{2.60}$$

设 $I(v|\beta)$和 $Q(v|\beta)$分别是在 β 给定条件下的 $I(v)$和 $Q(v)$。由于 $n(t)$是高斯随机过程，故$I(v|\beta)$和 $Q(v|\beta)$也是高斯随机变量，并且可以证明

$$\mathrm{E}\{I(v\mid\beta)Q(v\mid\beta)\}=\mathrm{E}\{I(v\mid\beta)\}\mathrm{E}\{Q(v\mid\beta)\}$$

即 $I(v|\beta)$和 $Q(v|\beta)$是相互独立的随机变量。也可以证明 $I(v|\beta)$和 $Q(v|\beta)$都具有方差 $N_0k^2\varepsilon/2$，并且均值分别为

$$\begin{cases}\mathrm{E}\{I(v\mid\beta)\}=kA\varepsilon\cos\beta\\\mathrm{E}\{Q(v\mid\beta)\}=-kA\varepsilon\sin\beta\end{cases} \tag{2.61}$$

则在观测区间 τ 上，分辨单元中存在信号时，随机变量 $I(v)$和 $Q(v)$在给定 β 条件下的联合概率密度函数为

$$f_{\mathrm{I,Q}}(x,y\mid\beta)=\frac{1}{\pi N_0k^2\varepsilon}\exp\left\{-\frac{1}{N_0k^2\varepsilon}[(x-kA\varepsilon\cos\beta)^2+(y+kA\varepsilon\sin\beta)^2]\right\} \tag{2.62}$$

通过式(2.43)的变换可得，当观测区间 τ 上分辨单元存在信号时，随机变量 $D_1(v)$和 $\Phi(v)$在给定 β 条件下的联合概率密度函数为

$$f_{D_1,\alpha}(r,\phi\mid\beta)=\frac{r}{\pi N_0k^2\varepsilon}\exp\left\{-\frac{1}{N_0k^2\varepsilon}[r^2-2kA\varepsilon r\cos(\phi+\beta)+(kA\varepsilon)^2]\right\} \tag{2.63}$$

对于给定的初始相位 β，$D_1(v)$的概率密度函数为式(2.63)的边缘积分，即

$$f_{D_1}(r\mid\beta)=\int_0^{2\pi}f_{D_1,\alpha}(r,\phi\mid\beta)\mathrm{d}\phi=\frac{2r}{N_0k^2\varepsilon}\exp\left[-\left(\lambda+\frac{r^2}{N_0k^2\varepsilon}\right)\right]I_0\left(\frac{2A}{kN_0}r\right) \tag{2.64}$$

其中，$\lambda=E/N_0$。假设 β 是在区间$[0,2\pi]$上均匀分布的随机变量，则

$$f_{D_1}(r)=\frac{1}{2\pi}\int_0^{2\pi}f_{D_1}(r\mid\beta)\mathrm{d}\beta=\frac{2r}{N_0k^2\varepsilon}\exp\left[-\left(\lambda+\frac{r^2}{N_0k^2\varepsilon}\right)\right]I_0\left(\frac{2A}{kN_0}r\right)\tag{2.65}$$

这是莱斯分布，也称为广义瑞利分布。因此

$$P_\mathrm{d}=\int_S^\infty f_{D_1}(x)\mathrm{d}x=\frac{2}{N_0k^2\varepsilon}\int_S^\infty r\exp\left[-\left(\lambda+\frac{r^2}{N_0k^2\varepsilon}\right)\right]I_0\left(\frac{2A}{kN_0}r\right)\mathrm{d}r$$

设 $z=r\sqrt{\dfrac{2}{N_0k^2\varepsilon}}$，并利用式(2.56)得

$$P_\mathrm{d}=\int_{\sqrt{2\ln(P_\mathrm{fa}^{-1})}}^\infty z\exp\left[-\left(\lambda+\frac{z^2}{2}\right)\right]I_0(z\sqrt{2\lambda})\,\mathrm{d}z\tag{2.66}$$

式(2.66)表明，P_d 是 P_fa 和 λ 的函数。

当接收机匹配于反射信号 $s(t)$时，可以证明 λ 等于匹配滤波器输出的单脉冲信号的平均功率与噪声的平均功率之比。对由式(2.33)给出的具有未知非起伏幅度和均匀分布初始相位的单脉冲信号检测来说，通过对式(2.66)进行数值求解，可以获得在确知恒定的白高斯噪声环境下的(P_d，P_fa，SNR)关系曲线，该曲线被称为对非起伏目标进行单脉冲检测的 Swerling 曲线，或者称为单脉冲检测的 Swerling 0 型曲线。图 2.2 给出了几种 P_fa 值的单脉冲检测的 Swerling 0 型曲线。

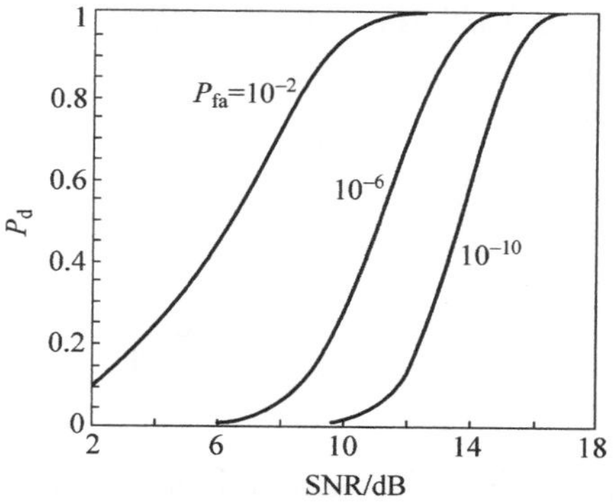

图 2.2 单脉冲检测的 Swerling 0 型曲线

2.3.2 对 Swerling 起伏目标的单脉冲线性检测

由式(2.45)可知，式(2.35)中的似然比可以表示为

$$l(v)=\mathrm{e}^{-\lambda}I_0\left[\frac{2A}{kN_0}D(v)\right]\tag{2.67}$$

式(2.67)是在确知恒定的白高斯噪声中对式(2.34)的具有恒定幅值 A 和均匀分布初始相位信号进行检测的贝叶斯判决准则的似然比表达式。当 A 是一个起伏的随机变量，并且概率密度函数为 $f_A(a)$时，$l(v)$可以表示为

$$l(v)=\int_A\mathrm{e}^{-\varepsilon a^2/N_0}I_0\left[\frac{2a}{kN_0}D(v)\right]f_A(a)\mathrm{d}a\tag{2.68}$$

由于 $I_0(\cdot)$是单调递增函数，且 $f_A(a)$是非负的，因此 $l(v)$是 $D(v)$的单调递增函数，故存在阈值 S 满足 $D(v)\geqslant S$ 与 $l(v)\geqslant l$ 等价，以及 $D(v)<S$ 与 $l(v)<l$ 等价。于是，贝叶斯判决准则可以简化为

$$\begin{cases}D(v)\geqslant S, & \text{判决目标存在}\\ D(v)<S, & \text{判决目标不存在}\end{cases}\tag{2.69}$$

该检测的框图与图 2.1 所示的框图相同。

给定 P_fa，对于信号 $s(t)$进行检测的阈值可以通过求解式(2.70)得到

$$P_\mathrm{fa}=\int_S^\infty f_{D_0}(x)\mathrm{d}x\tag{2.70}$$

在确定 S 以后，(P_d，P_fa，SNR)关系曲线由式(2.71)确定

$$P_{\mathrm{d}}=\int_{S}^{\infty} f_{D_1}(x)\mathrm{d}x \tag{2.71}$$

前面已经说明,$D_0(v)$服从瑞利分布,其概率密度函数见式(2.54)。代入式(2.70)可得

$$P_{\mathrm{fa}}=\int_{S}^{\infty}\frac{2x}{N_0 k^2\varepsilon}\exp\left(-\frac{x^2}{N_0 k^2\varepsilon}\right)\mathrm{d}x=\exp\left(-\frac{S^2}{N_0 k^2\varepsilon}\right)$$

由上式可以解得

$$S=\sqrt{N_0 k^2\varepsilon\ln(P_{\mathrm{fa}}^{-1})}=\sigma\sqrt{2\ln(P_{\mathrm{fa}}^{-1})} \tag{2.72}$$

由于$D_0(v)$和$f_{D_0}(x)$与$s(t)$无关,所以此结果与2.3.1节中获得的对非起伏信号检测的阈值是一致的。

1. 幅值按瑞利分布起伏的情况

下面假设由式(2.34)给出的反射信号$s(t)$的幅值A按瑞利分布起伏,且初始射频相位是在$[0,2\pi]$上均匀分布的随机变量,在此假设下研究在确知恒定的白高斯噪声中对上述形式信号的检测,建立(P_{d},P_{fa},SNR)关系。

当信号$s(t)$在观测区间τ上存在于分辨单元中时,由式(2.65)可知,在给定A条件下$D_1(v)$的概率密度函数为

$$f_{D_1}(r\mid A)=\frac{2r}{N_0 k^2\varepsilon}\exp\left[-\left(\lambda+\frac{r^2}{N_0 k^2\varepsilon}\right)\right]I_0\left(\frac{2A}{kN_0}r\right) \tag{2.73}$$

其中,$\lambda=E/N_0=A^2\varepsilon/N_0$。因此,$f_{D_1}(r)$可以通过计算式(2.73)关于$A$的数学期望而得到,即

$$f_{D_1}(r)=\mathrm{E}_A\{f_{D_1}(r\mid A)\}$$

前面已经假设幅度A是瑞利分布的随机变量,即A的概率密度函数为

$$f_A(a)=\frac{a}{A_0^2}\exp\left(-\frac{a^2}{2A_0^2}\right),\quad a\geqslant 0$$

其中,A_0是常数。因此

$$f_{D_1}(r)=\frac{2r}{N_0 k^2\varepsilon}\exp\left(-\frac{r^2}{N_0 k^2\varepsilon}\right)\int_0^{\infty}\exp\left(-\frac{a^2\varepsilon}{N_0}\right)I_0\left(\frac{2a}{kN_0}r\right)\frac{a}{A_0^2}\exp\left(-\frac{a^2}{2A_0^2}\right)\mathrm{d}a$$

设$y=\dfrac{a}{A_0}$,$\lambda'=\dfrac{2A_0^2\varepsilon}{N_0}$,则

$$f_{D_1}(r)=\frac{2r}{N_0 k^2\varepsilon}\exp\left(-\frac{r^2}{N_0 k^2\varepsilon}\right)\int_0^{\infty}\exp\left(-\frac{\lambda' y^2}{2}\right)I_0\left(\frac{2A_0}{kN_0}ry\right)y\exp\left(-\frac{y^2}{2}\right)\mathrm{d}y \tag{2.74}$$

把式(2.74)代入式(2.71)得检测概率为

$$P_{\mathrm{d}}=\int_{S}^{\infty}\frac{2r}{N_0 k^2\varepsilon}\exp\left(-\frac{r^2}{N_0 k^2\varepsilon}\right)\left[\int_0^{\infty}\exp\left(-\frac{\lambda' y^2}{2}\right)I_0\left(\frac{2A_0}{kN_0}ry\right)y\exp\left(-\frac{y^2}{2}\right)\mathrm{d}y\right]\mathrm{d}r$$

设$x=\dfrac{r}{\varepsilon A_0 k}$,则

$$P_{\mathrm{d}}=\int_{S/(\varepsilon A_0 k)}^{\infty}\lambda' x\exp\left(-\frac{\lambda' x^2}{2}\right)\left\{\int_0^{\infty}y\exp\left[-\frac{(1+\lambda')y^2}{2}\right]I_0(\lambda' xy)\mathrm{d}y\right\}\mathrm{d}x \tag{2.75}$$

可以证明

$$\int_0^{\infty}z^{u-1}\exp(-p^2z^2)I_{\zeta}(\alpha z)\mathrm{d}z=\frac{\Gamma\left(\dfrac{u+\zeta}{2}\right)\left(\dfrac{\alpha}{2p}\right)^{\zeta}}{2p^u\Gamma(\zeta+1)}\exp\left(\frac{\alpha^2}{4p^2}\right)$$

$$_1F_1\left(\frac{\zeta-u}{2}+1,\zeta+1;-\frac{\alpha^2}{4p^2}\right) \tag{2.76}$$

其中，$\Gamma(\cdot)$是伽马函数，$_1\mathrm{F}_1(a,b;c)$是合流超几何函数，它的展开式为

$$_1F_1(a,b;c)=1+\frac{a}{b}c+\frac{a(a+1)}{2!b(b+1)}c^2+\cdots \tag{2.77}$$

将式(2.76)用于式(2.75)的第一重积分，并进行如下变换

$$z=y,\quad u=2,\quad \zeta=0,\quad \alpha=\lambda' x,\quad p=\sqrt{\frac{1+\lambda'}{2}}$$

得到

$$\int_0^{\infty} y\exp\left[-\frac{(1+\lambda')y^2}{2}\right]I_0(\lambda' xy)\mathrm{d}y=\frac{1}{1+\lambda'}\exp\left[\frac{(\lambda' x)^2}{2(1+\lambda')}\right]{}_1F_1\left[0,1;-\frac{(\lambda' x)^2}{2(1+\lambda')}\right]$$

由于$_1F_1(0,1;z)=1$。因此式(2.75)的检测概率为

$$\begin{aligned}P_{\mathrm{d}}&=\int_{S/(\varepsilon A_0 k)}^{\infty}\lambda' x\exp\left(-\frac{\lambda' x^2}{2}\right)\frac{1}{1+\lambda'}\exp\left[\frac{(\lambda' x)^2}{2(1+\lambda')}\right]\mathrm{d}x\\&=\exp\left[-\frac{\lambda' S^2/(\varepsilon A_0 k)^2}{2(1+\lambda')}\right]=\exp\left[\frac{\ln(P_{\mathrm{fa}})}{1+\lambda'}\right]=(P_{\mathrm{fa}})^{1/(1+\lambda')}\end{aligned} \tag{2.78}$$

式(2.78)表明，P_{d}是P_{fa}和λ'的函数，λ'见式(2.74)，由此可以确定(P_{d}，P_{fa}，SNR)关系。

当接收机与接收信号相匹配时，SNR等于λ。其中，λ最初的定义见式(2.35)，即

$$\lambda=\frac{E}{N_0}=\frac{A^2\varepsilon}{N_0}$$

由于幅值A是随机变量，所以λ也是一个随机变量。λ的均值为

$$\mathrm{E}\{\lambda\}=\mathrm{E}_A\left\{\frac{A^2\varepsilon}{N_0}\right\}=\frac{\varepsilon}{N_0}\mathrm{E}_A\{A^2\}=\frac{\varepsilon}{N_0}\int_0^{\infty}\frac{a^3}{A_0^2}\exp\left(-\frac{a^2}{2A_0^2}\right)\mathrm{d}a=\frac{2A_0^2\varepsilon}{N_0}=\lambda' \tag{2.79}$$

因此，式(2.78)中的λ'是检波前的信号与噪声能量之比的均值，它可以作为对λ的估计。因此对式(2.78)进行数值求解，可以获得在确知恒定的白高斯噪声环境中对由式(2.34)给出的具有瑞利起伏幅值和在$[0,2\pi]$上均匀分布的初始相位的反射信号进行检测的典型的(P_{d}，P_{fa}，SNR)曲线，即对瑞利起伏目标进行单脉冲检测的Swerling曲线，或称为单脉冲检测的Swerling Ⅰ或Swerling Ⅱ型曲线。对于几种P_{fa}值，单脉冲检测的Swerling Ⅰ或Swerling Ⅱ型曲线在图2.3中给出。

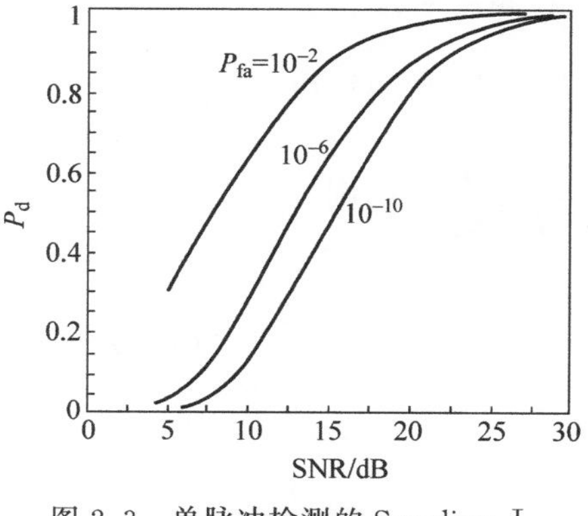

图2.3 单脉冲检测的Swerling Ⅰ或Swerling Ⅱ型曲线

2. 幅值按一主加瑞利分布起伏的情况

假设由式(2.34)给出的反射信号的幅值A按一主加瑞利分布起伏，且射频初始相位是在$[0,2\pi]$上均匀分布的随机变量，在此假设下研究在确知恒定的白高斯噪声中对上述形式的信号所进行的检测，建立(P_{d}，P_{fa}，SNR)关系。

由式(2.73)可知，$D_1(v)$以A为条件的概率密度函数为

$$f_{D_1}(r \mid A)=\frac{2r}{N_0 k^2 \varepsilon}\exp\left[-\left(\lambda+\frac{r^2}{N_0 k^2 \varepsilon}\right)\right] I_0\left(\frac{2A}{kN_0}r\right)$$

假设幅值 A 是一主加瑞利分布的随机变量,其概率密度函数为

$$f_A(a)=\frac{9a^3}{2A_0^4}\exp\left(-\frac{3a^2}{2A_0^2}\right),\quad a \geqslant 0 \tag{2.80}$$

其中,A_0 是常数。因此,$D_1(v)$的概率密度函数为

$$f_{D_1}(r)=\frac{2r}{N_0 k^2 \varepsilon}\exp\left(-\frac{r^2}{N_0 k^2 \varepsilon}\right)\int_0^{\infty}\exp\left(-\frac{a^2 \varepsilon}{N_0}\right) I_0\left(\frac{2a}{kN_0}r\right)\frac{9a^3}{2A_0^4}\exp\left(-\frac{3a^2}{2A_0^2}\right)\mathrm{d}a$$

设 $y=\sqrt{3}a/A_0$,$\lambda'=4A_0^2\varepsilon/(3N_0)$,则

$$f_{D_1}(r)=\frac{r}{N_0 k^2 \varepsilon}\exp\left(-\frac{r^2}{N_0 k^2 \varepsilon}\right)\int_0^{\infty} y^3 \exp\left[-\frac{(1+\lambda'/2)y^2}{2}\right] I_0\left(\frac{2A_0}{\sqrt{3}kN_0}ry\right)\mathrm{d}y$$

利用式(2.76),并且令

$$z=y,\quad u=4,\quad \zeta=0,\quad \alpha=\frac{2A_0}{\sqrt{3}kN_0}r,\quad p=\sqrt{\frac{1}{2}\left(1+\frac{\lambda'}{2}\right)}$$

可得

$$f_{D_1}(r)=\frac{r}{N_0 k^2 \varepsilon}\exp\left(-\frac{r^2}{N_0 k^2 \varepsilon}\right)\frac{2}{(1+\lambda'/2)^2}\exp\left[\frac{\lambda' r^2}{(\lambda'+2)N_0 k^2 \varepsilon}\right]{}_1F_1\left[-1,1;-\frac{\lambda' r^2}{(\lambda'+2)N_0 k^2 \varepsilon}\right]$$

由式(2.77)可得

$${}_1F_1(-1,1;z)=1-z$$

因此

$$f_{D_1}(r)=\frac{2r}{(1+\lambda'/2)^2 N_0 k^2 \varepsilon}\left[1+\frac{\lambda' r^2}{(\lambda'+2)N_0 k^2 \varepsilon}\right]\exp\left[-\frac{2r^2}{(\lambda'+2)N_0 k^2 \varepsilon}\right] \tag{2.81}$$

所以,检测概率为

$$P_{\mathrm{d}}=\int_S^{\infty} f_{D_1}(r)\mathrm{d}r=\frac{2}{(1+\lambda'/2)^2 N_0 k^2 \varepsilon}\int_S^{\infty} r\left[1+\frac{\lambda' r^2}{(\lambda'+2)N_0 k^2 \varepsilon}\right]\exp\left[-\frac{2r^2}{(\lambda'+2)N_0 k^2 \varepsilon}\right]\mathrm{d}r$$

设 $x=\dfrac{r}{k\sqrt{N_0\varepsilon/2}}$,则

$$\begin{aligned}P_{\mathrm{d}}&=\frac{1}{(1+\lambda'/2)^2}\int_{\frac{S}{k\sqrt{N_0\varepsilon/2}}}^{\infty} x\left(1+\frac{\lambda' x^2/2}{\lambda'+2}\right)\exp\left(-\frac{x^2}{\lambda'+2}\right)\mathrm{d}x\\&=\left[1+\frac{\lambda' S^2/2}{(1+\lambda'/2)^2 N_0 k^2 \varepsilon}\right]\exp\left[-\frac{S^2}{(1+\lambda'/2)N_0 k^2 \varepsilon}\right]\end{aligned}$$

将式(2.72)的 $S=\sqrt{N_0 k^2 \varepsilon \ln(P_{\mathrm{fa}}^{-1})}$ 代入上式得

$$\begin{aligned}P_{\mathrm{d}}&=\left[1+\frac{2\lambda'}{(2+\lambda')^2}\ln(P_{\mathrm{fa}}^{-1})\right]\exp\left[\frac{\ln(P_{\mathrm{fa}})}{1+\lambda'/2}\right]\\&=\left[1+\frac{2\lambda'}{(2+\lambda')^2}\ln(P_{\mathrm{fa}}^{-1})\right](P_{\mathrm{fa}})^{\frac{1}{1+\lambda'/2}}\end{aligned} \tag{2.82}$$

式(2.82)表明 P_{d} 是 P_{fa} 和 λ'的函数。由于幅值 A 是随机变量,所以

$$\lambda=\frac{E}{N_0}=\frac{A^2\varepsilon}{N_0}$$

也是一个随机变量。λ 的均值为

$$\mathrm{E}\{\lambda\}=\mathrm{E}_A\left\{\frac{A^2\varepsilon}{N_0}\right\}=\frac{\varepsilon}{N_0}\int_0^{\infty}\frac{9a^5}{2A_0^4}\exp\left(-\frac{3a^2}{2A_0^2}\right)\mathrm{d}a=\frac{4A_0^2\varepsilon}{3N_0}=\lambda' \tag{2.83}$$

因此，λ'可以作为对 λ 的估计。前面已经说明，当接收机与接收信号相匹配时，关系(P_d，P_{fa}，SNR)与关系(P_d，P_{fa}，λ)是等价的，因此式(2.82)可以用来研究(P_d，P_{fa}，SNR)关系。

对式(2.82)进行数值求解，可以获得在确知恒定的白高斯噪声环境中对式(2.34)给出的具有一主加瑞利起伏幅值和在$[0,2\pi]$上均匀分布的初始相位的反射信号进行检测的(P_d，P_{fa}，SNR)曲线，即对一主加瑞利起伏目标进行单脉冲检测的 Swerling 曲线，或称为单脉冲检测的 Swerling Ⅲ或 Swerling Ⅳ型曲线。对于几种 P_{fa} 值，单脉冲检测的 Swerling Ⅲ或 Swerling Ⅳ型曲线在图 2.4 中给出。对于 Swerling 起伏目标的单脉冲检测的结论在很多文献中都有论述，如文献[10]。

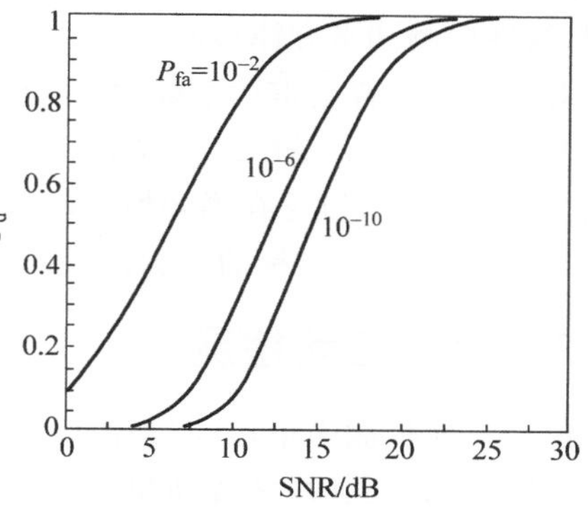

图 2.4　单脉冲检测的 Swerling Ⅲ或 Swerling Ⅳ型曲线

2.4　多脉冲检测

Marcum[4]于 1947 年开始了对多脉冲检测的研究。在低信噪比情况下，他证明了在接收机噪声中检测非起伏非相参脉冲串的准最优检测策略具有平方律检测器的形式，并且得到了相应的 Swerling 曲线。

Swerling[5]将研究扩展到了在确知恒定的接收机噪声中检测 Swerling 起伏非相参脉冲串的准最优检测策略。在一般情况下，至少对于低信噪比，检测非起伏或 Swerling 起伏非相参脉冲串的准最优检测策略具有平方律检测器形式。

在确知恒定的接收机噪声中和大信噪比情况下检测非起伏非相参脉冲串，多脉冲线性检测器是准最优的[8]。对于在接收机噪声中对非起伏或 Swerling 起伏目标的多脉冲检测，线性检测和平方律检测的性能非常接近，差别大约为 0.2dB[4-5]，并且当脉冲积累数 n 较小时，平方律检测实际上只比线性检测稍差。

在接收机噪声中，对于检测与 n 个非相参脉冲 $s_i(t)(i=1,2,\cdots,n)$ 相对应的信号，多脉冲的线性检测和平方律检测都具有单脉冲线性检测的一般形式，即

$$\begin{cases}S(v)\geqslant S, & \text{判决目标存在}\\ S(v)<S, & \text{判决目标不存在}\end{cases}$$

对于多脉冲检测，$\boldsymbol{v}=\{v_1,v_2,\cdots,v_n\}$是从各个单脉冲观测区间取得的一组观测，检测统计量 $S(\boldsymbol{v})$是各个单脉冲检测统计量的函数。设 $D_i(v_i)(i=1,2,\cdots,n)$是 n 个由式(2.43)定义的单脉冲检测统计量，它们均由相应信号 $s_i(t)$的观测 $v_i(t)$形成。因此对于多脉冲平方律检测，总的检测统计量为

$$S(\boldsymbol{v})=\sum_{i=1}^{n}D_i^2(v_i)$$

对于多脉冲线性检测，总的检测统计量为

$$S(\boldsymbol{v})=\sum_{i=1}^{n}D_i(v_i)$$

对于所有 Swerling 模型，虽然平方律检测是准最优的多脉冲检测策略，但是实际上以前很少使用它，这主要是由于处理平方值困难。若在接收机噪声中检测非起伏或 Swerling 起伏目标，对于单脉冲检测，线性检测和平方律检测具有相同的性能，对于多脉冲检测，它们之间的特性也非常相似。

对于脉冲多普勒雷达系统，线性检测要求的数据存储量可能很大。因此，线性检测通常不用于 CPI 间的非相参积累，而是用二元检测，因为准最优的二元检测在非瑞利包络杂波环境中通常能提供更好的性能。2.4.1 节将讨论在确知恒定的接收机噪声中对非起伏或 Swerling 起伏目标检测的二元检测。对各种 Swerling 起伏目标的多脉冲线性检测将在 2.4.2 节中讨论。

2.4.1 二元检测

由于二元检测简单，且与其他检测策略相比具有相对较好的性能，以及具有处理大量分辨单元的能力，因此它成为在脉冲多普勒和 MTD 雷达系统中广泛应用的检测策略。使其广泛应用的另一个特性是它在非高斯环境中的性能。

在每个分辨角度上发射一组脉冲(或相参脉冲串)$s_i(t)(i=1,2,\cdots,n)$的情况下，通常使用二元检测在脉冲或脉冲串间做积累。为了检测信号 $s_i(t)$，在每个分辨单元上按照式(2.43)形成检测统计量 $D_i(v_i)(i=1,2,\cdots,n)$，其中 $v_i(t)$是在该分辨单元中对第 i 个脉冲 $s_i(t)$的观测。二元检测在每个分辨单元上对接收信号进行两级检验，其原理框图见图 2.5。第一级检验与 2.3 节中讨论的在接收机噪声中对非起伏目标或 Swerling 起伏目标的单脉冲线性检测相同。后者在贝叶斯意义上是最优的，因此第一级检验也是最优的。第二级检验在每个分辨单元上将第一级检验超过第一级阈值的数目与第二级阈值(一个整数 M，$1\leqslant M\leqslant n$)进行比较。

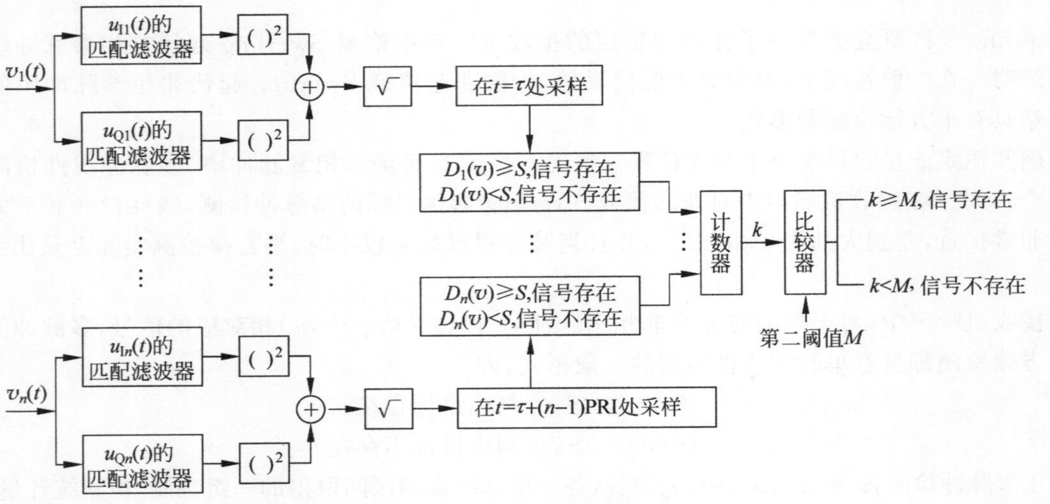

图 2.5 二元检测的原理框图

将两级检验结合，就得到利用某个分辨单元上一组 n 个非相参脉冲信号对应的 n 个观测的集合，在接收机噪声中对非起伏或 Swerling 起伏目标进行检测的二元检测策略可以表示为

$$\begin{cases} \sum_{i=1}^{n} \operatorname{sgn}[D_i(v_i) - S_i] \geqslant M, & \text{判决目标存在} \\ \sum_{i=1}^{n} \operatorname{sgn}[D_i(v_i) - S_i] < M, & \text{判决目标不存在} \end{cases} \tag{2.84}$$

其中，$\operatorname{sgn}(\cdot)$表示符号函数。当分辨单元中不存在目标信号时，检测统计量 $D_i(v_i)(i=1,2,\cdots,n)$即为杂波包络，用 $D_{0i}(v_i)(i=1,2,\cdots,n)$表示，通常可以假设它们是独立同分布的。因此，设阈值 $S_1=S_2=\cdots=S_n=S$。用 $P_{\mathrm{fa,sp}i}$ 表示第 i 个观测在第一级检验产生的虚警概率，相应的检测概率用 $P_{\mathrm{d,sp}i}$ 表示。所以

$$P_{\mathrm{fa,sp1}} = P_{\mathrm{fa,sp2}} = \cdots = P_{\mathrm{fa,sp}n} = P_{\mathrm{fa,sp}} = \int_S^{\infty} f_{D_0}(x)\mathrm{d}x$$

其中，$f_{D_0}(x)$是 $D_{0i}(v_i)(i=1,2,\cdots,n)$的概率密度函数。$P_{\mathrm{fa,sp}i}$ 和 $P_{\mathrm{d,sp}i}$ 以及单脉冲信噪比之间的关系可以直接利用 2.3 节中的结论，即在接收机噪声中对非起伏或 Swerling 起伏目标进行单脉冲检测的结论，可以参考图 2.2～图 2.4 所示的 Swerling 曲线。

二元检测的虚警概率 P_{fa} 和检测概率 P_{d} 对应的检测是指对一组 n 个观测 $s_i(t)(i=1,2,\cdots,n)$的检测。下面建立 P_{fa} 和 P_{d} 与单个观测的虚警概率($P_{\mathrm{fa,sp}}$)和检测概率($P_{\mathrm{d,sp}}$)之间的关系。对于二元检测，通常做以下两个假设。

(1) 事件$\{D_1(v_1)\geqslant S_1\}$，$\{D_2(v_2)\geqslant S_2\}$，…，$\{D_n(v_n)\geqslant S_n\}$是独立事件。

(2) $P_{\mathrm{fa,sp1}}=P_{\mathrm{fa,sp2}}=\cdots=P_{\mathrm{fa,sp}n}=P_{\mathrm{fa,sp}}$ 和 $P_{\mathrm{d,sp1}}=P_{\mathrm{d,sp2}}=\cdots=P_{\mathrm{d,sp}n}=P_{\mathrm{d,sp}}$。

因此，在分辨单元中不存在目标的条件下，n 个第一级检验中至少有 M 个使$\{D_i(v_i)\geqslant S_i\}$事件发生的概率，即二元检测的虚警概率为

$$\begin{aligned} P_{\mathrm{fa}} &= \Pr\left\{ \sum_{i=1}^{n} \operatorname{sgn}[D_i(v_i) - S_i] \geqslant M \mid \text{目标不存在} \right\} \\ &= 1 - \sum_{k=0}^{M-1} \Pr\left\{ \sum_{i=1}^{n} \operatorname{sgn}[D_i(v_i) - S_i] = k \mid \text{目标不存在} \right\} \end{aligned} \tag{2.85}$$

由于

$$P_{\mathrm{fa,sp}} = \Pr[D_i(v_i) \geqslant S \mid \text{目标不存在}]$$

并且根据假设(1)可得，在分辨单元中不存在目标的条件下，在 n 次观测中$\{D_i(v_i)\geqslant S_i\}$事件发生 k 次的概率为

$$\frac{n!}{k!(n-k)!}(P_{\mathrm{fa,sp}})^k(1-P_{\mathrm{fa,sp}})^{n-k}$$

因此二元检测的虚警概率为

$$P_{\mathrm{fa}} = 1 - \sum_{k=0}^{M-1} \frac{n!}{k!(n-k)!}(P_{\mathrm{fa,sp}})^k(1-P_{\mathrm{fa,sp}})^{n-k} \tag{2.86}$$

类似地，在分辨单元中存在目标的条件下，n 个第一级检验中至少有 M 个使$\{D_i(v_i)\geqslant S_i\}$事件发生的概率，即二元检测的检测概率为

$$P_{\mathrm{d}} = 1 - \sum_{k=0}^{M-1} \frac{n!}{k!(n-k)!}(P_{\mathrm{d,sp}})^k(1-P_{\mathrm{d,sp}})^{n-k} \tag{2.87}$$

给定 M、n、虚警概率和检测概率，解式(2.86)和式(2.87)可以获得相应的 $P_{\mathrm{fa,sp}}$ 和 $P_{\mathrm{d,sp}}$ 值。

2.4.2 线性检测

假设 $s_i(t)(i=1,2,\cdots,n)$是式(2.34)定义的 n 个持续时间为 τ 的接收信号,$v_i(t)(i=1,2,\cdots,n)$是式(2.7)定义的 n 个观测,其中 $n_i(t)(i=1,2,\cdots,n)$是噪声。定义向量

$$\boldsymbol{s}(t)=[s_1(t),\cdots,s_n(t)]$$

$$\boldsymbol{v}(t)=[v_1(t),\cdots,v_n(t)]$$

$$\boldsymbol{n}(t)=[n_1(t),\cdots,n_n(t)]$$

在有目标情况下

$$\boldsymbol{v}(t)=\boldsymbol{s}(t)+\boldsymbol{n}(t)=\boldsymbol{v}_1(t)$$

在没有目标情况下

$$\boldsymbol{v}(t)=\boldsymbol{n}(t)=\boldsymbol{v}_0(t)$$

因为观测向量 $\boldsymbol{v}(t)$是 $\boldsymbol{v}_0(t)$和 $\boldsymbol{v}_1(t)$二者之一,则由式(2.43)可知,由 $v_i(t)$形成的检测统计量 $D_i(v_i)$可以被认为是两个随机变量 $D_{0i}(v_i)$和 $D_{1i}(v_i)$之一。其中,$D_{0i}(v_i)$是 $D_i(v_i)$在分辨单元中没有目标情况下的样本,$D_{1i}(v_i)$是 $D_i(v_i)$在分辨单元中有目标情况下的样本。

定义

$$LD(\boldsymbol{v})=D_1(v_1)+D_2(v_2)+\cdots+D_n(v_n)$$

$LD(\boldsymbol{v})$是两个随机变量 $LD_0(\boldsymbol{v})$和 $LD_1(\boldsymbol{v})$之一。其中,$LD_0(\boldsymbol{v})$是 $LD(\boldsymbol{v})$在所有 $D_i(v_i)(i=1,2,\cdots,n)$中都不存在 $s_i(t)$条件下的样本,$LD_1(\boldsymbol{v})$是 $LD(\boldsymbol{v})$在所有 $D_i(v_i)(i=1,2,\cdots,n)$中都存在$s_i(t)$条件下的样本。$LD(\boldsymbol{v})$被定义为线性检测的检测统计量。因此,线性检测的判决准则为

$$\begin{cases} LD(\boldsymbol{v})\geqslant S, & \text{对于所有 } i,\text{判决 } s_i(t) \text{ 在 } D_i(v_i) \text{ 中存在} \\ LD(\boldsymbol{v})<S, & \text{对于所有 } i,\text{判决 } s_i(t) \text{ 在 } D_i(v_i) \text{ 中不存在} \end{cases} \tag{2.88}$$

其中,S 是检测阈值,可以通过解如下方程获得

$$P_{\mathrm{fa}}=\int_S^{\infty} f_{LD_0}(x)\mathrm{d}x$$

其中,$f_{LD_0}(x)$是随机变量 $LD_0(\boldsymbol{v})$的概率密度函数。图 2.6 给出了多脉冲线性检测器的原理框图。

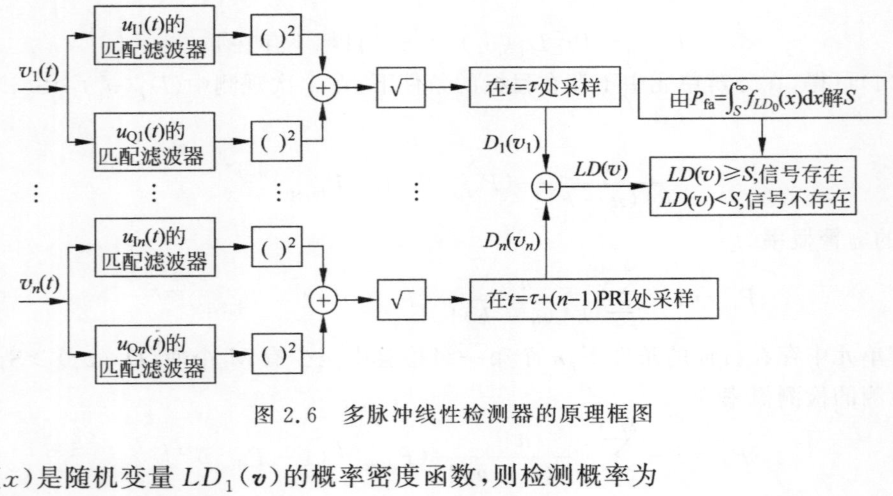

图 2.6 多脉冲线性检测器的原理框图

设 $f_{LD_1}(x)$是随机变量 $LD_1(\boldsymbol{v})$的概率密度函数,则检测概率为

$$P_{\mathrm{d}}=\int_{S}^{\infty} f_{LD_1}(x)\,\mathrm{d}x$$

进而可以建立(P_{d},P_{fa},SNR)关系。

概率密度函数 $f_{LD_1}(x)$的形式取决于对 $n(t)$和 $s_i(t)(i=1,2,\cdots,n)$的统计特性的假设。一方面,由于统计特性对每种 Swerling 模型都是不同的,所以 $f_{LD_1}(x)$的形式也不相同,因此需要对每种模型建立(P_{d},P_{fa},SNR)关系。另一方面,$f_{LD_0}(x)$仅取决于 $n(t)$的统计特性,因此对于在接收机噪声中的检测,$f_{LD_0}(x)$对于所有 Swerling 模型都是相同的。

在本章中,若称信号存在于检测单元中,意指对于所有 $i\in\{i=1,2,\cdots,n\}$,$s_i(t)$都存在于 $D_i(v_i)(i=1,2,\cdots,n)$中。相反,若称信号不存在于检测单元中,意指对于所有 $i\in\{i=1,2,\cdots,n\}$,$s_i(t)$都不存在于 $D_i(v_i)(i=1,2,\cdots,n)$中。下面将在不同 Swerling 模型条件下推导概率密度函数 $f_{LD_1}(x)$和 $f_{LD_0}(x)$的表达式,研究检测阈值和(P_{d},P_{fa},SNR)关系。

1. 确定检测阈值

设 $f_{D_{0i}}(x)$和 $C_{D_{0i}}(t)$分别表示随机变量 $D_{0i}(v_i)$的概率密度函数和特征函数,$C_{LD_0}(t)$表示 $LD_0(\boldsymbol{v})$的特征函数。由于 $D_{0i}(v_i)(i=1,2,\cdots,n)$是统计独立同分布的,因此由 $LD_0(\boldsymbol{v})$的定义可知

$$C_{LD_0}(t)=[C_{D_{0i}}(t)]^n=\left[\int_0^{+\infty} f_{D_{0i}}(x)\mathrm{e}^{\mathrm{j}tx}\,\mathrm{d}x\right]^n$$

因此

$$f_{LD_0}(x)=\frac{1}{2\pi}\int_{-\infty}^{+\infty} C_{LD_0}(t)\mathrm{e}^{-\mathrm{j}tx}\,\mathrm{d}t \tag{2.89}$$

对于式(2.88)的线性检测策略,虚警概率为

$$P_{\mathrm{fa}}=\int_S^{\infty}\frac{1}{2\pi}\int_{-\infty}^{+\infty}\left[\int_0^{+\infty} f_{D_{0i}}(r)\mathrm{e}^{\mathrm{j}tr}\,\mathrm{d}r\right]^n\mathrm{e}^{-\mathrm{j}tx}\,\mathrm{d}t\,\mathrm{d}x$$

由式(2.54)可知

$$f_{D_{0i}}(x)=\frac{2x}{N_0k^2\varepsilon}\exp\left[-\frac{x^2}{N_0k^2\varepsilon}\right],\quad x\geqslant 0 \tag{2.90}$$

因此

$$P_{\mathrm{fa}}=\int_S^{\infty}\frac{1}{2\pi}\int_{-\infty}^{+\infty}\left[\int_0^{+\infty}\frac{2r}{N_0k^2\varepsilon}\exp\left(-\frac{r^2}{N_0k^2\varepsilon}\right)\mathrm{e}^{\mathrm{j}tr}\,\mathrm{d}r\right]^n\mathrm{e}^{-\mathrm{j}tx}\,\mathrm{d}t\,\mathrm{d}x \tag{2.91}$$

令

$$y=\frac{r}{k\sqrt{N_0\varepsilon/2}},\quad u=tk\sqrt{N_0\varepsilon/2},\quad z=\frac{x}{k\sqrt{N_0\varepsilon/2}}$$

则

$$\begin{aligned}P_{\mathrm{fa}}&=\int_{\frac{S}{k\sqrt{N_0\varepsilon/2}}}^{\infty}\frac{1}{2\pi}\int_{-\infty}^{+\infty}\left[\int_0^{+\infty} y\exp\left(-\frac{y^2}{2}\right)\mathrm{e}^{\mathrm{j}uy}\,\mathrm{d}y\right]^n\mathrm{e}^{-\mathrm{j}uz}\,\mathrm{d}u\,\mathrm{d}z\\&=q_n\left(\frac{S}{k\sqrt{N_0\varepsilon/2}}\right)\end{aligned} \tag{2.92}$$

所以

$$S=q_n^{-1}(P_{\mathrm{fa}})k\sqrt{N_0\varepsilon/2} \tag{2.93}$$

其中,$q_n^{-1}(\cdot)$是 $q_n(\cdot)$的逆函数。

由式(2.90)可得

$$\mathrm{E}\{D_{0i}(v_i)\}=\int_0^{\infty}\frac{2r^2}{N_0k^2\varepsilon}\exp\left(-\frac{r^2}{N_0k^2\varepsilon}\right)\mathrm{d}r=k\sqrt{\frac{\pi N_0\varepsilon}{4}}$$

因此,式(2.93)可以表示为

$$S=\mathrm{E}\{D_{0i}(v_i)\}\sqrt{\frac{2}{\pi}}q_n^{-1}(P_{\mathrm{fa}}) \tag{2.94}$$

这个结果与在接收机噪声中最优单脉冲检测的结果类似(见式(2.57)),检测阈值也等于检测统计量$D(\boldsymbol{v})$的期望值乘以P_{fa}的一个函数。该结果对于在瑞利包络杂波条件下建立多脉冲线性检测的CFAR检测是有用的。

2. Swerling 0 型目标的(P_{d},P_{fa},SNR)关系

若目标的一组n个反射信号服从Swerling 0型模型,就是假设每个信号具有相同的未知非起伏幅值A和均匀分布的初始相位。如果$D_i(\boldsymbol{v}_i)$是对脉冲串中单个脉冲的检测统计量,那么对于所有$i\in\{1,2,\cdots,n\}$,由式(2.65)可得

$$f_{D_{1i}}(r)=\frac{2r}{N_0k^2\varepsilon}\exp\left[-\left(\lambda+\frac{r^2}{N_0k^2\varepsilon}\right)\right]I_0\left(\frac{2A}{kN_0}r\right) \tag{2.95}$$

其中,λ、N_0、ε和k的定义同2.3节。

设$f_{D_{1i}}(x)$和$C_{D_{1i}}(t)$分别是随机变量$D_{1i}(v_i)$的概率密度函数和特征函数,$C_{LD_{1i}}(t)$是$LD_1(\boldsymbol{v})$的特征函数。由于$D_{1i}(v_i)(i=1,2,\cdots,n)$是独立同分布的,因此

$$C_{LD_1}(t)=[C_{D_{1i}}(t)]^n=\left[\int_0^{+\infty}f_{D_{1i}}(x)\mathrm{e}^{\mathrm{j}tx}\mathrm{d}x\right]^n$$

因此

$$f_{LD_1}(x)=\frac{1}{2\pi}\int_{-\infty}^{+\infty}C_{LD_1}(t)\mathrm{e}^{-\mathrm{j}tx}\mathrm{d}t$$

则多脉冲线性检测的检测概率P_{d}为

$$P_{\mathrm{d}}=\int_S^{\infty}\frac{1}{2\pi}\int_{-\infty}^{+\infty}\left[\int_0^{+\infty}f_{D_{1i}}(r)\mathrm{e}^{\mathrm{j}tr}\mathrm{d}r\right]^n\mathrm{e}^{-\mathrm{j}tx}\mathrm{d}t\,\mathrm{d}x \tag{2.96}$$

因此

$$P_{\mathrm{d}}=\int_S^{\infty}\frac{1}{2\pi}\int_{-\infty}^{+\infty}\left[\int_0^{+\infty}\frac{2r}{N_0k^2\varepsilon}\exp\left[-\left(\lambda+\frac{r^2}{N_0k^2\varepsilon}\right)\right]I_0\left(\frac{2A}{kN_0}r\right)\mathrm{e}^{\mathrm{j}tr}\mathrm{d}r\right]^n\mathrm{e}^{-\mathrm{j}tx}\mathrm{d}t\,\mathrm{d}x$$

令

$$y=\frac{r}{k\sqrt{N_0\varepsilon/2}},\quad u=tk\sqrt{N_0\varepsilon/2},\quad z=\frac{x}{k\sqrt{N_0\varepsilon/2}}$$

则

$$\begin{aligned}P_{\mathrm{d}}&=\int_{\frac{S}{k\sqrt{N_0\varepsilon/2}}}^{\infty}\frac{1}{2\pi}\int_{-\infty}^{+\infty}\left[\int_0^{+\infty}y\exp\left[-\left(\lambda+\frac{y^2}{2}\right)\right]I_0(y\sqrt{2\lambda})\mathrm{e}^{\mathrm{j}uy}\mathrm{d}y\right]^n\mathrm{e}^{-\mathrm{j}uz}\mathrm{d}u\,\mathrm{d}z\\&=g_n\left(\lambda,\frac{S}{k\sqrt{N_0\varepsilon/2}}\right)\end{aligned} \tag{2.97a}$$

因此由式(2.93)得

$$P_{\mathrm{d}}=g_n[\lambda,q_n^{-1}(P_{\mathrm{fa}})] \tag{2.97b}$$

在评价多脉冲线性检测器性能时通常用数值方法求解式(2.92)和式(2.97a)。如上所述，当接收机与每个接收脉冲相匹配时，λ 等于信噪比。

Swerling 曲线是在平方律检测器的情况下推导出的，上述关于线性检测的(P_d，P_{fa}，SNR)关系不完全等价于在接收机噪声中对非起伏目标进行多脉冲检测的 Swerling 0 型曲线。然而有文献证明[4-5]，平方律检测和线性检测间的信噪比差别低于 0.2dB。对于几个主要的 P_{fa} 和 n 值，图 2.7 和图 2.8 给出了对 Swerling 0 型目标的多脉冲线性检测的性能曲线。

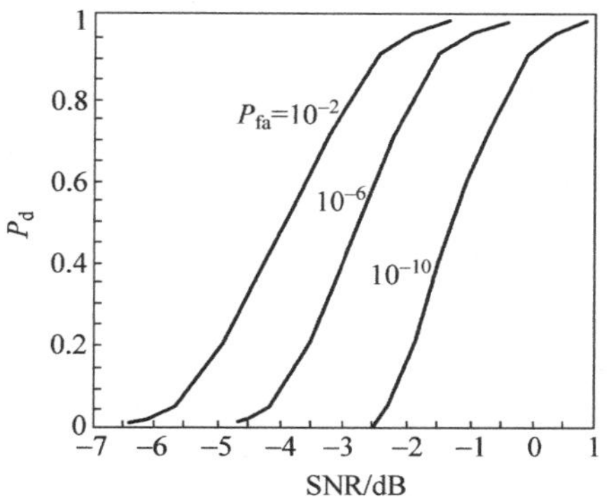

图 2.7 Swerling 0 型目标——多脉冲线性检测($n=100$)的检测性能曲线

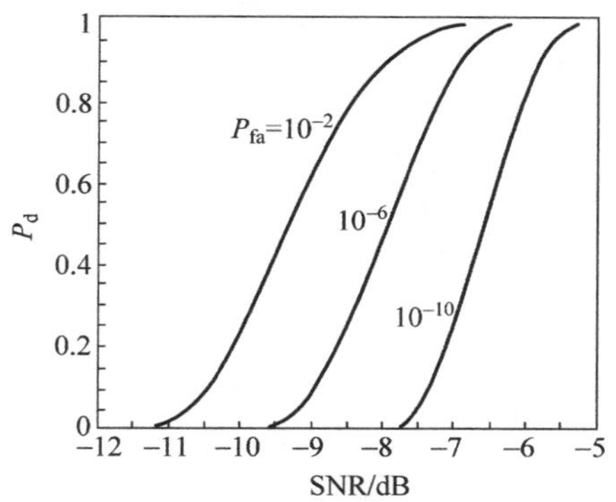

图 2.8 Swerling 0 型目标——多脉冲线性检测($n=1000$)的检测性能曲线

3. Swerling Ⅰ型目标的(P_d，P_{fa}，SNR)关系

若目标的一组 n 个反射信号服从 SwerlingⅠ型模型，则脉冲串中每个脉冲都具有相等的幅值 A，并且 A 是服从瑞利分布的随机变量。由于式(2.97)不依赖于关于 A 的假设，因此此式在 A 是瑞利分布的随机变量时同样成立。并且由于 $\lambda=E/N_0=A^2\varepsilon/N_0$ 是 A 的函数，因此 λ 也是一个随机变量。与 2.3 节中讨论的方法相同，λ 的期望值 λ' 可以作为对 λ 的估计，于是式(2.97a)和式(2.97b)可以用来建立(P_d，P_{fa}，λ')关系。如前所述，当接收机与每个接收的脉冲信号相匹配时，λ 等于信噪比(SNR)，即(P_d，P_{fa}，SNR)关系与(P_d，P_{fa}，λ)关系等价。因此，可以利用(P_d，P_{fa}，λ')关系来研究(P_d，P_{fa}，SNR)关系。

当目标存在时，随机变量 $LD(\boldsymbol{v})$ 的概率密度函数是以 A 为条件的条件概率密度函数，用 $f_{LD_1}(x|A)$ 表示。当 A 是瑞利分布的随机变量时，$LD(\boldsymbol{v})$ 的概率密度函数 $f_{LD_1}(x)$ 等于 $f_{LD_1}(x|A)$ 关于 A 的期望值。于是，对 Swerling Ⅰ型起伏目标的多脉冲线性检测的检测概率为

$$P_d=\int_S^{\infty}\int_A f_{LD_1}(x\mid a)f_A(a)\,\mathrm{d}a\,\mathrm{d}x=\int_A\int_S^{\infty}f_{LD_1}(x\mid a)\,\mathrm{d}x f_A(a)\,\mathrm{d}a$$

其中，$\int_S^{\infty}f_{LD_1}(x\mid a)\mathrm{d}x$ 由式(2.97)给出。于是

$$P_d=\mathrm{E}_A\left\{g_n\left(\lambda,\frac{S}{k\sqrt{N_0\varepsilon/2}}\right)\right\}=\int_0^{\infty}g_n\left(\frac{\varepsilon}{N_0}a^2,\frac{S}{k\sqrt{N_0\varepsilon/2}}\right)\frac{a}{A_0^2}\exp\left(-\frac{a^2}{2A_0^2}\right)\mathrm{d}a$$

令 $x=\dfrac{a}{A_0}$，得

$$\begin{aligned}P_d&=\int_0^{\infty}g_n\left(\frac{\varepsilon}{N_0}A_0^2x^2,\frac{S}{k\sqrt{N_0\varepsilon/2}}\right)x\exp\left(-\frac{x^2}{2}\right)\mathrm{d}x\\&=h_n\left(\lambda',\frac{S}{k\sqrt{N_0\varepsilon/2}}\right)=h_n[\lambda',q_n^{-1}(P_{fa})]\end{aligned}\tag{2.98}$$

其中，$\lambda' = \mathrm{E}_A\{\lambda\} = 2A^2\varepsilon/N_0$。

在给定的关于目标幅值起伏和背景噪声的假设下，通常用数值方法求解式(2.98)，以获得根据式(2.88)进行检测的(P_d，P_fa，SNR)关系。这个关系是关于线性检测的，近似等于平方律检测。对于Swerling Ⅰ型目标，图2.9和图2.10给出了几个主要P_fa和n值情况下的多脉冲平方律检测的检测性能曲线。

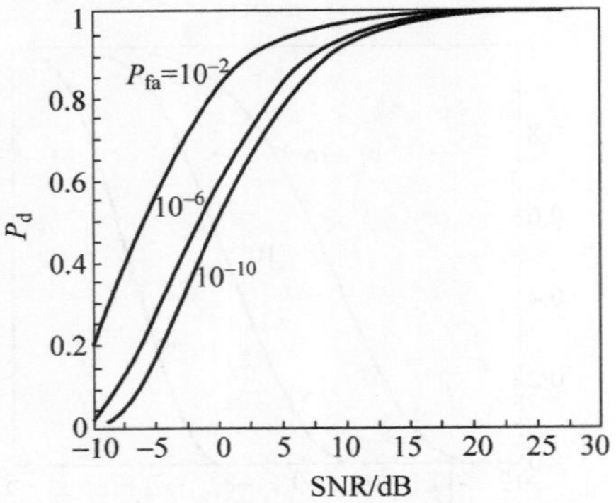

图2.9 Swerling Ⅰ型目标——多脉冲平方律检测($n=100$)的检测性能曲线

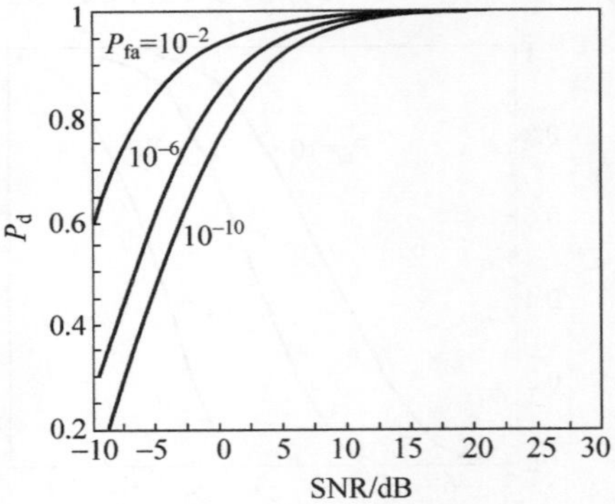

图2.10 Swerling Ⅰ型目标——多脉冲平方律检测($n=1000$)的检测性能曲线

4. Swerling Ⅱ型起伏目标的(P_d，P_fa，SNR)关系

若目标的一组n个反射信号服从Swerling Ⅱ型模型，则非相参脉冲串中每个脉冲的幅值$A_i(i=1,2,\cdots,n)$可以假设是相互统计独立的瑞利分布的随机变量。

设$D_i(v_i)$是对脉冲串中第i个脉冲进行检测的单脉冲检测统计量，见式(2.43)的定义。由于$A_i(i=1,2,\cdots,n)$是瑞利分布的随机变量，则$D_{1i}(v_i)$的概率密度函数$f_{D_{1i}}(r)$由式(2.74)给出，利用式(2.76)可得

$$f_{D_{1i}}(r) = \frac{\lambda' r}{(1+\lambda')A_0^2k^2\varepsilon^2}\exp\left[-\frac{\lambda' r^2}{2(1+\lambda')A_0^2k^2\varepsilon^2}\right] \tag{2.99}$$

其中

$$\lambda' = \frac{2A_0^2\varepsilon}{N_0} \tag{2.100}$$

因此，检测统计量$D_{1i}(v_i)(i=1,2,\cdots,n)$是具有参数$A_0k\varepsilon\sqrt{(1+\lambda')/\lambda'}$的独立同分布的瑞利分布的随机变量。

由式(2.96)得多脉冲线性检测的检测概率为

$$\begin{aligned}P_\mathrm{d} &= \int_S^\infty \frac{1}{2\pi}\int_{-\infty}^{+\infty}\left[\int_0^{+\infty} f_{D_{1i}}(r)\mathrm{e}^{\mathrm{j}tr}\,\mathrm{d}r\right]^n \mathrm{e}^{-\mathrm{j}tx}\,\mathrm{d}t\,\mathrm{d}x \\ &= \int_S^\infty \frac{1}{2\pi}\int_{-\infty}^{+\infty}\left[\int_0^{+\infty} \frac{\lambda' r}{(1+\lambda')A_0^2k^2\varepsilon^2}\exp\left(-\frac{\lambda' r^2}{2(1+\lambda')A_0^2k^2\varepsilon^2}\right)\mathrm{e}^{\mathrm{j}tr}\,\mathrm{d}r\right]^n \mathrm{e}^{-\mathrm{j}tx}\,\mathrm{d}t\,\mathrm{d}x\end{aligned}$$

用$\dfrac{1+\lambda'}{\lambda'}A_0^2k^2\varepsilon^2$置换式(2.91)中的$N_0k^2\varepsilon/2$时，则上式与式(2.91)相同。因此，采用式(2.92)中定义的函数$q_n(\cdot)$可以将检测概率表示为

$$P_{\mathrm{d}}=q_n\left(\frac{S}{A_0 k\varepsilon}\sqrt{\frac{\lambda'}{1+\lambda'}}\right)=q_n\left[\frac{1}{\sqrt{1+\lambda'}}q_n^{-1}(P_{\mathrm{fa}})\right] \tag{2.101}$$

当接收机与脉冲串中的每个脉冲匹配时，检波前脉冲串中第 i 个脉冲与噪声的能量之比 λ_i 等于检波输出的平均信号能量与平均噪声能量之比。而且式(2.79)已经证明 λ'是 λ 的均值。由于非相参脉冲串中每个脉冲都具有同分布的瑞利起伏幅值，因此对于脉冲串中的每个脉冲，λ'是检波前的单脉冲信号与噪声的能量之比的均值，并且对所有 i，λ'是 λ_i 的估计，λ'同时也是 SNR 的估计。所以，由式(2.101)的(P_{d}，P_{fa}，λ')关系得到(P_{d}，P_{fa}，SNR)关系。这个关系是关于线性检测的，近似等于平方律检测。对于 Swerling Ⅱ型目标，以及几个不同 P_{fa} 和 n 值，图 2.11 和图 2.12 给出了多脉冲平方律检测的检测性能曲线。

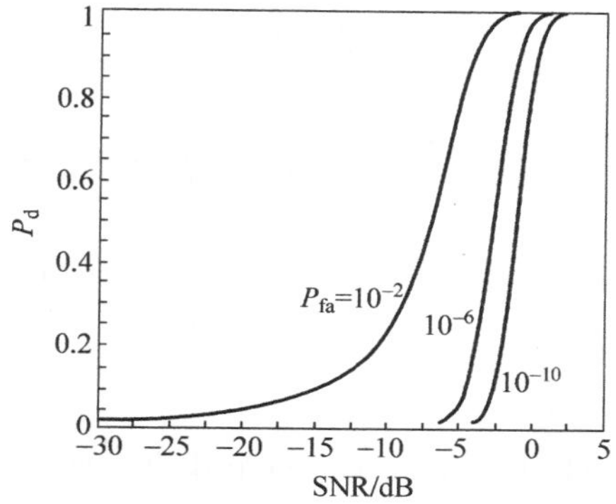

图 2.11 Swerling Ⅱ型目标——多脉冲平方检测($n=100$)的检测性能曲线

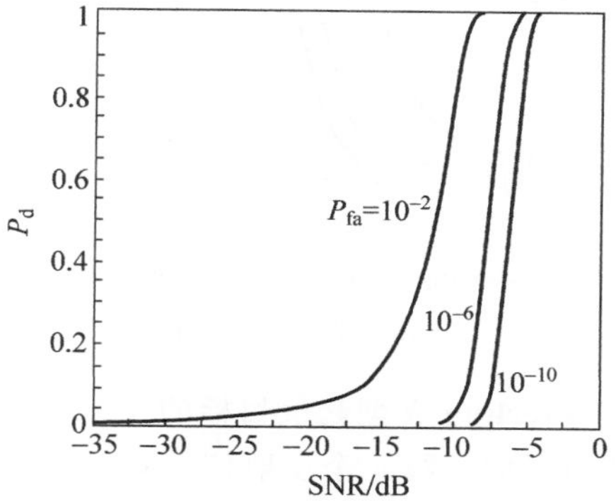

图 2.12 Swerling Ⅱ型目标——多脉冲平方律检测($n=1000$)的检测性能曲线

5. Swerling Ⅲ型起伏目标的(P_{d}，P_{fa}，SNR)关系

若目标的一组 n 个反射信号服从 Swerling Ⅲ型模型，则可以假设脉冲串中每个脉冲具有相等的幅值 A，并且 A 是服从一主加瑞利分布的随机变量。式(2.83)已经证明 $\lambda'=4A_0^2\varepsilon/(3N_0)$是 λ 的均值，于是 λ'可以作为 λ 的估计。如前所述，式(2.97)仍然可以用来建立(P_{d}，P_{fa}，λ')关系，以获得(P_{d}，P_{fa}，SNR)关系。

当目标存在时，随机变量 $LD(\boldsymbol{v})$的概率密度函数是以 A 为条件的条件概率密度函数，用 $f_{LD_1}(x|A)$表示。当 A 是一主加瑞利分布的随机变量时，$LD(\boldsymbol{v})$的概率密度函数 $f_{LD_1}(x)$等于 $f_{LD_1}(x|A)$关于 A 的期望值。因此，对 Swerling Ⅲ型起伏目标的多脉冲线性检测的检测概率为

$$P_{\mathrm{d}}=\int_S^\infty\int_A f_{LD_1}(x\mid a)f_A(a)\mathrm{d}a\,\mathrm{d}x=\int_A\int_S^\infty f_{LD_1}(x\mid a)\mathrm{d}x f_A(a)\mathrm{d}a$$

其中，$\int_S^\infty f_{LD_1}(x\mid a)\mathrm{d}x$ 由式(2.97)给出。因此

$$P_{\mathrm{d}}=\mathrm{E}_A\left\{g_n\left(\lambda,\frac{S}{k\sqrt{N_0\varepsilon/2}}\right)\right\}=\int_0^\infty g_n\left(\frac{\varepsilon}{N_0}a^2,\frac{S}{k\sqrt{N_0\varepsilon/2}}\right)\frac{9a^3}{2A_0^4}\exp\left(-\frac{3a^2}{2A_0^2}\right)\mathrm{d}a$$

令 $x=\dfrac{\sqrt{3}a}{A_0}$，得

$$P_{\mathrm{d}}=\int_0^\infty g_n\left(\frac{A_0^2\varepsilon x^2}{3N_0},\frac{S}{k\sqrt{N_0\varepsilon/2}}\right)\frac{x^3}{2}\exp\left(-\frac{x^2}{2}\right)\mathrm{d}x=w_n\left(\lambda',\frac{S}{k\sqrt{N_0\varepsilon/2}}\right)=w_n[\lambda',q_n^{-1}(P_{\mathrm{fa}})] \tag{2.102}$$

在给定的关于目标幅度起伏和背景噪声的假设下，通常采用数值方法求解式(2.102)以获得根据式(2.88)进行检测的(P_d,P_{fa},SNR)关系。这个线性检测的关系近似等于平方律检测的关系。对于Swerling Ⅲ型目标，以及几个不同P_{fa}和n值，图2.13和图2.14给出了多脉冲平方律检测的检测性能曲线。

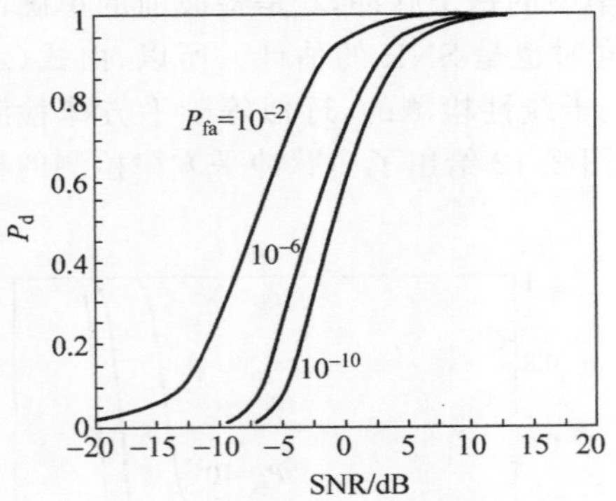

图2.13 Swerling Ⅲ型目标——多脉冲平方律检测($n=100$)的检测性能曲线

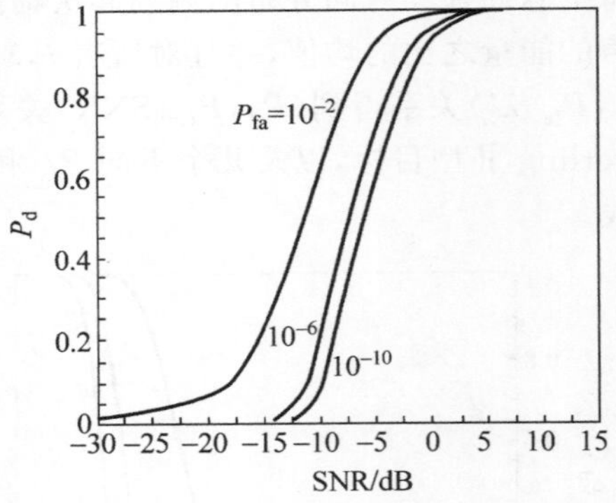

图2.14 Swerling Ⅲ型目标——多脉冲平方律检测($n=1000$)的检测性能曲线

6. Swerling Ⅳ型起伏目标的(P_d,P_{fa},SNR)关系

若目标的一组n个反射信号服从Swerling Ⅳ型模型，则可以假设脉冲串中每个脉冲的幅度A是统计独立的一主加瑞利分布的随机变量。确定(P_d,P_{fa},SNR)关系的方法与前面讨论的Swerling Ⅱ型的情况相似，仅在幅度分布的假设上不同。

设$D_i(v_i)$是对脉冲串中第i个脉冲进行检测的单脉冲检测统计量，其定义见式(2.43)。由于A_i($i=1,2,\cdots,n$)服从一主加瑞利分布，则由式(2.81)可知$D_{1i}(v_i)$的概率密度函数为

$$f_{D_{1i}(r)}=\frac{2r}{N_0k^2\varepsilon(1+\lambda'/2)^2}\left[1+\frac{\lambda' r^2/2}{N_0k^2\varepsilon(1+\lambda'/2)}\right]\exp\left[-\frac{r^2}{(1+\lambda'/2)N_0k^2\varepsilon}\right] \tag{2.103}$$

其中

$$\lambda'=\frac{4A_0^2\varepsilon}{3N_0} \tag{2.104}$$

由式(2.96)可知多脉冲线性检测策略的检测概率为

$$P_d=\int_S^{\infty}\frac{1}{2\pi}\int_{-\infty}^{+\infty}\left[\int_0^{+\infty}f_{D_{1i}}(r)\mathrm{e}^{\mathrm{j}tr}\,\mathrm{d}r\right]^n\mathrm{e}^{-\mathrm{j}tx}\,\mathrm{d}t\,\mathrm{d}x$$

于是，把式(2.103)代入上式得到检测概率为

$$P_d=\int_S^{\infty}\frac{1}{2\pi}\int_{-\infty}^{+\infty}\left\{\int_0^{+\infty}\frac{2r}{N_0k^2\varepsilon(1+\lambda'/2)^2}\left[1+\frac{\lambda' r^2/2}{N_0k^2\varepsilon(1+\lambda'/2)}\right]\times\right.$$
$$\left.\exp\left[-\frac{r^2}{(1+\lambda'/2)N_0k^2\varepsilon}\right]\mathrm{e}^{\mathrm{j}tr}\,\mathrm{d}r\right\}^n\mathrm{e}^{-\mathrm{j}tx}\,\mathrm{d}t\,\mathrm{d}x$$

令

$$y=\frac{r}{k\sqrt{N_0\varepsilon/2}},\quad u=tk\sqrt{N_0\varepsilon/2},\quad z=\frac{x}{k\sqrt{N_0\varepsilon/2}}$$

则由式(2.93)得

$$P_{\mathrm{d}}=\int_{\frac{S}{k\sqrt{N_0\varepsilon/2}}}^{\infty}\frac{1}{2\pi}\int_{-\infty}^{+\infty}\left\{\int_0^{+\infty}\frac{y}{(1+\lambda'/2)^2}\left[1+\frac{\lambda'/2}{(1+\lambda'/2)}\frac{y^2}{2}\right]\times\exp\left(-\frac{y^2/2}{1+\lambda'/2}\right)\mathrm{e}^{juy}\mathrm{d}y\right\}^n\mathrm{e}^{-juz}\mathrm{d}u\,\mathrm{d}z$$

$$=v_n\left(\lambda',\frac{S}{k\sqrt{N_0\varepsilon/2}}\right)=v_n[\lambda',q_n^{-1}(P_{\mathrm{fa}})] \tag{2.105}$$

式(2.83)已经证明，对于具有一主加瑞利起伏幅值的单脉冲，式(2.104)定义的λ'等于λ(检波前脉冲串中单个脉冲与噪声的能量之比)的均值。因此，对于所有i，λ'可以作为脉冲串中第i个脉冲的λ_i的估计。当接收机匹配于每个接收的脉冲信号时，λ_i等于SNR。因此，式(2.105)的(P_{d}，P_{fa}，λ')关系可以用来研究(P_{d}，P_{fa}，SNR)关系。

在给定关于目标幅度起伏和背景噪声的假设下，可以用数值方法求解式(2.105)，以获得多脉冲线性检测的(P_{d}，P_{fa}，SNR)关系。这个关系是关于线性检测的，近似等于平方律检测的关系。对于Swerling Ⅳ型目标，以及几个不同P_{fa}和n值，图2.15和图2.16给出了多脉冲平方律检测的检测性能曲线。

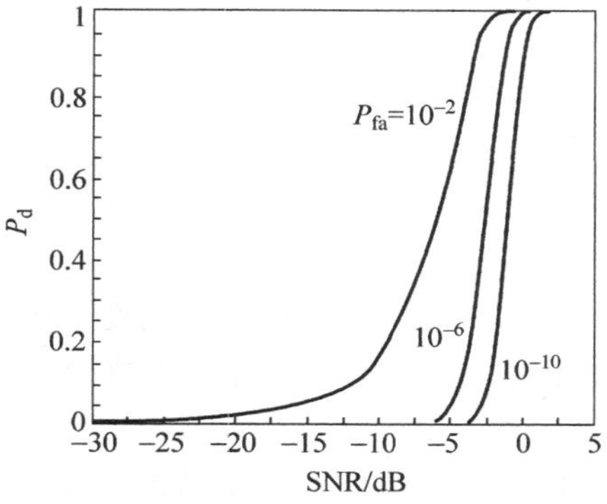

图2.15　Swerling Ⅳ型目标——多脉冲平方律检测($n=100$)的检测性能曲线

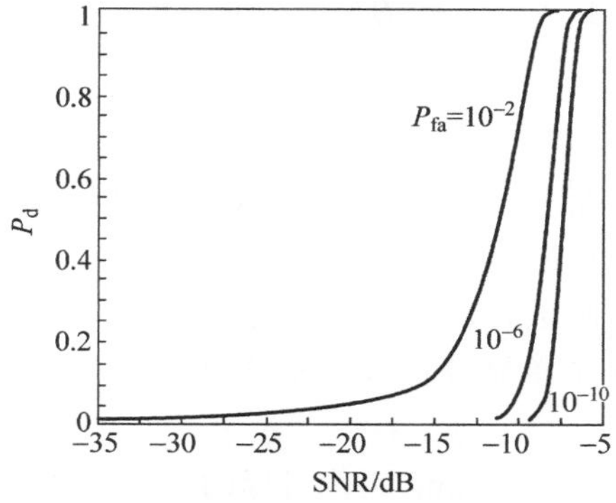

图2.16　Swerling Ⅳ型目标——多脉冲平方律检测($n=1000$)的检测性能曲线

2.4.3　相参脉冲串检测

2.3节中已经提到了相参脉冲串的最优检测问题，其处理方式是将相参脉冲串当作一个波形来对待。因此，对于相参脉冲串的检测来说，图2.1最优单脉冲线性检测器框图中的“匹配滤波器”指的是整个相参脉冲串的匹配滤波器。由于相参脉冲串的频谱是梳齿形的，故它的匹配滤波器可以由单个脉冲的匹配滤波器再串接上积累器组成。这样得到的最佳检测系统在形式上将有别于图2.1给出的检测系统，但检测效果在本质上是一样的，如图2.17所示。

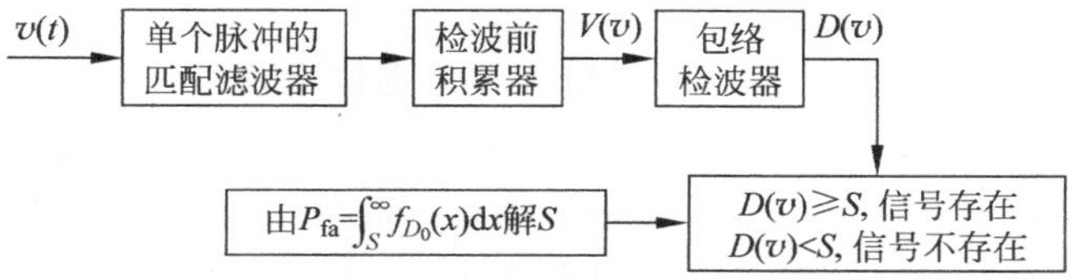

图2.17　相参脉冲串的最佳检测系统的原理框图

图 2.17 中积累器的频谱是梳齿状的，完成对脉冲串的匹配滤波。这个积累器也称为相参积累器，它的积累作用是在检波前完成的，积累后的信噪比是单个脉冲信噪比的 n 倍，相当于在噪声能量不变的情况下，信号能量变为脉冲串中各脉冲能量之和。因此，对相参脉冲串来说，将能量集中在一个脉冲上还是分散到 n 个脉冲上，最佳检测系统的检测能力不变。相参积累的优势在于在检波前就将脉冲串能量集中起来，这样可以减少检波过程中小信号的损失。当然，相参积累在以前实现较困难。

相参脉冲串的最佳检测是在检波前完成积累而在检波后进行门限判决。这是因为实际工作中碰到的相参脉冲串，其各脉冲之间的相位具有确定的关系，因而可以利用相位关系实现相参积累，但脉冲串的初相往往是未知的，因此判决应在包络检波以后进行。

相参脉冲串经过检波前积累，再经过包络检波后，噪声的概率密度函数为

$$f_{D_0}(x)=\frac{2x}{N_0k^2\varepsilon}\exp\left(-\frac{x^2}{N_0k^2\varepsilon}\right),\quad x\geqslant 0 \tag{2.106}$$

这是参数为$\dfrac{N_0k^2\varepsilon}{2}$的瑞利分布。而信号加噪声的概率分布为

$$f_{D_1}(r)=\frac{2r}{N_0k^2\varepsilon}\exp\left[-\left(\lambda+\frac{r^2}{N_0k^2\varepsilon}\right)\right]I_0\left(\frac{2A}{kN_0}r\right) \tag{2.107}$$

这是莱斯分布，其中 $\lambda=E/N_0$。信号能量 E 是脉冲串内能量的总和，$E=nE_0$。n 为脉冲数，而 E_0 为单个脉冲的能量。

实际上，式(2.106)就是式(2.54)，而式(2.107)在形式上与式(2.65)相同，信噪比变量中信号能量是 n 个单脉冲信号能量之和，因此可以将相参脉冲串的检测性能分析归结到单脉冲检测性能分析中。

对于 Swerling 起伏目标的相参脉冲串回波的检测问题，只需研究扫描间起伏的情况。如果存在脉冲间起伏，那么相参脉冲串中脉冲间的相参性就会被破坏，从而转化为非相参脉冲串的检测问题。与非起伏目标的相参脉冲串回波的检测问题一样，Swerling Ⅰ和 Swerling Ⅲ型目标的相参脉冲串回波检测的检测器结构与图 2.17 相同，其(P_d,P_{fa},SNR)关系分别与式(2.78)和式(2.82)相同。

2.5 小结

本章介绍了雷达在接收机噪声中对非起伏和 Swerling 起伏目标进行自动检测的经典理论。主要内容有：雷达距离方程与内部噪声环境、区域杂波和入射角、空间杂波环境、干扰环境的关系、虚警率、目标雷达截面积起伏的 Swerling 模型、匹配滤波、单脉冲检测、多脉冲检测。其中，多脉冲检测又分为非相参脉冲串的二元检测、线性检测及相参脉冲串的最优检测 3 部分。此外，文献[11-14]对经典自动检测理论也进行了讨论，这些内容是进行 CFAR 处理研究[15]的基础。

在解决雷达"四抗"问题的推动下，雷达体制和信号处理技术持续发展，在观测维度、处理域和目标特征利用方面不断突破。例如，相控阵为雷达信号处理提供了更高的自由度，可以做空时频联合处理，还有极化、波形等维度可以利用[16-18]。尤其对于隐身目标和强杂波背景中低信杂比信号检测问题，在时域和频域中单纯依赖幅度特征的检测将不再具有优势，需要在新的变换域中寻求可用于检测的新特征[19-22]，因此雷达目标检测呈现多维、多域、多特征的趋势，相应的 CFAR 处理方法也发生了很大的变化。

参考文献

[1] Swerling P. Detection of fluctuating pulsed signal in the presence of noise[J]. IRE Transactions on IT, 1957, 3(3): 175-178.

[2] Swerling P. More on detection of fluctuating targets[J]. IRE Transactions on IT, 1965, 11(3): 269-308.

[3] Skolnik M I. Radar handbook[M]. 2nd ed. New York: McGraw-Hill Book Company, 1990.

[4] Marcum J I. A statistical theory of detection by pulsed radar and mathematical appendix[J]. IRE Transactions on IT, 1960, 6(2): 209-211.

[5] Swerling P. Probability of detection for fluctuating targets[J]. IRE Transactions on IT, 1960, 6(2): 269-308.

[6] Rice S O. Mathematical analysis of random noise[J]. Bell System Technical Journal, 1944, 23(3): 282-332 and 1945, 24(1): 45-156.

[7] North D O. An analysis of the factors which determine signal/noise discrimination in pulsed carrier systems[J]. Proceedings of the IEEE, 2005,51(7): 1015-1028.

[8] Di Franco J V, Rubin W L. Radar detection[M]. Englewood Cliffs, NJ: Prentice-Hall, 1968; Norwood, MA: Artech House Publisher, 1980; Raleigh, NC: SciTech Publishing Inc. , 2004.

[9] Schleler D C. Automatic detection and radar data processing[M]. Deham Massachusetts: Artech House, 1980.

[10] Meyer D P, Meyer H A. Radar target detection-handbook of theory and practice[M]. New York: Academic Press, 1973.

[11] Levanon N. Radar principles[M]. New York: John Wiley & Sons, 1988.

[12] Minkler G, Minkler J. The principles of automatic radar detection in clutter-CFAR[M]. Baltimore: MD Magellan Book Company, 1990.

[13] De Maio A, Greco M S. Modern Radar Detection Theory[M]. Institution of Engineering and Technology, 2016.

[14] De Maio A, Greco M S, Orlando D. Introduction to Radar Detection [M]. Institution of Engineering and Technology, 2016.

[15] 何友，关键，孟祥伟，等. 雷达自动检测和CFAR处理方法综述[J]. 系统工程与电子技术，2001，23(1)：9-14.

[16] 关键，黄勇，何友. 基于自适应脉冲压缩-Capon滤波器的MIMO阵列雷达CFAR检测器[J]. 中国科学，2011，41(10)：1268-1282.

[17] 简涛，廖桂生，何友，等. 非高斯杂波下基于子空间的距离扩展目标检测器[J]. 电子学报，2017，45(6)：1342-1348.

[18] 王作珍. 自适应子空间信号检测理论和技术研究[D]. 成都：电子科技大学，2020.

[19] 陈小龙，关键，何友，等. 高分辨稀疏表示及其在雷达动目标检测中的应用[J]. 雷达学报，2017，6(3)：239-251.

[20] 许述文，白晓惠，郭子薰，等. 海杂波背景下雷达目标特征检测方法的现状与展望[J]. 雷达学报，2020，9(4)：684-714.

[21] 刘宁波，关键，宋杰，等. 海杂波频谱的多充分性特性分析[J]. 中国科学，2013，43(6)：768-783.

[22] 丁昊，刘宁波，董云龙，等. 雷达海杂波测量实验回顾与展望[J]. 雷达学报，2019，8(3)：281-302.

第3章

CHAPTER 3

均值类 CFAR 处理方法

3.1 引言

从本章开始将讨论一些在雷达目标检测恒虚警处理方面比较重要的典型恒虚警率(Constant False Alarm Rate,CFAR)处理方法,给出它们的基本数学模型,并在 3 种典型背景中分析其性能。根据所基于杂波背景的分布,CFAR 处理方法大体上分为两大部分:高斯背景中的 CFAR 处理方法和非高斯背景中的 CFAR 处理方法,前者包括第 3~6 章的内容,后者包括第 7~9 章的内容。

本章讨论的是均值(Mean Level,ML)类 CFAR 处理方法。它们的共同特点是在局部干扰功率水平估计中采用了取均值的方法。最经典的均值类 CFAR 方法是单元平均(Cell Averaging,CA)[1]方法,后为改善非均匀杂波背景中的检测性能,又相继出现了选大(Greatest Of,GO)[2]、选小(Smallest Of,SO)[3]和加权单元平均(Weighted Cell Averaging,WCA)[4-5]等方法。本章对它们进行综合的评价和比较。

3.2 基本模型描述

假设 $v(t)$是某个分辨单元的一个观测,$D(v)$是由 $v(t)$形成的检测统计量。对于平方律检测,$D(v)$具有如下形式

$$D(v)=I^2(v)+Q^2(v) \tag{3.1}$$

其中,$I(v)$和 $Q(v)$分别是信号的同相分量和正交分量。在检测单元中无目标信号时,$D(v)$对应于杂波回波,是一个随机变量,用 D_0 表示。对于线性检测,$D(v)$是式(3.1)的平方根,即 $D(v)=\sqrt{I^2(v)+Q^2(v)}$。在一般的杂波环境中,可以假设杂波包络服从瑞利分布,但是瑞利分布参数随时间和空间常常是变化的。这个模型常用于描述脉冲宽度大于 0.5μs 和入射余角大于 5°的海杂波,以及在未开发地带观测到的入射余角大于 5°的地杂波等杂波背景。

在确知恒定的接收机噪声中对 Swerling 起伏和非起伏目标进行单脉冲线性和多脉冲检测的最重要的三种典型检测方案是第 2 章中描述的单脉冲线性、多脉冲线性和二元检测方案。这些检测方案也同样适用于在确知不变的瑞利包络杂波环境中对 Swerling 起伏和非起伏目标的线性和二元检测。对于在这种杂波环境中的单脉冲线性检测和二元检测的第一级检测,与前面类似,其检测阈值 S 可以由式(3.2)解得

$$P_{fa}=\Pr[D(v)\geqslant S\mid \mathrm{H}_0]=\int_S^{\infty} f_{D_0}(x)\mathrm{d}x \tag{3.2}$$

其中，H_0 表示检测单元中不存在目标的假设。对于多脉冲检测，S 值则由式(3.3)解得：

$$P_{fa}=\Pr[LD(v)\geqslant S\mid H_0]=\int_S^{\infty}f_{LD_0}(x)\mathrm{d}x \tag{3.3}$$

其中，$LD(v)$ 代表由 $v(t)$ 得到的多脉冲线性检测的检测统计量，在检测单元中没有信号时用 LD_0 表示。

由式(2.55)可知，对于在接收机噪声中的固定门限的单脉冲线性检测，由于单边噪声功率谱密度 N_0 和 P_{fa} 之间的指数关系，N_0（或瑞利参数 $\delta=k\sqrt{N_0\varepsilon/2}$）的微小变化或不确定性将引起 P_{fa} 大的变化或不确定性，图 3.1 展示了 N_0 的 3dB 变化使虚警概率由 10^{-6} 变化到 10^{-3}，因此必须改进检测策略以保持相对恒定的虚警率。在参数未知或时变的瑞利包络杂波中，CFAR 处理的基本特征是提供把背景参数变化考虑在内的检测阈值 S，使检测具有相对恒定的 P_{fa}。

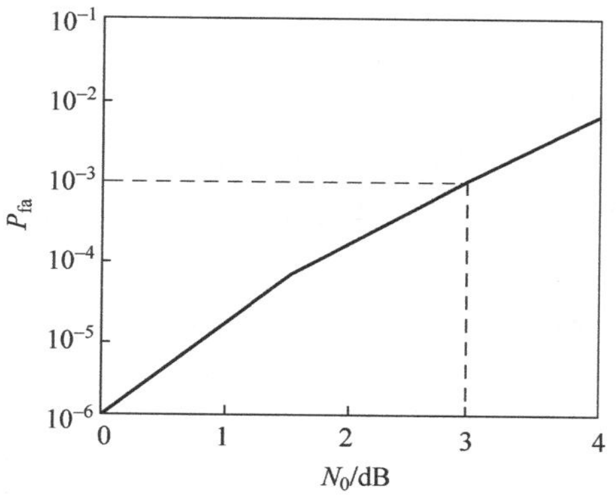

图 3.1 固定阈值检测的虚警概率

3.3 节将介绍在瑞利包络杂波环境中广泛使用的 CA 处理方法。在均匀的瑞利杂波背景条件下，CA 方法利用与检测单元相邻的一组独立同分布(Independent and Identically Distributed，IID)的参考单元样本的均值来估计杂波功率水平(见图 3.2)，它提供了对非起伏和 Swerling 起伏目标的最优或准最优检测。

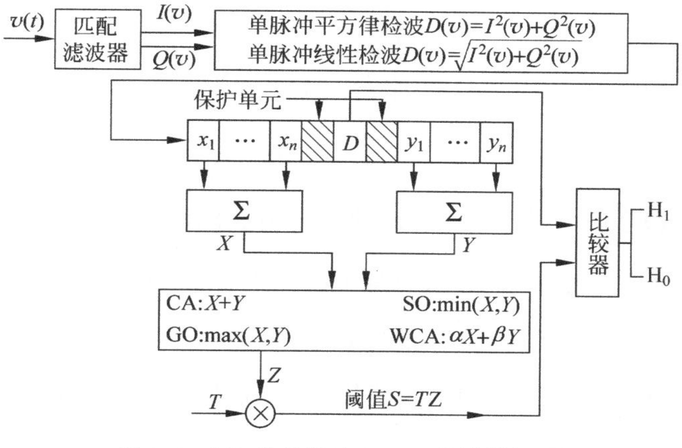

图 3.2 ML 类单脉冲 CFAR 检测器框图

当杂波包络样本间的空间距离很近时，样本间会不独立。即使保证了样本空间距离使样本相互独立，但是杂波的非均匀性也常使同分布条件遭到破坏，并且这种影响可能会随着采样间距的增加而加重。因此，在参考单元样本的独立性和平稳性之间要有一个权衡。最优参考单元距离要根据相关区域中不同气象条件或干扰条件下背景杂波空间起伏特性来确定，所要求的参考单元样本的 IID 条件应该贯穿于均匀背景条件中。瑞利参数阶跃变化(杂波边缘)的影响在 3.11 节中分析讨论。

ML 类单脉冲 CFAR 检测器结构可以用图 3.2 描述。除检测单元样本的表示方法外，其余按照大多数 CFAR 文献的表示方法，分别用 $x_i(i=1,2,\cdots,n)$ 和 $y_i(i=1,2,\cdots,n)$ 表示检测单元两侧参考单元(也称作前沿、后沿参考滑窗)样本，参考滑窗长度 $R=2n$，n 为前沿和后沿参考滑窗长度，X 和 Y 分

别是前沿和后沿滑窗对杂波强度的局部估计，此时自适应判决准则为

$$D \underset{H_0}{\overset{H_1}{\gtrless}} TZ \tag{3.4}$$

其中，H_1 表示存在目标的假设，H_0 表示不存在目标的假设，Z 是由 X 和 Y 形成的参考滑窗中的杂波强度估计，T 是标称化因子，D 表示检测单元中的检测统计量 $D(v)$。与检测单元最邻近的是两个保护单元，用来防止目标能量泄漏到参考单元中，影响 ML 类检测器对杂波强度的两个局部估计值。

高斯分布杂波的包络服从瑞利分布，经平方律检波后，每个参考单元样本服从指数分布，其概率密度函数为

$$f_D(x)=\frac{1}{\lambda'}\exp\left(-\frac{x}{\lambda'}\right),\quad x\geqslant 0 \tag{3.5}$$

在参考单元中不存在目标的假设 H_0 下，λ' 是背景杂波加热噪声总的平均功率水平，用 μ 表示；在存在目标的 H_1 假设下，λ' 是 $\mu(1+\lambda)$。其中，λ 是目标信号与杂波加噪声的平均功率比值。于是有

$$\lambda'=\begin{cases}\mu, & H_0\\ \mu(1+\lambda), & H_1\end{cases} \tag{3.6}$$

下面主要考虑杂波背景的影响。在均匀杂波背景中，参考单元样本 $x_i(i=1,2,\cdots,n)$ 和 $y_i(i=1,2,\cdots,n)$ 是 IID 的，并且它们的 λ' 都是 μ；由于门限 $S=TZ$ 是一个随机变量，因而需要对 Z 求统计平均，将虚警概率表示为

$$\begin{aligned}P_{\mathrm{fa}}&=\mathrm{E}_S\{\Pr[D(v)\geqslant S\mid H_0]\}=\int_0^{\infty}f_Z(z)\int_{Tz}^{\infty}\frac{1}{\mu}\mathrm{e}^{-x/\mu}\,\mathrm{d}x\,\mathrm{d}z\\&=\int_0^{\infty}\mathrm{e}^{-Tz/\mu}f_Z(z)\,\mathrm{d}z=M_Z(u)\mid_{u=T/\mu}\end{aligned} \tag{3.7}$$

其中，$f_Z(z)$是 Z 的概率密度函数，$M_Z(u)$是 Z 的矩母函数(Moment Generating Function，MGF)，当 $u=\dfrac{T}{\mu(1+\lambda)}$时，式(3.7)便转化成在均匀杂波背景中的检测概率表达式

$$P_{\mathrm{d}}=M_Z(u)\mid_{u=\frac{T}{\mu(1+\lambda)}} \tag{3.8}$$

若在背景杂波功率 μ 确知的假设下进行最优检测，只需要一个固定阈值 S_0 来判定目标是否存在，这时的虚警概率 P_{fa} 为

$$P_{\mathrm{fa}}=\Pr[D\geqslant S_0\mid H_0]=\mathrm{e}^{-S_0/\mu} \tag{3.9}$$

其中，S_0 是固定的最优阈值，最优检测的检测概率 P_{d} 为

$$P_{\mathrm{d,opt}}=\Pr[D\geqslant S_0\mid H_1]=\mathrm{e}^{-S_0/[\mu(1+\lambda)]}$$

再结合式(3.9)得

$$P_{\mathrm{d,opt}}=P_{\mathrm{fa}}^{1/(1+\lambda)} \tag{3.10}$$

在非均匀杂波背景中，参考单元样本不再服从 IID 假设。例如，在杂波边缘环境中，杂波功率从一种水平变化到另一种水平。在参考滑窗中存在干扰目标的情况下，即多目标环境中，一些参考单元样本会出现峰值。对于杂波边缘情况，这里只考虑背景杂波功率水平从高到低阶跃变化的情况，即杂波边缘前沿情况，至于后沿情况可以由此类推。杂波边缘前沿情况也就是假设 $2n$ 个参考单元中前 N_C 个单元

样本服从分布

$$f_1(x)=\frac{1}{\mu_0\gamma}\exp\left(-\frac{x}{\mu_0\gamma}\right) \tag{3.11}$$

其余$(2n-N_C)$个参考单元样本服从分布

$$f_2(x)=\frac{1}{\mu_0}\exp\left(-\frac{x}{\mu_0}\right) \tag{3.12}$$

其中，γ 是两种杂波功率强度之比。

对任意一个应用判决准则式(3.4)的 CFAR 检测器，D 和 Z 是独立的随机变量，因此，单脉冲平方律检测假设下 CFAR 检测器的虚警概率在 $0\leqslant N_C\leqslant n$ 时为

$$P_{\mathrm{fa}}=\int_0^{\infty}f_Z(z)\left[\int_{Tz}^{\infty}\frac{1}{\mu_0}\exp\left(-\frac{x}{\mu_0}\right)\mathrm{d}x\right]\mathrm{d}z=\int_0^{\infty}f_Z(z)\exp\left(-\frac{Tz}{\mu_0}\right)\mathrm{d}z=M_Z(u)\Big|_{u=\frac{T}{\mu_0}} \tag{3.13}$$

在 $n\leqslant N_C\leqslant 2n$ 时为

$$P_{\mathrm{fa}}=\int_0^{\infty}f_Z(z)\left[\int_{Tz}^{\infty}\frac{1}{\mu_0\gamma}\exp\left(-\frac{x}{\mu_0\gamma}\right)\mathrm{d}x\right]\mathrm{d}z=\int_0^{\infty}f_Z(z)\exp\left(-\frac{Tz}{\mu_0\gamma}\right)\mathrm{d}z=M_Z(u)\Big|_{u=\frac{T}{\mu_0\gamma}} \tag{3.14}$$

其中，$M_Z(u)$是随机变量 Z 的 MGF，但需注意参考滑窗中的强杂波样本数目不同时，$M_Z(u)$的数学解析表达式是不同的。

3.3 CA-CFAR 检测器

在 CA-CFAR 检测器[1]中，背景杂波功率水平的估计由 $R=2n$ 个参考单元样本的均值得到。它是在参考单元样本服从指数分布的假设下对杂波功率水平的一个极大似然估计。为了便于计算，常把因子 $1/R$ 归到标称化因子中，取

$$Z=\sum_{i=1}^{n}x_i+\sum_{j=1}^{n}y_j \tag{3.15}$$

Z 称为总的杂波功率水平估计。

由于指数分布是 Γ 分布在 $\alpha=1$ 的特殊情况，Γ 分布的 PDF 为

$$f(x)=\beta^{-\alpha}x^{\alpha-1}\mathrm{e}^{-x/\beta}/\Gamma(\alpha),\quad x\geqslant 0,\alpha\geqslant 0,\beta\geqslant 0 \tag{3.16}$$

其中，α 和 β 是两个参数，$\Gamma(\alpha)$就是通常的 Γ 函数。对于整数 α，$\Gamma(\alpha)$等于$(\alpha-1)!$。式(3.16)对应的累积分布函数(Cumulative Distribution Function，CDF)用 $G(\alpha,\beta)$表示，对于服从 Γ 分布的随机变量 X，将它记为 $X\sim G(\alpha,\beta)$。X 的矩母函数为

$$M_X(u)=(1+\beta u)^{-\alpha} \tag{3.17}$$

根据 IID 的假设得到 $x_i\sim G(1,\mu)$和 $y_i\sim G(1,\mu)$。由于多个独立随机变量和的矩母函数等于各个随机变量矩母函数的积，因此由式(3.17)可知 CA-CFAR 检测器中对杂波功率水平的估计 $Z\sim G(2n,\mu)$。这样，把式(3.17)代入式(3.8)就得到 CA-CFAR 检测器的检测概率

$$P_{\mathrm{d}}=\left(1+\frac{T}{1+\lambda}\right)^{2n} \tag{3.18}$$

其中，T 为标称化因子，可由式(3.18)在 $\lambda=0$ 时得到标称化因子 T 与虚警概率间的关系，即

$$T=(P_{\mathrm{fa}})^{-1/2n}-1 \tag{3.19}$$

从式(3.18)和式(3.19)可以看到，检测概率和虚警概率不依赖 μ，因此 CA-CFAR 是具有恒虚警特

性的。更一般地说，对于尺度型分布(如指数分布)，均值类处理方法形成的检测统计量是关于尺度变换不变的统计量，因此具有 CFAR 性质[6]。

在 CFAR 检测器性能分析中，Rohling[7]定义了一个平均判决阈值 ADT，它是一个标称化的量，即

$$\mathrm{ADT}=\frac{\mathrm{E}\{TZ\}}{\mu}=\frac{T\mathrm{E}\{Z\}}{\mu}=\frac{T}{\mu}\int_0^{\infty} zf(z)\mathrm{d}z \tag{3.20}$$

ADT 是计算检测性能损失的一种可供选择的度量。对于某一给定的参考单元总数 R 和 P_{fa}，ADT 是不依赖检测概率的。ADT 越小，表示检测性能越好，即检测概率越高。

利用矩母函数和均值的关系可得

$$\mathrm{ADT}=-\frac{T}{\mu}\left.\frac{\mathrm{d}M_Z(u)}{\mathrm{d}u}\right|_{u=0} \tag{3.21}$$

对于 CA-CFAR 检测器，有

$$M_Z(u)=\frac{1}{(1+\mu u)^{2n}}$$

故

$$\mathrm{ADT}_{\mathrm{CA}}=-\frac{T}{\mu}\left.\frac{\mathrm{d}}{\mathrm{d}u}(1+\mu u)^{-2n}\right|_{u=0}=2nT \tag{3.22}$$

对于固定门限的最优检测器，有 $\mathrm{ADT}=\mathrm{E}\{S_0\}/\mu$，ADT 可以作为比较某种 CFAR 检测器在均匀背景中与最优检测器的差别的一种度量，也可以用它近似两种 CFAR 检测器在均匀背景中的信杂噪比差别(单位：dB)

$$\Delta=10\log\frac{\mathrm{E}\{T_1Z_1\}}{\mathrm{E}\{T_2Z_2\}}\mathrm{dB} \tag{3.23}$$

然而，ADT 只描述了一个随机变量 S/μ 的均值。对于一个随机变量，它的方差也是确定该随机变量的重要指标。而且在分析检测性能过程中已经发现，检测概率的大小不仅取决于 ADT 值，而且也取决于对杂波功率水平估计的方差。为此，文献[8-9]在分析韦布尔背景中的 OS-CFAR 性能时，引入了一个新的变量 SD_{ADT}(Standard Deviation of Average Decision Threshold)，来表示韦布尔分布尺度参数的标称化 ADT 的均方差，这个概念对瑞利包络杂波背景中的性能分析也具有指导意义。

3.4 GO 和 SO-CFAR 检测器

CA-CFAR 检测在杂波边缘中会引起虚警率的明显上升，而在多目标环境中将导致检测性能的下降，这些不足促进了对其他 CFAR 方案的寻求。作为 CA-CFAR 的修正方案，GO-CFAR[2] 和 SO-CFAR[3] 被提出。但是，它们各自只能解决其中一个问题，并且还带来了一定的附加检测损失，GO 在杂波边缘环境中能保持好的虚警控制性能，但在多目标环境中会出现“目标遮蔽”现象；当干扰目标只位于前沿滑窗或后沿滑窗时，SO 具有良好的多目标分辨能力，但是它的虚警控制能力又很差。

GO 主要是针对杂波边缘而设计的(见图 3.2)，它取两个局部估计的较大者作为总的杂波功率水平估计，即有

$$Z=\max(X,Y) \tag{3.24}$$

其中

$$X=\sum_{i=1}^{n}x_i,\quad Y=\sum_{j=1}^{n}y_j \tag{3.25}$$

它们是相互独立的随机变量。于是，Z 的 PDF 为

$$f_Z(z)=f_X(z)F_Y(z)+f_Y(z)F_X(z) \tag{3.26}$$

其中，f 和 F 分别是 X 和 Y 的 PDF 和 CDF。在均匀杂波背景中，$F_X=F_Y$，$X,Y\sim G(n,\mu)$。这样就可以求得 Z 的矩母函数，进而得到 GO-CFAR 检测器在均匀杂波背景中的虚警概率为

$$P_{\mathrm{fa,GO}}=2(1+T)^{-n}-2\sum_{i=0}^{n-1}\binom{n+i-1}{i}(2+T)^{-(n+i)} \tag{3.27}$$

其中，T 是依赖 P_{fa} 的设计值和参考滑窗尺寸 $R=2n$ 的常数，用 $T/(1+\lambda)$ 代替式(3.27)中的 T 可得到 GO-CFAR 检测器的检测概率 P_{d} 为

$$P_{\mathrm{d,GO}}=2\left(1+\frac{T}{1+\lambda}\right)^{-n}-2\sum_{i=0}^{n-1}\binom{n+i-1}{i}\left(2+\frac{T}{1+\lambda}\right)^{-(n+i)} \tag{3.28}$$

当雷达探测特定距离单元的目标时，需要降低临近干扰目标的影响。SO-CFAR 方案可以解决单边滑窗中出现多个干扰目标时引起的检测性能下降的问题。SO-CFAR 使用式(3.25)中 X 和 Y 的较小者作为总的杂波功率水平估计

$$Z=\min(X,Y) \tag{3.29}$$

于是，Z 的 PDF 为

$$f_Z(z)=f_X(z)+f_Y(z)-[f_X(z)F_Y(z)+f_Y(z)F_X(z)] \tag{3.30}$$

式(3.30)方括号中的式子就是在式(3.26)中给出的 GO-CFAR 检测器 Z 的 PDF。因此，SO-CFAR 检测器在均匀杂波背景中的虚警概率为

$$P_{\mathrm{fa,SO}}=M_X\left(\frac{T}{\mu}\right)+M_Y\left(\frac{T}{\mu}\right)-P_{\mathrm{fa,GO}}=2\sum_{i=0}^{n-1}\binom{n+i-1}{i}(2+T)^{-(n+i)} \tag{3.31}$$

其中，$M_X(\cdot)$和 $M_Y(\cdot)$分别是 X 和 Y 的 MGF，可以由式(3.17)计算得到。SO 的检测概率由 $T/(1+\lambda)$代替式(3.31)中的 T 得到，即

$$P_{\mathrm{d,SO}}=2\sum_{i=0}^{n-1}\binom{n+i-1}{i}\left(2+\frac{T}{1+\lambda}\right)^{-(n+i)} \tag{3.32}$$

将求得的 GO-CFAR 检测器关于杂波功率水平估计的 MGF 代入式(3.21)，即可得到 GO-CFAR 的 ADT 为

$$\mathrm{ADT}_{\mathrm{GO}}=2T\left[n-\sum_{i=0}^{n-1}\binom{n+i-1}{i}(i+n)\cdot 2^{-(n+i+1)}\right] \tag{3.33}$$

类似地，可以得到 SO-CFAR 的 ADT，即

$$\mathrm{ADT}_{\mathrm{SO}}=2T\sum_{i=0}^{n-1}\binom{n+i-1}{i}(i+2)\cdot 2^{-(n+i+1)} \tag{3.34}$$

3.5 WCA-CFAR 检测器

在 WCA(Weighted Cell-Averaging)-CFAR 检测器[4]中，总的杂波功率水平估计是通过对前沿滑窗和后沿滑窗的局部估计按照杂波强度加权得到的(见图 3.2)，即有

$$Z=\alpha X+\beta Y \tag{3.35}$$

α 和 β 的最优加权值是在保持虚警概率恒定的同时使检测概率最大的条件下得到的。根据两个相互独立的随机变量之和的 MGF 的关系得到 Z 的 MGF 为

$$M_Z(u) = M_{\alpha X}(u) \cdot M_{\beta Y}(u) \tag{3.36}$$

把式(3.36)代入式(3.7)得到 WCA-CFAR 检测器在均匀杂波背景中的虚警概率为

$$P_{fa} = M_{\alpha X}(u) \cdot M_{\beta Y}(u)\big|_{u=T/\mu} = M_X(\alpha u) \cdot M_Y(\beta u)\big|_{u=T/\mu} = (1+\alpha T)^{-n}(1+\beta T)^{-n} \tag{3.37}$$

其中,M_X 和 M_Y 分别是 X 和 Y 的 MGF,可以由式(3.17)得到,将 T 用 $T/(1+\lambda)$ 替代就得到 WCA-CFAR 检测器的检测概率 P_d,把式(3.36)代入式(3.21)得 WCA-CFAR 检测器的 ADT 表达式为

$$\text{ADT} = (\alpha + \beta)nT \tag{3.38}$$

3.6 采用对数检波的 CA-CFAR 检测器

采用对数检波的一类 CFAR 检测器通常由对数检波器和后面的减法电路组成[10]。CA-LOG/CFAR 与传统的 CA-CFAR 相似,用一组由对数检波器输出的样本的均值形成估计。图 3.3 是 CA-LOG/CFAR 检测器的原理框图。

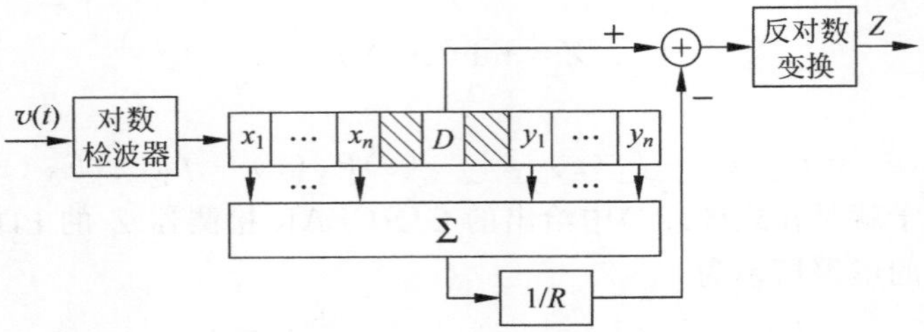

图 3.3 CA-LOG/CFAR 检测器的原理框图

CA-LOG/CFAR 有两个明显的优点:第一,与传统 CA-CFAR 相比,它可以在很大的背景杂噪动态范围上工作;第二,归一化过程由减法电路完成,比传统 CA-CFAR 使用的除法更易于实现。但是,这两个优点的代价是在相同样本数时具有较差的检测性能。相比于线性和二元检测器,检测损失可达 8dB。

使用对数检波器时很难像平方律检波那样得到 P_d 和 P_{fa} 的解析表达式,因此对它的性能分析需要借助 Monte Carlo 仿真的方法[11]。

3.7 单脉冲线性 CA-CFAR 检测器

式(3.5)~式(3.14)是在平方律检波条件下建立的,3.3 节对 CA-CFAR 的讨论也是在这个条件下进行的。使用线性检波器的 CA-CFAR 检测器利用的是回波信号的包络而不是包络的平方(平方律检波)进行检测的。除了一些简单的情况,求取 IID 随机变量之和的 PDF 闭型解很困难。如果杂波包络服从瑞利分布,平方律检波后为指数分布,参考单元样本的 PDF 闭型解是容易得到的,然而线性检波器却不是这样,这给性能评价带来了困难。但是,线性检测相对于平方律检测有一些优点,例如,易于实现及动态范围大等,这就吸引人们对它进行研究,但需要采用一些其他变通的方法来分析其性能。

Divito 和 Moretti 在文献[12]中得到了线性 CA-CFAR 检测器的虚警概率近似表达式

$$P_{fa} = \left\{1 + \frac{T^2}{2n\left[c - (c-1)e^{-(2n+1)}\right]}\right\}^{-2n} \tag{3.39}$$

其中，$c=4/\pi$，$2n$ 是滑窗长度，T 是标称化因子。

Raghavan[13]用多个 χ^2 概率密度的混合来近似瑞利分布，获得了瑞利杂波背景中的 P_{fa}，以及在此背景中的瑞利起伏目标的检测概率 P_{d} 的解析表达式，然后用 Monte Carlo 仿真分析对上述结果进行了验证，同时也给出了对瑞利杂波背景中的非起伏目标的仿真分析结果。Raghavan 还认为若适当选取混合型中的项数，该方法适用于非瑞利情形。

3.8 多脉冲 CA-CFAR 检测器

3.8.1 双门限 CA-CFAR 检测器

采用双门限检测的 CA-CFAR(见图 3.4)由于存储处理方便，易于实现，且具有良好的检测性能，成为实际雷达检测中常用的检测策略。假设 $v_i(t)(i=1,2,\cdots,N)$ 是第 i 个脉冲 $s_i(t)(i=1,2,\cdots,N)$ 发射后在同一分辨单元(检测单元)中的观测，$D_i(v_i)(i=1,2,\cdots,N)$ 是由每个观测 $v_i(t)(i=1,2,\cdots,N)$ 形成的检测统计量，假设 M 是双门限检测的第二门限；在 N 个观测中，若有 M 个观测对应的检测统计量超过第一门限，则判为目标存在，否则判为目标不存在。

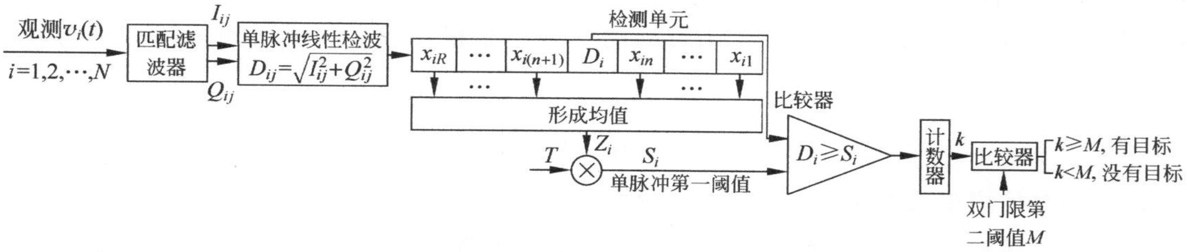

图 3.4 双门限检测的 CA-CFAR 检测器框图

第一门限检测的单脉冲虚警概率为

$$P_{\text{fa,sp}}=\int_0^{\infty}\left[\int_{S_i}^{\infty} f_{D_{0i}}(x)\mathrm{d}x\right] f_{S_i}(t)\mathrm{d}t \tag{3.40}$$

其中，$f_{D_{0i}}(x)$是第 i 个脉冲对应的随机变量 $D_{0i}(v_i)$ 的概率密度函数，S_i 是双门限检测的第一门限 $S_i=TZ_i$，$f_{S_i}(\cdot)$是随机变量 S_i 的概率密度函数。

总虚警概率和总检测概率是对 N 个脉冲的观测结果进行检测积累后的概率。它们仍然可以分别利用式(2.86)和式(2.87)进行计算。

3.8.2 多脉冲非相参积累 CA-CFAR 检测器

采用多脉冲非相参积累检测的 CA-CFAR 如图 3.5 所示，其检测统计量为

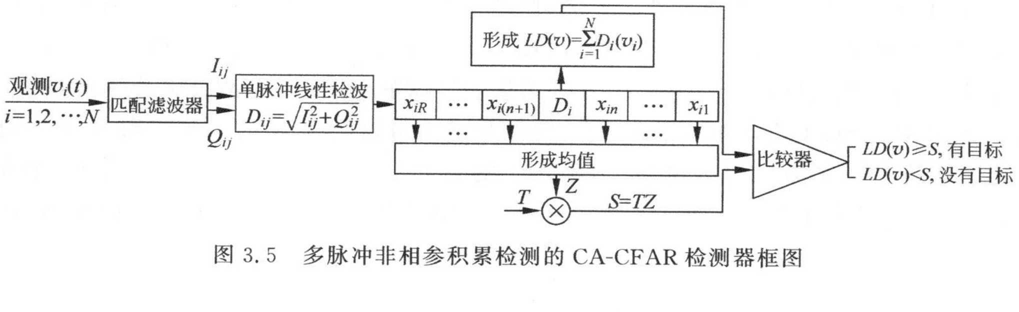

图 3.5 多脉冲非相参积累检测的 CA-CFAR 检测器框图

$$LD(v)=D_1(v_1)+D_2(v_2)+\cdots+D_n(v_n)$$

多脉冲非相参积累检测准则为

$$\begin{cases}LD(v)\geqslant S, & \mathrm{H}_1\\ LD(v)< S, & \mathrm{H}_0\end{cases}$$

因此,虚警概率为

$$P_{\mathrm{fa}}=\int_0^{\infty}\left[\int_S^{\infty} f_{LD_0}(x)\mathrm{d}x\right] f_S(t)\mathrm{d}t \tag{3.41}$$

其中,$f_{LD_0}(x)$是随机变量$LD(v)$在H_0假设下的概率密度函数,$f_S(x)$是$S(S=TZ)$的概率密度函数。

假定在H_0假设下检测单元的杂波包络服从瑞利分布,则多脉冲非相参积累CA-CFAR的检测统计量的概率密度函数为

$$f_{LD_0}(x)=\frac{1}{2\pi}\int_{-\infty}^{\infty}\left[\int_0^{\infty}\frac{r}{\sigma^2}\exp\left(\frac{r^2}{2\sigma^2}\right)\mathrm{e}^{\mathrm{j}tr}\,\mathrm{d}r\right]^n\exp(-\mathrm{j}tx)\mathrm{d}t \tag{3.42}$$

其中,σ是瑞利分布的参数。

在噪声中检测非起伏或Swerling起伏目标的多脉冲非相参积累检测概率采用如下的计算公式形式:

$$P_{\mathrm{d}}=\int_0^{\infty}\left[\int_S^{\infty} f_{LD_1}(x)\mathrm{d}x\right] f_S(t)\mathrm{d}t \tag{3.43}$$

其中,$f_{LD_1}(\cdot)$是检测单元中存在目标时检测统计量$LD(v)$的概率密度函数。对于线性检测,很难得到$f_{LD_0}(\cdot)$、$f_{LD_1}(\cdot)$及$f_S(t)$的解析表达式,因此也就很难得到P_{fa}和P_{d}的解析表达式,往往需要用仿真方法来分析其性能。

当采用平方律检波时,在多脉冲非相参积累检测情况下,文献[14]采用解析方法分析了CA-CFAR在均匀背景和参考单元中出现多个干扰目标时的检测性能。

3.9 ML类CFAR检测器在均匀杂波背景中的性能

3.3节和3.4节给出了经平方律检波后在均匀杂波背景中CA、GO和SO-CFAR检测器对Swerling Ⅱ型目标的检测概率P_{d}的计算公式,在$P_{\mathrm{fa}}=10^{-6}$,$R=16$和$R=32$时,它们的检测性能曲线如图3.6所示。

当参考滑窗长度R增加时,CA、GO和SO-CFAR检测器性能均向最优检测靠近。在上述假设下,CA-CFAR采用的是对背景杂波功率水平的极大似然估计,它在这种意义上是最优的。与CA-CFAR检测器相比,GO只表现出很小的检测性能下降,典型值在0.1~0.3dB[15]。上面这些对GO的分析都只局限于单脉冲处理和Swerling Ⅱ型起伏目标,Ritcey[16]把对GO检测器的分析扩展到了对非起伏和χ^2分布起伏目标的非相参脉冲积累检测。分析结果表明,CFAR损失随滑窗大小和脉冲积累数而变化,不依赖起伏参数。SO-CFAR的检测性能主要依赖参考滑窗长度R的大小。当R很小时,其检测性能损失比其他的CFAR方案要大得多,但是损失随着R增加而急剧减少[17]。当P_{fa}较低时,SO-CFAR的检测性能损失也很大。Weiss在文献[17]中给出了SO相对于CA的附加检测损失,见表3.1。

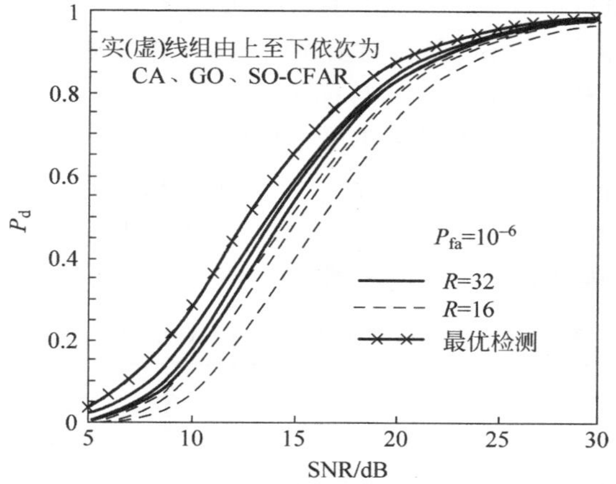

图 3.6　CA、GO 和 SO-CFAR 检测器的检测性能

表 3.1　SO-CFAR 相对于 CA-CFAR 的附加检测损失(无干扰目标, $P_d=0.5$)　(单位: dB)

P_{fa}	R			
	4	8	16	32
10^{-4}	6.63	2.58	0.99	0.41
10^{-6}	16.3	4.51	1.76	0.70
10^{-8}	16.2	6.69	2.64	1.05

采用对数检波的 CA-CFAR 检测器在动态范围和实现上相对于线性 CA-CFAR 检测器有优势,但这是以牺牲检测性能为代价的。在均匀杂波背景中,Hansen 和 Ward[11]在小样本数时采用基于重要采样技术的仿真方法,而在大样本数时用二阶矩方法进行性能分析。结果表明,当 $R>8$ 时,CA-LOG/CFAR 的 CFAR 损失比线性检波 CA-CFAR 高 65%;当 $R\leqslant 8$ 时,它们的 CFAR 损失差别变小;当 $R=1$ 时,两者具有相同的检测性能。这个结论可以用下面的经验公式表示,即

$$N_{LOG}=1.65N_{LIN}-0.65$$

其中,N_{LOG} 和 N_{LIN} 分别是 CA-LOG/CFAR 和线性检波 CA-CFAR 具有相同检测性能时的样本数。

文献[13]用多个 χ^2 分布的混合代替瑞利 PDF 解析式的近似分析结果与仿真分析结果都表明,线性检波 CA 和平方律检波 CA 在均匀瑞利杂波背景中检测起伏和非起伏目标时具有几乎相同的性能。

除上述结果外,Dillard 在文献[18]中把 Steenson 的工作[19]进行了延伸,分析了单脉冲 CA-CFAR 检测器对非起伏目标的检测,得到了检测概率 P_d 的计算式。结果表明,检测损失随着单脉冲虚警概率下降而上升,并且随着参考单元数 R 增加而下降。当单脉冲虚警概率很大时,由双门限检测带来的检测性能损失很小,Nitzberg[20]还得到了对起伏信号更通用的检测概率计算式。

3.10　ML 类 CFAR 检测器在多目标环境中的性能

在对检测单元进行目标(称为主目标)检测时,若在参考滑窗中还出现其他的目标(称为干扰目标),CA-CFAR 的检测阈值就会上升,CA-CFAR 对主目标的检测性能会严重下降[17],这就是所谓的"目标遮蔽"现象。通过适当调整参考滑窗长度 R 和 T 值可以在一定程度上缓解这个问题,但是不能真正解

决问题。并且当 R 较大时,虽然可以减小干扰目标在杂波功率水平估计中的比重,但是干扰目标和杂波尖峰进入参考滑窗的机会也多了。SO 是针对上述问题对 CA 的修正型,当干扰目标只是出现在前沿滑窗或后沿滑窗中时,它对分辨空间上邻近的目标十分奏效[21]。当强干扰目标造成的遮蔽效应对于 CA 和 GO 很严重时,SO 在这种情况下却几乎不受干扰目标强度影响。实际上,当 $R\leqslant 16$ 时,GO 几乎不能检测在检测单元和参考滑窗中的一对目标[17]。Weiss 在文献[17]中给出了有一个干扰目标时 CA、GO 和 SO 检测器 P_d 的闭型解,以及 SO 相对于没有干扰目标时 CA 的检测损失,见表 3.2。表中的环境为一个干扰目标,且其强度趋于无穷,检测概率 $P_d=0.5$。

表 3.2 SO-CFAR 相对于 CA-CFAR 的附加损失 (单位:dB)

P_{fa}	R			
	4	8	16	32
10^{-4}	8.62	3.90	1.92	1.02
10^{-6}	13.2	5.84	2.71	1.35
10^{-8}	18.1	7.96	3.50	1.69

在干扰目标同时分布在前沿和后沿滑窗中时,Gandhi[21] 的分析结果表明,SO 的检测性能虽然优于 CA 和 GO,但是也严重恶化。

Weiss 还指出,GO 应结合一些删除参考单元中大回波样本的方法来使用,因此他提出一种在两个局部估计中分别先剔除几个最大的样本,再用 GO 逻辑形成 Z 的修正方案,尽管他没有分析它的性能,但相信它在由杂波边缘和空间上邻近目标组成的复杂非均匀环境中会起适当作用,能使虚警概率或遮蔽效应减小。

Al-Hussaini 在文献[22]中分析了在 L 个干扰目标的环境中具有 M 个非相参脉冲积累的 CA-CFAR 检测性能,并在主目标和干扰目标均为 Swerling Ⅱ 型和平方律检波的假设下,给出了 P_d 的解析表达式。结果表明,当 L 增加时,P_{fa} 急剧下降,P_d 也相应恶化。因此,他建议使用阈值补偿技术[23-24]。

阈值补偿技术基于雷达跟踪系统提供的干扰目标的信息修正标称化因子 T 来达到补偿阈值的目的。Mclane[23] 等提出的是一种具有二级阈值控制过程的修正的 CA-CFAR,利用随机出现的干扰目标的先验信息修正阈值,使虚警概率保持在一定水平的前提下使检测概率尽可能得以恢复。Al-Hussaini 和 Ibrahim[24] 把这种技术扩展到 GO-CFAR 和 SO-CFAR。当有一个干扰目标出现在参考滑窗中时,采用阈值补偿技术可使 CA 和 GO 的检测性能获得明显的提高。其中,GO 的改善最明显,特别是在参考单元数较小时,而 SO 的性能改善却不明显。

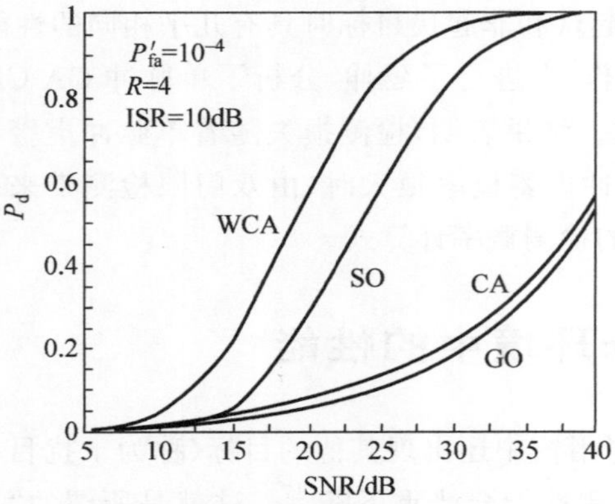

图 3.7 GO、CA、SO、WCA-CFAR 的检测概率

WCA 也可以被认为是一种阈值补偿方法,它作为 CA 的一种修正型,主要是针对多目标环境的。当参考滑窗中有一个干扰目标,且主目标和干扰目标均为 Swerling Ⅱ 型目标时,Barkat[5] 给出了 WCA 及 CA、GO 和 SO 的 P_d 表达式,并利用这些结果分析了它们在上述环境中的检测性能。结果表明,WCA 的检测性能比其他三个的检测性能都好,见图 3.7。图 3.7 中 ISR 是干扰与目标信号功率比,P'_{fa} 是虚警概率设计的指定值。在高信杂比时,WCA 对存在强、弱干扰目标

时的检测性能几乎相同，并且随着参考单元数的增加而提高[5]，WCA 的加权值基于干扰的先验信息以使检测概率极大并保持恒定的虚警概率。

3.11 ML 类 CFAR 检测器在杂波边缘环境中的性能

杂波边缘(Clutter Edge)描述的是检测不同背景特性区域间的过渡区情况，这种情况的典型例子是降雨区的边缘、海洋陆地交界处等。如果检测单元处于弱杂波区，而参考滑窗中其他一些参考单元处于强杂波区，那么即使信噪比很大也会对目标检测产生遮蔽效应，P_d 和 P_{fa} 都会下降。如果检测单元处于强杂波区，而其他一些参考单元处于弱杂波区，那么虚警概率会急剧上升。对于 $R=32$，设定虚警概率 $P'_{fa}=10^{-6}$ 的 CA-CFAR 检测器，一个 20dB 杂波强度的变化可使 P_{fa} 上升近 3 个数量级。这个问题是搜索雷达设计中应考虑的一个重要问题，GO 作为 CA 修正型，是专门针对杂波边缘情况的解决方案。

在杂波边缘环境中一般只分析检测器的虚警性能。假设背景杂波包络服从瑞利分布，经过平方律检波器检波后服从指数分布。杂波边缘的数学模型一般考虑为参考滑窗内杂波强度呈现一个阶跃变化，也就是参考滑窗中一部分单元服从式(3.11)而另一些单元服从式(3.12)的分布。此时各参考单元样本不再是同分布的。这种假设是一种理想情况，实际上杂波功率经常是有起伏的，并且幅度间部分相关。但是，这种简单的假设有利于对 CFAR 处理方法在杂波边缘的性能进行分析。

文献[25]给出了 GO-CFAR 在杂波边缘环境中虚警控制能力的近似分析，而文献[21]给出了 CA、GO、SO-CFAR 在杂波边缘环境中虚警概率的解析表达式。根据式(3.11)和式(3.12)的假设，当杂波边缘处于前沿滑窗中时，CA-CFAR 的杂波功率水平估计为

$$Z=\sum_{i=1}^{N_C} x_i+\left(\sum_{i=N_C+1}^{n} x_i+\sum_{j=1}^{n} y_j\right)=Z_1+Z_2 \tag{3.44}$$

其中，$Z_1 \sim G(N_C, \mu_0 \gamma)$，$Z_2 \sim G(2n-N_C, \mu_0)$，$\gamma$ 是两种杂波强度之比。因为 Z_1 和 Z_2 是统计独立的，所以 Z 的 MGF 是 Z_1 和 Z_2 各自的 MGF 的积。因此，当 $N_C \leqslant n$ 时，CA-CFAR 检测器的虚警概率为

$$P_{fa}=\frac{1}{(1+\gamma T)^{N_C}(1+T)^{2n-N_C}} \tag{3.45}$$

但是，应该指出 T 是由均匀杂波背景中的虚警概率设计值决定的阈值因子。在非均匀杂波背景中(包括杂波边缘和多目标环境)，实际的虚警概率将偏离其设计指定值。因此，今后用 P'_{fa}和 P_{fa} 分别表示虚警概率的设计值和实际值。在均匀杂波背景中，$P'_{fa}=P_{fa}$，所以不加区分。

当杂波边缘扫过检测单元，更多的强杂波将进入参考滑窗。此时，$n \leqslant N_C \leqslant 2n$，则 CA-CFAR 检测器的虚警概率为[18]

$$P_{fa}=\frac{1}{(1+T)^{N_C}(1+T/\gamma)^{2n-N_C}} \tag{3.46}$$

对于 GO 和 SO-CFAR 检测器，推导它们在杂波边缘环境中的虚警概率公式较为复杂，在这里只给出结果，文献[21]得到了这个结果，但是表达式极为复杂，而文献[26]得到的结果相对比较简单。文献[26]只给出了杂波边缘处于前沿滑窗时，即 $0 \leqslant N_C \leqslant n$ 时，GO-CFAR 检测器的虚警概率

$$P_{\mathrm{fa}}=\frac{[1/(1+\gamma\theta)]^{N_C}}{(1+\theta)^{n-N_C}}+\sum_{r=0}^{n-1}\binom{n+r-1}{r}\left[(1+\theta)^{r-n}-1\right](2+\theta)^{N_C-r-n}\times \tag{3.47}$$

$$(1+\gamma\theta+\gamma)^{-N_C}{}_2F_1\left(N_C,-r,n;\frac{1-\gamma}{1+\gamma+\gamma\theta}\right)$$

其中，$\theta=T/n$，${}_2F_1\left(N_C,-r,n;\frac{1-\gamma}{1+\gamma+\gamma\theta}\right)$是超几何函数，其展开式为

$${}_2F_1(a,-b,N_C;z)=\sum_{i=0}^{b}\frac{\Gamma(a+i)\Gamma(b+1)\Gamma(N_C)}{\Gamma(a)\Gamma(b+1-i)\Gamma(N_C+i)}\frac{(-1)^i}{i!}z^i \tag{3.48}$$

杂波边缘处于后沿滑窗中，即 $n\leqslant N_C\leqslant 2n$ 时，GO-CFAR 检测器的虚警概率为

$$P_{\mathrm{fa}}=\frac{[1/(1+\gamma\theta)]^{N_C-n}}{(1+\theta)^{2n-N_C}}+\gamma^{2n-N_C}\sum_{r=0}^{n-1}\binom{n+r-1}{r}\left[(1+\gamma\theta)^{r-n}-1\right]\times \tag{3.49}$$

$$(1+\gamma+\gamma\theta)^{N_C-r-2n}(\gamma\theta+2)^{n-N_C}{}_2F_1\left(N_C-n,-r;\ n;\ \frac{1-\gamma}{2+\gamma\theta}\right)$$

并且本书还推导出了 SO-CFAR 检测器在杂波边缘环境中的虚警概率解析表达式，在 $0\leqslant N_C\leqslant n$ 时为

$$P_{\mathrm{fa,SO}}=(1+T)^{-n}\left[1+\left(\frac{1+T}{1+\gamma T}\right)^{N_C}\right]-P_{\mathrm{fa,GO}} \tag{3.50}$$

其中，$P_{\mathrm{fa,GO}}$ 由式(3.47)给出，在 $n\leqslant N_C\leqslant 2n$ 时为

$$P_{\mathrm{fa,SO}}=(1+T)^{-n}+(1+T/\gamma)^{N_C-2n}(1+T)^{n-N_C}-P_{\mathrm{fa,GO}} \tag{3.51}$$

其中，$P_{\mathrm{fa,GO}}$ 也由式(3.47)给出。

图 3.8 给出的是 CA、GO、SO 检测器在杂波边缘环境中当 $P'_{\mathrm{fa}}=10^{-6}$，$R=32$ 时 3 组 γ 值的性能曲线，横坐标是被强杂波占据的参考单元数。随着 N_C 的增加，即杂波边缘由参考滑窗左边向右边滑动时，P_{fa} 先下降，在杂波边缘扫过检测单元时，P_{fa} 出现一个跳跃，形成一个尖峰，称之为"虚警尖峰"。"虚警尖峰"低说明 CFAR 检测器的虚警控制能力强。从图 3.8 中可以看出，GO 的虚警控制能力优于 CA 和 SO，并且对 γ 的变化不敏感，然而 CA 在 $\gamma=15\mathrm{dB}$ 时尖峰升高了 3 个数量级，SO 的尖峰则升高了 5 个数量级。

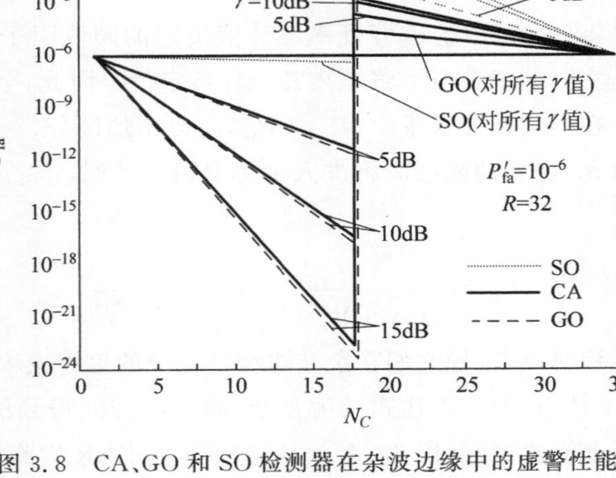

图 3.8　CA、GO 和 SO 检测器在杂波边缘中的虚警性能

3.12 比较与总结

在均匀杂波背景中，采用平方律检波的CA与线性检波CA的性能几乎相同，都具有较好的检测性能，而采用对数检波的CA在滑窗长度大于8时的CFAR损失比线性检波CA上升65%，在R较小时CFAR损失也变小。与CA相比，GO只表现出很小的检测性能下降，典型差值都在0.1～0.3dB，而SO-CFAR的检测性能十分依赖于参考单元数，当R很小时，它的检测性能损失比CA、GO-CFAR要大得多，但随着R增加而急剧减少；当P_{fa}较小时，SO-CFAR的检测性能损失也很大。文献[27]针对距离-多普勒二维CA-CFAR检测器，提出了一种圆形参考滑窗，相比于其他参考滑窗(矩形滑窗、交叉滑窗和列滑窗)，圆形参考滑窗能够获得更优的检测概率和更低的CFAR损失；而文献[28]在SAR图像目标CFAR检测中研究了变滑窗技术，滑窗尺寸的自适应变化与待检测的目标尺寸有关。

在多目标环境中，当干扰目标只分布在前、后沿滑窗中的一个滑窗中时，SO表现出很强的检测性能，且对干扰目标强度变化很不敏感，然而CA和GO的检测性能却严重下降。实际上，当$R\leqslant 16$时，GO在参考滑窗中存在干扰目标时几乎检测不到目标。在多个干扰目标同时分布在前沿和后沿滑窗中时，虽然SO还能保持相对于CA和GO的优势，但是其检测性能也严重下降。WCA在多目标环境中的检测性能比其他三个的都好，在高信杂比时，WCA对干扰目标的强度变化很不敏感。随着参考单元数的增加，WCA的检测性能也得以提高，并且与干扰目标的分布位置无关。但是它需要关于干扰目标分布情况的先验信息。

在杂波边缘环境中，GO的虚警控制能力明显强于CA和SO，它的虚警尖峰几乎不受γ影响。在$\gamma=15$dB时，CA的虚警尖峰上升了3个数量级，SO的虚警尖峰则上升了5个数量级(见式(3.44)～式(3.51))。

总之，这几种均值类CFAR检测器各有利弊。CA在均匀杂波背景中的检测性能最好，然而在非均匀背景中性能严重下降；GO具有很好的杂波边缘保护能力且在均匀杂波背景中相比CA检测性能下降不多，但是它在多目标环境中的检测性能下降到了令人不能接受的地步；SO具有较好的抗干扰目标能力，但是它在均匀杂波背景中的检测性能和在杂波边缘中的虚警性能都很差；虽然WCA的性能比较全面，但是它需要关于干扰的先验信息。

为了方便读者深入研究，本书将本章介绍的几种CFAR方法的有关文献总结如下。对于单脉冲平方律检波CA方法，除文献[1]外，还有文献[6,17,25,29,30]在不同的目标类型和背景下研究了CA。对于多脉冲检测情况，文献[31-32]分析了CA在没有干扰目标情况下对各种起伏目标的检测性能，而文献[14,22]分析了在参考单元中出现多个干扰目标时的性能。文献[10-11]分析了对数检波的CA。文献[12-13]分析了线性检波情况。对GO的研究，除文献[2]外，还有文献[5-6,14-17,21,24,26,33]。对SO进行研究的文献有[3,5,14,17,21,24]。

为了提高CA-CFAR检测器在多目标环境和杂波边缘中的性能，WCA-CFAR对两个子滑窗的局部估计进行加权获得对杂波的全局估计。GO-CFAR和SO-CFAR可以看作WCA-CFAR加权系数取特定值时的特例。为了提高利用参考滑窗内样本对杂波估计的有效性，文献[34-35]提出了对参考滑窗内样本进行准最佳加权的方法(QBW-CFAR)。为减少QBW-CFAR的样本排序时间并改善在杂波边缘的性能，基于子滑窗技术又提出了准最佳加权有序统计最大选择(QBWGO-CFAR)算法[36]和修正的准最佳加权(MQBW-CFAR)算法[37]。文献[38-39]又进一步提出了最佳线性无偏最大选择(BLUGO-CFAR)算法和修正的最佳线性无偏(IBLU-CFAR)算法，并进行了性能分析和比较。文献[40]提出先对参考滑窗样本进行拟合优度检验,若确认是 IID 的,则直接采用 CA-CFAR 进行检测,否则采用距离非均匀检测算法剔除非均匀样本,获取剩余的均匀样本,进而采用 CA-CFAR 进行检测。文献[41]提出了基于 Grubbs 准则剔除参考滑窗中的异常值,然后利用 CA-CFAR 进行检测的 CAG-CFAR 检测器。这些检测器的特点是都在参考滑窗中进行了某种方式的参考单元筛选,因此都应属于本书第 6 章所讨论的自适应 CFAR 检测的范畴。

在 CFAR 问题的研究中,通常假设接收的参考样本是统计独立和同分布的。然而,在实际雷达应用环境中,由于干扰目标回波或者杂波边缘的存在,不是所有样本都是同分布的。在非均匀背景环境中,CA-CFAR 检测器可能既得不到设计的虚警概率,也得不到高的检测概率。为缓解各参考样本不是同分布而导致的性能下降问题,前面章节中讨论了大量的检测器方案。不过在这些检测器中,几乎所有检测器都是假设杂波样本是统计独立的。在许多实际情况中,如气象杂波和金属箔片,杂波样本可能不是统计独立的,而是部分相关的。这时,再利用原来的假设已不能正确分析检测器的性能。因此,需要对参考单元样本相关条件下的 CFAR 检测进行研究。

Himonas 和 Barkat 还在文献[42]中研究了杂波的空间相关性对 CA-CFAR 检测器的虚警概率和检测概率的影响。他们假设杂波样本是瑞利分布且部分相关的,而热噪声样本是瑞利分布但不相关的,并假设杂波功率比热噪声功率高得多。为了研究杂波的空间相关性的影响,假设所有距离分辨单元的平均杂波功率是相同的。对于这种环境,他们得到了 CA-CFAR 检测器的实际虚警概率的准确表达式,它是杂波协方差矩阵的一个函数,也就是 CA-CFAR 检测器的 CFAR 参数 T 依赖于可能随时间变化的杂波协方差矩阵。同时,他们提出了一个可以估计杂波协方差矩阵的广义 CA-CFAR 检测器 GCA,阈值标称化因子可利用杂波协方差矩阵估计来计算。他们用计算机仿真方法分析了 GCA-CFAR 检测器的性能,结果表明它能保持与杂波样本相关性无关的虚警率。当杂波回波是部分相关时,GCA-CFAR 检测器获得了优于传统 CA-CFAR 检测器的检测性能,它是一种推广了的 CA-CFAR 检测器。该检测器适用于杂波回波的相关程度未知的情况。GCA-CFAR 检测器不仅能适应于杂波功率的变化,也能适应于杂波的相关程度的变化。文献[43]研究了相关噪声对 CA-CFAR 检测器虚警概率的影响,对比分析了各噪声样本(参考单元样本)统计独立与部分相关两种情况下的检测性能,结果表明,两者检测性能非常接近,这意味着各噪声样本统计独立的假设在工程实践中是合理且可行的。文献[44]研究表明,距离维和多普勒维加窗处理会在参考距离单元样本中引入相关性,从而导致基于参考距离单元样本 IID 假设的 CA-CFAR 检测器的虚警概率出现较大误差。Armstrong 和 Griffiths[45],以及 Watts[46]等研究了 CA、CAGO 和 OS 等 CFAR 检测器在空间相关 K 分布杂波中的检测性能,给出了杂波尖峰导致的检测损失及 CFAR 处理带来的附加损失,分析了杂波形状参数的估计对检测性能的影响,并提出了一种简便的用于分析部分相关杂波中检测性能的数值方法。关于相关杂波背景中的 CFAR 检测问题还可参考文献[47-48]。

本章研究的 CFAR 检测器适用于尺度型分布背景,即只有在指数分布等尺度型分布杂波背景下才能保持 CFAR 特性,而对于其他杂波分布背景,则不是 CFAR 的。文献[49-50]分析了 CA-CFAR 检测器在均匀 Weibull 杂波中的检测性能,其中文献[49]给出了检测概率与虚警概率的精确表达式。文献[51-54]则提出一种变换处理,将这些适用于指数分布背景的 CFAR 检测器映射为可工作于其他分布类型的检测器,这些分布类型包括 Pareto 分布、Weibull 分布等。这种变换处理方法的问题在于它依赖于杂波参数,进而导致与杂波参数有关的 CFAR 损失。

在基于神经网络的雷达目标检测技术研究中,文献[55-57]提出采用 CA-CFAR、GO-CFAR、SO-CFAR 的检测结果对神经网络进行训练,进而在保持高检测概率的同时获得较低的虚警概率。文献

[58]则以 CA、OS、GO、SO-CFAR 门限为基础，结合待检测单元数据，利用神经网络来形成新的检测门限。

参考文献

[1] Finn H M, Johnson R S. Adaptive detection mode with threshold control as a function of spatially sampled clutter-level estimates[J]. RCA Review, 1968, 29: 414-464.

[2] Hansen V G. Constant false alarm rate processing in search radars. IEEE International Radar Conference[C]. London: IEEE Radar Present and Future, 1973: 325-332.

[3] Trunk G V. Range resolution of targets using automatic detectors[J]. IEEE Transactions on AES, 1978, 14(5): 750-755.

[4] Barkat M, Varshney P K. A weighted cell-averaging CFAR detector for multiple target situation[C]. Baltimore: Proceedings of the 21st Annual Conference on Information Sciences and Systems, 1987: 118-123.

[5] Himonas S D, Barkat M, Varshney P V. CFAR detection for multiple target situations[J]. IEEE Proceedings, 1989, 136(5): 193-209.

[6] Guan jian, Peng Yingning, He You. Proof of CFAR by the use of the invariant test[J]. IEEE Transactions on AES, 2000, 36(1): 336-339.

[7] Rohling H. Radar CFAR thresholding in clutter and multiple target situations[J]. IEEE Transactions on AES, 1983, 19(4): 608-621.

[8] 何友，Rohling H. 有序统计恒虚警(OS-CFAR)检测器在 Weibull 干扰背景中的性能[J]. 电子学报，1995,23(1): 79-84.

[9] He Y. Leistungsfähigkeit der order statistics constant false alarm rate(OS-CFAR) schaltung vor einom Weibull-Storhintergrund[J]. Ortung and Navigation, 1993: 133-155.

[10] Hansen V G. Studies of logarithmic radar receiver using pulse-length discrimination[J]. IEEE Transactions on AES, 1965, 1(4): 246-253.

[11] Hansen V G, Ward H R. Detection performance of the cell averaging LOG/CFAR receiver[J]. IEEE Transactions on AES, 1972, 8(5): 648-652.

[12] Di Vito A, Moretti G. Probability of false alarm in CA-CFAR device downstream from linear-law detector[J]. Electronics Letters, 1989, 25(24): 1692-1693.

[13] Raghavan R S. Analysis of CA-CFAR processors for linear-law detection[J]. IEEE Transactions on AES, 1992, 28(3): 661-665.

[14] Ei-Mashade M B, Al-Hussaini E K. Performance of CFAR detectors for M-sweeps in the presence of interfering targets[J]. Signal Processing, 1994, 38(2): 211-222.

[15] Ritcey J A, Hines J L. Performance of max-mean level detector with and without censoring[J]. IEEE Transactions on AES, 1989, 25(2): 213-223.

[16] Ritcey J A. Detection analysis of the MX-MLD with noncoherent integration[J]. IEEE Transactions on AES, 1990, 26(3): 569-576.

[17] Weiss M. Analysis of some modified cell-averaging CFAR processors in multiple-target situations[J]. IEEE Transactions on AES, 1982, 18(1): 102-114.

[18] Dillard G M. Mean-level detection of nonfluctuating signals[J]. IEEE Transactions on AES, 1974, 10(6): 795-799.

[19] Steenson B O. Detection performance of a mean-level threshold[J]. IEEE Transactions on AES, 1968, 4(3): 529-534.

[20] Nitzberg R. Analysis of the arithmetic mean CFAR normalizer for fluctuating targets[J]. IEEE Transactions on

AES, 1978, 14(1): 44-47.

[21] Gandhi P P, Kassam S A. Analysis of CAFR processors in nonhomogeneous background[J]. IEEE Transactions on AES, 1988, 24(4): 427-445.

[22] Al-Hussaini E K. Performance of a cell averaging radar detector in the presence of interfering targets[J]. Frequenz, 1989, 43(1): 21-23.

[23] Mclane P J, Wittke P H, Sip C K. Threshold control for automatic detection in radar systems[J]. IEEE Transactions on AES, 1982, 18(2): 242-247.

[24] Al-Hussaini E K, Ibrahim B M. Comparison of adaptive cell-averaging detectors for multiple-target situations[J]. IEE Proceedings, 1986, 123(3): 217-223.

[25] Moore J D, Lawrence N B. Comparison of two CFAR methods used with square law detection of Swerling I targets: IEEE International Radar Conference[C]. Arlington VA: IEEE International Radar Conference, 1980: 403-409.

[26] Wilson S L. Two CFAR algorithms for interfering targets and nonhomogeneous clutter[J]. IEEE Transactions on AES, 1993, 29(1): 57-72.

[27] Wang W J, Wang R Y, Jiang R K, et al. Modified reference window for two-dimensional CFAR in radar target detection[J]. IET International Radar Conference(IRC 2018), The Journal of Engineering, 2019(21): 7924-7927.

[28] Chen S Y, Li X J. A new CFAR algorithm based on variable window for ship target detection in SAR images[J]. Signal, Image and Video Processing, 2019(13): 779-786.

[29] Helstrom C W, Ritcey J A. Evaluating radar detection probability by steepest descent integration[J]. IEEE Transactions on AES, 1984, 20(3): 624-634.

[30] Rohling H, Schilrmam J. Zar Entdeckunfsleistang storadapliver Radarsignal-Verarbeitcmg-ssysteme(CFAR)[J]. NteArchiv, 1981:169-177.

[31] Al-Hussaini Emad K, Al-Hussaini Essam K. Performance of a mean level detector processing M-correlated sweeps[J]. IEEE Transactions on AES, 1981, 17(2): 329-334.

[32] Hou X Y, Morinaga N T. Direct evaluation of radar detection probability[J]. IEEE Transactions on AES, 1987, 23(2): 418-423.

[33] Pace P E, Taylor L L. False alarm analysis of the envelope detection GO-CFAR processor[J]. IEEE Transactions on AES, 1994, 30(3): 848-864.

[34] 孟祥伟,何友. 准最佳加权有序统计恒虚警检测器[J]. 系统工程与电子技术,1997, 19(5): 14-17.

[35] Meng X W, Guan J, He Y. A discussion of linear weighted order statistics CFAR algorithm[J]. Journal of System Engineering and Electronics, 2004, 15(3): 232-236.

[36] Meng X W, He Y. 基于准最佳加权有序统计的最大选择 CFAR 检测算法[J]. 电子学报, 1997, 25(12): 74-78.

[37] 孟祥伟, 何友. 一种改进的准最佳加权有序统计恒虚警检测器[J]. 现代雷达,1997, 19(2): 57-62.

[38] 孟祥伟,关键,何友. 基于最佳线性无偏检测算法的最大选择恒虚警检测器[J]. 系统工程与电子技术, 2003, 25(5): 564-580.

[39] Meng X W, He Y. The best linear unbiased with greatest of selection(BLUGO) CFAR algorithm: Proceedings of IEEE Aerospace Conference[C]. Big Sky, USA: IEEE Aerospace Conference Proceedings,2004, 1980-1985.

[40] Zaimbashi A. An adaptive cell averaging-based CFAR detector for interfering targets and clutter-edge situations[J]. Digital Signal Processing, 2014,31: 59-68.

[41] Zhou W, Xie J H, Xi K, et al. Modified cell averaging CFAR detector based on Grubbs criterion in non-homogeneous background[J]. IET Radar, Sonar & Navigation, 2019, 13(1): 104-112.

[42] Himonas S D, Barkat M. Adaptive CFAR detection in partially correlated clutter[J]. IEE Proceedings, 1990, 137(5): 387-394.

[43] He M, Jia K X, Cheng T. False alarm probability of the digital channelized receiver based CA-CFAR detector[J]. WISM 2011, Part I, LNCS 6987, 2011: 86-91.

[44] Melebari A, Alomar W, Gaffar M Y A, et al. The effect of windowing on the performance of the CA-CFAR and OS-CFAR algorithms[C]. Johannesburg: 2015 IEEE Radar Conference, 2015: 249-254.

[45] Armstrong B C, Griffiths H D. CFAR detection of fluctuating targets in spatially correlated K-distributed clutter[J]. IEE Proceedings, 1991, 138(2): 139-152.

[46] Watts S. Cell-averaging CFAR gain in spatially correlated K-distributed clutter[J]. IEE Proceedings, 1996, 143(5): 321-327.

[47] 简涛,何友,苏峰,等. 非高斯杂波下自适应雷达目标检测新方法[J].航空学报, 2010, 31(3): 579-586.

[48] Aysin C H, Altunkan H. CA-CFAR detection in spatially correlated K-distributed sea clutter[C]. Antalya, Turkey: 2009 IEEE 17th Signal Processing and Communications Applications Conference,2009: 840-843.

[49] Fernando D A G, Andrea C F R, Gustavo F, et al. CA-CFAR detection performance in homogeneous Weibull clutter[J]. IEEE Geoscience and Remote Sensing Letters, 2019: 1-5.

[50] Mohamed B E M. Binary integration performance analysis of CA family of CFAR strategies in homogeneous Weibull clutter[J]. Radio-electronics and Communications Systems, 2020, 63(1): 24-41.

[51] Weinberg G V. Constant false alarm rate detectors for Pareto clutter models[J]. IET Radar Sonar Navigation, 2013(7): 153-163.

[52] Weinberg G V. General transformation approach for constant false alarm rate detector development[J]. Digital Signal Process, 2014,30: 15-26.

[53] Weinberg G V. The constant false alarm rate property in transformed noncoherent detection processes[J]. Digital Signal Process, 2016,51: 1-9.

[54] Shuji S. CFAR processing by converting Weibull to Rayleigh distributions[J]. Electronics and Communications in Japan, 2020,103(5-6): 3-10.

[55] Jabran A, Karl E O. A neural network target detector with partial CA-CFAR supervised training[C]. Brisbane: In Proceedings of International Conference on Radar, 2018.

[56] Jabran A, Karl E O. GO-CFAR trained neural network target detectors[C]. Boston: IEEE Radar Conference, 2019: 1-5.

[57] Jabran A. Training of neural network target detectors mentored by SO-CFAR[C]. Amsterdam: IEEE EUSIPCO, 2020: 1522-1526.

[58] Budiman P A R, Dayat K. Neural network-based adaptive selection CFAR for radar target detection in various environments[J]. International Journal of Intelligent Systems Technologies and Applications, 2019, 18(4): 377-390.

第4章 有序统计类 CFAR 处理方法

CHAPTER 4

4.1 引言

基于对参考单元样本处理方式的不同，参量型雷达目标 CFAR 处理方法可分为两大类：一类是第 3 章介绍的均值类，是最早被提出并被广泛研究的检测器；另一类很有代表性的是 OS 类 CFAR 处理方法，建立在 Rohling[1] 提出的 OS(Ordered Statistics)-CFAR 处理方法基础之上，利用参考滑窗内的有序样本进行处理。有序统计方法的思想来源于数字图像处理中的中值滤波。在多目标环境中，相比均值类方法，OS 类方法具有明显的性能优势；同时，在均匀杂波背景中，相比均值类方法，OS 类方法的性能下降也是适度的、可以接受的。

本章将讨论 OS-CFAR 检测器及由 OS 方法派生出来的一些基于排序的 CFAR 检测器，并分析它们在 3 种典型背景环境中的性能。

4.2 基本模型描述

假设检测单元中的目标回波幅值、接收机噪声和背景杂波三者统计独立且都服从零均值复高斯分布，则三者之和的包络服从瑞利分布，经过平方律检波器之后，检测单元样本 x 服从指数分布，其 PDF 和 CDF 分别为

$$f_1(x)=\frac{1}{\lambda'}\mathrm{e}^{-x/\lambda'},\quad x\geqslant 0 \tag{4.1}$$

$$F_1(x)=1-\mathrm{e}^{-x/\lambda'},\quad x\geqslant 0 \tag{4.2}$$

式中

$$\lambda'=\begin{cases}\mu, & \mathrm{H}_0\\ \mu(1+\lambda), & \mathrm{H}_1\end{cases} \tag{4.3}$$

μ 表示“杂波加噪声”的平均功率，即背景功率水平，λ 表示信号与“杂波加噪声”平均功率的比值，即信杂噪比，H_0 表示“目标不存在”的假设，H_1 表示“目标存在”的假设。在均匀背景中，参考单元样本 $x_i(i=1,2,\cdots,R)$ 只包含“杂波加噪声”，且 $x_i(i=1,2,\cdots,R)$ 是统计独立同分布的，服从参数 $\lambda'=\mu$ 的指数分布。在多目标环境中，统计独立的参考单元样本 $x_i(i=1,2,\cdots,R)$ 可分为两大类：一类只包含“杂波加噪声”，服从参数 $\lambda'=\mu$ 的指数分布；另一类包含“干扰目标+杂波+噪声”，服从参数 $\lambda'=\mu(1+\lambda_i)$ 的

指数分布。在杂波边缘环境中，统计独立的参考单元样本 $x_i(i=1,2,\cdots,R)$ 也分为两类，都是只包含"杂波加噪声"，区别在于，一类服从参数 $\lambda'=\mu$ 的指数分布，另一类服从参数 $\lambda'=\gamma\mu$ 的指数分布，γ 表示这两种背景功率水平之比。

OS 类 CFAR 检测器的共同点是对参考单元样本由小到大进行排序。在均匀背景情况下，R 个样本中的第 k 个有序样本的 PDF 为

$$f_{(k)}(x)=k\binom{R}{k}[1-F_0(x)]^{R-k}[F_0(x)]^{k-1}f_0(x) \tag{4.4}$$

它的 CDF 为

$$F_{(k)}(x)=\sum_{i=k}^{R}\binom{R}{i}[1-F_0(x)]^{R-i}[F_0(x)]^{i} \tag{4.5}$$

式中，$f_0(x)$ 和 $F_0(x)$ 分别表示均匀背景中参考单元样本 $x_i(i=1,2,\cdots,R)$ 的 PDF 和 CDF

$$f_0(x_i)=\frac{1}{\mu}\mathrm{e}^{-x_i/\mu},\quad F_0(x)=1-\mathrm{e}^{-x_i/\mu},\quad x_i\geqslant 0, i=1,2,\cdots,R \tag{4.6}$$

4.3　OS-CFAR 检测器

OS-CFAR 检测器框图如图 4.1 所示，其中 D 是检测单元样本，$x_i(i=1,2,\cdots,R)$ 是参考单元样本，R 是参考单元数(为了方便，通常取偶数)。

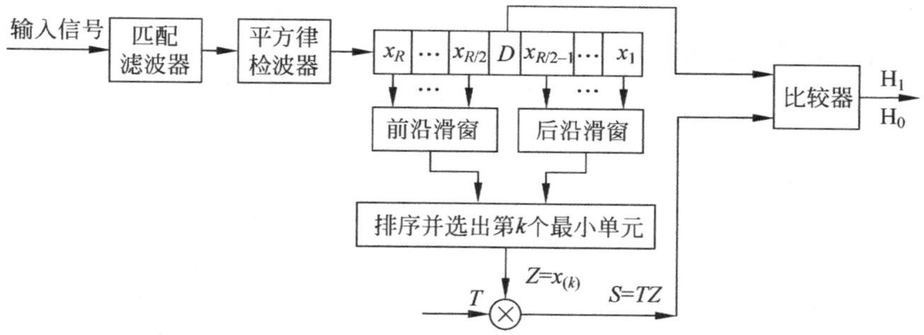

图 4.1　OS-CFAR 检测器框图

OS-CFAR 检测器首先对参考单元样本由小到大进行排序，有

$$x_{(1)}\leqslant x_{(2)}\leqslant\cdots\leqslant x_{(R)} \tag{4.7}$$

然后，取第 k 个排序样本 $x_{(k)}$ 作为对背景功率水平的估计 Z，即

$$Z=x_{(k)} \tag{4.8}$$

其中，序值 k 被称为"代表序值"[2]。那么由式(4.4)可知，均匀杂波背景中 Z 的 PDF 为

$$f_Z(z)=\frac{k}{\mu}\binom{R}{k}\mathrm{e}^{-(R-k+1)z/\mu}(1-\mathrm{e}^{-z/\mu})^{k-1} \tag{4.9}$$

Z 的 MGF(Moment Generating Function，矩母函数、矩生成函数)为

$$M_Z(u)=k\binom{R}{k}\frac{\Gamma(R-k+1+u\mu)\Gamma(k)}{\Gamma(R+u\mu+1)} \tag{4.10}$$

所以，由式(3.7)和式(3.8)得 OS-CFAR 检测器在均匀杂波背景中的检测概率和虚警概率分别为

$$P_{\mathrm{d}}=k\binom{R}{k}\frac{\Gamma[R-k+1+T/(1+\lambda)]\Gamma(k)}{\Gamma[R+T/(1+\lambda)+1]} \tag{4.11}$$

$$P_{\mathrm{fa}}=k\binom{R}{k}\frac{\Gamma(R-k+1+T)\Gamma(k)}{\Gamma(R+T+1)} \tag{4.12}$$

由式(4.9)可得 Z 的均值为

$$\mathrm{E}(Z)=\mu\sum_{i=1}^{k}\frac{1}{R-k+i} \tag{4.13}$$

所以，OS-CFAR 检测器的平均判决阈值 ADT 为

$$\mathrm{ADT}=\frac{\mathrm{E}(TZ)}{\mu}=T\sum_{i=1}^{k}\frac{1}{R-k+i} \tag{4.14}$$

在多目标环境中，若仅分析强干扰目标的影响，也就是假定干扰目标信号与“杂波加噪声”的功率比 INR 是无限的，这时，干扰目标回波总是占据参考滑窗中排序样本的最高位置，这在某种意义上是一种最糟情况。对于有限的 INR，检测损失将会变小。对于这种多目标环境，可以通过用($R-\mathrm{IL}$)代替式(4.11)中的 R 来评估(IL 代表参考滑窗中出现的干扰目标数)OS-CFAR 在多目标环境中的性能，即

$$P_{\mathrm{d}}=k\binom{R-\mathrm{IL}}{k}\frac{\Gamma[R-\mathrm{IL}-k+1+T/(1+\lambda)]\Gamma(k)}{\Gamma[R-\mathrm{IL}+T/(1+\lambda)+1]} \tag{4.15}$$

在非均匀杂波背景中，参考单元样本不再服从 IID 假设。在杂波边缘环境中，杂波功率水平由一种水平急剧变化到另一种水平。这里仅考虑从低杂波功率水平到高功率水平的过渡区情况，也就是假设 R 个参考单元中有 N_C 个单元服从分布

$$f_1(x)=\frac{1}{\gamma\mu_0}\mathrm{e}^{-\frac{x}{\gamma\mu_0}} \tag{4.16}$$

而其余 $R-N_C$ 个单元服从分布

$$f_2(x)=\frac{1}{\mu_0}\mathrm{e}^{-x/\mu_0} \tag{4.17}$$

其中，γ 是两种杂波功率水平之比。

由文献[3]可知，Z 的 CDF 为

$$F_Z(z)=\sum_{i=k}^{R}\sum_{j=\max(0,i-N_C)}^{\min(i,R-N_C)}\binom{N_C}{i-j}\binom{R-N_C}{j}(1-\mathrm{e}^{-z/\mu_0})^j\mathrm{e}^{-z(R-N_C-j)/\mu_0}\times (1-\mathrm{e}^{-z/\gamma\mu_0})^{i-j}\mathrm{e}^{-z(N_C-i+j)/\gamma\mu_0} \tag{4.18}$$

对 $F_Z(z)$ 求导可得 Z 的 PDF，当 $0\leqslant N_C\leqslant R/2$ 时，在杂波边缘环境中 OS-CFAR 检测器的虚警概率为

$$\begin{aligned}P_{\mathrm{fa}}&=\int_0^{\infty}F'_Z(z)\left[\int_{Tz}^{\infty}\frac{1}{\mu_0}\mathrm{e}^{-x/\mu_0}\mathrm{d}x\right]\mathrm{d}z=\frac{T}{\mu_0}\int_0^{\infty}F'_Z(z)\mathrm{e}^{-Tz/\mu_0}\mathrm{d}z\\&=\frac{T}{\mu_0}\sum_{i=k}^{R}\sum_{j=\max(0,i-N_C)}^{\min(i,R-N_C)}\binom{N_C}{i-j}\binom{R-N_C}{j}\sum_{n_1=0}^{j}\binom{j}{n_1}(-1)^{n_1}\sum_{n_2=0}^{i-j}\binom{i-j}{n_2}(-1)^{n_2}\times\\&\quad\int_0^{\infty}\exp\left\{-\frac{z}{\mu_0}\left[T+n_1+R-N_C-j+\frac{1}{\gamma}(n_2+N_C-i+j)\right]\right\}\mathrm{d}z\end{aligned} \tag{4.19}$$

为便于书写，这里用一个函数 Q 来表示式(4.19)中的和式，即设

$$Q(q;k,R,N_C)=\sum_{i=k}^{R}\sum_{j=\max(0,i-N_C)}^{\min(i,R-N_C)}\binom{N_C}{i-j}\binom{R-N_C}{j}\sum_{n_1=0}^{j}\binom{j}{n_1}(-1)^{n_1}\sum_{n_2=0}^{i-j}\binom{i-j}{n_2}(-1)^{n_2}q \tag{4.20}$$

这样，就可以把式(4.19)写为

$$P_{\mathrm{fa}}=\frac{T}{\mu_0}Q\left[\int_0^{\infty}\mathrm{e}^{-\frac{z}{\mu_0}(T+a)}\mathrm{d}z;k,R,N_C\right] \tag{4.21}$$

其中

$$a=n_1+R-N_C-j+\frac{1}{\gamma}(n_2+N_C-i+j) \tag{4.22}$$

因此，当 $0\leqslant N_C\leqslant R/2$ 时，OS-CFAR 检测器的虚警概率为

$$P_{\mathrm{fa}}=Q\left(\frac{T}{T+a};k,R,N_C\right) \tag{4.23}$$

当 $R/2\leqslant N_C\leqslant R$ 时，OS-CFAR 检测器的虚警概率为

$$\begin{aligned}P_{\mathrm{fa}}&=\int_0^{\infty}F'_Z(z)\left(\int_{Tz}^{\infty}\frac{1}{\gamma\mu_0}\mathrm{e}^{-x/\gamma\mu_0}\mathrm{d}x\right)\mathrm{d}z=\frac{T}{\gamma\mu_0}Q\left[\int_0^{\infty}\mathrm{e}^{-\frac{x}{\gamma\mu_0}(T+\gamma a)}\mathrm{d}x;k,R,N_C\right]\\&=Q\left(\frac{T}{T+\gamma a};k,R,N_C\right)\end{aligned} \tag{4.24}$$

4.4　CMLD-CFAR 检测器

Rickard 和 Dillard 提出的删除均值(Censored Mean Level Detector，CMLD)恒虚警率检测器[4]也是一种 OS 类 CFAR 检测器。它首先将参考滑窗中的参考单元样本按幅值大小进行排序，筛除掉 r 个较大的参考单元样本，取剩余样本的线性组合作为对检测单元杂波功率水平的估计 Z，即

$$Z=\sum_{i=1}^{R-r}x_{(i)} \tag{4.25}$$

$x_{(i)}(i=1,2,\cdots,R-r)$是对参考单元样本进行排序后的第 i 个有序样本，它们之间不再是统计独立的。为了求 Z 的 PDF，需要作一个线性变换，即

$$Z=\sum_{i=1}^{R-r}x_{(i)}=\sum_{i=1}^{R-r}\nu_i \tag{4.26}$$

其中 $\nu_i=(R-r-i+1)[x_{(i)}-x_{(i-1)}]$，$x_{(0)}\equiv 0$。$\nu_i(i=1,2,\cdots,R-r)$之间统计独立，且 ν_i 的 PDF 为

$$f(\nu_i)=\frac{c_i}{\mu}\mathrm{e}^{-c_i\nu_i/\mu} \tag{4.27}$$

其中

$$c_i=\frac{R-i+1}{R-r-i+1} \tag{4.28}$$

于是，ν_i 的 MGF 为

$$M_{\nu_i}(u)=\frac{c_i}{\mu u+c_i} \tag{4.29}$$

式(4.26)中 Z 的 MGF 为各个 ν_i 的 MGF 的乘积，即

$$M_Z(u)=\prod_{i=1}^{R-r}\frac{c_i}{\mu u+c_i} \tag{4.30}$$

那么，Z 的 PDF 为

$$f_Z(z)=L^{-1}\left\{\prod_{i=1}^{R-r}\frac{c_i}{\mu u+c_i}\right\}=L^{-1}\left\{\sum_{i=1}^{R-r}\frac{c_i}{\mu u+c_i}\left(\prod_{\substack{j\neq i\\ j=1}}^{R-r}\frac{c_j}{c_j-c_i}\right)\right\} \tag{4.31}$$

其中，L^{-1} 表示拉普拉斯反变换。定义系数 a_i 为

$$a_i=\frac{\prod\limits_{j=1}^{R-r}c_j}{\prod\limits_{\substack{j\neq i\\ j=1}}^{R-r}(c_j-c_i)}=\binom{R}{r}\binom{R-r}{i-1}(-1)^{i-1}\left(\frac{R-i+1-r}{r}\right)^{R-r-1} \tag{4.32}$$

可得 Z 的 PDF 和 MGF 为

$$f_Z(z)=\sum_{i=1}^{R-r}\frac{a_i}{\mu}\mathrm{e}^{-c_i z/\mu} \tag{4.33}$$

和

$$M_Z(u)=\sum_{i=1}^{R-r}\frac{a_i}{c_i+\mu u} \tag{4.34}$$

把式(4.34)代入式(3.7)和式(3.8)，得到 CMLD-CFAR 检测器的检测概率和虚警概率分别为

$$P_{\mathrm{d}}=\sum_{i=1}^{R-r}\frac{a_i}{c_i+T/(1+\lambda)} \tag{4.35}$$

$$P_{\mathrm{fa}}=\sum_{i=1}^{R-r}\frac{a_i}{c_i+T} \tag{4.36}$$

文献[5]中，Ritcey 给出了 CMLD-CFAR 检测器在 Swerling Ⅱ型干扰目标环境中的检测概率表达式。Barkat 等[6]推导了干扰目标为 Swerling Ⅳ型时的检测概率表达式。

在杂波边缘环境中，由于很难得到 $f_Z(z)$ 的解析表达式，因此无法导出 P_{fa} 的闭式解，需要借助 Monte Carlo 仿真来分析 CMLD-CFAR 检测器在杂波边缘环境中的检测性能。

4.5 TM-CFAR 检测器

作为 OS-CFAR 检测器的推广，Gandhi 和 Kassam 提出了剔除平均 TM-CFAR(Trimmed-Mean-CFAR)检测器[7-8]。该检测器先对参考滑窗内的样本进行排序，然后剔除掉最小的 r_1 个样本和最大的 r_2 个样本，最后对余下的样本取平均作为检测单元杂波功率水平的估计 Z，即

$$Z=\sum_{i=r_1+1}^{R-r_2}x_{(i)} \tag{4.37}$$

CA-CFAR、OS-CFAR 和 CMLD-CFAR 检测器都可看作 TM-CFAR 检测器的特例。尽管排序前参考滑窗中样本 $x_1,x_2,\cdots,x_R$ 是 IID 的，但是经过排序后所得的有序样本 $x_{(1)},x_{(2)},\cdots,x_{(R)}$ 不再服

从 IID 假设，需做如下线性变换

$$\begin{cases} W_1 = x_{(r_1+1)} \\ W_2 = x_{(r_1+2)} - x_{(r_1+1)} \\ \vdots \\ W_i = x_{(r_1+i)} - x_{(r_1+i-1)} \\ \vdots \\ W_{R-r_1-r_2} = x_{(R-r_2)} - x_{(R-r_2-1)} \end{cases} \tag{4.38}$$

所得变量 $W_1, W_2, \cdots, W_{R-r_1-r_2}$ 是统计独立的随机变量序列。W_i 的 PDF 为

$$f_{W_1}(x) = \frac{1}{\mu}(R-r_1)\binom{R}{r_1}\mathrm{e}^{-(R-r_1)x/\mu}(1-\mathrm{e}^{-x/\mu})^{r_1}, \quad x \geqslant 0 \tag{4.39}$$

$$f_{W_i}(x) = \frac{1}{\mu}(R-i+1-r_1)\mathrm{e}^{-(R-i+1-r_1)x/\mu}, \quad x \geqslant 0, \quad 2 \leqslant i \leqslant R-r_1-r_2 \tag{4.40}$$

再定义一个变换

$$V_i = (R-r_1-r_2-i+1)W_i, \quad i = 1, 2, \cdots, R-r_1-r_2 \tag{4.41}$$

因为 W_i 是独立的，所以 V_i 也是独立的随机变量。杂波功率水平估计 Z 可表示为

$$Z = \sum_{i=1}^{R-r_1-r_2} V_i \tag{4.42}$$

Z 的 MGF 是各个 V_i 的 MGF 的积。因此虚警概率可表示为

$$P_{\mathrm{fa}} = \prod_{i=1}^{R-r_1-r_2} M_{V_i}(T) \tag{4.43}$$

式中

$$\begin{cases} M_{V_1}(T) = \dfrac{R!}{r_1!(R-r_1-1)!(R-r-r_2)}\displaystyle\sum_{j=0}^{r_1}\frac{\binom{r_1}{j}(-1)^{r_1-j}}{\dfrac{R-j}{R-r_1-r_2}+T} \\ M_{V_i}(T) = \dfrac{b_i}{b_i+T}, \quad i = 1, 2, \cdots, R-r_1-r_2 \end{cases} \tag{4.44}$$

其中，$b_i = (R-r_1-i+1)/(R-r_1-r_2-i+1)$。用 $T/(1+\lambda)$ 代替式(4.43)中的 T 就得到 TM-CFAR 检测器的检测概率 P_{d}，把式(4.43)代入式(3.21)可得到 TM-CFAR 检测器的 ADT，即

$$\mathrm{ADT} = \frac{T(R!)}{(R-r_1-1)!}\sum_{j=0}^{r_1}\frac{(-1)^{r_1-j}}{(R-j)j!(r_1-j)!}\left(\frac{R-r_1-r_2}{R-j} + \sum_{i=2}^{R-r_1-r_2}\frac{1}{b_i}\right) \tag{4.45}$$

表 4.1 给出了 $P_{\mathrm{fa}} = 10^{-6}$，$R = 24$ 时对称和非对称剔除的 TM-CFAR 检测器的 T 和 ADT 的几组数值。当 $r_1 = r_2 = 0$ 时，T 和 ADT 也就是 CA-CFAR 检测器的 T 和 ADT 的数值。

文献[9]分析了 r_1 值与 r_2 值对 TM-CFAR 在不同环境下检测性能的影响，得出 r_2 值影响更大的结论，然后基于 KL 散度（Kullback-Leibler Divergence，KLD）与 Ostu 改进了 TM-CFAR 检测器，使之

能够根据背景自适应地剔除可能存在的干扰目标,并且不需要干扰目标的位置或者数目这一类先验信息。

表 4.1 TM-CFAR 的阈值因子 T 和 ADT 值

对称剔除				非对称剔除			
r_1	r_2	T	ADT	r_1	r_2	T	ADT
0	0	0.778	18.67	2	4	1.548	20.28
1	1	0.941	18.99	2	7	2.566	22.05
2	2	1.121	19.40	2	10	4.590	24.80
3	3	1.329	19.81	2	15	17.60	35.33
4	4	1.585	20.27	2	17	40.50	46.75
5	5	1.907	20.74	2	20	408.0	126.0
6	6	2.338	21.27	4	2	1.140	19.39
7	7	2.941	21.87	7	2	1.200	19.40
8	8	3.841	22.50	10	2	1.313	19.42
9	9	5.340	23.19	14	2	1.643	19.58
10	10	8.351	23.98	17	2	2.280	19.89
11	11	17.45	24.93	20	2	4.900	20.67

4.6 MX-CMLD CFAR 检测器

Ritcey[10] 把 Weiss 在文献[11]中提出的设想付诸实现,就是将 CMLD 和 MX-MLD(GO-CFAR 检测器)结合起来,进而提出 MX-CMLD CFAR 检测器,它先是采用 CMLD 算法分别估计前沿、后沿滑窗中杂波功率水平 X 和 Y,然后采用选大逻辑选取 X 和 Y 中的最大者,作为检测器对杂波功率水平的估计

$$Z = \max(X, Y) \tag{4.46}$$

其中

$$\begin{cases} X = \sum_{i=1}^{n-r-1} x_{(i)} + (r+1)x_{(n-r)} \\ Y = \sum_{i=1}^{n-r-1} y_{(i)} + (r+1)y_{(n-r)} \end{cases} \tag{4.47}$$

这种检测器用 MX-CMLD(R, r)表示。R 为参考单元总数;$n=R/2$ 为前、后沿滑窗中参考单元的个数,而 $\{x_{(1)} \leqslant x_{(2)} \leqslant \cdots \leqslant x_{(n)}\}$ 和 $\{y_{(1)} \leqslant y_{(2)} \leqslant \cdots \leqslant y_{(n)}\}$ 是分别来自 $\{x_1, x_2, \cdots, x_n\}$ 和 $\{y_1, y_2, \cdots, y_n\}$ 的有序统计量。MX-CMLD 检测器能够抑制前、后沿滑窗中干扰目标对杂波功率水平估计的影响。可以证明,式(4.47)中对杂波功率水平的估计 $X/(n-r)$ 和 $Y/(n-r)$ 是采用余下杂波样本对杂波功率水平估计的无偏最小方差估计(Unbiased Minimum Variance Estimator,UMVE),该估计达到了基于$(n-r)$个杂波样本的无偏估计的克拉美-罗界[8]。

图 4.2 给出了 MX-CMLD 检测器的框图,其中 X 和 Y 的计算如式(4.47)所示。该检测器的基本思想是将前、后沿滑窗中的大样本值视为异常值,并将这些异常值从式(4.47)中剔除,而只保留$(n-r)$个最小的杂波样本参与式(4.47)的计算。关于式(4.47)中权系数$(r+1)$的讨论请参见文献[12]。

在均匀背景或干净环境的假设条件下,X 与 Y 是统计独立同分布的,且都服从 Gamma 分布,这与

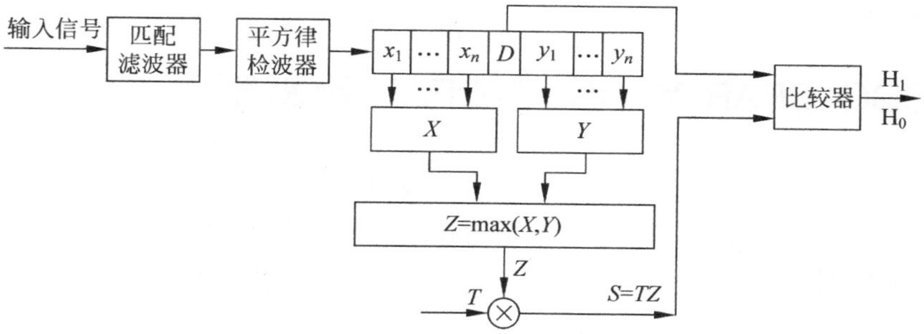

图 4.2　MX-CMLD 检测器框图

GO-CFAR 检测器中的估计结果是一样的[12]。此时，X 与 Y 的概率密度函数可写为

$$f_X(z)=f_Y(z)=z^{n-r-1}\mathrm{e}^{-z}/\Gamma(n-r),\quad z>0 \tag{4.48}$$

其中，Γ(·)是 Gamma 函数。相比于 GO-CFAR 检测器，删除大异常值的代价主要是减少了参考样本数以及由此导致的 SNR 损失增加，但这种损失并不大。图 4.3 给出了非均匀环境中 GO-CFAR、MX-CMLD CFAR 和 CMLD-CFAR 检测器的检测门限曲线，其可以清楚地看到非均匀环境对不同检测器的检测门限的影响。仿真中，强弱杂波区的杂波功率水平相差 30dB，且在强弱杂波区都可能存在干扰目标。

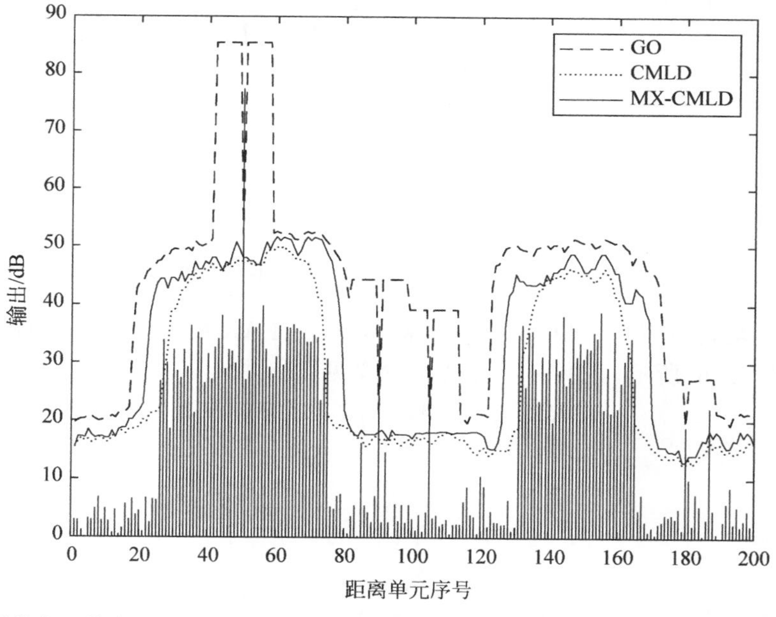

图 4.3　非均匀环境中 GO-CFAR、MX-CMLD CFAR 和 CMLD-CFAR 检测器的检测门限曲线

在文献[13]中，Ritcey 和 Hines 又将 MX-CMLD 归于 MAX 类有序统计检测器。所谓 MAX 类有序统计检测器就是选用前、后沿滑窗中两个局部估计中的最大者作为检测单元杂波功率水平的估计 Z。而且，在前、后沿滑窗的局部估计器(例如 MX-CMLD)中都使用了排序的方法。他们把这种 MAX 类有序统计检测器推广为，在两个局部估计中采用有序统计量的任意线性组合，即

$$X=\sum_{i=1}^{n}c_i x_{(i)},\quad Y=\sum_{i=1}^{n}c_i y_{(i)} \tag{4.49}$$

其中,c_i 是非负加权值。

4.7 OSGO-CFAR 和 OSSO-CFAR 检测器

Elias 等提出了 OSGO 和 OSSO[14] 两种恒虚警率检测器方法,它们都是先对前沿、后沿滑窗中参考单元样本按幅值大小排序,然后分别取前沿、后沿滑窗中第 k 个最小的样本 $x_{(k)}$ 和 $y_{(k)}$ 作为杂波功率水平的局部估计 X 和 Y,即

$$X=x_{(k)},\quad Y=y_{(k)} \tag{4.50}$$

OSGO-CFAR 和 OSSO-CFAR 检测器的杂波功率水平估计 Z 分别取为

$$Z_{\mathrm{OSGO}}=\max(X,Y) \tag{4.51}$$

和

$$Z_{\mathrm{OSSO}}=\min(X,Y) \tag{4.52}$$

图 4.4 给出了 OSGO-CFAR 和 OSSO-CFAR 检测器的框图,其中 X 和 Y 的计算如式(4.50)所示。

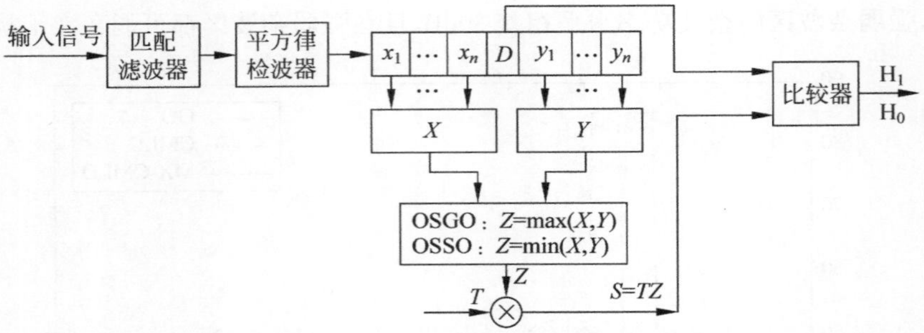

图 4.4 OSGO-CFAR 和 OSSO-CFAR 检测器框图

文献[14]得到了均匀背景中 OSGO-CFAR 和 OSSO-CFAR 的检测概率和虚警概率的解析表达式,如式(4.53)~式(4.56)所示,其中 $R=2n$,λ 的含义见式(4.3)。

$$P_{\mathrm{fa,OSGO}}=2k^2\binom{R/2}{k}^2\sum_{j=0}^{R/2-k}\sum_{i=0}^{R/2-k}\binom{R/2-k}{j}\binom{R/2-k}{i}\frac{(-1)^{R-2k-j-i}}{R/2-i}\frac{\Gamma(R-j-i)\Gamma(T+1)}{\Gamma(R-j-i+T+1)} \tag{4.53}$$

$$P_{\mathrm{d,OSGO}}=2k^2\binom{R/2}{k}^2\sum_{j=0}^{R/2-k}\sum_{i=0}^{R/2-k}\binom{R/2-k}{j}\binom{R/2-k}{i}\frac{(-1)^{M-2k-j-i}}{R/2-i}\frac{\Gamma(R-j-i)\Gamma[T/(1+\lambda)+1]}{\Gamma[R-j-i+T/(1+\lambda)+1]} \tag{4.54}$$

$$P_{\mathrm{fa,OSSO}}=2k\binom{R/2}{k}\left[\frac{\Gamma(k)\Gamma(T+R/2-k+1)}{\Gamma(T+R/2+1)}-k\binom{R/2}{k}\times\right.$$
$$\sum_{j=0}^{R/2-k}\sum_{i=0}^{R/2-k}\binom{R/2-k}{j}\binom{R/2-k}{i}$$

$$\left.\frac{(-1)^{R-2k-j-i}}{R/2-i}\frac{\Gamma(R-j-i)\Gamma(T+1)}{\Gamma(R-j-i+T+1)}\right] \tag{4.55}$$

$$P_{\mathrm{d,OSSO}}=2k\binom{R/2}{k}\left\{\frac{\Gamma(k)\Gamma[T/(1+\lambda)+R/2-k+1]}{\Gamma[T/(1+\lambda)+R/2+1]}-k\binom{R/2}{k}\times\right.$$
$$\sum_{j=0}^{R/2-k}\sum_{i=0}^{R/2-k}\binom{R/2-k}{j}\binom{R/2-k}{i}$$
$$\left.\frac{(-1)^{R-2k-j-i}}{R/2-i}\frac{\Gamma(R-j-i)\Gamma[T/(1+\lambda)+1]}{\Gamma[R-j-i+T/(1+\lambda)+1]}\right\} \tag{4.56}$$

表 4.2 给出了 $R=32$、$P_{\mathrm{fa}}=10^{-6}$ 以及平方律检波条件下 OS-CFAR、OSGO-CFAR、OSSO-CFAR 检测器取不同 k 值时的门限因子 T；表 4.3 中 $P_{\mathrm{fa}}=10^{-8}$，其他条件与表 4.2 相同。

表 4.2 取不同 k 值时的门限因子 T($R=32$、$P_{\mathrm{fa}}=10^{-6}$、平方律检波)

k	$1\leqslant k\leqslant R$	$1\leqslant k\leqslant R/2$	
	OS-CFAR	OSGO-CFAR	OSSO-CFAR
32	5.440460	—	—
31	6.579733	—	—
30	7.606999	—	—
29	8.615643	—	—
28	9.644402	—	—
27	10.71759	—	—
26	11.85498	—	—
25	13.07535	—	—
24	14.39853	—	—
23	15.84684	—	—
22	17.44654	—	—
21	19.22939	—	—
20	21.23478	—	—
19	23.51253	—	—
18	26.12690	—	—
17	29.16228	—	—
16	32.73206	—	—
15	36.99163	7.263981	11.79037
14	42.16010	9.073313	14.83076
13	48.55500	11.05236	18.40066
12	56.65329	13.32843	22.80241
11	67.19936	16.04618	28.46141
10	81.41584	19.40623	36.05826
9	101.4347	23.71711	46.77891
8	—	29.49240	62.87313
7	—	37.66004	89.07619
6	—	50.07664	—
5	—	71.00034	—

表 4.3 取不同 k 值时的门限因子 T($R=32$、$P_{fa}=10^{-8}$、平方律检波)

k	$1\leqslant k\leqslant R$	$1\leqslant k\leqslant R/2$	
	OS-CFAR	OSGO-CFAR	OSSO-CFAR
32	8.169095	—	—
31	9.778580	—	—
30	11.26185	—	—
29	12.73983	—	—
28	14.26541	—	—
27	15.87409	—	—
26	17.59633	—	—
25	19.46286	—	—
24	21.50714	—	—
23	23.76814	—	—
22	26.29271	—	—
21	29.13885	—	—
20	32.37988	—	—
19	36.11030	—	—
18	40.45426	—	—
17	45.57830	—	—
16	51.71125	—	—
15	59.17445	10.81250	18.97285
14	68.43278	13.50601	24.07937
13	80.18070	16.51451	30.29332
12	95.49571	20.04474	38.23946
11	—	24.34975	48.86809
10	—	29.79788	63.79190
9	—	36.97913	85.98831
8	—	46.91550	121.49221
7	—	61.54199	—
6	—	84.95542	—
5	—	127.2706	—

Han 和 Lee[15] 提出了两种改进的 OS-CFAR 检测器，被称为 OSCA-CFAR 和 OSGO-CFAR，这里的 OSGO-CFAR 与文献[14]中的相同。文献[15]分析了它们在多脉冲非相参积累条件下对 χ^2 分布起伏目标的检测性能。图 4.5 给出了非均匀环境中 CA、OS、OSGO、OSSO 检测器的检测门限曲线，其仿真条件与图 4.3 相同。由图 4.5 可知，CA 检测器对杂波边缘及干扰目标的适应能力较差，导致杂波边缘和大目标附近的检测门限偏高，容易漏警。OS、OSGO 同样面临杂波边缘小目标的漏警问题。OSSO 能适应干扰目标环境，但在杂波边缘环境中易产生虚警。

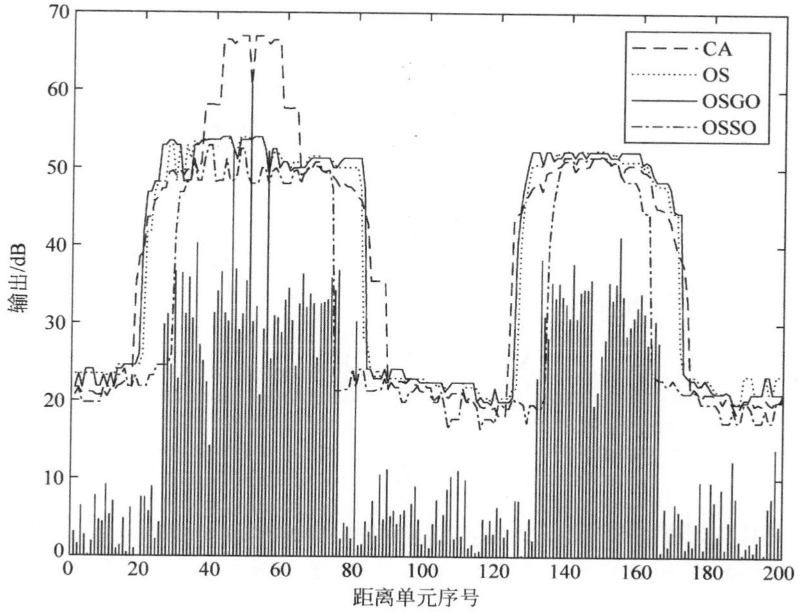

图 4.5 非均匀环境中 CA、OS、OSGO、OSSO 检测器的检测门限曲线

4.8 S-CFAR 检测器

Cao[16]提出了另一种基于转换的恒虚警率检测方案(Switching CFAR,S-CFAR),它基于检测单元的信息来选择合适的参考单元样本从而形成背景功率水平的估计,其实质是在 CA-CFAR 和 CMLD-CFAR 间进行切换来估计杂波功率水平。分析结果表明,S-CFAR 在均匀背景下性能接近 CA-CFAR,在非均匀背景下也具有较好的鲁棒性。S-CFAR 算法的步骤如下(为方便读者查阅原文,这里采用与原文相一致的符号)。

(1) $2N$ 个参考单元样本按照下面的原则被分到两个集合 S_0 和 S_1 中:

$$z_k \underset{S_0}{\overset{S_1}{\gtrless}} \alpha z \tag{4.57}$$

其中,$z_k(k=1,2,\cdots,2N)$为参考单元样本,若小于门限 αz,则归到集合 S_0 中,反之归到集合 S_1 中,z 为检测单元样本,$\alpha<1$ 是标称因子。

(2) 计算集合 S_0 中参考单元样本的个数 n_0,满足下列条件时,认为检测单元中存在目标:

$$z > \frac{\beta_0}{n_0}\sum_{z_k \in S_0} z_k, \quad n_0 > N_T \tag{4.58}$$

$$z > \frac{\beta_1}{2N}\sum_{k=1}^{2N} z_k, \quad n_0 \leqslant N_T \tag{4.59}$$

其中,β_0 和 β_1 为常量,N_T 为整数门限。

文献[16]中很大一部分内容是关于 S-CFAR 的虚警概率 P_{fa} 和检测概率 P_{d} 解析表达式的推导。文献[17]利用有序统计量的理论推导了 P_{fa} 和 P_{d} 另外一种形式的表达式,如式(4.60)和式(4.61)所示,它们具有较为简单的形式,因此容易较精确地计算出 S-CFAR 的门限参数。

$$P_{\mathrm{d}}(N,N_T,\alpha,\beta_0,\beta_1,\lambda)=\frac{\beta_0}{1+\lambda}\sum_{n_0=N_T+1}^{2N}\left[\binom{2N}{n_0}\sum_{j=0}^{n_0}\binom{n_0}{j}(-1)^j\frac{\bar{\Phi}_{V_r}(u)}{u}\bigg|_{u=\beta_0\left(A_j+\frac{1}{1+\lambda}\right)}\right]+ \frac{\beta_1}{1+\lambda}\sum_{n_0=0}^{N_T}\left[\binom{2N}{n_0}\sum_{j=0}^{n_0}\binom{n_0}{j}(-1)^j\frac{\bar{\Phi}_{V_0}(u)}{u}\bigg|_{u=\beta_1\left(A_j+\frac{1}{1+\lambda}\right)}\right] \tag{4.60}$$

其中，$\bar{\Phi}_{V_r}(u)=\prod_{i=1}^{2N-r}\frac{c_i}{u+c_i}$，$c_i=\frac{(2N-r)(2N-i+1)}{2N-r-i+1}$，$r=2N-n_0$，$A_j=(2N-n_0+j)\alpha$，$\lambda$ 的含义见式(4.3)，令 $\lambda=0$ 得到 P_{fa} 的表达式

$$P_{\mathrm{fa}}(N,N_T,\alpha,\beta_0,\beta_1)=\beta_0\sum_{n_0=N_T+1}^{2N}\left[\binom{2N}{n_0}\sum_{j=0}^{n_0}\binom{n_0}{j}(-1)^j\frac{\bar{\Phi}_{V_r}(u)}{u}\bigg|_{u=\beta_0(A_j+1)}\right]+ \beta_1\sum_{n_0=0}^{N_T}\left[\binom{2N}{n_0}\sum_{j=0}^{n_0}\binom{n_0}{j}(-1)^j\frac{\bar{\Phi}_{V_0}(u)}{u}\bigg|_{u=\beta_1(A_j+1)}\right] \tag{4.61}$$

图 4.6 给出了非均匀环境中 S-CFAR 检测器的检测门限曲线。该曲线受参数 N、N_T、α、β_0、β_1 的影响较大，且由于参数较多，因此在给定虚警概率条件下，如何确定一组合适的参数使得检测概率达到最大，是一件很困难的事情。

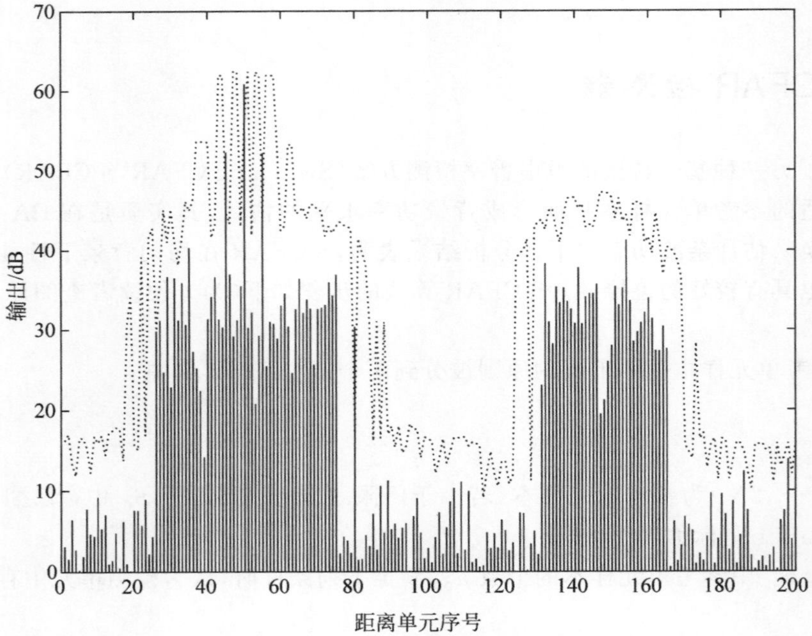

图 4.6 非均匀环境中 S-CFAR 检测器的检测门限曲线

4.9 其他 OS 类 CFAR 检测器

4.3 节～4.8 节讨论了 6 种比较有代表性的 OS 类 CFAR 检测器。其他新的 CFAR 检测器也大多基于有序统计方法对参考样本进行筛选，以估计检测单元处的杂波功率水平。

4.9.1　CATM-CFAR 检测器

文献[18]将 CA-CFAR 与 TM-CFAR 结合起来，提出了一种 CATM-CFAR 检测器，图 4.7 给出了其原理框图。

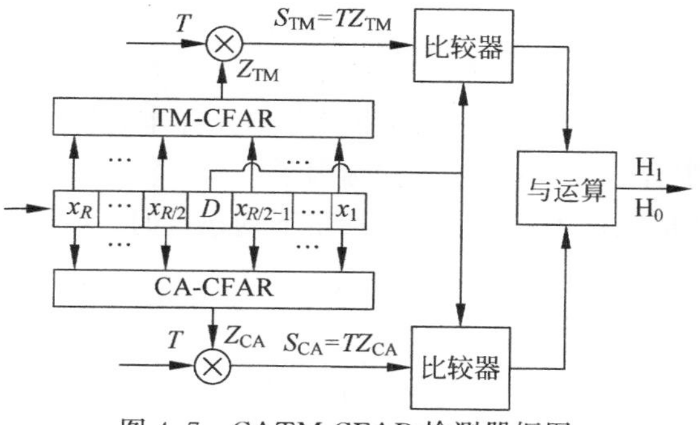

图 4.7　CATM-CFAR 检测器框图

CA-CFAR 检测器与 TM-CFAR 检测器同时独立工作，且采用相同的门限因子 T。首先，CA-CFAR 检测器与 TM-CFAR 检测器分别形成各自的杂波功率水平估计 Z_{CA} 与 Z_{TM}；然后，利用 T 计算各自的检测门限 S_{CA} 与 S_{TM}，并分别与检测单元样本 Y 进行判决得到各自关于目标存在与否的判决结果；最后，对这两个判决结果进行与运算，得到 CATM-CFAR 检测器的最终判决结果。文献[18]推导了 CATM-CFAR 检测器的检测概率、虚警概率和平均判决阈值的表达式

$$P_{fa,CATM}=(1+T)^{-N}\prod_{i=1}^{N-T_1-T_2}M_i(T) \tag{4.62}$$

$$P_{d,CATM}=\left(1+\frac{T}{1+\lambda}\right)^{-N}\prod_{i=1}^{N-T_1-T_2}M_i\left(\frac{T}{1+\lambda}\right) \tag{4.63}$$

$$\mathrm{ADT}_{CATM}=TN+\frac{TN!}{(N-T_1-1)!}\sum_{j=0}^{T_1}\frac{(-1)^{T_1-j}}{(N-j)j!(T_1-j)!}\left(\frac{N-T_1-T_2}{N-j}+\sum_{i=2}^{N-T_1-T_2}\frac{N-T_1-T_2-i+1}{N-T_1-i+1}\right) \tag{4.64}$$

其中，λ 表示信噪比。

$$M_1(T)=\frac{N!}{T_1!\ (N-T_1-1)!\ (N-T_1-T_2)}\sum_{j=0}^{T_1}\frac{\binom{T_1}{j}(-1)^{T_1-j}}{\dfrac{N-j}{N-T_1-T_2}+T} \tag{4.65}$$

$$M_i(T)=\frac{\dfrac{N-T_1-i+1}{N-T_1-T_2-i+1}}{\dfrac{N-T_1-i+1}{N-T_1-T_2-i+1}+T},\quad i=2,3,\cdots,N-T_1-T_2 \tag{4.66}$$

4.9.2　SOSGO-CFAR 与 MS-CFAR 检测器

S-CFAR 检测器通过在 CA-CFAR 和 CMLD-CFAR 间进行切换来估计杂波功率水平。文献[19-20]借

鉴了 S-CFAR 检测器的这种切换选择的思想,在 CA-CFAR 与 OSGO-CFAR 之间进行选择,形成了 SOSGO-CFAR 检测器(文献[20]中称为 IOSGO-CFAR 检测器),以提升杂波边缘环境中的检测性能。

图 4.8 给出了 IOSGO-CFAR 检测器框图。算法步骤如下:

(1) $R=2n$ 个参考单元样本按照下面的原则被分到两个集合 S_0 和 S_1 中:

$$x_k \underset{S_0}{\overset{S_1}{\gtrless}} \alpha D, \quad y_k \underset{S_0}{\overset{S_1}{\gtrless}} \alpha D \tag{4.67}$$

其中,D 为检测单元样本,$x_k, y_k(k=1,2,\cdots,n)$为参考单元样本,若小于门限 αD,则归到集合 S_0 中,其中元素记为 z_k,反之归到集合 S_1 中,$\alpha<1$ 是标称因子。

(2) 计算集合 S_0 中参考单元样本的个数 n_0,满足下列条件时,认为检测单元中存在目标:

$$D > \beta_0 \cdot \frac{1}{n_0}\sum_{z_k\in S_0} z_k, \quad n_0 > N_T \tag{4.68}$$

$$X = x_{(k)}, \quad Y = y_{(k)}, \quad D > \beta_1 \cdot \max(X,Y), \quad n_0 \leqslant N_T \tag{4.69}$$

其中,β_0 与 β_1 为门限因子,N_T 为整数门限。

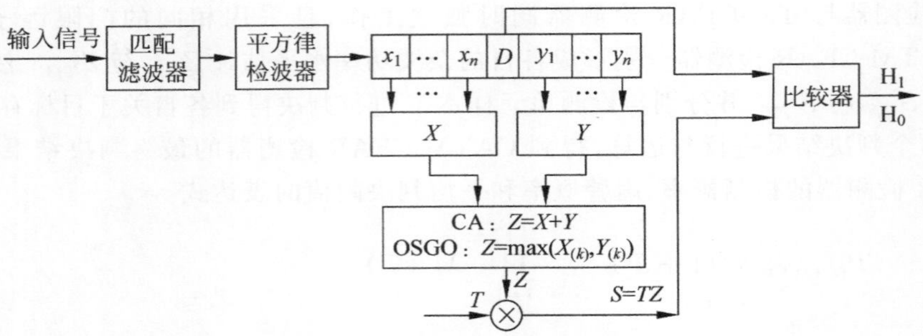

图 4.8 IOSGO-CFAR 检测器框图

文献[21]讨论了另一种修正的 S-CFAR 检测器(记为 MS-CFAR),即在 CA-CFAR 与 GO-CFAR 之间进行选择,目的是改善 S-CFAR 检测器在杂波边缘环境中的检测性能。文献[21]还给出了检测概率与虚警概率表达式,不过该方法没有考虑多目标环境。

4.10 OS 类 CFAR 检测器的性能分析

4.10.1 在均匀杂波背景中的性能

OS-CFAR 检测器在均匀背景和非均匀背景中的性能分析已经出现在很多文献中,其中,Gandhi 和 Kassam 在文献[7]中的分析比较全面。他们详细比较了 CA、GO、SO 和 OS 在均匀杂波背景中的检测性能,见图 4.9。由图 4.9 可见,OS 在均匀背景中的检测性能介于 GO 和 SO 之间,与 GO 十分接近,明显优于 SO。在 $R=24$,$P_{\text{fa}}=10^{-6}$,$P_{\text{d}}=0.5$ 时,表 4.4 给出了一组关于 Swerling Ⅱ 型目标 CFAR 检测损失的数据。

表 4.4 CA、GO、SO 和 OS-CFAR 检测器的检测损失

CFAR 检测器	CA	GO	SO	OS($k=18$)	OS($k=21$)
CFAR 损失/dB	1.31	1.65	2.47	1.93	1.82

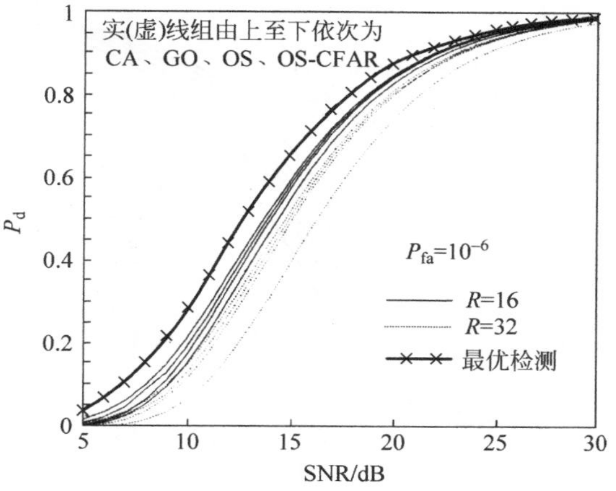

图 4.9 CA、GO、SO 和 OS-CFAR 检测器在均匀背景中的检测性能

Levanon 在文献[22]中给出了 OS 对强干扰目标的附加检测损失计算方法,并在文献[23]中与 Shor 一起分析了 OS-CFAR 在脉冲积累情况下的检测性能。

在均匀背景中,由于 CMLD 删除了一些参考单元样本,所以它相对于 CA 有一定的 CFAR 损失;并且当参考单元数 R 固定时,基本上是 r 越大,CFAR 损失越大。因为 r 增加,CMLD 可以利用的参考单元样本数也就减小,所以 CFAR 损失增加。

对于 TM-CFAR 检测器,r_1 和 r_2 值较小情况下的 TM 在均匀杂波背景中具有较好的检测性能。当 $r_1=r_2=0$ 时,TM 就变成了 CA,因此此时 TM 的检测性能最好。一般情况下,r_2 对 TM-CFAR 检测器检测性能的影响大,而 r_1 的影响小;并且,r_2 较小时 TM 的 ADT 值优于 OS-CFAR 的 ADT 值,也就是说,TM 在均匀背景中的检测性能要稍好于 OS。但是,考虑到要应对多目标情况,r_2 应该较大,而 r_1 应该较小,以便在均匀杂波背景中获得好的检测性能,若考虑第 5 章讨论的自动筛选技术[24-25],那么 r_2 也可以取得小些。

当样本筛除个数 $r=0$ 时,MX-CMLD 检测器退化为 MX-MLD 检测器。一般情况下,由于 MX-CMLD 删除了一些参考单元样本,所以在均匀杂波背景中的检测损失要比相同滑窗长度的 MX-MLD 大一些[10]。OSGO 要比 MX-CMLD 简单,但这是以稍高的检测损失为代价的。

文献[14]的分析表明,OSGO 在均匀杂波背景中与 OS 的检测性能相近,只有一个很小的可以忽略的 CFAR 损失。然而 OSSO 的 CFAR 损失要比 OS 高得多。OSGO 和 OSSO 相对于 OS 有一个共同的优点:它的样本排序处理时间只有 OS 的一半。文献[15]的分析结果表明,随着非相参脉冲积累数的增加,OSCA 和 OSGO 的检测性能明显增强。当选择适当的 k 值时,OSCA 在均匀杂波背景中的检测性能优于 OS 和 OSGO。

4.10.2 在多目标环境中的性能

尽管 OS 在均匀杂波背景中与 CA 和 GO 相比有一定的检测损失,但是在多目标环境中,OS 具有明显的优势。由图 4.10 可知,当 R、k 固定,随着干扰目标数 IN 的增加,OS 的检测性能会有所下降。较小的 k 值在其他条件相同时具有较好的应对干扰目标的能力,然而这是以牺牲均匀背景中的检测性能为代价的。当 IN$>R-k$ 时,OS 的检测性能将严重下降。

GO 在多目标环境中几乎检测不到有用目标,当干扰目标同时分布在前、后沿两个滑窗中时,SO 的

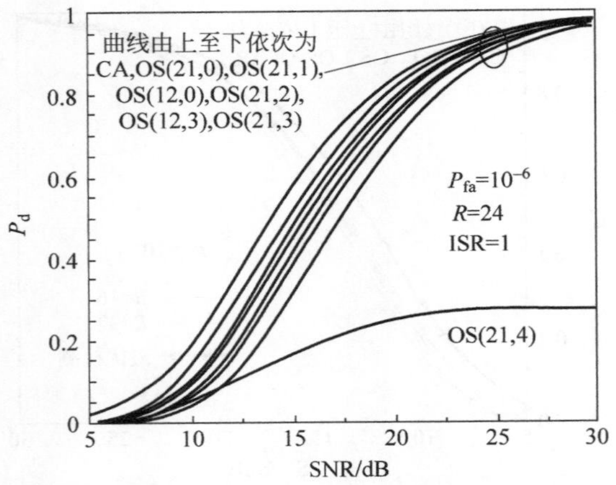

图 4.10 OS-CFAR 在多目标环境中的检测性能(OS(21,1)分别表示代表序值 $k=21$ 和干扰目标数 IN=1)

检测性能也会严重下降。但 OS 不受这些影响,只要干扰目标数 IN$<R-k$,OS 就可以较好地检测出目标。

Levanon 在文献[22]中给出了强干扰目标环境中 OS-CFAR 检测瑞利起伏目标的 CFAR 损失的计算方法。结果表明,当干扰目标数 IN$<R-k$ 时只带来很小的 CFAR 损失。在均匀杂波背景中 CA 略优于 OS,而在多目标环境中,OS 应对干扰目标的能力明显优于 CA 方法。

Ritcey[26]通过分析发现 OS 检测器的检测过程与双门限检测的过程有相同之处。因此,可以利用双门限检测来实现 OS 检测器,这样就无须排序处理,进而缩短了处理时间,增加了实时性,适于硬件实现。为了进一步提高实时性,他又提出了更适于硬件实现的 OS 检测器的脉动结构,并把这种结构移植到 MX-OSD[13]上。分析结果表明,当干扰目标数为随机数时,所带来的 CFAR 损失随干扰目标数量的随机性增大而急剧增大,远大于干扰目标数固定时的 CFAR 损失。这表明,在分析多目标环境中 OS 检测器的检测性能时,通常所采用的固定干扰目标数模型与实际干扰目标数模型不相符会造成显著的性能差异。

Nagle 和 Saniie[2]用有序样本的渐近效率(Asymptotic Efficiency,AE)来优化不同杂波分布条件下的 OS-CFAR 检测器,这样对于不同杂波分布类型来说,可以利用渐近分析的结果来选择最有效的代表序值。因此,渐近分析结果提供了未知目标信息情况下使 OS-CFAR 检测器具有通常意义上最优性能的代表序值。渐近效率通过有序统计量的渐近特性和对应的渐近标称化因子来估计。

对于 CMLD 在多目标环境中的性能也有大量的文献可以参考,文献[4]给出了 CMLD(用 D_r 表示,其中 r 表示删除样本的个数,且 $r=1,2,\cdots,R-1$)在不同类型 Swerling 主目标和干扰目标环境中的检测性能曲线。结果表明,当有一个干扰目标时,$D_r(r=1,2)$的性能明显优于 CA,并且 D_1 和 D_2 的检测性能几乎不受干扰目标强度影响。当有两个干扰目标时,D_2 明显优于 D_1,且 D_1 明显优于 D_0。针对干扰目标数固定情况下的 Swerling Ⅱ型和非起伏型干扰目标环境,Ritcey[5]研究了 CMLD 对 Swerling Ⅱ型目标的检测性能。当其他条件相同时,随着干扰目标数的增加,CMLD 的 CFAR 损失也随之增加,但是 CFAR 损失并不随 r 单调变化。CA 受干扰目标强度影响很大,而 CMLD 在干扰目标强度增长到一定程度时,其 CFAR 损失的变化就很小了。在强干扰目标环境中,CMLD 未删除单元数等于 OS 的序值时,CMLD 的检测性能与 OS 相差无几[23]。

Barkat 等[6]比较了 CMLD 对 Swerling Ⅱ型和 Swerling Ⅳ型目标的检测性能,给出了存在一个

Swerling Ⅱ型和 Swerling Ⅳ型干扰目标时检测概率的渐近公式(即干扰目标与噪声功率比 INR→∞),并由此公式计算得到不同 R 值时的若干条检测曲线。在 $r=R-\text{IN}$(IN 为干扰目标数)时,对于不同 IN 值,CMLD 的检测性能变化不大,即当 CMLD 删除的干扰目标数正确时它具有良好的检测性能,但是当干扰目标数未知时,删除干扰目标数的不同会使检测性能发生十分明显的变化。

Gandhi 和 Kassam[7] 对 TM、CA、GO、SO 和 OS 方法在非均匀背景中的性能进行了比较全面的分析和比较。为对付干扰目标,r_2 需取较大的值,而较大的 r_2 值会影响 TM 在均匀背景的检测性能。在 $R=24$ 时,TM($r_1=20,r_2=1$ 或 2)可以分别对抗 1 个或 2 个干扰目标,而 OS($k=21$)可以对抗 3 个干扰目标。但是,TM($r_1=20,r_2=1$ 或 2)在均匀杂波背景中的检测性能在一定程度上要比 OS($k=21$)好。

像 CMLD 一样,MX-CMLD 和 OSGO 这两种检测器也都是可以对抗干扰目标的 CFAR 检测器。假设干扰目标强度为无限大,且分布在一个滑窗中,那么在干扰目标数 $\text{IN}>\max(R-k,r)$(R 为滑窗长度,r 为 CMLD 的删除点,k 为 OS 的代表序值)时,这两种检测器均失效。当删除的单元过多时,CFAR 损失将很大。若强干扰目标同时分布在两个滑窗中,当前沿滑窗中干扰目标数 IL 固定,后沿滑窗中干扰目标数 IR 增加时,CFAR 损失也随之增加。并且不同 IL 值对应的曲线在 IR 达到允许的最大值时都重合在一起,这说明一个局部估计在一定程度上恶化时,另一个局部估计器很小程度的恶化是可以忽略的,这是采用选大逻辑的结果。

在多个干扰目标分布在前沿、后沿滑窗的条件下,Wilson[27] 还将 OS、GOOSE(MX-OSD,OSGO)、CML(CMLD)和 CGO(MX-CMLD)进行了比较。当干扰目标数小于容许限度时,CMLD 和 MX-CMLD 的 CFAR 损失比 OS 和 MX-OSD 的要小些。这是因为 CMLD 和 MX-CMLD 用多个杂波样本来估计杂波功率水平,而不是像 OS 和 MX-OSD 那样只使用一个有序样本作为杂波功率水平的估计。当干扰目标同时分布在前沿、后沿滑窗中时,GO 检测器的性能严重恶化,这是 GO 检测器应对干扰目标情况时存在的一个严重缺陷。若在局部估计中使用 OS 和 CMLD 方法可以在一定程度上克服这个缺点。当干扰目标数超过容许限度时,这四种 CFAR 检测器的性能均下降,MX-CMLD 和 CMLD 的性能下降要比 MX-OSD 和 OS 轻缓些。

OSGO 具有 OS 在非均匀杂波背景和多目标环境中的优点,只附加一个很小的 CFAR 损失,而 OSSO 的 CFAR 损失就比较大,见表 4.5。由表 4.5 可知,干扰目标数达到最大容许限度时,OSSO 的 CFAR 损失的变化比 OSGO 小。

表 4.5 OSGO、OSSO 和 OS 在多目标环境中的 CFAR 损失

IL,IR	0.0	1.0	1.1	2.0	2.1	3.0	3.1	3.2	4.0	4.2	5.0	6.0
OSGO	1.5892	2.0081	2.3225	2.7086	2.9003	4.0246	4.0958	4.2818	—	—	—	—
OSSO	2.3268	2.6246	—	2.8694	—	3.0384	—	—	—	—	—	—
OS	1.3925	1.7564	2.1597	2.1597	2.6404	2.6404	3.2109	3.9759	3.2109	5.1926	3.9759	5.1926

OSGO 和 OSSO: $R=2n=32,k=13$; OS: $R=32,k=26$。

CATM 通过与运算融合了 CA 与 TM 的检测结果,这给非均匀背景中的 CFAR 检测器设计提供了新思路,文献[18]的仿真结果表明,CATM 具有优于 CA 与 TM 的检测性能。

4.10.3 在杂波边缘背景中的性能

OS-CFAR 检测器在杂波边缘背景中的虚警控制能力随 k 值的变化而显著变化。图 4.11 给出了

$R=24$ 时的 OS 和 GO 在杂波边缘中取三组不同 k 值时虚警概率随参考滑窗中强杂波单元数 N_C 变化的曲线。

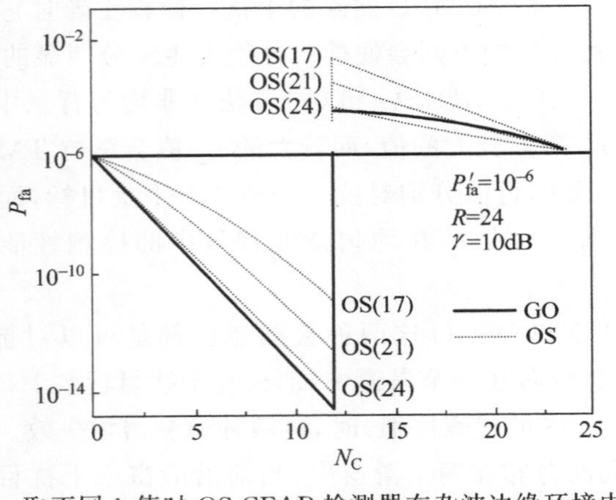

图 4.11　取不同 k 值时 OS-CFAR 检测器在杂波边缘环境中的性能

由图 4.11 可见，当 $k=R$ 时，OS 的性能稍优于 GO。然而 $k=R$ 在实际中是不可行的，因为此时它在参考滑窗中没有为应对干扰目标留有余地，从而会产生严重的目标遮蔽效应。随着 k 值的减小，OS 的虚警尖峰明显上升。例如，$k=17$ 比 $k=21$ 的虚警尖峰高出一个数量级还多，即使是在 $k=21$ 时，OS 的性能还是比 GO 差很多。文献[7]对 OS 抗边缘杂波性能不如人意的原因进行了详细分析。

Rohling 在文献[1]中提出有关杂波边缘"膨胀"和"收缩"的概念。当 $k>R/2$ 时，估计的强杂波区宽度大于实际杂波区宽度，称为"膨胀"；当 $k<R/2$ 时则相反，称为"收缩"。通过选择高 k 值而采用"膨胀"的方法对杂波功率水平进行估计可以较好地对抗边缘杂波。

Wilson 在文献[27]中对比了 CA、GO、OS、CMLD(CML)、MX-CMLD(CGO)和 MX-OSD(GOOSE、OSGO)抗边缘杂波的性能。通过分析发现，包含 GO 逻辑的 CFAR 方法均具有较好的抗边缘杂波能力，如 GO、MX-CMLD 和 OSGO 等方法。

在文献[8]中，Blake 通过对比分析 TM、OS 与 CA 在多目标和非均匀杂波中的性能，认为使用多分位数的 TM-CFAR 是不值得的，因为它只是在检测性能上相对于单分位数 OS-CFAR(通常的 OS)有一定程度的提高，但是在复杂性和抗边缘杂波能力方面都不如 OS。

OSGO 具有 OS 在非均匀背景中的优点，并且 CFAR 损失也较小[28]，处理时间也只是 OS 的一半。而 OSSO 除具有与 OSGO 相同的信号处理速度外，它的抗边缘杂波能力和在均匀背景中的检测性能都比 OS 差很多。

4.11 比较与总结

基于有序统计 CFAR 检测器的几种比较有代表性的检测方案包括：OS、CMLD、TM、MX-CMLD(也称作 CGO)、MX-OSD(也称作 GOOSE 或 OSGO)、OSCA、OSSO。

OS 在均匀杂波背景中的检测性能介于 GO 和 SO 之间[29]，与 GO 比较接近，明显优于 SO。对于 CMLD 与 OS，当两者参考单元数相同，且高端不参与杂波功率水平估计的有序样本数相同时，它们的检测性能十分相近。TM 在均匀杂波背景中的检测性能相对于 OS 有一定程度的提高，但是算法实现方面较 OS 复杂，并且其抗边缘杂波能力也要比 OS 差。CMLD 与 OS 的抗边缘杂波能力接近，但都相对 GO 有着明显的差距。

MX-CMLD 和 MX-OSD 与 GO 同属于采用选大逻辑的 MAX 类有序统计检测器，但是前两种方案删除了一些参考单元样本，所以它们在均匀杂波背景中的检测损失要比 GO 大些。同理，MX-OSD 的检测损失比 MX-CMLD 的还要大些。OSGO 与 MX-OSD 是同一种方案，而 OSSO 在均匀背景和杂波边缘中的性能都要比 OSGO 差很多。

由于 MX-CMLD 和 MX-OSD 同属于 MAX 类检测器，都采用了选大逻辑，所以它们的抗边缘杂波能力要比 CMLD 和 OS 强，具有 GO-CFAR 检测器在杂波边缘时的一些性能优势。

在多目标环境中，OS 类 CFAR 检测器相对于 ML 类 CFAR 检测器具有一定的优势。因为 OS 类 CFAR 检测器删除了一些可能是干扰目标信号的参考单元样本，在一定程度上减小了干扰目标回波进入杂波功率水平估计的概率，使杂波功率水平估计更趋于合理。当干扰目标数超过容许的限度时，OS、MX-OSD、CMLD 和 MX-CMLD 这 4 种 CFAR 检测器的性能均严重下降，其中 CMLD 和 MX-CMLD 的下降程度要比 OS 和 MX-OSD 轻一些。这是由于 CMLD 比 OS 采用更多参考单元样本进行杂波强度估计的缘故。同理，TM 在多目标环境中的检测性能也要比 OS 好些。

总之，采用选大逻辑的 MAX 类 CFAR 检测器的抗边缘杂波性能较好，而用 CMLD 方法形成局部估计的 CFAR 检测器要比用单一有序统计量作为局部估计的 CFAR 检测器的检测性能好些，但是 CMLD 会带来相对较差的抗边缘杂波性能。因此，综合考虑雷达目标检测领域 3 种典型背景下的检测性能，MX-CMLD 是 OS 类 CFAR 检测器中综合性能较优的检测器。在 OS 类 CFAR 检测器中，Barkat 和 Nagle 等还提出了两种 CFAR 检测器，分别是移动有序统计(Moving OS，MOS)[30]和线性组合有序统计(Linearly Combined OS，LCOS)[31] CFAR 检测器。文献[30-31]分别分析了这两种检测器，并将其与其他一些 OS 类检测器进行了比较。文献[32]为兼顾检测器在均匀背景和多目标环境中的性能，对参考滑窗中的有序样本采取了加窗处理。相比直接剔除掉高端和低端有序样本的 TM-CFAR 和 CMLD-CFAR 方法，加窗处理只要对高端和低端有序样本加较小的权值，就可比较明显地克服检测器在均匀背景中性能下降的问题，且具有有效应对其他干扰目标的能力。文献[33]提出采用模拟退火技术来优化分布式 OS-CFAR 系统的检测门限，优化参数包括统计阶数与尺度因子，结果表明这种优化技术能够搜索整个解空间而避免陷入局部优化值中。

多脉冲线性检测主要包括以下流程：首先是对每个参考单元和检测单元的多个脉冲回波样本进行线性积累，然后采用与单脉冲情况相同的处理方法对杂波功率水平进行估计，最后进行比较与判决。由于难以获得线性检波后随机变量之和的 PDF 表达式，因此很难进行检测性能的解析分析，往往需要借助仿真分析。并且，如果一种检测器在单脉冲情况下的检测性能很好的话，那么其在多脉冲情况下的检测性能会更好，但具体情况如何，还需进行详细的分析。文献[15，34-37]分析了几种 OS 类 CFAR 检测器在多脉冲非相参积累情况下的检测性能。对于双门限检测，每种 OS 类 CFAR 检测器和 ML 类 CFAR 检测器都可以先在每个脉冲上进行相应的 CFAR 处理，再根据积累准则进行目标是否存在的最终判决。针对采用双门限检测的 OSGO-CFAR 和 OS-CFAR 检测器，文献[38-39]系统和详细地分析它们在均匀背景以及多目标和杂波边缘引起的非均匀背景中的性能，并将杂波背景考虑为韦布尔非高斯背景。文献[40]比较了线性、平方律和贝塞尔三种不同检波方式下 CA-CFAR 与 OS-CFAR 的检测性能，结果表明，平方律检波为 OS-CFAR 提供了优于线性检波与贝塞尔检波的性能。对于 CA-CFAR，当 Weibull 形状参数数值较大时，平方律检波为其提供了最佳检测性能，而当 Weibull 形状参数较小时，线性检波为其提供了最佳检测性能。文献[41]针对广义 Swerling-Chi 起伏目标条件下的 OS-CFAR 检测性能进行了分析，并推导了非均匀背景中的检测概率解析表达式。文献[42]分析了 Weibull 背景中结合二值积累的 OS-CFAR 的检测性能，并推导了虚警概率表达式。分析表明，在单脉冲处理情况下，结合二值积累的 OS-CFAR 的检测性能明显优于 OS-CFAR，而且对杂波边缘虚警率的控制更加有效。

参考文献

[1] Rohling H. Radar CFAR thresholding in clutter and multiple target situations[J]. IEEE Transactions on AES, 1983, 19(4): 608-621.

[2] Nagle D T, Saniie J. Asymptotic analysis of OS-CFAR detectors for general clutter distributions[J]. Illinois Institute of Technology, Chicago, 1991,1(6): 25-30.

[3] Wilson S L. Two CFAR algorithms for interfering targets and nonhomogeneous clutter[J]. IEEE Transactions on AES. 1993, 29(1): 57-72.

[4] Rickard J T, Dillard G M. Adaptive detection algorithms for multiple target situations[J]. IEEE Transactions on AES, 1977, 13(4): 338-343.

[5] Ritcey J A. Performance analysis of the censored mean-level detector[J]. IEEE Transactions on AES, 1986, 22(4): 443-454.

[6] Himonas S D, Barkat M, Varshney P V. CFAR detection for multiple target situations[J]. IEE Proceedings, 1989, 136(5): 193-209.

[7] Gandhi P P, Kassam S A. Analysis of CFAR processors in nonhomogeneous background[J]. IEEE Transactions on AES, 1988, 24(4): 427-445.

[8] Blake S. OS-CFAR theory for multiple targets and nonuniform clutter[J]. IEEE Transactions on AES, 1988, 24(6): 785-790.

[9] 郭辰锋. 复杂背景目标检测技术研究[D]. 哈尔滨：哈尔滨工业大学，2020.

[10] Ritcey J A, Hines J L. Performance of max-mean-level detector with and without censoring[J]. IEEE Transactions on AES, 1989, 25(2): 213-223.

[11] Weiss M. Analysis of some modified cell-averaging CFAR processors in multiple-target situations[J]. IEEE Transactions on AES, 1982, 18(1): 102-114.

[12] Ritcey J A. Performance analysis of the censored mean-level detector[J]. IEEE Transaction on Aerospace and Electronic Systems, July 1986, 22(4): 44-53.

[13] Ritcey J A, Hines J L. Performance of max family of order statistic CFAR detectors[J]. IEEE Transactions on AES, 1991, 27(1): 48-57.

[14] Elias A R, De Mercad M G, Davo E R. Analysis of some modified order statistic CFAR: OSGO and OSSO CFAR[J]. IEEE Transactions on AES, 1990, 26(1): 197-202.

[15] Lee H. Performance of modified order statistics CFAR detectors with noncoherent integration[J]. Signal Processing, 1993, 31(1): 31-42.

[16] Cao T V. Constant false alarm rate algorithm based on test cell information[J]. IET Radar Sonar Navigation, 2008, 2(3): 200-213.

[17] Meng X W. Comments on constant false alarm rate algorithm based on test cell information[J]. IET Radar Sonar Navigation, 2009, 3(6): 646-649.

[18] Dejan I, Milenko A, Bojan Z. A new model of CFAR detector[J]. De Gruyter, 2014, 68(3-4): 125-136.

[19] Song Y Z, Meng X W, Qu F Y. A new CFAR method based on test cell statistics[C]. Guilin: Proc. IET International Radar Conference, 2009: 1-4.

[20] 李园园. 基于杂波分类识别的自适应 IOSGO-CFAR 算法研究及实现[D]. 西安：西安电子科技大学，2019.

[21] 曲付勇，孟祥伟，宋玉珍. 基于检测统计量的修正转换恒虚警检测算法[J]. 火力与指挥控制，2009, 34(10):

149-152.

[22] Levanon N. Detection loss due to interfering targets in ordered statistics CFAR[J]. IEEE Transactions on AES, 1988, 4(6): 678-681.

[23] Shor M, Levanon N. Performances of order statistics CFAR[J]. IEEE Transactions on AES, 1991, 27(2): 214-224.

[24] He Y. Performance of some generalized modified order statistics CFAR detectors with automatic censoring technique in multiple target situations[J]. IEE Proceedings, 1994,141(4): 205-212.

[25] 何友，Rohling H. 两种具有自动筛选技术的广义有序统计恒虚警检测器及其在多目标情况下的性能[J]. 电子科学学刊，1994，16(6)：582-590.

[26] Ritcey J A, Hwang J N. Detection performance and systolic architectures for OS-CFAR detectors: IEEE International Radar Conference[C]. Arlington: IEEE International Radar Conference,1990: 112-116.

[27] Wilson S L. Two CFAR algorithms for interfering targets and nonhomogeneous clutter[J]. IEEE Transactions on AES. 1993, 29(1): 57-72.

[28] Di Vito A, Galati G, Mura R. Analysis and comparison of two order situations CFAR systems [J]. IEE Proceedings, 1994, 141(2): 109-115.

[29] Rohling H. Analyse neuer methoden zar storadaptiven zielerkennung in einem radarsignalprozessor mit konstanter falschalarmwahrscheinlkhkeit(CFAR)[J]. ntzArchiv, 1983, 5(4): 101-111.

[30] Cho C M, Barkat M. Moving ordered statistics CFAR detection for nonhomogeneous backgrounds [J]. IEE Proceedings, 1993, 140(5): 284-290.

[31] Nagle D T, Saniie J. Performance analysis of linearly combined order statistic CFAR detectors [J]. IEEE Transactions on AES, 1995, 31(2): 522-532.

[32] Meng X W, Guan J, He Y. Performance analysis of the weighted window CFAR algorithms [C]. Adelaide: Proceedings of International Conference on Radar, 2003.

[33] Abdou L, Taibaoui O, Moumen A, et al. Threshold optimization in distributed OS-CFAR system by using simulated annealing technique [C]. Sousse: 2015 4th International Conference on Systems and Control (ICSC), 2015.

[34] Ei Mashade M B, Al Hussaini E K. Performance of CFAR detectors for M-sweeps in the presence of interfering targets[J]. Signal Processing, 1994, 38(2): 211-222.

[35] Ei Mashade M B. Detection performance of the trimmed-mean CFAR processor with noncoherent integration[J]. IEE Proceedings, 1995, 142(1): 18-24.

[36] Kim C J, Han D S, Lee H S. Generalized OS-CFAR detector with noncoherent integration[J]. Signal Processing, 1993, 31(1): 43-46.

[37] Lim C J, Lee H S. Performance of order-statistics CFAR detector with noncoherent integration in homogeneous situations[J]. IEE Proceedings, 1993, 140(5): 291-296.

[38] Meng X W. Performance analysis of ordered-statistic greatest of-constant false alarm rate with binary integration for M-sweeps[J]. IET Radar, Sonar and Navigation, 2010,4(1): 37-48.

[39] Meng X W. Performance analysis of OS-CFAR with binary integration for Weibull background [J]. IEEE Transactions on AES, 2013, 49(2): 1357-1366.

[40] Melebari A, Mishra A K, Gaffar M Y A. Comparison of square law, linear and Bessel detectors for CA and OS CFAR algorithms[C]. Johannesburg: 2015 IEEE Radar Conference, 2015. 10, 10. 1109/ Radar Conference, 2015.

[41] Kong L J, Wang B, Cui G L, et al. Performance prediction of OS-CFAR for generalized Swerling-Chi fluctuating targets[J]. IEEE Transactions on Aerospace and Electronic Systems, 2016, 52(1): 492-500.

[42] Meng X W. Performance analysis of OS-CFAR with binary integration for Weibull background [J]. IEEE Transactions on Aerospace and Electronic Systems, 2013, 49(2): 1357-1366.

第5章 采用自动筛选技术的GOS类CFAR检测器

CHAPTER 5

5.1 引言

任何一种方法都有其优缺点，这取决于检测器的工作环境与假定的杂波和目标统计模型。从雷达目标恒虚警检测器的发展历程可以看出，研究人员一直试图利用不同方法的优点并将它们组合起来，以兼顾检测器在均匀背景和各种非均匀背景中的性能。文献[1]将OS和GO与SO结合起来，提出了OSGO和OSSO-CFAR检测器，文献[2]对OSGO的性能进行了分析。文献[3-4]又提出了3种广义的有序统计检测器，即广义有序统计单元平均(Generalized Ordered Statistic Cell Averaging，GOSCA)、广义有序统计最大选择(Generalized Ordered Statistic Greatest Of，GOSGO)和广义有序统计最小选择(Generalized Ordered Statistic Smallest Of，GOSSO)，以及一种自动筛选技术。文献[5-16]对这方面的研究更推进了一步，用CA、CM、TM方法分别替代这3种检测器后沿滑窗中OS的局部估计，就又派生出了MOSCA-CFAR[5]、OSCAGO-CFAR[6-7]、OSCASO-CFAR[7]、MOSCM-CFAR[8-9]、OSCMGO-CFAR[10-11]、OSCMSO-CFAR[12]、MOSTM-CFAR[13-14]、OSTMGO-CFAR[15-16]、OSTMSO-CFAR检测器。文献[17-23]对这些检测器进行了推广和统一，分别提出了MTM、TMGO、TMSO-CFAR检测器。由于这些新的CFAR检测器都采用了文献[3]提出的一种自动筛选技术，因此就将这些CFAR检测器统称为采用自动筛选技术的GOS类CFAR检测器。5.2节将分别给出OS-OS类、OS-CA类和TM-TM类CFAR检测器的检测原理框图和基本数学模型，5.3～5.5节将分别给出OS-OS类、OS-CA类和TM-TM类CFAR检测器在均匀杂波背景中的检测概率、虚警概率及平均判决阈值ADT的表达式。5.6节将研究它们在均匀背景和多目标情况下的性能，并给出数值分析结果。5.7节将研究它们在杂波边缘中的性能，并进行性能分析和比较。

5.2 基本模型描述

5.2.1 OS-OS类CFAR检测器的模型描述

GOSCA，GOSGO，GOSSO-CFAR 3种检测器的共同特点是在两个局部估计中都采用了基于OS的杂波功率水平估计方法，所以称之为OS-OS类CFAR检测器，它们的基本结构如图5.1所示。其中，D为检测单元，两侧为参考单元，也称作参考滑窗。参考滑窗长度$R=R_1+R_2$，R_1和R_2分别为前沿和后沿参考滑窗长度。自适应判决准则为

$$D \underset{H_0}{\overset{H_1}{\gtrless}} TZ \tag{5.1}$$

其中，H_1 表示有目标，H_0 表示没有目标；Z 是杂波功率水平估计，T 是标称化因子，D 代表检测单元样本值；$x_i(i=1,2,\cdots,R_1)$ 和 $y_i(i=1,2,\cdots,R_2)$ 分别是前沿和后沿参考单元样本。

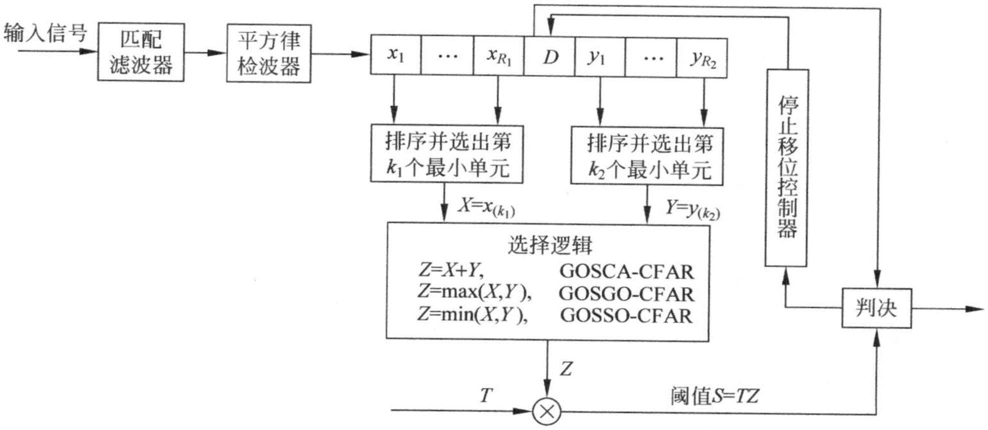

图 5.1 OS-OS 类 CFAR 检测器的基本结构

在图 5.1 中设置了停止移位控制器，它与参考单元移位寄存器一起实现了自动筛选技术[3-4]。如果根据式(5.1)做出检测单元中目标存在的判决(H_1 假设成立)，则右边参考滑窗(称为后沿滑窗)停止移位。这样，所有通过左边参考滑窗(称为前沿滑窗)并在检测单元被检测到的目标，将不再进入后沿滑窗。一般情况下，通过上述自动筛选处理后，后沿滑窗中的干扰目标数总是少于前沿滑窗中的干扰目标数。当目标的检测概率 $P_d \geqslant 0.9$ 时，后沿滑窗中的干扰目标数以概率 1 等于前沿滑窗中干扰目标数的 1/10。在恒虚警率检测算法中，这是一种新的自动筛选技术。另外，这种新的自动筛选技术不需要附加其他处理时间，并且可应用到所有的参量 CFAR 方案中。通过它可明显降低干扰目标的影响，进而提高探测性能。

1. 均匀杂波背景的基本模型

假设接收机噪声服从复高斯分布，经平方律检波，其输出结果服从指数分布。在对 Swerling Ⅱ 型目标进行检测的假设下，检测统计量 D 的 PDF 为

$$f_D(x)=\frac{1}{\lambda'}\exp\left(-\frac{x}{\lambda'}\right),\quad x\geqslant 0 \tag{5.2}$$

其中

$$\lambda'=\begin{cases}\mu, & H_0\\ \mu(1+\lambda), & H_1\end{cases} \tag{5.3}$$

λ 代表信号与"杂波加噪声"功率之比，μ 代表"杂波加噪声"的功率水平。不失一般性，在这里的分析中令 $\mu=1$。由于在均匀杂波噪声背景中得到的参考单元样本 $x_i(i=1,2,\cdots,R_1)$ 和 $y_i(i=1,2,\cdots,R_2)$ 与 D 是 IID 的，那么它们具有相同的 PDF，即

$$f_{x_i}(x)=f_{y_i}(x)=e^{-x},\quad x>0, H_0 \tag{5.4}$$

它们共同的 CDF 为

$$F_{x_i}(x)=F_{y_i}(x)=1-e^{-x},\quad x>0, H_0 \tag{5.5}$$

OS-OS 类 CFAR 检测器的前沿和后沿滑窗的局部估计器均采用 OS 方法，于是

$$X = x_{(k_1)}, \quad Y = y_{(k_2)} \tag{5.6}$$

$x_{(k_1)}$ 是来自前沿滑窗 R_1 个参考样本中的第 k_1 个排序样本。因此，X 的 PDF 为

$$f_{x_{(k_1)}}(x) = k_1 \binom{R_1}{k_1} (1 - \mathrm{e}^{-x})^{k_1 - 1} \mathrm{e}^{-(R_1 - k_1 + 1)x} \tag{5.7}$$

Y 的 PDF 为

$$f_{y_{(k_2)}}(x) = k_2 \binom{R_2}{k_2} (1 - \mathrm{e}^{-x})^{k_2 - 1} \mathrm{e}^{-(R_2 - k_2 + 1)x} \tag{5.8}$$

它们的 CDF 分别为

$$F_{x_{(k_1)}}(x) = \sum_{i=k_1}^{R_1} \binom{R_1}{k_1} (1 - \mathrm{e}^{-x})^{i} \mathrm{e}^{-(R_1 - i)x} \tag{5.9}$$

$$F_{y_{(k_2)}}(x) = \sum_{i=k_2}^{R_2} \binom{R_2}{i} (1 - \mathrm{e}^{-x})^{i} \mathrm{e}^{-(R_2 - i)x} \tag{5.10}$$

对于应用判决准则式(5.1)的任意一类的 CFAR 检测器，D 和 Z 是统计独立的随机变量。这样，在均匀杂波背景中，CFAR 检测器的检测概率为

$$\begin{aligned} P_{\mathrm{d}} &= \int_0^\infty f_Z(z) \Pr(D \geqslant TZ \mid \mathrm{H}_1) \mathrm{d}z = \int_0^\infty f_Z(z) \left[\int_{TZ}^\infty \frac{1}{1+\lambda} \exp\left(-\frac{x}{1+\lambda}\right) \mathrm{d}x \right] \mathrm{d}z \\ &= \int_0^\infty \mathrm{e}^{-[T/(1+\lambda)]z} f_Z(z) \mathrm{d}z = M_Z(u) \big|_{u = T/(1+\lambda)} \end{aligned} \tag{5.11}$$

虚警概率为

$$P_{\mathrm{fa}} = \int_0^\infty f_Z(z) \Pr(D \geqslant TZ \mid H_0) \mathrm{d}z = M_Z(u) \big|_{u=T} \tag{5.12}$$

$M_Z(u)$是 Z 的矩母函数(MGF，亦称矩生成函数)。

2. 杂波边缘情况下的基本模型

在杂波边缘环境中衡量的是虚警概率，它与目标模型无关。将杂波边缘考虑为参考滑窗内杂波功率发生一个阶跃，那么，参考单元样本的 PDF 为

$$f(x) = \frac{1}{\gamma'} \mathrm{e}^{-x/\gamma'} \tag{5.13}$$

其中

$$\gamma' = \begin{cases} \mu, & \mathrm{H_N} \\ \gamma\mu, & \mathrm{H_C} \end{cases} \tag{5.14}$$

式中，$\mathrm{H_N}$ 代表纯噪声情形，即弱杂波区；$\mathrm{H_C}$ 代表杂波加噪声情形，即强杂波区。设强杂波区占据的参考单元数为 N_C，其余的$(R - N_\mathrm{C})$个单元是弱杂波区。因为虚警概率只取决于两种杂波功率水平之比，而与其中任何一种杂波的绝对强度无关，所以，不失一般性，仍可取 $\mu = 1$，于是 γ 便是强、弱杂波功率水平之比。$\mathrm{H_N}$ 和 $\mathrm{H_C}$ 代表的杂波区情形的 PDF 分别为

$$f_1(x) = \mathrm{e}^{-x}, \quad \mathrm{H_N} \tag{5.15}$$

$$f_2(x) = \frac{1}{\gamma} \mathrm{e}^{-x/\gamma} \quad \mathrm{H_C} \tag{5.16}$$

OS-OS 类 CFAR 检测器使用 OS 方法来计算局部估计。当 $0 \leqslant N_\mathrm{C} \leqslant R_1$ 时，为杂波边缘进入前沿滑窗的情况，而后沿滑窗仍是充满幅值服从 e^{-x} 分布的弱杂波区情况。式(5.6)中的 X 和 Y 的概率密

度函数在 $0\leqslant N_C\leqslant R_1$ 时分别为[2]

$$f_X(x)=\sum_{i=k_1}^{R_1}\sum_{j=\max(0,i-N_C)}^{\min(i,R_1-N_C)}\binom{N_C}{i-j}\binom{R_1-N_C}{j}\sum_{n_1=0}^{j}\binom{j}{n_1}(-1)^{n_1}\sum_{n_2=0}^{i-j}\binom{i-j}{n_2}(-1)^{n_2}(-b)\mathrm{e}^{-bx} \tag{5.17}$$

$$f_Y(y)=k_2\binom{R_2}{k_2}\sum_{r=0}^{k_2-1}\binom{k_2-1}{r}(-1)^r\mathrm{e}^{-ay} \tag{5.18}$$

其中

$$a=r+R_2-k_2+1 \tag{5.19}$$

$$b=n_1+R_1-N_C-j+(n_2+N_C-i+j)/\gamma \tag{5.20}$$

为便于书写，用一个函数 Q 代替式(5.17)的右边，即有

$$Q(q;k_1,R_1,N_C)=\sum_{i=k_1}^{R_1}\sum_{j=\max(0,i-N_C)}^{\min(i,R_1-N_C)}\binom{N_C}{i-j}\binom{R_1-N_C}{j}\sum_{n_1=0}^{j}\binom{j}{n_1}(-1)^{n_1}\times\sum_{n_2=0}^{i-j}\binom{i-j}{n_2}(-1)^{n_2}q \tag{5.21}$$

则式(5.17)变为

$$f_X(x)=Q(-b\mathrm{e}^{-bx};k_1,R_1,N_C) \tag{5.22}$$

X 和 Y 的 CDF 为

$$F_X(x)=Q(\mathrm{e}^{-bx};k_1,R_1,N_C) \tag{5.23}$$

$$F_Y(y)=k_2\binom{R_2}{k_2}\sum_{r=0}^{k_2-1}\binom{k_2-1}{r}\frac{(-1)^r}{a}(1-\mathrm{e}^{-ay}) \tag{5.24}$$

在 $R_1\leqslant N_C\leqslant R$ 时，前沿滑窗已是充满幅值服从$\frac{1}{\gamma}\mathrm{e}^{-x/\gamma}$ 分布的强杂波情况，而后沿滑窗中出现的是杂波功率水平发生跳变的情况。前沿滑窗的第 k_1 个排序样本和后沿滑窗的第 k_2 个排序样本的 PDF 在 $R_1\leqslant N_C\leqslant R$ 时分别为

$$f_X(x)=\frac{k_1}{\gamma}\binom{R_1}{k_1}\sum_{r=0}^{k_1-1}\binom{k_1-1}{r}(-1)^r\mathrm{e}^{-xd/\gamma} \tag{5.25}$$

$$f_Y(y)=Q(-f\mathrm{e}^{-fy};k_2,R_2,N_C-R_1) \tag{5.26}$$

其中

$$d=r+R_1-k_1+1 \tag{5.27}$$

$$f=n_1+R-N_C-j+(n_2+N_C-R_1-i+j)/\gamma \tag{5.28}$$

它们的累积分布函数 CDF 分别为

$$F_X(x)=k_1\binom{R_1}{k_1}\sum_{r=0}^{k_1-1}\binom{k_1-1}{r}(-1)^r\left(\frac{1-\mathrm{e}^{-xd/\gamma}}{d}\right) \tag{5.29}$$

$$F_Y(y)=Q(\mathrm{e}^{-fy};k_2,R_2,N_C-R_1) \tag{5.30}$$

对于应用判决准则式(5.1)的任意一类 CFAR 检测器，D 和 Z 是独立的随机变量。因此，CFAR 检测器在 $0\leqslant N_C\leqslant R_1$ 时的虚警概率为

$$P_{\mathrm{fa}} = \int_0^{\infty} f_Z(z) \Pr\{D \geqslant Tz \mid \mathrm{H_N}\} \, \mathrm{d}z$$
$$= \int_0^{\infty} f_Z(z) \left(\int_{Tz}^{\infty} \mathrm{e}^{-x} \, \mathrm{d}x \right) \mathrm{d}z = \int_0^{\infty} \mathrm{e}^{-Tz} f_Z(z) \mathrm{d}z = M_Z(u) \big|_{u=T} \tag{5.31}$$

在 $R_1 \leqslant N_C \leqslant R$ 时

$$P_{\mathrm{fa}} = \int_0^{\infty} f_Z(z) \Pr\{D \geqslant Tz \mid \mathrm{H_C}\} \, \mathrm{d}z = M_Z(u) \big|_{u=T/\gamma} \tag{5.32}$$

5.2.2 OS-CA 类检测器的模型描述

5.2.1 节已经建立了 OS-OS 类 CFAR 检测器在均匀杂波背景和杂波边缘环境中的基本模型。下面将建立 OS-CA 类 CFAR 检测器在这两种环境中的基本模型。OS-CA 类 CFAR 检测器的共同特点是前沿滑窗的局部估计 X 采用了 OS 方法,而后沿滑窗的局部估计 Y 采用了 CA 方法,所以称之为 OS-CA 类 CFAR 检测器,其基本结构如图 5.2 所示。

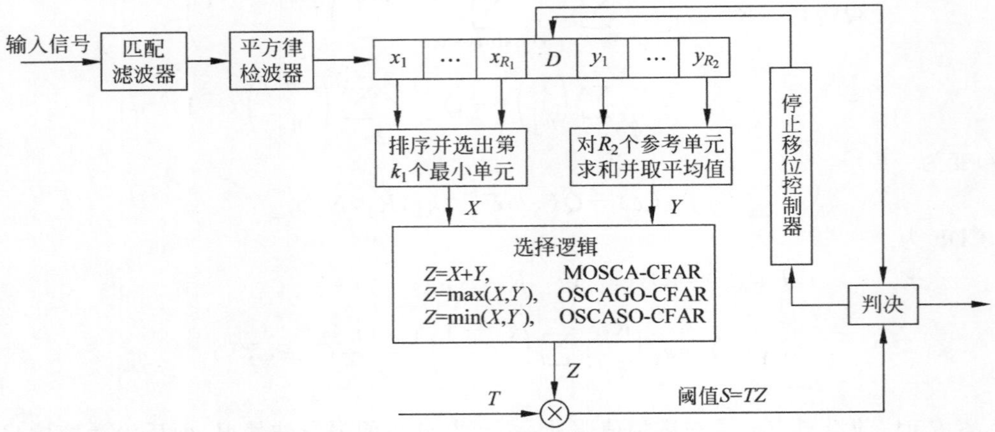

图 5.2 OS-CA 类 CFAR 检测器的基本结构

下面将建立 OS-CA 类 CFAR 检测器的基本模型。

1. 均匀背景中的基本模型

$$X = x_{(k)}, \quad Y = \frac{1}{R_2} \sum_{i=1}^{R_2} y_i \tag{5.33}$$

式(5.33)中定义的 X 与在式(5.6)中的 X 定义相同,所以在均匀杂波背景中,OS-CA 类 CFAR 检测器的局部估计 X 的 PDF 和 CDF 分别为式(5.7)和式(5.9)。由 CA 方法产生的后沿滑窗局部估计 Y 的 PDF 和 CDF 分别为

$$f_Y(y) = R_2^{R_2} y^{R_2 - 1} \frac{\mathrm{e}^{-R_2 y}}{\Gamma(R_2)} \tag{5.34}$$

$$F_Y(y) = 1 - \mathrm{e}^{-R_2 y} \sum_{i=0}^{R_2 - 1} \frac{(R_2 y)^i}{i!} \tag{5.35}$$

2. 杂波边缘情况下的基本模型

在杂波边缘环境中,杂波分布模型与 5.2.1 节中的相同。所以,当 $0 \leqslant N_C \leqslant R_1$ 时,前沿滑窗为存

在杂波边缘的情况，而后沿滑窗是充满幅值服从 e^{-x} 分布的均匀背景杂波情况。X 和 Y 的 PDF 在 $0\leqslant N_C\leqslant R_1$ 时分别为[2]

$$f_X(x)=Q(-g\mathrm{e}^{-gx};k,R_1,N_C) \tag{5.36}$$

$$f_Y(y)=R_2^{R_2}y^{R_2-1}\frac{\mathrm{e}^{-R_2y}}{\Gamma(R_2)} \tag{5.37}$$

它们的 CDF 分别为

$$F_X(x)=Q(\mathrm{e}^{-gx};k,R_1,N_C) \tag{5.38}$$

$$F_Y(y)=\int_0^y f_Y(t)\mathrm{d}t \tag{5.39}$$

其中，Q 函数的定义见式(5.21)。g 的定义如下

$$g=n_1+R_1-N_C-j+(n_2+N_C-i+j)/\gamma \tag{5.40}$$

当 $R_1\leqslant N_C\leqslant R$ 时，前沿滑窗已是充满幅值服从 $\frac{1}{\gamma}\mathrm{e}^{-x/\gamma}$ 分布的均匀杂波情况，而后沿滑窗中出现的是杂波边缘情况。X 和 Y 的 PDF 在 $R_1\leqslant N_C\leqslant R$ 时分别为[2]

$$f_X(x)=\frac{k}{\gamma}\binom{R_1}{k}(\mathrm{e}^{-x/\gamma})^{R_1-k+1}(1-\mathrm{e}^{-x/\gamma})^{k-1} \tag{5.41}$$

$$f_Y(y)=R_2^{R_2}y^{R_2-1}\frac{\mathrm{e}^{-R_2y}}{\Gamma(R_2)}\gamma^{R_1-N_C}{}_1F_1\left[N_C-R_1,R_2;R_2y\left(1-\frac{1}{\gamma}\right)\right] \tag{5.42}$$

其中，${}_1F_1(a,b;c)$是合流超几何函数。X 和 Y 的 CDF 分别为

$$F_X(x)=Q(\mathrm{e}^{-gx};k,R_1,N_C)\mid_{N_C=R_1}=\mathrm{e}^{-R_1x/\gamma}\sum_{i=k}^{R_1}\binom{R_1}{i}(\mathrm{e}^{x/\gamma}-1)^i \tag{5.43}$$

$$F_Y(y)=\int_0^y f_Y(t)\mathrm{d}t \tag{5.44}$$

其中，g 见式(5.40)。

因为 OS-CA 类 CFAR 检测器的判决准则和背景模型与 OS-OS 类 CFAR 检测器相同，所以它的虚警概率也可以引用式(5.31)和式(5.32)来计算。至此，OS-OS 类和 OS-CA 类 CFAR 检测器的基本模型已经建立。下面将给出 TM-TM 类 CFAR 检测器的基本模型，而 OS-OS 类、OS-CA 类、OS-CM 类、CM-CM 类、OS-TM 类等 CFAR 检测器作为 TM-TM 类 CFAR 检测器的特例，都可以包含在其中。

5.2.3　TM-TM 类检测器的模型描述

前面已经建立了 OS-OS 类和 OS-CA 类 CFAR 检测器在均匀杂波背景和杂波边缘环境中的基本模型。下面，将建立 TM-TM 类 CFAR 检测器的基本模型。TM-TM 类 CFAR 检测器的共同特点是前沿滑窗中的局部估计 X 采用了 TM 方法。后沿滑窗中的局部估计 Y 采用了 TM 方法。所以，称之为 TM-TM 类 CFAR 检测器。TM 方法是指在局部估计中剔除掉最大的 p 个和最小的 q 个样本，对剩下的样本求平均，作为杂波平均功率水平的局部估计；CA 方法等价于 $p=q=0$ 条件下的 TM 方法；OS 方法等价于 $p=R-k$、$q=k-1$ 条件下的 TM 方法，k 为 OS 方法中的“代表序值”；CM 方法等价于 $q=0$ 条件下的 TM 方法。因此可以说，CA、OS 和 CM 都是 TM 检测器的特例，故提出的采用自动筛选技术的 OS-OS 类、OS-CA 类、OS-CM 类、CM-CM 类和 OS-TM 类 CFAR 检测器都可以作为 TM-TM 类 CFAR 检测器的特例进行分析。TM-TM 类 CFAR 检测器的基本结构如图 5.3 所示。

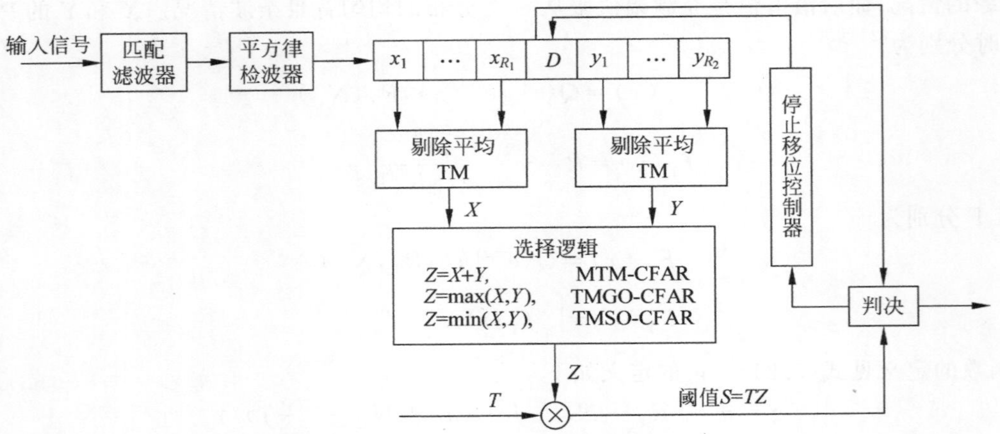

图 5.3 TM-TM 类 CFAR 检测器的基本结构

下面将建立 TM-TM 类 CFAR 检测器的基本模型[17-23]，TM-TM 类 CFAR 检测器前沿和后沿滑窗均采用剔除平均(TM)方法来产生杂波功率水平的局部估计 X 和 Y。对于 TM-TM 类 CFAR 检测器的前沿滑窗来说，前沿滑窗共有 R_1 个参考单元样本 $x_i(i\in\{1,2,\cdots,R_1\})$，先剔除掉最大的 R_1'' 个和最小的 R_1' 个样本，对剩下的 $R_1-R_1'-R_1''$ 个样本求平均，作为前沿滑窗对杂波平均功率水平的局部估计

$$X=\frac{1}{R-R'_1-R''_1}\sum_{i=R'_1+1}^{R_1-R''_1}x_{(i)} \tag{5.45}$$

式中，$x_{(i)}$ 为前沿滑窗的第 i 个有序样本。

尽管参考单元变量 $x_i(i\in\{1,2,\cdots,R_1\})$ 是统计独立的，但对它们排序后得到的有序统计量 $x_{(i)}$ $(i\in\{1,2,\cdots,R_1\})$ 不是统计独立的，不能简单地对 $x_{(i)}$ 的 PDF 进行卷积来求 X 的 PDF。故引入辅助变量[24]

$$w_i=(R_1+1-i)[x_{(i)}-x_{(i-1)}],\quad (1\leqslant i\leqslant R_1,x_{(0)}=0) \tag{5.46}$$

它们是统计独立的，且有相同的概率密度函数 $f(x)=\mathrm{e}^{-x}$，通过简单的代数运算，便有

$$X=\sum_{i=1}^{R_1-R''_1}\beta_i w_i \tag{5.47}$$

其中

$$\beta_i=\begin{cases}\dfrac{1}{R_1+1-i} & i=1,2,\cdots,R'_1+1\\[2ex] \dfrac{R_1-R''_1-i+1}{(R_1+1-i)(R_1-R'_1-R''_1)} & i=R'_1+2,\cdots,R_1-R''_1\end{cases} \tag{5.48}$$

故 X 的 MGF 为

$$M_X(u)=\prod_{i=1}^{R_1-R''_1}\frac{c'_i}{\mu u+c'_i} \tag{5.49}$$

式中，$c_i'=\beta_i^{-1}$。若 $R_1'=R_1''=0$，前沿滑窗便是单元平均(CA)的情形，若 $R_1'=0$，前沿滑窗便是删除平均(CM)的情形，若 $R_1''=R_1-R_1'-1$，前沿滑窗便是 OS 的情形。

将互不相同的 c'_i 记为 $s'_i(i=1,2,\cdots,m')$，r'_i 为 s'_i 的重数，有 $\sum\limits_{i=1}^{m'}r'_i=R_1-R''_1$，式(5.49)变为

$$M_X(u)=\prod_{i=1}^{m'}\left(\frac{s'_i}{\mu u+s'_i}\right)^{r'_i} \tag{5.50}$$

对式(5.50)取拉普拉斯反变换,得到 X 的概率密度函数

$$f_X(t)=\sum_{i=1}^{m'}\sum_{j=1}^{r'_i}\frac{a'_{ij}}{\mu}\left(\frac{t}{\mu}\right)^{j-1}\mathrm{e}^{-\frac{s'_i}{\mu}t}u(t) \tag{5.51}$$

其中

$$a'_{ij}=\frac{1}{(j-1)!(r'_i-j)!}\left\{\frac{\mathrm{d}^{r'_i-j}}{\mathrm{d}u^{r'_i-j}}\left[M_X(u)(\mu u+s'_i)^{r'_i}\right]\right\}_{u=-\frac{s'_i}{\mu}} \tag{5.52}$$

X 的累积分布函数为

$$F_X(t)=\sum_{i=1}^{m'}\sum_{j=1}^{r'_i}a'_{ij}\frac{\Gamma(j)}{(s'_i)^j}\left[1-\mathrm{e}^{-\frac{s'_i}{\mu}t}\sum_{k=0}^{j-1}\frac{\left(\frac{s'_i}{\mu}t\right)^k}{k!}\right]u(t) \tag{5.53}$$

对于 TM-TM 类 CFAR 检测器的后沿滑窗来说,后沿滑窗共有 R_2 个参考单元样本 $y_i(i\in\{1,2,\cdots,R_2\})$,先剔除掉最大的 R''_2 个样本和最小的 R'_2 个样本,对剩下的 $R_2-R'_2-R''_2$ 个样本求平均作为后沿滑窗对杂波平均功率水平的局部估计

$$Y=\frac{1}{R_2-R'_2-R''_2}\sum_{j=R'_2+1}^{R_2-R''_2}y_{(j)} \tag{5.54}$$

式中,$y_{(j)}$ 为后沿滑窗的第 j 个有序样本。

类似地,可求出局部估计 Y 的 MGF 为

$$M_Y(u)=\prod_{j=1}^{R_2-R''_2}\frac{c''_j}{\mu u+c''_j} \tag{5.55}$$

式中

$$c''_j=\begin{cases}R_2+1-j, & j=1,2,\cdots,R'_2+1\\ \dfrac{(R_2+1-j)(R_2-R''_2-R'_2)}{R_2-R''_2-j+1}, & j=R'_2+2,\cdots,R_2-R''_2\end{cases} \tag{5.56}$$

若 $R'_2=R''_2=0$,后沿滑窗便是 CA 的情形,若 $R''_2=0$,后沿滑窗便是删除平均(CM)的情形,若 $R''_2=R_2-R'_2-1$,后沿滑窗便是 OS 的情形。

采用与前沿滑窗相同的方法,可以求出局部估计 Y 的 PDF 和 CDF 分别为

$$f_Y(t)=\sum_{e=1}^{n'}\sum_{f=1}^{r''_e}\frac{a''_{ef}}{\mu}\left(\frac{t}{\mu}\right)^{f-1}\mathrm{e}^{-\frac{s''_e}{\mu}t}u(t) \tag{5.57}$$

$$F_Y(t)=\sum_{e=1}^{n'}\sum_{f=1}^{r''_e}a''_{ef}\frac{\Gamma(f)}{(s''_e)^f}\left[1-\mathrm{e}^{-\frac{s''_e}{\mu}t}\sum_{g=0}^{f-1}\frac{\left(\frac{s''_e}{\mu}t\right)^g}{g!}\right]u(t) \tag{5.58}$$

式中

$$a''_{ef}=\frac{1}{(f-1)!(r''_e-f)!}\left\{\frac{\mathrm{d}^{r''_e-f}}{\mathrm{d}u^{r''_e-f}}\left[M_Y(u)(\mu u+s''_e)^{r''_e}\right]\right\}_{u=-\frac{s''_e}{\mu}} \tag{5.59}$$

因为 TM-TM 类 CFAR 检测器的判决准则和背景模型与 OS-OS 类 CFAR 检测器相同,所以它在均匀背景中的检测概率和虚警概率仍可按照式(5.11)和式(5.12)来计算。在杂波边缘环境中,求局部

估计 X 或 Y 的 PDF,需要对多个来自非均匀背景的有序统计量的 PDF 进行卷积积分,以至于难以得到它的解析解,故后面将用虚警尖峰和仿真的方法对 TM-TM 类 CFAR 检测器在杂波边缘环境中的性能进行分析。

5.3 GOSCA、GOSGO、GOSSO-CFAR 检测器

GOSCA-CFAR、GOSGO-CFAR 和 GOSSO-CFAR 检测器属于 OS-OS 类 CFAR 检测器,其基本数学模型已在 5.2 节给出。本节将在均匀背景条件下,推导它们的检测概率、虚警概率及 ADT,并对它们进行分析。

5.3.1 GOSCA-CFAR 检测器

在 GOSCA-CFAR 检测器中,取两个局部估计的和作为总的杂波功率估计,即

$$Z = X + Y \tag{5.60}$$

Z 的 PDF 为

$$f_Z(z) = \int_0^z f_X(x) f_Y(z-x)\mathrm{d}x, \quad z > 0 \tag{5.61}$$

在均匀杂波背景下随机变量 Z 的 MGF 为

$$M_Z(u) = M_X(u) \cdot M_Y(u) \tag{5.62}$$

其中

$$\begin{aligned} M_X(u) &= k_1 \binom{R_1}{k_1} \int_0^\infty \mathrm{e}^{-uz} [\exp(-z)]^{R_1-k_1+1} [1-\exp(-z)]^{k_1-1} \mathrm{d}z \\ &= k_1 \binom{R_1}{k_1} \frac{\Gamma(R_1-k_1+1+u)\Gamma(k_1)}{\Gamma(R_1+u+1)} \end{aligned} \tag{5.63}$$

$$\begin{aligned} M_Y(u) &= k_2 \binom{R_2}{k_2} \int_0^\infty \mathrm{e}^{-uz} [\exp(-z)]^{R_2-k_2+1} [1-\exp(-z)]^{k_2-1} \mathrm{d}z \\ &= k_2 \binom{R_2}{k_2} \frac{\Gamma(R_2-k_2+1+u)\Gamma(k_2)}{\Gamma(R_2+u+1)} \end{aligned} \tag{5.64}$$

把式(5.63)和式(5.64)分别代入式(5.11)和式(5.12),可以得到 GOSCA-CFAR 的检测概率和虚警概率分别为

$$\begin{aligned} P_{\mathrm{d}} = &\frac{\Gamma(R_1+1)}{\Gamma(R_1-k_1+1)} \frac{\Gamma[R_1-k_1+1+T/(1+\lambda)]}{\Gamma[R_1+T/(1+\lambda)+1]} \frac{\Gamma(R_2+1)}{\Gamma(R_2-k_2+1)} \times \\ &\frac{\Gamma[R_2-k_2+1+T/(1+\lambda)]}{\Gamma[R_2+T/(1+\lambda)+1]} \end{aligned} \tag{5.65}$$

$$P_{\mathrm{fa}} = \frac{\Gamma(R_1+1)\Gamma(R_1-k_1+1+T)}{\Gamma(R_1-k_1+1)\Gamma(R_1+T+1)} \frac{\Gamma(R_2+1)\Gamma(R_2-k_2+1+T)}{\Gamma(R_2-k_2+1)\Gamma(R_2+T+1)} \tag{5.66}$$

对任意指定的 P_{fa},用数值方法解方程式(5.66)可获得 GOSCA-CFAR 检测器的阈值因子 T。当 R_1 和 R_2 固定时,T 是 k_1 和 k_2 的函数。当 $R_1=R_2$ 时,有 $T[k_1,k_2]=T[k_2,k_1]$、$P_{\mathrm{d}}[k_1,k_2]=P_{\mathrm{d}}[k_2,k_1]$。表 5.1 给出了 $P_{\mathrm{fa}}=10^{-6}$,三种 R_1、R_2 条件下的 GOSCA-CFAR 检测器的阈值因子 T。

表 5.1(a) GOSCA-CFAR 检测器的阈值因子 $T(P_{fa}=10^{-6}, R_1=R_2=12)$

k_2	k_1					
	7	8	9	10	11	12
7	14.6136	12.5990	10.8338	9.2397	7.7371	6.1944
8	12.5990	11.0229	9.6057	8.2944	7.0288	5.6967
9	10.8338	9.6057	8.4743	7.4032	6.3455	5.2044
10	9.2397	8.2944	7.4032	6.5403	5.6687	4.7046
11	7.7371	7.0288	6.3455	5.6687	4.9689	4.1737
12	6.1944	5.6967	5.2044	4.7046	4.1737	3.5514

表 5.1(b) GOSCA-CFAR 检测器的阈值因子 $T(P_{fa}=10^{-6}, R_1=R_2=16)$

k_2	k_1						
	10	11	12	13	14	15	16
10	10.885	9.843	8.865	7.932	7.019	6.090	5.055
11	9.843	8.964	8.129	7.332	6.522	5.696	4.764
12	8.865	8.129	7.421	6.728	6.032	5.303	4.467
13	7.932	7.322	6.728	6.138	5.538	4.902	4.160
14	7.019	6.522	6.032	5.538	5.030	4.483	3.833
15	6.090	5.696	5.303	4.902	4.483	4.024	3.469
16	5.055	4.764	4.467	4.160	3.833	3.469	3.0184

表 5.1(c) GOSCA-CFAR 检测器的阈值因子 $T(P_{fa}=10^{-6}, R_1=R_2=32)$

k_2	k_1												
	20	21	22	23	24	25	26	27	28	29	30	31	32
20	8.7342	8.3393	7.9520	7.5704	7.1928	6.8172	6.4410	6.0609	5.6726	5.2695	4.8405	4.3636	3.7783
21	8.3393	7.9763	7.6189	7.2656	6.9150	6.5649	6.2131	5.8565	5.4909	5.1099	4.7028	4.2483	3.6874
22	7.9520	7.6189	7.2899	6.9636	6.6387	6.3132	5.9850	5.6512	5.3078	4.9485	4.5631	4.1308	3.5942
23	7.5704	7.2656	6.9636	6.6632	6.3629	6.0611	5.7558	5.4442	5.1224	4.7845	4.4205	4.0104	3.4983
24	7.1928	6.9150	6.6387	6.3629	6.0863	5.8075	5.5244	5.2345	4.9339	4.6172	4.2745	3.8864	3.3990
25	6.8172	6.5649	6.3132	6.0611	5.8075	5.5509	5.2895	5.0208	4.7412	4.4454	4.1239	3.7580	3.2956
26	6.4410	6.2131	5.9850	5.7558	5.5244	5.2895	5.0493	4.8015	4.5427	4.2677	3.9675	3.6240	3.1870
27	6.0609	5.8565	5.6512	5.4442	5.2345	5.0208	4.8015	4.5744	4.3362	4.0821	3.8034	3.4828	3.0718
28	5.6726	5.4909	5.3078	5.1224	4.9339	4.7412	4.5427	4.3362	4.1189	3.8859	3.6290	3.3318	2.9479
29	5.2695	5.1099	4.9485	4.7845	4.6172	4.4454	4.2677	4.0821	3.8859	3.6745	3.4402	3.1673	2.8118
30	4.8405	4.7028	4.5631	4.4205	4.2745	4.1239	3.9675	3.8034	3.6290	3.4402	3.2296	2.9826	2.6576
31	4.3636	4.2483	4.1308	4.0104	3.8864	3.7580	3.6240	3.4828	3.3318	3.1673	2.9826	2.7640	2.4733
32	3.7783	3.6874	3.5942	3.4983	3.3990	3.2956	3.1870	3.0718	2.9479	2.8118	2.6576	2.4733	2.2242

根据式(3.21)可以推出 GOSCA-CFAR 检测器的平均判决门限 ADT 为

$$\mathrm{ADT}=T\left[\sum_{i=1}^{k_1}\frac{1}{R_1-k_1+i}+\sum_{j=1}^{k_2}\frac{1}{R_2-k_2+j}\right] \tag{5.67}$$

表 5.2(a)～(c)给出了 GOSCA-CFAR 检测器的 ADT 值,分析条件与表 5.1 相同。

表 5.2(a) GOSCA-CFAR 检测器的平均判决门限 ADT($R_1=R_2=12$)

k_2	k_1					
	7	8	9	10	11	12
7	23.963	23.179	22.640	22.389	22.616	24.301
8	23.179	22.484	21.995	21.757	21.952	23.488
9	22.640	21.995	21.523	21.270	21.404	22.759
10	22.389	21.757	21.270	20.971	21.011	22.142
11	22.616	21.952	21.404	21.011	20.901	21.730
12	24.301	23.488	22.759	22.142	21.730	22.041

表 5.2(b) GOSCA-CFAR 检测器的平均判决门限 ADT($R_1=R_2=16$)

k_2	k_1						
	10	11	12	13	14	15	16
10	20.262	19.963	19.752	19.656	19.734	20.167	21.794
11	19.963	19.674	19.467	19.392	19.423	19.811	21.334
12	19.752	19.467	19.256	19.140	19.170	19.505	20.897
13	19.656	19.365	19.140	18.996	18.985	19.256	20.501
14	19.734	19.423	19.170	18.985	18.920	19.104	20.167
15	20.167	19.811	19.505	19.256	19.104	19.160	19.986
16	21.794	21.334	20.897	20.501	20.167	19.986	20.409

表 5.2(c) GOSCA-CFAR 检测器的平均判决门限 ADT($R_1=R_2=32$)

k_2	k_1												
	20	21	22	23	24	25	26	27	28	29	30	31	32
20	16.687	16.628	16.578	16.540	16.514	16.504	16.513	16.549	16.623	16.759	17.008	17.515	18.944
21	16.628	16.569	16.519	16.479	16.453	16.440	16.447	16.479	16.548	16.678	16.917	17.406	18.795
22	16.578	16.519	16.468	16.428	16.399	16.384	16.387	16.415	16.479	16.601	16.829	17.300	18.647
23	16.540	16.479	16.428	16.385	16.354	16.336	16.335	16.358	16.416	16.529	16.745	17.197	18.499
24	16.514	16.453	16.399	16.354	16.319	16.297	16.292	16.310	16.360	16.464	16.667	17.097	18.352
25	16.504	16.440	16.384	16.336	16.297	16.271	16.261	16.271	16.314	16.407	16.595	17.002	18.205
26	16.513	16.447	16.387	16.335	16.292	16.261	16.244	16.247	16.279	16.361	16.533	16.913	18.061
27	16.549	16.479	16.415	16.358	16.310	16.271	16.247	16.241	16.262	16.330	16.483	16.835	17.920
28	16.623	16.548	16.479	16.416	16.360	16.314	16.279	16.262	16.271	16.322	16.453	16.771	17.787
29	16.759	16.678	16.601	16.529	16.464	16.407	16.361	16.330	16.322	16.353	16.457	16.735	17.668
30	17.008	16.917	16.829	16.745	16.667	16.595	16.533	16.483	16.453	16.457	16.526	16.753	17.585
31	17.515	17.406	17.300	17.197	17.097	17.002	16.913	16.835	16.771	16.735	16.753	16.907	17.602
32	18.944	18.795	18.647	18.499	18.352	18.205	18.061	17.920	17.787	17.668	17.585	17.602	18.054

5.3.2 GOSGO-CFAR 检测器

在 GOSGO-CFAR 检测器中,取两个杂波强度局部估计 X 和 Y 中的较大者作为杂波功率水平的估计 Z,即

$$Z=\max(X,Y) \tag{5.68}$$

式(5.68)中 Z 的 PDF 为

$$f_Z(z)=f_X(z)F_Y(z)+f_Y(z)F_X(z) \tag{5.69}$$

设 Z 的 PDF 为

$$M_Z(u)=M_1(u)+M_2(u) \tag{5.70}$$

其中

$$\begin{aligned}M_1(u)&=k_1\binom{R_1}{k_1}\sum_{i=k_2}^{R_2}\binom{R_2}{i}\int_0^{\infty}\mathrm{e}^{-z(R-k_1+1-i+u)}(1-\mathrm{e}^{-z})^{k_1-1+i}\mathrm{d}z\\&=k_1\binom{R_1}{k_1}\sum_{i=k_2}^{R_2}\binom{R_2}{i}\frac{\Gamma(R-k_1+1-i+u)\Gamma(k_1+i)}{\Gamma(R+1+u)}\end{aligned} \tag{5.71}$$

$$\begin{aligned}M_2(u)&=k_2\binom{R_2}{k_2}\sum_{i=k_1}^{R_1}\binom{R_1}{i}\int_0^{\infty}\mathrm{e}^{-z(R-k_2+1-i+u)}(1-\mathrm{e}^{-z})^{k_2-1+i}\mathrm{d}z\\&=k_2\binom{R_2}{k_2}\sum_{i=k_1}^{R_1}\binom{R_1}{i}\frac{\Gamma(R-k_2+1-i+u)\Gamma(k_2+i)}{\Gamma(R+1+u)}\end{aligned} \tag{5.72}$$

把式(5.71)～式(5.72)分别代入式(5.11)和式(5.12)，可以得到 GOSGO-CFAR 检测器的检测概率为

$$\begin{aligned}P_{\mathrm{d}}=&k_1\binom{R_1}{k_1}\sum_{i=k_2}^{R_2}\binom{R_2}{i}\frac{\Gamma[R-k_1+1-i+T/(1+\lambda)]\Gamma(k_1+i)}{\Gamma[R+1+T/(1+\lambda)]}+\\&k_2\binom{R_2}{k_2}\sum_{i=k_1}^{R_1}\binom{R_1}{i}\frac{\Gamma[R-k_2+1-i+T/(1+\lambda)]\Gamma(k_2+i)}{\Gamma[R+1+T/(1+\lambda)]}\end{aligned} \tag{5.73}$$

虚警概率为

$$\begin{aligned}P_{\mathrm{fa}}=&k_1\binom{R_1}{k_1}\sum_{i=k_2}^{R_2}\binom{R_2}{i}\frac{\Gamma(R-k_1+1-i+T)\Gamma(k_1+i)}{\Gamma(R+1+T)}+\\&k_2\binom{R_2}{k_2}\sum_{i=k_1}^{R_1}\binom{R_1}{i}\frac{\Gamma(R-k_2+1-i+T)\Gamma(k_2+i)}{\Gamma(R+1+T)}\end{aligned} \tag{5.74}$$

对于任意一个给定的 P_{fa}，解式(5.74)可以得到相应的阈值因子 T，见表 5.3，其中 $P_{\mathrm{fa}}=10^{-6}$，取三种 R_1、R_2 的组合，分别对应表 5.3(a)～表 5.3(c)。

表 5.3(a)　GOSGO-CFAR 检测器的阈值因子 $T(P_{\mathrm{fa}}=10^{-6}, R_1=16, R_2=8)$

k_1	k_2				
	3	4	5	6	7
10	28.1635	25.36127	22.32724	19.16493	15.88194
11	23.48164	21.62279	19.45496	17.05749	14.43926
12	19.58476	18.38765	16.87274	15.08662	13.02852
13	16.26951	15.53537	14.5145	13.2201	11.63692
14	13.38022	12.96435	12.31603	11.41873	10.24125

表 5.3(b) GOSGO-CFAR 检测器的阈值因子 $T(P_{fa}=10^{-6}, R_1=24, R_2=8)$

k_1	k_2				
	3	4	5	6	7
18	15.900	15.377	14.568	13.450	11.985
19	14.045	13.694	13.103	12.235	11.040
20	12.352	12.128	11.713	11.105	10.099
21	10.786	10.652	10.375	9.8974	9.1512
22	9.3073	9.2350	9.2350	8.7354	8.1770

表 5.3(c) GOSGO-CFAR 检测器的阈值因子 $T(P_{fa}=10^{-6}, R_1=48, R_2=16)$

k_1	k_2				
	10	11	12	13	14
36	12.13769	11.76988	11.27386	10.63072	9.814000
37	11.51132	11.20650	10.78240	10.21818	9.485880
38	10.89328	10.64495	10.28711	9.797180	9.145760
39	10.28269	10.08448	9.787340	9.367100	8.793050
40	9.678450	9.524120	9.282110	8.926960	8.426790
41	9.079090	8.962530	8.770070	8.475380	8.045570
42	8.482640	8.397860	8.249320	8.010360	7.647340
43	7.886230	7.827430	7.717100	7.528960	7.229070
44	7.285510	7.247170	7.169210	7.026750	6.786170
45	6.673450	6.650400	6.598930	6.496680	6.311300
46	6.037480	6.025120	5.994260	5.926350	5.791660

这样,根据式(3.21)得到 GOSGO-CFAR 检测器的平均判决门限 ADT 为

$$\text{ADT}=T\left[\sum_{i=k_2}^{R_2}\frac{k_1\binom{R_1}{k_1}\binom{R_2}{i}}{(k_1+i)\binom{R_1+R_2}{k_1+i}}\sum_{j=1}^{k_1+i}\frac{1}{R-k_1-i+j}+\sum_{i=k_1}^{R_1}\frac{k_2\binom{R_2}{k_2}\binom{R_1}{i}}{(k_2+i)\binom{R_1+R_2}{k_2+i}}\sum_{j=1}^{k_2+i}\frac{1}{R-k_2-i+j}\right] \tag{5.75}$$

对于特定的 P_{fa}、R_1 和 R_2 值,解式(5.75)可以得到 GOSGO-CFAR 检测器的 ADT 值,见表 5.4,其中 $P_{fa}=10^{-6}$,取三种 R_1、R_2 的组合,分别对应表 5.4(a)~表 5.4(c)。

表 5.4(a) GOSGO-CFAR 检测器的平均判决门限 ADT($R_1=16, R_2=8$)

k_1	k_2				
	3	4	5	6	7
10	26.790	25.315	24.736	25.497	27.943
11	26.035	24.636	23.744	23.964	25.930
12	25.520	24.294	23.223	22.919	24.228

续表

k_1	k_2				
	3	4	5	6	7
13	25.216	24.225	23.120	22.390	22.936
14	25.177	24.448	23.439	22.438	22.209

表 5.4(b) GOSGO-CFAR 检测器的平均判决门限 ADT($R_1=24, R_2=8$)

k_1	k_2				
	3	4	5	6	7
18	21.139	20.654	20.251	20.482	22.248
19	20.991	20.580	20.117	20.043	21.271
20	20.919	20.569	20.132	19.828	20.516
21	20.095	20.722	20.304	19.857	20.038
22	21.184	21.028	20.691	20.189	19.938

表 5.4(c) GOSGO-CFAR 检测器的平均判决门限 ADT($R_1=48, R_2=16$)

k_1	k_2				
	10	11	12	13	14
36	16.828	16.821	17.071	17.745	18.998
37	16.822	16.759	16.902	17.435	18.560
38	16.836	16.734	16.783	17.174	18.148
39	16.868	16.742	16.714	16.968	17.774
40	16.916	16.780	16.693	16.822	17.447
41	16.979	16.847	16.721	16.740	17.180
42	17.060	16.941	16.796	16.727	16.986
43	17.166	17.067	16.920	16.789	16.882
44	17.311	17.234	17.101	16.935	16.886
45	17.523	17.468	17.358	17.186	17.031
46	17.864	17.830	17.749	17.593	17.376

5.3.3 GOSSO-CFAR 检测器

在 GOSSO-CFAR 检测器中,取两个局部估计中的较小者作为杂波功率水平估计 Z(见图 5.1),即

$$Z=\min(X,Y) \tag{5.76}$$

Z 的 PDF 为

$$f_Z(z)=f_X(z)+f_Y(z)-[f_X(z)F_Y(z)+f_Y(z)F_X(z)] \tag{5.77}$$

Z 的 MGF 为

$$M_Z(u)=M_1(u)+M_2(u)-M_3(u) \tag{5.78}$$

其中

$$M_1(u)=k_1\binom{R_1}{k_1}\frac{\Gamma(R_1-k_1+1+u)\Gamma(k_1)}{\Gamma(R_1+u+1)},$$

$$M_2(u)=k_2\binom{R_2}{k_2}\frac{\Gamma(R_2-k_2+1+u)\Gamma(k_2)}{\Gamma(R_2+u+1)}$$

$M_3(u)$就是式(5.62)的$M_Z(u)$。这样,利用式(5.11)和式(5.12)就得到GOSSO-CFAR检测器的检测概率为

$$\begin{aligned}P_d=&k_1\binom{R_1}{k_1}\left\{\frac{\Gamma[R_1-k_1+1+T/(1+\lambda)]\Gamma(k_1)}{\Gamma[R_1+T/(1+\lambda)+1]}-\right.\\&\left.\sum_{i=k_2}^{R_2}\binom{R_2}{i}\frac{\Gamma[R-k_1+1-i+T/(1+\lambda)]\Gamma(k_1+i)}{\Gamma[R+1+T/(1+\lambda)]}\right\}+\\&k_2\binom{R_2}{k_2}\left\{\frac{\Gamma[R_2-k_2+1+T/(1+\lambda)]\Gamma(k_2)}{\Gamma[R_2+T/(1+\lambda)+1]}-\right.\\&\left.\sum_{i=k_1}^{R_1}\binom{R_1}{i}\frac{\Gamma[R-k_2+1-i+T/(1+\lambda)]\Gamma(k_2+i)}{\Gamma[R+1+T/(1+\lambda)]}\right\}\end{aligned}\tag{5.79}$$

虚警概率为

$$\begin{aligned}P_{fa}=&k_1\binom{R_1}{k_1}\left[\frac{\Gamma(R_1-k_1+1+T)\Gamma(k_1)}{\Gamma(R_1+T+1)}-\sum_{i=k_2}^{R_2}\binom{R_2}{i}\frac{\Gamma(R-k_1+1-i+T)\Gamma(k_1+i)}{\Gamma(R+T+1)}\right]+\\&k_2\binom{R_2}{k_2}\left[\frac{\Gamma(R_2-k_2+1+T)\Gamma(k_2)}{\Gamma(R_2+T+1)}-\sum_{i=k_1}^{R_1}\binom{R_1}{i}\frac{\Gamma(R-k_2+1-i+T)\Gamma(k_2+i)}{\Gamma(R+T+1)}\right]\end{aligned}\tag{5.80}$$

对于任意一个给定的P_{fa},解式(5.80)可以得到相应的阈值因子T,见表5.5,其中$P_{fa}=10^{-6}$,取三种R_1、R_2的组合,分别对应表5.5(a)~表5.5(c)。

表5.5(a) GOSSO-CFAR检测器的阈值因子$T(P_{fa}=10^{-6},R_1=16,R_2=8)$

k_1	k_2			
	5	6	7	8
10	86.37536	47.23851	34.35014	32.93261
11	86.36912	46.76193	29.82262	26.24488
12	86.36871	46.70356	28.19715	21.52767
13	86.36869	46.69776	27.85391	18.53081
14	86.36869	46.69730	27.80601	17.17190
15	86.36869	46.69727	27.80114	16.83758
16	86.36869	46.69727	27.80080	16.79525

表 5.5(b) GOSSO-CFAR 检测器的阈值因子 $T(P_{fa}=10^{-6}, R_1=22, R_2=10)$

k_1	k_2			
	7	8	9	10
11	45.8481	42.04449	41.73404	41.71925
12	42.58262	35.51278	34.6700	34.61993
13	41.5411	31.23843	29.30299	29.15482
14	41.30058	28.91079	25.21741	24.82728
15	41.2549	27.98157	22.24251	21.32816
16	41.24727	27.71659	20.34385	18.47166
17	41.24614	27.65767	19.41099	16.17266
18	41.2460	27.64669	19.09116	14.44528
19	41.24598	27.64496	19.01285	13.36677

表 5.5(c) GOSSO-CFAR 检测器的阈值因子 $T(P_{fa}=10^{-6}, R_1=48, R_2=16)$

k_1	k_2			
	12	13	14	15
30	21.45412	19.18121	18.57805	18.48175
31	21.20340	18.41445	17.52571	17.36690
32	21.07024	17.84116	16.59094	16.33567
33	21.00516	17.44921	15.78136	15.38258
34	20.97544	17.20769	15.10808	14.50535
35	20.96261	17.07379	14.58119	13.70555
36	20.95735	17.00632	14.20233	12.98929
37	20.95530	16.97497	13.95741	12.36747
38	20.95454	16.96139	13.81671	11.85427

利用式(3.21)可以推出 GOSSO-CFAR 检测器的平均判决门限 ADT 为

$$\mathrm{ADT}=T\left\{\left[\sum_{i=1}^{k_1}\frac{1}{R_1-k_1+i}-\sum_{i=k_2}^{R_2}\frac{k_1\binom{R_1}{k_1}\binom{R_2}{i}}{(k_1+i)\binom{R_1+R_2}{k_1+i}}\sum_{j=1}^{k_1+i}\frac{1}{R-k_1-i+j}\right]+\right.$$
$$\left.\left[\sum_{i=1}^{k_2}\frac{1}{R_2-k_2+i}-\sum_{i=k_1}^{R_1}\frac{k_2\binom{R_2}{k_2}\binom{R_1}{i}}{(k_2+i)\binom{R_1+R_2}{k_2+i}}\sum_{j=1}^{k_2+i}\frac{1}{R-k_2-i+j}\right]\right\} \tag{5.81}$$

对于 $P_{fa}=10^{-6}$，三种 R_1、R_2 的组合，表 5.6(a)～(c)给出了 GOSSO-CFAR 检测器的 ADT 的一些典型值。

表 5.6(a) GOSSO-CFAR 检测器的平均判决门限 ADT($R_1=16, R_2=8$)

k_1	k_2			
	5	6	7	8
10	61.098	38.650	30.542	30.371
11	65.765	42.571	30.403	28.317
12	69.575	46.520	32.586	27.090
13	72.469	50.043	36.051	27.158
14	74.463	52.934	39.763	29.307
15	75.658	55.057	43.217	33.495
16	76.227	56.349	46.037	39.009

表 5.6(b) GOSSO-CFAR 检测器的平均判决门限 ADT($R_1=22, R_2=10$)

k_1	k_2			
	7	8	9	10
11	29.036	27.649	27.863	27.973
12	29.861	26.223	26.192	26.343
13	31.894	25.681	24.904	25.071
14	34.337	26.244	23.990	24.059
15	36.741	27.813	23.574	23.253
16	38.930	29.904	23.908	22.644
17	40.822	32.091	25.166	22.299
18	42.371	34.155	27.146	22.429
19	43.555	35.977	29.439	23.416

表 5.6(c) GOSSO-CFAR 检测器的平均判决门限 ADT($R_1=48, R_2=16$)

k_1	k_2			
	12	13	14	15
30	28.859	30.069	35.063	44.028
31	28.847	29.010	33.131	41.386
32	29.073	28.292	31.435	38.949
33	29.489	27.908	29.996	36.704
34	30.071	27.827	28.841	34.650
35	30.816	28.001	28.001	32.794
36	31.734	28.387	27.492	31.155
37	32.848	28.964	27.308	29.766
38	34.186	29.733	27.417	28.672

5.4 MOSCA、OSCAGO、OSCASO-CFAR 检测器

MOSCA、OSCAGO、OSCASO-CFAR 检测器属于 OS-CA 类 CFAR 检测器，其基本模型已在 5.2 节给出。本节将在均匀背景条件下，推导它们的检测概率和虚警概率的数学模型，计算门限参数 T 和 ADT。

5.4.1 MOSCA-CFAR 检测器

在 MOSCA-CFAR 检测器中，取两个局部估计的和作为总的杂波功率估计(见图 5.2)，即

$$Z = X + Y \tag{5.82}$$

在式(5.82)中 Z 的 PDF 为

$$f_Z(z) = \int_0^z f_X(x) f_Y(z-x)\mathrm{d}x \tag{5.83}$$

它的 MGF 为

$$M_Z(u) = M_X(u) \cdot M_Y(u) \tag{5.84}$$

其中

$$M_X(u) = k\binom{R_1}{k}\frac{\Gamma(R_1 - k + 1 + u)\Gamma(k)}{\Gamma(R_1 + u + 1)}, \quad M_Y(u) = \int_0^\infty \mathrm{e}^{-uz} f_Y(y)\mathrm{d}y = \frac{1}{(1 + u/R_2)^{R_2}}$$

把式(5.84)代入式(5.11)和式(5.12)，可以得到 MOSCA-CFAR 的检测概率和虚警概率表达式，分别为

$$P_\mathrm{d} = k\binom{R_1}{k}\frac{\Gamma[R_1 - k + 1 + T/(1+\lambda)]\Gamma(k)}{\Gamma[R_1 + T/(1+\lambda) + 1]}\frac{1}{\{1 + T/[(1+\lambda)R_2]\}^{R_2}} \tag{5.85}$$

$$P_\mathrm{fa} = k\binom{R_1}{k}\frac{\Gamma(R_1 - k + 1 + T)\Gamma(k)}{\Gamma(R_1 + T + 1)(1 + T/R_2)^{R_2}} \tag{5.86}$$

对任意指定的 P_fa，用数值方法解方程式(5.86)可获得 MOSCA-CFAR 检测器的阈值因子 T。当 R_1、R_2 固定时，T 是 k 的函数。对于 $P_\mathrm{fa}=10^{-6}$，表 5.7(a)～表 5.7(c)给出了 R 分别为 24、32 和 64 条件下 MOSCA-CFAR 检测器的阈值因子 T。当 $R_2=0$ 时，MOSCA 退化成 OS-CFAR；同理，当 $R_1=0$ 时 MOSCA 变成了 CA-CFAR。

表 5.7(a) MOSCA-CFAR 检测器的阈值因子 $T(R=24)$

k	R_1		
	8	16	22
1	19.152	31.288	436.228
2	16.756	26.964	195.816
3	14.664	23.559	116.822
4	12 805	20.798	80.370
5	11.122	18.504	60.045
6	9.559	16.56	47.277
7	8.051	14.883	38.575
8	6.475	13.413	32.284

续表

k	R_1		
	8	16	22
9	—	12.105	27.527
10	—	10.926	23.802
11	—	9.848	20.801
12	—	8.845	18.326
13	—	7.895	16.242
14	—	6.973	14.455
15	—	6.039	12.899
16	—	5.005	11.523
17	—	—	10.285
18	—	—	9.1550
19	—	—	8.1030
20	—	—	7.0960
21	—	—	6.0950
22	—	—	5.0060

表 5.7(b) MOSCA-CFAR 检测器的阈值因子 $T(R=32)$

k	R_1				
	8	16	24	28	30
1	16.7	20.1	32.4	85.2	482.1
2	14.9	18.4	28.8	65.0	227.5
3	13.3	16.9	25.9	52.3	140.0
4	11.8	15.6	23.5	43.7	98.5
5	10.4	14.4	21.4	37.4	74.9
6	9.1	13.3	19.6	32.6	59.9
7	7.7	12.3	18.1	28.9	49.6
8	6.3	11.3	16.7	25.8	42.1
9	—	10.4	15.5	23.3	36.3
10	—	9.6	14.4	21.1	31.9
11	—	8.8	13.4	19.3	28.2
12	—	8.0	12.4	17.7	25.2
13	—	7.3	11.6	16.3	22.7
14	—	6.5	10.8	15.0	20.6
15	—	5.7	10.1	13.9	18.8
16	—	4.8	9.4	12.9	17.1
17	—	—	8.7	11.9	15.7
18	—	—	8.1	11.1	14.4
19	—	—	7.5	10.3	13.3
20	—	—	6.9	9.5	12.2
21	—	—	6.3	8.8	11.3

续表

k	R_1				
	8	16	24	28	30
22	—	—	5.7	8.2	10.4
23	—	—	5.0	7.5	9.6
24	—	—	4.3	6.9	8.8
25	—	—	—	6.3	8.0
26	—	—	—	5.7	7.3
27	—	—	—	5.0	6.6
28	—	—	—	4.2	5.9
29	—	—	—	—	5.2
30	—	—	—	—	4.4

表 5.7(c)　MOSCA-CFAR 检测器的阈值因子 $T(R=64)$

k	R_1					
	8	16	24	28	32	40
1	14.365	15.128	15.792	16.176	16.637	18.022
2	13.121	14.289	15.115	15.542	16.027	17.40
3	11.928	13.489	14.469	14.936	15.444	16.809
4	10.774	12.725	13.851	14.356	14.887	16.245
5	9.644	11.991	13.258	13.801	14.353	15.708
6	8.519	11.285	12.689	13.267	13.841	15.194
7	7.358	10.604	12.141	12.754	13.348	14.702
8	6.064	9.944	11.612	12.259	12.874	14.231
9	—	9.300	11.101	11.781	12.417	13.778
10	—	8.669	10.606	11.319	11.975	13.342
11	—	8.047	10.125	10.872	11.548	12.922
12	—	7.426	9.657	10.437	11.133	12.516
13	—	6.798	9.201	10.014	10.732	12.125
14	—	6.148	8.753	9.602	10.341	11.746
15	—	5.448	8.313	9.199	9.960	11.378
16	—	4.625	7.879	8.805	9.589	11.022
17	—	—	7.448	8.417	9.226	10.675
18	—	—	7.018	8.035	8.870	10.338
19	—	—	6.585	7.657	8.520	10.009
20	—	—	6.144	7.282	8.176	9.689
21	—	—	5.689	6.907	7.837	9.375
22	—	—	5.207	6.530	7.501	9.068
23	—	—	4.675	6.149	7.167	8.767
24	—	—	4.028	5.758	6.833	8.471
25	—	—	—	5.351	6.498	8.180
26	—	—	—	4.916	6.160	7.893
27	—	—	—	4.433	5.816	7.610

续表

k	R_1					
	8	16	24	28	32	40
28	—	—	—	3.838	5.461	7.328
29	—	—	—	—	5.090	7.049
30	—	—	—	—	4.691	6.771
31	—	—	—	—	4.244	6.492
32	—	—	—	—	3.689	6.212
33	—	—	—	—	—	5.929
34	—	—	—	—	—	5.641
35	—	—	—	—	—	5.346
36	—	—	—	—	—	5.039
37	—	—	—	—	—	4.716
38	—	—	—	—	—	4.365
39	—	—	—	—	—	3.968
40	—	—	—	—	—	3.470

根据式(3.21)可推出 MOSCA-CFAR 检测器的平均判决门限 ADT 为

$$\mathrm{ADT}=T\left(\sum_{i=1}^{k}\frac{1}{R_1-k+i}+1\right) \tag{5.87}$$

表 5.8(a)～(c)给出了 $P_{\mathrm{fa}}=10^{-6}$，R 分别为 24、32 和 64 条件下 MOSCA-CFAR 检测器的 ADT 值。

表 5.8(a) MOSCA-CFAR 检测器的平均判决门限 ADT($R=24$)

k	R_1		
	8	16	22
1	21.546	33.244	454 404
2	21.244	30.447	212.489
3	21.036	28.285	132.079
4	20.93	26.57	94.693
5	20.96	25.181	73.748
6	21.200	24.041	60.555
7	21.881	23.095	51.552
8	24.073	22.304	45.044
9	—	21.642	40.127
10	—	21.095	36.284
11	—	20.655	33.195
12	—	20.32	30.655
13	—	20.112	28.522
14	—	20.087	26.698
15	—	20.416	25.114
16	—	21.926	23.715
17	—	—	22.453

续表

k	R_1		
	8	16	22
18	—	—	21.294
19	—	—	20.198
20	—	—	19.107
21	—	—	17.935
22	—	—	16.399

表 5.8(b)　MOSCA-CFAR 检测器的平均判决门限 ADT($R=32$)

k	R_1				
	8	16	24	28	30
1	18.787	21.356	33.750	88.243	498.17
2	18.891	20.777	31.252	69.729	242.928
3	19.079	20.290	29.283	58.116	154.494
4	19.287	19.929	27.688	50.308	112.346
5	19.599	19.596	26.284	44.614	88.309
6	20.182	19.308	25.105	40.305	73.020
7	20.928	19.087	24.189	37.044	62.531
8	23.422	18.790	23.300	34.299	54.906
9	—	18.594	22.595	32.141	48.992
10	—	18.535	21.951	30.217	44.572
11	—	18.457	21.384	28.711	40.812
12	—	18.379	20.742	27.372	37.797
13	—	18.596	20.371	26.226	35.308
14	—	18.725	19.947	25.134	33.254
15	—	19.270	19.665	24.284	31.523
16	—	21.027	19.346	23.529	29.813
17	—	—	18.993	22.697	28.493
18	—	—	18.840	22.180	27.242
19	—	—	18.695	21.611	26.269
20	—	—	18.579	20.988	25.205
21	—	—	18.539	20.542	24.476
22	—	—	18.673	20.313	23.682
23	—	—	18.880	19.829	23.060
24	—	—	20.537	19.622	22.396
25	—	—	—	19.491	21.693
26	—	—	—	19.535	21.255
27	—	—	—	19.636	20.867
28	—	—	—	20.694	20.620
29	—	—	—	—	20.774
30	—	—	—	—	21.978

表 5.8(c) MOSCA-CFAR 检测器的平均判决门限 ADT($R=64$)

k	R_1					
	8	16	24	28	32	40
1	16.161	16.074	16.450	16.754	17.157	18.473
2	16.636	16.135	16.402	16.673	17.045	18.281
3	17.111	16.195	16.359	16.597	16.940	18.103
4	17.610	16.256	16.320	16.527	16.842	17.934
5	18.174	16.318	16.284	16.463	16.751	17.778
6	18.894	16.383	16.253	16.403	16.666	17.630
7	19.998	16.455	16.225	16.348	16.585	17.492
8	22.545	16.536	16.201	16.297	16.511	17.363
9	—	16.627	16.182	16.251	16.443	17.240
10	—	16.737	16.168	16.210	16.378	17.125
11	—	16.878	16.158	16.173	16.319	17.017
12	—	17.060	16.154	16.140	16.263	16.914
13	—	17.317	16.158	16.112	16.213	16.818
14	—	17.711	16.167	16.089	16.167	16.728
15	—	18.418	16.185	16.071	16.125	16.641
16	—	20.261	16.216	16.060	16.088	16.561
17	—	—	16.260	16.054	16.056	16.485
18	—	—	16.324	16.056	16.027	16.414
19	—	—	16.414	16.066	16.004	16.347
20	—	—	16.543	16.088	15.986	16.285
21	—	—	16.741	16.123	15.977	16.226
22	—	—	17.058	16.176	15.974	16.172
23	—	—	17.653	16.257	15.979	16.122
24	—	—	19.238	16.375	15.994	16.076
25	—	—	—	16.555	16.022	16.035
26	—	—	—	16.848	16.068	15.999
27	—	—	—	17.409	16.140	15.969
28	—	—	—	18.910	16.247	15.941
29	—	—	—	—	16.416	15.921
30	—	—	—	—	16.693	15.909
31	—	—	—	—	17.224	15.903
32	—	—	—	—	18.661	15.907
33	—	—	—	—	—	15.923
34	—	—	—	—	—	15.956
35	—	—	—	—	—	16.012
36	—	—	—	—	—	16.101
37	—	—	—	—	—	16.248
38	—	—	—	—	—	16.493
39	—	—	—	—	—	16.977
40	—	—	—	—	—	18.317

5.4.2 OSCAGO-CFAR 检测器

在 OSCAGO-CFAR 检测器中，取两个局部估计 X 和 Y(见图 5.2)中的较大者作为杂波功率水平估计 Z，即

$$Z=\max(X,Y) \tag{5.88}$$

式(5.88)中 Z 的 PDF 为

$$f_z(z)=f_x(z)F_Y(z)+f_Y(z)F_x(z) \tag{5.89}$$

设 Z 的 MGF 为

$$M_z(u)=M_1(u)+M_2(u) \tag{5.90}$$

其中

$$\begin{aligned}M_1(u)&=\int_0^{\infty}\mathrm{e}^{-uz}f_x(z)F_Y(z)\mathrm{d}z\\&=k\binom{R_1}{k}\int_0^{\infty}\mathrm{e}^{-(R_1-k+1-u)z}(1-\mathrm{e}^{-z})^{k-1}\left[1-\mathrm{e}^{-R_2z}\sum_{i=0}^{R_2-1}\frac{(R_2z)^i}{i!}\right]\mathrm{d}z\\&=k\binom{R_1}{k}\left[\frac{\Gamma(R_1-k+1+u)\Gamma(k)}{\Gamma(R_1+u+1)}-\frac{1}{R_2}\sum_{i=0}^{R_2-1}\sum_{j=0}^{k-1}\binom{k-1}{j}(-1)^j\times\right.\\&\left.\left(\frac{R_2}{R+1-k+j+u}\right)^{i+1}\right]\end{aligned} \tag{5.91}$$

$$\begin{aligned}M_2(u)&=\int_0^{\infty}\mathrm{e}^{-uz}f_Y(z)F_x(z)\mathrm{d}z=\frac{R_2^{R_2}}{\Gamma(R_2)}\int_0^{\infty}z^{R_2-1}\mathrm{e}^{-(R_2+u)z}\sum_{i=k}^{R_1}\binom{R_1}{i}(1-\mathrm{e}^{-z})^i\mathrm{e}^{-z(R_1-i)}\mathrm{d}z\\&=\sum_{i=k}^{R_1}\binom{R_1}{i}\sum_{i=0}^{i}\binom{i}{j}(-1)^j\times\left(\frac{R_2}{R+j-i+u}\right)^{R_2}\end{aligned} \tag{5.92}$$

将式(5.90)～式(5.92)代入式(5.11)和式(5.12)可得 OSCAGO-CFAR 检测器的检测概率为

$$\begin{aligned}P_{\mathrm{d}}=&\frac{R_1!\Gamma[R_1-k+T/(1+\lambda)+1]}{(R_1-k)!\Gamma[R_1+T/(1+\lambda)+1]}-k\binom{R_1}{k}\frac{1}{R_2}\sum_{i=0}^{R_2-1}\sum_{j=0}^{k-1}\binom{k-1}{j}(-1)^j\times\\&\left[\frac{R_2}{R+1-k+j+T/(1+\lambda)}\right]^{i+1}+\sum_{i=k}^{R_1}\binom{R_1}{i}\sum_{j=0}^{i}\binom{i}{j}(-1)^j\times\\&\left[\frac{R_2}{R+j-i+T/(1+\lambda)}\right]^{R_2}\end{aligned} \tag{5.93}$$

虚警概率为

$$\begin{aligned}P_{\mathrm{fa}}=&\frac{R_1!\Gamma(R_1-k+T+1)}{(R_1-k)!\Gamma(R_1+T+1)}-k\binom{R_1}{k}\frac{1}{R_2}\sum_{i=0}^{R_2-1}\sum_{j=0}^{k-1}\binom{k-1}{j}(-1)^j\times\\&\left(\frac{R_2}{R+1-k+j+T}\right)^{i+1}+\sum_{i=k}^{R_1}\binom{R_1}{i}\sum_{j=0}^{i}\binom{i}{j}(-1)^j\left(\frac{R_2}{R+j-i+T}\right)^{R_2}\end{aligned} \tag{5.94}$$

对于任意一个给定的 P_{fa}，解式(5.94)可以得到相应的阈值因子 T，见表 5.9(a)～表 5.9(c)，其中 $P_{\mathrm{fa}}=10^{-6}$，R 分别为 24、32 和 64。

表 5.9(a) OSCAGO-CFAR 检测器的阈值因子 $T(R=24)$

k	R_1		
	8	16	22
1	21.86	36.641	139.754
2	21.532	35.687	140.352
3	20.837	34.166	140.94
4	19.742	32.255	127.547
5	18.269	30.128	94.563
6	16.444	27.914	73.706
7	14.251	25.694	59.49
8	11.534	23.515	49.245
9	—	21.401	41.54
10	—	19.362	35.538
11	—	17.396	30.755
12	—	15.494	26.815
13	—	13.64	23.486
14	—	11.806	20.796
15	—	9.941	18.276
16	—	7.914	16.209
17	—	—	14.352
18	—	—	12.58
19	—	—	11.018
20	—	—	9.531
21	—	—	8.072
22	—	—	6.519

表 5.9(b) OSCAGO-CFAR 检测器的阈值因子 $T(R=32)$

k	R_1			
	8	16	22	24
1	18.658	21.937	29.783	36.876
2	18.546	21.907	29.658	36.49
3	18.235	21.81	29.361	35.746
4	17.635	21.598	28.849	34.665
5	16.693	21.235	28.122	33.326
6	15.376	20.701	27.208	31.817
7	13.63	20	26.149	30.211
8	11.271	19.144	24.985	28.56
9	—	18.15	23.749	26.914
10	—	17.035	22.467	25.249
11	—	15.813	21.115	23.7
12	—	14.492	19.846	22.318
13	—	13.072	18.495	20.321
14	—	11.543	17.237	19.07

续表

k	R_1			
	8	16	22	24
15	—	9.868	15.855	17.865
16	—	7.93	14.688	16.586
17	—	—	13.357	15.123
18	—	—	12.075	14.066
19	—	—	10.815	12.555
20	—	—	9.493	11.213
21	—	—	8.123	10.14
22	—	—	6.591	8.935
23	—	—	—	7.698
24	—	—	—	6.277

表 5.9(c) OSCAGO-CFAR 检测器的阈值因子 $T(R=64)$

k	R_1			
	8	16	22	24
1	15.667	16.009	16.359	16.502
2	15.646	16.009	16.359	16.501
3	15.56	16.007	16.359	16.501
4	15.329	16.001	16.358	16.501
5	14.86	15.979	16.356	16.5
6	14.062	15.925	16.348	16.496
7	12.824	15.811	16.329	16.484
8	10.916	15.608	16.289	16.457
9	—	15.289	16.206	16.415
10	—	14.831	16.078	16.299
11	—	14.215	15.866	16.154
12	—	13.426	15.61	16.171
13	—	12.445	15.178	16.186
14	—	11.249	14.707	16.294
15	—	9.791	14.125	15.407
16	—	7.951	13.448	14.201
17	—	—	12.761	16.529
18	—	—	11.691	21.494
19	—	—	10.649	12.671
20	—	—	9.504	11.034
21	—	—	8.175	10.104
22	—	—	6.632	9.025
23	—	—	—	7.753
24	—	—	—	6.318

于是,根据式(3.21)得到 OSCAGO-CFAR 的平均判决门限 ADT 为

$$\mathrm{ADT}=T\left[\sum_{i=1}^{k}\frac{1}{R_1-k+i}-\frac{1}{R_2^2}k\binom{R_1}{k}\sum_{i=0}^{R_2-1}(i+1)\sum_{j=0}^{k-1}\binom{k-1}{j}(-1)^j\times\right.$$

$$\left.\left(\frac{R_2}{R+1-k+j}\right)^{i+2}+\sum_{i=k}^{R_1}\binom{R_1}{i}\sum_{j=0}^{i}\binom{i}{j}(-1)^j\left(\frac{R_2}{R+j-i}\right)^{R_2+1}\right] \tag{5.95}$$

对于特定的 P_{fa}、R_1 和 R_2 值，解式(5.95)可以得到 OSCAGO-CFAR 检测器的 ADT 值，见表 5.10(a)～表 5.10(c)，其中 $P_{fa}=10^{-6}$，R 分别为 24、32 和 64。

表 5.10(a) OSCAGO-CFAR 检测器的平均判决门限 ADT(R=24)

k	R_1		
	8	16	22
1	21.864	36.641	139.798
2	21.577	35.69	140.533
3	21.086	34.18	141.407
4	20.658	32.303	128.414
5	20.802	30.258	95.719
6	22.105	28.212	75.189
7	25.121	26.297	61.34
8	31.446	24.617	51.504
9	—	23.255	44.251
10	—	22.273	38.744
11	—	21.702	34.505
12	—	21.548	31.155
13	—	21.803	28.46
14	—	22.478	26.496
15	—	23.746	24.708
16	—	26.767	23.497
17	—	—	22.584
18	—	—	21.812
19	—	—	21.461
20	—	—	21.439
21	—	—	21.978
22	—	—	24.138

表 5.10(b) OSCAGO-CFAR 检测器的平均判决门限 ADT(R=32)

k	R_1			
	8	16	22	24
1	18.66	21.937	29.783	36.876
2	18.575	21.907	29.658	36.49
3	18.416	21.812	29.362	35.747
4	18.369	21.607	28.852	34.669
5	18.882	21.269	28.131	33.337
6	20.549	20.804	27.233	31.844

续表

k	R_1			
	8	16	22	24
7	23.957	20.263	26.209	30.27
8	30.711	19.733	25.113	28.677
9	—	19.324	23.997	27.128
10	—	19.149	22.913	25.615
11	—	19.295	21.865	24.291
12	—	19.801	21.04	23.233
13	—	20.658	20.292	21.636
14	—	21.854	19.833	20.948
15	—	23.527	19.424	20.459
16	—	26.813	19.476	20.04
17	—	—	19.502	19.531
18	—	—	19.763	19.688
19	—	—	20.241	19.325
20	—	—	20.85	19.279
21	—	—	21.871	19.835
22	—	—	24.328	20.386
23	—	—	—	21.383
24	—	—	—	23.704

表 5.10(c)　OSCAGO-CFAR 检测器的平均判决门限 ADT($R=64$)

k	R_1			
	8	16	22	24
1	15.668	16.009	16.359	16.502
2	15.663	16.009	16.359	16.501
3	15.68	16.007	16.359	16.501
4	15.881	16.003	16.358	16.501
5	16.675	15.987	16.356	16.5
6	18.661	15.958	16.349	16.496
7	22.466	15.921	16.333	16.485
8	29.726	15.916	16.302	16.462
9	—	16.026	16.245	16.43
10	—	16.366	16.179	16.34
11	—	17.042	16.101	16.254
12	—	18.103	16.103	16.395
13	—	19.517	16.113	16.646
14	—	21.23	16.335	17.171
15	—	23.323	16.738	16.867
16	—	26.881	17.352	16.418
17	—	—	18.276	20.539
18	—	—	18.916	29.198

续表

k	R_1			
	8	16	22	24
19	—	—	19.819	19.11
20	—	—	20.831	18.747
21	—	—	21.999	19.65
22	—	—	24.478	20.545
23	—	—	—	21.523
24	—	—	—	23.857

5.4.3 OSCASO-CFAR 检测器

在 OSCASO-CFAR 检测器中，取两个局部估计 X 和 Y 中的较小者作为杂波功率水平估计 Z(见图 5.2)，即

$$Z=\min(X,Y) \tag{5.96}$$

式(5.96)中 Z 的 PDF 为

$$f_Z(z)=f_X(z)+f_Y(z)-[f_X(z)F_Y(z)+f_Y(z)F_X(z)] \tag{5.97}$$

它的 MGF

$$M_z(u)=M_1(u)+M_2(u)-M_3(u) \tag{5.98}$$

其中

$$M_1(u)=\int_0^{\infty}\mathrm{e}^{-uz}f_x(z)\mathrm{d}z=\frac{\Gamma(R_1-k+u+1)\Gamma(k)}{\Gamma(R_1+u+1)}k\binom{R_1}{k}$$

$$M_2(u)=\int_0^{\infty}\mathrm{e}^{-uz}f_Y(z)\mathrm{d}z=\frac{1}{(1+u/R_2)^{R_2}}$$

$M_3(u)$就是式(5.90)的 $M_z(u)$。这样，利用式(5.11)就得到 OSCASO-CFAR 检测器的检测概率为

$$\begin{aligned}P_{\mathrm{d}}=&\frac{1}{[R_1+T/R_2(1+\lambda)]^{R_2}}+k\binom{R_1}{k}\frac{1}{R_2}\sum_{i=0}^{R_2-1}\sum_{j=0}^{k-1}\binom{k-1}{j}(-1)^j\times\\&\left[\frac{R_2}{R+1-k+j+T/(1+\lambda)}\right]^{i+1}-\\&\sum_{i=k}^{R_1}\binom{R_1}{i}\sum_{j=0}^{i}\binom{i}{j}(-1)^j\left[\frac{R_2}{R+j-i+T/(1+\lambda)}\right]^{R_2}\end{aligned} \tag{5.99}$$

虚警概率为

$$\begin{aligned}P_{\mathrm{fa}}=&\frac{1}{(1+T/R_2)^{R_2}}+k\binom{R_1}{k}\frac{1}{R_2}\sum_{i=0}^{R_2-1}\sum_{j=0}^{k-1}\binom{k-1}{j}(-1)^j\left(\frac{R_2}{R+1-k+j+T}\right)^{i+1}-\\&\sum_{i=k}^{R_1}\binom{R_1}{i}\sum_{j=0}^{i}\binom{i}{j}(-1)^j\left(\frac{R_2}{R+j-i+T}\right)^{R_2}\end{aligned} \tag{5.100}$$

对于任意一个给定的 P_{fa}，解式(5.100)可以得到相应的 OSCASO-CFAR 检测器的阈值因子 T，见表 5.11(a)～表 5.11(c)($P_{\mathrm{fa}}=10^{-6}$,$R=24,32,64$)。另外，利用式(3.21)可以推出 OSCASO-CFAR 的 ADT 为

$$\mathrm{ADT}=T\left[1+\frac{1}{R_2^2}k\binom{R_1}{k}\sum_{i=0}^{R_2-1}(i+1)\sum_{j=0}^{k-1}\binom{k-1}{j}(-1)^j\left(\frac{R_2}{R+1-k+j}\right)^{i+2}-\right.$$
$$\left.\sum_{i=k}^{R_1}\binom{R_1}{i}\sum_{j=0}^{i}\binom{i}{j}(-1)^j\left(\frac{R_2}{R+j-i}\right)^{R_2+1}\right] \tag{5.101}$$

表 5.11(a)　OSCASO-CFAR 检测器的阈值因子 $T(R=24)$

k	R_1		
	8	16	19
2	7475.8	15476.4	18474.7
3	688.2	1482.8	1780.1
4	196	442.7	534.8
5	86.4	206.8	251.7
6	46.7	120.4	148.6
7	28.2	79.5	102.4
8	22.3	57.1	82
9	—	44.5	75.6
10	—	38.9	74.4
11	—	37.3	74.3
12	—	15476.4	74.3
13	—	1482.8	74.2
14	—	442.7	74.2
15	—	206.8	74.2
16	—	120.4	74.2
17	—	—	74.2
18	—	—	74.2
19	—	—	74.2

表 5.11(b)　OSCASO-CFAR 检测器的阈值因子 $T(R=32)$

k	R_1			
	8	16	19	24
2	7475.8	15476.4	18474.7	23471.2
3	688.2	1482.8	1780.1	2275.5
4	196	442.7	534.8	688.1
5	86.4	206.8	251.5	326
6	46.7	120.4	147.6	192.8
7	27.8	79.5	98.3	129.5
8	19.5	56.6	70.7	94.1
9	—	42.4	53.6	72.2
10	—	33	42.2	57.7
11	—	26.6	34.2	48
12	—	23.3	28.9	41.5

续表

k	R_1			
	8	16	19	24
13	—	22.2	26	38.6
14	—	22	24.9	38
15	—	21.9	24.7	37.2
16	—	21.9	24.6	38.5
17	—	—	24.6	37
18	—	—	24.6	36.6
19	—	—	24.6	37.6
20	—	—	—	36.4
21	—	—	—	37
22	—	—	—	36.9
23	—	—	—	37
24	—	—	—	37

表 5.11(c) OSCASO-CFAR 检测器的阈值因子 $T(R=64)$

k	R_1			
	8	16	19	24
2	7475.8	15476.4	18474.7	23471.2
3	688.2	1482.8	1780.1	2275.5
4	196	442.7	534.8	688.1
5	86.4	206.8	251.5	326
6	46.7	120.4	147.6	192.8
7	27.8	79.5	98.3	129.5
8	17.5	56.6	70.7	94.1
9	—	42.4	53.6	72.1
10	—	32.9	42.1	57.4
11	—	26.1	34	46.9
12	—	21	27.9	39.1
13	—	17.6	23.2	33
14	—	16.3	19.6	28.8
15	—	16	17.3	25.2
16	—	16	16.4	21.5
17	—	—	16.2	21.6
18	—	—	16.2	18.5
19	—	—	16.2	16.7
20	—	—	—	16.9
21	—	—	—	16.4
22	—	—	—	16.5
23	—	—	—	16.5
24	—	—	—	16.5

对于 $P_{fa}=10^{-6}$, $R=24,32,64$，表 5.12(a)～表 5.12(c)给出了 ADT 的一些典型值。

表 5.12(a) OSCASO-CFAR 检测器的平均判决门限 ADT($R=24$)

k	R_1		
	8	16	19
2	1986.706	1997.71	1996.126
3	290.811	296.818	296.322
4	115.274	122.193	121.889
5	64.442	73.729	73.573
6	40.796	53.105	53.398
7	26.935	42	43.878
8	22.11	35.175	40.917
9	—	31.204	43.076
10	—	30.357	47.587
11	—	31.7	52.527
12	—	33.546	57.23
13	—	35.111	61.427
14	—	36.14	65.166
15	—	36.707	68.277
16	—	36.943	70.701
17	—	—	72.426
18	—	—	73.5
19	—	—	74.034

表 5.12(b) OSCASO-CFAR 检测器的平均判决门限 ADT($R=32$)

k	R_1			
	8	16	19	24
2	1990.643	1998.838	1998.627	1998.315
3	292.205	297.311	297.234	297.111
4	116.213	122.668	122.658	122.556
5	65.091	74.293	74.345	74.291
6	41.163	53.795	54.005	53.983
7	26.693	42.819	43.254	43.31
8	19.364	35.776	36.575	36.804
9	—	30.663	31 967	32.473
10	—	26.619	28.488	29.422
11	—	23.333	25.664	27.402
12	—	21.693	23.672	26.218
13	—	21.469	22.829	26.688
14	—	21.724	23.031	28.444
15	—	21.825	23.684	29.826
16	—	21.889	24.124	32.72
17	—	—	24.419	32.99

续表

k	R_1			
	8	16	19	24
18	—	—	24.55	33.901
19	—	—	24.593	35.849
20	—	—	—	35.427
21	—	—	—	36.501
22	—	—	—	36.691
23	—	—	—	36.934
24	—	—	—	36.988

表 5.12(c) OSCASO-CFAR 检测器的平均判决门限 ADT($R=64$)

k	R_1			
	8	16	19	24
2	1994.475	1999.014	1998.722	1998.453
3	293.716	297.422	297.293	297.179
4	117.305	122.814	122.735	122.632
5	65.873	74.518	74.47	74.398
6	41.601	54.141	54.209	54.144
7	26.854	43.31	43.575	43.554
8	17.407	36.401	37.042	37.163
9	—	31.362	32.592	32.937
10	—	27.216	29.176	29.957
11	—	23.452	26.335	27.655
12	—	19.93	23.644	25.764
13	—	17.232	21.039	24.014
14	—	16.194	18.619	22.843
15	—	15.978	16.903	21.476
16	—	15.998	16.259	19.392
17	—	—	16.16	20.315
18	—	—	16.192	17.9
19	—	—	16.199	16.44
20	—	—	—	16.793
21	—	—	—	16.365
22	—	—	—	16.491
23	—	—	—	16.498
24	—	—	—	16.5

5.5 MTM、TMGO、TMSO-CFAR 检测器

MTM、TMGO、TMSO-CFAR 检测器属于 TM-TM 类 CFAR 检测器，其基本模型已在 5.2 节给出。本节将在均匀背景条件下，推导出它们检测概率和虚警概率的数学模型以及 ADT 的解析表达式。

5.5.1　MTM-CFAR 检测器

MTM-CFAR 检测器[17-18]选取前沿、后沿滑窗两个局部估计 X 和 Y 的和，作为总的杂波平均功率水平估计，有

$$Z=X+Y \tag{5.102}$$

Z 的 PDF 为

$$f_Z(z)=\int_0^z f_X(x)f_Y(z-x)\mathrm{d}x,\quad z>0 \tag{5.103}$$

利用卷积定理，由式(5.49)、式(5.55)和式(5.103)，可得 Z 的矩母函数

$$M_z(u)=M_X(u)M_Y(u)=\prod_{i=1}^{R_1-R_1''}\frac{c_i'}{\mu u+c_i'}\prod_{j=1}^{R_2-R_2''}\frac{c_j''}{\mu u+c_j''} \tag{5.104}$$

把式(5.104)代入式(5.11)和式(5.12)，可求出 MTM-CFAR 检测器的虚警概率 P_{fa} 和检测概率 P_{d} 的表达式[17-18]

$$P_{\mathrm{fa}}=\prod_{i=1}^{R_1-R''}\frac{c_i'}{T+c_i'}\prod_{j=1}^{R_2-R_2''}\frac{c_j''}{T+c_j''} \tag{5.105}$$

$$P_{\mathrm{d}}=\prod_{i=1}^{R_1-R_1''}\frac{c_i'}{T/(1+\lambda)+c_i'}\prod_{j=1}^{R_2-R_2''}\frac{c_j''}{T/(1+\lambda)+c_j''} \tag{5.106}$$

对于任意一个给定的 P_{fa}，解式(5.105)可以得到相应的 MTM-CFAR 检测器的阈值因子 T。另外，利用式(3.21)可推得 MTM-CFAR 检测器的平均判决门限 ADT 为[17-18]

$$\mathrm{ADT}_{\mathrm{MTM}}=T\left(\sum_{i=1}^{R_1-R'}\frac{1}{c_i'}+\sum_{j=1}^{R_2-R_2''}\frac{1}{c_j''}\right) \tag{5.107}$$

在均匀背景中，对于任意给定的 P_{fa}，解式(5.105)可得 MTM 的阈值因子 T。对于 $P_{\mathrm{fa}}=10^{-6}$，前、后沿滑窗长度 $R_1=R_2=16$ 时，图 5.4 和图 5.5 分别给出了前、后沿滑窗剔除平均(TM)同时从高端剔除($R_1'=R_2'=0$、$R_1''=R_2''$增加)和同时从低端剔除($R_1''=R_2''=2$、$R_1'=R_2'$增加)时，阈值因子 T 和平均判决门限 ADT 变化的曲线。实线为从高端剔除随 $R_1''(R_2'')$变化的曲线，虚线为从低端剔除随 $R_1'(R_2')$变化的曲线。随着高有序样本的减少，MTM 的 ADT 是逐渐增加的，而逐渐剔掉低有序样本，MTM 的 ADT 并不直接上升，这是因为高有序样本估计的方差小，而低有序样本估计的方差大的缘故[25-26]。

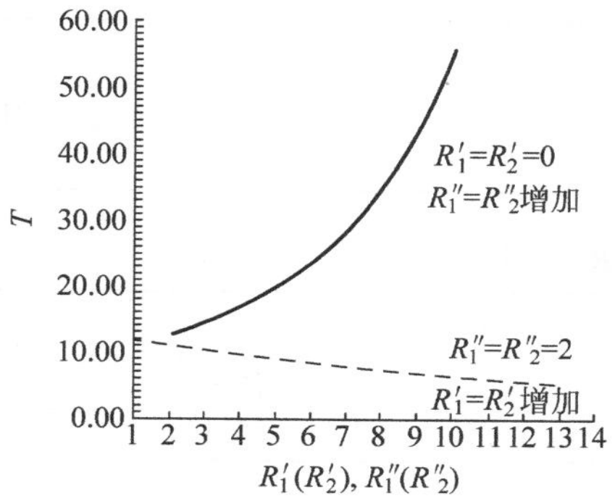

图 5.4　MTM 的阈值因子 T 随 $R_1''(R_2'')$ 或 $R_1'(R_2')$变化的曲线

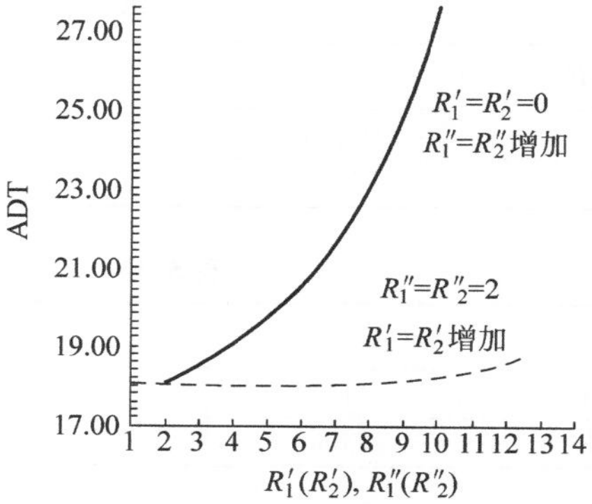

图 5.5　MTM 的平均判决门限 ADT 随 $R_1''(R_2'')$ 或 $R_1'(R_2')$变化的曲线

5.5.2 TMGO-CFAR 检测器

TMGO-CFAR 检测器选取 X 和 Y 中的最大者作为杂波功率水平估计 Z，即

$$Z=\max\{X,Y\} \tag{5.108}$$

于是，Z 的 PDF 为

$$f_Z(t)=f_X(t)F_Y(t)+f_Y(t)F_X(t) \tag{5.109}$$

其中，$f_X(t)$、$F_X(t)$、$f_Y(t)$、$F_Y(t)$ 分别为 X 和 Y 的 PDF 和 CDF。利用式(5.51)～式(5.53)和式(5.57)～式(5.59)，对式(5.109)取拉普拉斯变换得 Z 的 MGF

$$M_z(u)=\sum_{i=1}^{m'}\sum_{j=1}^{r_i'}\sum_{e=1}^{n'}\sum_{f=1}^{r_e''}a'_{ij}a''_{ef}\left\{\frac{\Gamma(f)}{(s''_e)^f}\left[\frac{\Gamma(j)}{(\mu u+s'_i)^j}-\sum_{g=0}^{f-1}\frac{(s''_e)^g}{g!}\frac{\Gamma(j+g)}{(\mu u+s'_i+s''_e)^{j+g}}\right]+\right.$$
$$\left.\frac{\Gamma(j)}{(s'_i)^j}\left[\frac{\Gamma(f)}{(\mu u+s''_e)^f}-\sum_{k=0}^{j-1}\frac{(s'_i)^k}{k!}\frac{\Gamma(f+k)}{(\mu u+s'_i+s''_e)^{f+k}}\right]\right\} \tag{5.110}$$

将式(5.110)代入式(5.11)和式(5.12)，就可分别求出 TMGO-CFAR 检测器的 P_{fa} 和 P_{d} 解析表达式[19-21]。

基于平均判决门限 ADT 的计算式(3.21)，可求得 TMGO-CFAR 检测器的平均判决门限 ADT 为[19-21]

$$\mathrm{ADT_{TMGO}}=T\sum_{i=1}^{m'}\sum_{j=1}^{r_i'}\sum_{e=1}^{n'}\sum_{f=1}^{r_e''}a'_{ij}a''_{ef}\left\{\frac{\Gamma(f)}{(s''_e)^f}\left[\frac{\Gamma(j+1)}{(s'_i)^{j+1}}-\sum_{g=0}^{f-1}\frac{(s''_e)^g}{g!}\frac{\Gamma(j+g+1)}{(s'_i+s''_e)^{j+g+1}}\right]+\right.$$
$$\left.\frac{\Gamma(j)}{(s'_i)^j}\left[\frac{\Gamma(f+1)}{(s''_e)^{f+1}}-\sum_{k=0}^{j-1}\frac{(s'_i)^k}{k!}\frac{\Gamma(f+k+1)}{(s'_i+s''_e)^{f+k+1}}\right]\right\} \tag{5.111}$$

在均匀背景中，对于任意给定的 P_{fa}，解式(5.12)和式(5.110)可得 TMGO 检测器的阈值因子 T。对于 $P_{\mathrm{fa}}=10^{-6}$，前沿、后沿滑窗长度 $R_1=R_2=10$ 时，表 5.13 和表 5.14 分别给出了前、后沿滑窗同时从高端剔除($R_1'=R_2'=0$、$R_1''=R_2''$增加)和同时从低端剔除($R_1''=R_2''=2$、$R_1'=R_2'$增加)时，阈值因子 T 和平均判决门限 ADT 变化的部分数值结果。可以看出，随着高有序样本的减少，TMGO 的 ADT 是逐渐增加的，而逐渐剔除掉低有序样本，TMGO 的 ADT 并不直接上升，这也是因为高有序样本估计的方差小，而低有序样本估计的方差大的缘故。

表 5.13　TMGO 的 T 和 ADT 随 $R_1''(R_2'')$ 变化的情况

$R_1''(R_2'')$	2	3	4	5	6	7
T	31.06	39.95	53.54	75.49	116.6	216.4
ADT	23.98	25.78	28.85	33.64	42.24	61.89

表 5.14　TMGO 的 T 和 ADT 随 $R_1'(R_2')$ 变化的情况

$R_1'(R_2')$	2	3	4	5	6	7
T	24.45	21.84	19.51	17.49	15.70	14.18
ADT	23.63	23.56	23.50	23.59	23.85	24.53

5.5.3 TMSO-CFAR 检测器

TMSO-CFAR[22-23] 检测器选取前沿、后沿滑窗两个局部估计 X 和 Y 中的最小者作为杂波平均功

率水平估计 Z，即

$$Z=\min\{X,Y\} \tag{5.112}$$

于是，Z 的 PDF 为

$$f_Z(t)=f_X(t)+f_Y(t)-[f_X(t)F_Y(t)+f_Y(t)F_X(t)] \tag{5.113}$$

同样，计算 TMSO-CFAR 检测器的 P_{fa} 及 P_d 的关键就在于求 Z 的矩母函数 $Mz(u)$。对式(5.113)取拉普拉斯变换，得 Z 的矩母函数

$$\begin{aligned}M_z(u)=&\prod_{i=1}^{m'}\left(\frac{s'_i}{\mu u+s'_i}\right)^{r'_i}+\prod_{j=1}^{n'}\left(\frac{s''_j}{\mu u+s''_j}\right)^{r''_j}-\sum_{i=1}^{m'}\sum_{j=1}^{r'_i}\sum_{e=1}^{n'}\sum_{f=1}^{r''_e}a'_{ij}a''_{ef}\\&\left\{\frac{\Gamma(f)}{(s''_e)^f}\left[\frac{\Gamma(j)}{(\mu u+s'_i)^j}-\sum_{g=0}^{f-1}\frac{(s''_e)^g}{g!}\frac{\Gamma(j+g)}{(\mu u+s'_i+s''_e)^{j+g}}\right]+\frac{\Gamma(j)}{(s'_i)^j}\right.\\&\left.\left[\frac{\Gamma(f)}{(\mu u+s''_e)^f}-\sum_{k=0}^{j-1}\frac{(s'_i)^k}{k!}\frac{\Gamma(f+k)}{(\mu u+s'_i+s''_e)^{f+k}}\right]\right\}\end{aligned} \tag{5-114}$$

将式(5.114)代入式(5.11)和式(5.12)，就可分别求出 TMSO-CFAR 检测器的 P_{fa} 和 P_d 解析表达式[22-23]。

利用平均判决门限 ADT 的计算式(3.21)，可求得 TMSO-CFAR 检测器平均判决门限 ADT 的解析表达式[22-23]

$$\begin{aligned}\mathrm{ADT}_{\mathrm{TMSO}}=&T\sum_{i=1}^{m'}\frac{r'_i}{s'_i}+T\sum_{j=1}^{n'}\frac{r''_j}{s''_j}-T\sum_{i=1}^{m'}\sum_{j=1}^{r'_i}\sum_{e=1}^{n'}\sum_{f=1}^{r''_e}a'_{ij}a''_{ef}\\&\left\{\frac{\Gamma(f)}{(s''_e)^f}\left[\frac{\Gamma(j+1)}{(s'_i)^{j+1}}-\sum_{g=0}^{f-1}\frac{(s''_e)^g}{g!}\frac{\Gamma(j+g+1)}{(s'_i+s''_e)^{j+g+1}}\right]+\right.\\&\left.\frac{\Gamma(j)}{(s'_i)^j}\left[\frac{\Gamma(f+1)}{(s''_e)^{f+1}}-\sum_{k=0}^{j-1}\frac{(s'_i)^k}{k!}\frac{\Gamma(f+k+1)}{(s'_i+s''_e)^{f+k+1}}\right]\right\}\end{aligned} \tag{5-115}$$

对于 $P_{fa}=10^{-6}$，前沿和后沿滑窗长度 $R_1=R_2=10$ 时，表 5.15 和表 5.16 分别给出了前、后沿滑窗剔除平均(TM)同时从高端剔除($R'_1=R'_2=0$、$R''_1=R''_2$增加)和同时从低端剔除($R''_1=R''_2=2$、$R'_1=R'_2$增加)时，阈值因子 T 和平均判决门限 ADT 变化的部分数值结果。可以看出，随着高有序样本的减少，TMSO 的 ADT 是逐渐增加的，而逐渐剔除掉低有序样本，TMSO 的 ADT 并不直接上升，这也是因为高有序样本的方差小，而低有序样本的方差大的缘故。

表 5.15 TMSO 的 T 和 ADT 随 $R''_1(R''_2)$ 变化的情况

$R''_1(R''_2)$	2	3	4	5	6	7
T	64.92	95.59	148.5	258.8	556.2	1848.5
ADT	50.12	61.70	80.02	115.3	201.4	528.6

表 5.16 TMSO 的 T 和 ADT 随 $R'_1(R'_2)$ 变化的情况

$R'_1(R'_2)$	2	3	4	5	6	7
T	52.10	46.66	41.64	37.36	33.68	30.72
ADT	50.35	50.33	50.16	50.41	51.18	53.15

5.6 GOS 类 CFAR 检测器在均匀背景和多目标环境中的性能

5.6.1 GOS 类 CFAR 检测器在均匀背景中的性能

对于 GOSCA-CFAR 检测器，当 $R_1=R_2$ 时有 $\mathrm{ADT}(k_1,k_2)=\mathrm{ADT}(k_2,k_1)$。表 5.2 给出了 $R_1=R_2=16$、$P_{fa}=10^{-6}$ 时 GOSCA 的 ADT 值。在前面的推导过程中，即使杂波功率水平参数 μ 不假设为 1，也可证明式(5.67)右侧总是不依赖于 μ 的，即 ADT 也是不依赖于参数 μ 的。一般来说，这对任意 CFAR 方案也都是成立的。为了便于把 GOSCA 与 OS-CFAR 检测器进行比较，表 5.17 给出了 $R=32$、$P_{fa}=10^{-6}$ 时 OS-CFAR 的部分 ADT 值。从对比结果可以看出，在均匀背景下 GOSCA 的性能要比 OS 好一些。因此，ADT_{GOSCA} 普遍比 ADT_{OS} 小。另外，在 GOSCA 中，序值 k_1 和 k_2 的选择要比 OS 具有更大的自由度。这一特性对处理多目标情况是有利的[27]。实际上，参考单元数 $R(R=R_1+R_2)$ 的值越大，GOSCA 与 OS 检测性能的差别越小。在参考单元数 R 固定的情况下，$R_1=R_2$ 时的 GOSCA 的检测性能最好。当 $R_1=R_2=16$ 时，最优的 k_1、k_2 值分别为 $k_1=14$、$k_2=14$。在这种情况下，它们将极大化检测概率 P_d。

表 5.17 OS-CFAR 检测器的 ADT

k	19	20	21	22	23	24	25	26	27	28	29	30	31	32
ADT	20.65	20.28	19.97	19.70	19.48	19.30	19.16	19.06	19.02	19.04	19.17	19.46	20.12	22.08

当 $R_1=R_2$ 时，GOSGO 也存在 $\mathrm{ADT}(k_1,k_2)=\mathrm{ADT}(k_2,k_1)$的结论，见表 5.4。由平均判决门限的定义可知，ADT 越小，则检测概率 P_d 越大。当 R 固定时，较大的差值 R_1-R_2 总是使 GOSGO 具有较好的检测性能。当 $R_2=0$，即不考虑后沿滑窗的影响时，GOSGO 就退化成 OS-CFAR。为了在杂波边缘环境中保持好的性能并抑制过高的虚警概率，GOSGO 的 R_1 和 R_2 的选择应采取折中处理。如果取 $R_1=R_2$、$k_1=k_2$，则 GOSGO 便退化成 OSGO。在 $R_1=R_2=16$ 的 GOSGO 或 OSGO 中，最优的 ADT 值位于 $k=14$，而不是文献[1]的 $k=13$。显然，就 R_1、R_2、k_1、k_2 的选择来说，GOSGO 比 OSGO 具有更大的灵活性。因为在 OSGO 中，$R_1=R_2$、$k_1=k_2$ 只有两个参数可变，而在 GOSGO 中有四个参数可变。

从对 ADT_{GOSSO} 的分析中可以发现，当 R 固定时，GOSSO-CFAR 的 R_1 和 R_2 间存在一个最优组合。例如，在 $R_1=22(k_1=17)$ 且 $R_2=10(k_2=10)$ 时，ADT_{GOSSO} 的值最小。取 $R_1=R_2$、$k_1=k_2$，则 GOSSO 便退化成 OSSO。$R_1=R_2=16$ 时，OSSO 的最小 ADT 值位于 $k_1=k_2=15$，与文献[1]的结论不一致。从表 5.2 和表 5.4 与表 5.6 的平均判决门限的比较来看，在均匀背景中，GOSGO 的检测性能与 OS 相近，GOSSO 要比 OS 差些，但是也能被接受。

从表 5.8 的 ADT_{MOSCA} 值和表 5.17 的 ADT_{OS} 值的比较来看，在均匀背景中，MOSCA 的检测性能明显优于 OS[27]，它基本上位于 CA($\mathrm{ADT}_{CA}=17.278$)和 OS 之间，是对 CA 和 OS 的折中，这一点也可以从对杂波功率水平的估计方式上看出。

从表 5.10 的 ADT 值来看，OSCAGO 在均匀背景中的检测性能不仅比 GOSGO 要好，甚至比 OS 还好。例如，$R_1=8$、$R_2=24$、$k_1=3$ 或 4 时，OSCAGO 的 ADT=18.4。然而，$R=32$、$k_1=28$ 时，OS 的 ADT=19.049。从表 5.12 的 ADT 值来看，OSCASO 的性能与 GOSSO 相近。例如，$R_1=19$、$k_1=13$、$R_2=13$ 时，$\mathrm{ADT}_{OSCASO}=22.829$，而 $R_1=22$、$k_1=16$、$R_2=10$、$k_2=10$ 的 $\mathrm{ADT}_{GOSSO}=22.644$(见表 5.6)。

对于 $P_{fa}=10^{-6}$，前沿、后沿滑窗长度 $R_1=R_2=10$ 的情况，TMGO 在 $R_1'=R_2'=4$、$R_1''=R_2''=2$ 时的平均判决门限 ADT=23.50，而 GOSGO 或 OSGO 在 $k_1=k_2=8$ 时的 ADT 值为 24.53，这时它们具有相同的抗干扰目标能力，故 TMGO-CFAR 在均匀背景中的性能是优于 GOSGO 或 OSGO 的。在相同条件下，TMSO 在 $R_1'=R_2'=4$、$R_1''=R_2''=2$ 时的平均判决门限 ADT 值为 50.16，而 GOSSO 或 OSSO 在 $k_1=k_2=8$ 时的 ADT 值为 53.15，这时它们具有相同的抗干扰目标能力，故 TMSO-CFAR 在均匀背景中的性能是优于 GOSSO 或 OSSO 的。这是因为 TMGO-CFAR 和 TMSO-CFAR 相比，OSGO 和 OSSO 采用了较多杂波样本参与杂波功率水平估计的缘故。

文献[28]分析了 GOSSO 与 GOSGO 在 Pareto 分布混响背景下的检测性能，证明了在尺度参数已知情况下，这两个检测器对形状参数具有恒虚警特性，且在均匀混响背景下，GOSGO 的检测性能与 OS-CFAR 相近，在混响边缘情况下具有最好的虚警控制能力；而对于多目标干扰情况，GOSSO 比其他两种检测器的检测性能更优。文献[29]则针对 Pearson 分布混响背景分析了 GOSGO 与 GOSSO 的检测性能。文献[30]针对非均匀韦布尔分布和非均匀 K 分布背景，对比分析了 GOSCA、GOSGO、GOSSO 的检测性能，得到的结论与上述分析类似。文献[31]针对 OS、GOSSO 和 GOSGO 在非均匀环境下检测性能严重下降的问题，基于韦布尔分布模型和模糊量化的软决策方法，研究了一种加权有序统计量的模糊恒虚警(记为 WOSF-CFAR)检测算法。通过计算前沿和后沿滑窗对应的模糊隶属函数值，采用代数积、代数和、最大选择和最小选择 4 种融合规则对 2 个滑窗的模糊输出量进行融合，并与门限进行比较，判别目标有无。仿真表明，提出的检测方法与 GOSGO、GOSSO 算法相比，在均匀噪声、杂波边缘干扰和多目标干扰环境下均具有较好的检测性能，尤其是采用代数积融合规则时，检测性能最优，且提出的检测算法在均匀环境下也具有最佳的检测性能。

5.6.2 GOS 类 CFAR 检测器在多目标环境中的性能

这里研究参考滑窗中存在干扰目标时对 GOS 类 CFAR 检测器检测性能的影响。对与 GOSCA-CFAR、GOSGO-CFAR 和 GOSSO-CFAR 检测器，当左边滑窗中的干扰目标数小于或等于 R_1-k_1，右边的小于或等于 R_2-k_2 时，这些干扰目标的影响可以在下面统计假设条件下进行分析。当没有干扰目标时，得到来自 R_1 个独立同分布样本的第 k_1 个有序样本和来自 R_2 个独立同分布样本的第 k_2 个有序样本。假设干扰目标回波与背景杂波功率比 INR 是无限的，以至于来自干扰目标回波的样本总是占据最高的有序样本。在某种意义上，这是一种最糟的情况。对于左边滑窗中有 IL 个干扰目标的情况，实际上所利用的是 $R_1-\mathrm{IL}$ 个样本中的第 k_1 个有序样本。而当右边滑窗中有 IR 个干扰目标时，所获得的是 $R_2-\mathrm{IR}$ 个样本中的第 k_2 个有序样本。这种强干扰目标的影响可以利用 5.3 节中推出的检测概率表达式通过代换来计算。例如，对于 GOSCA-CFAR 检测器，在式(5.65)中，通过用 $R_1-\mathrm{IL}$ 和 $R_2-\mathrm{IR}$ 分别代替 R_1 和 R_2 就可实现。这就产生了一个新的信噪比 λ 与 P_d 间的关系。应该注意的是，阈值因子 T 是利用 R_1 和 R_2 得到的，而不是用 $R_1-\mathrm{IL}$ 和 $R_2-\mathrm{IR}$ 获得的。对于有限的 INR，检测损失将会变小。

为了便于比较，这里同时也研究了 OS-CFAR 在多目标情况下的性能。对于 $R_1=16(k_1=11)$ 且 $R_2=16(k_2=13)$ 时的 GOSCA 和 $R=32(k=24)$ 时的 OS，它们最多允许处理 8 个干扰目标。由于在检测中采用了自动筛选技术，这使得干扰目标一般不会等数量地分布在前、后沿滑窗中。在分析中仅考虑与固定阈值方案相比较的 CFAR 损失，并只处理 Swerling Ⅱ 型目标。其他起伏模型的 CFAR 损失与此是相近的，这是因为 CFAR 处理对 Swerling 模型不敏感。GOSCA 和 OS 检测器的 CFAR 损失被总结在表 5.18 中。在 OS-CFAR 中，干扰目标数为 IL+IR。从表 5.18 可以看出，GOSCA 的 CFAR 损失比 OS 小。这是因为 GOSCA-CFAR 在代表序值 k_1 和 k_2 的选择上比 OS 具有更大的灵活性。图 5.6 给出了这两个 CFAR 检测器的检测性能曲线。

表 5.18 GOSCA 和 OS 检测器在多目标环境下的 CFAR 损失

检测器	IL,IR								
	0,0	1,0	2,0	3,0	3,1	4,1	4,2	5,2	5,3
GOSCA	1.4606dB	1.6905dB	1.9739dB	2.3483dB	2.7108dB	3.2058dB	3.6757dB	4.4973dB	5.2013dB
OS	1.4472dB	1.7516dB	2.0793dB	2.4498dB	2.8672dB	3.3488dB	3.9394dB	4.7163dB	5.9564dB

GOSCA：$R_1=R_2=16,k_1=11,k_2=13$；OS：$R=32,k=24$

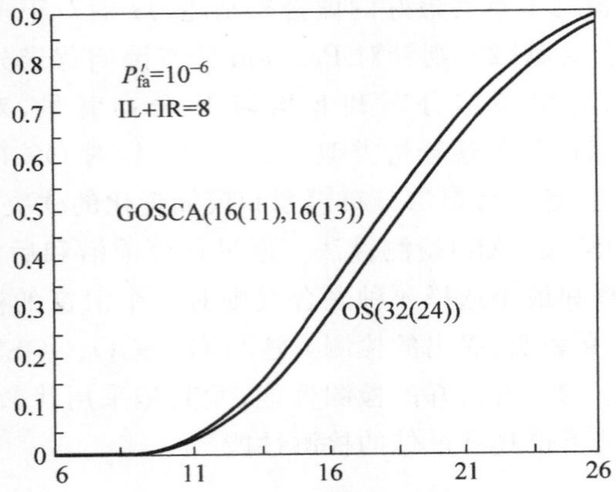

图 5.6 GOSCA 和 OS-CFAR 检测器的检测性能曲线

用 R_1-IL 和 R_2-IR 分别代替 R_1 和 R_2，同样可以得到 GOSGO 和 GOSSO 在多目标环境中的检测概率 P_d。表 5.19 给出了在 Swerling Ⅱ型目标条件下 5 种检测器的 CFAR 损失。从表 5.19 的结果可知：GOSGO 的损失比 OSGO 小，GOSSO 和 OSSO 的性能相近。当左边滑窗中干扰目标数 IL 小于或等于 1 时，GOSSO 比 OSSO 性能好；但是，当 IL≥2 时，OSSO 的损失小于 GOSSO。如果考虑自动筛选技术的作用，则 GOSSO 较 OSSO 具有更大的鲁棒性。一般来说，OS 的损失比 GOSGO 和 GOSSO 都小。但是，当干扰目标数等于可被允许的最大数时，例如，在表 5.19 中 IL+IR=6 时，在这 3 种 CFAR 检测器中，OS 的性能最差。为了更充分地说明这个问题，图 5.7 给出了这种情况下的检测性能曲线。

表 5.19 5 种检测器在多目标环境下的 CFAR 损失

检测器	IL,IR										
	0,0	1,0	1,1	2,0	2,1	3,0	3,1	3,2	4,0	4,2	5,0	6,0
GOSGO	1.55dB	1.97dB	2.27dB	2.53dB	2.73dB	3.29dB	3.41dB	3.95dB	4.58dB	4.89dB	—	—
OSGO	1.59dB	2.01dB	2.32dB	2.71dB	2.90dB	4.02dB	4.10dB	4.28dB	—	—	—	—
GOSSO	2.13dB	2.52dB	—	2.95dB	—	3.42dB	—	—	3.94dB	—	4.52dB	5.16dB
OSSO	2.33dB	2.63dB	—	2.87dB	—	3.04dB	—	—	—	—	—	—
OS	1.39dB	1.76dB	2.16dB	2.16dB	2.64dB	3.64dB	3.21dB	3.97dB	3.21dB	5.19dB	3.98dB	5.19dB

GOSGO：$R_1=24,k_1=20,R_2=8,k_2=6$；GOSSO：$R_1=22,k_1=16,R_2=10,k_2=10$；OSGO 和 OSSO：$R_1=R_2=16,k_1=k_2=13$；OS：$R=32,k=26$

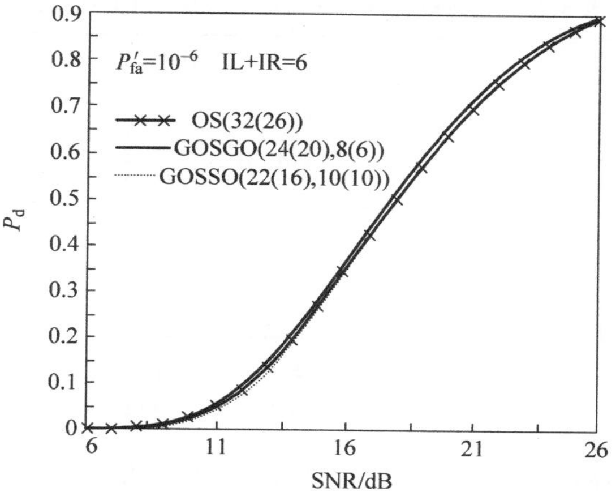

图 5.7　GOSGO-CFAR、GOSSO-CFAR 和 OS-CFAR 检测器的检测性能曲线

对于 MOSCA-CFAR 检测器，同样可通过用 R_1-IL 代替式(5.85)中的 R_1 来获得多目标环境中的 P_d。由于 MOSCA 采用了自动筛选技术，当检测概率较高时，目标回波一般不会进入右边的参考滑窗。因此，这里仅研究左边参考滑窗内的干扰目标对检测性能的影响。表 5.20 给出了在 Swerling Ⅱ 型目标条件下的 CFAR 损失。对于 $R_1=20(k_1=14)$ 且 $R_2=12$ 的 MOSCA 和 $R=32(k=26)$ 的 OS，它们最多可以抗击 6 个干扰目标。

表 5.20　MOSCA 和 OS 检测器在多目标环境下的 CFAR 损失

检测器	IL						
	0	1	2	3	4	5	6
MOSCA	1.2509dB	1.4934dB	1.7758dB	2.1156dB	2.5613dB	3.1719dB	4.2261dB
OS	1.3925dB	1.7564dB	2.1597dB	2.6404dB	3.2109dB	3.9759dB	5.1926dB

MOSCA：$R_1=20, R_2=12, k_1=14$；OS：$R=32, k=26$

从表 5.20 的结果可以看出，在多目标情况下，MOSCA 的检测性能明显好于 OS。当 IL≥5 时，MOSCA 的 CFAR 损失比 OS 大约小 1dB。而且，MOSCA 在工程实现上也比 OS 容易，在参考单元样本幅值的排序过程中，MOSCA 的处理时间大约是 OS 的一半。另外，在杂波边缘环境下，MOSCA 也具有与 OS 相近的性能。

当干扰目标数小于或等于容许的最大干扰目标数时，同样可以通过用 R_1-IL 代替式(5.93)或式(5.99)中的 R_1 来获得 OSCAGO 或 OSCASO 在多目标环境中的检测性能。表 5.21 给出了这几个检测器的 CFAR 损失。

从表 5.21 的结果可以看出，OSCAGO-CFAR 在均匀杂波背景和多目标情况下的检测性能都比 OS 和 GOSGO 明显好，并且随着干扰目标数的变化，这种优势更为明显。当干扰目标数等于 4 时，OSCAGO 的 CFAR 损失比 GOSGO 减少了约 3dB。由此可见，OSCAGO 在检测性能方面的明显优势。在均匀背景中，OSCAGO 的检测性能与 GOSSO 相近。但是，在多目标情况下，当干扰目标数小于容许的最大数时，随着干扰目标数的增加，OSCAGO 的检测性能更显得比 GOSSO 好。并且，在表 5.21 给出的条件下，当干扰目标数为 4～6 时，OSCAGO 的检测性能甚至超过了 OS 和 GOSGO。

表 5.21　5 种检测器在多目标环境中的 CFAR 损失　(单位：dB)

检测器	IL,IR											
	0,0	1,0	1,1	2,0	2,1	3,0	3,1	3,2	4,0	4,2	5,0	6,0
OSCAGO	1.28	1.29	—	1.31	—	1.37	—	—	1.52	—	1.91	3.18
GOSGO	1.55	1.97	2.27	2.53	2.73	3.29	3.41	3.95	4.57	4.89	—	—
OSCASO	2.32	2.47	—	2.61	—	2.71	—	—	2.79	—	2.83	2.85
GOSSO	2.13	2.52	—	2.95	—	3.42	—	—	3.94	—	4.52	5.16
OS	1.39	1.76	2.16	2.16	2.64	2.64	3.21	3.98	3.21	5.19	3.98	5.19

OSCAGO：$R_1=8, k_1=2, R_2=24$；GOSGO：$R_1=24, k_1=20, R_2=8, k_2=6$；OS：$R=32, k=26$

OSCASO：$R_1=19, k_1=13, R_2=13$；GOSSO：$R_1=22, k_1=16, R_2=10, k_2=10$

对于 MTM-CFAR 检测器，当左边滑窗内的干扰目标数目 IL 小于或等于 R_1''，右边滑窗内的干扰目标数目 IR 小于或等于 R_2''时，同样可以通过用(R_1-IL)，(R_2-IR)，($R_1''-$IL)，($R_2''-$IR)代替前、后沿滑窗数学模型中的参数 R_1、R_2、R_1''、R_2''来获得 MTM 在多目标环境中的检测性能，对弱干扰目标，检测损失将变小。表 5.22 给出了在 Swerling Ⅱ型目标条件下，几种检测器在出现多目标时的 CFAR 损失。MTM 选取 $R_1=R_2=16$、$R_1'=5$、$R_1''=3$、$R_2'=5$、$R_2''=3$，MCM 选取 $R_1=R_2=16$、$R_1'=0$、$R_1''=3$、$R_2'=0$、$R_2''=3$，GOSCA 选取 $R_1=R_2=16$、$k_1=k_2=13$，TM 选取 $R=32$、$R'=10$、$R''=6$，CM 选取 $R=32$、$R'=0$、$R''=6$，OS 选取 $R=32$、$k=26$。从表 5.22 中的结果可以看出，与 TM 相比，MTM 在均匀背景及多目标环境中的性能均获得了改善，MCM[32] 作为 MTM 的特例也相比 CM 获得了性能改善，MTM 的性能明显优于 GOSCA 和 OS，而从这几种检测器在均匀背景中的性能表现来看，MTM 最好，从多目标环境中的性能表现来看，MCM 最好。

表 5.22　MTM 等几种检测器在多目标情况下的 CFAR 损失($P_{fa}=10^{-6}, P_d=0.5$)　(单位：dB)

检测器	IL,IR									
	0,0	1,0	1,1	2,0	2,1	2,2	3,0	3,1	3,2	3,3
MTM	1.248	1.525	1.786	1.884	2.125	2.441	2.419	2.633	2.916	3.345
MCM	1.264	1.528	1.777	1.866	2.098	2.397	2.365	2.572	2.842	3.247
GOSCA	1.372	1.747	2.093	2.281	2.590	3.038	3.221	3.475	3.848	4.540
CM	1.282	1.532	1.804	1.804	2.103	2.440	2.103	2.440	2.838	3.364
TM	1.264	1.526	1.811	1.811	2.126	2.481	2.126	2.481	2.901	3.463
OS	1.388	1.750	2.158	2.158	2.633	3.211	2.633	3.211	3.971	5.192

MTM：$R_1=R_2=16$、$R_1'=5$、$R_1''=3$、$R_2'=5$、$R_2''=3$；MCM：$R_1=R_2=16$、$R_1'=0$、$R_1''=3$、$R_2'=0$、$R_2''=3$；GOSCA：$R_1=R_2=16$、$k_1=k_2=13$；TM：$R=32$、$R'=10$、$R''=6$；CM：$R=32$、$R'=0$、$R''=6$；OS：$R=32$、$k=26$

在 Swerling Ⅱ型目标条件下，表 5.23 给出了 TMGO、TMSO 和 GOSGO 或 OSGO、GOSSO 或 OSSO 检测器在出现多目标时的 CFAR 损失。由此可见，由于 TMGO 采用了更多的参考单元样本参与杂波功率水平的估计，TMGO 在均匀背景和多目标环境中的性能均比 GOSGO 或 OSGO 有所改善，而 TMSO 在均匀背景及多目标环境中的性能均相比 GOSSO 或 OSSO 获得了改善。对于采用选小逻辑的 CFAR 检测器，它的 CFAR 损失对参考单元数目比较敏感，参考单元数目较少时，它的 CFAR 损失较大，这里参考单元数目取 $R=R_1+R_2=20$，但这并不影响检测器性能的比较，若参考单元数目 $R=32$，它们引起的 CFAR 损失是可以接受的。

表 5.23　TMGO 和 TMSO 等检测器在多目标情况下的 CFAR 损失($P_{fa}=10^{-6}$, $P_d=0.5$)(单位: dB)

检测器	IL,IR								
	0,0	1,0	1,1	2,0	2,1	2,2	3,0	3,1	3,2
TMGO	2.449	2.770	3.259	3.309	3.660	4.383	4.328	4.506	4.948
GOSGO	2.621	2.950	3.655	3.567	4.081	5.253	4.906	5.159	5.898
TMSO	4.424	4.834	5.204	5.227	5.734	6.227	5.570	6.228	6.991
GOSSO	4.589	5.099	5.475	5.597	6.151	6.675	6.029	6.826	7.754

TMGO: $R_1=R_2=10$、$R_1'=3$、$R_1''=3$、$R_2'=3$、$R_2''=2$
GOSGO: $R_1=R_2=10$、$k_1=7$、$k_2=8$
TMSO: $R_1=R_2=10$、$R_1'=3$、$R_1''=3$、$R_2'=3$、$R_2''=2$
GOSSO: $R_1=R_2=10$、$k_1=7$、$k_2=8$

5.7　GOS 类 CFAR 检测器在杂波边缘环境中的性能

5.6 节已经分析了在 $P_{fa}=10^{-6}$ 条件下 GOS 类 CFAR 检测器在均匀杂波和多目标环境中的检测性能。本节将分别讨论它们在杂波边缘环境中的性能。因此，首先给出 GOS 类 CFAR 检测器在杂波边缘中的虚警概率计算公式，然后分析它们在杂波边缘环境中的性能。

5.7.1　GOSCA-CFAR 检测器在杂波边缘环境中的性能

当 $0\leqslant N_C\leqslant R_1$ 时，GOSCA-CFAR 检测器的虚警概率为[33-34]

$$P_{fa}=M_Z(u)\big|_{u=T}=k_2\binom{R_2}{k_2}\frac{\Gamma(R_2+T+1-k_2)\Gamma(k_2)}{\Gamma(R_2+T+1)}$$

$$\sum_{i=k_1}^{R_1}\sum_{j=\max(0,i-N_C)}^{\min(i,R_1-N_C)}\binom{N_C}{i-j}\binom{R_1-N_C}{j}\sum_{n_1=0}^{j}\binom{j}{n_1}\times(-1)^{n_1}\gamma T\frac{\Gamma(g_1)\Gamma(i-j+1)}{\Gamma[g_1+(i-j)+1]}\tag{5.116}$$

其中

$$g_1=\gamma(T+n_1+R_1-N_C-j)+N_C-i+j\tag{5.117}$$

当 $R_1<N_C\leqslant R$ 时，GOSCA-CFAR 检测器的虚警概率为[33-34]

$$P_{fa}=M_Z(u)\big|_{u=T/\gamma}=k_1\binom{R_1}{k_1}\frac{\Gamma(R_1+T+1-k_1)\Gamma(k_1)}{\Gamma(R_1+T+1)}$$

$$\sum_{i=k_2}^{R_2}\sum_{j=\max(0,i-N_C+R_1)}^{\min(i,R-N_C)}\binom{N_C-R_1}{i-j}\binom{R-N_C}{j}\times$$

$$\sum_{n_1=0}^{j}\binom{j}{n_1}(-1)^{n_1}T\frac{\Gamma(h)\Gamma(i-j+1)}{\Gamma[h+(i-j)+1]}\tag{5.118}$$

其中

$$h=T+\gamma(n_1+R_2-N_C+R_1-j)+N_C-R_1-i+j\tag{5.119}$$

图 5.8 给出了在 $\gamma=15$dB 的杂波边缘环境中，取给定的几组 k_1 和 k_2 值时 GOSCA 的虚警概率随进入参考滑窗中强杂波单元数 N_C 变化的曲线。当 $N_C=R_1$ 时出现尖峰，称之为虚警尖峰。总的看来，k_1 增大，k_2 减小，则虚警尖峰降低。因为虽然大的 k_1 值对应的 T 值较小，但是当强杂波充满前沿滑窗时，大

的 k_1 值所带来的高的局部估计 X 可以补偿 T 值的下降，使之仍具有较低的虚警尖峰。选择小的 k_2 值具有较低的虚警尖峰，是因为 k_2 越小 T 值越大，则 T 值可以补偿 Z 的降低，使 TZ 较大，虚警尖峰也就随之降低。总之，对于 GOSCA-CFAR 检测器的虚警尖峰控制来说，大 k_1 值和小 k_2 值比较有效。

图 5.9 给出了 GOSCA-CFAR 检测器在强弱杂波比 $\gamma=15\text{dB}$ 时，虚警尖峰随 k_1 和 k_2 值变化的三维图。由图 5.9 可见，$k_1=16$ 且 $k_2=1$ 时 GOSCA-CFAR 检测器的虚警尖峰最低，即虚警控制能力最强。考虑到在多目标情况中发挥 GOSCA 所具有的 OS 的优点，k_1 的选择还应该留有余量。并且，为了提高检测性能，k_2 也应该选得恰当，以降低 ADT 值。

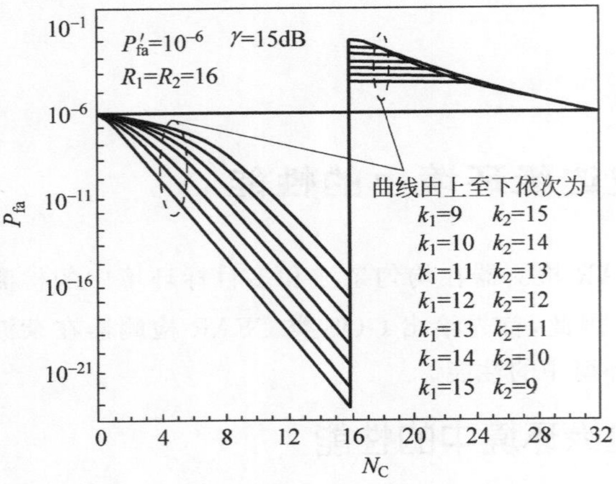

图 5.8 杂波边缘环境下 GOSCA-CFAR 检测器的虚警性能

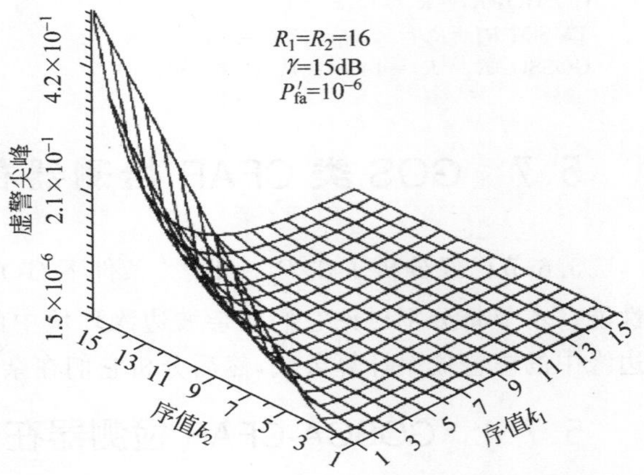

图 5.9 GOSCA-CFAR 检测器的虚警尖峰与 k_1 和 k_2 值关系

图 5.10 给出 $k_1=11$、$k_2=13$ 和 $k_1=13$、$k_2=11$ 时 GOSCA 与 CA 等方法抗杂波边缘性能的比较。GOSCA(k_1, k_2)与 GOSCA(k_2, k_1)在均匀背景下的检测性能相同。所以，在抗干扰目标能力允许的情况下，且干扰目标总数固定时，分配 k_1 和 k_2 值应该主要考虑其在杂波边缘环境中的虚警控制能力。上述的分析都是指对抗杂波边缘前沿。此时的 γ 为非负数。当 γ 为负值时表示杂波边缘后沿。当然，大 k_1 值和小 k_2 值的 GOSCA 的抗杂波边缘性能较强的结论是有条件的，是针对杂波边缘前沿而言的。然而，OS 和 CA 的抗边缘杂波性能是不分前沿和后沿的。因此，从综合对抗杂波边缘前沿和后沿的能力来看，GOSCA 的抗杂波边缘性能与 CA 和 $k_{OS}=(k_1+k_2)_{GOSCA}$ 的 OS 相近，比抗杂波边缘性能好的 GO 要差。

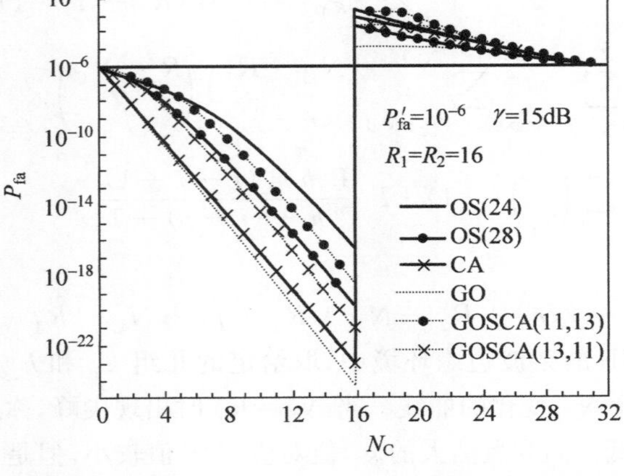

图 5.10 杂波边缘环境下 GOSCA、GO、CA 和 OS 检测器的虚警性能比较

5.7.2 GOSGO-CFAR和GOSSO-CFAR检测器在杂波边缘环境中的性能

当 $0\leqslant N_C\leqslant R_1$ 时，利用式(5.17)～式(5.24)和式(5.31)可以得到GOSGO-CFAR检测器的虚警概率为[35-36]

$$P_{fa}=k_2\binom{R_2}{k_2}\sum_{i=0}^{k_2-1}\binom{k_2-1}{i}(-1)^iQ\left[\frac{T}{(T+b)(T+a+b)};k_1,R_1,N_C\right] \tag{5.120}$$

式(5.120)包括 $N_C=R_1$ 时，检测单元幅值仍服从 e^{-x} 分布的情况，其中 a 和 b 分别见式(5.19)和式(5.20)。

当 $R_1\leqslant N_C\leqslant R$ 时，利用式(5.25)～式(5.30)和式(5.32)可得到GOSGO-CFAR检测器的虚警概率为[35-36]

$$P_{fa}=\frac{k_1}{\gamma}\binom{R_1}{k_1}\sum_{i=0}^{k_1-1}\left\{\binom{k_1-1}{i}(-1)^iQ\left[\frac{T}{(T+\gamma f)(T+d+\gamma f)};k_2,R_2,N_C-R_1\right]\right\} \tag{5.121}$$

式(5.121)包括 $N_C=R_1$ 时，检测单元幅值服从 $\frac{1}{\gamma}e^{-x/\gamma}$ 分布的情况，其中 d 和 f 分别见式(5.27)和式(5.28)。

当 $0\leqslant N_C\leqslant R_1$ 时，利用式(5.17)～式(5.24)及式(5.31)可得到GOSSO-CFAR检测器的虚警概率为[37]

$$\begin{aligned}P_{fa}=&Q\left(\frac{T}{T+b};k_1,R_1,N_C\right)+k_2\binom{R_2}{k_2}\sum_{i=0}^{k_2-1}\binom{k_2-1}{i}(-1)^i\frac{1}{T+a}-\\&k_2\binom{R_2}{k_2}\sum_{i=0}^{k_2-1}\left\{\binom{k_2-1}{i}\frac{(-1)^i}{a}Q\left[\left(\frac{T}{T+b}-\frac{T}{T+a+b}\right);k_1,R_1,N_C\right]\right\}\end{aligned} \tag{5.122}$$

式(5.122)包括 $N_C=R_1$ 时，检测单元幅值仍服从 e^{-x} 分布的情况，其中 a 和 b 分别见式(5.19)和式(5.20)。

当 $R_1\leqslant N_C\leqslant R$ 时，由式(5.25)～式(5.30)和式(5.32)可得到GOSSO-CFAR检测器的虚警概率为[37]

$$\begin{aligned}P_{fa}=&k_1\binom{R_1}{k_1}\sum_{i=0}^{k_1-1}\left[\binom{k_1-1}{i}(-1)^i\frac{1}{d+T}+Q\left(\frac{T}{T+\gamma f};k_1,R_2,N_C-R_1\right)\right]-\\&\frac{k_1}{\gamma}\binom{R_1}{k_1}\sum_{i=0}^{k_1-1}\left\{\binom{k_1-1}{i}(-1)^iQ\left[\frac{T}{(T+\gamma f)(T+d+\gamma f)};k_2,R_2,N_C-R_1\right]\right\}\end{aligned} \tag{5.123}$$

式(5.123)包括 $N_C=R_1$ 时，且检测单元幅值仍服从 $\frac{1}{\gamma}e^{-x/\gamma}$ 分布的情况，其中 d 和 f 分别见式(5.27)和式(5.28)。

下面根据已经获得的GOSGO和GOSSO-CFAR检测器在杂波边缘环境中的虚警概率计算公式，分析它们在杂波边缘环境中的性能[36-37]。图5.11给出了GOSGO-CFAR检测器在 $R_1=24$、$R_2=8$、$P_{fa}'=10^{-6}$，强弱杂波比 $\gamma=15$dB时，虚警尖峰随 k_1 和 k_2 变化的三维图。由图5.11可见，在上述条件下，$k_1=24$、$k_2=1$ 时，GOSGO的虚警尖峰最低，即虚警控制能力最强。考虑到在多目标情况中应发挥GOSGO所具有的OS的优点，k_1 的选择还应该留有余量。并且，为了提高检测性能，k_2 也应该选得适当，以降低ADT值。总之，对于GOSGO-CFAR检测器的虚警尖峰控制来说，大的 k_1

值和小的 k_2 值比较有效。在 k_1 较大时，虚警尖峰已经很低，它对 γ 的变化不是很敏感。例如，$R_1=24$、$k_1=18$、$R_2=8$、$k_2=8$ 时，有下面一组虚警尖峰数据：

$$\gamma=5\text{dB 时}, P_{\text{fa}}=3.1656\times10^{-5}$$

$$\gamma=15\text{dB 时}, P_{\text{fa}}=5.9997\times10^{-5}$$

$$\gamma=25\text{dB 时}, P_{\text{fa}}=5.9996\times10^{-5}$$

这说明 GOSGO-CFAR 检测器在强杂波环境中有良好的虚警控制性能。

GOSSO-CFAR 检测器的虚警尖峰随 k_1 和 k_2 变化的情况与 GOSGO-CFAR 检测器正好相反，即 k_1 越小，k_2 越大，则 GOSSO-CFAR 检测器的虚警尖峰越低。这是因为虚警尖峰考虑的是前沿滑窗充满强杂波样本而后沿滑窗仍是弱杂波样本的情况，但 GOSSO-CFAR 检测器在出现杂波边缘的情况下，几乎总是选择后沿滑窗中的杂波样本去设置门限，由于前述的原因，小的 k_1 值对杂波边缘的变化会迟钝一些，而大的 k_2 值会带来较好的虚警控制能力。图 5.12 给出了 GOSSO 的虚警概率随进入参考滑窗中强杂波样本数 N_C 变化的几组曲线。总体来看，GOSSO 的虚警尖峰很高，达到了 10^{-1} 数量级，也就是说，GOSSO-CFAR 检测器在杂波边缘环境中失去了虚警控制能力。

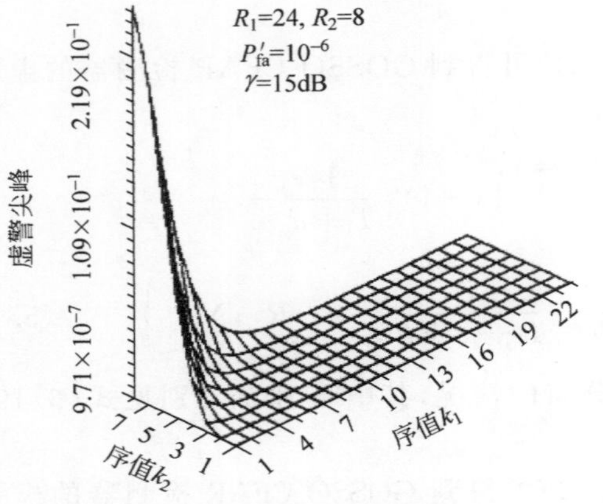

图 5.11 GOSGO-CFAR 检测器的虚警尖峰与 k_1 和 k_2 的关系

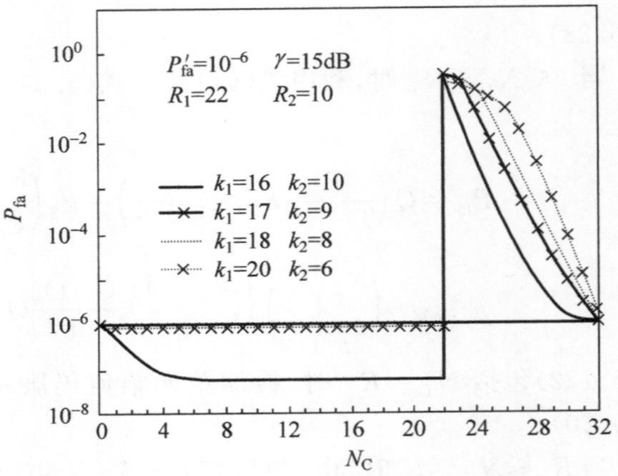

图 5.12 GOSSO 在杂波边缘环境中的几组性能曲线

图 5.13 比较了 GOSGO(24(20),8(6))、GOSSO(22(16),10(10))与 CA 等经典方法在杂波边缘环境中的虚警控制性能曲线。由图 5.13 可见，GO 和 GOSGO 都表现出了很强的虚警控制能力，在图 5.13 中所示参数的情况下，GOSGO 的性能甚至超过了 GO。而 SO 和 GOSSO 在杂波边缘环境中的虚警控制能力很差，但是它们在多目标环境中具有优势。

5.7.3 MOSCA-CFAR 检测器在杂波边缘环境中的性能

当 $0\leqslant N_C\leqslant R_1$ 时，MOSCA-CFAR 检测器的虚警概率为

$$P_{\text{fa}}=Q\left(\frac{T}{T+g};k,R_1,N_C\right)\left(1+\frac{T}{R_2}\right)^{-R_2} \tag{5.124}$$

其中，$g=n_1+R_1-N_C-j+(n_2+N_C-i+j)/\gamma$，函数 Q 由式(5.21)给出。式(5.124)包括 $N_C=R_1$ 时检测单元幅值仍服从 e^{-x} 分布的情况。

当 $R_1\leqslant N_C\leqslant R$ 时，可得到 MOSCA-CFAR 检测器的虚警概率为

$$P_{\mathrm{fa}}=\frac{\Gamma(R_1+1)\Gamma(T+R_1-k+1)}{\Gamma(R_1-k+1)\Gamma(T+R_1+1)}\left(1+\frac{T}{\gamma R_2}\right)^{N_C-R}\left(1+\frac{T}{R_2}\right)^{R_1-N_C} \tag{5.125}$$

式(5.125)包括 $N_C=R_1$ 时,检测单元幅值服从 $\frac{1}{\gamma}\mathrm{e}^{-x/\gamma}$ 分布的情况。

下面分析 MOSCA-CFAR 检测器在参考滑窗中出现杂波边缘时的虚警控制性能[38]。图 5.14 给出的是在 $R_1=R_2=16$、$\gamma=15\text{dB}$ 的杂波边缘环境中,k 取不同值时的几组虚警控制性能曲线。大 k 值的 P_{fa} 在杂波边缘扫过参考滑窗的全程中都低于小 k 值的 P_{fa}。大 k 值对 MOSCA 在杂波边缘环境中的虚警尖峰控制比较有效。k 值的选择范围应该在具有较低 ADT 值的 k 值和 R_1 之间。对于 $R_1=20$、$R_2=12$ 的 MOSCA,k 在 14～20 选择是比较合适的,此时的抗杂波边缘性能与检测性能均比较好。如果考虑多目标环境中的检测性能,k 的选择应该留有一定的余量。这时,$k=14$ 更为合适。

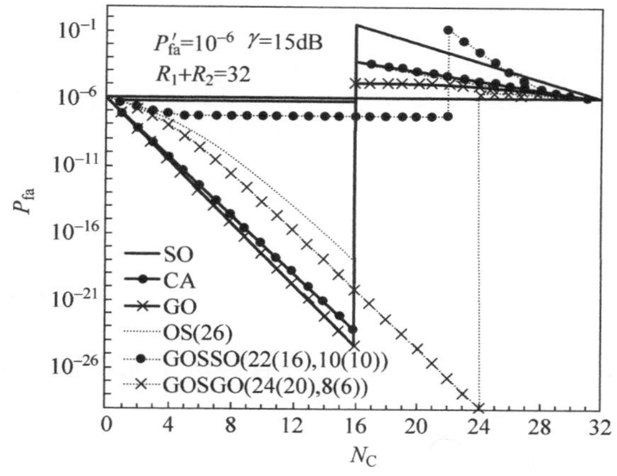

图 5.13 GOSGO、GOSSO 与 CA 等经典方法在杂波边缘环境中的性能曲线

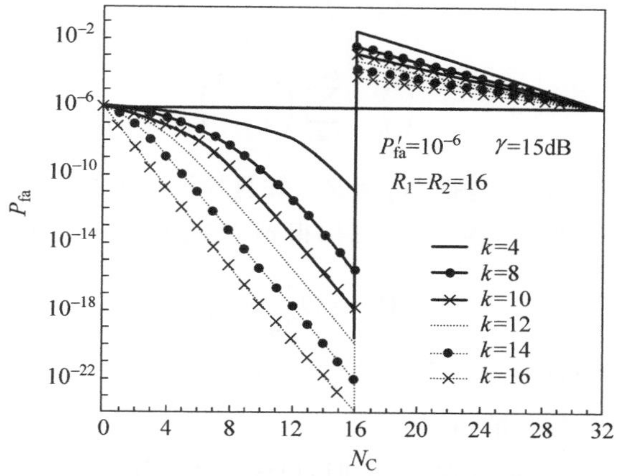

图 5.14 MOSCA 随 k 取不同值时的几组抗杂波边缘性能曲线

图 5.15 给出了 MOSCA 和其他几种 CFAR 方法的抗杂波边缘性能的比较,由图 5.15 可见,当 $(R_1-k)_{\mathrm{MOSCA}}=(R-k_1-k_2)_{\mathrm{GOSCA}}$ 时 MOSCA 的虚警尖峰比 GOSCA 略高,而当杂波边缘处于后沿参考滑窗时,其 P_{fa} 水平与 GOSCA 相当。与 OS 相比,当 $(R_1-k)_{\mathrm{MOSCA}}=(R-k)_{\mathrm{OS}}$ 时,MOSCA 的虚警尖峰与 OS 相当,但是当杂波边缘处于后沿参考滑窗时,其 P_{fa} 水平却比 OS 略高。因此,可以说,MOSCA 在获得均匀背景和多目标环境中良好检测性能的同时也保持了良好的抗杂波边缘性能。

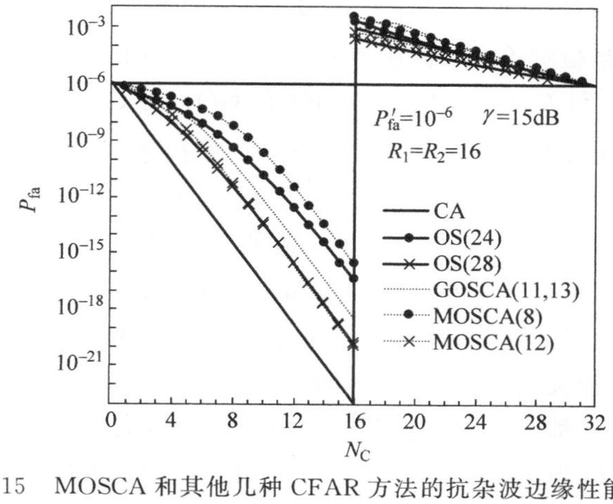

图 5.15 MOSCA 和其他几种 CFAR 方法的抗杂波边缘性能比较

5.7.4 OSCAGO、OSCASO-CFAR 检测器在杂波边缘环境中的性能

当 $0 \leqslant N_C \leqslant R_1$ 时，可得到 OSCAGO-CFAR 检测器的虚警概率为[33]

$$P_{fa} = Q\left[\frac{T}{g+T}\left(\frac{R_2}{R_2+g+T}\right)^{R_2}; k, R_1, N_C\right] \tag{5.126}$$

其中，$g = n_1 + R_1 - N_C - j + (n_2 + N_C - i + j)/\gamma$。式(5.126)包括 $N_C = R_1$，并且检测单元幅值仍然服从 e^{-x} 分布的情况。

当 $R_1 \leqslant N_C \leqslant R$ 时，可得到 OSCAGO-CFAR 检测器的虚警概率为[33]

$$P_{fa} = \sum_{i=k}^{R_1}\binom{R_1}{i}\sum_{j=0}^{i}\binom{i}{j}(-1)^{i-j}(R+T-j)^{R_1-N_C}\frac{R_2^{R_2}}{[R_2+(T+R_1-j)/\gamma]^{R_1+R_2-N_C}} +$$

$$k\binom{R_1}{k}\sum_{i=0}^{k-1}\binom{k-1}{i}\frac{(-1)^i}{i+T+R_1-k+1}\frac{(R_2+i+T-k+1)^{R_1-N_C}R_2^{R_2}}{[R_2+(i+T+R_1-k+1)/\gamma]^{R_1+R_2-N_C}} \tag{5.127}$$

式(5.127)包括 $N_C = R_1$，并且检测单元幅值服从 $e^{-x/\gamma}/\gamma$ 分布的情况。

当 $0 \leqslant N_C \leqslant R_1$ 时，得到 OSCASO-CFAR 检测器的虚警概率为

$$P_{fa} = Q\left\{\frac{T}{g+T}\left[1-\left(\frac{R_2}{g+T+R_2}\right)^{R_2}\right]; k, R_1, N_C\right\} + \left(1+\frac{T}{R_2}\right)^{-R_2} \tag{5.128}$$

其中，$g = n_1 + R_1 - N_C - j + (n_2 + N_C - i + j)/\gamma$。式(5.128)包括 $N_C = R_1$，并且检测单元幅值服从 e^{-x} 分布的情况。

当 $R_1 \leqslant N_C \leqslant R$ 时，OSCASO-CFAR 检测器的虚警概率为

$$P_{fa} = \frac{R_2^{R_2}(R_2+T)^{R_1-N_C}}{(R_2+T/\gamma)^{R-N_C}} + k\binom{R_1}{k}\sum_{i=0}^{k-1}\left\{\binom{k-1}{i}\frac{(-1)^j}{i+T+R_1-k+1}\times\right.$$

$$\left.\left[1-\frac{R_2^{R_2}(R+i+T-k+1)^{R_1-N_C}}{[R_2+(i+T+R_1-k+1)/\gamma]^{R-N_C}}\right]\right\} - R_2^{R_2}\sum_{i=k}^{R_1}\binom{R_1}{i}$$

$$\sum_{j=0}^{i}\binom{i}{j}\frac{(-1)^{i-j}(R+T-j)^{R_1-N_C}}{[R_2+(T+R_1-j)/\gamma]^{R-N_C}} \tag{5.129}$$

式(5.129)包括 $N_C = R_1$，并且检测单元幅值服从 $e^{-x/\gamma}/\gamma$ 分布的情况。

图 5.16 给出了 $R_1 = 24$ 和 $R_2 = 8$ 的 OSCAGO 和 GO、OS-CFAR 检测器的抗杂波边缘性能曲线[39-41]。由图 5.16 可以看出，大 k 值时的 P_{fa} 在杂波边缘扫过参考滑窗的全程都低于小 k 值时的 P_{fa}，并且随着 R_1 的增大，虚警尖峰对 k 值变化的敏感程度下降。所以，k 较大时，应主要从均匀背景和多目标环境中的检测性能方面考虑 k 的选择。k 值的选择范围应该在具有较低 ADT 值对应的 k' 值与 R_1 之间。对于 $R_1 = 24$ 和 $R_2 = 8$ 的 OSCAGO，k 在 18～22 选择是比较合适的，此时的抗杂波边缘性能和检测性能均较好。虽然 k 取得最大可能值($k = R_1$)时的 OSCAGO 与图 5.16 中给出的几种检测器相比具有最好的抗杂波边缘性能，并且明显超过了所有其他的检测器，但是考虑到在均匀杂波和多目标环境中的检测性能，$k = R_1$ 是不可取的。而 OSCAGO(24(18)，8)的抗杂波边缘性能与抗杂波边缘性能较好的 GO 相比又有明显的提升，并且它在均匀杂波背景和多目标环境中也保持

了良好的检测性能。

图 5.17 给出了 $R_1=R_2=16$ 的 OSCASO 和 SO、GOSSO 几种检测器的抗杂波边缘性能比较[39-41]。从图 5.17 中可以看出，$R_1=R_2=16$ 时的 OSCASO 抗杂波边缘性能与 SO、GOSSO 相当。$k=12\sim16$ 时的 OSCASO 几乎与 SO 重合，但它的虚警尖峰达到了 10^{-1} 数量级，可见其抗杂波边缘性能是很不理想的。虽然，OSCASO(8)的抗杂波边缘性能较 OSCASO(12)有所提高，即在大 N_C 值段，它的 P_{fa} 低于图 5.17 中给出的所有其他检测器。但是，考虑到均匀杂波环境中的检测性能，小 k 是不可取的。前面已提到，OSCAGO 检测器存在均匀杂波环境中的检测性能和杂波边缘环境中的抗杂波边缘性能的矛盾，而这个矛盾在 OSCASO-CFAR 检测器中显得更加突出。

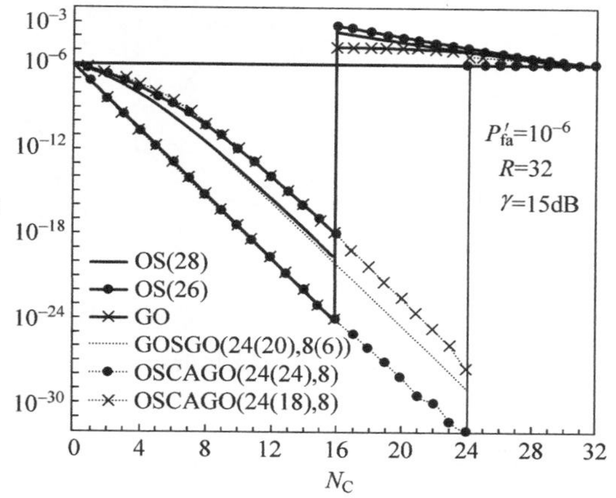

图 5.16　OSCAGO 和其他几种检测器的抗杂波边缘性能曲线

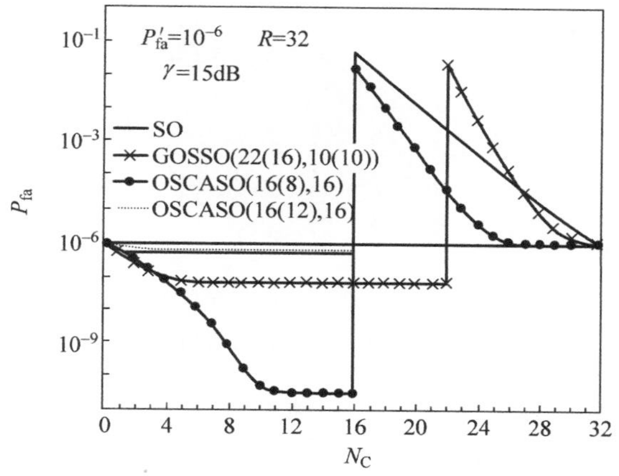

图 5.17　OSCASO 和其他几种检测器的抗杂波边缘性能

5.7.5　MTM、TMGO-CFAR 检测器在杂波边缘环境中的性能

当强杂波区从前沿滑窗进入，检测单元也进入强杂波区，而后沿滑窗仍处于弱杂波区时，此时的虚警率上升最为严重，达到虚警率上升的峰值，称为虚警尖峰，以此衡量 MTM-CFAR 和 TMGO-CFAR 检测器在杂波边缘环境中虚警率上升的程度。若前沿滑窗进入强杂波区，而后沿滑窗仍处于弱杂波区，那么虚警尖峰的计算公式为

$$P_{fa}=\int_0^{\infty} f_Z(z)\left[\int_{Tz}^{\infty}\frac{1}{\gamma}\exp\left(-\frac{x}{\gamma}\right)\mathrm{d}x\right]\mathrm{d}z=\int_0^{\infty} f_Z(z)\exp\left(-\frac{Tz}{\gamma}\right)\mathrm{d}z \tag{5.130}$$

式中，T 为均匀背景中预先设置好的阈值因子，$f_Z(z)$仍通过式(5.103)求解，但 $f_X(z)$为参数 $\lambda'=1$ 的指数分布，$f_Y(z)$为参数 $\lambda'=\gamma$ 的指数分布，利用卷积定理，得到 MTM-CFAR 检测器的虚警尖峰[17-18]

$$P_{fa}=\prod_{i=1}^{R_1-R''_1}\frac{c'_i}{T+c'_i}\prod_{j=1}^{R_2-R''_2}\frac{c''_j}{T/\gamma+c''_j} \tag{5.131}$$

图 5.18 给出了 MTM-CFAR 检测器在前、后沿滑窗长度 $R_1=R_2=16$，杂波强度比 $\gamma=15$dB 的条件下，前、后沿滑窗从高端剔除($R'_1=R'_2=0$，R''_1和 R''_2变化)时虚警尖峰随 R''_1和 R''_2变化的情况。图 5.19 给出了前、后沿滑窗从低端剔除($R''_1=R''_2=0$，R'_1和 R'_2变化)时虚警尖峰随 R'_1和 R'_2变化的情况。从图 5.19 中可以看出，R''_1增大，虚警尖峰升高，R''_2增大，虚警尖峰降低，R'_1增大，虚警尖峰降低，R'_2增大，虚警尖峰升高；还可

以看出，虚警尖峰对前沿滑窗参数 R_1' 和 R_1'' 的变化，要比对后沿滑窗参数 R_2' 和 R_2'' 的变化敏感，这是由于高有序样本对杂波功率水平的变化反应大，低有序样本对杂波功率水平的变化反应小，前沿滑窗中杂波功率水平强($\gamma=15$dB)的缘故。对于前、后沿滑窗长度 $R_1=R_2=16$，杂波强度比为 $\gamma=15$dB 的情况，认为几种检测器有相同的抗干扰目标能力，MTM 选取 $R_1'=10$、$R_1''=2$、$R_2'=2$、$R_2''=2$，GOSCA 选取 $k_1=14$、$k_2=14$，OS 选取 $k=26$，表 5.24 给出了它们各自的阈值因子 T 和虚警尖峰的大小。从表 5.24 中的情况来看，MTM 通过合适的参数选择，可以获得比 GOSCA 和 OS 还低的虚警尖峰。

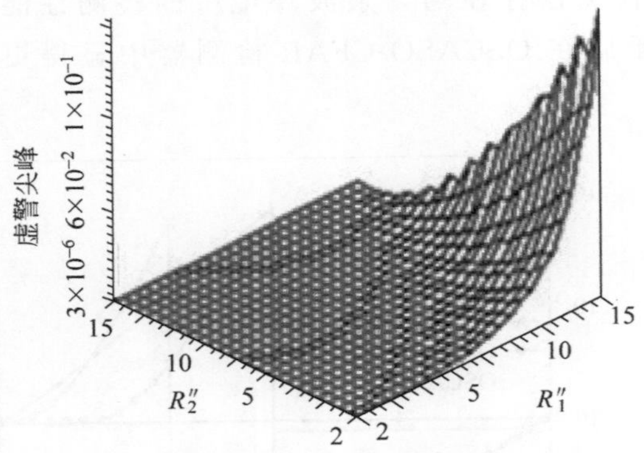

图 5.18　前、后沿滑窗从高端剔除时虚警尖峰随 R_1'' 和 R_2'' 变化的情况

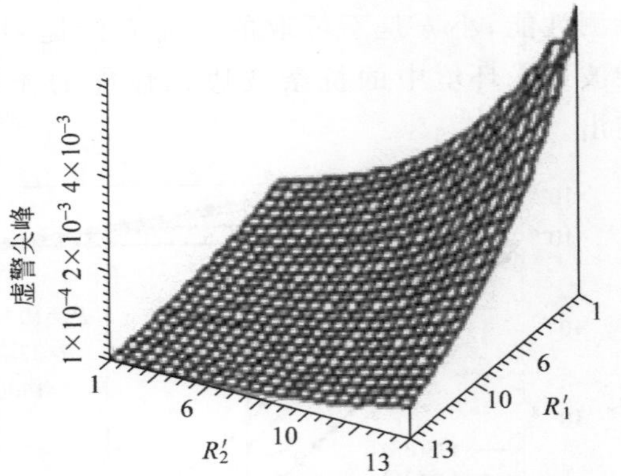

图 5.19　前、后沿滑窗从低端剔除时虚警尖峰随 R_1' 和 R_2' 变化的情况

表 5.24　$\gamma=15$dB 时几种检测器虚警概率 P_{fa} 上升程度的比较

CFAR 检测器	$P_{fa}=10^{-6}$ 时的阈值因子 T	虚警尖峰
MTM	8.0246($R_1'=10,R_1''=2,R_2'=2,R_2''=2$)	1.8962×10^{-4}
GOSCA	5.0300($k_1=12,k_2=12$)	7.4452×10^{-4}
OS	11.8550($k=26$)	6.4938×10^{-4}
CA	17.2776	7.6271×10^{-4}

对于 TMGO-CFAR 检测器，当前沿滑窗进入强杂波区而后沿滑窗仍处于弱杂波区时，由式(5.130)可得到 TMGO-CFAR 检测器的虚警尖峰计算公式为[19-21]

$$P_{fa}=\sum_{i=1}^{m'}\sum_{j=1}^{r_i'}\sum_{e=1}^{n'}\sum_{f=1}^{r_e''}a'_{ij}a''_{ef}\left\{\frac{\Gamma(f)}{(s''_e)^f}\left[\frac{\Gamma(j)}{(T+s'_i)^j}-\sum_{g=0}^{f-1}\frac{(\gamma s''_e)^g}{g!}\frac{\Gamma(j+g)}{(T+s'_i+\gamma s''_e)^{j+g}}\right]+\right.$$
$$\left.\frac{\Gamma(j)}{(s'_i)^j}\left[\frac{\Gamma(f)}{(T/\gamma+s''_e)^f}-\sum_{k=0}^{j-1}\frac{(s'_i/\gamma)^k}{k!}\frac{\Gamma(f+k)}{(T/\gamma+s'_i/\gamma+s''_e)^{f+k}}\right]\right\} \tag{5.132}$$

给定 $R_1=R_2=10$、$R_1'=R_2'=3$、$P_{fa}=10^{-6}$，杂噪比 $\gamma=10$dB。图 5.20 给出了 TMGO-CFAR 的虚警尖峰，相对于从高端剔除的样本个数 R_1'' 和 R_2'' 的变化曲线。图 5.21 给出了 TMGO-CFAR 的虚警尖峰，相对于从低端剔除的样本个数 R_1' 和 R_2' 的变化曲线。可以注意到，当 R_1'' 增加时，虚警尖峰上升，当 R_2'' 增加时，虚警尖峰下降；当 R_1' 增加时，虚警尖峰下降，当 R_2' 增加时，虚警尖峰上升。这主要是因为高端有序统计量对杂波功率水平的变化，比低端有序统计量对杂波功率水平的变化更为敏感。而且，前沿滑窗处于强杂波区时，为了保持虚警概率的恒定，背景功率水平的估计应该反应检测单元中杂波功率的变化，且当后沿滑窗处于弱杂波区时，它应该不受杂波功率变化的影响。

在 TMGO-CFAR 检测器的前、后沿滑窗长度 $R_1=R_2=10$、$R_1''=R_2''=2$、$R_1'=6$、$R_2'=2$，杂波功率水平之比为 $\gamma=15(\mathrm{dB})$，并认为检测器有相同的抗干扰目标能力的条件下，表 5.25 给出了 TMGO 和 GOSGO 的阈值因子 T 和虚警尖峰的大小。从表 5.25 中的情况来看，通过合适的参数选择，TMGO 可以获得比 GOSGO 或 OSGO 还低的虚警尖峰。

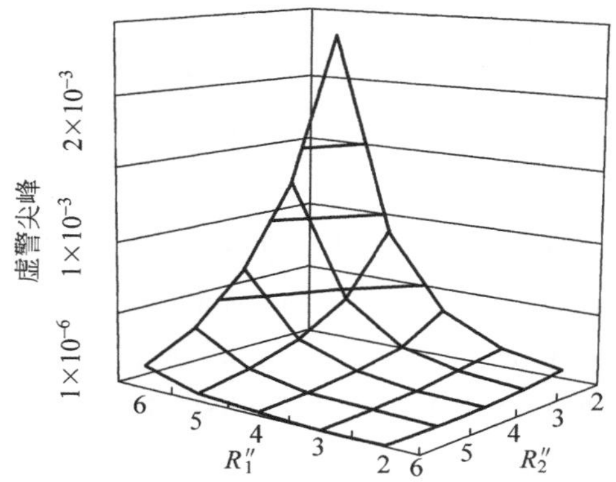

图 5.20 TMGO 的虚警尖峰相对于 $R_1''(R_2'')$ 的变化情况

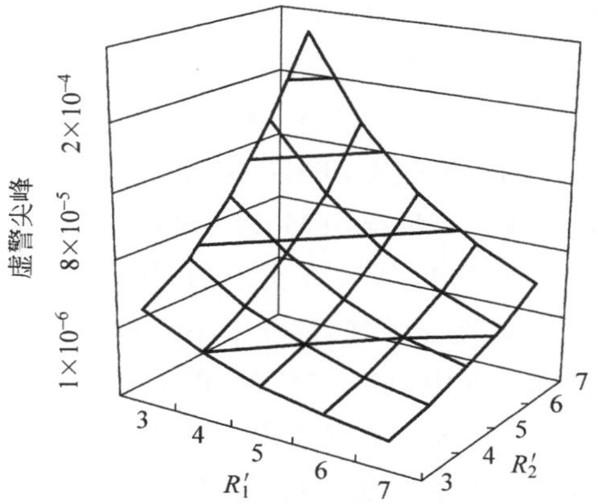

图 5.21 TMGO 的虚警尖峰相对于 $R_1'(R_2')$ 的变化情况

表 5.25 $\gamma=15\mathrm{dB}$ 时 TMGO 和 GOSGO 检测器虚警概率 P_{fa} 上升程度的比较

检 测 器	$P_{\mathrm{fa}}=10^{-6}$ 时的阈值因子 T	P_{fa} 上升的峰值
TMGO	19.2485($R_1''=R_2''=2$、$R_1'=6$、$R_2'=2$)	1.7087×10^{-5}
GOSGO	14.1772($k_1=k_2=8$)	5.7064×10^{-5}

由于采用选小逻辑的 CFAR 检测器的优势主要在于多目标环境中，而杂波边缘环境是它的劣势，在这里不再讨论 TMSO 在杂波边缘环境中的性能。

5.8 比较与总结

衡量一种检测器性能的好坏主要考虑其在 3 种典型背景情况中的性能，其一是参考滑窗内的杂波样本是统计独立同分布的，即均匀背景的情况；其二是参考滑窗内的杂波功率水平突然发生变化，即杂波边缘环境的情况，当检测单元处于强杂波区时会发生虚警概率上升，而当检测单元处于弱杂波区时又会发生目标遮掩效应；其三是参考滑窗内出现多目标，即多目标环境的情况，由于落入参考滑窗的其他目标会抬高检测门限，因此通常会检测出较强的目标信号而漏检较弱的目标，致使检测概率降低。本章研究了 GOSCA、GOSGO、GOSSO、MOSCA、OSCAGO、OSCASO、MTM、TMGO 和 TMSO 检测器在以上 3 种杂波背景中的性能，给出了用于分析这些检测器性能的数学模型和数值分析结果。同时指出其他诸如 MOSCM、OSCMGO、OSCMSO、MOSTM、OSTMGO 和 OSTMSO-CFAR 等检测器都可以作为 MTM、TMGO 和 TMSO 的特例进行分析。在这些 CFAR 检测器中，由于都采用了文献[3-4]提出的一种自动筛选技术，因此这些 CFAR 检测器统称为采用自动筛选技术的 GOS 类 CFAR 检测器。

基于前沿、后沿滑窗中杂波功率水平估计方式的不同，将对前沿、后沿滑窗中两个局部估计取平均的检测器 GOSCA、MOSCA、MOSCM、MOSTM、MCM、MTM，称为 TM-TM-CA 类检测器；将对前沿、后沿滑窗中杂波局部估计采用选大逻辑的检测器 GOSGO、OSCAGO、OSCMGO、OSTMGO、TMGO，称为 TM-TM-GO 类检测器；将对前沿、后沿滑窗中两个局部估计采用选小逻辑的检测器 OSSO、GOSSO、OSCASO、OSCMSO、OSTMSO、TMSO，称为 TM-TM-SO 类检测器。在均匀背景中，TM-TM-CA 类检测器的 CFAR 损失比 TM-TM-GO 类和 TM-TM-SO 类要小一些，而 TM-TM-GO 类比 TM-TM-SO 类要小一些，对于多目标情况也有类似的结论。对于杂波边缘情况，TM-TM-GO 类的虚警控制能力最强，TM-TM-SO 类的虚警控制能力极差，TM-TM-CA 类介于二者之间。但对于单边密集干扰目标情况，TM-TM-SO 类具有优势。

对于 TM-TM-CA 类中的 GOSCA、MOSCA、MOSCM、MOSTM 和 MTM 方法来说，在均匀背景中，MTM 性能最好，MOSCA 比 GOSCA 好，MOSCM 和 MOSTM 的性能次于 MTM 和 MOSCA。在多目标情况中，也是 MTM 最好，其次分别是 MOSCM、MOSTM 和 GOSCA，而 MOSCA 在后沿滑窗出现干扰目标时性能恶化。在杂波边缘情况中，它们的虚警控制能力不是太强，MOSCM 和 MOSCA 略优于其他的方法。可见，当对杂波边缘环境中的性能要求不是太高时，MTM 方法是值得考虑的方法。

对于 TM-TM-GO 类中的 GOSGO、OSCAGO、OSCMGO、OSTMGO 和 TMGO 方法来说，在均匀背景中，TMGO 性能最好，OSCAGO 比 GOSGO 好，OSCMGO 和 OSTMGO 的性能次于 TMGO 和 OSCAGO。在多目标情况中，也是 TMGO 最好，其次分别是 OSCMGO、OSTMGO 和 GOSGO，而 OSCAGO 在后沿滑窗出现干扰目标时性能会恶化。在杂波边缘情况中，突出的虚警控制能力是 TM-TM-GO 类方法的优势，其中 OSCMGO 和 OSCAGO 优于 GOSGO 和 TMGO。

对于 TM-TM-SO 类中的 GOSSO、OSCASO、OSCMSO、OSTMSO 和 TMSO 方法来说，在均匀背景中，TMSO 性能最好，OSCASO 比 GOSSO 好，OSCMSO 和 OSTMSO 的性能次于 TMSO 和 OSCASO。在多目标情况中，也是 TMSO 最好，其次分别是 OSCMSO、OSTMSO 和 GOSSO，而 OSCAGO 在后沿滑窗出现干扰目标时性能会恶化。在杂波边缘情况中，TM-TM-SO 类检测器丧失了虚警控制能力，但当干扰目标密集地分布于单边参考滑窗时，TM-TM-SO 类检测器比 TM-TM-CA 类和 TM-TM-GO 类性能要好。

任何方法都有其优缺点，而在实际的雷达检测环境中，均匀背景、多目标和杂波边缘出现的情况也会不断变化，这就需要针对具体情况对雷达检测方案做出合适的选择。从近些年来提出的一些新的 CFAR 检测方法来看，其思想是，综合利用不同方法的优点并将它们组合起来，以兼顾检测器在均匀背景和各种非均匀背景中的性能。例如，E-CFAR[42](Ensemble CFAR)将 CA、GO、SO、TM、GM(Geometric Mean)方法组合起来形成检测统计量；Smith 和 Varshney[43]提出的 VI-CFAR(Variable Index CFAR)基于 VI 和前、后沿滑窗中杂波样本均值的比值将检测器自动切换成 CA、GO 和 SO；Cao 提出的 S-CFAR[44-45](Switching CFAR)技术基于检测单元统计量对检测方法进行切换。文献[46]提出使用神经网络来切换 CA-CFAR 与 OS-CFAR，以同时获得均匀与非均匀环境下的最佳检测性能。文献[47]提出了一种新的非线性融合 CFAR 检测器，该检测器基于 CA、OS、TM 3 种 CFAR 检测器的并行处理，以消除尽可能多的虚假目标，同时尽可能减少对真实目标检测概率的影响。文献[48]在 FMCM 雷达系统中，将 OS-CFAR 与 CA-CFAR 结合起来形成二维

CFAR，应用于距离-多普勒二维数据矩阵中，从而可以在不增加计算复杂度的情况下避免多目标检测时的遮蔽效应，此外，还可以通过调整参考窗口的大小来优化检测性能。文献[49]则将GOS-CFAR与CA-CFAR结合起来形成二维CFAR。另外，还有一些研究将OS类、GOS类CFAR检测器从通常的SISO(单输入单输出)雷达推广到MIMO(多输入多输出)雷达上，推导了检测性能闭式表达式[50-51]。文献[52]将GOS检测器应用于频谱感知领域，并结合OR规则，形成了基于GOS-OR检测器的协作频谱感知方案。

参考文献

[1] Elias A R, De Mercard M G, Davo E R. Analysis of some modified order statistic CFAR: OSGO and OSSO CFAR[J]. IEEE Transactions on AES, 1990, 26(1): 197-202.

[2] Wilson S L. Two CFAR algorithms for interfering targets and nonhomogeneous clutter[J]. IEEE Transactions on AES. 1993, 29(1): 57-72.

[3] 何友, Rohling H. 两种具有自动筛选技术的广义有序统计恒虚警检测器及其在多目标情况下的性能[J]. 电子科学学刊, 1994, 16(6): 582-590.

[4] He Y. Performance of some generalized modified order statistics CFAR detectors with automatic censoring technique in multiple target situations[J]. IEEE Proceedings, 1994, 141(4): 205-212.

[5] He Y, Guan J, Peng Y N, et al. A new CFAR detector based on ordered statistics and cell averaging[C]. Beijing: CIE 1996 International Radar Conference, 1996: 106-108.

[6] He Y, Guan J. A new CFAR detector with greatest of selection[C]. Alexandria: IEEE 1995 International Conference on Radar, 1995: 589-591.

[7] 何友, 关键, Rohling H. 基于最大和最小选择的两种新的恒虚警检测器[J]. 系统工程与电子技术, 1995, 17(7): 6-16.

[8] Meng X W, He Y. A new CFAR detector based on order statistics and censored mean[C]. Nanjing: Proc. of IEEE International Conference on Neural Networks and Signal Processing, 1995: 1178-1181.

[9] 何友, 孟祥伟. 一种采用自动筛选技术的鲁棒恒虚警检测器[J]. 宇航学报, 1998, 19(2): 19-23.

[10] He Y, Meng X W. Performance of a new greatest of selection CFAR detector based on order statistics and censored mean[C]. Beijing: Proc. of 1996 Third International Conference on Signal Processing(ICSP'96), 1996: 565-567.

[11] 孟祥伟, 何友. 一种新的基于有序统计和筛选平均的最大选择恒虚警检测器[J]. 信号处理, 1996, 12(4): 316-321.

[12] 孟祥伟, 何友. 一种新的最小选择恒虚警检测器[J]. 系统工程与电子技术, 1997, 19(1): 25-32.

[13] 孟祥伟, 何友. 一种新的基于有序统计和剔除平均的恒虚警检测器[J]. 现代雷达, 1996, 18(1): 96-104.

[14] Meng X W, He Y. A new CFAR detector with automatic censoring[C]. Beijing: Proc. of 1996 Third International Conference on Signal Processing(ICSP'96), 1996: 1632-1635.

[15] Meng X W, He Y. Performance analysis of a new greatest of selection CFAR detector[C]. Beijing: Proc. of CIE International Conference on Radar(CIE'ICR), 1996, 401-404.

[16] 何友, 孟祥伟. 一种基于有序统计和剔除平均的最大选择恒虚警检测器[J]. 电子学报, 1998, 26(3): 75-79.

[17] 孟祥伟, 何友, 关键. 采用子滑窗技术的修正剔除平均恒虚警检测算法[J]. 仪器与仪表学报, 2001, 22(3): 231-239.

[18] He Y, Meng X W. Performance of a new CFAR detector based on trimmed mean[C]. Beijing: Proceedings of IEEE International Conference on System, Men and Cybernetics(IEEE'SMC), 1996, 702-706.

[19] Meng X W, He Y. A unified model of greatest of selection logic CFAR algorithm-TMGO[C]. New York:

Proceedings of 1998 International Radar Symposium(IRS'98)，1998：613-620.

[20] Meng X W，He Y. Two generalized greatest of selection CFAR algorithms[C]. Beijing：CIE International Conference on Radar，2001，359-362.

[21] 何友，孟祥伟. 剔除平均最大选择 CFAR 检测[J]. 电子科学学刊，1999，21(6)：779-785.

[22] 孟祥伟，何友. 最小选择恒虚警检测器的统一模型——TMSO[J]. 仪器仪表学报，1997，18(5)：481-485.

[23] Meng X W，Guan J，He Y. A generalized smallest of selection CFAR algorithm[C]. Australia：2003 International Conference on Radar，2003.

[24] David H A. Order Statistics[M]. New York：Wiley，1981.

[25] 孟祥伟. 负指数分布有序统计量的统计分析及其在恒虚警处理中的应用[C]. 北京：《信号处理》杂志社，2001：299-302.

[26] Meng X W，Guan J，He Y. Order statistics for negative exponential distribution and its application[C]. Nanning：Proceedings of the International Conference on Neural Networks and Signal processing，2003.

[27] 孙艳丽，李建海，陈贻焕. 基于自动筛选技术的广义有序统计类 CFAR 检测性能分析[J]. 计算机与数字工程，2016，44(9)：1677-1680.

[28] 魏嘉，徐达，闫晟，等. OSSO 和 OSGO 恒虚警检测器在 Pareto 分布混响背景下的性能分析[J]. 信号处理，2019，35(9)：1599-1606.

[29] 徐振国，蔡龙，冯常慧. Pearson 分布混响背景下 OSGO 和 OSSO 检测器设计[J]. 舰船科学技术，2013，35(10)：88-91.

[30] 马江彦. 非均匀杂波环境下恒虚警的研究[D]. 大连：大连海事大学，2012.

[31] 王陆林，刘贵如，邹姗. 基于威布尔分布杂波模型的加权有序统计模糊 CFAR 检测算法[J]. 重庆邮电大学学报(自然科学版)，2019，31(2)：245-252.

[32] 孟祥伟，何友. 两种改进的适用于多目标情况的恒虚警检测算法[J]. 电子科学学刊，1997，19(5)：592-595.

[33] 何友，关键，孟祥伟，等. 雷达目标检测与恒虚警处理[M]. 2 版. 北京：清华大学出版社，2011.

[34] 关键. 几种新的 CFAR 检测器在干扰边缘环境中的性能[D]. 烟台：海军航空工程学院，1994.

[35] 何友，Rohling H. 一个新的组合恒等式及其证明[J]. 海军航空工程学院学报，1993，3：18-21.

[36] 关 键，何友. GOSGO-CFAR 检测器在干扰边缘环境下的性能分析[J]. 现代雷达，1994，16(3)：94-97.

[37] 关键，何友. 两种恒虚警检测器在干扰边缘中的性能分析[J]. 电子科学学刊，1996，18(3)：243-248.

[38] 关键，何友. MOSCA-CFAR 检测器在干扰边缘中的性能分析[J]. 信号处理，1995，11(4)：237-244.

[39] 关键，何友. OSCAGO-CFAR 检测器在干扰边缘中的性能分析[J]. 电子学报，1996，24(3)：56-60.

[40] Gandhi P，Kassam S A. Analysis of CFAR processors in nonhomogeneous background[J]. IEEE Transactions on AES，1988，24(4)：427-445.

[41] Rohling H. Radar CFAR thresholding in clutter and multiple target situations[J]. IEEE Transactions on AES，1983，19(4)：608-621.

[42] Srinivasan R. Robust radar detection using ensemble CFAR processing[J]. IEEE Proceedings Radar，Sonar and Navigation，2000，147(6)：291-297.

[43] Smith M E，Varshney P K. Intelligent CFAR processor based on data variability[J]. IEEE Transactions on Aerospace and Electronic Systems，2000，36(3)：837-847.

[44] Cao T V. Constant false alarm rate algorithm based on test cell information[J]. IET Radar Sonar Navigation，2008，2(3)：200-213.

[45] Meng Xiangwei. Comments on 'constant false alarm rate algorithm based on test cell information'[J]. IET Radar Sonar Navigation，2009，3(6)：646-649.

[46] Budiman P A，Rohman，Kurniawan D，et al. Switching CA/OS CFAR using neural network for radar target detection in non-homogeneous environment[C]. New York：2015 International Electronics Symposium (IES)，2015.

[47] Ivković D，Andrić M，Zrnić B. Nonlinear fusion CFAR detector[C]. New York：2015 16th International Radar

Symposium(IRS)，2015.

[48] Kronauge M，Rohling H. Fast two-dimensional CFAR procedure[J]. IEEE Transactions on Aerospace and Electronic Systems，2013，49(3)：1817-1823.

[49] Li S，Bi X，Huang L B，et al. 2-D CFAR procedure of multiple target detection for automotive radar[J]. SAE International Journal of Passenger Cars-Electronic and Electrical Systems，2018，11(1)：65-70.

[50] Baadeche M，Soltani F. Performance analysis of ordered CFAR detectors for MIMO radars[J]. Digital Signal Processing，2015，44：47-57.

[51] Baadeche M，Soltani F，Mezache A. Generalization of some CFAR detectors for MIMO radars[J]. International Journal of Signal Processing Systems，2016，4(1)：22-26.

[52] Kim Chang-joo，Jin E S，Cheon Kyung-yul，et al. Robust spectrum sensing under noise uncertainty for spectrum sharing[J]. ETRI Journal，2019，41(2)：176-183.

第6章 自适应 CFAR 检测器

CHAPTER 6

6.1 引言

雷达目标自动检测技术通常利用与检测单元邻近的距离或多普勒单元中的样本来估计检测单元背景的平均功率。这种方法在许多噪声和杂波条件下提供了接近最优的目标检测性能,同时也保持了一定的虚警率。然而,当参考滑窗中出现干扰目标时,可能会导致检测阈值升高,进而使得对主目标的检测性能严重下降,例如 CA、GO 等 CFAR 检测器就是如此。

干扰目标的统计特性不同于背景杂波的统计特性。为得到背景杂波的平均功率水平估计,自然需要将干扰目标从参考单元样本序列中剔除出去,然后基于处理后的参考单元样本序列计算检测阈值。例如,CMLD 和 OS 等方法都是采用此类做法的典型。然而,这些方法有时需要关于干扰目标数的先验信息。在第 3～5 章所讨论的那些 CFAR 检测器中,Z 的形成方法所涉及的选择逻辑、算法和有关参数都是固定的,如 GO 逻辑、SO 逻辑、代表序值、删除点等。实际上,干扰目标的数目和空间分布以及杂波边缘的位置都是随机的,因此需要一类既能自适应于干扰目标数和空间分布情况的变化,又能自适应于杂波边缘位置变化的 CFAR 检测器,这类检测器被称为自适应 CFAR 检测器。很多学者在自适应确定选择逻辑、算法和参数方面做了大量研究,本章将论述其中的一些典型例子。

6.2 CCA-CFAR 检测器

针对经典的 CA 方法在多目标环境中检测性能下降的问题,Barboy 提出了 CCA(Censored Cell-Averaging)-CFAR 检测器[1],即删除单元平均 CFAR 检测器。它摆脱了以往 CFAR 方法使用所有参考单元样本或完全依赖干扰目标先验信息(如干扰目标数)来形成杂波平均功率水平估计的思想,提出了一种多步删除方案,使干扰目标逐一被删除,剩下的杂波和噪声样本就是与检测单元杂波和噪声特性相近的样本,以此来形成检测门限。

CCA-CFAR 检测器对参考单元样本中可能存在的尖峰信号(参考单元样本中可能含有干扰目标的样本或者其他与检测单元背景统计特性不一致的样本)进行剔除的基本过程如下。

(1) 假设 $x_1,x_2,\cdots,x_R$ 是 R 个参考单元样本值,T_0 是与 R 及给定的 $P_{\text{fa}}^{(0)}$ 对应的门限因子。求第一个和值,即

$$Z_{R_0}=\sum_{k_0=1}^{R_0}x_{k_0},\quad R_0=R \tag{6.1}$$

然后将每个参考单元样本与如下门限进行比较

$$S_1 = T_0 Z_{R_0} \tag{6.2}$$

将超过这个门限的参考单元样本从和值中剔除，用剩余的参考单元样本形成一个新的和值

$$Z_{R_1} = \sum_{k_1=1}^{R_1} x_{k_1}, \quad R_1 = R_0 - j_1 \tag{6.3}$$

其中，j_1 是被剔除的参考单元样本数，k_1 是剩余的参考单元样本的下标。

(2) 同样引入门限因子 T_1，它对应于 R_1 和 $P_{\mathrm{fa}}^{(1)}$。然后将剩余的参考单元样本与如下门限进行比较

$$S_2 = T_1 Z_{R_1} \tag{6.4}$$

剔除超过这个门限的参考单元样本值，j_2 是本次操作中被剔除的参考单元样本数。利用剩余的 $R_2 = R_0 - j_1 - j_2$ 个参考单元样本形成一个新的和值

$$Z_{R_2} = \sum_{k_2=1}^{R_2} x_{k_2} \tag{6.5}$$

上述过程继续下去，直到参考单元样本中检测不到超过门限的样本为止。显然，这种算法总是收敛的。因此，在若干次操作之后，参考单元样本中不出现过门限样本就是CCA-CFAR检测方法的终止准则。

概率 $P_{\mathrm{fa}}^{(i)}$ 是每个筛选层的虚警率。总的平均虚警数为 $\sum_i P_{\mathrm{fa}}^{(i)} R_i$，因此总的虚警概率为

$$P_{\mathrm{fa}} = \frac{\sum_i P_{\mathrm{fa}}^{(i)} R_i}{\sum_i R_i} \tag{6.6}$$

要保持虚警率恒定的最简单方法就是使所有的 $P_{\mathrm{fa}}^{(i)}$ 等于预定的 P_{fa} 值。

图6.1给出了CCA-CFAR检测过程的原理框图。每级的尖峰信号剔除包括3步：①尖峰功率的求和；②尖峰信号计数；③零值插入(插入缓冲器中取代尖峰信号功率)。当缓存器中所有参考单元样本都通过比较器时，就完成了第一个周期。如果尖峰信号计数器的输出不为零，那么下一个周期一开始就从参考单元样本值之和中减去尖峰信号之和，然后重复第一个周期的方法，唯一不同的是该周期的门限因子由剔除尖峰之后剩余的参考单元数决定。若在当前周期中没有尖峰信号过门限，那么剔除过程就结束了。于是，再用最后一个周期得到的参考单元样本之和与该周期的门限因子所形成的检测门限，对检测单元样本进行门限判决，从而完成CCA-CFAR的检测过程。

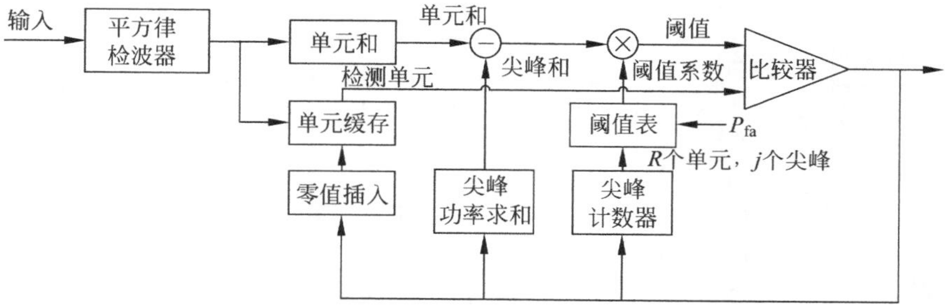

图6.1 CCA-CFAR检测过程的原理框图

文献[1]通过仿真分析将CCA与CMLD和OS方法进行了比较。结果表明，在已知干扰目标数的条件下，CMLD具有较好的性能。而在其他情况下，CCA表现出更好的性能。CCA具有较大的检测损

失，这是由于它没有先验信息可以利用，且不能完全删掉强干扰目标造成的。虽然对强目标来说，OS 比 CCA 的性能更好，但是对于弱目标检测，CCA 更可取。在密集目标环境中，CCA 的优势就更明显了，并且 CCA 可以容纳的干扰目标数不像 OS 方法那样受指定 k 值的限制。

6.3 HCE-CFAR 检测器

针对杂波边缘问题，Finn 提出了自适应于杂波边缘位置的 HCE(Heterogeneous Clutter Estimate)方法[2]，其基本原理是，先确定杂波边缘位置，进而判断检测单元所处的杂波区域，然后再选择杂波功率水平估计方法。图 6.2 给出了 HCE-CFAR 的流程框图。

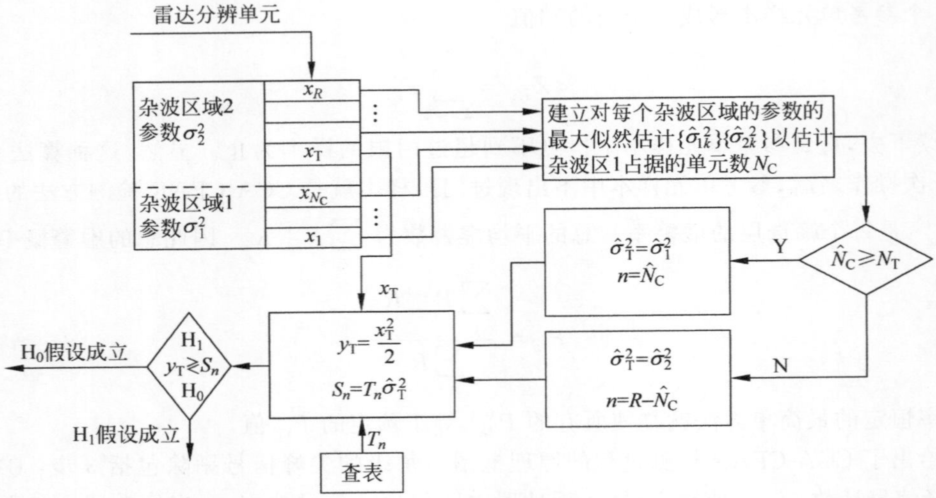

图 6.2 HCE-CFAR 的流程框图

HCE-CFAR 的关键在于估计杂波区域 1 的长度 N_C 以及检测单元杂波功率水平 σ_T^2。文献[2]提供了无偏 HCE-CFAR 检测器的方案。该方案首先形成关于两个杂波区域分离点位置的 $R-1$ 个假设 $H_{Ck}(k=1,2,\cdots,R-1)$，并用最大似然法确定接受哪个假设，以获得杂波边缘位置的最大似然估计。图 6.3 给出了适用于瑞利分布杂波的无偏 HCE-CAFR 的流程图。

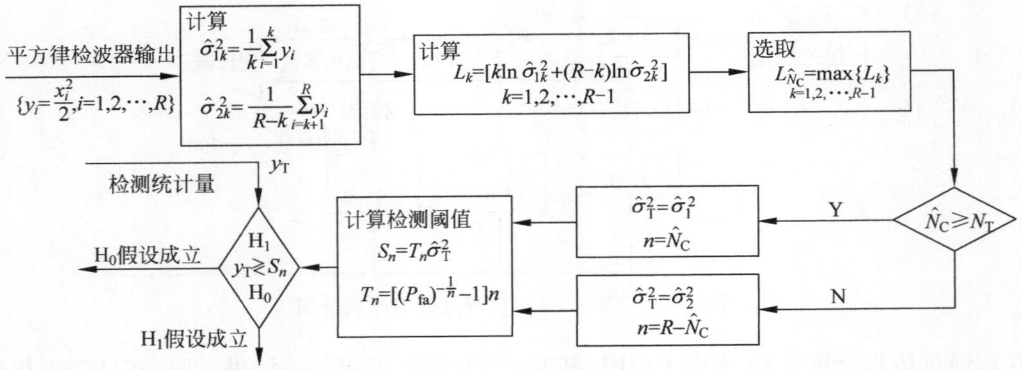

图 6.3 瑞利杂波中的无偏 HCE-CFAR 的流程图

在瑞利分布杂波条件下，假设参考单元中杂波区域 1 的参数为 σ_1^2，杂波区域 2 的参数为 σ_2^2，则判决准则为

$$y_T \underset{H_0}{\overset{H_1}{\gtrless}} T_n \hat{\sigma}_T^2 \tag{6.7}$$

如果$\{H_{Ck}\}$中的一个 $k=\hat{N}_C$ 的假设满足最大似然条件，对应的 σ_1^2 和 σ_2^2 估计为

$$\hat{\sigma}_1^2 = \frac{1}{\hat{N}_C} \sum_{i=1}^{\hat{N}_C} y_i \tag{6.8}$$

$$\hat{\sigma}_2^2 = \frac{1}{R-\hat{N}_C} \sum_{i=\hat{N}_{C+1}}^{R} y_i \tag{6.9}$$

其中，$\hat{N}_C$ 是由最大似然估计得到的区域分离点。将 $\hat{N}_C$ 与检测单元位置 N_T 进行比较，就可以确定检测单元所处的杂波区域。$\hat{\sigma}_T^2$ 是检测单元所处杂波区域的平均功率水平估计。如果 $\hat{N}_C$ 是准确的，那么可以用式(6.10)来确定 T_n

$$P_{fa} = \left(\frac{1}{1+T_n/n}\right)^n \tag{6.10}$$

其中，n 的取值为 $\hat{N}_C$ 或 $R-\hat{N}_C$，根据流程中间结果确定，如图 6.2 所示。

图 6.4 和图 6.5 分别给出了无偏 HCE-CFAR 方案在预先设定的虚警概率 $P'_{fa}=10^{-4}$ 和 $R=30$ 条件下的检测概率与虚警概率关于实际 N_C 值的关系曲线。其中，γ 是检测单元中杂波与另一区域杂波的功率水平之比，图 6.4 中还给出了与 CA 的性能比较。由图 6.4 可知，HCE-CFAR 的检测概率 P_d 作为 N_C 的函数保持了相对的恒定，避免了因遮蔽效应造成的像 CA-CFAR 那样剧烈变化的检测性能。然而，无偏 HCE-CFAR 的虚警率上升了，通常让人难以接受，无偏 HCE-CFAR 的这种 P_{fa} 失控在区域分离点接近检测单元和 γ 不大时更为严重。这是由于对 N_C 估计的偏差造成的。当 γ 较小时，这种偏差较大。因此，文献[2]又提出了有偏 HCE-CFAR 方案，并详细研究了有偏 HCE-CFAR 在杂波边缘环境中的性能以及相对于 CA 和无偏 HCE-CFAR 的性能改善。此外，Finn 还讨论了对数正态杂波中的 HCE-CFAR 检测器。

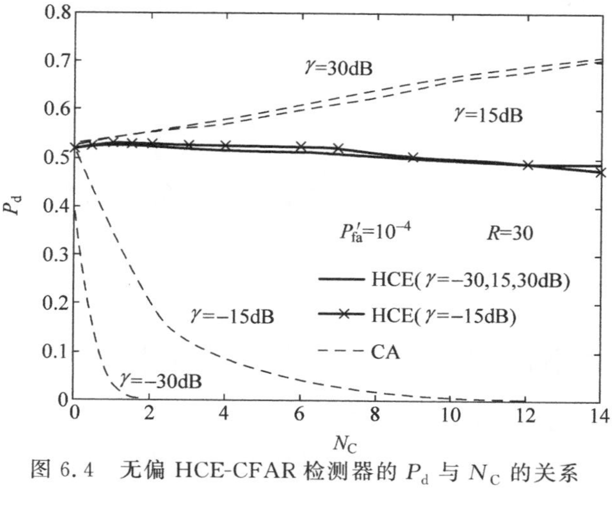

图 6.4　无偏 HCE-CFAR 检测器的 P_d 与 N_C 的关系

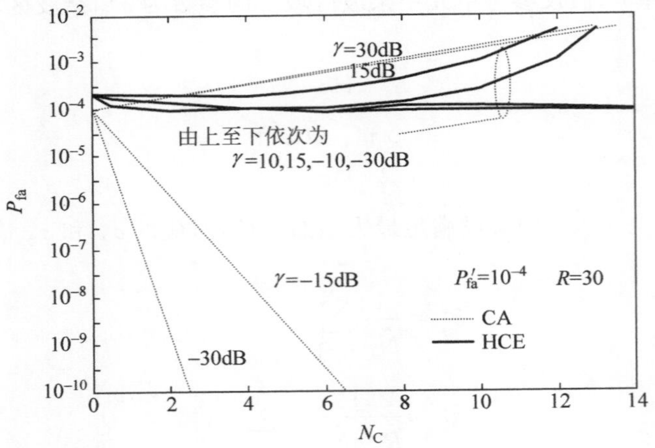

图 6.5 无偏 HCE-CFAR 检测器的 P_{fa} 与 N_C 的关系

6.4 E-CFAR 检测器

E-CFAR 检测器是由 Perry 和 Urkowitz 在一项专利(U. S. Patent 3995270)中首先提出来的,但没有提供任何数学分析,其中 E 是 excision(删除)的缩写。此后 Goldman 和 Bar-David 在文献[3]中对 E-CAFR 检测器进行了数学分析。E-CFAR 是针对多目标情况的,它将超过删除阈值的强干扰样本从参考单元样本集合中删除,然后对剩余样本做平均处理,由此形成检测阈值,删除了强干扰样本后的样本集合能够代表检测单元中的杂波功率水平。如果删除阈值设置得当,干扰目标的影响是可以克服的。文献[4]提出的 TS-CFAR(截断统计量 CFAR 检测器)也采用了这种预先设置固定门限以剔除可能包含的干扰目标的做法,但该文献没有分析如何设置该固定门限。

6.4.1 E-CFAR 检测器结构

图 6.6 给出了 E-CFAR 检测器的原理框图。删除器从参考单元样本集合$\{x_i\}$中删除超过删除阈值 B_E 的样本,然后对通过删除器的样本集合$\{y_i\}$取平均,再乘上一个门限因子 T,就得到检测阈值 S。

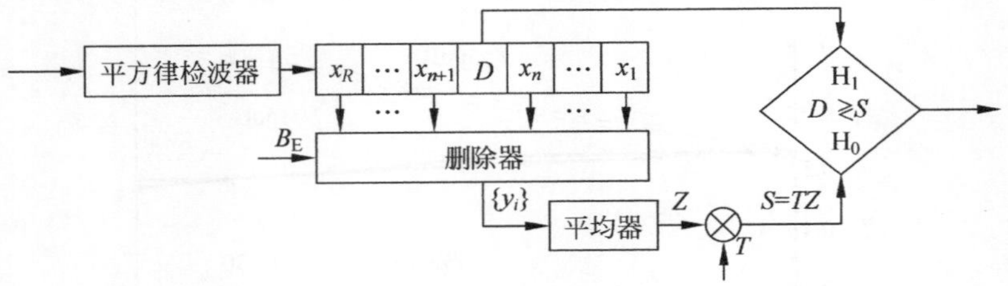

图 6.6 E-CFAR 检测器的原理框图

6.4.2 E-CFAR 检测器在均匀杂波背景中的性能

假设在高斯白噪声背景中,对瑞利起伏目标(Swerling Ⅰ型)进行检测,那么,平方律检波器输出的杂波加噪声样本服从参数为 σ^2 的指数分布。在上述假设下,可得 P_d 和 P_{fa} 的解析表达式为[5-7]

$$P_{\mathrm{fa}}=\sum_{i=1}^{R}\binom{R}{i}P_{\mathrm{E}}^{i}(1-P_{\mathrm{E}})^{R-i}Q^{i}\left(\mathrm{j}\frac{T}{\lambda_0 i};\lambda_0,B_{\mathrm{E}}\right)\tag{6.11}$$

$$P_{\mathrm{d}}=\sum_{i=1}^{R}\binom{R}{i}P_{\mathrm{E}}^{i}(1-P_{\mathrm{E}})^{R-i}Q^{i}\left(\mathrm{j}\frac{T}{\lambda_{\mathrm{S}} i};\lambda_0,B_{\mathrm{E}}\right)\tag{6.12}$$

其中

$$P_{\mathrm{E}}=1-\exp\left(-\frac{B_{\mathrm{E}}}{\lambda}\right)\tag{6.13}$$

$$\lambda=\begin{cases}\lambda_0=2\sigma^2, & \text{对于杂波采样}\\ \lambda_{\mathrm{S}}=2\sigma^2(1+\mathrm{SNR}), & \text{对于目标信号的采样}\end{cases}\tag{6.14}$$

$$Q(u;\lambda,B_{\mathrm{E}})=\int_0^{B_{\mathrm{E}}}\exp(\mathrm{j}uy)f_y(y)\mathrm{d}y=\frac{1-\exp\left(-\frac{B_{\mathrm{E}}}{\lambda}+\mathrm{j}B_{\mathrm{E}}u\right)}{(1-\mathrm{j}\lambda u)\left[1-\exp\left(-\frac{B_{\mathrm{E}}}{\lambda}\right)\right]}\tag{6.15}$$

其中

$$f_y(y)=\frac{\exp(-y/\lambda)}{\lambda\left[1-\exp(-B_{\mathrm{E}}/\lambda)\right]},\quad 0\leqslant y\leqslant B_{\mathrm{E}}$$

$f_y(y)$是杂波幅值低于B_{E}时的条件概率密度，$Q(u;\lambda,B_{\mathrm{E}})$是其对应的条件特征函数。

文献[3]和文献[5]还分别给出了E-CFAR检测器对非起伏目标(Swerling 0型)和瑞利起伏目标(Swerling Ⅰ型)的检测概率表达式。定义删除系数$\alpha=B_{\mathrm{E}}/\lambda_0$，则CA-CFAR相当于$\alpha\to\infty$的情况。在均匀杂波背景中，E-CFAR对起伏目标的检测损失要比非起伏目标大些，两种情况下检测损失的差值与CA-CFAR相应的差值相当。更重要的是，E-CFAR的检测损失随α的增加而急剧下降。例如，$P_{\mathrm{d}}=0.9$时，对于非起伏目标情况，$\alpha=0.2, 0.3, 0.5$时E-CFAR相对于CA-CFAR的检测损失分别为1.2，0.5和0.2dB[3]。而对于起伏目标情况也有相同的趋势[5]。较大的α值在实际应用中是可能出现的，所以E-CFAR在这两种情况下的检测损失都不大。实际上，与CA相比，$\alpha\geqslant 1$时E-CFAR的检测损失可以忽略。SO与$\alpha=0.75$时E-CFAR的检测性能相近。而当$\alpha\geqslant 1$时，E-CFAR的检测性能优于SO-CFAR[6]。

6.4.3 E-CFAR检测器在多目标环境中的性能

在用于估计噪声功率水平的样本集合中，强干扰信号的存在是一些典型ML类检测器性能下降的主要原因。为了分析E-CFAR检测器在多目标环境中的性能，假设R个样本构成的集合中有IN个样本是干扰目标，且平方律检波器输出的干扰目标样本服从期望为λ_{J}的指数分布。与式(6.14)中λ_{S}的定义相似，$\lambda_{\mathrm{J}}=\lambda_0(1+\mathrm{INR})$，这里INR是干扰目标与噪声的平均功率之比。平均器输出为

$$Z=\frac{1}{n+m}\left(\sum_{i=1}^{n}y_{\mathrm{N}_i}+\sum_{i=1}^{m}y_{\mathrm{J}_i}\right)\tag{6.16}$$

其中，n是通过删除器的杂波加噪声样本y_{N_i}的个数，m是通过删除器的干扰目标样本y_{J_i}的个数，$0\leqslant m\leqslant \mathrm{IN}$，$1\leqslant n\leqslant R-\mathrm{IN}$。在上述条件下，E-CFAR的虚警概率和检测概率分别为

$$P_{\mathrm{fa}}=\sum_{n=1}^{R-\mathrm{IN}}q_{\mathrm{N}}(n)\sum_{m=0}^{\mathrm{IN}}q_{\mathrm{J}}(m)Q^{n}\left[\mathrm{j}\frac{T}{\lambda_0(n+m)};\lambda_0,B_{\mathrm{E}}\right]Q^{m}\left[\mathrm{j}\frac{T}{\lambda_0(n+m)};\lambda_{\mathrm{J}},B_{\mathrm{E}}\right]\tag{6.17}$$

$$P_{\mathrm{d}}=\sum_{n=1}^{R-\mathrm{IN}} q_{\mathrm{N}}(n) \sum_{m=0}^{\mathrm{IN}} q_{\mathrm{J}}(m) Q^{n}\left[\mathrm{j} \frac{T}{\lambda_{\mathrm{S}}(n+m)} ; \lambda_{0}, B_{\mathrm{E}}\right] Q^{m}\left[\mathrm{j} \frac{T}{\lambda_{\mathrm{S}}(n+m)} ; \lambda_{\mathrm{J}}, B_{\mathrm{E}}\right] \tag{6.18}$$

其中，$q_{\mathrm{N}}(n)$和$q_{\mathrm{J}}(m)$分别是n个杂波加噪声样本和m个干扰目标样本通过删除器的概率，且

$$q_{\mathrm{N}}(n)=\binom{R-\mathrm{IN}}{n} P_{\mathrm{EN}}^{n}(1-P_{\mathrm{EN}})^{R-\mathrm{IN}-n} \tag{6.19}$$

$$q_{\mathrm{J}}(m)=\binom{\mathrm{IN}}{m} P_{\mathrm{EJ}}^{m}(1-P_{\mathrm{EJ}})^{\mathrm{IN}-m} \tag{6.20}$$

P_{EN}和P_{EJ}分别是一个杂波加噪声样本和一个干扰目标样本通过删除器的概率，且

$$P_{\mathrm{EN}}=1-\exp(-B_{\mathrm{E}}/\lambda_{0}) \tag{6.21}$$

$$P_{\mathrm{EJ}}=1-\exp(-B_{\mathrm{E}}/\lambda_{\mathrm{J}}) \tag{6.22}$$

函数Q的定义见式(6.15)。需要注意的是，$\lim\limits_{B_{\mathrm{E}}\to\infty} P_{\mathrm{d}}$与文献[8]给出的CA的$P_{\mathrm{d}}$是一致的。

图6.7给出了E-CFAR检测器在存在两个干扰目标的环境中对主目标的检测能力与删除系数α的关系。由图6.7可知，在α很宽的取值范围内，曲线较为平坦，与没有干扰情况下的CA接近。对于大α值，E-CFAR的P_{d}逐渐下降到CA在相同条件下的水平。

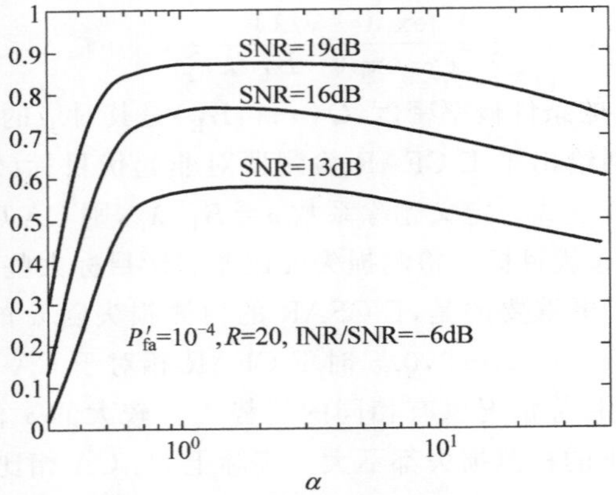

图6.7 E-CFAR检测器的检测概率与删除系数α的关系(两个干扰目标)

在多目标环境中，为了使虚警概率相对恒定，应该根据干扰目标数(IN值)来调整T。文献[7]给出了这种方案的E-CFAR的P_{d}曲线。对于高λ_{S}值，即使IN值较大，检测性能的下降也很小。在实际应用中，干扰目标数不太可能超过R的1/4。对于$R=20$，实用的α将超过3。因此，即使不能根据实际IN来调整T，使用IN=0或其他预定的IN值对应的T值进行检测，检测性能变化也很小。

如果已知检测器的输入噪声功率水平，那么就可确定删除系数α，并根据式(6.19)计算出通过删除器的样本个数的期望值。如果删除的样本数超过期望值，就假设存在干扰目标，然后查表更换门限因子T，这样也可以减小检测损失。

6.5 OSTA-CFAR检测器

6.5.1 OSTA-CFAR检测器基本原理

为了提高OS在杂波边缘环境中的性能，文献[9]提出了一种阈值可调的OS-CFAR检测器，

即 OSTA-CFAR 检测器(Threshold Adjustable,TA)。OS 中把排序后的第 k 个样本值 $x_{(k)}$ 作为杂波功率水平估计。在杂波边缘环境中,从对杂波平均功率水平的总体估计效果来看,若估计值 $x_{(k)}$ 比实际值小,则 P_{fa} 就会上升,导致检测器性能下降,所以,需要事先判断杂波边缘是否存在。如果杂波边缘存在,可以用比 $x_{(k)}$ 大些的 $x_{(k+1)}$ 作为杂波功率水平估计。OSTA 的方法如下,用 $x_{(l)}$($1<l<k$)与均匀杂波功率水平估计 a(在检测杂波边缘前获得)进行比较。如果 $x_{(l)}>a$,说明强杂波区大于 $R-l$,则 $Z=x_{(k)}$;如果 $x_{(l)}<a$,说明弱杂波区大于 l,则 $Z=x_{(k+1)}$,以此降低杂波边缘的影响。仿真分析结果表明,在杂波边缘环境中,OSTA 的 CFAR 性能要比 OS 和 CA 好得多。

OSTA 的关键是选择 l。如果 l 太小,虚警率会下降,P_d 也同时下降;如果 l 太大,P_{fa} 将急剧增加。因此,l 要选得适当。但是,很难从理论上说明怎样选择 l 值,需要用仿真的方法来选择 l。

6.5.2　OSTA-CFAR 检测器在杂波边缘环境中的性能

在瑞利杂波条件下,表 6.1 给出了 CA-CFAR、OS-CFAR、OSTA-CFAR 在杂波边缘环境中的几个 P_{fa} 数据。当预先设定的虚警概率 $P'_{fa}=10^{-6}$ 和 $R=14$ 以及积累脉冲数 $M=10$ 时,$k=12$ 的 OS 具有较好的结果;对于 OSTA,通过仿真选择得到 $k=12$ 和 $l=3$ 具有较好的结果。当强杂波区参考单元数 $N_C>R/2$ 时,较小的 N_C 会导致较低的杂波功率估计,这就是 P_{fa} 随着 N_C 减小而上升的原因。由表 6.1 可知,当 $N_C=R/2$ 并且检测单元处于强杂波区时,CA 的 P_{fa} 在数量级上由 10^{-5} 跳跃到 10^{-1},变化了大约 4 个数量级;OS 的 P_{fa} 由 10^{-6} 跳跃到 10^{-2},变化了大约 4 个数量级。OS 的虚警控制性能相对于 CA 只有一个很小的提高,与理想值还相差很远。OSTA 的 P_{fa} 由 10^{-6} 变化到 10^{-5},只变化了一个数量级。因此,在上述条件下,这 3 种 CFAR 检测器中,OSTA 在杂波边缘中的虚警控制性能是最好的一个。

表 6.1　3 种 CFAR 检测器在杂波边缘环境中的虚警控制性能 P_{fa}

N_C	检测器		
	CA-CFAR	OS-CFAR($k=12$)	OSTA-CFAR($k=12,l=3$)
7	0.400×10^{-0}	0.122×10^{-1}	0.947×10^{-4}
11	0.223×10^{-2}	0.519×10^{-4}	0.261×10^{-5}
14	0.715×10^{-4}	0.227×10^{-5}	0.227×10^{-5}

6.5.3　OSTA-CFAR 检测器在多目标环境中的性能

表 6.2 给出了 CA-CFAR、OS-CFAR、OSTA-CFAR 在多目标环境中的 P_{fa} 值,其中多目标环境是指存在一个瑞利分布的干扰目标,其他条件包括:$P'_{fa}=10^{-6}$,$R=14$,积累脉冲数 $M=10$,干扰目标与噪声的平均功率之比(简称干噪比,INR)取 5dB、10dB 和 20dB 三个值。很明显,随着干噪比增加,CA 的 P_{fa} 急剧下降,当干杂比增加到 20dB 时,P_{fa} 在数量级上由 10^{-6} 变到 10^{-17},变化了大约 11 个数量级。这说明 CA 在多目标环境中的性能极不稳定,其检测性能也是如此。OS 和 OSTA 的 P_{fa} 只改变了 1 个数量级,几乎不受干噪比变化的影响,保持了稳定的性能。

表 6.2 3 种 CFAR 检测器在多目标环境中的虚警控制性能 P_{fa}

INR	检测器		
	CA-CFAR	OS-CFAR($k=12$)	OSTA-CFAR($k=12,l=3$)
5dB	0.132×10^{-6}	0.471×10^{-5}	0.471×10^{-5}
10dB	0.951×10^{-8}	0.364×10^{-6}	0.364×10^{-6}
20dB	0.940×10^{-16}	0.338×10^{-6}	0.338×10^{-6}

6.6 VTM-CFAR 检测器

6.6.1 VTM-CFAR 检测器基本原理

TM 是 OS 的一种改进型,其总体性能只是比设计得当的 OS 稍好一些,r_1 值($r_1<R-r_2$)较高的 TM 可以控制杂波边缘环境中的虚警率,并且检测损失不大。r_2 在 TM 的性能调整方面起着重要作用,但是,r_2 的选取往往面临多目标环境中的检测性能与杂波边缘环境中的虚警控制性能二者之间取舍的矛盾。

文献[10]提出了可变削减平均,即 VTM(Variably Trimmed Mean)-CFAR 检测器。它是 TM 的一种修正型,且其性能优于 TM 与 OS。因为固定的 r_2 是限制 TM 性能的一个重要原因,所以 VTM 采用基于数据的一套规则来自适应地调整 r_2,然后利用削减后剩余的有序样本值 $x_{(1)}\leqslant x_{(2)}\leqslant\cdots\leqslant x_{(R-r_2)}$ 的线性组合形成 Z。VTM-CFAR 检测器的原理框图如图 6.8 所示。

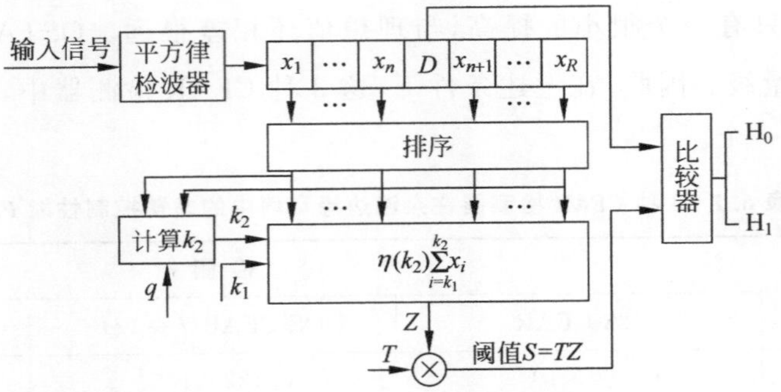

图 6.8 VTM-CFAR 检测器的原理框图

VTM 的设计原则是:①使用较小的 r_2 值或者不用 r_2,以获得均匀杂波背景中良好的检测性能和杂波边缘环境中增强的虚警控制能力;②使用适当大的 r_2 值以获得在多目标环境中对干扰目标不敏感的检测性能。为满足这两个条件,把统计量 Z 定义为

$$Z=\eta(K_2)\sum_{i=k_1}^{K_2}x_{(i)} \tag{6.23}$$

其中,K_2 是一个由相应规则确定的随机变量,$\eta(K_2)$是归一化系数。确定 K_2 的规则为

$$\text{如果 } x_{(k_2)}\leqslant(1+q)x_{(k_1)}<x_{(k_2+1)},\quad \text{取 } K_2=k_2 \tag{6.24}$$

q 是一个非负的实值参数。这是对 $x_i(i=1,2,\cdots,R)$的分离过程,以确定 K_2。定义 $K_2=k_2$ 为“E_{k_2}事件”。

6.6.2 VTM-CFAR 检测器在均匀杂波背景中的性能

在均匀杂波背景中，VTM 的虚警率可以表示为[10]

$$P_{\text{fa}} = \sum_{i=0}^{R-k_1} \frac{\binom{R-k_1}{i}}{\left(1+\frac{T}{i+1}\right)^i} \sum_{j=0}^{i} \binom{i}{j} (-1)^j M_{k_1}[B(i,j,T)] \tag{6.25}$$

其中

$$B(i,j,T) = q(R-k_1+j-i) + \frac{T}{i+1}(1+i+qj) \tag{6.26}$$

$$M_{k_1}(T) = \prod_{i=0}^{k_1-1} \frac{R-i}{R-i+T} \tag{6.27}$$

其中，$M_{k_1}(T)$是 OS(k_1)的虚警概率。很明显，式(6.25)表示的 P_{fa} 不依赖于噪声功率，因此 VTM 是 CFAR 的。给定 P_{fa}、k_1 和 q 的设计值，就可以计算出 T，若用 $T/(1+\text{SNR})$代替式(6.25)中的 T 就得到 VTM-CFAR 检测器在均匀杂波背景中的检测概率 P_{d}。由式(6.24)可知，当 $q=0$ 时，VTM 就是 OS(k_1)。当 $q\to\infty$时，VTM 的性能趋近于 $r_1=k_1-1$ 和 $r_2=0$ 的 TM。

VTM 的 ADT 可以由式(6.25)对 T 求微分并且令 $T=0$ 得到，即有

$$\text{ADT} = T\sum_{i=0}^{R-k_1} \frac{\binom{R-k_1}{i}}{i+1} \left\{ \sum_{j=0}^{i} \binom{i}{j} (-1)^j M_{k_1}[(R-k_1-i+j)q] \times \left[i + (i+jq+1) \sum_{n=0}^{k_1-1} \frac{1}{R-n+(R-k_1-i+j)q} \right] \right\} \tag{6.28}$$

图 6.9 给出了 VTM 的 ADT 随 q 和 k_1 变化的曲线。对于固定的 q，ADT 并不随 k_1 单调变化。在 $0\leqslant q<1$ 时，$k_1=21$ 的 ADT 达到了最小值。对于 OS 也有相似的性质。对于一个固定 k_1，ADT 随 q 变化有一个起伏，在 $q\approx1$ 处最高；在 $q\to\infty$时，VTM 的 ADT 值趋近 TM 的 ADT 值。

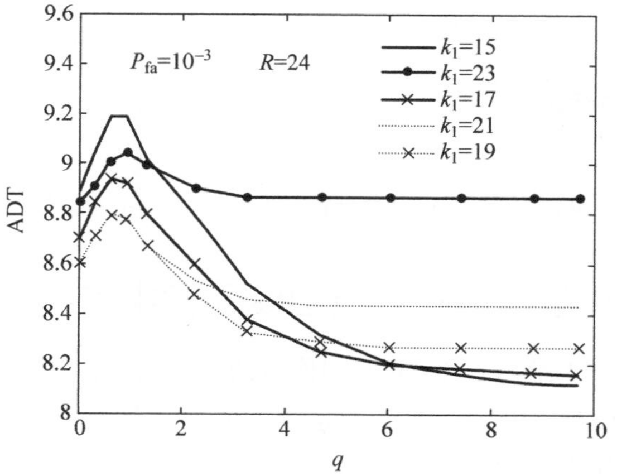

图 6.9 对于不同 k_1 值 VTM-CFAR 检测器的 ADT 值与 q 的关系曲线

ADT 是分析 CFAR 检测器在均匀杂波背景中检测性能的一种很方便的指标。与 OS 不同，VTM 不能直接确定最优 k_1 值，因为同时还要考虑到 q 的影响。k_1 通常由其他的约束条件来确定（见下面的结论），并且选择的 q 使 ADT 最小化。由图 6.9 可知，$q\geqslant4$ 的 ADT 有很宽的最小值范围。因此，q 可以在很宽范围内选择，且不会导致 VTM 在均匀背景中的检测性能有太大的损失。

6.6.3 VTM-CFAR 检测器在多目标环境中的性能

很明显，只要适当选择 q，VTM 在多目标环境中最多可以对抗 $R-k_1$ 个干扰目标。如果 q 太大，VTM 的检测性能损失也较大。这是因为对于大 q，k_2 不等于 k_1 的概率也增加了，导致形成较大的 Z。

在预设虚警概率为 $P'_{\mathrm{fa}}=10^{-3}$，$R=24$，$k_1=21$ 的条件下，在 3 个与主目标相同强度的干扰目标环境中，图 6.10 给出了 VTM 的检测性能曲线。由图 6.10 可知，对于固定的 k_1 值，在信噪比值较高时，$q=2$ 时的 VTM 相对于 OS(k_1)($q=2$)的检测性能有一定的增强。因为对于 $0\leqslant q\leqslant4$ 和大信噪比来说，包含干扰目标的样本通常以较大的概率大于 $(1+q)x_{(21)}$，并且不包含在式(6.23)的和式中。因此，VTM 显示出类似于 OS(21)的性能。但是，这时一个较小的 T 值会反过来导致较低的阈值，使 P_{d} 有轻微上升。对于大 q 值($q\geqslant4.0$)，式(6.23)的和式中包含至少一个干扰目标回波的概率增加，所以 P_{d} 通常随着 q 的增加而下降。对于其他的 k_1 值也可以得到类似的结果[10]。

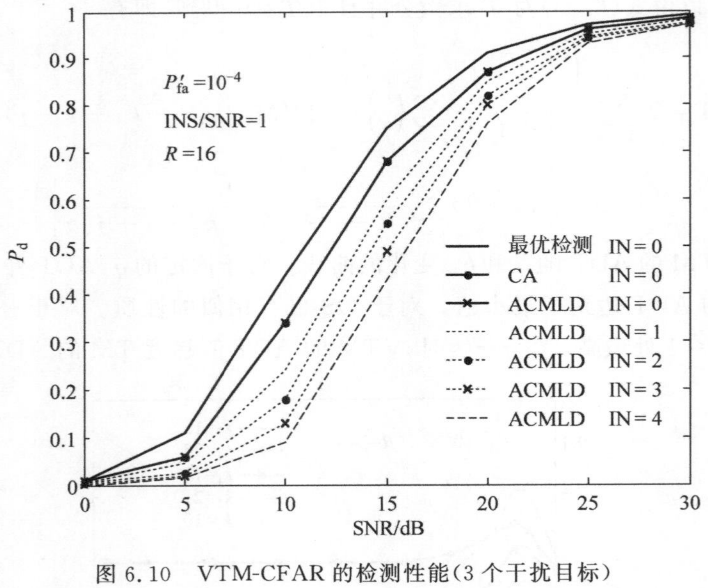

图 6.10 VTM-CFAR 的检测性能(3 个干扰目标)

6.6.4 VTM-CFAR 检测器在杂波边缘环境中的性能

在杂波边缘环境中，图 6.11 给出了 k_1 取不同值时的虚警概率。考虑最糟的情况，假设检测单元处于杂波边缘处，且处于强杂波区。很明显，对于固定的 q 值，P_{fa} 随着 k_1 增加而下降。类似地，对于固定的 k_1 且 $q\geqslant2$，VTM 的 P_{fa} 通常随着 q 的增大而下降，且低于 OS($q=0$)的 P_{fa}。在 $0<q\leqslant2$ 时，由于 VTM 和 OS(k_1)的 T 值不同，VTM 的 P_{fa} 有轻微的上升。

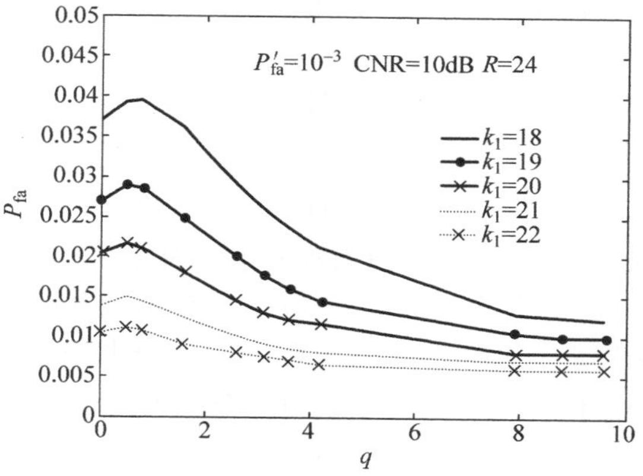

图 6.11 VTM-CFAR 的 P_{fa} 与 q 的关系

6.6.5 VTM-CFAR 检测器的参数选择

为了使 VTM 的总体性能优于最优 OS($R=24,k_1=21$)，可以考虑参数 k_1 和 q 的如下几种可能的选择，这也表明 VTM 在设计上具有灵活性。

(1) 在保持与 OS(21)相近检测性能的同时，如果要降低虚警概率，可以选择 $k_1=21$ 和 $q\approx 4$。此时，在 3 个干扰目标环境中，VTM 的 $P_d=0.6$ 与 OS(21)的 $P_d=0.64$ 十分接近。另外，在图 6.11 所示的最糟情况中，VTM 的 P_{fa} 从 OS(21)的 $P_{fa}=0.014$ 降到了 0.008。

(2) 如果要同时提高杂波边缘与多目标环境中的性能，可选择 $k_1=20$ 和 $q\approx 4$。在 3 个干扰目标环境中，VTM 的 $P_d=0.67$，在杂波边缘环境中，$P_{fa}=0.011$，相对于 OS(21)有轻微改善。而且，$k_1=20$ 的 VTM 可多对抗一个干扰目标。在 4 个干扰目标环境中，VTM 的 $P_d=0.57$，而此时 OS(21)的 P_d 只有 0.41。

(3) 如果要对抗更多的干扰目标，同时使虚警概率的要求也得到适当满足，可以选择 $k_1=19$ 和 $q\approx 5$，这样便可以对抗 5 个干扰目标，并且杂波边缘环境中的 $P_{fa}=0.013$，其他一些 k_1 和 q 也可以满足这个要求。因此，干扰目标数可自适应调整是 VTM 相对于 OS 的一个优点。

6.7 Himonas 的一系列 CFAR 检测器

6.7.1 GCMLD-CFAR 检测器

1. GCMLD-CFAR 检测器描述

广义删除均值检测器，即 GCMLD(Generalized Censored Mean-Level Detector)-CFAR 检测器是由 Himonas 和 Barkat 等在文献[11]中提出的。它利用迭代方法分别确定前后滑窗中的删除点，不需要任何关于干扰目标数的先验信息。

设前、后滑窗长度均为 n。操作过程是：先分别对前后滑窗中各单元样本值进行排序，$x_{(1)}\leqslant x_{(2)}\leqslant\cdots\leqslant x_{(n)}$ 和 $x_{(n+1)}\leqslant x_{(n+2)}\leqslant\cdots\leqslant x_{(2n)}$，然后分别在前后滑窗中实施下述操作以确定相应的样本删除数 r_1 和 r_2。以前沿滑窗为例，首先比较 $x_{(2)}$ 与 $T_1x_{(1)}$，T_1 是满足错误删除概率 P_{FC} 的因子；如果 $x_{(2)}$ 大于 $T_1x_{(1)}$，则判定 $x_{(2)}$ 是干扰目标回波，$r_1=n-1$；如果 $x_{(2)}$ 小于 $T_1x_{(1)}$，则判定 $x_{(2)}$ 是不

包含干扰目标回波的杂波样本；在这种情况下，令 $Z_2=x_{(1)}+x_{(2)}$；在第 k 步，比较 $x_{(k+1)}$ 与 T_kZ_k，并进行如下判决

$$x_{(k+1)} \underset{H_0}{\overset{H_1}{\gtrless}} T_kZ_k \tag{6.29}$$

其中，T_k 是第 k 步的标称化因子，Z_k 是 k 个较低有序样本之和。第 k 步中，H_1 表示 $x_{(k+1)},x_{(k+2)},\cdots,x_{(n)}$ 为干扰目标回波的假设；H_0 表示 $x_{(1)},x_{(2)},\cdots,x_{(k+1)}$ 为不包含干扰目标的杂波样本的假设。这个删除过程在 H_1 成立时即刻停止。也就是说，确定有 $n-k$ 个干扰目标回波样本，取 Z_k 作为前沿滑窗局部估计 X。图 6.12 是 GCMLD-CFAR 检测器的框图，图 6.13 是上述删除过程的流程框图。

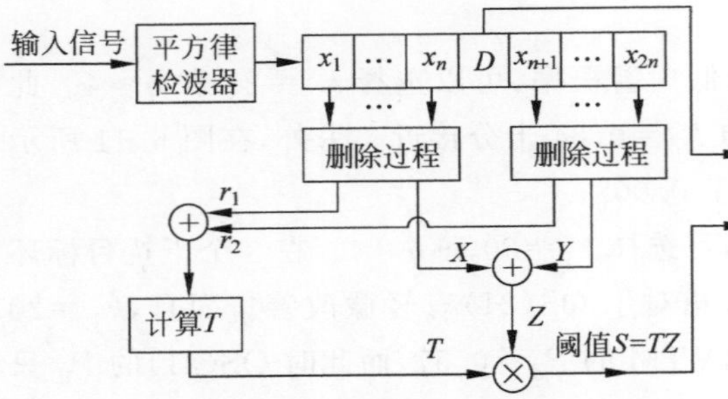

图 6.12　GCMLD-CFAR 检测器的框图

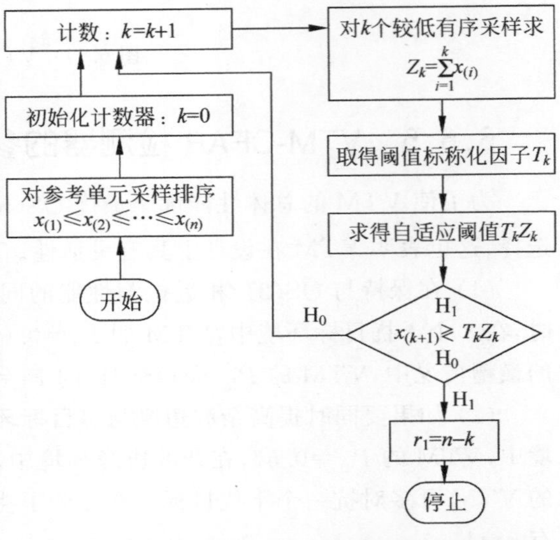

图 6.13　GCLMD-CFAR 中删除过程的流程框图

2. GCMLD-CFAR 检测器在多目标环境中的性能

假设第 i 个参考单元样本的 PDF 为

$$f(x_i)=\begin{cases} e^{-x_i}, & x_i\geqslant 0,\text{杂波采样} \\ \dfrac{1}{1+(\mathrm{INR})_i}\exp\left[-\dfrac{x_i}{1+(\mathrm{INR})_i}\right], & x_i\geqslant 0,\text{干扰目标与杂波之和的采样} \end{cases} \tag{6.30}$$

设

$$\gamma_i=\frac{1}{1+(\mathrm{INR})_i},\quad i=1,2,\cdots,k \tag{6.31}$$

与 $x_{(n-i+1)}(i=1,2,\cdots,k)$ 对应，并且 $(\mathrm{INR})_i$ 是第 i 个干扰目标的平均功率与杂波功率比。

在删除过程的第 $n-k$ 步，若 H_1 成立则有 k 个较大的有序样本值被判定为干扰目标样本。因此，删除 k 个干扰目标的概率为

$$P_{C,k}=\Pr\{D_{n-k}>0\mid H_1\} \tag{6.32}$$

其中，$D_{n-k}=x_{(n-k+1)}-T_{n-k}Z_{n-k}$ 是与式(6.29)定义的假设检验等价的检验统计量，由于在 $(\mathrm{INR})_1\geqslant(\mathrm{INR})_2\geqslant\cdots\geqslant(\mathrm{INR})_k$ 时，式(6.32)的计算极其烦琐，甚至得不到闭型解。为了简化计算，并且不失一

般性，假设$(\mathrm{INR})_1=(\mathrm{INR})_2=\cdots=(\mathrm{INR})_{k-1}=I$。相应地，令$\gamma_1=\gamma_2=\cdots=\gamma_{k-1}=\gamma$。那么，$\gamma$趋近于0时，$P_{\mathrm{C},k}$的解析表达式为[11]

$$P_{\mathrm{C},k}=\gamma_k\binom{n}{k}^{-1}\sum_{i=1}^{n-k}\left\{\frac{1}{(1+T_{n-k})^{n-k-i}}\left[\prod_{j=1}^{i}\frac{n+1-j}{\gamma_k+(n-k+1-j)(1+T_{n-k})-1}\right]\times\left(\prod_{j=i+1}^{n-k}\frac{n+1-j}{n-k+1-j}\right)\right\}+\frac{1}{(1+\gamma_k T_{n-k})^{n-k}} \tag{6.33}$$

在式(6.33)中，令$\gamma_k=1$，则得第$n-k$步的错误删除概率为

$$P_{\mathrm{FC}}=\frac{n-k+1}{(1+T_{n-k})^{n-k}} \tag{6.34}$$

在删除$r=r_1+r_2$个样本条件下，对于主目标的检测概率为

$$P_{\mathrm{d}}(r;\ \mathrm{SNR},I\to\infty)=\frac{1}{\left(1+\dfrac{T}{1+\mathrm{SNR}}\right)^{R-r}} \tag{6.35}$$

总的检测概率P_{d}为

$$P_{\mathrm{d}}(\mathrm{SNR},I\to\infty)=\sum_{r=1}^{R-2}P_{\mathrm{d}}(r;\ \mathrm{SNR},I\to\infty)\Pr\{E_r\} \tag{6.36}$$

其中，E_r表示r个样本已被删除的事件，且

$$\Pr\{E_r\}=\sum_{r_1,r_2\in B}P_{\mathrm{C},r_1}(I\to\infty)P_{\mathrm{C},r_2}\{I\to\infty\} \tag{6.37}$$

其中，P_{C,r_1}和P_{C,r_2}分别是从前沿滑窗删除r_1个样本和从后沿滑窗删除r_2个样本的概率。B表示满足$r=r_1+r_2$条件的整数对(r_1,r_2)的值域空间。

由计算机仿真结果可知，GCMLD对多目标的分辨能力明显优于CA。与干扰目标同时分布在前后两个滑窗时相比，当所有干扰目标都落在前、后滑窗中的一个时，GCMLD的检测性能稍差，但其总体性能还是增强的。GCMLD的优点是不需要任何关于干扰目标数的先验信息。与CCA相比，GCMLD在强或弱干扰目标环境中的检测性能都有所提高[11]。

6.7.2 GO/SO-CFAR 检测器

GO/SO-CFAR检测方法由Himonas和Barkat在文献[12]中提出，其检测器方框图如图6.14所示。这个方法的目的就是自适应地确定检测单元是处于强杂波区还是处于弱杂波区，然后确定用于形成Z的样本集。因此，GO/SO的优势在于对抗边缘杂波。

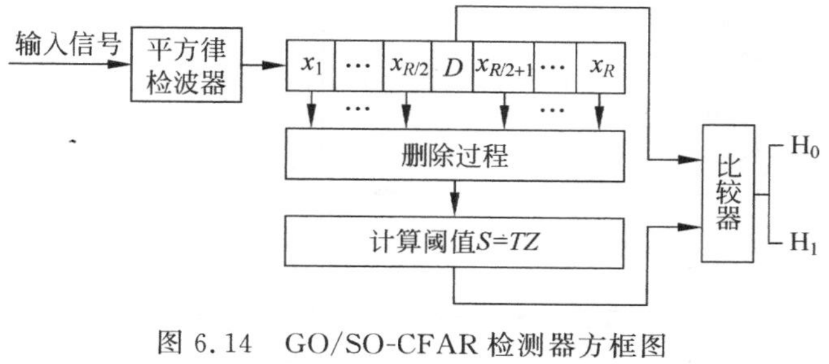

图6.14　GO/SO-CFAR检测器方框图

1. GO/SO-CFAR 检测器描述

对参考单元样本进行排序：$x_{(1)} \leqslant x_{(2)} \leqslant \cdots \leqslant x_{(R)}$，然后实施下面的操作。

第 1 步：用 $x_{(1)}$ 代表杂波样本。比较 $x_{(2)}$ 和 $T_1 x_{(1)}$，T_1 是满足给定的错误删除概率 P_{FC} 的因子。

(1) 如果 $x_{(2)} < T_1 x_{(1)}$，判定 $x_{(2)}$ 和 $x_{(1)}$ 是同一分布的样本，然后进行第二步。

(2) 如果 $x_{(2)} > T_1 x_{(1)}$，判定 $x_{(2)}$ 是来自强杂波区的样本，也就是 $x_{(2)}, x_{(3)}, \cdots, x_{(R)}$ 是强杂波区域的回波信号。

第 k 步：更新背景杂波功率水平 $Z_k = x_{(1)} + x_{(2)} + \cdots + x_{(k)}$。把 $x_{(k+1)}$ 与新阈值 $T_k Z_k$ 进行比较，Z_k 也是满足 P_{FC} 的。

(1) 如果 $x_{(k+1)} < T_k Z_k$，判决 $x_{(k+1)}$ 是弱杂波区样本，然后进行第 $k+1$ 步；

(2) 如果 $x_{(k+1)} > T_k Z_k$，判决 $x_{(k+1)}, x_{(k+2)}, \cdots, x_{(R)}$ 是强杂波区样本。

总之，这个删除过程是一个连续的二元假设检验，在第 k 步为

$$x_{(k+1)} \underset{H_N}{\overset{H_C}{\gtrless}} T_k Z_k \tag{6.38}$$

其中

$$\begin{cases} H_N: \gamma = 1, & x_1, x_2, \cdots, x_{(k+1)} \text{ 为弱杂波样本} \\ H_C: \gamma = \dfrac{1}{1 + \text{CNR}}, & x_{(k+1)}, x_{(k+2)}, \cdots, x_{(R)} \text{ 为强杂波样本} \end{cases} \tag{6.39}$$

当 H_C 成立或者 $k=R$ 时，删除过程停止。这样若判定 $x_{(1)}, x_{(2)}, \cdots, x_{(k)}$ 是弱杂波区样本，则其余的较高样本都是强杂波区样本。如果 $k > R/2$，则判定检测单元处于弱杂波区，由 $x_{(1)}, x_{(2)}, \cdots, x_{(k)}$ 形成 Z；如果 $k \leqslant R/2$，则判定检测单元处于强杂波区，删除 $x_{(1)}, x_{(2)}, \cdots, x_{(k)}$，用其余的较高样本来形成 Z。因此，称这种检测方法为 GO/SO。图 6.15 是这种方法的删除过程流程图。

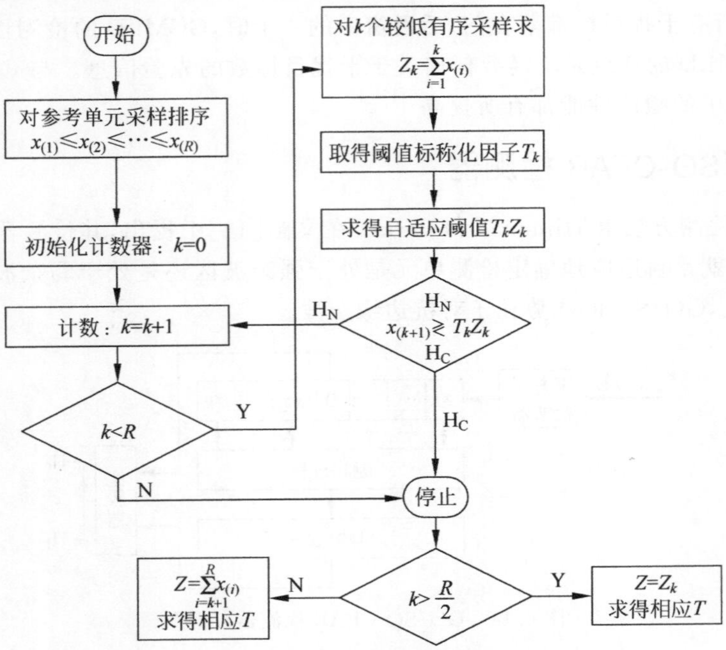

图 6.15 GO/SO-CFAR 删除过程流程图

2. GO/SO-CFAR 检测器的性能分析

在均匀杂波背景中，删除过程第 k 步的错误删除概率 P_{FC} 被定义为

$$P_{FC} = \Pr\{D_k > 0 \mid H_N\} \tag{6.40}$$

其中，检验统计量 $D_k = x_{(k+1)} - T_k Z_k$，$P_{FC}$ 的解析表达式为[12]

$$P_{FC} = \binom{R}{k} \frac{1}{[1 + T_k(R-k)]^k} \tag{6.41}$$

因子 T_k 见表 6.3。

表 6.3 GO/SO-CFAR 删除过程的标称化因子($R=16$, $P_{FC}=10^{-4}$)

步序	1	2	3	4	5	6	7	8	9	10	11	12	13	14	15
T_k	10666.6	78.15	13.50	5.360	2.976	1.976	1.463	1.165	0.980	0.862	0.790	0.757	0.768	0.859	1.223

在杂波边缘环境中，将杂波边缘的检测概率设为 P_{DC}，也就是 H_C 成立的概率。在相同的杂噪比条件下，弱杂波区长度增加使噪声功率水平估计的起伏减小，因此 P_{DC} 将随弱杂波区长度增加而增强，且 P_{DC} 也将随杂噪比增加而增加。

当删除过程判定检测单元处于弱杂波区(删除点 $k > R/2$)时，删除掉 $R-k$ 个较高的有序样本 $x_{(k+1)}, x_{(k+2)}, \cdots, x_{(R)}$，并用 k 个较低的有序样本形成 Z，即

$$Z = \sum_{i=1}^{k} x_{(i)} \tag{6.42}$$

如果检测单元被判定为处于强杂波区(删除点 $k \leqslant R/2$)，则删除掉 k 个较低的有序样本，把剩余的样本求和形成 Z，即

$$Z = \sum_{i=k+1}^{R} x_{(i)} \tag{6.43}$$

式(6.42)和式(6.43)各自对应的总的虚警概率分别为

$$P_{fa} = \binom{R}{k} \prod_{i=1}^{k} \left(T + \frac{R-i+1}{k-i+1}\right)^{-1}, \quad k > R/2 \tag{6.44}$$

$$P_{fa} = \binom{R}{k} (1+T)^{-(R-k-1)} \sum_{i=0}^{k} \binom{R}{i} (-1)^i \left(1 + \frac{i}{R-k} + T\right)^{-1}, \quad k \leqslant R/2 \tag{6.45}$$

其中，T 是标称化因子，可以预先计算制成表格。在应用中，根据估计的 k 值，查表确定 T 值。在 $R=16$，$P_{FC}=10^{-4}$，$P'_{fa}=10^{-4}$ 条件下，表 6.4 给出了一些典型的 T 值。

表 6.4 GO/SO-CFAR 检测器的标称化因子

步序	1	2	3	4	5	6	7	8	9	10	11	12	13	14	15	16
T	0.785	0.787	0.798	0.813	0.834	0.960	0.890	0.945	4.830	3.550	2.700	2.080	1.630	1.208	1.020	0.778

在检测单元处于强杂波区时，N_C 增加使检测阈值升高，对于给定的信噪比，GO/SO 的检测概率随着 N_C 的增加而略有下降。TM[8] 的检测概率严重依赖于删除点，在杂噪比较高时，如果没有删除掉强杂波区样本，那么 TM 就会出现遮蔽现象，然而 GO/SO 不受遮蔽效应的影响。实际上，GO/SO 在高杂噪比条件下的性能与适当选取 r_2 值条件下的 TM$(0, r_2)$ 的性能相同，但是 TM$(0, r_2)$ 需要关于杂波空

间分布的准确先验信息。当杂噪比较小时，TM(0,r_2)的检测性能优于 GO/SO，因为此时 GO/SO 经常检测不出杂波边缘。GO/SO 相对于 TM(0,r_2)的最大检测损失大约为 2dB。

6.7.3 ACMLD-CFAR 检测器

ACMLD(Automatic Censored Mean Level Detector)-CFAR 检测器也是在文献[12]中提出的。它是假设均匀杂波背景中包含干扰目标回波时，为实现自动删除参考滑窗中可能存在的干扰目标的问题而设计的。ACMLD 不需要关于干扰目标数的先验信息。

1. ACMLD-CFAR 检测器描述

ACMLD 采用与 GO/SO 相同的删除算法，但是 ACMLD 无须判断杂波边缘。在第 k 步，$x_{(k+1)}$ 与阈值 $T_k Z_k$ 比较，按如下检验进行判决

$$x_{(k+1)} \underset{H_0}{\overset{H_1}{\gtrless}} T_k Z_k \tag{6.46}$$

其中，T_k 和 Z_k 与 6.7.1 节中的定义相同。H_1 代表 $x_{(k+1)}$ 中包含干扰目标回波样本的假设，H_0 代表 $x_{(k+1)}$ 中不包含干扰目标回波样本的假设。如果 H_1 成立，那么删除 $x_{(k+1)}, x_{(k+2)}, \cdots, x_{(R)}$，并且算法停止。杂波功率水平估计为

$$Z = \sum_{i=1}^{k} x_{(i)} \tag{6.47}$$

式(6.47)与式(6.42)相同，因此虚警概率如式(6.44)所示。删除过程在第 k 步中，错误删除 $R-k$ 个有序样本的概率也与 GO/SO 的相同，见式(6.41)。$P_{fa}'=10^{-4}$，$P_{FC}=10^{-4}$ 和 $R=16$ 时的标称因子如表 6.5 所示。

表 6.5 ACMLD 的标称化因子

K	1	2	3	4	5	6	7	8	9	10	11	12	13	14	15	16
T	160000	1080	168.5	58.00	27.72	15.85	10.02	6.75	4.830	3.550	2.700	2.080	1.630	1.280	1.020	0.778

2. ACMLD-CFAR 检测器的性能分析

在存在 IN 个与主目标强度相同的干扰目标的条件下，图 6.16 给出了 ACMLD 的检测性能曲线，同时也给出了最优固定阈值检测的检测概率。当 IN=0 时，即在均匀杂波背景中，CA 与 ACMLD 完全重合，从 ACMLD 不受遮蔽效应影响的角度来说，ACMLD 的检测性能是占优的。

在有 IN 个干扰目标的环境中，文献[12]比较了 TM(0,r_2)和 ACMLD 对强和弱干扰目标的检测性能。当 r_2<IN 时，TM 的检测性能受遮蔽效应的影响较严重，而且这种影响依赖于干扰目标的强度。然而 ACMLD 不受遮蔽效应的影响。但是当 r_2>IN 时，ACMLD 相对于 TM(0,r_2)有一定的检测损失，这可能是由于删除过程没有删除掉参考滑窗中所有的干扰目标造成的。

6.7.4 GTL-CMLD-CFAR 检测器

GTL-CMLD(Generalized Two-Level Censored Mean Level Detector)也是在文献[12]中提出的。它结合了 6.7.2 节和 6.7.3 节中的自动删除算法，对不希望的样本进行有效删除。GTL-CMLD 包括两级删除过程。在第一级删除过程中，确定检测单元是处于弱杂波区还是强杂波区，并删除不包含检测单

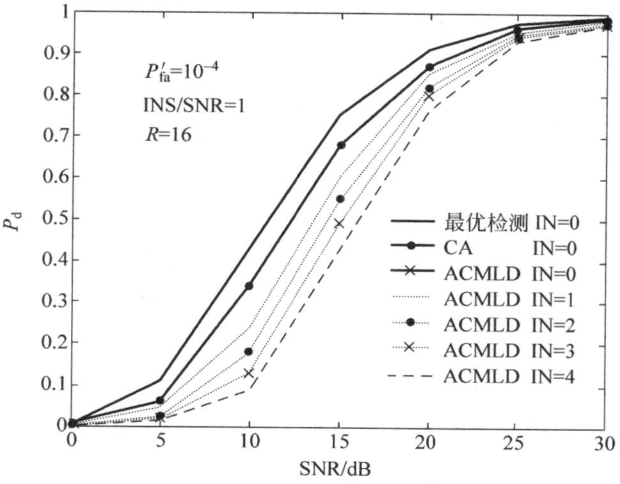

图 6.16　ACMLD 的检测性能曲线

元的杂波区样本；在第二级删除过程中，删除参考滑窗中出现的干扰目标。

1. GTL-CMLD 检测器的描述

前面讨论的 GO/SO 和 ACMLD 是分别为对抗杂波边缘和干扰目标而设计的。GLT-CMLD 是为同时对抗干扰目标和杂波边缘而设计的。与 GO/SO 和 ACMLD 删除过程一样，GLT-CMLD 首先对参考样本进行排序，然后实施 GO/SO 的删除过程，确定杂波边缘位置，其中的删除过程流程图如图 6.17 所示。

具体步骤如下。在第 k 步，进行如下的检验

$$x_{(k+1)} \underset{H_B}{\overset{H_A}{\gtrless}} T(k \mid 0)Z(k \mid 0) \tag{6.48}$$

其中，$T(k|0)$是在没有较低有序样本被删除的条件下的第 k 个标称化因子，$Z(k|0)$是该条件下 k 个较低的有序样本的和

$$Z(k \mid 0)=\sum_{i=1}^{k} x_{(i)} \tag{6.49}$$

H_B 假设表示 $x_{(k+1)}$ 与 $x_{(1)},x_{(2)},\cdots,x_{(k)}$ 来自于同一分布的情形，H_A 假设表示 $x_{(k+1)}$ 的分布参数大于 k 个较低有序样本的分布参数的情形。如果 H_A 成立，那么停止第一级删除。如果 $k>R/2$，那么判定检测单元中的杂波与 $k_{C1}=k$ 个较低有序样本的分布相同，算法停止。检测单元杂波功率水平估计为

$$Z=\sum_{i=1}^{k} x_{(i)} \tag{6.50}$$

如果 $k \leqslant R/2$，那么判定检测单元中的杂波样本与 k_{C1} 个较低有序样本服从不同的分布。删除 k_{C1} 个较低的有序样本，更新背景杂波功率水平估计 $Z(1|k_{C1})=x_{(k_{C1}+1)}$，也就是让 $x_{(k_{C1}+1)}$ 代表杂波功率水平，然后转入第二级删除过程。在第二级删除过程中，首先将样本 $x_{(k_{C1}+2)}$ 与 $T(1|k_{C1})Z(1|k_{C1})$进行比较。其中，$T(1|k_{C1})$是 k_{C1} 个较低的有序样本被删除后的第二级删除过程第一步的标称化因子。第二级删除算法与 ACMLD 相同，也就是在第 k 步进行下面的假设检验

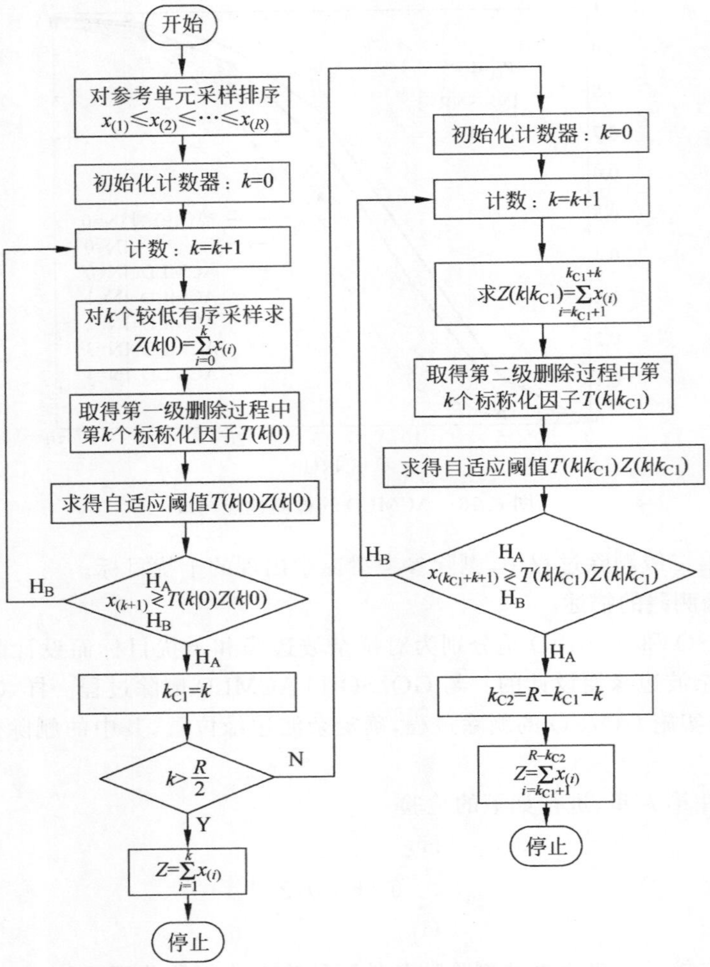

图 6.17 GLT-CMLD 删除过程流程图

$$x_{(k_{C1}+k+1)}\underset{H_B}{\overset{H_A}{\gtrless}}T(k\mid k_{C1})Z(k\mid k_{C1})\tag{6.51}$$

其中，$Z(k|k_{C1})$是第k步的背景噪声功率水平，且有

$$Z(k\mid k_{C1})=\sum_{i=k_{C1}+1}^{k_{C1}+k}x_{(i)}\tag{6.52}$$

如果H_B成立，那么判定$x_{(k_{C1}+k+1)}$与$x_{(k_{C1}+1)},x_{(k_{C1}+2)},\cdots,x_{(k_{C1}+k)}$来自于同一分布，更新背景噪声功率水平估计，然后进行下一步假设检验。如果H_A成立，那么删除$k_{C2}=R-k_{C1}-k$个较高的有序样本并停止删除过程。由未被删除的有序样本形成Z，即

$$Z=\sum_{i=k_{C1}+1}^{R-k_{C2}}x_{(i)}\tag{6.53}$$

最后判决检测单元中是否存在目标

$$D \underset{H_0}{\overset{H_1}{\gtrless}} TZ \tag{6.54}$$

其中，H_1 代表检测单元中有目标，H_0 代表没有目标。

2. GTL-CMLD 检测器的性能分析

在上述第二级删除过程中，式(6.51)的检验与如下的检验是等价的

$$D(k \mid k_{C1}) = x_{(k_{C1}+k+1)} - T(k \mid k_{C1}) Z(k \mid k_{C1}) \underset{H_B}{\overset{H_A}{\gtrless}} 0 \tag{6.55}$$

第二级删除过程中第 k 步的错误删除概率 P_{FC} 为

$$\begin{aligned} P_{FC} &= \Pr\{D(k \mid k_{C1}) > 0 \mid H_B\} \\ &= \frac{R!}{k_{C1}!(R-k_{C1}-k)!(k-1)!} \frac{1}{[1+(R-k_{C1}-k)T(k \mid k_{C1})]^{k-1}} \times \\ &\quad \sum_{i=0}^{k_{C1}} \left\{ \binom{k_{C1}}{i} \frac{(-1)^i}{i+k[1+(R-k_{C1}-k)T(k \mid k_{C1})]} \right\} \end{aligned} \tag{6.56}$$

每步的 $T(k|k_{C1})$ 由式(6.56)迭代计算得出。对于 $P_{FC}=10^{-4}$，$R=16$，$T(k|k_{C1})$ 值在表 6.6 中给出。

表 6.6 GTL-CMLD 删除过程的标称化因子 T

k_{C1}	k														
	1	2	3	4	5	6	7	8	9	10	11	12	13	14	15
0	1066 6.6	78.15	13.50	5.360	2.976	1.976	1.463	1.165	0.980	0.862	0.790	0.757	0.768	0.859	1.223
1	110.0	15.50	5.730	3.110	2.009	1.495	1.182	0.991	0.869	0.795	0.761	0.771	0.863	1.227	—
2	24.50	7.030	3.480	2.188	1.570	1.228	1.020	0.890	0.810	0.773	0.782	0.872	1.238	—	—
3	11.72	1.430	2.540	1.730	1.310	1.070	0.920	0.835	0.793	0.797	0.887	1.258	—	—	—
4	7.707	3.301	2.030	1.463	1.162	0.980	0.878	0.823	0.826	0.912	1.286	—	—	—	—
5	5.860	2.750	1.750	1.312	1.073	0.938	0.870	0.961	0.945	1.325	—	—	—	—	—
6	4.875	2.360	1.583	1.223	1.033	0.932	0.911	0.990	1.374	—	—	—	—	—	—
7	4.301	2.150	1.484	1.182	1.035	0.980	1.050	1.450	—	—	—	—	—	—	—
8	3.950	2.020	1.445	1.185	1.095	1.143	1.545	—	—	—	—	—	—	—	—

总的虚警概率为

$$\begin{aligned} P_{fa} &= \frac{R!}{k_{C1}!k_{C2}!} \prod_{i=2}^{R-k_{C1}-k_{C1}} \frac{1}{(R-k_{C1}-k_{C2}-i+1)T+(R-k_{C1}-i+1)} \times \\ &\quad \sum_{j=0}^{k_{C1}} \left[\binom{k_{C1}}{j} (-1)^j \frac{1}{(R-k_{C1}-k_{C2})T+(R-k_{C1}+j)} \right] \end{aligned} \tag{6.57}$$

其中，$R=16$，预设的 $P'_{fa}=10^{-4}$ 且 $P_{FC}=10^{-4}$ 的标称化因子 $T(k_{C1},k_{C2})$ 由表 6.7 给出。

表 6.7 GTL-CMLD 的标称化因子 $T(k_{C1},k_{C2})$

k_{C2}	k_{C1}							
	0	1	2	3	4	5	6	7
0	0.778	1.010	1.280	1.630	2.080	2.700	3.550	4.830
1	0.781	1.016	1.295	1.645	2.104	2.730	3.610	4.910
2	0.787	1.028	1.310	1.668	2.142	2.790	3.710	5.070
3	0.798	1.044	1.335	1.710	2.205	2.890	3.875	5.337
4	0.813	1.065	1.375	1.770	2.300	3.045	4.108	5.852
5	0.834	1.102	1.425	1.850	2.425	3.250	4.565	6.680
6	0.860	1.148	1.502	1.975	2.640	3.630	5.240	8.130
7	0.890	1.210	1.605	2.150	2.938	4.205	6.445	11.18
8	0.945	1.290	1.750	2.405	3.420	5.200	8.920	20.65

当检测单元处于强杂波区的杂波边缘环境中时，在 $P'_{fa}=10^{-4}$，$R=16$，杂波区域 1 的长度为 10 的条件下，图 6.18 比较 TM(r_1,4)和 GTL-CMLD 的虚警控制能力。除 CNR=8～16dB 的区间外，GTL-CMLD 的虚警性能优于所有的 TM(r_1,4)($r_1=0,3,6,9$)。在与图 6.18 相同的条件下，图 6.19 给出了在 CNR=20dB 的杂波边缘环境中有一个干扰目标(处于弱杂波区)时的虚警概率与干杂比的关系，图 6.19 中假设了干扰目标小于或等于强杂波平均功率水平，即 INR≤CNR。由图 6.19 可知，当干杂比很小时，P_{fa} 与没有干扰目标时的 P_{fa} 几乎相同。随着 INR 增加，干扰目标幅度与强杂波区样本差别减小，杂波边缘变得模糊，导致虚警率增加。当 INR=CNR 时，随着弱杂波样本数减少，对弱杂波功率水平估计的起伏增加，因此对杂波边缘的检测概率减小，P_{fa} 也比没有干扰目标时的高。

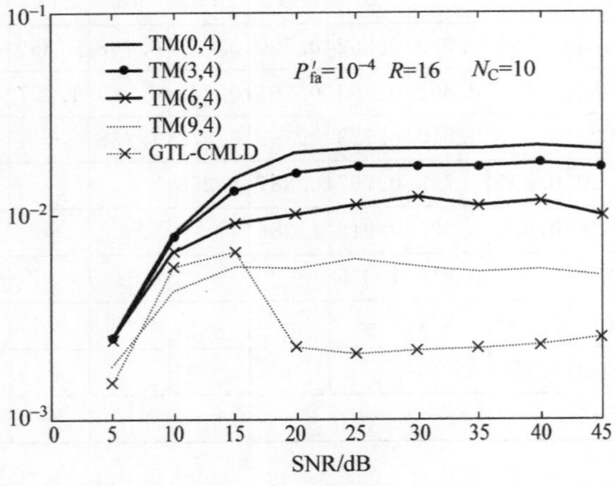

图 6.18 TM(r_1,4)和 GTL-CMLD 的 P_{fa} 与 CNR 的关系曲线

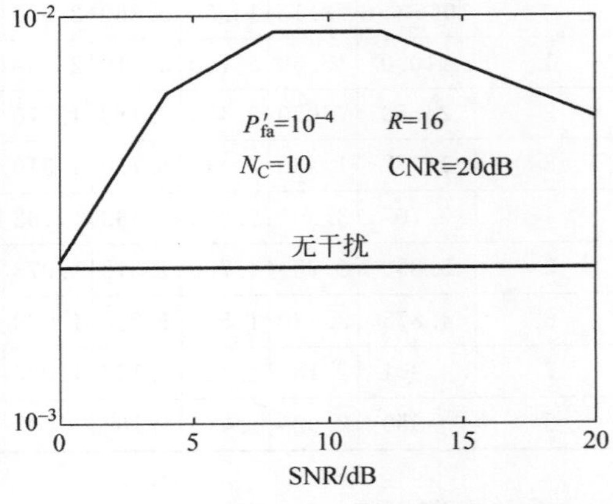

图 6.19 GTL-CMLD 的 P_{fa} 与 INR 的关系曲线

当检测单元处于弱杂波区时，只有当包含干扰的参考单元数小于 $R/2$ 时，GTL-CMLD 才能获得较优的检测性能。否则，删除算法将判定检测单元处于强杂波区，从而过高地估计了检测单元杂波功率水平，导致检测性能严重下降。

在存在 4 个干扰目标和 $R=16$ 的条件下，TM(r_1,4)具有较优的检测性能，而 TM(r_1,2)受到了遮蔽效应的影响严重，这是由与干扰环境有关的错误先验信息导致的。然而，GTL-CMLD 不受遮蔽效应的影响，它的性能是较优的。当干扰目标同时分布在强弱杂波区时，对于不同的边缘杂波，GTL-CMLD的检测性能虽然随杂噪比增加会有所下降，但是总体性能还是较优的。

6.7.5 ACGO-CFAR 检测器

1. ACGO-CFAR 检测器描述

ACGO(Adaptive Censored Greatest Of)-CFAR 检测器是由 Himonas 在文献[13]中提出的，它利用了与 GCMLD 相同的删除算法。设 $x_{(i)}$ 和 $x_{(n+i)}$ $(i=1,2,\cdots,n)$ 分别是前、后滑窗中经过排序的样本值，则 ACGO 删除算法的假设检验为

$$x_{(k+1)} \underset{\mathrm{H}_n}{\overset{\mathrm{H}_i}{\gtrless}} T_k Z_k \tag{6.58}$$

与 GCMLD 不同的是，ACGO 选取前后参考滑窗的局部估计 X 和 Y 中的较大者，作为检测单元杂波功率水平估计，即

$$Z=\max(X,Y) \tag{6.59}$$

其中

$$X=\frac{1}{r_1}\sum_{i=1}^{r_1}x_{(i)},\quad Y=\frac{1}{r_2}\sum_{i=1}^{r_2}x_{(n+i)} \tag{6.60}$$

其中，r_1 和 r_2 是利用删除算法分别在前后滑窗参考样本中确定的删除点。删除过程第 k 步的错误删除概率为

$$P_{\mathrm{FC}}=\binom{n}{k}\frac{1}{[1+T_k(n-k)]^k} \tag{6.61}$$

式(6.61)与式(6.34)中 GCMLD 的 P_{FC} 表达式不同，它是在没有干扰目标条件下，即相当于干杂比趋于零的条件下得到的，而式(6.34)是在干杂比为无限大的假设下得到的。换句话说，式(6.34)是 $k-1$ 个有序样本对应平均功率非常大的回波时在第 $n-k$ 步被错误删除的概率。然而此处的 P_{FC} 是在第 k 步中当所有 $n-k$ 个较高有序样本对应不包括干扰的回波时被全部错误地删除的概率。因此，为得到式(6.61)，假设所有参考单元样本都是 IID 的。总的虚警概率为

$$P_{\mathrm{fa}}=\sum_{r_1=1}^{n}\sum_{r_2=1}^{n}P_{\mathrm{fa}}(r_1,r_2)\Pr\{r_1,r_2\} \tag{6.62}$$

其中，$P_{\mathrm{fa}}(r_1,r_2)$表示前、后滑窗中删除点分别为 r_1 和 r_2 时的条件虚警概率；r_1 和 r_2 是随机变量，它们的联合概率分布函数为 $\Pr\{r_1,r_2\}$，且有

$$\Pr\{r_1,r_2\}=\Pr\{r_1\}\Pr\{r_2\} \tag{6.63}$$

P_{fa} 和 $\Pr\{r_1,r_2\}$都依赖于特定的背景环境，求它们的解析表达式是很困难的。因此，需要利用计算机仿真来分析 ACGO 的虚警控制性能。

2. ACGO-CFAR 检测器的性能分析

在 ACGO-CFAR 检测器设计中，P_{FC} 是一个重要参数。当 $P_{\mathrm{FC}}>P'_{\mathrm{fa}}$时，会产生过度删除样本的结果，导致 P_{fa} 上升。当 $P_{\mathrm{FC}}<P'_{\mathrm{fa}}$时，删除的样本不足，并且干扰目标被删除的概率较低，因此影响了检测性能。为了避免这种不稳定性，可采用 $P_{\mathrm{FC}}=P'_{\mathrm{fa}}$。当 $P_{\mathrm{FC}}=10^{-4}$ 时，实际虚警概率比设计值大约高一个数量级[13]。

在不包含干扰目标的杂波边缘环境中,当杂噪比较小时,ACGO 的虚警控制能力与 GO 相同,并且优于 TM。随着杂噪比增加,GO 的虚警控制能力要比 ACGO 和 TM 好。在 N_C 值的某个范围内,ACGO 优于 TM,而在另一范围内,TM 优于 ACGO。与 GCMLD 相比,ACGO 也表现出了良好的虚警控制能力。从检测性能方面来看,在均匀杂波背景中,ACGO 与 GO 的检测性能相同,CFAR 损失在 0.1～0.3dB($R=16$,$P_{FC}=P'_{fa}=10^{-4}$)[13]。

干扰目标均匀分布于前后两个滑窗时,ACGO 的检测性能较好,而干扰目标集中在前、后沿滑窗的一个滑窗中时,ACGO 的检测损失较大。这是由于在相同干杂比条件下,在若干个样本中,删除较少的干扰目标的概率大于删除较多的干扰目标的概率。TM 的 P_d 不依赖干扰目标分布,而依赖于事先指定的删除点。与 TM 不同,从不受遮蔽效应影响的角度来说,ACGO 在几种干扰目标分布情况下都是稳健的[13]。

在杂波边缘和干扰目标同时存在的环境中,检测单元处于强杂波区时,ACGO 也表现出了稳健性。当检测单元处于弱杂波区,且 CNR=30dB,前沿滑窗中不包含干扰目标时,ACGO 具有最优的性能,最大 CFAR 损失为 2.5dB;当前沿滑窗中有 1 个干扰目标时,损失为 4dB[13]。较小的杂噪比对应了删除概率的下降,这将导致检测损失增加。

6.8 VI-CFAR 检测器

VI(Variability Index)-CFAR 检测器[14],即变化指数恒虚警检测器,是由 Michael 等提出的。它是基于 CA-CFAR、SO-CFAR、GO-CFAR 的一种综合检测方法,该算法能通过计算参考单元的二阶统计量(变化指数 V_{VI})及前后滑窗均值之比,动态地调整杂波功率水平的估计方法。VI-CFAR 的框图如图 6.20 所示。

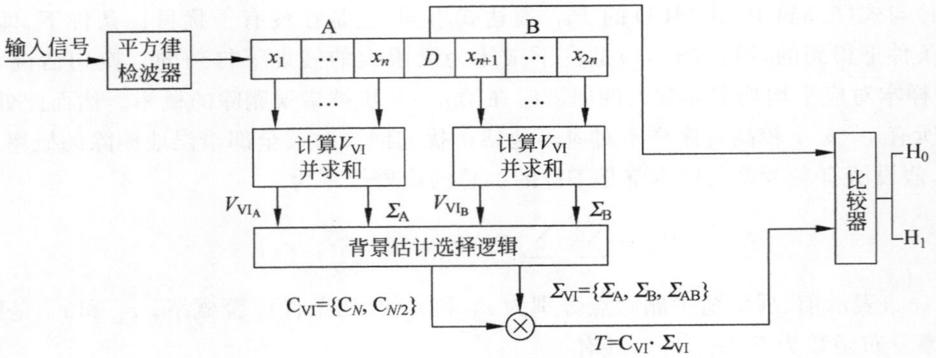

图 6.20 VI-CFAR 检测器的框图

V_{VI} 是一个二阶统计量,与形状参数的估计非常相似。对于每个滑窗(A 或 B)

$$V_{VI}=1+\frac{\hat{\sigma}^2}{\hat{\mu}^2}=1+\frac{1}{n-1}\sum_{i=1}^{n}\frac{(x_i-\bar{x})^2}{(\bar{x})^2} \tag{6.64}$$

式中,$\hat{\sigma}^2$ 是方差的估计值,$\hat{\mu}^2$ 是均值的估计值,$\bar{x}$ 是 n 个参考距离单元样本的算术平均值。通过对 V_{VI} 与门限 K_{VI} 比较

$$V_{VI}\leqslant K_{VI}\Rightarrow \text{均匀杂波}$$

$$V_{\mathrm{VI}} > K_{\mathrm{VI}} \Rightarrow \text{非均匀杂波} \tag{6.65}$$

可以判别 V_{VI} 是来自均匀杂波还是非均匀杂波。

前滑窗 A 和后滑窗 B 的均值比 V_{MR} 定义为

$$V_{\mathrm{MR}} = \bar{x}_{\mathrm{A}} / \bar{x}_{\mathrm{B}} = \sum_{i \in \mathrm{A}} x_i \Big/ \sum_{i \in \mathrm{B}} x_i \tag{6.66}$$

其中，$\bar{x}_{\mathrm{A}}$ 表示前滑窗的均值，$\bar{x}_{\mathrm{B}}$ 表示后滑窗的均值。通过将 V_{MR} 与门限 K_{MR} 进行比较

$$\begin{gathered} K_{\mathrm{MR}}^{-1} \leqslant V_{\mathrm{MR}} \leqslant K_{\mathrm{MR}} \Rightarrow \text{均值相同} \\ V_{\mathrm{MR}} < K_{\mathrm{MR}}^{-1} \text{ 或 } V_{\mathrm{MR}} > K_{\mathrm{MR}} \Rightarrow \text{均值不同} \end{gathered} \tag{6.67}$$

可以确定前后滑窗的均值是否相同。而 VI-CFAR 的检测门限是根据 VI 和 MR 假设检验的结果确定的，确定方法如表 6.8 所示。其中，背景乘积常数 C_N 或 $C_{N/2}$ 中的 N 表示参考单元数目 $2n$。当两个滑窗都用上时，采用 C_N，当只利用前滑窗或后滑窗时，采用 $C_{N/2}$。

表 6.8　自适应门限生成方法

序号	前滑窗杂波是否非均匀	后滑窗杂波是否非均匀	均值是否有差异	VI-CFAR 自适应门限	等价的 CFAR 处理方法
1	否	否	否	$C_N \Sigma_{\mathrm{AB}}$	CA-CFAR
2	否	否	是	$C_{N/2} \max(\Sigma_{\mathrm{A}}, \Sigma_{\mathrm{B}})$	GO-CFAR
3	是	否	—	$C_{N/2} \Sigma_{\mathrm{B}}$	CA-CFAR
4	否	是	—	$C_{N/2} \Sigma_{\mathrm{A}}$	CA-CFAR
5	是	是	—	$C_{N/2} \min(\Sigma_{\mathrm{A}}, \Sigma_{\mathrm{B}})$	SO-CFAR

6.8.1　VI-CFAR 检测器在不同背景中的应用

1. 均匀环境

假设 K_{VI} 和 K_{MR} 已经确定，杂波为均匀杂波，即前滑窗和后滑窗都是非易变的，并且具有相同的均值，这相当于表 6.8 中的第一种情况。此时 VI-CFAR 检测性能近似于 CA-CFAR，具有较少的 CFAR 损失。

2. 多目标环境

当有一个或多个目标存在于前滑窗或后滑窗时，相应的 VI 表明它是一个非均匀的杂波背景。当这种情况只出现在一个滑窗内时，则用另一个滑窗的数据作为 CA-CFAR 算法的背景估计，这是表 6.8 中的第 3 和第 4 种情况。当所选滑窗的杂波均匀时，则存在轻微的检测性能损失，这是因为只采用了 $N/2$ 个距离单元作为背景的估计，而没有用所有 N 个距离单元。

如果 VI 表明两个滑窗都是非均匀的，则选择较小的滑窗均值作为背景的估计，相当于 SO-CFAR 方法，这就是表 6.8 中的第 5 种情况。

3. 杂波边缘环境

杂波边缘位置及待检测单元在弱杂波区域还是在强杂波区域都是无法先验已知的。当杂波最初进入滑窗 A 的参考单元时，有一个或多个单元包含较高功率水平的杂波，这种情况与存在一个干扰目标的情况非常相似，从而可以采用滑窗 B 作为背景的估计。当杂波占据整个滑窗 A 时，包含杂波的滑窗

和只有噪声的滑窗相似，也是均匀的，但是两个滑窗的均值不同。为了不产生过多的虚警，选择较大的均值作为背景估计。这就是表 6.8 中的第 2 种情况。当杂波继续向前进入滑窗 B 时，滑窗 B 就变成非均匀背景，而滑窗 A 包含均匀的杂波。对于这种情况，选择滑窗 A 估计背景杂波，相当于表 6.8 中的第 3 和第 4 种情况。最后，当两个滑窗都充满杂波时，每个都似乎是均匀环境，这时选用表 6.8 中的第 1 种算法。

6.8.2 VI-CFAR 检测器的性能分析

图 6.21～图 6.23 分别为均匀环境、存在一个干扰目标的环境和杂波边缘环境下不同检测器的性能曲线。由图 6.21 可以看出，在均匀环境下，所有的 CFAR 检测器性能相似，但与最优检测器相比，都存在一定的 CFAR 损失，VI-CFAR 与 CA-CFAR 和 GO-CFAR 相比要稍差一些，但要优于 OS-CFAR 和 SO-CFAR。由图 6.22 可以看出，当参考单元中存在一个干扰目标时，CA-CFAR 和 GO-CFAR 性能明显变差，而 VI-CFAR、OS-CFAR 和 SO-CFAR 几乎不受影响。图 6.23 给出了 CA-CFAR、OS-CFAR、GO-CFAR、SO-CFAR 及 VI-CFAR 在杂波边缘环境下的虚警概率。由图 6.23 可以看出，在杂波边缘环境中，VI-CFAR 具有较优的虚警控制能力。

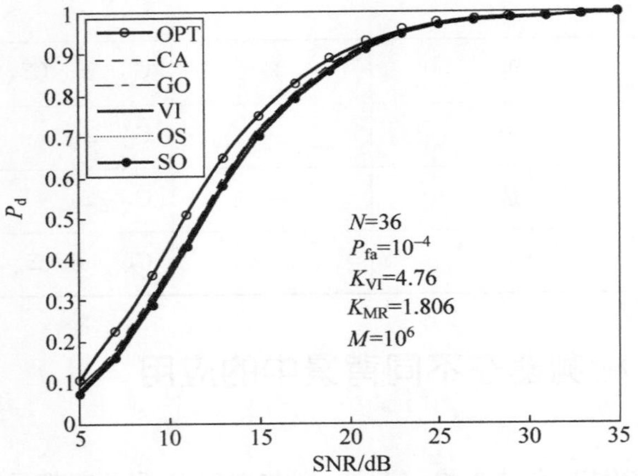

图 6.21　均匀环境下 CA、OS、GO、SO 和 VI-CFAR 的检测性能比较

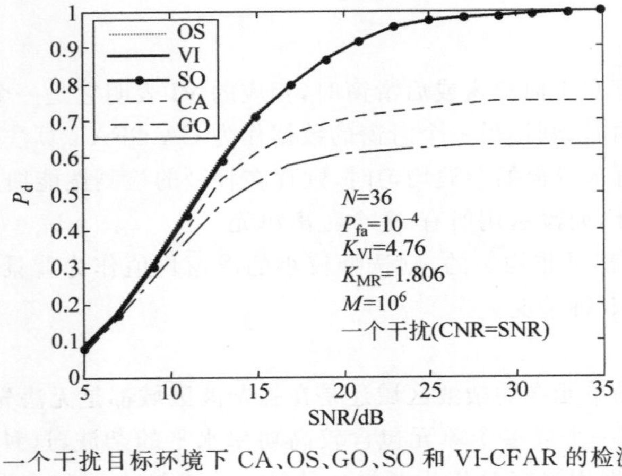

图 6.22　一个干扰目标环境下 CA、OS、GO、SO 和 VI-CFAR 的检测性能比较

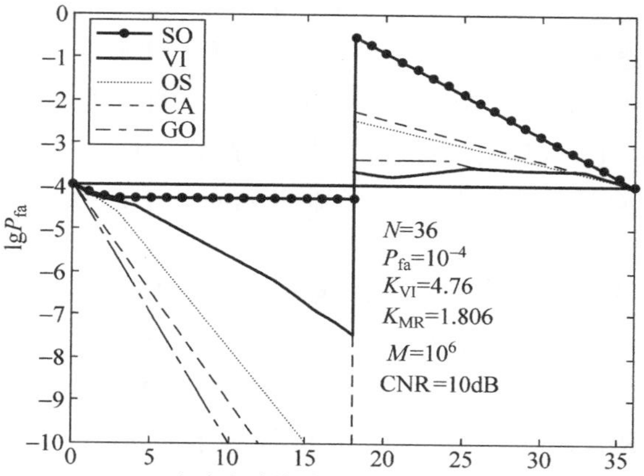

图 6.23 杂波边缘环境下(CNR=10dB)CA、OS、GO、SO 和 VI-CFAR 的虚警概率比较

总体来说，VI-CFAR 检测器综合了 CA-CFAR、GO-CFAR 及 SO-CFAR 的优点。在均匀环境中具有较低的 CFAR 损失，对于存在干扰和杂波边缘的非均匀背景，具有一定的鲁棒性。另外，VI-CFAR 的复杂性要低于 OS-CFAR，并且在杂波边缘中具有更优的虚警控制能力，其虚警性能也要优于其他自适应检测器，如 AOS(Adaptive Order Statistic)-CFAR 检测器[15]和 ET(Estimation Test)-CFAR 检测器[16]。但是当两个滑窗中都存在干扰时，VI-CFAR 的检测性能有明显下降，与 SO-CFAR 性能接近。文献[17]对 VI-CFAR 进行了改进，使其能在 CA-CFAR、改进的 CA-CFAR[18]及 S-CFAR[19]之间进行选择，而文献[20-21]的改进使其能在 CA-CFAR、GO-CFAR、SO-CFAR 及 OS-CFAR 之间进行选择。文献[22]则通过评估相邻两样本之间的差异情况来确定选择 CA-CFAR 还是 OS-CFAR。

6.9 基于回波形状信息的删除单元平均 CFAR 检测器

CFAR 检测方法在形成检测门限时一般包括两个步骤[23]，一是估计背景功率水平，二是计算门限因子，且这两个步骤都很大程度地依赖于对背景杂波类型的假设。其中，在 CFAR 要求下，门限因子的计算依赖于对杂波统计分布类型的假设，在这里假设杂波的统计分布类型为指数分布。在估计背景功率水平时，传统 CFAR 检测方法总是基于背景类型(背景类型分为三类：均匀背景、杂波边缘背景和多目标环境)的某个假设来获取足够的独立同分布样本[23]。例如，常用的 CA-CFAR 方法基于均匀背景假设，相应的检测单元背景功率水平是利用邻近距离单元的样本均值来估计的；GO-CFAR 方法则基于杂波边缘背景假设，相应的检测单元背景功率水平通过选择两侧邻近距离单元样本均值中的较大者来估计；而 SO-CFAR 方法则基于单边有多目标的背景假设，相应的检测单元背景功率水平通过选择两侧邻近距离单元样本均值中的较小者来估计。然而，在实际雷达工作环境中，只有单一背景类型的情况是很难出现的，更常见的是同时由地面、海面、建筑物、树林、湿地、湖泊、其他目标、强散射点距离旁瓣等形成的、涵盖三类背景类型的复杂非均匀环境。这种复杂非均匀环境使得基于单一背景类型假设而设计的 CFAR 检测器难以获得足够的独立同分布样本来进行背景功率水平估计，同时保护距离单元数和参考距离单元数的设置往往面临着两难问题。

针对上述问题,本章讨论了多种自适应 CFAR 检测器,其通常的解决办法是,设计复杂的样本删除逻辑、CFAR 算法选择逻辑及算法参数调整方案来适应实际的复杂非均匀环境。这些方法在一定程度上改善了 CFAR 检测器在复杂非均匀环境中的检测性能,但还存在如下一些问题:①样本删除逻辑对背景类型仍存在一定的依赖性;②样本删除逻辑只利用了回波幅度信息;③部分算法复杂度高,特别是多次删除过程中的门限因子难以确定;④部分样本删除逻辑仍存在较强的主观性。

本节针对常规低分辨率雷达海面目标 CFAR 检测问题,提供一种环境适应能力强、可工程实现、易操作的背景功率水平估计方法,进而结合门限因子的计算形成有效的检测门限,得到的检测器称为基于回波形状信息的删除单元平均 CFAR 检测器[24],简写为 ESECA(Envelope Shape-based Excision Cell Average)-CFAR 检测器,相应的二元假设检验的形式与式(6.54)相同,不同之处是,ESECA-CFAR 检测器中的 Z 是通过 6.9.2 节所述的基于回波形状信息的删除单元平均方法来获得的。

6.9.1 基于回波形状信息的删除单元平均方法

基于回波形状信息的删除单元平均方法[24]的主要工作是解决复杂非均匀背景下估计背景功率水平的参考单元选取问题。

该类方法的出发点是,如果一个点是目标点,那么它至少应该具备如下两个特征:①它应该属于某个凸起;②在一定的检测概率和虚警概率要求下,它所在的那个突起应当高于两侧距离单元的背景均值。下面从这两个特征出发细化具体的操作。

第一个特征是从目标的回波形状进行考虑的。经过匹配滤波、杂波抑制、平方律检波等一系列处理后,在进行 CFAR 处理时,距离维上,目标的包络形状往往具有图 6.24 中右侧椭圆所示的形状,这里将这种类似钟形的形状称为凸起。这种凸起的规则性取决于目标的强弱、目标尺寸的大小、目标周围的环境、雷达的距离分辨率、雷达的距离采样率等因素。在雷达参数都确定的条件下,通过分析大量的实测回波数据,来总结目标回波包络的特点。目标回波包络的特点掌握得越准确、越精细,则越有利于抑制虚警。这部分研究是开放的。

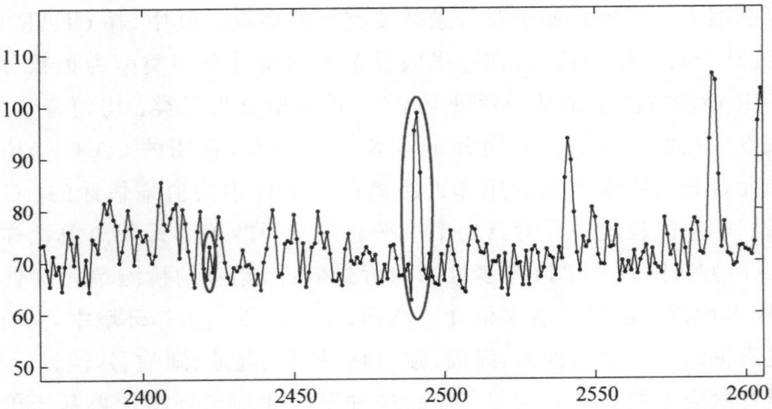

图 6.24 疑似目标的示意图

当受限于数据量较小,不能准确掌握典型目标的包络特征,而无法总结出较精细的规律时,这里采用了一种较粗的描述方式:"在距离维上,极大值点左右往下,至少一边有 3 个点(含极大值点),则认为是可疑目标",如图 6.24 中右侧椭圆指示的部分。

针对上述描述方式,有如下几点讨论。

(1) 由于此类方法需要考虑目标距离维回波包络的形状,因此需要通过预处理来消除毛刺对目标回波包络形状的影响。毛刺的产生来源可能包括噪声、杂波及采样等因素,预处理方法主要是距离维平滑。此处,距离维平滑滤波器设置为[0.3,0.4,0.3]。后续的背景均值估计过程都是在平滑后的距离维数据上进行的。

(2) 极大值点左右往下,两边含两个点的情况,就不认为是目标(如图 6.24 中左侧椭圆指示的部分),原因在于,在当前给定的雷达参数条件下,这种情况下的极大值点不会具有较高的幅值。

(3) 这种描述方式考虑了某些大目标距离维回波包络顶部出现凹口的情况,以及两个目标相距较近使得回波连在一起导致其顶部也出现类似凹口的形状,其凹口深度也许只下降 1 个点,如图 6.25 中椭圆所指示的部分。

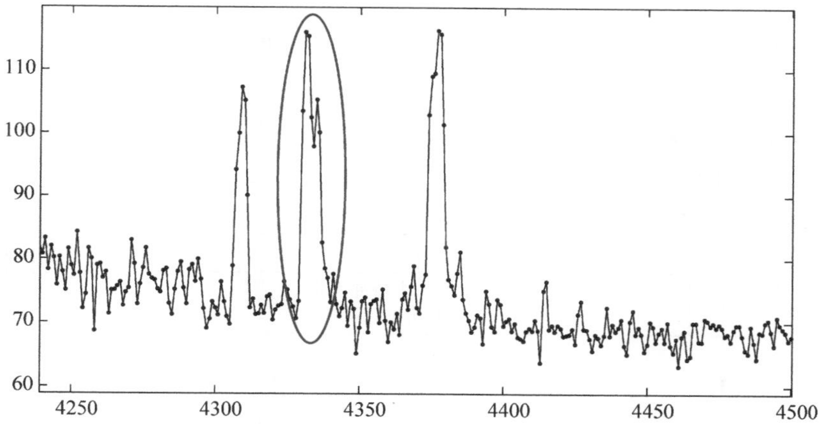

图 6.25　目标顶部出现凹口的示意图

(4) 根据这种描述,在后续估计背景功率水平时,就不需要考虑保护单元的设置问题,同时也为有效参考单元的选取设定了参照物。具体做法是:首先分别计算可疑目标左右两边极大值与极小值的均值(极大值与极小值之间 1/2 的位置,这一点也是第二个特征的具体体现。既然要求凸起比较明显的高于周边,则可认为周边的大部分背景值都应在凸起的一半之下),分别记为左删除门限和右删除门限;然后从极小值往左数 N 个参考单元,取这些参考单元值小于左删除门限的单元计算均值,记为左均值;同样的做法计算右均值。

若想进一步压制邻近距离单元中强点干扰的影响,在计算左删除门限和右删除门限时,还可取极大值与极小值之间 1/3 的位置。

显然,上述做法可降低多目标干扰的影响。

(5) 根据左均值和右均值的大小,可判断目标是在均匀杂波区还是在杂波边缘区,或是在大目标脉压旁瓣的"肩膀"位置,如图 6.26 中椭圆所指示的部分。具体做法是:如果左右两边的均值相差小于 V_T,则认为左右两边是均匀背景,二者的均值就作为当前目标单元的背景功率水平估计;如果左右两边

的均值相差大于 V_T，则认为左右两边是杂波边缘，取二者的大者作为当前目标单元的背景功率水平估计。

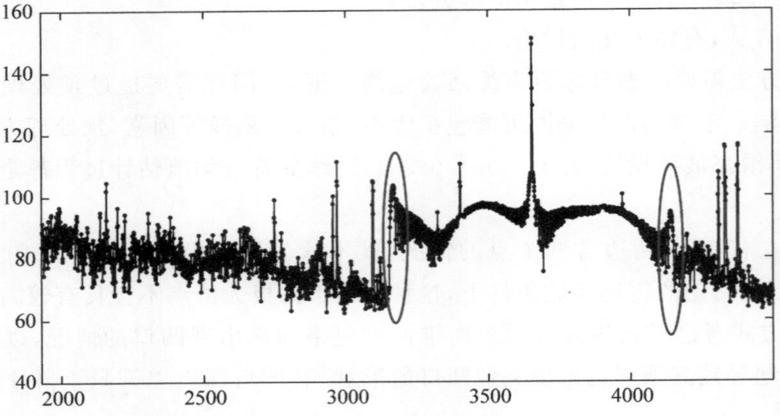

图 6.26　大目标脉压旁瓣的"肩膀"问题

基于上述讨论，图 6.27 给出了 ESECA-CFAR 检测器的处理框图。

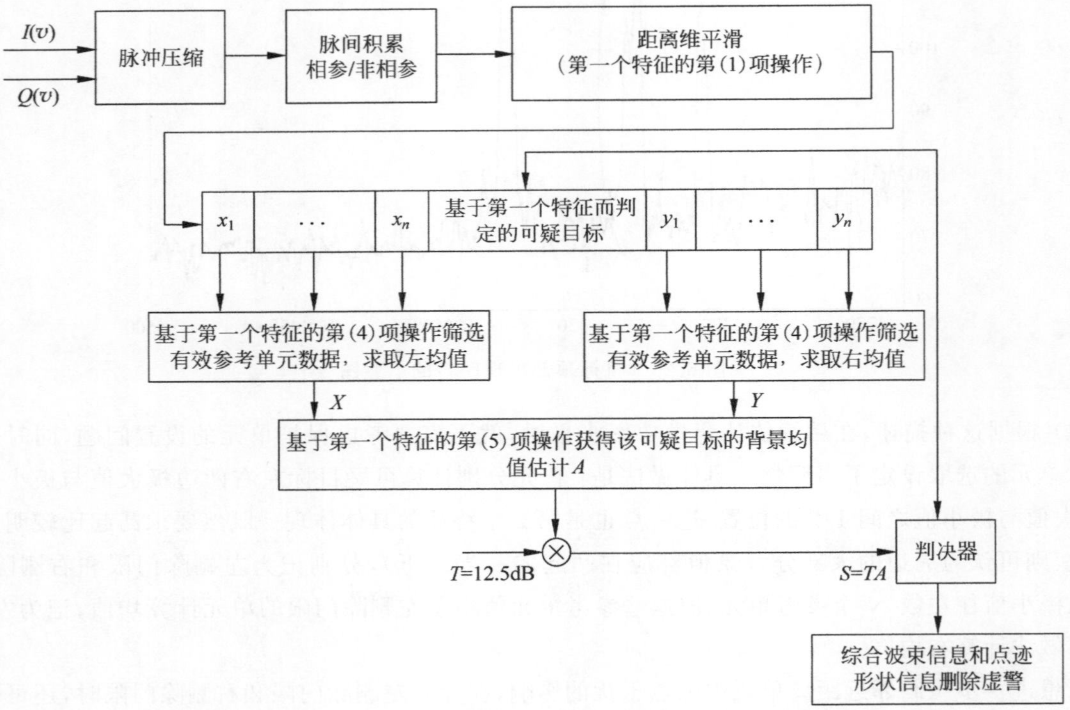

图 6.27　利用回波形状信息的单参数的自适应 CFAR 检测方法的处理框图

相比于已有的删除类自适应 CFAR 方法[23]，ESECA-CFAR 检测器具备以下特点。

(1) 疑似目标的判定是一个开放的问题，可根据对雷达回波数据的仔细分析给出更准确、更精细的描述，进而较大限度地提高检测性能。而本节中给出的判定规则是很粗略的，可适用于几乎所有雷达类型。

(2) 利用疑似目标这一参照物，可不用设置保护单元，同时也给有效参考单元选择规则及杂波边缘识别规则的设计带来了便利。

(3) 相比于常规 CFAR 方法，本方法带来的性能改善体现在对复杂非均匀背景的适应能力方面，而在均匀背景中，其性能与 CA-CFAR 相当。

6.9.2 检测性能仿真分析

下面对均匀背景和非均匀背景下的 CA-CFAR、GO-CFAR、ESECA-CFAR 检测器的性能进行对比分析，背景杂波服从指数分布。

图 6.28 给出了虚警概率为 10^{-4} 时，均匀背景条件下 CA-CFAR 检测器、GO-CFAR 检测器与 ESECA-CFAR 检测器的 SCR-P_d 曲线。由图 6.28 可知，在均匀背景下，这三种检测器的检测性能相当。

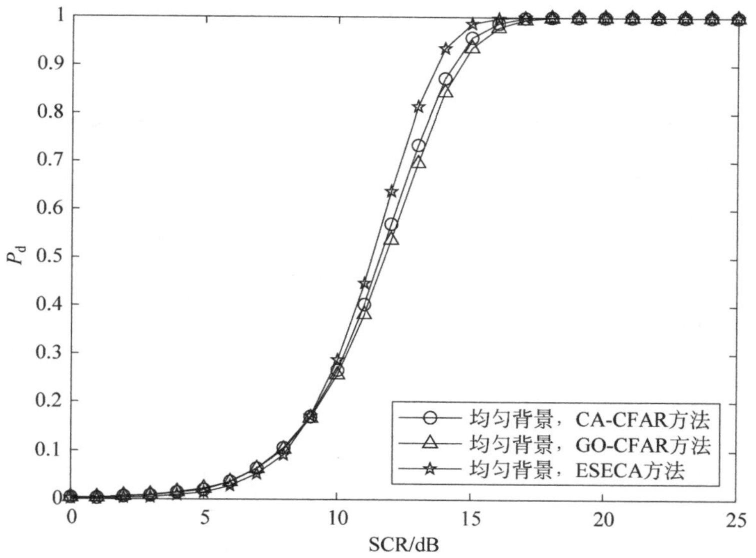

图 6.28 均匀背景下 CA-CFAR、GO-CFAR 与 ESECA-CFAR 检测器的性能对比

图 6.29 给出了非均匀背景条件下 CA-CFAR、GO-CFAR 与 ESECA-CFAR 检测器的 SCR-P_d 曲线，其中非均匀背景是通过修改距离单元中杂波功率水平(设置强弱两种杂波功率水平的差距约为 23dB)和增加干扰目标(在参考距离单元范围内，设置了 4 个干扰目标，其功率水平分别为 2 个 13dB 和 2 个 15dB)两种方式来仿真的。由图 6.29(a)可知，在多目标背景下，ESECA-CFAR 检测器的检测性能优于 CA-CFAR、GO-CFAR 检测器，其优势与设置的干扰目标数量和强度等因素有关，在本仿真条件下，达到了 8dB 以上；而且干扰目标强度越大，ESECA-CFAR 检测器的优势越大。由图 6.29(b)可知，在杂波边缘背景下(目标处在弱杂波区)，ESECA-CFAR 检测器优于 CA-CFAR、GO-CFAR 检测器 2～4dB 以上，其优势与设置的杂波边缘强度对比度以及目标相对于杂波边缘的位置等因素有关；而且杂波边缘强度对比度越大，ESECA-CFAR 检测器的优势越大。

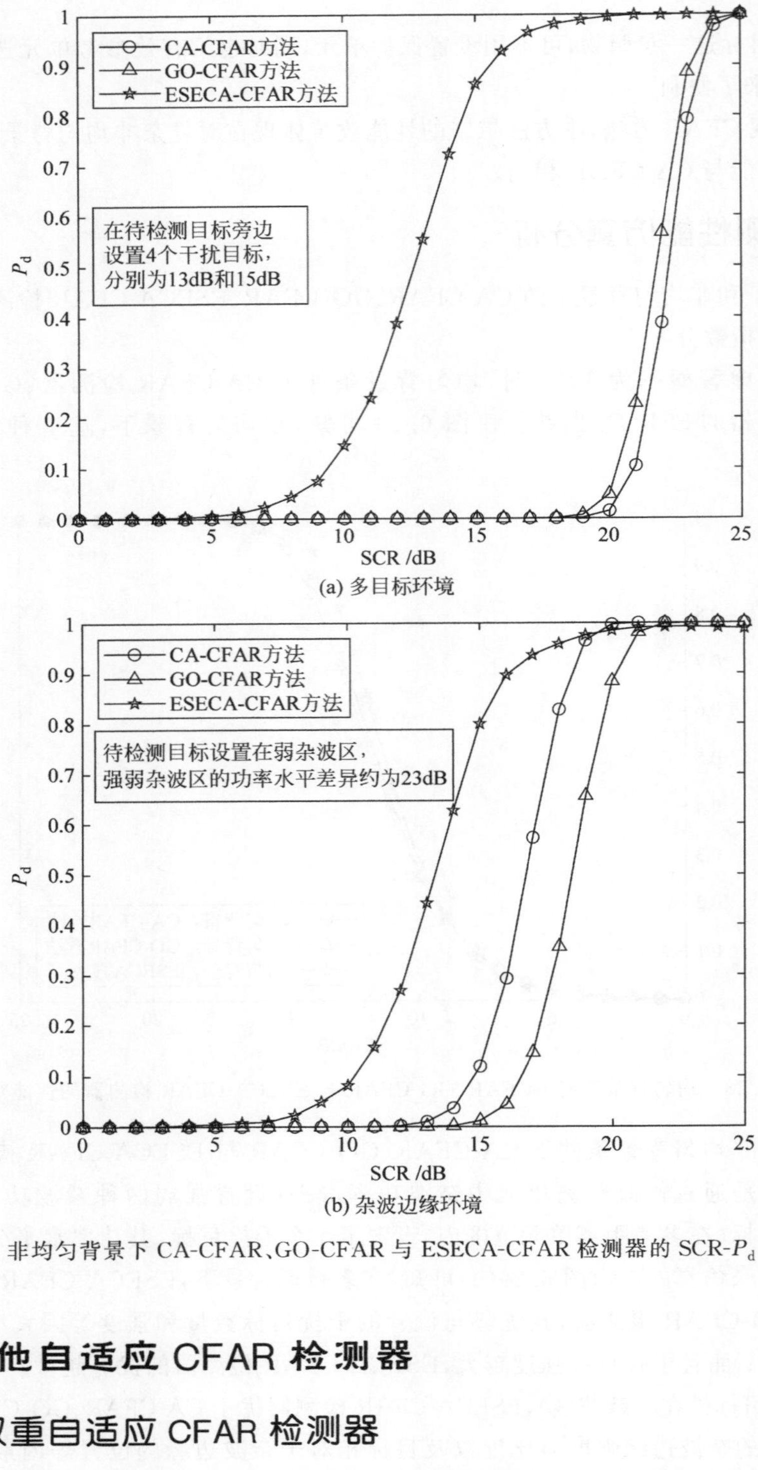

图 6.29 非均匀背景下 CA-CFAR、GO-CFAR 与 ESECA-CFAR 检测器的 SCR-P_d 曲线对比

6.10 其他自适应 CFAR 检测器

6.10.1 双重自适应 CFAR 检测器

Cole 和 Chen 提出的双重自适应 CFAR 检测器[25]框图如图 6.30 所示。它利用了一个与主检测并行的辅助检测来测量背景杂波的非均匀程度，即通过移位操作将背景杂波样本逐一地与 $S_1=T_1Z$

进行比较。大于 T_1Z 计为 1,否则计为 0,然后输入移位寄存器。T_1Z 是较低的阈值,因此通过辅助检测确定的虚警数可以反映背景杂波的分布情况。用辅助检测的虚警数,也就是移位寄存器各单元的和,控制标称化因子 T_2,并且使和值越高,T_2 值越高,这样可以在杂波边缘环境中较好地控制虚警率。移位寄存器位数的选择要适中,过小时不能反映背景杂波分布情况,过大时覆盖的杂波区会过多。

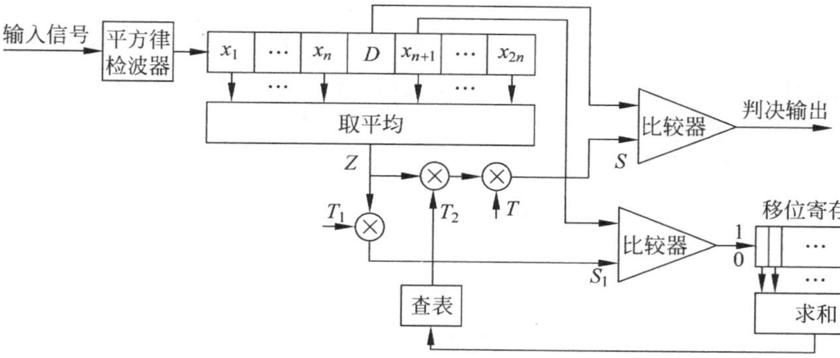

图 6.30 双重自适应 CFAR 检测器框图

6.10.2 AC-CFAR 检测器

如何形成 CFAR 的门限是 CFAR 技术的一个关键。文献[26-27]提出了一种新的形成 Z 的方法,它从两个由均值法产生的局部估计中选出一个与被检测单元样本更相近的估计作为总的杂波功率水平估计,并与标称化因子 T 相乘,由此形成 CFAR 处理的检测阈值。称这种方法为 AC-CFAR 方法(Approach Cell,AC),其框图如图 6.31 所示。

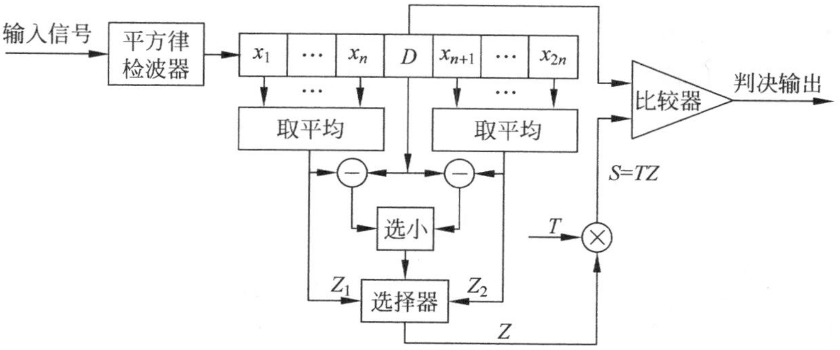

图 6.31 AC-CFAR 检测器框图

就检测损失而言,相比于 OS 和 GO,AC 的损失最小[26-27],其性能优于这两种常见的 CFAR 方法。在杂波边缘环境中,GO 检测不到目标,OS 在滑窗长度 R 较小时尚可检测目标,但是随着 R 的增大,OS 对目标信号的反应越来越不灵敏,从而丢失目标。而 AC-CFAR 方法可以有效地工作。因此,AC-CFAR 方法抗杂波边缘的性能比 GO 和 SO 优越。当两个目标回波信号幅度相差较大时,GO 会造成大目标遮蔽小目标,而 AC 和 OS 不会出现这种情况,它们都能有效地抑制这种遮蔽现象。在多目标环境中,GO 的检测性能很弱,当干杂比较大时,几乎检测不到目标,而 AC 和 OS 都具有很好的抗多目标能力。

6.10.3 改进的 CA-CFAR 检测器

CA-CFAR 检测器实现 CFAR 的前提是杂波环境为均匀背景。当杂波边缘或干扰目标出现在参考滑窗中时，这一前提不再成立，CA 也不再能维持 CFAR 性能。因为在这种情况下，有一侧参考单元的杂波样本值与检测单元的样本值并不来自同一分布，这些单元的样本值并不反映检测单元背景杂波的强度，不能用来计算检测阈值。但另一侧参考单元样本值仍与检测单元处于同一平稳的干扰区域。若能判断出这一侧参考单元并用来计算检测阈值，则在杂波边缘环境中或一侧滑窗内出现干扰目标时，仍能保持较好的 CFAR 性能。基于上述思想，文献[18]提出了一种改进的 CA-CFAR 检测器，如图 6.32 所示。其中，Z_b 是杂波边缘检测门限。对杂波边缘的检测可能出现两种错误，第一种是检测单元处于弱杂波区，而判断为处于强杂波区，第二种错误则相反。Z_b 是根据给定的其中一种错误概率的设计值而预先设定的，通过比较 Z_a 和 Z_b 确定检测单元处于强杂波区还是弱杂波区。

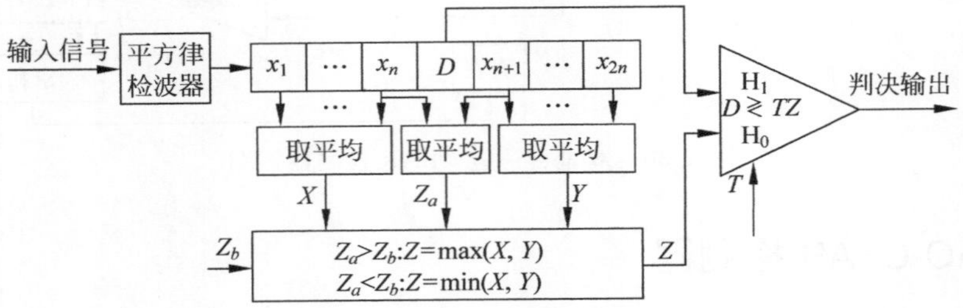

图 6.32 改进的 AC-CFAR 检测器框图

6.10.4 自适应长度 CFAR 检测器

为了对抗边缘杂波和干扰目标，文献[28]中提出了自适应长度 CFAR 检测器。它能够分辨出杂波边缘并且自适应地修正它的滑窗长度，且它能删除不止一个干扰目标。自适应长度 CFAR 检测器，即 AL(Adaptive Length)-CFAR 检测器的框图如图 6.33 所示。

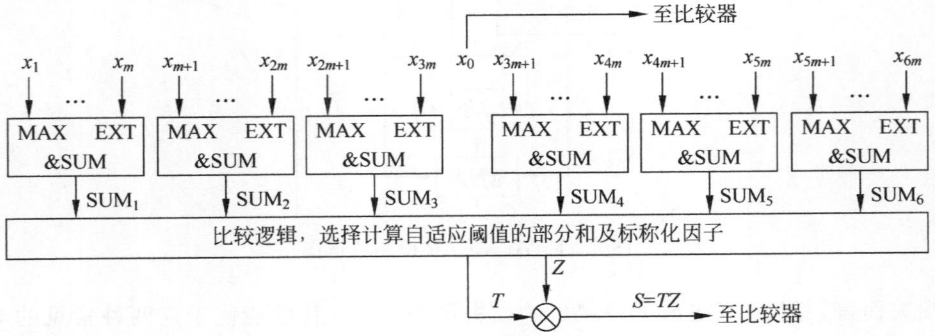

图 6.33 AL-CFAR 检测器的框图

AL-CFAR 检测器处理参考单元样本的方法如下。

(1) 把参考滑窗分成几个部分(在图 6.33 中是 6 个部分)，每部分包含 m 个样本，$m>4$。

(2) 对于每个部分，提取最大的 4 个样本值，以便删除可能的干扰目标，用 MAX EXT 表示；选择的 m 能使每个部分出现多于一个目标的概率达到可以忽略的程度。

(3) 对每部分的剩余样本求和,用 SUM 表示,得到各部分的求和结果 $SUM_1,\cdots,SUM_L$。

(4) 对这些部分和进行一系列的比较,检测杂波边缘和干扰目标。

文献[28]的仿真分析表明,AL 的抗边缘杂波性能和在均匀杂波背景中的检测性能均介于 CA 和 GO 之间,且检测性能更接近 CA,虚警控制能力接近 GO。而在参考滑窗中出现干扰目标时,由于 AL 采用了删除最大四个样本的机制,所以 AL 的检测性能明显优于 CA 和 GO。

6.10.5 ACCA-ODV-CFAR 检测器

基于有序数据可变性(Ordered Data Variability,ODV)的自动删除平均(Automatic Censored Cell Averaging,ACCA)CFAR 检测器是由 Farrouki 等针对非均匀背景环境提出的[29]。其原理是通过一系列假设检验,在排列好的单元中动态选择合适的集合来估计背景功率水平,然后利用可变指数统计量作为形状参数来判断序列单元的取舍。ACCA-ODV-CFAR 的框图如图 6.34 所示。

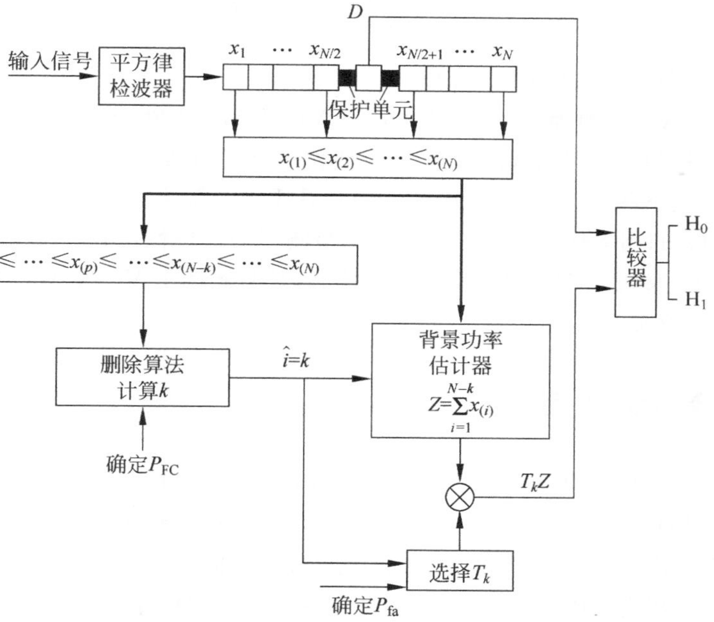

图 6.34 ACCA-ODV-CFAR 检测器框图

对参考单元进行排序,采用删除算法删除最大的 k 个,并将剩下的单元用来形成背景杂波估计,即

$$Z=\sum_{i=1}^{N-k}x_{(i)} \tag{6.68}$$

根据虚警概率 P_{fa} 选择门限因子 T_k,因此目标存在与否的假设检验为

$$D \mathop{\gtrless}_{H_0}^{H_1} T_k Z \tag{6.69}$$

其中的删除算法的基本过程如下。

(1) 选取 $x=x_{(N-k)}$。

(2) 形成有序样本 $E_x=\{x_{(1)},x_{(2)},\cdots,x_{(p)},\cdots\}$。

(3) 基于 E_x 计算 ODV 统计量 V_k，并将其作为形状参数。

(4) 执行基于 ODV 的假设检验(其中假设检验的错误概率预设为 P_{FC})，重复步骤(1)～(4)，其中 $k=0,1,\cdots$，直到 $d_k=0$ 或者 $k=N-p$ 停止。

(5) $\hat{i}=k$ 就是删除单元的数目估计。

一旦删除单元的数目估计 $\hat{i}$ 确定，那么就用$(N-\hat{i})$个最小的样本来估计统计量 Z。在均匀环境中，从 N 个样本中删除最大的 k 个时，Swerling 起伏目标的检测概率 $P_{\mathrm{d}}(k)$可以表示为[30-31]

$$P_{\mathrm{d}}(k)=\binom{N}{N-k}\prod_{j=1}^{N-k}\left(\frac{T_k}{1+\mathrm{SNR}}+\frac{N-j+1}{N-k-j+1}\right)^{-1} \tag{6.70}$$

对于给定的虚警概率 P_{fa}，门限 T_k 可以由式(6.70)计算得到，其中令 SNR=0。由于 k 是随机的，因此目标的检测概率为

$$P_{\mathrm{d}}=\sum_{k=0}^{N-p}\Pr\{\hat{i}=k\}\,P_{\mathrm{d}}(k) \tag{6.71}$$

其中，$\Pr\{\hat{i}=k\}$表示在均匀环境中的删除概率。

ACCA-ODV-CFAR 检测器由 OS-CFAR 和 VI-CFAR 合成，在多个干扰目标环境下，它不像 VI-CFAR 那样受干扰位置的影响，因此其性能要优于 VI-CFAR。对于杂波边缘环境，OS-CFAR 和 VI-CFAR 的性能明显下降，而 ACCA-ODV-CFAR 几乎不受影响。

总之，基于 ODV 的删除技术增强了单元平均检测器的鲁棒性，当多个干扰目标不止存在于单个滑窗内时表现出较好的、接近于 OS-CFAR 的检测性能，且明显优于 VI-CFAR。对于杂波边缘环境，同样也表现出较好的检测性能。

6.11 比较与小结

本章讨论了一些自适应 CFAR 处理方法。实际上，前几章讨论的 CFAR 检测器形成检测阈值的方法也是自适应的，之所以又在本章归纳出一类自适应 CFAR 处理方法，是因为它们在形成检测阈值过程中不需要任何关于干扰(如杂波边缘和干扰目标)的先验信息，可以自动适应于干扰的变化。由于自适应 CFAR 方法没有利用先验信息，所以它们在某些干扰环境中相对非自适应 CFAR 方法有一定的检测损失。在实际应用中，先验信息不是总能得到的，或者是不准确的，因此在先验信息与实际干扰环境不符时，自适应 CFAR 方法就显示出了优越性。

CCA 是为了改善 CFAR 检测器对抗多个干扰目标的性能而提出的。多目标环境中，在未知干扰目标数的条件下，CCA 表现出比 CMLD 更好的性能。与 OS 相比，在强干扰目标环境中，CCA 稍有逊色；而对于弱干扰目标，CCA 更可取。在密集目标环境中，CCA 的优势就更加明显，并且 CCA 可以容纳的干扰目标数不会像 OS 方法那样受指定 k 值的限制，CCA 还可以检测本身可能是目标的干扰。

HCE-CFAR 方法的含义就是非均匀杂波参数估计法。顾名思义，它是为应对边缘杂波而设计的，分为无偏结构和有偏结构两种情况。无偏 HCE 的检测概率 P_{d} 在干扰区分离点位置变化时保持了相对恒定，避免了遮蔽效应的影响。然而它在某些情况下的虚警率上升很严重。有偏 HCE 则克服了这个问题。与 CA 相比，在 CA 的检测性能严重下降，目标完全被遮蔽时，HCE 却保持了相对较好的性能。

E-CFAR 是针对多目标环境的。在均匀杂波背景中，删除系数 $\alpha>1$ 的 E-CFAR 与 CA 相比的检测

损失几乎可以忽略。在多目标环境中，与抗多个干扰目标性能较好的 SO 相比，E-CFAR 具有不受干扰目标在滑窗中的分布位置影响的优点。而且 α 较小的 E-CFAR 在有一个强度无限大的干扰目标的环境中，检测损失几乎可以忽略。在干扰目标数较大时，它的检测性能下降也较小。

OSTA 是为改善 OS 的抗边缘杂波性能而设计的。它获得了优于 CA 和 OS 的虚警控制性能。在多目标环境中，OSTA 与 OS 的性能几乎相同。

VTM 是削减点可以自适应地随干扰环境变化的 TM-CFAR 检测器。因此，当参数 q 取某些值时，VTM 可以转变成 OS 或 TM。VTM 具有不亚于 OS 的抗干扰目标的性能。当 $q \geqslant 2$ 时，VTM(k,q) 还具有优于 OS(k) 的抗边缘杂波性能。

基于逐个参考样本的多步删除规则，Himonas 等提出了一系列自适应 CFAR 检测方案：GCMLD，GO/SO，ACMLD，GTL-CMLD，ACGO。与 CCA 相比，GCMLD 在强或弱干扰目标环境中的检测性能都有所改善。即使当参考滑窗长度较小时，GCMLD 也保持了较优的性能，然而 CCA 的性能却急剧下降。GO/SO 是针对杂波边缘环境的，当杂噪比较高时，能获得与 TM$(r_1,0)$ 相同的 P_{fa}。当杂噪比较低时，GO/SO 的 P_{fa} 比 TM 的 P_{fa} 要高些。TM$(0,r_2)$$(r_2<N_C)$ 受到遮蔽效应的严重影响，从不受遮蔽效应影响的角度来说，GO/SO 的性能是占优的。ACMLD 是为多目标环境设计的，它在各种强度干扰目标环境中都不受遮蔽效应的影响。在杂波边缘和干扰目标同时出现的环境中，当检测单元处于强杂波区时，GTL-CMLD 能获得较优的检测性能。当检测单元处于杂波边缘的强杂波区中时，ACGO 表现出与 GO 相近的虚警控制能力；当干扰目标出现在杂波边缘环境中时，它也可以有效地删除干扰目标回波，从而不受遮蔽效应的影响。

VI-CFAR 检测器综合了 CA-CFAR、GO-CFAR 及 SO-CFAR 的优点，在均匀环境中具有较低的 CFAR 损失，对于存在干扰和杂波边缘的非均匀背景，具有一定的鲁棒性。但是当两个滑窗中都存在干扰时，VI-CFAR 的检测性能下降，与 SO-CFAR 的性能接近。

ESECA-CFAR 检测器从实测数据中目标回波包络形状的特点出发，来设计参考单元删除规则，因此具有较强的复杂非均匀环境的适应能力，在均匀背景中，其性能与 CA-CFAR 相当；对于存在干扰和杂波边缘的非均匀背景，其性能显著优于 CA-CFAR 与 GO-CFAR，且背景的非均匀程度越大，其性能优势越大。

6.10 节中讨论的几种自适应 CFAR 检测器也在不同程度上克服了以往一些典型 CFAR 方法在非均匀干扰背景中性能下降的问题。另外，Viswanathan[32] 等提出了一种 CFAR 检测器，称作 SE-CFAR。Himonas[33] 又提出了检波后积累的自适应 CFAR 检测器(API)。当存在随机脉冲干扰时，API-CFAR 检测器能够获得稳健的虚警控制性能，而当没有干扰存在时能够获得最优检测性能。文献[34]针对随机到达的二项分布脉冲干扰背景，研究了 EXC(Excision)-CFAR 和 EXC-CFAR-BI(Binary Integration) 检测器，并推导了检测概率和虚警概率的解析式。文献[35]针对非均匀背景，提出了一种基于切换算法和 OS-CFAR 的 SOS-CFAR-I(Switching Ordered Statistic CFAR type I)检测器。在均匀环境和非均匀环境中对 Swerling Ⅰ型目标的检测结果表明，与 CA-CFAR、GO-CFAR、SO-CFAR 和 OS-CFAR 相比，SOS-CFAR-I 在接近杂波边缘时可以显著减少虚警率。文献[36]针对杂波边缘的非均匀环境，采用两步法 CFAR 检测结构，获得了对非平稳环境的鲁棒性。文献[37]提出了一种基于加权有序统计和模糊规则的 CFAR 处理器，该处理器能利用基于模糊逻辑的软规则来克服杂波边缘和干扰目标的问题。文献[38]提出了一种基于自动删除算法的最小选择 CFAR 检测方法，该方法的排序时间为自动删除均值检测器的一半，且在多目标环境中，当强干扰目标较多时，该方法优于自动删除均值检测器。文献[39]针对非均匀环境，先用小波检查非均匀位置，并应用小波进行去噪，再结合 CA-CFAR，提出了 WT-CA-CFAR

检测器，这种检测器在均匀环境中具有与 CA-CFAR 相似的性能，且对存在干扰的环境具有鲁棒性。另外，这种自适应技术在 GOSCA、GOSGO、MTM、TMGO 等 CFAR 检测器[40-43]上的推广还有待于深入研究。文献[19]针对 CA-CFAR 在多目标环境中性能下降的问题，利用指数分布条件下，方差均值平方比等于 1 的特点，自适应删除使方差均值平方比偏离指定置信区间的参考单元样本，然后利用剩余样本估计背景功率水平，进而实现 CFAR 检测。文献[44-45]利用二阶统计差异假设与 Shapiro-Wilk 指数检验来筛选有效参考单元，进而估计背景功率水平。文献[46]采用 Grubbs 准则来删除参考单元中的异常样本，进而形成基于 Grubbs 准则的修正的单元平均 CFAR 检测器，即 CAG-CFAR 检测器。文献[47]针对毫米波雷达在非均匀环境下的目标检测问题，提出利用有序样本一阶差分(FOD)结果的假设检验来删除异常样本，进而形成新的自动删除单元平均 CFAR 检测器。文献[48]提出了一种智能多策略融合(Intelligent Multi-Strategy Fusion，IMSF)CFAR 检测算法，通过结合 FOCA-CFAR、BOCA-CFAR 和 OSVI-CFAR 的预处理结果，IMSF-CFAR 可以利用比传统 CFAR 方法更合适的独立同分布参考单元，这意味着更好的检测性能。文献[49]利用拟合优度检验来验证参考单元的 IID 假设，当满足 IID 假设时，采用 CA-CFAR 进行检测，否则采用距离非均匀检测算法为 CA-CFAR 提供均匀样本。

总之，本章讨论的各种自适应 CFAR 方法都有各自的优势，能在特定的环境中获得性能改善。但是这些方法还不能在所有干扰环境中都具有稳健性，并且算法往往还很复杂，因此在工程实现上难度较大，缺乏对这些自适应 CFAR 方法进行全面系统的比较。考虑到自适应 CFAR 方法不需要关于干扰的先验信息，且能自适应于干扰变化，有必要对此类方法做进一步深入研究。

参考文献

[1] Barboy B，Lomes A，Perkalski E. Cell-averaging CFAR for multiple target situations[J]. IEEE Proceedings，1986，133(2)：176-186.

[2] Finn H M. A CFAR design for a window spanning two clutter fields[J]. IEEE Transactions on AES，1986，22(1)：155-169.

[3] Goldman H，Bar David D I. Analysis and application of the excision CFAR detector[J]. IEEE Proceedings，1988，135(6)：563-575.

[4] 王元恺，肖泽龙，吴礼，等. 调频连续波雷达的二维截断统计量恒虚警检测方法[J]. 西安交通大学学报，2017，51(10)：113-119.

[5] Conto E，Longo M，Lops M. Analysis of the excision CFAR detector in the presence of fluctuating targets[J]. IEEE Proceedings，1989，136(6)：290-291.

[6] Conto E，Longo M，Lops M. Analysis of the excision CFAR in multiple target situations[C]. New York：Proceedings of the International Symposium on Noise and Clutter Rejection in Radars and Imaging Sensors，1989：566-571.

[7] Goldman H. Performance of excision CFAR detector in the presence of interferes[J]. IEEE Proceedings，1990，137(3)：163-171.

[8] Gandhi P P，Kassam S A. Analysis of CAFR processors in nonhomogeneous background[J]. IEEE Transactions on AES，1988，24(4)：427-445.

[9] Lei S，Li W. Research of ordered statistic CFAR detector[C]. New York：Proceedings of the International Symposium on Noise and Clutter Rejection in Radars and Imaging Sensors，1989：560-565.

[10] Ozgunes I，Gandhi P P，Kassam S A. A variably trimmed mean CFAR radar detector[J]. IEEE Transactions on AES，1992，28(1)：1002-1014.

[11] Barkat M, Himonas S D, Varshney P V. CFAR detection for multiple target situations[J]. IEEE Proceedings, 1989, 136(5): 193-209.

[12] Himonas S D, Barkat M. Automatic censored CFAR detection for nonhomogeneous Environments[J]. IEEE Transactions on AES, 1992, 28(1): 286-304.

[13] Himonas S D. Adaptive censored greatest-of CFAR detection[J]. IEEE Proceedings, 1992, 139(3): 247-255.

[14] Smith M E, Varshney P K. Intelligent CFAR processor based on data variability[J]. IEEE Transactions on AES, 2000, 36(3): 837-847.

[15] Gandhi P P, Kassam S A. Analysis of CFAR processors in nonhomogeneous background[J]. IEEE Transactions on AES, 1998, 24(4): 427-445.

[16] Viswanathan R, Eftekhari A. A selection and estimation test for multiple target detection[J]. IEEE Transactions on AES, 1992, 28(4): 505-519.

[17] Cao T T V. Constant false-alarm rate algorithm based on test cell information[J]. IET Radar, Sonar & Navigation, 2008, 2(3): 200-213.

[18] 齐国青. 单元平均恒虚警率检测器性能的改善[C]. CCSP D3-9, 1992: 517-520.

[19] 徐从安, 简涛, 何友, 等. 方差均值平方比恒虚警检测器[J]. 电光与控制, 2012, 19(9): 59-62.

[20] 王陆林, 刘贵如, 邹姗, 等. 非均匀噪声环境下分布式 IVI-CFAR 检测算法[J]. 重庆邮电大学学报(自然科学版), 2019, 31(4): 509-516.

[21] Zhao Likai, Li Sen, Hu Guangzhao. An improved VI-CFAR detector based on GOS[C]. 2016 7th International Conference on Mechatronics and Manufacturing, 2016,45: 101-104.

[22] Hong S W, Han D S. Performance Analysis of an Environmental Adaptive CFAR Performance Analysis of an Environmental Adaptive CFAR Detector[J]. Mathematical Problems in Engineering, 2014.

[23] 何友, 关键, 孟祥伟, 等. 雷达目标检测与恒虚警处理[M]. 2 版. 北京: 清华大学出版社, 2011.

[24] 黄勇, 关键, 张林, 等. 结合回波形状特征与迭代删除的背景功率水平估计方法: 中国, 201418009644. 8[P]. 2017-12-29.

[25] Cole L G, Chen P W. Constant false alarm rate detector for a pulse radar in a maritime environment[J]. IEEE Transactions on AES, 1978, 14(6): 1110-1113.

[26] 吉书龙, 皇甫堪, 周良柱, 等. 一种新的雷达恒虚警(CFAR)处理器[J]. 国防科技大学学报, 1990, 12(4): 116-121.

[27] Ji Shulong, Huang Fukan, et al. The studies of radar new CFAR processor[C]. In: Proceedings of ICSP, 1990: 657-658.

[28] Basile M, Di Vito A, Falessi C, et al. An adaptive length CA-CFAR device for an ATC radar[C]. Peking: Proceedings of CIE International Conference on Radar, 1991: 419-422.

[29] Farrrouki A, Barkat M. Automatic censoring CFAR detector based on ordered data variability for nonhomogeneous environments[J]. IEEE Proceedings Radar Sonar Navig. , 2005, 152(1): 43-51.

[30] Rickard J T, Dillard G M. Adaptive detection algorithms for multiple target situations[J]. IEEE Transactions on AES, 1977, 13(4): 338-343.

[31] Himonas S D, Barkat M. Automatic censored CFAR detection for nonhomogeneous environments[J]. IEEE Transactions on AES, 1992, 28(1): 286-304.

[32] Viswanathan R, Eftekhari A. A selection and estimation test for multiple target detection[J]. IEEE Transactions on AES, 1992, 28(2): 505-518.

[33] Himonas S D. CFAR integration processors in randomly arriving impulse interference[J]. IEEE Transactions on AES, 1994, 30(3): 809-817.

[34] Ivan G, Christo A K. Excision CFAR BI detector in randomly arriving impulse interference[C]. 2005 IEEE International Radar Conference, 2005: 950-955.

[35] Erfanian S, Vakili V T. Introducing switching ordered statistic CFAR type I in different radar environments[J].

EURASIP Journal on Advances in Signal Processing，2009，89(6)：1023-1031.

[36] Lombardo P. Adaptive CFAR detection for clutter-edge heterogeneity using Bayesian inference[J]. IEEE Transactions on AES，2003，39(4)：1462-1470.

[37] Zaimbashi A，Taban M R，Nayebi M M，et al. Weighted order statistic and fuzzy rules CFAR detector for Weibull clutter[J]. Signal Processing(Elsevier)，2008，88(3)：558-570.

[38] 许江湖，张明敏，王平波. 基于自动删除算法的最小选择恒虚警检测器[J]. 武汉理工大学学报，2006，30(16)：1065-1068.

[39] Alamdari M，Hashemi M M. An improved CFAR detector using wavelet shrinkage in multiple target environments[C]. New York：9th International Symposium Signal Processing and its Applications，ISSPA 2007.

[40] 何友，Rohling H. 两种具有自动筛选技术的广义有序统计恒虚警检测器及其在多目标情况下的性能[J]. 电子科学学刊，1994，16(6)：582-590.

[41] He You. Performance of some generalized modified order statistics CFAR detectors with automatic censoring technique in multiple target situations[J]. IEEE Proceedings，1994，141(4)：205-212.

[42] He You，Meng Xiangwei. Performance of a new CFAR detector based on trimmed mean. Proc[C]. Beijing：IEEE International conference on System，Men and Cybernetics(IEEE'SMC)，1996：702-706.

[43] Meng Xiangwei，He You. Two generalized greatest of selection CFAR algorithms[C]. Beijing：CIE International Conference on Radar，2001：359-362.

[44] 芦永强，韩壮志，张宏伟. 一种非平稳复杂环境下的自适应恒虚警算法[J]. 雷达科学与技术，2018，16(3)：333-337.

[45] Ali A，Abderrazak A，Bencheikh M L，et al. A new adaptive CFAR processor in multiple target situations[J]. IEEE 2017 Seminar on Detection Systems Architectures and Technologies，2017，3：1-4.

[46] Zhou Wei，Xie Junhao，Xi Kun，et al. Modified cell averaging CFAR detector based on Grubbs criterion in non-homogeneous background[J]. IET Radar，Sonar & Navigation，2019，13(1)：104-112.

[47] Jiang Wen，Huang Yulin，Yang Jianyu. Automatic censoring CFAR detector based on ordered data difference for low-flying helicopter safety[J]. Sensors，2016，16(7)：1055-1058.

[48] Ouyang Siyuan，Tang Jun，Yang Wenming，et al. An intelligent CFAR algorithm based on multistrategy fusion[C]. New York：Twelfth International Conference on Digital Image Processing，2020.

[49] Amir Z. An adaptive cell averaging-based CFAR detector for interfering targets and clutter edge situations[J]. Digital Signal Processing，2014，31：59-68.

第 7 章 经典非高斯杂波背景中的 CFAR 检测器

CHAPTER 7

7.1 引言

在前面几章讨论中，均假设杂波包络服从瑞利分布，然而在许多应用中，实验数据表明，杂波包络的概率密度函数(PDF)与瑞利分布相比有一个长的拖尾，此时雷达将面临非瑞利包络杂波或非高斯杂波[1]。当用高分辨率雷达(脉冲宽度一般小于 0.5μs)或在低入射角(一般小于 5°)的地/海杂波中进行目标检测时，目标往往包含在非高斯杂波中[2-4]。

描述非瑞利包络杂波的经典统计模型是韦布尔和对数正态分布[5]，另外杂波尖峰更多的 α 稳定分布也得到了越来越多的关注[6]。与瑞利模型相比，韦布尔模型能在很宽范围内很好地与实验数据相匹配[7]。韦布尔分布杂波包络的 PDF 表示为

$$f(x)=\frac{c}{b}\left(\frac{x}{b}\right)^{c-1}\exp\left[-\left(\frac{x}{b}\right)^{c}\right],\quad x\geqslant 0 \tag{7.1}$$

其中，b 是尺度参数，表示分布的强度；c 是形状参数，表示分布的偏斜度。表 7.1 给出了一些典型的地杂波和海杂波形状参数 c 的观测值[8]。

表 7.1　不同条件下韦布尔分布的形状参数

地形或海况	频率	波束宽度/°	视角/°	脉冲宽度/μs	c
岩石山脉	S	1.5	—	2	0.512
有树林的山坡	L	1.7	约 0.5	3	0.626
森林	X	1.4	0.7	0.17	0.506～0.531
耕地	X	1.4	0.7～5.0	0.17	0.606～2.0
海况 1	X	0.5	4.7	0.02	1.452
海况 2	K_u	5	1.0～8.30	0.1	1.160～1.783

把式(7.1)与瑞利分布比较可得，$c=2$ 时的韦布尔分布即为瑞利分布，即瑞利分布是韦布尔分布的一个特例。因此，韦布尔分布比瑞利分布能适应更宽的杂波范围。

对数正态分布杂波包络的 PDF 为

$$f(x)=\frac{1}{\sqrt{2\pi}\sigma x}\exp\left[-\frac{(\ln x-\mu)^2}{2\sigma^2}\right],\quad x\geqslant 0 \tag{7.2}$$

其中，μ 是尺度参数，取值范围为$(-\infty,+\infty)$；σ 是形状参数，取值范围为$(0,+\infty)$，有时也被称作对数标准差。另外，参数 $\rho=\mathrm{e}^{\sigma^2/2}$ 为均值与中值比，也可以用来控制 PDF 的形状。实际杂波数据的 ρ 范围

是 1.065～1.93[9]，对应的 σ 值为 0.355～1.147。

当杂波包络服从非瑞利分布时，最优检测问题变得比前几章中讨论的内容更复杂[10]。由于得不到背景杂波加目标的统计特征的闭型表示，因此难以获得似然比的精确表达式。鉴于这种原因，目前尚没有一种非高斯杂波中最优接收机的通用形式。

与单参数的瑞利分布不同，韦布尔和对数正态分布都是双参数分布。一个参数是表示强度的尺度参数，另一个是表示分布偏斜度的形状参数[11]。而常规高斯背景中的 CFAR 检测方法是单参数估计方法，在韦布尔或对数正态包络杂波中若仍采用高斯背景中的 CFAR 检测方法，一般假设某个参数(通常是形状参数)已知，这实际上仍然是单参数 CFAR 检测方法。然而，实际杂波形状参数的时变性也会引起系统虚警概率的变化。若同时考虑到形状和尺度参数的变化，则要使用双参数 CFAR 检测方法。对于在韦布尔或对数正态杂波背景中的单参数 CFAR 检测方法及双参数 CFAR 检测方法，将在本章的后续内容中进行讨论，另外，本章也将对 α 稳定分布杂波下目标 CFAR 检测问题进行简要分析。

7.2 Log-t CFAR 检测器

在形状和尺度参数均未知的韦布尔或对数正态杂波背景下，Log-t 检测器[12]为 CFAR 检测提供了一种准最优单脉冲检测策略。具体来说，它允许由参考单元估计形状和尺度参数，以便在形状和尺度参数都变化的环境中进行检测并保持 CFAR。当形状和尺度参数变化时，Log-t 检测器保持 CFAR 的有效性依赖于参考单元样本的先验条件，即参考单元杂波样本是 IID 的。

7.2.1 对数正态分布中的 Log-t CFAR 检测器

对于服从对数正态分布的雷达杂波回波 $X_0, X_1, \cdots, X_R$，若对其取对数变换，即

$$Y_i = \log X_i, \quad i = 0, 1, \cdots, R \tag{7.3}$$

所得到的随机变量 $Y_0, Y_1, \cdots, Y_R$ 将变成易于处理的正态分布，Log-t 检测器将 Y_0 作为检测单元，而将检测单元邻近的 R 个参考单元的回波样本 $Y_1, Y_2, \cdots, Y_R$ 作为参考样本，并假定回波样本 $Y_0, Y_1, \cdots, Y_R$ 是 IID 的。Log-t 检测器的检测统计量 t 为

$$t = \frac{Y_0 - \dfrac{1}{R}\sum_{i=1}^{R} Y_i}{\left[\dfrac{1}{R}\sum_{i=1}^{R}\left(Y_i - \dfrac{1}{R}\sum_{j=1}^{R} Y_j\right)^2\right]^{1/2}} \tag{7.4}$$

在杂波环境中，当 $X_i (i=0,1,\cdots,R)$ 是 IID 的对数正态随机变量时，即 $Y_0, Y_1, \cdots, Y_R$ 服从正态分布时，H_0 假设下检测统计量 t_0 具有 Student-t 分布，即它的 PDF 具有如下形式：

$$f_{t_0}(x) = \frac{\Gamma[(R+1)/2]}{\sqrt{R\pi}\,\Gamma(R/2)}(1 + x^2/R)^{-(R+1)/2} \tag{7.5}$$

式(7.5)表明，$f_{t_0}(x)$ 不依赖于杂波包络的尺度参数和形状参数。因此，在对数正态杂波中，当参考单元样本 $Y_i (i=1,2,\cdots,R)$ 满足 IID 的条件时，Log-t 检测器提供了对于尺度参数和形状参数的双参数 CFAR 检测策略[13]。

在对数正态杂波环境下，对于给定的 P_{fa} 和 R 值，检测阈值 T 可以通过式(7.6)计算：

$$P_{fa} = \int_T^{\infty} \frac{\Gamma[(R+1)/2]}{\sqrt{R\pi}\,\Gamma(R/2)}(1 + x^2/R)^{-(R+1)/2}\,dx \tag{7.6}$$

并且对于给定的 P_{fa} 和 R 值，可以用标准 Student-t 分布来确定要求的门限 T。

7.2.2 韦布尔分布中的 Log-t CFAR 检测器

值得一提的是，Log-t 检测器对于韦布尔分布杂波下目标检测仍然有效，其在韦布尔分布杂波中仍然能维持虚警概率的恒定。对于式(7.4)表示的检测统计量 t，将它用未经对数变换前的 $X_0, X_1, \cdots, X_R$ 表示出来，即

$$t=\frac{\ln\left[\prod_{j=1}^{R}\left(\frac{X_0}{X_j}\right)\right]}{\sqrt{\frac{1}{R}\sum_{k=1}^{R}\left\{\ln\left[\prod_{\ell=1}^{R}\left(\frac{X_k}{X_\ell}\right)\right]\right\}^2}} \tag{7.7}$$

通过观察式(7.7)可得，若对服从任意韦布尔分布杂波回波$(X_0, X_1, \cdots, X_R)$进行指数变换 $Z_i=\alpha X_i^{\beta}$，通过适当选取 α 和 β，X_i 均可变成标准的瑞利分布。经过 $Z_i=\alpha X_i^{\beta}$ 的指数变换，式(7.7)的分子和分母中的 α 和 β 均被抵消了，这就意味着式(7.7)与 α 和 β 无关，也就证明了 Log-t 检测器在韦布尔分布杂波中仍具有 CFAR 特性。

在韦布尔杂波下单脉冲检测场景中，Log-t 检测器的检测阈值 T 更难获得。对于较大的 R 值，可利用样本 $\{Y_i\}$ 的均值 m' 和方差 σ' 的联合 PDF[12] 确定检测阈值 T。设

$$m'=\frac{1}{R}\sum_{i=1}^{R}Y_i,\quad \sigma'=\left[\frac{1}{R}\sum_{i=1}^{R}\left(Y_i-\frac{1}{R}\sum_{j=1}^{R}Y_j\right)^2\right]^{1/2}=\left[\frac{1}{R}\sum_{i=1}^{R}(Y_i-m')^2\right]^{1/2} \tag{7.8}$$

且检测阈值 T 可通过式(7.9)求解：

$$P_{fa}=\int_{m'}\int_{\sigma'}\left[\int_{e^{\sigma' T+m'}}^{\infty}f_{X_0}(x)\mathrm{d}x\right]f_{m',\sigma'}(m',\sigma')\mathrm{d}\sigma'\mathrm{d}m' \tag{7.9}$$

其中，$f_{m',\sigma'}(m',\sigma')$是 m' 和 σ' 的联合 PDF，$f_{X_0}(x)$是 X_0 的 PDF。

现有研究表明[12]，当 $R\gg 20$，且 $P_{fa}\geqslant 10^{-4}$ 时，式(7.9)可以写成

$$P_{fa}=\frac{\exp\left[-\frac{1}{2}\left(1+\frac{1}{4}\sigma_z^2 e^{z_0}\right)e^{z_0}\right]}{\sqrt{1+\frac{1}{2}\sigma_z^2 e^{z_0}}} \tag{7.10}$$

其中

$$\sigma_z^2=4\left[\mathrm{var}(m')+T^2\mathrm{var}(\sigma')+2T\rho(m',\sigma')\right]$$

式中

$$\mathrm{var}(m')=\frac{\pi^2}{24R},\quad \mathrm{var}(\sigma')=\frac{0.45}{R},\quad \rho(m',\sigma')=-\frac{0.234}{R}$$

z_0 是如下超越方程的解：

$$\begin{cases} z_0+\frac{1}{2}\sigma_z^2 e^{z_0}=2(m_0+\sigma_0 T) \\ m_0=\mathrm{E}(m')=0.058 \\ \sigma_0=\sqrt{\mathrm{var}(\ln D_0)}=\frac{\pi}{\sqrt{24}} \end{cases}$$

在可接受的 R 和 P_{fa} 范围内，文献[12]基于式(7.10)求解检测阈值 T，结果如图 7.1 所示。其中，

T_∞是已知尺度参数和形状参数条件下的检测阈值，在 $P_{fa}=10^{-3}$、10^{-4} 和 10^{-6} 时分别为 1.96、2.18 和 2.50。

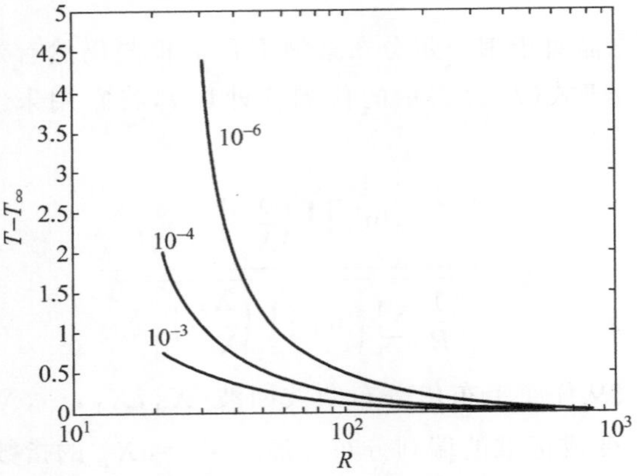

图 7.1 韦布尔杂波中的 Log-t 检测器检测阈值特性

当检测单元同时包含韦布尔或对数正态分布杂波和目标信号时，很难得到关于检测统计量的 PDF 解析表达式，因而难以采用解析方法分析 Log-t 检测器的检测性能。文献[12]指出，对于高信杂比，$P_d(P_{fa}, \mathrm{SCR})$的表达式可以在如下的近似基础上建立：即对于起伏目标模型，信号加杂波包络的 PDF 可以用信号包络的概率密度较好地近似。Goldstein[12] 发现，在 P_d 大于 0.5 时，这种近似解与文献[9]的仿真结果之间没有明显区别。

由于 Log-t 检测器采取的是单脉冲检测策略，因而它在现代脉冲多普勒或 MTD 雷达系统中的应用受到限制，因为在进行距离和频率模糊分辨时，需要用多脉冲检测策略来实现。但在这个过程中，Log-t 检测器可以用在单脉冲线性检测器中以提供在韦布尔或对数正态杂波中的双门限检测的第一级。

7.3 韦布尔分布中有序统计类 CFAR 检测器

许多实测数据已表明，陆地、森林、海浪、雨雪等杂波在高分辨雷达中或低入射角的情况下，其幅度分布会明显偏离瑞利分布而呈现长拖尾情况。在许多情况下，韦布尔分布可以很好地拟合实际环境获取的杂波数据[3]。韦布尔背景下 CFAR 检测的研究可划分为两类：一类是双参数 CFAR 检测方案，其假定尺度参数和形状参数均未知，需要利用参考滑窗中的样本同时对两个参数进行估计；另一类是假定形状参数已知的单参数 CFAR 检测方案。例如，7.2 节介绍的 Goldstein 提出的 Log-t 检测器[7]；本节将要介绍的 Weber-Haykin 恒虚警算法[14-15]，其门限由两个有序样本来形成；7.4 节将要介绍的韦布尔背景下最大似然 CFAR 检测器[16]，同时考虑了形状参数已知和未知的两种情况。还有一种双参数 CFAR 处理方法[17]，通过对数变换将韦布尔分布转化为 Gumbel 分布，利用 local-scale 分布参数的最佳线性无偏估计(Best Linear Unbiased Estimation，BLUE)来调整检测门限，将在 7.5 节中讨论。

本节将先研究韦布尔分布下的单参数 CFAR 检测器，即 OS-CFAR 和 OSGO-CFAR 检测器。值得注意的是，韦布尔分布下的单参数 CFAR 检测方法有其限制条件，即假定形状参数是已知的。

7.3.1 OS-CFAR 检测器在韦布尔背景中的检测性能

文献[18]首先介绍了 OS-CFAR 在韦布尔背景中的检测性能。假定背景杂波服从韦布尔分布，且

杂波样本 $x_i(i=1,2,\cdots,R)$ 间是统计独立的，其中 R 表示参考单元的个数。韦布尔分布具有两个参数，第一个参数与平均功率有关，称之为尺度参数；第二个参数和分布的偏斜度有关，称之为形状参数[15]。韦布尔分布杂波经平方律检波后输出的 PDF 和 CDF 分别为

$$f(x)=\rho^{-c/2}\frac{c}{2}x^{(c/2)-1}\exp\left[-\left(\frac{x}{\rho}\right)^{c/2}\right],\quad x\geqslant 0 \tag{7.11}$$

和

$$F(x)=1-\exp\left[-\left(\frac{x}{\rho}\right)^{c/2}\right],\quad x\geqslant 0 \tag{7.12}$$

对比式(7.1)中线性检波后韦布尔分布杂波包络的 PDF 可知，功率尺度参数 $\rho=b^2$。

将参考单元样本 $x_j(j=1,2,\cdots,R)$ 按幅值大小排序，选择第 k 个有序样本 $x_{(k)}$ 作为对杂波功率水平的估计 Z，即

$$Z=x_{(k)},\quad 1\leqslant k\leqslant R \tag{7.13}$$

$x_{(k)}$ 的 PDF 和 CDF 分别表示为

$$f_k(x)=k\binom{R}{k}[1-F(x)]^{R-k}[F(x)]^{k-1}f(x) \tag{7.14}$$

和

$$F_k(x)=\sum_{i=k}^{R}\binom{R}{i}[1-F(x)]^{R-i}[F(x)]^{i} \tag{7.15}$$

令 $Z=x_{(k)}$ 与门限因子 T 相乘，得到的积 TZ 作为检测门限。检测单元样本 D 与之进行比较，若 D 超过门限值，则判定目标存在。其中 T 值由设定的虚警概率 α 所决定。在假设 H_0 下，韦布尔背景下 OS-CFAR 的虚警概率 P_{fa} 为[18]

$$P_{\mathrm{fa}}=\mathrm{E}_Z\{\Pr\{D\geqslant TZ\mid \mathrm{H}_0,Z\}\}=\int_0^{\infty}\exp\left[-\left(\frac{Tx}{\rho}\right)^{c/2}\right]f_k(x)\mathrm{d}x=\frac{R!\,\Gamma(R-k+1+T^{c/2})}{(R-k)!\,\Gamma(R+1+T^{c/2})} \tag{7.16}$$

从式(7.16)可见，OS-CFAR 在韦布尔背景中的虚警概率与形状参数有关，因而 OS-CFAR 只能在韦布尔分布形状参数已知的情况下，维持对功率尺度参数变化的 CFAR 特性，即它是一种单参数 CFAR 方法。这里有必要评估形状参数背离设定值对 OS-CFAR 检测器虚警概率的影响。门限参数 T 根据设定的虚警概率 $\alpha=10^{-6}$ 和 $c=2.0$ 求得。图 7.2 给出了 OS-CFAR 检测器的虚警概率相对形状参数 c 的变化曲线，取 $R=32$，$k=28$ 和 $k=24$。由图 7.2 可知，当 c 从 2.0 变到 1.0 时，OS-CFAR 的虚警概率的上升超过了 3 个数量级。另外可以看出，OS-CFAR 检测器采用高序值的有序统计量有助于对虚警概率的控制。

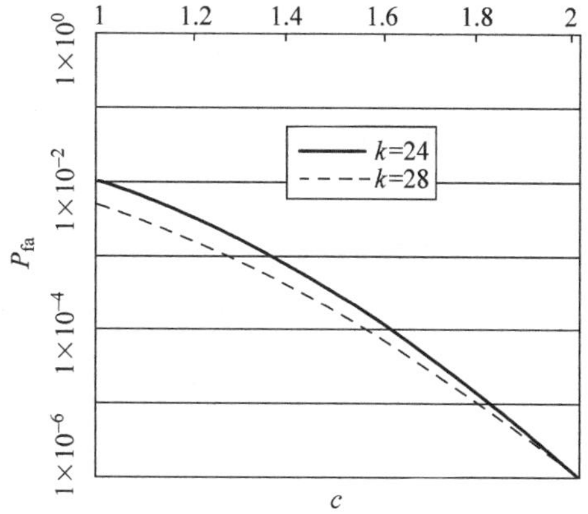

图 7.2 OS-CFAR 检测器的虚警概率相对形状参数 c 的变化曲线

与高斯背景不同，在韦布尔背景中 OS-CFAR 检测器不存在检测概率解析表达式，此时可采用计算机仿真方法进行分析[14]。表 7.2 和表 7.3 分别给出了 $c=0.8$ 和 $c=0.6$ 时 OS-CFAR 检测器在韦布尔背景中的 CFAR 损失(CFAR loss)。从表 7.2 和表 7.3 中的结果可以看出，OS-CFAR 在韦布尔背景中的 CFAR 损失比

其在高斯背景中大得多，主要原因在于，为了控制韦布尔分布的长拖尾，检测器需要采用较高的检测阈值，也就导致了较大的 CFAR 损失。

表 7.2 OS-CFAR 检测器在韦布尔背景中的 CFAR 损失($c=0.8, P_d=0.9, P_{fa}=10^{-6}$)

杂波样本数 R	代表序值 k	门限因子 T	CFAR 损失
12	11	29.1	9dB
24	20	23.45	5dB
48	40	17.44	2.5dB
72	60	15.78	1.5dB

表 7.3 OS-CFAR 检测器在韦布尔背景中的 CFAR 损失($c=0.6, P_d=0.9, P_{fa}=10^{-6}$)

杂波样本数 R	代表序值 k	门限因子 T	CFAR 损失
12	11	89.6	13dB
24	20	67.1	7dB
48	40	45.2	4dB
72	60	39.6	2.5dB

7.3.2 OSGO-CFAR 检测器在韦布尔背景中的检测性能

假定背景杂波服从韦布尔分布，并考虑平方律检波。OSGO-CFAR 采用前沿、后沿滑窗中的第 k 个有序样本 $x_1(k)$ 和 $x_2(k)$ 的较大者作为对杂波功率水平的估计，即

$$Z=\max\{x_1(k), x_2(k)\} \tag{7.17}$$

Z 与 T 相乘得到自适应门限 TZ，T 为门限因子，取决于均匀背景中设定的虚警概率。检测单元的输出 D 与 TZ 相比较，做出有无目标的判决；R 表示参考滑窗中杂波样本的个数。前沿、后沿滑窗中第 k 个有序样本的 PDF 和 CDF 分别为

$$f_k(t)=k\binom{R/2}{k}\left[1-F(t)\right]^{R/2-k}\left[F(t)\right]^{k-1}f(t) \tag{7.18}$$

和

$$F_k(t)=\sum_{i=k}^{R/2}\binom{R/2}{i}\left[1-F(t)\right]^{R/2-i}\left[F(t)\right]^{i} \tag{7.19}$$

检测统计量 Z 的 PDF 为

$$f_Z(t)=2f_k(t)F_k(t) \tag{7.20}$$

检测单元的虚警概率为

$$\begin{aligned}P_{fa}&=\mathrm{E}_Z\{\Pr(D\geqslant TZ\mid \mathrm{H}_0, Z)\}\\&=\int_0^{\infty}f_Z(t)\exp\left[-\left(\frac{Tt}{\rho}\right)^{c/2}\right]\mathrm{d}t\\&=2k\binom{R/2}{k}\sum_{i=k}^{N/2}\binom{R/2}{i}\int_0^{\infty}\left[1-F(t)\right]^{R-k-i}\left[F(t)\right]^{k+i-1}f(t)\cdot\\&\quad\exp\left[-\left(\frac{Tt}{\rho}\right)^{c/2}\right]\mathrm{d}t\end{aligned} \tag{7.21}$$

将式(7.18)和式(7.19)代入式(7.21)中,作变量代换 $x=\left(\frac{t}{\rho}\right)^{c/2}$ 进行积分后,将自变量 x 换回 t,可得

$$\begin{aligned}P_{\mathrm{fa}} &= 2k\binom{R/2}{k}\sum_{i=k}^{R/2}\binom{R/2}{i}\int_0^{\infty}\exp[-(R-k-i+1+T^{c/2})t]\,[1-\exp(-t)]^{k+i-1}\mathrm{d}t \\ &= 2k\binom{R/2}{k}\sum_{i=k}^{R/2}\binom{R/2}{i}\frac{\Gamma(k+i)\Gamma(R-k-i+1+T^{c/2})}{\Gamma(R+1+T^{c/2})}\end{aligned} \tag{7.22}$$

从式(7.22)可见,OSGO-CFAR 在韦布尔背景中的虚警概率取决于形状参数,因而 OSGO-CFAR 在韦布尔背景中也是一种单参数 CFAR 方法,其只能对尺度参数的变化保持 CFAR 特性。图 7.3 给出了 OSGO-CFAR 检测器的虚警概率相对形状参数 c 的变化曲线,取 $R=32$,$k=14$ 和 $k=10$。门限参数 T 根据设定的虚警概率 $\alpha=10^{-6}$ 和 $c=2.0$ 求得。从图 7.3 中可看出,当 c 从 2.0 变到 1.0 时,OSGO-CFAR 的虚警概率的上升也超过了 3 个数量级。另外可以看出,OSGO-CFAR 检测器采用高序值的有序统计量也有助于对虚警概率的控制。

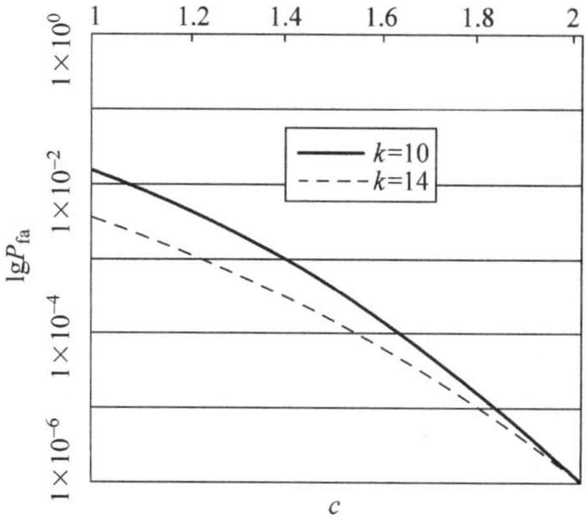

图 7.3 OSGO-CFAR 检测器的虚警概率相对形状参数 c 的变化曲线

在韦布尔背景中 OSGO-CFAR 检测概率的解析表达式也不存在,此时可采用计算机仿真进行分析。表 7.4 和表 7.5 分别给出了 $c=0.8$ 和 $c=0.6$ 时 OSGO-CFAR 检测器在韦布尔背景中的 CFAR 损失。从表 7.4 和表 7.5 中可知,OSGO-CFAR 检测器在韦布尔背景中的 CFAR 损失比它在高斯背景中大得多。由于多个韦布尔分布随机变量之和的统计分布解析表达式难以获取,相应的性能分析需借助于仿真方法。

表 7.4 OSGO-CFAR 检测器在韦布尔背景中 CFAR 损失($c=0.8$,$P_{\mathrm{d}}=0.9$,$P_{\mathrm{fa}}=10^{-6}$)

杂波样本数 R	代表序值 k	门限因子 T	CFAR 损失
12	5	38.8	10dB
24	10	21.5	5.5dB
48	20	16.1	3dB
72	30	14.6	2dB

表 7.5 OSGO-CFAR 检测器在韦布尔背景中 CFAR 损失($c=0.6$,$P_{\mathrm{d}}=0.9$,$P_{\mathrm{fa}}=10^{-6}$)

杂波样本数 R	代表序值 k	门限因子 T	CFAR 损失
12	5	131.3	14dB
24	10	59.8	8dB
48	20	40.7	4.5dB
72	30	35.8	3dB

7.3.3 韦布尔背景中 Weber-Haykin 恒虚警检测算法

Weber 和 Haykin 在文献[14]中提出了一种利用两个有序统计样本设置检测门限的方法,可实现

韦布尔背景中的双参数 CFAR 检测,常称为 Weber-Haykin 恒虚警检测算法[15],简称 WH 算法。为便于理解和说明,下文将从单参数 OS-CFAR 方法引申出双参数 WH 算法[15]。

对 R 个 IID 的韦布尔随机变量按照从小到大进行排序得 $x_{(1)}, x_{(2)}, \cdots, x_{(R)}$。对韦布尔背景中单参数 OS-CFAR 来说,其检测门限为 $S_Z = Tx_{(k)}$,其虚警概率的解析表达式如式(7.16)所示。但式(7.16)为平方律检波结果,若对回波信号采用线性检波[1],则虚警概率变为

$$P_{\mathrm{fa}} = \frac{R!\,\Gamma(R-k+1+T^c)}{(R-k)!\,\Gamma(R+1+T^c)} \tag{7.23}$$

通过观察式(7.23)可得,若对门限因子 T 利用形状参数进行如下修正:

$$S = T^{1/c} x_{(k)} \tag{7.24}$$

则虚警概率将变为

$$P_{\mathrm{fa}} = \frac{R!\,\Gamma(R-k+1+T)}{(R-k)!\,\Gamma(R+1+T)} \tag{7.25}$$

所得到的虚警概率将与韦布尔分布的形状参数无关,即可实现韦布尔分布中双参数 CFAR 处理。注意到检测门限式(7.24)需要预先获得形状参数 c,可利用两个有序统计样本进行形状参数估计,即

$$\hat{c} = \frac{\ln[-\ln(1-h_j)] - \ln[-\ln(1-h_i)]}{\ln x_{(j)} - \ln x_{(i)}} \tag{7.26}$$

式中,$h_j = \dfrac{j}{R+1}$。将式(7.24)中的形状参数 c 用它的估计值 $\hat{c}$ 来代替,并令 $k=i$,就得到 WH 算法的检测门限

$$S = x_{(i)} \left(\frac{x_{(j)}}{x_{(i)}}\right)^{\beta} = x_{(i)}^{1-\beta} x_{(j)}^{\beta} \tag{7.27}$$

式中,$\beta = \dfrac{\ln T}{\ln[-\ln(1-h_j)] - \ln[-\ln(1-h_i)]}$,$T$ 由设定的虚警概率根据式(7.25)确定。

以上的推导考虑了对雷达回波采用线性检波的情况。若考虑平方律检波,检测门限式(7.27)的形式及 β 的计算公式仍将不变。这是因为在韦布尔背景中双参数 WH 恒虚警检测算法与形状参数是无关的。

接下来将证明 WH 算法的 CFAR 特性,并给出由 β 确定 P_{fa} 的方法。令第 i 个排序样本的值为随机变量 x,第 j 个排序样本值为随机变量 y,且保持 $i<j$,则 $x_{(i)}$ 和 $x_{(j)}$ 的联合概率密度 $f_{ij}(x,y)$ 为

$$f_{ij}(x,y) = \begin{cases} Q_2 [F(x)]^{i-1} f(x) [F(y)-F(x)]^{j-i-1} f(y) [1-F(y)]^{R-j}, & x \leqslant y \\ 0, & \text{其他} \end{cases} \tag{7.28}$$

其中,$f(x)$、$f(y)$ 均为排序前的 PDF,$F(x)$、$F(y)$ 均为排序前的 CDF,且

$$Q_2 = \frac{\Gamma(R+1)}{\Gamma(i)\Gamma(j-i)\Gamma(R-j+1)}$$

WH 算法的检测门限为 $S = x_{(i)}^{1-\beta} x_{(j)}^{\beta}$,在 $x_{(i)}$ 和 $x_{(j)}$ 为确定值的情况下虚警概率为 $P_{\mathrm{fa}} = \exp\left[-\left(\dfrac{S}{b}\right)^c\right]$,但衡量检测器性能的虚警概率需要对 $x_{(i)}$ 和 $x_{(j)}$ 取统计平均,即

$$P_{\mathrm{fa}} = \int_0^{\infty} \left\{ \int_0^{y} \exp\left\{ -\left[\left(\frac{x}{b}\right)^{1-\beta} \left(\frac{y}{b}\right)^{\beta} \right]^c \right\} f_{ij}(x,y)\,\mathrm{d}x \right\} \mathrm{d}y$$

$$=Q_2\int_0^{\infty}\left\{\int_0^{v}\exp(-u^{1-\beta}v^{\beta}-u)[\exp(-v)]^{R-j+1}\times\right.$$
$$\left.[1-\exp(-u)]^{i-1}[\exp(-u)-\exp(-v)]^{j-i-1}du\right\}dv \quad (7.29)$$

其中，$u=\left(\frac{r}{b}\right)^c$，$v=\left(\frac{y}{b}\right)^c$。式(7.29)表明，$P_{fa}$ 不依赖于 b 或 c。因此，WH 算法在韦布尔杂波背景中是双参数 CFAR 的。由式(7.29)可知，β 值越小则 P_{fa} 越低。对于平方律检测器，对在瑞利背景中($c=2$ 时)检测瑞利起伏目标的特殊情况，检测概率 P_d 和 β 之间的关系式为

$$P_d=Q_2\int_0^{\infty}\left\{\int_0^{v}\exp\left(-\frac{1}{1+\lambda}u^{1-\beta}v^{\beta}\right)[1-\exp(-u)]^{i-1}\exp(-u)\times\right.$$
$$\left.[\exp(-v)]^{R-j+1}[\exp(-u)-\exp(-v)]^{j-i-1}du\right\}dv \quad (7.30)$$

其中，λ 是信号与噪声平均功率比。图 7.4 在 $R=16$，$i=3$，$j=16$，$P_{fa}=10^{-5}$ 时给出了 WH 算法、$k=12$ 的 OS 以及固定阈值 3 种检测器的检测性能曲线。

由图 7.4 可见，WH 算法的 CFAR 损失为 9.6dB($P_d=0.5$)，而 OS 的 CFAR 损失只有 2.3dB。通过这组最佳序值($i=3$，$j=16$)可达到 WH 算法的最小损失，但失去了 OS 抗干扰目标的能力。为了使 WH 算法获得与 OS 相近的抗干扰目标能力，需设置 $j<R$。然而，降低 j 值也将增加 CFAR 损失，见表 7.6。

表 7.6　$R=16$，$P_{fa}=10^{-5}$，$P_d=0.5$，$c=2.0$ 时 WH 算法的 CFAR 损失　　（单位：dB）

i	j					
	16	15	14	13	12	11
3	9.6	11.0	13.5	17.0	21.5	27.5
4	9.9	12.0	14.9	18.9	24.5	32.7
5	10.8	13.2	17.0	22.0	29.3	41.2

总之，从上述的结果可以看出：①当 i 和 j 间的差距增加时，CFAR 损失下降；②WH 算法的 CFAR 损失要比已知形状参数的单参数 CFAR 检测器的大。第一个特性说明要尽可能使用两个间距较大的有序样本。然而，使用有序统计检测器的主要优势就是对抗杂波边缘和干扰目标。因此，在二者之间需要一个内在的协调，既要使两个有序样本间距较大以获得较小的 CFAR 损失，又要使它们尽量接近以保持抗干扰目标和杂波边缘的性能。第二个特性说明在选择某种方法之前，应该把它与在最差形状参数条件下设计的单参数 CFAR 方法进行比较。

图 7.5 给出了 WH 算法和 Hansen[19] 方法的 CFAR 损失比较。图 7.5 中虚线表示 Hansen 方法，而“*”表示 WH 算法，其中，$R=32$，$i=5$，$j=31$，$P_d=0.5$。结果表明，WH 算法的 CFAR 损失小于 Hansen 方法。Hansen 方法的基本思想是，用经过对数变换后的参考单元样本值的均值估计尺度参数，用样本值的方差估计形状参数。

值得注意的是，若采用多个有序样本来设置门限[15]并不会带来检测性能的明显改善。在尺度参数和形状参数均未知的韦布尔背景中，WH 算法的性能优于 Hansen 方法的双参数检测器。然而，WH 算法与已知形状参数的韦布尔背景中单参数 OS 方法相比仍有较大损失。例如，由图 7.4 可知，在该条件下估计形状参数的代价是 7.3dB 的附加损失，如此大的损失需重新考虑形状参数估计的必要性。一种方案是不估计 c，而是设定 c，如果 c 的不确定性范围不是很大，则在可能的最大 c 值处的损失可能比采

用估计 c 时的损失小。

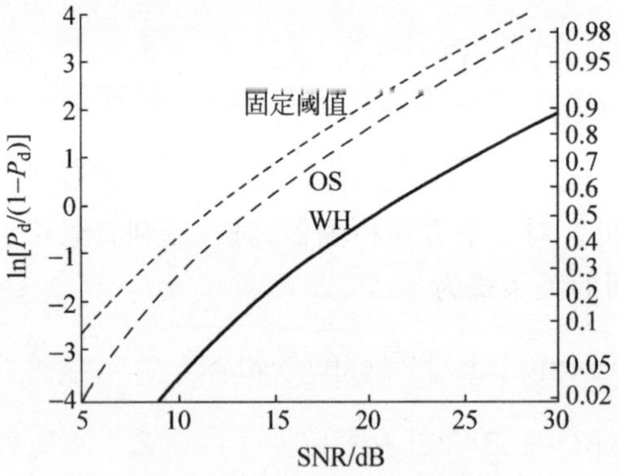

图 7.4　3 种检测器的检测性能曲线

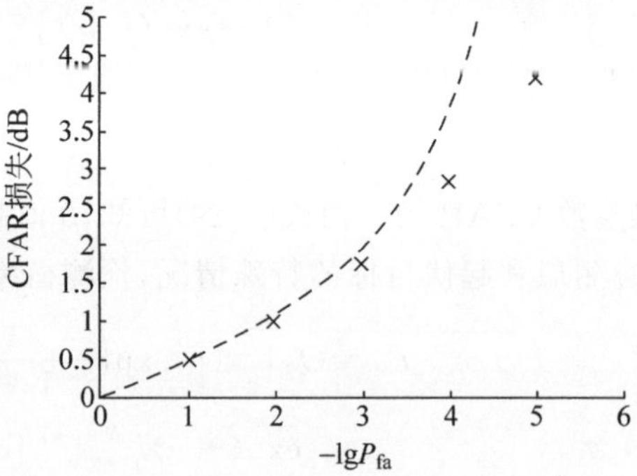

图 7.5　WH 算法和 Hansen 方法的 CFAR 损失

文献[15]给出了如下建议：当形状参数的不确定范围为 $1.5\leqslant c\leqslant 2$ 时，假定 $c=1.5$ 要比估计 c 的方法好。然而，如果 $1\leqslant c\leqslant 2$ 时，采用估计 c 的方法比较好。但在后一种情况，这个结论忽略了一个重要因素，即用 WH 算法估计 c 时，最优代表序值 $j=R$（见表 7.6），这将牺牲抗多干扰目标的性能。但是，由表 7.6 可见，降低 j 又会使损失急剧上升。因此，在干扰目标存在时，即便 c 的不确定性范围很大（如 $1\leqslant c\leqslant 2$），仍建议使用韦布尔背景中的单参数 OS 方法。

7.3.4　用参考单元样本的期望和中值估计 c 的方法

文献[20]和[21]中提出了利用参考单元样本的数学期望和中值估计形状参数 c，它是一种双参数 OS 方法。

韦布尔分布的数学期望和中值分别为

$$\mathrm{E}(x)=b\Gamma\left(1+\frac{1}{c}\right) \tag{7.31}$$

$$\mathrm{M}(x)=b(\ln 2)^{\frac{1}{c}} \tag{7.32}$$

另外，可以从有序统计序列中获得如下两个估计值

$$\hat{\mathrm{E}}(x)=\frac{1}{R-\mathrm{IN}}\sum_{j=1}^{R-\mathrm{IN}}x_{(j)} \tag{7.33}$$

$$\hat{\mathrm{M}}(x)=\begin{cases}x_{[(R-\mathrm{IN}+1)/2]}, & R-\mathrm{IN}\text{ 是奇数}\\ \dfrac{1}{2}\left[x_{[(R-\mathrm{IN})/2]}+x_{[(R-\mathrm{IN})/2+1]}\right], & R-\mathrm{IN}\text{ 是偶数}\end{cases} \tag{7.34}$$

其中，IN 代表参考滑窗中的干扰目标数，R 代表参考单元样本的数目。在多目标情况下，形状参数 c 的估计过程中应该从有序统计序列值中删除 IN 个较大的值。从式(7.31)～式(7.34)可得到如下结果：

$$\frac{\hat{\mathrm{M}}(x)}{\hat{\mathrm{E}}(x)}=\frac{(\ln 2)^{1/\hat{c}}}{\Gamma\left(1+\dfrac{1}{\hat{c}}\right)}=F(\hat{c}) \tag{7.35}$$

$$\hat{c}=F^{-1}\left(\frac{\hat{\mathrm{M}}(x)}{\hat{\mathrm{E}}(x)}\right) \tag{7.36}$$

式(7.36)即为形状参数 c 的估计值。因为形状参数 c 为 0.5～2，所以 $\hat{c}$ 和函数 $F(\hat{c})$ 的关系可以事先计算出并形成表格[20-21]。在自适应检测中可先由式(7.33)和式(7.34)的估值计算出函数 $F(\hat{c})$ 的值，再从表 7.7 中获得形状参数 c 的估计值。此处只考虑 $\hat{c}$ 为 0.45～2.3 的估计值，即数据表中只需要包含与 $\hat{c}$ 为 0.45～2.3 相对应的 $F(\hat{c})$ 的函数值。

由于形状参数 c 已不再是确定的了，因此必须重新计算并分析自适应检测性能。为了便于与双参数的 WH 检测算法进行比较，进行 40000 次 Monte Carlo 仿真，结果列于表 7.7。标称化因子 T 是在满足 $P'_{fa}=10^{-2}$ 条件下设置的。仿真条件为均匀背景，参考滑窗中干扰目标数 IN 为零。其中，f 表示形状参数 c 是已知固定的，s 表示 c 是估计的。从表 7.7 可看出，P_{fa} 要比设计值略高一点。对于某一固定形状参数 c 来说，滑窗长度 R 越小，这两个虚警概率值的差越大。一般地说，这种变化是可以接受的。此外，表 7.7 不仅给出了形状参数估计值 $\hat{c}$ 的均值 $E(\hat{c})$ 和均方差 $\sigma_{\hat{c}}$，而且包含了相应的 ADT 和 SD_{ADT} 值。后两个指标将在下面讨论。

表 7.7 检测性能的仿真结果

检测器参数	形状参数		P_{fa}	ADT	SD_{ADT}	$E(\hat{c})$	$\sigma_{\hat{c}}$
$R=16$ $k=16$	$c=0.6$	f	1.0025×10^{-2}	21.607	14.288	—	—
		s	1.6550×10^{-2}	22.106	18.647	0.7153	0.2755
	$c=1.2$	f	1.0025×10^{-2}	4.4435	1.3648	—	—
		s	1.6024×10^{-2}	4.4759	1.8130	1.4071	0.4898
	$c=2.0$	f	1.0025×10^{-2}	2.4207	0.4389	—	—
		s	1.1099×10^{-2}	2.5018	0.5648	1.9438	0.4096
$R=32$ $k=31$	$c=0.6$	f	1.0150×10^{-2}	16.658	7.3357	—	—
		s	1.3024×10^{-2}	16.872	9.2947	0.6490	0.1490
	$c=1.2$	f	1.0150×10^{-2}	3.9926	0.8468	—	—
		s	1.3550×10^{-2}	3.9942	1.1009	1.3303	0.3861
	$c=2.0$	f	1.0150×10^{-2}	2.2828	0.2880	—	—
		s	1.0700×10^{-2}	2.3288	0.3672	1.9660	0.3613
$R=64$ $k=60$	$c=0.6$	f	1.0374×10^{-2}	14.416	4.1588	—	—
		s	1.2175×10^{-2}	14.539	5.3864	0.6219	0.0927
	$c=1.2$	f	1.0374×10^{-2}	3.7591	0.5341	—	—
		s	1.2649×10^{-2}	3.7580	0.7288	1.2684	0.2685
	$c=2.0$	f	1.0374×10^{-2}	2.2080	0.1876	—	—
		s	1.0900×10^{-2}	2.2374	0.2566	1.9886	0.3142

对于韦布尔分布样本，在固定形状参数情况下，采用代表序值为 k 的有序统计量时，其度量 ADT 计算如下：

$$\mathrm{ADT}=\frac{E(TZ)}{b}=Tk\binom{R}{k}\sum_{i=0}^{k-1}(-1)^i\binom{k-1}{i}\frac{\Gamma(1+1/c)}{(R-k+i+1)^{1+1/c}} \tag{7.37}$$

对于固定的 R、预先给定的 P'_{fa} 及确定参数 k，度量 ADT 是依赖于形状参数的。形状参数 c 越大，ADT 与最优值 $[-\ln P_{fa}]^{1/c}$ 间的差则越小[21]。在上述条件下，CFAR 处理的性能取决于形状参数 c 的大小。即 c 越大，CFAR 检测器的性能越好。当 P'_{fa} 取不同值时，对于一个固定的 c，R 和 k 越大，CFAR

检测器的性能亦越好。给定的 P'_{fa} 越大,阈值估计精度越好。另外,在 P'_{fa} 设定的情况下,对任意形状参数 c,阈值的估计精度取决于参考滑窗长度 R。对于一个较大的 R 将导致较高精度的阈值估计。对于 $R=16$,当 $k=14$ 时,ADT 的值最小;对于 $R=24$,最小 ADT 的值位于 $k=20$;当 $R=32$ 和 64 时,最小的 ADT 分别位于 $k=27$ 和 $k=53$。使 ADT 最小的 k 值与形状参数 c 无关。也就是说,当 R 给定后,无论实际的 c 取何值,使 ADT 最小的 k 值是固定的。

度量模式 SD_{ADT}(Standard Deviation of Average Decision Threshold)利用的是 b 标称化的检测阈值 S 的均方差[21]。对于 OS-CFAR 和固定 c 值,韦布尔背景下 SD_{ADT} 表示为

$$
\begin{aligned}
SD_{ADT} &= \frac{[\mathrm{var}(S)]^{1/2}}{b} \\
&= T\left\{\left[\sum_{i=0}^{k-1}(-1)^i k\binom{R}{k}\binom{k-1}{i}\frac{\Gamma\left[1+\left(\frac{2}{c}\right)\right]}{(R-k+i+1)^{1+\frac{2}{c}}}\right]-\right. \\
&\quad \left.\left[\sum_{i=0}^{k-1}(-1)^i k\binom{R}{k}\binom{k-1}{i}\frac{\Gamma\left[1+\left(\frac{2}{c}\right)\right]}{(R-k+i+1)^{1+\frac{1}{c}}}\right]^2\right\}^{1/2}
\end{aligned}
\tag{7.38}
$$

在分析检测性能过程中发现,对于一个固定的参数 k、R 和信杂比,P_d 的大小不仅取决于 ADT 值,而且也取决于其方差 SD_{ADT}。即对于参数 k,较小的 ADT 值并不总是对应于较高的检测概率。例如,对于 $R=16$,$c=1.2$,ADT(15)=14.9725 小于 ADT(13)=15.1223,但是在同一信杂比条件下,$k=13$ 时的检测性能好于 $k=15$。这是因为阈值估计的方差 $SD_{ADT}(13)=3.84$ 比 $SD_{ADT}(15)=3.985$ 小,鉴于此,引入 SD_{ADT} 这种度量。

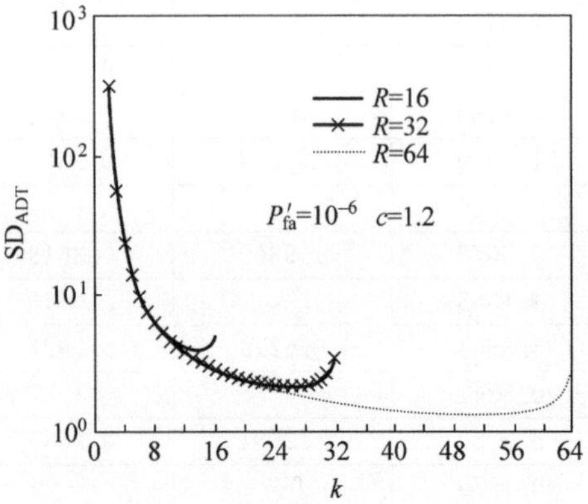

图 7.6 由理论计算得到的 SD_{ADT} 曲线

根据式(7.38),对于给定的 $P'_{fa}=10^{-6}$ 和形状参数 c,图 7.6 给出了一组依赖于参数 k 和 R 的 SD_{ADT} 曲线。对于 $R=16$,最小的 SD_{ADT} 值位于 $k=14$,且 ADT 也最小。但是对于 $R=32$ 和 64,在 $k=26$ 和 $k=52$ 时分别得到最小的 SD_{ADT} 值,而在 $k=27$ 和 $k=53$ 时 ADT 分别为最小。只有当 ADT 和 SD_{ADT} 同时达到最小时,阈值估计才是最优的。该特性既不取决于杂波背景分布的形状参数 c 也不依赖于预先给定的 P'_{fa}。

表 7.7 给出的结果表明,较大的 R 总是具有较好的 c 和 S(阈值)估计。对于 $R=16$,$c=2$ 来说,用本节的估计方法得出的 c 的估计值和度量 ADT 均好于 WH 算法,同时其 CFAR 损失也远低于后者。在上述估计过程中,虚警概率 P_{fa} 的变化是可以忽略的。

仿真的 ADT 值(表 7.7 给出的)与固定形状参数韦布尔背景中 OS 的 ADT 值基本相当[20-21],但是 SD_{ADT} 则产生了明显的变化,图 7.7 给出了 SD_{ADT} 对 k 的曲线。这一特性再次表明,SD_{ADT} 从对检测性能的描述来看是一个重要的度量。

对不同的参数 k 和 R 值,图 7.8 在瑞利目标条件下给出了 CFAR 损失曲线。由图 7.8 可知,与固定形状参数相比,用这种新的估计方法对 c 进行估计所带来的附加损失是不大的。例如,当 $R=16$,$c=2$,$k=10$ 时,其附加损失大约只有 4.3dB(对固定的 c,其 CFAR 损失是 3.38dB[22])。同时这一新结果

比以前的双参数 WH 算法好得多。在图 7.8 中，对于 $R=16, c=1.2, k=12$，其损失大约是 4dB，但 WH 算法的 CFAR 损失则达到了 21.5dB(需要说明的是，当 $P'_{fa}=10^{-5}$ 和 $P'_{fa}=10^{-6}$ 时，就对 CFAR 损失的影响来说，其差别很小，可以忽略不计)。实际上，像这样大的损失是很难被接受的，本节描述的对形状参数 c 的估计方案是可行的，因为它表现出很低的 CFAR 损失和很小的虚警率变化。当滑窗长度 R 较大时，这种方法将产生更好的检测性能。

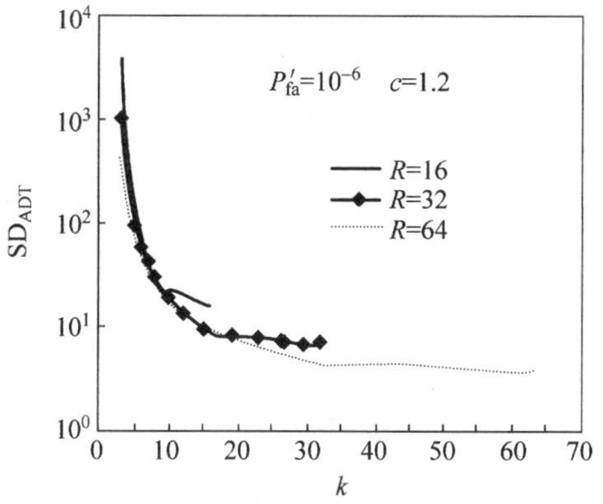

图 7.7　由 Monte Carlo 模拟得到的 SD_{ADT} 曲线

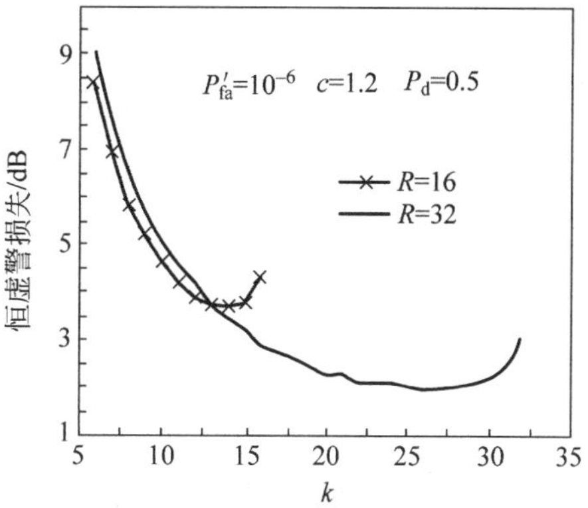

图 7.8　瑞利目标下的一组 CFAR 损失曲线

7.3.5　多脉冲二进制积累下 OS-CFAR 的检测性能

多脉冲积累是提高实际雷达系统检测性能的常用方式之一，根据积累方式可分为相参和非相参积累。除了回波信号直接相加，二进制积累也是一种常用的非相参积累方式[23]，其依据连续的 M 次单脉冲检测结果，当至少有 S 次检测出目标时判定目标存在。通常情况下，相参积累比非相参积累效果更佳，然而，非相参积累也存在容易实现、无须相参基准等优点，可以作为相参积累方式的有益补充。二进制积累是一种典型的非相参积累方法[23]，其在 M 次连续脉冲中至少有 S 次检测出目标，才判定目标存在，因此也称为重合检验。在已知韦布尔分布的尺度和形状参数条件下，文献[8]评估了二进制积累的检测性能。但实际应用中，韦布尔分布的参数难以预先完全已知，本节在功率尺度参数未知的韦布尔背景下，基于平方律检波器，分别针对均匀背景和非均匀背景，分析基于二进制积累的 OS-CFAR 检测器性能。

1. 均匀背景情况

假定连续 M 个扫描脉冲，单个脉冲检测对应的 OS-CFAR 虚警概率 P_{fa1} 可由式(7.16)获得，假定参考单元样本 $x_{ij}(i=1,2,\cdots,M;\ j=1,2,\cdots,R)$ 是 IID 的，因此单脉冲 OS-CFAR 的虚警概率 P_{fa1} 是相等的[23]。在平方律检波后，若利用 M 个脉冲进行非相参积累，基于二进制积累 OS-CFAR 的虚警概率为：

$$P_{fa}=\sum_{i=S}^{M}\binom{M}{i}(P_{fa1})^{i}(1-P_{fa1})^{M-i} \tag{7.39}$$

由式(7.16)和式(7.39)可知，虚警概率只与形状参数 c 有关，即韦布尔杂波下 OS-CFAR 是单参数 CFAR 的。在实际目标检测应用中，对于双参数韦布尔杂波，可以根据形状参数对杂波数据进行区域分割，每个区域内形状参数相同但杂波强度变化，区域内可采用单参数 CFAR 方法。形状参数的估计，可采用区域内较多的杂波样本进行估计，而杂波强度则基于滑窗内适量样本进行估计。如果区域内杂波样本足够多,则形状参数估计可获得较高精度,进而可等效为杂波形状参数已知的情况[24]。

假定目标为 Swerling Ⅱ型,由于目标加韦布尔杂波的 PDF 难以获得闭型表达式,可用相同平均功率 ρ_c 的瑞利分布来近似检测单元中杂波分布[16],即

$$\rho_c = \rho\Gamma\left(1+\frac{2}{c}\right) \tag{7.40}$$

式中,假定 ρ_c 与参考单元中韦布尔杂波平均功率相同,参照式(7.12)有 $\rho=b^2$。这种近似处理在低信杂比时估算出的 P_{d} 偏大,而中高信杂比时近似可信度高[25]。经过瑞利近似,目标加杂波的 PDF 为

$$f_t(x)=\frac{1}{\rho_c(1+\lambda)}\exp\left[-\frac{1}{\rho_c(1+\lambda)}x\right],\quad x\geqslant 0 \tag{7.41}$$

其中,λ 表示信杂比。因此,单脉冲检测时 OS-CFAR 的检测概率为

$$P_{\mathrm{d1}}=\int_0^{\infty} f_t(x)F_k\left(\frac{x}{T}\right)\mathrm{d}x=\frac{T}{\rho_c(1+\lambda)}\Phi(s)\bigg|_{s=\frac{T}{\rho_c(1+\lambda)}} \tag{7.42}$$

其中,$F_k(x)$可参见式(7.15),而 $F_k(x)$的拉氏变换 $\Phi(s)$可表示为

$$\Phi(s)=\rho\sum_{i=k}^{R}\binom{R}{i}\sum_{l=0}^{i}(-1)^l\binom{i}{l}\sum_{f=0}^{\infty}(-1)^f\frac{(R-i+l)}{f!}\frac{\Gamma\left(\frac{fc}{2}+1\right)}{(\rho s)^{fc/2+1}} \tag{7.43}$$

当形状参数 $c=2$ 时,式(7.43)简化为

$$\Phi(s)=\sum_{i=k}^{R}\binom{R}{i}\sum_{l=0}^{i}(-1)^l\binom{i}{l}\frac{1}{s+(R-i+l)/\rho} \tag{7.44}$$

因此,对 M 个脉冲非相参积累时,基于二进制积累的 OS-CFAR 的检测概率为[25]

$$P_{\mathrm{d}}=\sum_{i=S}^{M}\binom{M}{i}(P_{\mathrm{d1}})^i(1-P_{\mathrm{d1}})^{M-i} \tag{7.45}$$

2. 非均匀背景情况

下面分析杂波边缘下的虚警性能。杂波边缘中杂波功率变化建模为阶跃函数[26]。单个脉冲检测时假定有 L 个强杂波区参考单元,服从 $W(\gamma\rho,c)$,即功率尺度参数为 $\gamma\rho$ 且形状参数为 c 的韦布尔分布;弱杂波区的 $R-L$ 个参考单元服从 $W(\rho,c)$,γ 为强/弱杂波功率比[27]。假定前沿滑窗中有 L 个单元进入强杂波区,而后沿滑窗仍在弱杂波区,则第 k 个有序样本的 CDF 为

$$\begin{aligned}F'_k(x)=&\sum_{i=k}^{R}\sum_{j=\max(0,i-L)}^{\min(i,R-L)}\binom{R-L}{j}\binom{L}{i-j}\\&F^j(x)[1-F(x)]^{R-L-j}\bar{F}^{i-j}(x)[1-\bar{F}(x)]^{L-i+j}\end{aligned} \tag{7.46}$$

其中,$\bar{F}(x)$和 $F(x)$分别表示强杂波和弱杂波的 CDF。

若检测单元处于强杂波区,则单脉冲 OS-CFAR 的虚警概率为

$$P_{\mathrm{fa2}}=\int_0^{\infty}F'_k\left(\frac{x}{T}\right)\bar{f}(x)\mathrm{d}x=\int_0^{\infty}F'_k\left(\frac{x}{T}\right)(\gamma\rho)^{-c/2}\frac{c}{2}x^{(c/2)-1}\exp\left[-\left(\frac{1}{\gamma\rho}\right)^{c/2}\right]\mathrm{d}x \tag{7.47}$$

其中,$\bar{f}(x)$表示强杂波的 PDF。利用积分变量代换和 Gamma 函数组合恒等式,可得

$$P_{\mathrm{fa2}}=T^{c/2}\sum_{i=K}^{R}\sum_{j=\max(0,i-L)}^{\min(i,R-L)}\binom{R-L}{j}\binom{L}{i-j}\sum_{n_1=0}^{j}\binom{j}{n_1}(-1)^{n_1}\times$$

$$\frac{\Gamma[\gamma^{c/2}(n_1+R-L-j)+T^{c/2}+L-i+j]\Gamma(i-j+1)}{\Gamma[\gamma^{c/2}(n_1+R-L-j)+T^{c/2}+L+1]} \tag{7.48}$$

当检测单元处于弱杂波区时，根据对称性，将式(7.48)中 $\gamma^{c/2}$ 替换为 $1/\gamma^{c/2}$，即得到相应的虚警概率表达式。

根据不同脉冲检测间杂波区杂波样本的 IID 假设，可得杂波边缘下基于二进制积累 OS-CFAR 的虚警概率为

$$P_{\mathrm{fa}}=\sum_{i=S}^{M}\binom{M}{i}(P_{\mathrm{fa2}})^{i}(1-P_{\mathrm{fa2}})^{M-i} \tag{7.49}$$

3. 性能分析结果

首先分析均匀背景下的虚警性能，在虚警概率 $\alpha=10^{-6}$ 条件下，图 7.9 给出了二进制积累取 6/8 时 OS-CFAR 的虚警性能。由图 7.9 可知，当形状参数较小即杂波尖峰增多时，达到相同虚警概率所需 T 值增大。研究结果还表明[25]，为达到相同虚警概率，当二进制积累(S/M)的阈值 S 值变大时，T 值需相应减小；当门限因子保持不变时，杂波尖峰的增加会导致更高的虚警。值得注意的是，图 7.9 不同的曲线在 $T=1$ 处有一交叉点，此时虚警概率不仅与形状参数无关，亦与功率尺度参数无关，即二进制积累 OS-CFAR 检测器变成自由分布系统。

接着，评估形状参数估计值出现偏差时的虚警和检测性能。图 7.10 给出了基于二进制积累 OS-CFAR 检测器在不同二进制检测门限 S/M 下的虚警概率随形状参数的变化，其中单脉冲 OS-CFAR 标记为 OS，设定 $c=1.2$。可以看出，随着二进制门限 S 的增大，检测器虚警概率对形状参数的鲁棒性有所提高，研究结果还表明[48]，秩 k 值和 S 值具有类似的效果。与单脉冲 OS-CFAR 相比，二进制积累后检测器虚警概率随形状参数的变化幅度明显降低。图 7.11 分析了形状参数估计值出现偏差时的检测性能，其中，设定 $c=1.2$。由图 7.11 可知，形状参数在[0.9,2.0]区间时，检测性能基本不变，但当杂波尖峰进一步增加时，即 $c=0.6$ 时，检测概率反而提高。然而，这并不意味着杂波尖峰越多检测性能越好，因为检测门限是根据 $c=1.2$ 设定的，当 $c=0.6$ 时过多的杂波尖峰将带来更高的虚警概率。研究结果还表明[25]，对较小的形状参数，杂波尖峰更为严重，此时需要较大的二进制积累门限 S 来抑制尖峰杂波。

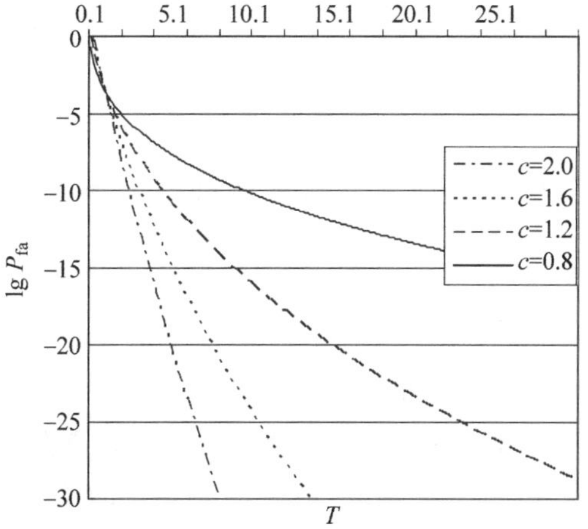

图 7.9　基于二进制积累(6/8)OS-CFAR 检测器虚警概率随门限因子 T 的变化曲线（$c=2.0,1.6,1.2,0.8,R=32,k=28$）

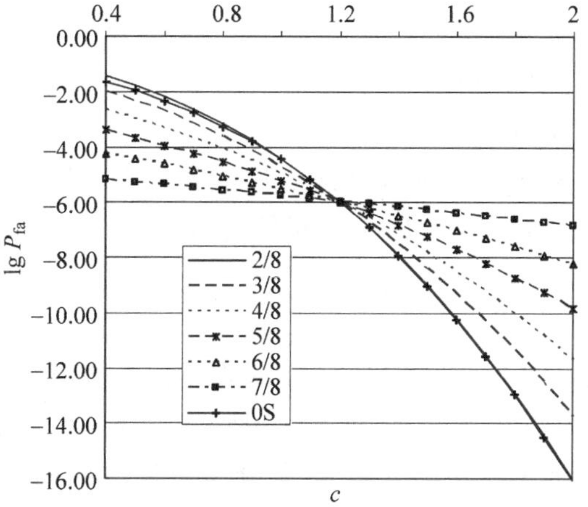

图 7.10　基于二进制积累 OS-CFAR 检测器在不同二进制检测门限 S/M 下虚警概率随形状参数 c 的变化曲线

图 7.12 进一步分析了不同数量($r=0,1,2,3,4$)强目标干扰下的检测性能。图 7.12 中结果与单脉冲 OS-CFAR 情形类似,即基于二进制积累 OS-CFAR 最多可对抗 $R-k$ 个干扰目标。对杂波边缘情况的仿真分析表明[25],基于二进制积累 OS-CFAR 检测器采用较高秩 k 值时虚警率控制能力更强;鉴于杂波的稀疏性质,给定强弱杂波功率比时,虚警率上升幅度未随形状参数 c 的减小而增加。另外,相比单脉冲检测情况,OS-CFAR 检测器在二进制积累时具有更好的虚警控制能力。

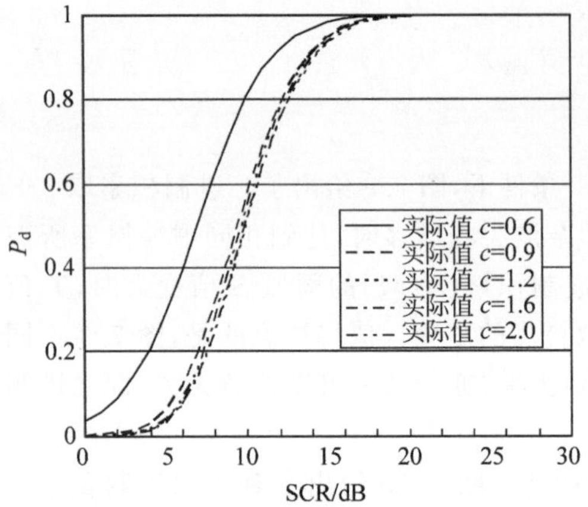

图 7.11 当 $c=0.6,0.9,1.2,1.6,2.0$ 时基于二进制积累(6/8) OS-CFAR 的检测性能曲线

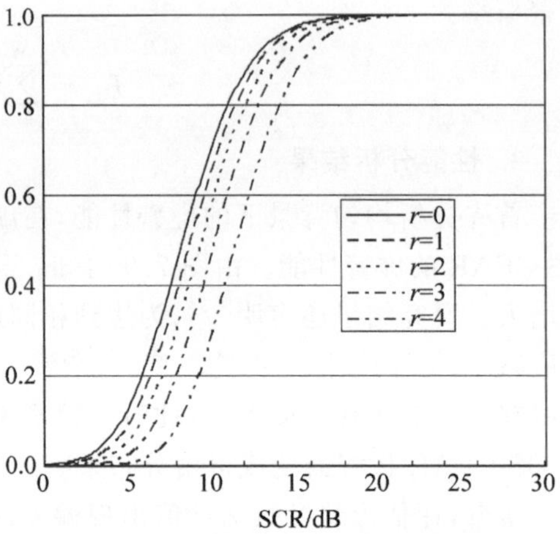

图 7.12 多目标情况下基于二进制积累(6/8)OS-CFAR 的检测性能曲线($r=0,1,2,3,4$,设定 $c=2.0,R=32,k=28,\alpha=10^{-6}$)

因幅值直接相加的非相参积累方式对强干扰非常敏感,强尖峰干扰往往导致虚警。考虑到强干扰目标对信号幅值的影响,相比于幅值直接相加的非相参积累方式,基于二进制积累(S/M)的 OS-CFAR 方法在对抗干扰目标方面具有明显优势。

7.3.6 多脉冲二进制积累下 OSGO-CFAR 的检测性能

本节在形状参数已知的韦布尔背景下,分别针对均匀背景和非均匀背景,分析基于二进制积累的 OSGO-CFAR 检测器性能[28]。

1. 均匀背景情况

利用 M 个脉冲进行二进制积累,假定脉冲间和参考单元间杂波样本间都是 IID 的[23],则基于二进制积累 OSGO-CFAR 的虚警概率为

$$P_{\mathrm{fa}}=\sum_{i=S}^{M}\binom{M}{i}(P_{\mathrm{fa1}})^{i}(1-P_{\mathrm{fa1}})^{M-i} \tag{7.50}$$

式中,单脉冲 OSGO-CFAR 检测器的虚警概率 P_{fa1} 可由式(7.22)获得。

假定目标为 Swerling Ⅱ型,由于目标加韦布尔杂波的 PDF 难以获得闭式解,与 7.3.5 节类似,可用相同平均功率 ρ_c 的瑞利分布来近似检测单元中杂波分布[16]。此时,OSGO-CFAR 的检测概率 P_{d1} 为

$$P_{\mathrm{d1}}=\int_0^{\infty}\exp\left[-\frac{Tt}{\rho_c(1+\lambda)}\right]f_z(t)\mathrm{d}t=\Phi_z(s)\Big|_{s=\frac{T}{\rho_c(1+\lambda)}} \tag{7.51}$$

其中,$\Phi_z(s)$ 为式(7.20)中 $f_z(t)$ 的拉普拉斯变换,可表示为

$$\Phi_z(s)=c\cdot k\binom{r/2}{k}\sum_{i=k}^{r/2}\sum_{j=0}^{k+i-1}\binom{r/2}{i}(-1)^j\sum_{f=0}^{\infty}(-1)^f\frac{(r+j-k-i+1)^f}{f!\rho^{(f+1)c/2}}\frac{\Gamma[(f+1)c/2]}{s^{(f+1)c/2}} \tag{7.52}$$

对于 $c=2$ 的特殊情况，式(7.51)简化为

$$P_{d1}=2k\binom{R/2}{k}\sum_{i=k}^{R/2}\binom{R/2}{i}\frac{\Gamma(k+i)\Gamma[R-k-i+1+T/(1+\lambda)]}{\Gamma[R+1+T/(1+\lambda)]} \tag{7.53}$$

因此，对 M 个脉冲非相参积累时，利用单脉冲 OSGO-CFAR 的检测概率 P_{d1} 的表达式(7.51)，基于二进制积累 OSGO-CFAR 的检测概率仍可参照式(7.45)。

2. 非均匀背景情况

下面讨论杂波边缘对虚警概率的影响。假定前沿滑窗中有 L 个单元进入强杂波区，而后沿滑窗仍在弱杂波区[26]。对于后沿滑窗，第 k 个有序样本的 CDF 为

$$F'_k(x)=\sum_{i=k}^{R/2}\sum_{j=\max(0,i-L)}^{\min(i,R/2-L)}\binom{R/2-L}{j}\binom{L}{i-j}F^j(x)[1-F(x)]^{R/2-L-j}\bar{F}^{i-j}(x)[1-\bar{F}(x)]^{L-i+j} \tag{7.54}$$

其中，$\bar{F}(x)$和 $F(x)$分别表示强杂波和弱杂波的 CDF，杂波功率水平估计 z 的 CDF 为

$$F_z(x)=F'_k(x)F_k(x) \tag{7.55}$$

式中，$F'_k(x)$和 $F_k(x)$分别参见式(7.54)和式(7.19)。如果检测单元处于弱杂波区，则单脉冲 OSGO-CFAR 的虚警概率为

$$P_{fa2}=\int_0^{\infty}F'_k\left(\frac{x}{T}\right)F_k\left(\frac{x}{T}\right)\cdot\rho^{-c/2}\frac{c}{2}x^{(c/2)-1}\exp\left[-\left(\frac{x}{\rho}\right)^{c/2}\right]dx \tag{7.56}$$

通过变量代换 $y=\left(\frac{x}{\rho}\right)^{c/2}$，并利用恒等式变换后化简，可得 P_{fa2} 为[28]

$$\begin{aligned}P_{fa2}=&T^{c/2}k\binom{R/2}{k}\sum_{i_1=0}^{k-1}(-1)^{i_1}\binom{k-1}{i_1}\sum_{i_2=k}^{R/2}\sum_{j_1=\max(0,i_2-L)}^{\min(i_2,R/2-L)}\binom{R/2-L}{j_1}\binom{L}{i_2-j_1}\sum_{j_2=0}^{j_1}(-1)^{j_2}\binom{j_1}{j_2}\\&\frac{\gamma^{c/2}}{a}\left\{\frac{\Gamma[(T^{c/2}+b_1)\gamma^{c/2}+b_2]\Gamma(i_2-j_1+1)}{\Gamma[(T^{c/2}+b_1)\gamma^{c/2}+L+1]}-\right.\\&\left.\frac{\Gamma[(T^{c/2}+a+b_1)\gamma^{c/2}+b_2]\Gamma(i_2-j_1+1)}{\Gamma[(T^{c/2}+a+b_1)\gamma^{c/2}+L+1]}\right\}\end{aligned} \tag{7.57}$$

其中

$$a=R/2-k+i_1+1 \tag{7.58}$$

$$b_1=R/2-L+j_2-j_1 \tag{7.59}$$

$$b_2=L-i_2+j_1 \tag{7.60}$$

对 M 个脉冲非相参积累时，利用单脉冲 OSGO-CFAR 的虚警概率 P_{fa2} 的表达式(7.57)，杂波边缘下基于二进制积累 OSGO-CFAR 的虚警概率 P_{fa} 仍可参照式(7.49)。

3. 性能分析结果

在均匀背景下虚警性能方面，与 7.3.5 节二进制积累 OS-CFAR 检测器相比，形状参数和二进制积累门限对二进制积累 OSGO-CFAR 检测门限 T 的影响是相似的[28]。随着形状参数 c 的减小，为了获得较好的检测性能，二进制积累门限 S 取值应稍大。

在多干扰目标情况下，假定感兴趣目标与干扰目标具有相同的信杂比。图7.13给出了基于二进制积累(6/8)OSGO-CFAR的检测性能曲线，其中OSGOB(m,n)表示二进制积累过程中前沿滑窗和后沿滑窗各有m个和n个干扰目标。由图7.13可知，当干扰目标的个数不超过$N/2-k$时，基于二进制积累OSGO-CFAR检测器均能有效检测目标。事实上，当参考滑窗扫过密集干扰目标时，二进制积累或单脉冲条件下OSGO-CFAR能对抗的干扰目标数最大为$N/2-k$。

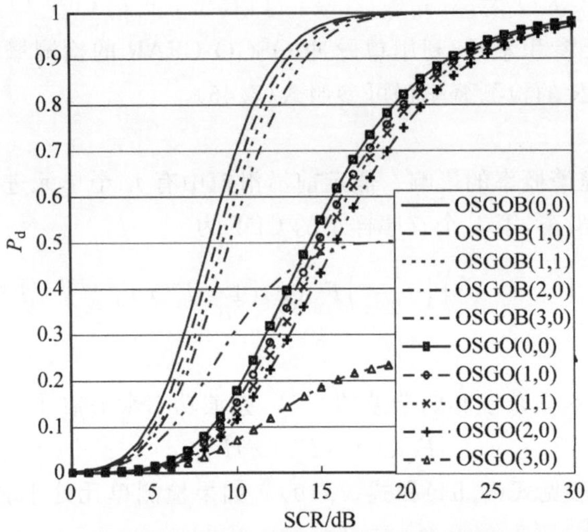

图7.13 多目标情况下基于二进制积累(6/8)OSGO-CFAR的检测性能曲线($c=2.0$,$R=32$,$k=14$,$\alpha=10^{-6}$)

针对杂波边缘情况的虚警性能分析表明[28]，二进制积累情况下OSGO-CFAR的虚警概率的上升程度小于单脉冲情况；对于OSGO-CFAR检测器，二进制积累不仅在检测性能方面明显优于单脉冲检测，且在杂波边缘情况下也具有更强的虚警控制能力。

综合7.3.5节和7.3.6节的分析结果，与单脉冲检测时OSGO-CFAR和OS-CFAR间的对比结果相似[29]：同样基于二进制积累检测，OSGO-CFAR在杂波边缘的虚警和检测能力均优于OS-CFAR；但在多目标情况下，后者对抗干扰目标的能力更强；另外，基于二进制积累检测，OSGO-CFAR的样本排序时间也是OS-CFAR的一半左右，但以牺牲抗干扰目标性能为代价。因此，实际雷达应用中需根据具体检测需求设置相应的CFAR处理方法[30]。

7.4 MLH-CFAR检测器

现有文献分析表明[15]，虽然WH算法比文献[19]所提方法具有更小CFAR损失，但是WH算法的损失仍然很大。因此，文献[15]又提出了基于两个以上有序样本来设置检测门限的方法。但相应分析表明，这几种方法[15,19]都有较大的CFAR损失，且这种损失与形状参数估计的方差有关。为了减小形状参数估计方差所导致的CFAR损失，文献[16]提出了用最大似然(Maximum Likelihood,MLH)方法估计韦布尔分布参数的MLH-CFAR算法。

7.4.1 形状参数已知时韦布尔分布背景中的MLH-CFAR检测器

1. 不删除的MLH-CFAR检测器(已知c)

在$c=2$时，线性检波输出的杂波包络服从瑞利分布，此时b的MLH估计为[31]

$$\hat{b}=\left(\frac{1}{R}\sum_{i=1}^{R}x_i^2\right)^{1/2} \tag{7.61}$$

检测阈值为

$$S=T\hat{b} \tag{7.62}$$

这是一个对于给定的虚警概率可使检测概率最大化的最优检测器。这个检测器被称为一致最优势(Uniformly Most Powerful,UMP)检测器。虚警概率可表示为[32]

$$P_{\mathrm{fa}}=\left(1+\frac{T^2}{R}\right)^{-R} \tag{7.63}$$

与 $c=2$ 的情形相似,当 c 已知但不一定等于 2 时,相应的 MLH 估计器为[33]

$$\hat{b}=\left(\frac{1}{R}\sum_{i=1}^{R}x_i^c\right)^{1/c} \tag{7.64}$$

检测阈值为

$$S=T\hat{b}=T\left(\frac{1}{R}\sum_{i=1}^{R}x_i^c\right)^{1/c} \tag{7.65}$$

虚警概率为

$$P_{\mathrm{fa}}=\prod_{i=1}^{R}\int_0^{\infty}\frac{c}{b}\left(\frac{x_i}{b}\right)^{c-1}\exp\left[-\left(1+\frac{T^c}{R}\right)\left(\frac{x_i}{b}\right)^c\right]\mathrm{d}x_i=\left(1+\frac{T^c}{R}\right)^{-R} \tag{7.66}$$

其中,$f_{D_0}(x)$是不存在目标时检测统计量的 PDF。式(7.66)表明,P_{fa} 不依赖于 b,这种算法确实是具有 CFAR 特性的。把式(7.66)中的 T 代入式(7.65)可得一个简洁的阈值表达式

$$S=\left[(P_{\mathrm{fa}}^{-1/R}-1)\sum_{i=1}^{R}x_i^c\right]^{1/c} \tag{7.67}$$

2. 删除的 MLH-CFAR 检测器(已知 c)

为了降低参考单元中干扰目标对 CFAR 检测器性能的影响,第 4 章中的 CMLD 方法删除 $R-k$ 个较大的有序样本。在瑞利背景中,利用 k 个较小的有序样本 $x_{(1)},x_{(2)},\cdots,x_{(k)}$ 可获得 b 的 MLH 估计[34]

$$\hat{b}=\left\{\frac{1}{k}\left[(R-k)x_{(k)}^2+\sum_{i=1}^{k}x_{(i)}^2\right]\right\}^{1/2} \tag{7.68}$$

其中,$x_{(i)}(i=1,2,\cdots,R)$是从小到大的有序样本。这个估计式与基于 k 个未删除有序样本估计 b 时的最大似然估计器具有相同的分布。

在韦布尔背景中,使用基于 R 个有序样本中 k 个较小有序样本估计 b 的 MLH 估计器为[33]

$$\hat{b}=\left\{\frac{1}{k}\left[(R-k)x_{(k)}^c+\sum_{i=1}^{k}x_{(i)}^c\right]\right\}^{1/c} \tag{7.69}$$

式(7.69)的估计与基于 k 个未删除有序样本估计 b 的 MLH 估计器具有相同的分布。如果保持式(7.62)的检测阈值,以及式(7.66)的 P_{fa} 和 T 之间的关系,并且用 k 代替 R,就得到 P_{fa} 和 S 之间的关系式

$$S=\left\{(P_{\mathrm{fa}}^{-1/k}-1)\left[(R-k)x_{(k)}^c+\sum_{i=1}^{k}x_{(i)}^c\right]\right\}^{1/c} \tag{7.70}$$

至此已经分析了线性检波器的情况,但是它也直接适用于平方律检波器情况。如果随机变量 X 服从具有形状参数 c 的韦布尔分布,则平方律检波器输出 $Y=X^2$ 也是韦布尔分布的,其形状参数为 $c/2$。

此时,不删除的 MLH-CFAR 的阈值为

$$S=\left[(P_{\text{fa}}^{-1/R}-1)\sum_{i=1}^{R}y_i^{c/2}\right]^{2/c} \tag{7.71}$$

而删除的 MLH-CFAR 的检测阈值为

$$S=\left\{(P_{\text{fa}}^{-1/k}-1)\left[(R-k)x_{(k)}^{c/2}+\sum_{i=1}^{k}x_{(i)}^{c/2}\right]^{2/c}\right\} \tag{7.72}$$

7.4.2 形状参数未知时韦布尔分布背景中的 MLH-CFAR 检测器

1. 形状和尺度参数均未知时的 MLH 估计器

在 b 和 c 均未知时为获得 MLH 估计,首先应给出 R 个参考单元的联合 PDF,然后将其对数对 b 和 c 分别求导,并且令导数等于零,即可得 $\hat{b}$ 和 $\hat{c}$。

假设相互独立的 R 个参考单元的联合 PDF 为

$$f(x_1,x_2,\cdots,x_R)=\left(\frac{c}{b^c}\right)^R\prod_{i=0}^{R}\left[x_i^{c-1}\exp\left(-\frac{x_i^c}{b^c}\right)\right] \tag{7.73}$$

基于式(7.73)分别对 b 和 c 进行求导并令导数为零,可得二者的 ML 估计分别为

$$\hat{b}^{\hat{c}}=\frac{1}{R}\sum_{i=1}^{R}x_i^{\hat{c}} \tag{7.74}$$

$$\frac{\sum_{i=1}^{R}x_i^{\hat{c}}\ln x_i}{\sum_{i=1}^{R}x_i^{\hat{c}}}-\frac{1}{R}\sum_{i=1}^{R}\ln x_i=\frac{1}{\hat{c}} \tag{7.75}$$

用迭代方法解式(7.75)即可得 $\hat{c}$,再代入式(7.74)得到 $\hat{b}$。

由式(7.1)可知,当 b 和 c 完全已知时,虚警概率为

$$P_{\text{fa}}=\exp\left[-\left(\frac{S}{b}\right)^c\right] \tag{7.76}$$

因此

$$S=b(-\ln P_{\text{fa}})^{1/c}$$

当 b 和 c 未知时,用它们的估计值代替,这样得到的 P_{fa} 要比式(7.76)确定的 P_{fa} 高一些。为了补偿,用参数 T 代替 $-\ln P_{\text{fa}}$,可以证明,这样可以得到要求的 P_{fa},阈值为

$$S=\hat{b}T^{1/\hat{c}} \tag{7.77}$$

对于大 R 值和较高的 P_{fa},T 只比 $-\ln P_{\text{fa}}$ 稍高,文献[16]的分析表明,由式(7.77)得到的是一个 CFAR 检测器,也是唯一可能的基于 b 和 c 的 MLH 估计的 CFAR 阈值。图 7.14 是利用 Monte Carlo 仿真得到的 T 和 P_{fa} 之间的关系。图 7.14 中 16、32 和∞分别表示 $R=16$、$R=32$ 和固定门限检测即 $R=\infty$ 时的数值结果。

2. 形状参数未知时的删除 MLH-CFAR 检测器

为了使 MLH-CFAR 具有抗干扰目标的能力,仍然可以利用删除 $R-k$ 个较大有序样本的方法。但是,这是以牺牲估计精度为代价的,且增加了 CFAR 损失。在删除情况下 b 和 c 的 MLH 估计为

$$\begin{cases} \dfrac{(R-k)x_{(k)}^{\hat{c}}\ln x_{(k)}+\sum\limits_{i=1}^{k}x_{(i)}^{\hat{c}}\ln x_{(i)}}{(R-k)x_{(k)}^{\hat{c}}+\sum\limits_{i=1}^{k}x_{(i)}^{\hat{c}}}-\dfrac{1}{k}\sum\limits_{i=1}^{k}\ln x_{(i)}=\dfrac{1}{\hat{c}} \\ \hat{b}=\left\{\dfrac{1}{k}\left[(R-k)x_{(k)}^{\hat{c}}+\sum\limits_{i=1}^{k}x_{(i)}^{\hat{c}}\right]\right\}^{1/\hat{c}} \end{cases} \tag{7.78}$$

与已知形状参数的情形相比，这里的 T 和 P_{fa} 的关系不仅依赖于 R，而且还依赖于 k。图 7.15 是在形状参数未知条件下利用 Monte Carlo 和解析计算[16]得到的结果，其中 $R=32$。图 7.15 中的曲线从上至下分别对应 $k=20,24,26,28,32$。对于给定的 R 和 P_{fa}，减小 k 意味着增加 T。

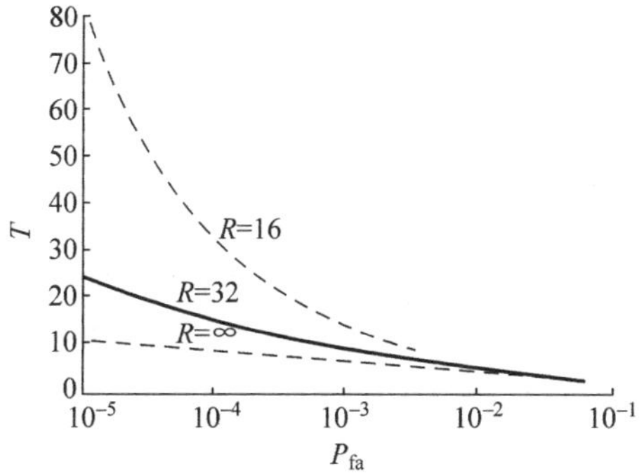

图 7.14 P_{fa} 和 T 之间的关系

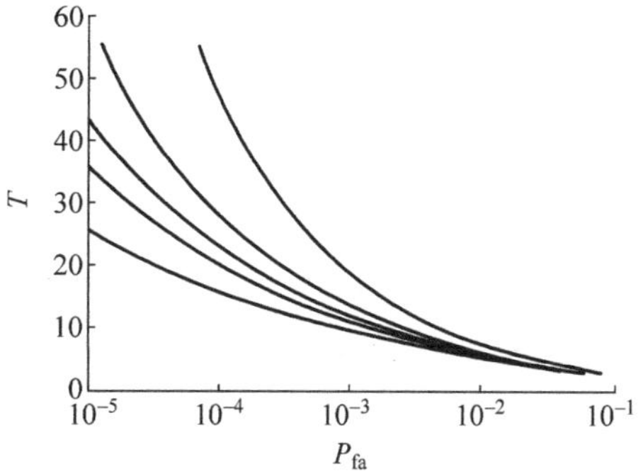

图 7.15 删除 MLH-CFAR 的 P_{fa} 和 T 之间的关系

7.4.3 检测概率和 CFAR 损失

针对韦布尔背景下 Swerling Ⅰ和 Swerling Ⅱ型目标的检测问题，尚难以取得检测单元中包含目标加杂波时 PDF 的闭型解。文献[16]提出一种容易计算的近似方法。当信杂比较高，即检测单元中的杂波成分较小时，可以假设检测单元中包含的是瑞利目标回波且具有与韦布尔杂波相同的平均功率。

1. 已知形状参数时的检测概率

根据上述假设得到检测统计量 D 的 PDF 近似为[16]

$$f_{D_1}(x)=\frac{2x}{b^2\Gamma\left(1+\dfrac{2}{c}\right)+b_t^2}\exp\left[-\frac{x^2}{b^2\Gamma\left(1+\dfrac{2}{c}\right)+b_t^2}\right] \tag{7.79}$$

信杂比定义为

$$\mathrm{SCR}=\frac{b_t^2}{b^2\Gamma\left(1+\dfrac{2}{c}\right)} \tag{7.80}$$

b_t^2 是瑞利起伏目标回波的平均功率，$b^2\Gamma\left(1+\dfrac{2}{c}\right)$是韦布尔杂波的平均功率。当形状参数已知时，式(7.64)中 MLH 估计 $\hat{b}$ 的 PDF 为

$$f_{\hat{b}}(x)=\left(\frac{R}{b^c}\right)^R\frac{c}{(R-1)!}x^{cR-1}\exp\left(-\frac{Rx^c}{b^c}\right) \tag{7.81}$$

因此，检测概率为

$$P_d = \int_0^\infty \left[\int_{Tx}^\infty f_{D_1}(t)\mathrm{d}t\right] f_{\hat{b}}(x)\mathrm{d}x$$

$$= \int_0^\infty \exp\left[\frac{-(Tx)^2}{b^2(1+\mathrm{SCR})\Gamma\left(1+\frac{2}{c}\right)}\right]\left(\frac{R}{b^c}\right)^R \frac{c}{(R-1)!}x^{cR-1}\exp\left(-\frac{Rx^c}{b^c}\right)\mathrm{d}x \tag{7.82}$$

令

$$y = \frac{Rx^c}{b^c}$$

当 $SCR \gg 1$ 时，可得检测概率为

$$P_d = \frac{1}{(R-1)!}\int_0^\infty y^{R-1}\exp\left[\frac{-T^2}{(1+\mathrm{SCR})\Gamma\left(1+\frac{2}{c}\right)}\left(\frac{y}{R}\right)^{2/c} - y\right]\mathrm{d}y \tag{7.83}$$

为了衡量近似的精确程度，把式(7.83)的计算结果与 Monte Carlo 仿真结果进行对比。结论是：式(7.83)对瑞利起伏目标和 SCR≫1 的条件下精确地描述了任意形状参数已知的韦布尔背景中 MLH-CFAR 的检测概率。当 c 趋近于 2 时，对于任意 SCR 都是精确的。式(7.83)也可用于形状参数 c 已知时的删除 MLH-CFAR，其尺度参数由式(7.69)估计，而与式(7.83)的唯一区别是用 k 代替 R。

图 7.16 包含了 $P_{fa}=10^{-5}$，$R=16$ 时几种形状参数取值时式(7.83)的数值解。当 c 较小时，杂波的 PDF 有较长的拖尾，使检测阈值上升，这就解释了对应的 P_d 的下降原因。P_d 下降的另一个原因是随着 c 减小 b 的估计精度下降。下面将分析 CFAR 损失对 c 的依赖性。

2. 已知形状参数时的 CFAR 损失

在杂波功率水平已知且检测阈值固定的非 CFAR 情况(瑞利目标、韦布尔杂波)下，下面的两个关系成立：

$$P_{fa} = \exp\left(\frac{-S_\infty^c}{b^c}\right)$$

$$P_d = \exp\left[\frac{-\left(\frac{S_\infty}{b}\right)^2}{(1+\mathrm{SCR})\,\Gamma\left(1+\frac{2}{c}\right)}\right]$$

由上面两式得到 SCR、P_{fa} 和 P_d 三者间的关系为

$$\mathrm{SCR}_\infty = \frac{[\ln(1/P_{fa})]^{\frac{2}{c}}}{[\ln(1/P_d)]\,\Gamma\left(1+\frac{2}{c}\right)} - 1 \tag{7.84}$$

在使用 CFAR 检测器时，对于给定的 P_{fa}、P_d 和 k，首先利用式(7.66)由要求的 P_{fa} 值计算 T；然后利用这个 T，迭代求解式(7.83)得到 SCR(用 k 代替 R)，则 CFAR 损失为

$$\frac{\mathrm{SCR}(P_{fa}, P_d, c, k)}{\mathrm{SCR}_\infty(P_{fa}, P_d, c)} \tag{7.85}$$

图 7.17 给出了用上述过程计算的 $P_{fa}=10^{-5}$，$P_d=0.5$，$k=16$ 时的 CFAR 损失，结果表明，$c=0.8$ 时比 $c=2$ 时的 CFAR 损失增加了 2.4dB。比较图 7.16 和图 7.17，可以注意到，为保持 $P_d=0.5$，c 由 2 下降到 0.8 需增加 13dB 的信杂比(见图 7.16)，而这正是由上述 2.4dB 的附加 CFAR 损失(见图 7.17)

引起的。其中，前者指为了满足 $P_d=0.5$，$c=0.8$ 时所需 SCR 与 $c=2$ 时所需 SCR 相差 13dB，而后者是指 $c=0.8$ 时的 CFAR 损失比 $c=2$ 时多出 2.4dB。

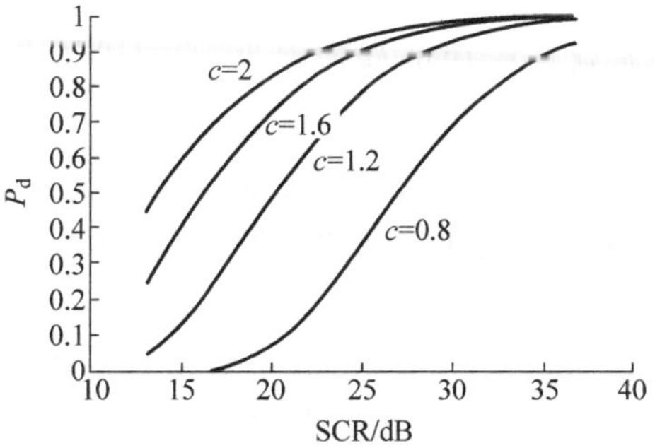

图 7.16 MLH-CFAR 的检测性能曲线

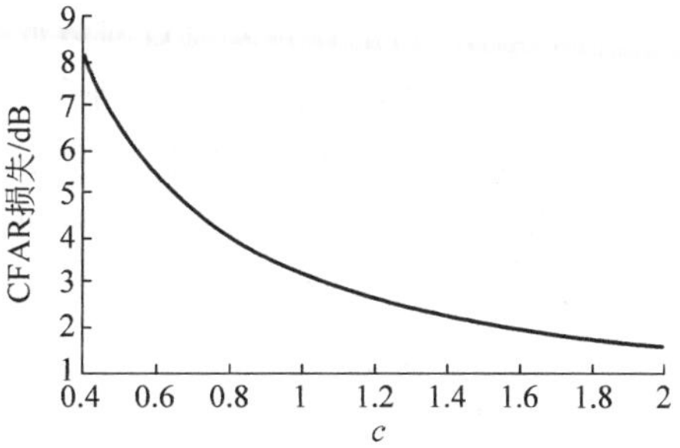

图 7.17 MLH 的 CFAR 损失和形状参数 c 的关系

3. 形状参数未知的情况

当 b 和 c 都由 MLH 方法估计时，得不到 $\hat{b}$ 和 $\hat{c}$ 分布的闭型解，为此给出如下的计算检测概率的近似方法：

$$P_d=\int_0^{\infty}\left[\int_S^{\infty} f_{D_1}(t)\mathrm{d}t\right] f_S(x)\mathrm{d}x \approx \Pr\{D>\mathrm{E}(S)\} \tag{7.86}$$

根据式(7.86)的假设，对于给定的 R、b、c 和 P_{fa}，通过仿真得到 $\mathrm{E}(S)$，然后用来计算 P_d。仍然假设检测单元中为瑞利目标回波加杂波，则检测概率为

$$P_d=\exp\left\{-\frac{\left[\dfrac{\mathrm{E}(S)}{b}\right]^2}{(1+\mathrm{SCR})\Gamma\left(1+\dfrac{2}{c}\right)}\right\} \tag{7.87}$$

其中 SCR 可表示为

$$\mathrm{SCR}=\frac{\mathrm{E}^2(S)}{b^2\ln\left(\dfrac{1}{P_d}\right)\Gamma\left(1+\dfrac{2}{c}\right)}-1$$

对于非 CFAR 情形，用 S_{∞} 代替 $\mathrm{E}(S)$ 也可以得到相同的表达式。因此有

$$\mathrm{SCR}_{\infty}=\frac{S_{\infty}^2}{b^2\ln\left(\dfrac{1}{P_d}\right)\Gamma\left(1+\dfrac{2}{c}\right)}-1$$

因此，CFAR 损失近似为

$$\frac{\mathrm{SCR}}{\mathrm{SCR}_{\infty}}\approx\frac{\mathrm{E}^2(S)}{S_{\infty}^2} \tag{7.88}$$

当 $R=16$，$P_{fa}=10^{-5}$，$P_d=0.5$，$c=1$ 时，由仿真得到的 SCR 为 37.8dB，非 CFAR 情况要求的 SCR 为 19.8dB，因此精确的 CFAR 损失为 18dB，而由式(7.88)得到的近似 CFAR 损失为 18.8dB，表明式(7.86)中的近似处理是可行的。

在形状参数未知的情况下，文献[16]还比较了 MLH 方法和 WH 算法的性能，结果表明，MLH 方法的 CFAR 损失一般要比 WH 算法的 CFAR 损失小。

7.5 BLUE-CFAR 检测器

7.2 节中的 Log-t 检测器首先通过对数放大器把韦布尔分布 PDF 转换成 Gumbel 分布的 PDF,它是一个位置-尺度类型分布,然后通过处理所有参考样本,对 Gumbel 变量的均值和标准差进行联合估计形成自适应阈值以确保 CFAR 性能。然而这种自适应方法不服从任何最优准则,也不适用于设定删除点的处理,这就使其性能易受非均匀杂波的影响。7.3 节介绍的 WH 算法,通过处理由参考样本形成的两个有序统计量来估计检测阈值,同样没有采用最优准则估计检测阈值。

为此,文献[17]和文献[35]分别提出了两种针对韦布尔和对数正态杂波背景的 CFAR 处理方案,它们通过由固定个数的参考样本有序统计量的线性组合估计检测阈值,并且允许从有序样本的高端和低端进行删除。组合权值是在估计的方差最小条件下确定的,且对位置和尺度参数的估计均是 BLUE。如果权值满足了这个条件,它的性能将超过 Log-t,即使在估计阈值之前有适当删除的情况下也是如此。

7.5.1 韦布尔背景中的 BLUE 检测器

1. 针对变化的尺度和形状参数的 CFAR 检测器

设$\{x_i\}$表示在包络检波中观测到的随机变量。在没有目标存在的 H_0 假设下,x 服从形状参数为 c、尺度参数为 b 的韦布尔分布(见式(7.1))。

在估计 b 和 c 的方法中,7.4 节中描述的 MLH 准则是一种可能的选择,然而,由于 MLH 的参数估计性能直接与样本个数有关,因此在小样本数时 MLH 估计不能获得最小方差和最优性能,且相应的 MLH 估计器对 b 和 c 的估计均没有闭型表达式,需要迭代求解。这样就产生了寻求新估计器的需求,这种估计器应该在较小计算复杂性的条件下获得与 MLH 估计器相当的性能。引入变量 $Y=\ln X$ 会使估计过程更加方便,实际中可在包络检波后插入对数放大器。对数的底数可以任意选择,一般采用自然对数。对数转换已经在雷达系统中广泛采用,它能在较大的杂波水平动态范围上工作,且能把假设的对数正态 PDF 变为更常用的正态 PDF。这里介绍对数转换的另一种重要作用,即在 H_0 假设下,转换后变量 Y 的 PDF 是位置-尺度类型的,即具有 Gumbel 函数形式。对式(7.1)的韦布尔分布作变量置换,设$(x/b)^c=\exp[(y-\alpha)/\beta]$,且使参数 $\alpha=\ln b$,$\beta=1/c$,即使 $y=\ln x$,可得

$$f_Y(y)=\frac{1}{\beta}\exp\left(\frac{y-\alpha}{\beta}\right)\exp\left[-\exp\left(\frac{y-\alpha}{\beta}\right)\right]$$

其中,$-\infty<\alpha<+\infty$是 Gumbel 分布位置参数,$\beta>0$ 是 Gumbel 分布尺度参数。在 H_0 假设下,表示为 $Y\sim\mathrm{Gu}(\alpha,\beta)$。Gumbel 分布的期望和方差为

$$\mathrm{E}(Y)=\alpha-\lambda\beta,\quad \mathrm{var}(Y)=\frac{\pi^2}{6}\beta^2$$

其中,$\gamma=\int_0^{\infty}(\ln x)\mathrm{e}^{-x}\,\mathrm{d}x\approx 0.5772$ 是 Euler 常数。如果 α 和 β 是已知参数,那么对于一个给定的 P_{fa},固定阈值 S_{∞} 为

$$S_{\infty}=\alpha+g_{\infty}\beta \tag{7.89}$$

S_{∞} 等于 Y 概率分布的$(1-P_{\mathrm{fa}})$分位点,$g_{\infty}=\ln(-\ln P_{\mathrm{fa}})$是标准 Gumbel 分布 $\mathrm{Gu}(0,1)$的$(1-P_{\mathrm{fa}})$分位点。

当参数未知时,需要一个统计量 S 满足式

$$\Pr\{Y > S \mid \mathrm{H}_0\} = P_{\mathrm{fa}} \tag{7.90}$$

并且不依赖于 α 和 β 值。由式(7.89)可知，对 S 的一个自然选择为

$$S = \hat{\alpha} + g\hat{\beta} \tag{7.91}$$

其中，$\hat{\alpha}$ 和 $\hat{\beta}$ 是 α 和 β 的估计，g 是阈值系数。

对数转换的优点是 α 和 β 可享有同变(equivariant)估计[17]。通常使用的位置和尺度参数估计器，包括 MLH 和线性估计器，如最优线性无偏估计器(BLUE)和最优线性不变估计器(BLIE)，实际上都是"同变"估计。如果在式(7.91)中使用同变估计 $\hat{\alpha}$ 和 $\hat{\beta}$，则[13]

$$\Pr\{Y > S \mid \mathrm{H}_0\} = \Pr\left\{\frac{Y - \hat{\alpha}}{\hat{\beta}} > g \mid \mathrm{H}_0\right\} = \text{常数} \tag{7.92}$$

这样对任意位置和尺度函数都可获得 CFAR 性能。这里有如下两点需要说明。

(1) 如果选择 g_{∞} 作为阈值系数，只有在 $R \to \infty$ 时式(7.92)所示概率的常数值才等于 P_{fa}。

假设 $\hat{\alpha}$ 和 $\hat{\beta}$ 是渐近无偏和一致的。用 $F_Q(u \mid \mathrm{H}_0) = \exp(-\mathrm{e}^u)$ 代表变量 $Q = (y - \alpha)/\beta$ 在 H_0 假设下的概率分布函数(CDF)的补，并且用 $f_\tau(u)$ 表示 τ 的 PDF，其中 τ 表示为

$$\tau = \frac{S - \alpha}{\beta} = \frac{\hat{\alpha} - \alpha}{\beta} + \frac{g\hat{\beta}}{\beta}$$

令 $R \to \infty$，则 τ 的方差趋近于零，$f_\tau(u)$ 趋近于 δ 函数 $\delta(u - g_{\infty})$。因此有

$$\Pr\{Y > S \mid \mathrm{H}_0\} = \Pr\{Q > \tau \mid \mathrm{H}_0\} = \int_{-\infty}^{+\infty} f_\tau(u) F_Q(u \mid \mathrm{H}_0)\mathrm{d}u \overset{R\to\infty}{\Rightarrow} \int_{-\infty}^{+\infty} \delta(u - g_{\infty}) F_Q(u \mid \mathrm{H}_0)\mathrm{d}u = P_{\mathrm{fa}}$$

(2) 对于有限的 R，如果 $g = g_{\infty}$，式(7.92)的概率值通常比 P_{fa} 大。

如果 $\hat{\alpha}$ 和 $\hat{\beta}$ 是使 $\mathrm{E}(\tau) = g_{\infty}$ 的无偏估计器，且进一步假设 $f_\tau(u) \approx 0 (u < 0)$，后一个假设在 P_{fa} 很小时是合适的，而且这也正是雷达目标检测所关心的情况。在这些假设下可得

$$\begin{aligned} \int_0^{\infty} f_\tau(u) F_Q(u \mid \mathrm{H}_0)\mathrm{d}u &\approx \int_{-\infty}^{+\infty} f_\tau(u) F_Q(u \mid \mathrm{H}_0)\mathrm{d}u \\ &= \mathrm{E}\{F_Q(\tau \mid \mathrm{H}_0) > F_Q[\mathrm{E}(\tau \mid \mathrm{H}_0)]\} = P_{\mathrm{fa}} \end{aligned}$$

上述推导利用了关于严格凸函数 $F_Q(u \mid \mathrm{H}_0)$ 的 Jensen 不等式[36]。

下面将在给定有限样本 R 和期望的 P_{fa} 值条件下，讨论阈值系数 g 的选择问题。

2. Gumbel 位置-尺度参数的线性估计

令 $Y_{(1)} \leqslant Y_{(2)} \leqslant \cdots \leqslant Y_{(R)}$ 表示服从 $\mathrm{Gu}(\alpha, \beta)$ 分布的有序样本，基于删除样本的 α 和 β 的线性估计被定义为

$$\hat{\alpha} = \sum_{i=r_1+1}^{R-r_2} A_i(R, r_1, r_2) Y_{(i)} \tag{7.93}$$

$$\hat{\beta} = \sum_{i=r_1+1}^{R-r_2} B_i(R, r_1, r_2) Y_{(i)} \tag{7.94}$$

其中，$A_i(R, r_1, r_2)$ 和 $B_i(R, r_1, r_2)$ 是依赖于样本数 R、低端删除点 r_1 和高端删除点 r_2 的权值。这些权值必须按照合适的最优准则计算，一旦权值给定，式(7.93)和式(7.94)就可计算出，并一直使用。与线性估计器计算量相比，MLH 估计的计算量很大，而且尽管 MLH 能保证渐近效率，但是对于有限样本 MLH 并不能获得最小方差。

如果 $\hat{\alpha}$ 和 $\hat{\beta}$ 是 α 和 β 的线性估计，且权值满足

$$\sum_{i=r_1+1}^{R-r_2} A_i(R, r_1, r_2) = 1 \tag{7.95}$$

$$\sum_{i=r_1+1}^{R-r_2} B_i(R, r_1, r_2) = 0 \tag{7.96}$$

那么，$\hat{\alpha}$ 和 $\hat{\beta}$ 是同变的。

比较线性估计器与一致最小方差(Uniformly Minimum Variance, UMV)估计器也是有价值的。假设这样的 UMV 估计器存在，就效率而论，线性估计器相对于它具有一定损失，但是这个损失在实际应用中是可以忽略的。然而，UMV 估计器经常是不存在的。实际上，一个参数 θ 达到 Cramer-Rao 方差下界的估计器存在的必要条件，就是 θ 的 UMV 无偏估计器存在的必要条件，它要求属于指数类分布。当然，韦布尔分布不满足这个条件，因此 b 和 c 的最小方差无偏估计器不存在。对 Gumbel 分布的 α 和 β 也有同样的结论。

在线性估计器中，BLUE 就是在所有无偏估计器中具有最小方差的线性估计器。令 $\boldsymbol{Z}=(Z_{(r_1+1)}, Z_{(r_1+2)}, \cdots, Z_{(R-r_2)})^{\mathrm{T}}$ 表示服从 Gu(0,1)分布经过排序的有序统计量构成的向量，E($\boldsymbol{Z}$)是它的均值向量，$\boldsymbol{B}$ 是其协方差矩阵，$\boldsymbol{C}=(\mathbf{1}, \mathrm{E}(\boldsymbol{Z}))$是一个$(R-R_1-R_2)\times 2$ 阶辅助矩阵，其中 $\mathbf{1}$ 为 1 的列矢量，$\boldsymbol{Y}=[Y_{(r_1+1)}, Y_{(r_1+2)}, \cdots, Y_{(R-r_2)}]^{\mathrm{T}}$ 是服从未知参数 Gumbel 分布的有序统计量向量。注意对于 Gu(0,1)来说，E($\boldsymbol{Z}$)是定值。那么 BLUE 为[17]

$$\begin{bmatrix} \hat{\alpha} \\ \hat{\beta} \end{bmatrix} = (\boldsymbol{C}^{\mathrm{T}}\boldsymbol{B}^{-1}\boldsymbol{C})^{-1}\boldsymbol{C}^{\mathrm{T}}\boldsymbol{B}^{-1}\boldsymbol{Y} \tag{7.97}$$

式(7.97)左边矢量具有对称协方差矩阵

$$(\boldsymbol{C}^{\mathrm{T}}\boldsymbol{B}^{-1}\boldsymbol{C})^{-1}\beta^2 = \beta^2 \begin{bmatrix} A_{11} & A_{12} \\ A_{12} & A_{22} \end{bmatrix} \tag{7.98}$$

对于样本数小于或等于 25，且只从高端删除的情况，用式(7.93)和式(7.94)形式表示式(7.97)的系数，则 $A_i(R, r_1, r_2)$和 $B_i(R, r_1, r_2)$可以由文献[37]中的表得到。实际上，这些表格是指最优线性不变估计器(Best Linear Invariant Estimator, BLIE)，也就是在转换中估计器的最小均方误差不变。然而，BLIE 和 BLUE 的系数是相互联系的。令$(\hat{\alpha}_{\mathrm{BLI}}, \hat{\beta}_{\mathrm{BLI}})$和$(\hat{\alpha}_{\mathrm{BLU}}, \hat{\beta}_{\mathrm{BLU}})$分别表示 α 和 β 的 BLIE 和 BLUE，则下面的关系式成立：

$$\hat{\alpha}_{\mathrm{BLI}} = \hat{\alpha}_{\mathrm{BLU}} - \frac{\hat{\beta}_{\mathrm{BLU}}}{1+A_{22}}A_{12} \tag{7.99}$$

$$\hat{\beta}_{\mathrm{BLI}} = \frac{\hat{\beta}_{\mathrm{BLU}}}{1+A_{22}} \tag{7.100}$$

对于大样本值及由有序样本序列两端进行删除的情况，能够用上述的过程计算出对应的系数。文献[17]的作者编制了计算 BLUE 系数的程序。通过式(7.99)和式(7.100)的关系，上述程序的计算结果可与文献[37]中的结果很好地吻合(至少在 6 个小数位上)。

对应于上述两种线性估计准则，阈值可以分别估计为

$$S_{\mathrm{BLI}} = \hat{\alpha}_{\mathrm{BLI}} + g_{\mathrm{BLI}}\hat{\beta}_{\mathrm{BLI}} \tag{7.101}$$

$$S_{\mathrm{BLU}}=\hat{\alpha}_{\mathrm{BLU}}+g_{\mathrm{BLU}}\hat{\beta}_{\mathrm{BLU}} \tag{7.102}$$

其中，g_{BLI} 和 g_{BLU} 是按照要求的 P_{fa} 设置的阈值系数，有趣的是这两个阈值经常是重合的。实际上，由 CFAR 的条件

$$\Pr\left\{\frac{Y-\hat{\alpha}_{\mathrm{BLI}}}{\hat{\beta}_{\mathrm{BLI}}}>g_{\mathrm{BLI}}\mid \mathrm{H}_0\right\}=\Pr\left\{\frac{Y-\hat{\alpha}_{\mathrm{BLU}}}{\hat{\beta}_{\mathrm{BLU}}}>g_{\mathrm{BLU}}\mid \mathrm{H}_0\right\}=P_{\mathrm{fa}}$$

可知，它们服从关系

$$g_{\mathrm{BLI}}=(1+A_{22})g_{\mathrm{BLU}}+A_{12}$$

这样，通过代换式(7.102)，可得

$$S_{\mathrm{BLI}}=S_{\mathrm{BLU}}$$

因此，对 CFAR 性能和检测性能不会产生影响。

MLH、BLIE 和 BLUE 都是渐近有效的和渐近正态的。当 R 趋近于无穷时，它们都趋近于各自的 Cramer-Rao 界。对于有限样本，Monte Carlo 仿真表明 MLH 和线性估计器具有几乎相同的精度，但是后者在小样本数时更可取。

3. 阈值系数的选择

为了使 CFAR 检测器在给定的虚警概率 P_{fa} 上工作，选择的系数 g 一定要满足式(7.65)。由于 $\hat{\alpha}$ 和 $\hat{\beta}$ 是同变估计，因而归一化变量 τ 的概率密度函数 $f_\tau(u;g)$ 不依赖于 α 和 β。用这种表示方法在于强调阈值系数 g 的作用，g 由式(7.103)的解确定，即

$$\int_{-\infty}^{+\infty}F_Q(u\mid \mathrm{H}_0)f_\tau(u;g)\mathrm{d}u=P_{\mathrm{fa}} \tag{7.103}$$

如果 $f_\tau(u;g)$ 已知，问题就可以解决了。实际上 τ 的一、二阶矩是已知的，因为

$$\mathrm{E}(\tau)=g,\quad \sigma_\tau^2=A_{11}+g^2A_{22}+2gA_{12} \tag{7.104}$$

确定 τ 的 PDF 的方法有几种在原理上是可信的。基于 S 的渐进正态性和 τ 的渐进正态性，以及 $\hat{\beta}/\beta$ 是近似 Gamma 分布的结果，用正态分布 PDF 和 Gamma 分布 PDF 的混合形式近似 τ 的 PDF，即

$$f_\tau(u;g)=\xi f_{\mathrm{N}}(u)+(1-\xi)f_\Gamma(u) \tag{7.105}$$

其中，$\xi\in[0,1]$，$f_{\mathrm{N}}(u)$ 表示正态 PDF，$f_\Gamma(u)$ 表示 Gamma 分布 PDF，也就是

$$f_\Gamma(u)=\frac{\eta^{\upsilon}u^{\upsilon-1}}{\Gamma(\upsilon)}\exp(-\eta u),\quad u\geqslant 0$$

当且仅当两个 PDF 具有相同的数学期望 g 和相同的方差 σ_τ^2 时，式(7.104)的条件才能得以满足。因此，Gamma 分布的两个参数为

$$\eta=g/\sigma_\tau^2,\quad \upsilon=g\eta$$

另外，式(7.105)中的权值 ξ 依赖于多个因素。ξ 的一个经验公式为[17]

$$\xi=\exp(1.327-0.760g+0.120\sigma_\tau^2-0.106r_2+0.017R) \tag{7.106}$$

通过比较由式(7.103)、式(7.105)、式(7.106)计算的 g 和由 Monte Carlo 仿真得到的 g，证实这个方法是可行的。这样就可以通过计算得到具有足够精度的阈值系数，至少在 $8\leqslant R\leqslant 32$ 范围内可以满足精度要求。

4. 性能分析

在检测单元中同时存在 Swerling Ⅱ 目标回波和杂波情况下，因 BLUE 检测概率难以获得闭型表达

式,可以利用仿真和近似的方法分析检测性能。

图 7.18 和图 7.19 表示了 BLUE、CA 和 Log-t CFAR 的检测性能曲线,其中分别对应 $R=8$ 和 $R=32$。形状参数固定为 $c=2$,否则比较是没有意义的,因为 CA 的虚警率随着 c 发生变化。双参数的估计方法相对于 CA 的检测损失在 $0.5<P_d<0.9$ 范围变化很小。由于双参数估计方法利用的关于背景分布的先验信息较少,其相对于单参数的 CA 方法的检测损失是固有的。随着样本数 R 的增加,损失明显减少。需要强调的是,如果不考虑尺度和形状参数而只考虑保持 CFAR 特性,那么 CA 和其他的单参数方法在某些情况下将失去作用。在双参数方法中,BLUE 相对于 Log-t 具有 1~2dB 的检测优势,这个差别还取决于 P_{fa} 的设计值和样本数 R。

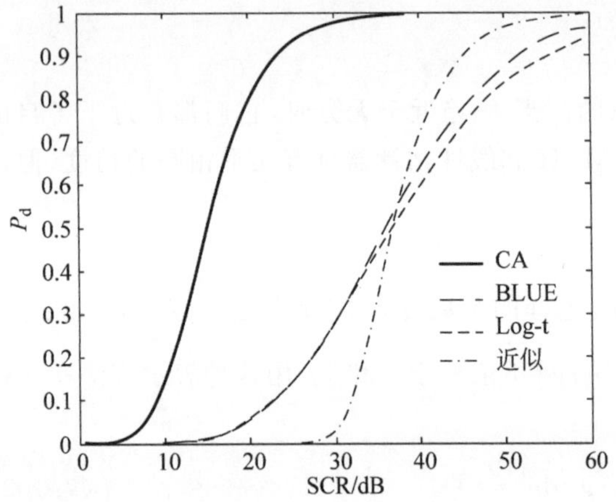

图 7.18 BLUE、CA 和 Log-t 检测器的检测性能曲线
($R=8, P_{fa}=10^{-5}$)

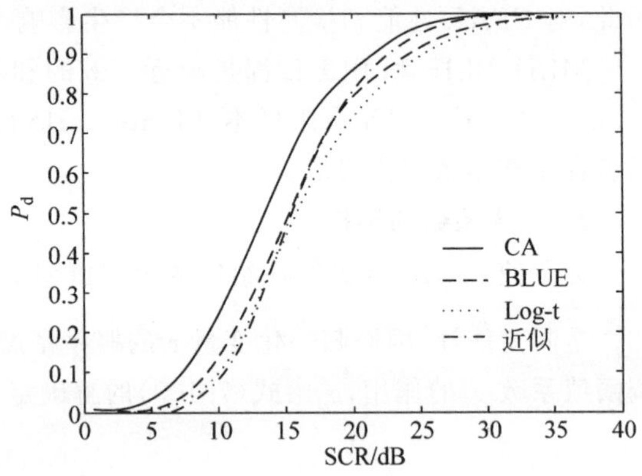

图 7.19 BLUE、CA 和 Log-t 检测器的检测性能曲线
($R=32, P_{fa}=10^{-5}$)

形状参数对检测性能的影响见图 7.20,它包含 $c=1,2,3$ 三种情况。很明显,形状参数具有很强的影响,在 c 由 3 变到 1 时,性能下降超过 10dB。在图 7.19 所示的条件下,$c=1$ 是最差的情况,实际上这对应有代表性的尖峰杂波情况,即与瑞利情况相比具有很高的概率拖尾。比较结果表明,BLUE 相比 Log-t 具有一定的检测优势,且随着杂波尖锐程度而增加,在 $c=1$ 时获得了 3dB 多的性能优势。可以推测,这个检测优势将随着 P_{fa} 下降而增加。

图 7.21 反映的是删除的影响。由图 7.21 可见,适度的删除既可以对抗多个干扰目标,又不使检测性能明显下降,例如 $r_1=0, r_2=1$ 和 $r_1=0, r_2=2$。过度的删除能对抗杂波边缘,例如,$r_1=6$,$r_2=6$,但却使检测性能明显下降,特别是对于低 P_{fa} 值。与不需要删除过程的 Log-t 相比,$r_2=2$ 的 BLUE 的性能仍然优于 Log-t。也就是说,在不损失均匀环境中检测能力的条件下,BLUE 获得了对抗干扰目标的能力。

上述分析表明,检测性能依赖于大量的系统参数和必要的条件(R, r_1, r_2, P_{fa})以及单个分布参数,也就是 c 或 β。通过由仿真得到的一个工作特性集合来选择系统参数是不实际的,因为这个集合要覆盖所有的参数组合情况,所以是十分庞大的。因此可使用式(7.107)来近似检测概率,即

$$P_d \approx \exp\left[-\frac{e^{2\beta g}}{\rho\Gamma(1+2\beta)}\right],\quad \rho \gg 1 \tag{7.107}$$

式中,ρ 的定义见文献[17],这个近似式至少可以在设计初始阶段使用。除了 R 较小的情况,式(7.107)可以在较宽范围内精确地预测性能,例如,$P_d>0.8$。图 7.18 和图 7.19 的"近似"就是指上述仿真的近

似结果，并且在 $R\to\infty$ 时有

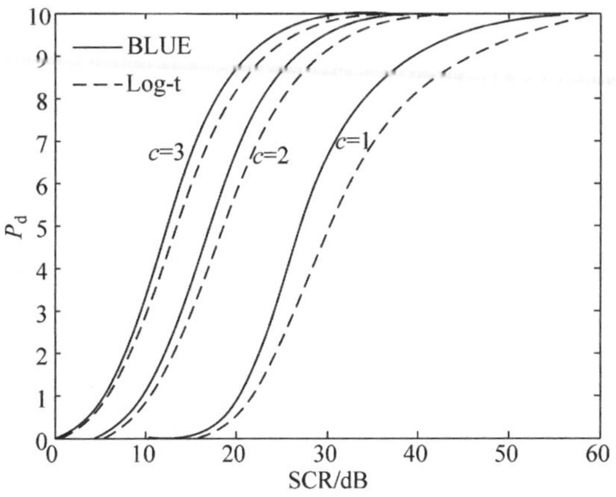

图 7.20　不同 c 值时的 BLUE 和 Log-t 检测器的检测性能曲线（$R=24, P_{fa}=10^{-5}$）

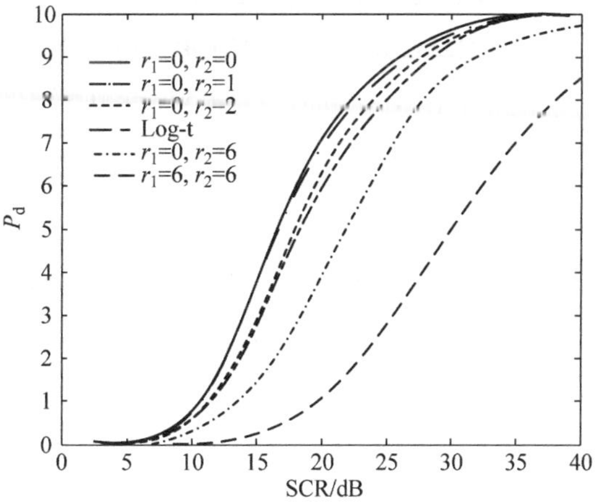

图 7.21　BLUE 检测器在几种删除情况下的检测性能曲线

$$\lim_{R\to\infty} P_d \approx \exp\left[-\frac{e^{2\beta g_{\infty}}}{\rho\Gamma(1+2\beta)}\right], \quad \rho \gg 1 \tag{7.108}$$

如果式(7.107)的 SCR 增长到原来的 $\exp[2\beta(g-g_{\infty})]$ 倍，它将获得与式(7.108)相同的检测概率，这也就是检测损失

$$\Delta = 10\log e^{2\beta(g-g_{\infty})} = \frac{8.68(g-g_{\infty})}{c} \quad \text{(dB)}$$

这个检测损失是 ATD(Average Threshold Deviation)的 8.68 倍。对于给定的系统参数，检测损失反比于 c，这与图 7.20 的结果一致。表 7.8 给出了 $c=1$ 时，对于不同样本数和删除数组合的归一化 ATD 值。4 处横线表示 ATD 太大，以至于不实用，这发生在删除后剩余的样本数很少的情况。

表 7.8　几种样本数和删除数时的归一化 ATD($P_{fa}=10^{-5}$)

R	$r_1=0$				$r_1=0.25R$			
	0	1	2	0.25R	0	1	2	0.25R
32	0.651	0.713	0.776	1.42	0.819	0.928	1.04	2.31
24	0.907	1.05	1.28	2.11	1.25	1.51	1.80	3.68
20	1.22	1.46	1.72	2.78	1.63	2.06	2.58	5.01
16	1.67	2.11	2.62	3.97	2.33	3.19	4.26	7.30
12	2.65	3.66	4.87	6.35	3.96	5.91	8.45	14.0
8	5.56	8.55	16.0	—	8.64	—	—	—

7.5.2　对数正态背景中的 BLUE-CFAR 检测器

对于对数正态背景，与 7.5.1 节相似，在经过对数变换 $Y=\ln X$ 之后，Y 的 PDF 也是位置-尺度型函数。下面的分析将表明，在对数正态背景中使用 BLUE 获取 CFAR 也是有利的。下面还将给出有关系

统参数设计的指导思想,以及利用检测性能曲线和 ATD 值的性能评价,并将与 Log-t 进行比较。BLUE 在多目标和杂波边缘环境中的检测性能比 Log-t 有所提高。

1. 对数正态杂波背景中的 BLUE-CFAR 检测器

用 X 代表包络检测中观测到的随机变量。在纯杂波假设(H_0 假设下),X 服从参数为 μ 和 σ 的对数正态分布,即

$$f_X(x)=\frac{1}{\sqrt{2\pi}\sigma x}\exp\left[-\frac{(\ln x-\mu)^2}{2\sigma^2}\right],\quad x\geqslant 0$$

将上述关系表示为 $H_0: X\sim L(\mu,\sigma)$。仍然令 $Y=\ln X$,Y 是对数放大器的输出,根据对数正态分布的定义可知,Y 服从均值和方差分别为 μ 和 σ 的高斯分布,表示为 $H_0: Y\sim N(\mu,\sigma)$。式(7.90)在这里仍然适用。如果 S 是一个不依赖于 μ 和 σ 的统计量,就可以获得 CFAR。如果 $\hat{\mu}$ 和 $\hat{\sigma}$ 是 μ 和 σ 的同变估计,则该统计量可表示为

$$S=\hat{\mu}+h\hat{\sigma} \tag{7.109}$$

其中,h 是阈值系数,可根据 P_{fa} 的设计值确定。

令 $Y_{(1)}\leqslant Y_{(2)}\cdots\leqslant Y_{(R)}$ 表示服从 $N(\mu,\sigma)$ 分布的有序样本。基于删除后的有序样本集,构成的 μ 和 σ 的线性估计分别为

$$\hat{\mu}=\sum_{i=r_1+1}^{R-r_2}A_i(R,r_1,r_2)Y_{(i)},\quad \hat{\sigma}=\sum_{i=r_1+1}^{R-r_2}B_i(R,r_1,r_2)Y_{(i)}$$

其中,$A_i(R,r_1,r_2)$和 $B_i(R,r_1,r_2)$是适当的权值。若 A_i 和 B_i 满足式(7.95)和式(7.96),则 $\hat{\mu}$ 和 $\hat{\sigma}$ 是 μ 和 σ 的同变估计。与韦布尔分布中的估计一样,这里的 A_i 和 B_i 也是计算一次后就可以一直使用。

令 $\boldsymbol{Y}=\left(Y_{(r_1+1)},Y_{(r_1+2)},\cdots,Y_{(R-r_2)}\right)^{\mathrm{T}}$ 表示删除后剩余的服从具有未知参数 $N(\mu,\sigma)$分布的有序样本向量,$\boldsymbol{\Omega}$ 是它的未知协方差矩阵。更进一步,令 $\boldsymbol{Z}=(Z_{(r_1+1)},Z_{(r_1+2)},\cdots,Z_{(R-r_2)})^{\mathrm{T}}$ 为服从 $N(0,1)$ 分布的有序统计量向量,$E(\boldsymbol{Z})$是它的均值向量,$\boldsymbol{B}$ 是它的协方差矩阵,$\boldsymbol{C}=[\boldsymbol{1},E(\boldsymbol{Z})]$是$(R-r_1-r_2)\times 2$ 阶辅助矩阵,则容易得到

$$E(\boldsymbol{Y})=\boldsymbol{C}(\mu,\sigma)^{\mathrm{T}},\quad \boldsymbol{\Omega}=\sigma^2\boldsymbol{B}$$

未知参数 μ 和 σ 的线性估计 $\hat{\mu}$ 和 $\hat{\sigma}$ 可以由广义最小二乘法得到,即

$$(\hat{\mu},\hat{\sigma})^{\mathrm{T}}=(\boldsymbol{C}^{\mathrm{T}}\boldsymbol{B}^{-1}\boldsymbol{C})^{-1}\boldsymbol{C}^{\mathrm{T}}\boldsymbol{B}^{-1}\boldsymbol{Y}$$

这些估计是无偏和同变的,并且在所有线性估计器中具有最小方差。更进一步,像 MLH 估计一样,BLUE 是渐进有效和渐进正态的。注意到 $\hat{\mu}$ 和 $\hat{\sigma}$ 通常是相关的,并且它们的协方差矩阵为

$$\boldsymbol{\Sigma}=\sigma^2(\boldsymbol{C}^{\mathrm{T}}\boldsymbol{B}^{-1}\boldsymbol{C})^{-1}$$

为方便起见,可改写成

$$\boldsymbol{\Sigma}=\sigma^2\begin{bmatrix}A_{11} & A_{12}\\ A_{12} & A_{22}\end{bmatrix} \tag{7.110}$$

对于样本数不大于 20 以及由有序样本序列两端进行删除的样本数情况,文献[38]提供了 $A_i(R,r_1,r_2)$和 $B_i(R,r_1,r_2)$的表格。为了考虑 R 直到 32 的较多样本,可以用前述的过程计算这些量。文献[35]的作者也编制了相应的计算机程序。

与上述线性估计准则对应的阈值由式(7.109)给出。其中,h 可以近似表示为[35]

$$h\approx\sqrt{\frac{1+A_{11}}{1+A_{22}}}\,t_{\eta,1-P_{fa}} \tag{7.111}$$

式(7.111)中,$t_{\eta,1-P_{\mathrm{fa}}}$ 是自由度为 η 的 Student-t 分布的$(1-P_{\mathrm{fa}})$分位点,η 表示为

$$\eta=0.230+\frac{1}{2A_{22}}$$

随着 R 趋近于无穷,由于 BLUE 的一致性,A_{11} 和 A_{22} 都趋近于零,并且 η 趋近于无穷。因此,式(7.111)的 h 趋近于极限值 h_{∞},也就是 N(0,1)的$(1-P_{\mathrm{fa}})$分位点。

2. 性能评价

假设目标回波信号的复包络具有瑞利分布的幅度和均匀分布的相位。就单脉冲检测而言,这对应于 Swerling Ⅰ型和 Swerling Ⅱ型起伏模型。

首先分析均匀杂波背景中的性能。若 Y 表示检测单元的观测值,则检测概率为

$$P_{\mathrm{d}}=\Pr\{Y\geqslant S\mid \mathrm{H}_1\}=\int_{-\infty}^{+\infty}P_Y(\mu\mid \mathrm{H}_1)f_S(u)\mathrm{d}u$$

其中,$P_Y(\mu\mid \mathrm{H}_1)$表示 Y 在 H_1 假设下 CDF 的补。由于 $P_Y(\mu\mid \mathrm{H}_1)$没有闭式表示,直接的解析分析是不可行的,因此需要用仿真或近似的方法。在高 SCR 和大样本数时,有[30]

$$P_{\mathrm{d}}\approx\exp\left[-\frac{\exp[2(h\sigma-\sigma^2)]}{\mathrm{SCR}}\right],\quad \mathrm{SCR}\geqslant 1 \tag{7.112}$$

这个表达式突出了不同的分布和设计参数的作用,表明一旦 SCR 给定,检测性能就不受尺度参数 μ 的影响,而只依赖于形状参数 σ。随着 σ 增加,也就是杂波变得尖锐,检测性能单调下降。

R 和 r_1,r_2 的作用隐含在阈值系数 h 中,不是很直观。为了定量地衡量这些影响,把 ATD 定义为

$$\mathrm{ATD}=\sigma(h-h_{\infty})=\sigma\alpha_L$$

其中,α_L 是归一化($\sigma=1$)的 ATD。ATD 代表阈值均值之间的差,即 $\mathrm{E}[S]=\mu+h\sigma$ 和已知杂波参数时的固定阈值 $S_{\infty}=\mu+h_{\infty}\sigma$ 之间的差。$h-h_{\infty}$ 越大,这个 CFAR 过程的检测损失越大。对式(7.112)取极限可得到理想固定阈值检测器的 P_{d},在相同 P_{d} 条件下,相对于已知杂波参数时固定阈值检测的 SCR,式(7.112)中 SCR 的增长量(检测损失)为

$$\Delta=8.68\sigma\alpha_L$$

对于 Log-t 也有相似的结论,此时有[35]

$$P_{\mathrm{d}}\approx\exp\left\{-\frac{\exp2[h_{\text{Log-t}}\mathrm{E}(\hat{\sigma}_{\text{Log-t}})-\sigma^2]}{\mathrm{SCR}}\right\},\quad \mathrm{SCR}\geqslant 1$$

这样,Log-t 的 ATD 为

$$\mathrm{ATD}_{\text{Log-t}}=\sigma\left[\sqrt{\frac{2}{R}}\frac{\Gamma\left(\frac{R}{2}\right)}{\Gamma\left(\frac{R-1}{2}\right)}h_{\text{Log-t}}-h_{\infty}\right]=\sigma\alpha_{\text{Log-t}}$$

在这种情况下,Log-t 检测器所对应的检测损失就是其 ATD 的 8.68 倍。因此,对于给定的系统参数,BLUE 和 Log-t 的检测损失都正比于 σ。表 7.9 给出了 BLUE 的归一化 ATD 值,同时也给出 Log-t 的归一化 ATD 值以供对比。

由前面的近似可以看出,尽管可以保证 P_{fa} 不依赖于形状参数 σ,但是该参数却对检测性能有很大影响,见图 7.22 和图 7.23。图 7.22 和图 7.23 中考虑了 $R=24$,以及 $P_{\mathrm{fa}}=10^{-3}$ 和 $P_{\mathrm{fa}}=10^{-5}$ 两种情形,并且包含了三种不同的处理:BLUE(0,0),即没有删除的 BLUE;Log-t;BLUE(2,2),指有序样本

序列的高端和低端各删除两个样本。图 7.22 和图 7.23 分别对应于 $\sigma=0.355$(适度的拖尾)和 $\sigma=1.147$(严重拖尾)两种情况。在由适度到严重拖尾杂波变化中,每种方法的性能都下降了 10～20dB 的量级。由图 7.22 和图 7.23 可知,BLUE(0,0)和 Log-t 实际上是重合的,这证实了 BLUE 的效率与 MLH 估计十分接近。另外,删除方法会带来一定的检测损失。

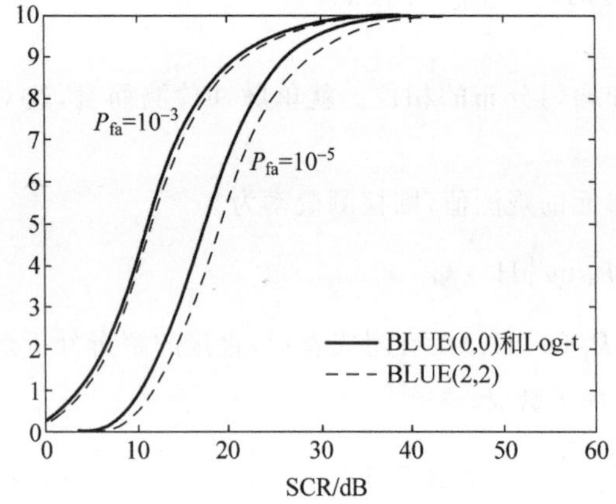

图 7.22 BLUE 和 Log-t 检测器在对数正态杂波中的检测性能曲线($\sigma=0.355$)

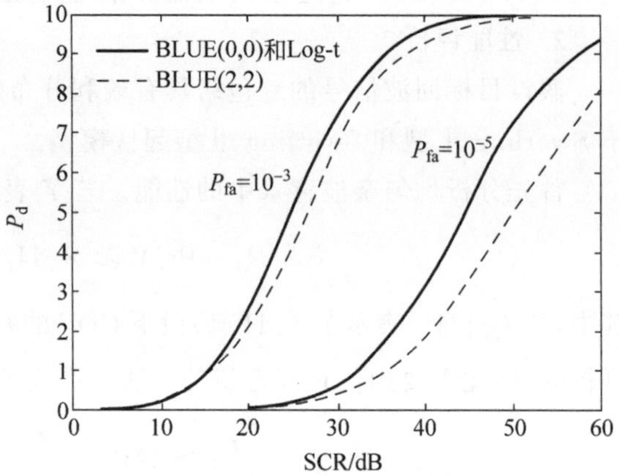

图 7.23 BLUE 和 Log-t 检测器在对数正态杂波中的检测性能曲线($\sigma=1.147$)

表 7.9 BLUE 在几种样本数和删除数的 ATD($P_{fa}=10^{-5}$)

R	$r_1=0$				$r_1=0.25R$				Log-t
	0	1	2	0.25R	0	1	2	0.25R	
32	0.806	0.849	0.896	1.27	1.27	1.38	1.50	2.69	0.793
28	0.945	1.00	1.07	1.51	1.51	1.67	1.84	3.31	0.940
24	1.14	1.23	1.33	1.86	1.86	2.09	2.36	4.29	1.13
20	1.44	1.58	1.74	2.42	2.42	2.81	3.30	6.00	1.43
16	1.95	2.21	2.53	3.43	3.43	4.22	5.32	9.60	1.92
12	3.00	3.63	4.49	5.73	5.73	8.01	12.0	20.5	2.96
8	6.19	9.03	14.6	14.6	14.6	31.2	102	102	6.07

由上述讨论可见,尽管式(7.85)的近似在理论上只有在 SCR→∞,P_d→1 时成立,但是在很宽的使用范围内,例如 $P_d>0.80$,都能得到较精确的性能预测,而且对于其他实用 R 值,这个近似也是正确的。

文献[35]还分析了 BLUE 在对数正态包络的杂波边缘非均匀环境中的性能。具体给出了杂噪比 CNR=10dB 的杂波边缘中,$\sigma=0.355$ 和 $\sigma=1.147$ 对应的 BLUE(0,0)、BLUE(2,2)、Log-t 及单参数方法的虚警控制性能。结果表明,单参数方法要比双参数方法的虚警控制能力差得多,这正是双参数方法的优势。考虑到单参数方法是在形状参数确知的条件下工作的,那么在参数变化的情况下,单参数方法性能下降将更严重。文献[35]在多目标环境中的性能分析表明,BLUE(0,0)和 BLUE(2,2)都优于 Log-t,即 BLUE 在一定程度上避免了遮蔽效应的影响。

7.6 Pearson 分布背景下的 CFAR 检测器

α 稳定分布的概念最早由 Levy 在研究广义中心极限定理时引入，其特征函数具有高斯分布特征函数的形式，因此其是高斯分布的推广，换言之，高斯分布是 α 稳定分布的特例($\alpha=2$)[39]。低阶 α 稳定分布的 PDF 相比高斯分布的 PDF 有着更长的拖尾，即远离均值或中值的尖峰样本数较多，α 值越小，其 PDF 的拖尾越长。α 稳定分布在雷达目标检测中的应用潜力已在部分实测数据中得到验证[40-41]，但其统计特性一般通过特征函数进行描述，且其相应的二阶矩并不存在；除少数特定情况外，α 稳定分布没有统一闭式的 PDF 表达式，这在一定程度上限制了其实际应用。特例中除高斯分布外，当 $\alpha=1$ 且对称参数 $\beta=0$ 时，α 稳定分布退化为特定的 Cauchy 分布；而当 $\alpha=0.5$ 且对称参数 $\beta=-1$ 时，α 稳定分布退化为特定的 Pearson 分布。本节将简要介绍 Pearson 分布背景下的 CFAR 检测器，主要讨论现有 CA-CFAR 和 OS-CFAR 等在 Pearson 分布背景下的有效性[42]。

7.6.1 Pearson 分布背景下的 CA-CFAR 检测器

假设平方律检波后，N 个参考单元的样本 $X_1, X_2, \cdots, X_N$ 服从 Pearson 分布，则其 PDF 可表示为[43]

$$p_X(x)=\begin{cases}\dfrac{\gamma}{\sqrt{2\pi}}\dfrac{1}{x^{3/2}}\mathrm{e}^{-\gamma^2/2x}, & x\geqslant 0\\ 0, & x<0\end{cases} \tag{7.113}$$

相应的 CDF 可表示为

$$P_X(x)=\Pr\{X_i\leqslant x\}=\begin{cases}2\left[1-\Phi\left(\dfrac{\gamma}{\sqrt{x}}\right)\right], & x\geqslant 0\\ 0, & x<0\end{cases} \tag{7.114}$$

式(7.114)中，γ 为 Pearson 分布的尺度参数，$\Phi(x)$ 表示标准高斯分布的 CDF。

对于 CA-CFAR 检测方法来说，杂波功率水平估计为 $Z=(1/N)\sum_{i=1}^{N}X_i$，由于其是 N 个 IID 的 Pearson 分布随机变量的平均值，因此 Z 仍服从 Pearson 分布，且相应的尺度参数为 $\gamma_Z=\sqrt{N}\gamma$，即 Z 的 PDF 可表示为

$$p_Z(z)=\begin{cases}\dfrac{\sqrt{N}\gamma}{\sqrt{2\pi}}\dfrac{1}{z^{3/2}}\mathrm{e}^{-N\gamma^2/2z}, & z\geqslant 0\\ 0, & z<0\end{cases} \tag{7.115}$$

此时，相应的虚警概率可表示为

$$\begin{aligned}P_{\mathrm{fa}}^{\mathrm{CA}}&=\int_0^{\infty}\Pr\{D\geqslant TZ\mid \mathrm{H}_0\}\,p_Z(z)\mathrm{d}z\\&=\sqrt{\frac{2N}{\pi}}\int_0^{\infty}\mathrm{erf}(y/\sqrt{2T})\,\mathrm{e}^{-Ny^2/2}\mathrm{d}y\end{aligned} \tag{7.116}$$

式中，T 为标称化因子，误差函数 $\mathrm{erf}(y)=\dfrac{2}{\sqrt{\pi}}\int_0^y \mathrm{e}^{-t^2}\mathrm{d}t$。由式(7.116)可以看出，虚警概率受比例因子 T 控制，且与 Pearson 分布的尺度参数 γ 无关，即 CA 方法在 Pearson 背景下仍是 CFAR 的[42]。

关于检测概率,可考虑瑞利起伏目标的情形,对于参数为 σ_s^2 的瑞利起伏目标,平方律检波器的输出服从指数分布[23],则相应的 CA-CFAR 检测概率可表示为

$$P_{\mathrm{d}}^{\mathrm{CA}}=\int_0^{\infty}\Pr\{D\geqslant TZ \mid \mathrm{H}_1\}\,p_Z(z)\mathrm{d}z=\int_0^{\infty}\mathrm{e}^{-TZ/2\sigma_s^2}\frac{\sqrt{N}\gamma}{\sqrt{2\pi}}\frac{1}{z^{3/2}}\mathrm{e}^{-N\gamma^2/2z}\mathrm{d}z$$

$$=\sqrt{\frac{2N}{\pi}}\int_0^{\infty}\mathrm{e}^{(-\gamma^2/\sigma_s^2)(T/2y^2)}\mathrm{e}^{-Ny^2/2}\mathrm{d}y \tag{7.117}$$

式(7.117)表明,检测概率和杂波尺度参数 γ 与瑞利起伏目标功率参数 σ_s 的比值有关。

7.6.2 Pearson 分布背景下的 OS-CFAR 检测器

对于 OS-CFAR 检测方法来说,杂波功率水平估计 $Z=X_{(k)}$ 是一个有序的统计值[20],其 PDF 可以根据参考单元样本的 $p_{X_i}(x)$ 和 $P_{X_i}(x)$ 来确定,即

$$p_{X_{(k)}}(x)=p_k(x)=\frac{1}{B(k,N-k+1)}[1-P_X(x)]^{N-k}\times[P_X(x)]^{k-1}p_X(x) \tag{7.118}$$

其中,$B(a,b)$ 表示 Beta 函数。

因此 Pearson 分布随机变量 $X_1,X_2,\cdots,X_N$ 的第 k 个有序统计量的 PDF 为

$$p_k(x)=\frac{[2\Phi(\gamma/\sqrt{x})-1]^{N-k}}{B(k,N-k+1)}[1-\Phi(\gamma/\sqrt{x})]^{k-1}\frac{\gamma}{\sqrt{2\pi}}\frac{2^{k-1}}{x^{3/2}}\mathrm{e}^{-\gamma^2/2x} \tag{7.119}$$

此时,OS-CFAR 的虚警概率可表示为

$$P_{\mathrm{fa}}^{\mathrm{OS}}=\int_0^{\infty}\Pr\{D\geqslant Tx \mid \mathrm{H}_0\}\,p_k(x)\mathrm{d}x$$

$$=\int_0^{\infty}[2\Phi(\gamma/\sqrt{Tx})-1]\frac{[2\Phi(\gamma/\sqrt{x})-1]^{N-k}}{B(k,N-k+1)}\times$$

$$[1-\Phi(\gamma/\sqrt{x})]^{k-1}\frac{\gamma\mathrm{e}^{-\gamma^2/2x}}{\sqrt{2\pi}}\frac{2^{k-1}}{x^{3/2}}\mathrm{d}x \tag{7.120}$$

利用变量代换 $y=\gamma/\sqrt{x}$,虚警概率可进一步表示为

$$P_{\mathrm{fa}}^{\mathrm{OS}}=\sqrt{\frac{2}{\pi}}\int_0^{\infty}\frac{\mathrm{erf}(y/\sqrt{2T})}{B(k,N-k+1)}[\mathrm{erf}(y/\sqrt{2})]^{N-k}[\mathrm{erfc}(y/\sqrt{2})]^{k-1}\mathrm{e}^{-y^2/2}\mathrm{d}y \tag{7.121}$$

其中,$\mathrm{erfc}(y)=1-\mathrm{erf}(y)$ 表示互补误差函数。由式(7.121)可以看出,虚警概率与 Pearson 分布参数 γ 无关,即 OS 方法在 Pearson 背景下仍是 CFAR 的[42]。

同样在目标幅度瑞利起伏的情形讨论检测概率,假定瑞利分布参数为 σ_s^2,利用变量代换 $y=\gamma/\sqrt{x}$,OS-CFAR 的检测概率可表示为

$$P_{\mathrm{d}}^{\mathrm{OS}}=\int_0^{\infty}\Pr\{D\geqslant Tx \mid \mathrm{H}_1\}\,p_k(x)\mathrm{d}x$$

$$=\sqrt{\frac{2}{\pi}}\int_0^{\infty}\frac{\mathrm{e}^{(-\gamma^2/\sigma_s^2)(T/2y^2)}}{B(k,N-k+1)}[\mathrm{erf}(y/\sqrt{2})]^{N-k}[\mathrm{erfc}(y/\sqrt{2})]^{k-1}\mathrm{e}^{-y^2/2}\mathrm{d}y \tag{7.122}$$

类似地,检测概率和杂波尺度参数 γ 与瑞利起伏目标功率参数 σ_s 的比值有关。

7.6.3 Pearson 分布背景下的 CMLD-CFAR 检测器

对于 4.4 节中所述筛选平均 CMLD-CFAR 检测方法来说,假定对 N 个参考单元中 IID 的 Pearson

分布数据样本 $X_1, X_2, \cdots, X_N$ 进行排序，对 p 个最大的 $X_k, k=1,2,\cdots,N$ 进行截尾，剩余变量记为 $W_1, W_2, \cdots, W_{N-p}$，CMLD-CFAR 检测器通过对样本值 $W_1, W_2, \cdots, W_{N-p}$ 进行平均来估计杂波功率水平 Z，即

$$Z=\frac{1}{N-p}\sum_{k=1}^{N-p} W_k \tag{7.123}$$

由于参考样本 $X_1, X_2, \cdots, X_N$ 是 IID 的，剩余样本 $W_1, W_2, \cdots, W_{N-p}$ 也是 IID 的[29]，为了确定 Z 的 PDF，需确定 $W_1, W_2, \cdots, W_{N-p}$ 的概率密度函数 $h_p(w)$。从前述 OS-CFAR 检测器分析的式(7.118)可知，第 k 个有序样本 $X_{(k)}$ 的 PDF 为

$$p_{X_{(k)}}(w)=p_k(w)=k\binom{N}{k}\left[1-P_X(w)\right]^{N-k}\left[P_X(w)\right]^{k-1} p_X(w) \tag{7.124}$$

其中，$p_X(w)$和 $P_X(w)$分别为排序之前 N 个 IID 随机变量的 PDF 和 CDF。由于单个 W_k 的秩取 1～$N-p$ 的可能性完全相同，$h_p(w)$可以通过对式(7.124)在秩 k 服从均匀分布的条件下取平均得到，即

$$h_p(w)=\frac{1}{N-p}\sum_{k=1}^{N-p} p_k(w) \tag{7.125}$$

注意到

$$p_X(w)=\frac{1}{N}\sum_{k=1}^{N} p_k(w)=\frac{N-p}{N} h_p(w)+\frac{1}{N}\sum_{k=N-p+1}^{N} p_k(w) \tag{7.126}$$

对式(7.126)求解 $h_p(w)$可得

$$\begin{aligned} h_p(w) &=\frac{N}{N-p} p_X(w)-\frac{1}{N-p}\sum_{k=N-p+1}^{N} p_k(w) \\ &=\frac{\gamma \mathrm{e}^{-\gamma^2/2w} w^{-3/2}}{\sqrt{2\pi}(N-p)}\left\{N-\sum_{k=N-p+1}^{N}\sum_{j=0}^{k-1}\frac{N!\left[\operatorname{erf}(\gamma/\sqrt{2w}\right]^{N-k+j}}{(N-k)!(k-1)!}\binom{k-1}{j}(-1)^j\right\} \end{aligned} \tag{7.127}$$

经过变量替换和化简，对应的特征函数为

$$\varphi_p(\tau)=g(\tau\gamma^2/2)=\int_0^{\infty}\frac{2\mathrm{e}^{j(\tau\gamma^2/2t^2)-t^2}}{\sqrt{\pi}(N-p)}\left\{N-\sum_{k=N-p+1}^{N}\sum_{j=0}^{k-1}\frac{N!\left[\operatorname{erf}(t)\right]^{N-k+j}}{(N-k)!(k-1)!}\binom{k-1}{j}(-1)^j\right\}\mathrm{d}t \tag{7.128}$$

将因子 $1/(N-p)$包含在标称化因子 T 中，Z 的 PDF 可由 $N-p$ 个特征函数为$\varphi_p(\tau)$的 IID 随机变量之和来确定，因此

$$p_Z(z)=\frac{1}{2\pi}\int_0^{\infty}\left[\varphi_p(\tau)\right]^{N-p}\mathrm{e}^{-j\tau z}\mathrm{d}\tau \tag{7.129}$$

此时，经过变量替换和化简，可得虚警概率为

$$\begin{aligned} P_{\mathrm{fa}}^{\mathrm{CMLD}} &=\int_0^{\infty}\Pr\{D\geqslant TZ \mid \mathrm{H}_0\} p_Z(z)\mathrm{d}z \\ &=\frac{1}{2\pi}\int_0^{\infty}\operatorname{erf}(1/\sqrt{Tq})\int_{-\infty}^{\infty}\left[g(v)\right]^{N-p}\mathrm{e}^{-jvq}\mathrm{d}v\mathrm{d}q \end{aligned} \tag{7.130}$$

由式(7.130)可知，虚警概率与 Pearson 分布的尺度参数 γ 无关，即 CMLD 方法在 Pearson 背景下仍是 CFAR 的[42]。

同样在目标幅度瑞利起伏的情形讨论检测概率，假定瑞利分布参数为 σ_s^2，CMLD-CFAR 的检测概率可表示为

$$
\begin{aligned}
P_{\mathrm{d}}^{\mathrm{CLMD}} &= \int_0^{\infty} \Pr\{D_1 \geqslant TZ \mid \mathrm{H}_1\} p_Z(z)\mathrm{d}z \\
&= \int_0^{\infty} \mathrm{e}^{-(T\gamma^2/4\sigma_s^2)} \int_{-\infty}^{\infty} [g(v)]^{N \quad p} \mathrm{e}^{-jvq} \mathrm{d}v\mathrm{d}q
\end{aligned}
\tag{7.131}
$$

类似地,检测概率和杂波尺度参数 γ 与瑞利起伏目标功率参数 σ_s 的比值有关。

在大部分场景中,随机过程的二阶矩 $\mathrm{E}(X^2)$ 一般作为信号强度的标准度量,并与功率和能量的物理概念相关联,但由于 α 稳定分布不存在有限的二阶矩,前述信号强度度量无法用于 α 稳定分布[6]。为了进行检测性能分析,依据 CA-CFAR、OS-CFAR 和 CMLD-CFAR 的检测概率表达式,文献[42]给出了广义信噪比(GSNR)定义,即

$$
\mathrm{GSNR} = 20\log\frac{\sigma_s}{\gamma}
\tag{7.132}
$$

仿真分析结果表明[42],对于 GSNR>75dB 的情况,排序秩值 k 越小,OS-CFAR 的检测性能越好;而对于低 GSNR 值,当排序秩值 k 值较大时,OS-CFAR 的检测性能更好。作为折中选择,$k=3N/4$ 的取值只会导致较小的额外 CFAR 损失。另外,由于 Pearson 分布的尖峰特征,在均匀 Pearson 分布背景杂波的情况下,通过选择合适的 k,OS-CFAR 的性能甚至可能优于 CA-CFAR。

7.7 Cauchy 分布背景下的 CFAR 检测器

如 7.6 节所述,α 稳定分布只在特定情况下具有闭型的 PDF,其中当 $\alpha=1$ 且对称参数 $\beta=0$ 时,α 稳定分布退化为特定的 Cauchy 分布。当雷达的同相分量和正交分量服从 IID 的 Cauchy 分布,即同相分量和正交分量构成的复样本服从 $\alpha=1$ 的复值各向同性对称 α 稳定分布[44]时,复值各向同性对称 α 稳定分布随机变量的相位均匀分布在 $[0,2\pi]$。当 $\alpha=2$ 时,其对应包络为瑞利分布;而当 $\alpha=1$ 时,其对应包络为 Cauchy-Rayleigh 分布(也称为广义 Rayleigh 分布)。本节将简要讨论 Cauchy-Rayleigh 分布下 OS-CFAR 检测器的性能[45]。

假定同相和正交分量服从联合各向同性对称 α 稳定分布,即线性检波后参考单元的包络服从 Cauchy-Rayleigh 分布,N 个参考单元中 IID 的 Cauchy-Rayleigh 分布样本记为 $Y_1, Y_2, \cdots, Y_N$,其 PDF 和 CDF 可分别表示为

$$
p_Y(y) = \begin{cases} \dfrac{y\gamma}{(y^2+\gamma^2)^{3/2}}, & y \geqslant 0 \\ 0, & y < 0 \end{cases}
\tag{7.133}
$$

$$
P_Y(y) = \begin{cases} 1 - \dfrac{\gamma}{\sqrt{y^2+\gamma^2}}, & y \geqslant 0 \\ 0, & y < 0 \end{cases}
\tag{7.134}
$$

其中,γ 为分布的尺度参数。

设 $\{Y(k), k=1,2,\cdots,N\}$ 表示参考单元样本排序后的有序样本,令 $Z=Y(k)$ 为杂波功率水平估计,其 PDF 可根据 $p_Y(y)$ 和 $P_Y(y)$ 按式(7.135)计算[42],即

$$
p_{Y_{(k)}}(y) = p_k(y) = \frac{1}{B(k, N-k+1)}[1-P_Y(y)]^{N-k} P_Y^{k-1}(y) p_Y(y)
\tag{7.135}
$$

对于 Cauchy-Rayleigh 分布,第 k 个值顺序统计量的 PDF 可表示为

$$p_k(y)=\frac{1}{B(k,N-k+1)}\left(\frac{\gamma}{\sqrt{y^2+\gamma^2}}\right)^{N-k}\left(1-\frac{\gamma}{\sqrt{y^2+\gamma^2}}\right)^{k-1}\frac{y\gamma}{(y^2+\gamma^2)^{3/2}} \tag{7.136}$$

其中，$B(a,b)$表示 Beta 函数。

此时，对于标称化因子 T，OS-CFAR 的虚警概率可表示为

$$\begin{aligned}P_{\mathrm{fa}}&=\int_0^{\infty}\Pr\{D\geqslant Ty\mid \mathrm{H}_0\}p_k(y)\mathrm{d}y\\&=\int_0^{\infty}\frac{\mathrm{e}^{-\frac{(T_D y)^2}{2\sigma_s^2}}}{B(k,N-k+1)}\left(\frac{\gamma}{\sqrt{y^2+\gamma^2}}\right)^{N-k}\left(1-\frac{\gamma}{\sqrt{y^2+\gamma^2}}\right)^{k-1}\frac{y\gamma}{(y^2+\gamma^2)^{3/2}}\mathrm{d}y\end{aligned} \tag{7.137}$$

式(7.137)表明，OS-CFAR 方法在 Cauchy-Rayleigh 分布未能保持 CFAR 特性。

同样在目标幅度瑞利起伏的情形讨论检测概率，假定瑞利分布参数为σ_s^2，则 OS-CFAR 的检测概率可表示为

$$\begin{aligned}P_{\mathrm{d}}&=\int_{-\infty}^{\infty}\Pr\{D\geqslant Ty\mid \mathrm{H}_1\}p_k(y)\mathrm{d}y\\&=\int_0^{\infty}\frac{\mathrm{e}^{-\frac{(Ty)^2}{2\sigma_s^2}}}{B(k,N-k+1)}\left(\frac{\gamma}{\sqrt{y^2+\gamma^2}}\right)^{N-k}\left(1-\frac{\gamma}{\sqrt{y^2+\gamma^2}}\right)^{k-1}\frac{y\gamma}{(y^2+\gamma^2)^{3/2}}\mathrm{d}y\end{aligned} \tag{7.138}$$

仿真分析结果表明[45]，在 Cauchy-Rayleigh 杂波背景下，选择更大的参考滑窗尺寸，OS-CFAR 方法检测性能并没有得到很大改善；随着 GSNR 的降低，ROC 性能曲线的形状基本保持不变，但整体向较高虚警概率方向移动；与 Pearson 杂波背景下类似，对于较高的 GSNR 值，排序秩值 k 越小，OS-CFAR 的检测性能越好，而对于较低的 GSNR 值，较大 k 值能在一定程度上改善 OS-CFAR 的检测性能。

7.8 比较与小结

对数正态和韦布尔杂波背景都具有两个变化的参数，因此适用于瑞利包络杂波背景的单参数方法在上述环境中不再具有 CFAR[46]。7.2 节介绍的 Log-t 就是为双参数的对数正态和韦布尔背景设计的，它利用全部参考单元样本估计均值和方差，以形成一个检测统计量，这是一种匹配对数正态杂波背景的最大似然(MLH)估计方法。因此，文献[16]关于韦布尔中的 MLH 方法在对数正态背景中的性能要比 Log-t 差。由于 Log-t 采用全部参考单元样本估计阈值，这必然导致在非均匀杂波背景中性能的下降。考虑到这一点，7.3 节介绍了韦布尔背景中各种 OS 方法[14-15,20-21]，相应的检测性能是比较好的；但 OS 和 OSGO 方法在韦布尔背景中的 CFAR 损失比其在高斯背景中大得多，通过与二进制积累相结合，OS 和 OSGO 在检测性能和虚警控制能力方面较单脉冲情况均有较大改善[25-28]。按照文献[15]中的分类法，WH 算法[14]是基于两个有序样本的方法，尽管 WH 算法比 Hansen 利用矩的双参数的估计方法的 CFAR 损失要小，但是文献[15]的分析表明它们的 CFAR 损失都很大，甚至是不能接受的。文献[21]提出的方法得到了实际可行的 CFAR 损失，例如在 $R=16$，$c=2$，$k=12$ 时，文献[20-21]中方法的 CFAR 损失大约是 4dB，WH 算法在相同条件下的 CFAR 损失却达到了 21.5dB。7.4 节的分析表明，与基于矩或有序统计的双参数估计方法相比，MLH-CFAR 表现出更小的 CFAR 损失，但这是以更复杂的计算为代价的[16]。这给实现上带来很大困难，有时几乎是不可实现的。

Log-t 和双参数 OS 阈值估计中都不服从任何最优准则。在使估计方差最小的约束条件下，7.5 节获得了 BLUE 产生阈值估计的有序统计量线性组合的权值。满足这个条件的 BLUE 将超过 Log-t 的性能，即使在对有序样本采取适当的删除情况下也是这样(见图 7.21)。由于 BLUE 用有序样本的线性组合估计阈值，因此它的计算量也是可行的。在韦布尔背景中，BLUE 相对于 Log-t 有 1～2dB 的检测优势(见图 7.18 和图 7.19)。这个差别还取决于样本数和 P_{fa} 设计值。形状参数 c 对 BLUE 和 Log-t 的影响也是很大的(见图 7.20)，在 c 由 3 变化到 1 过程中，它们的性能下降超过 10dB。在对数正态包络的均匀杂波或杂波边缘环境中，BLUE(0,0)总是具有与 Log-t 极其相近的性能。在干扰目标环境中，采用适当删除的 BLUE，如 $R=16$ 的 BLUE(2,2)，检测性能明显优于 Log-t，甚至 BLUE(0,0)检测性能也优于 Log-t。BLUE 不像 Log-t 那样受干扰目标遮蔽效应影响。

7.6 节分析了 α 稳定分布杂波下的雷达目标 CFAR 检测问题，由于 α 稳定分布一般不具有闭型的 PDF 表达式，现有文献一般针对其特例进行研究。对于 $\alpha=0.5$ 且对称参数 $\beta=-1$ 的 Pearson 分布，文献[42]深入分析了该背景下经典的 CA、OS 和 CMLD 等 CFAR 检测器的有效性，并推导了相应的虚警概率和检测概率解析表达式，为尖峰杂波下的 CFAR 检测提供了理论依据。由于 Pearson 分布的尖峰特征，在均匀 Pearson 分布背景杂波的情况下，通过选择合适的 k，OS-CFAR 的性能甚至可能优于 CA-CFAR。对于 $\alpha=1$ 且对称参数 $\beta=0$ 的 Cauchy 分布，7.7 节探讨了杂波包络服从 Cauchy-Rayleigh 分布时的目标 CFAR 检测问题[45]；结果表明，经典的 OS 在该杂波背景下丧失了 CFAR 特性，选择更大的参考滑窗尺寸时，OS 的检测性能并没有得到很大改善。关于 GO 和 SO 等 CFAR 方法在 Pearson 分布杂波下的虚警和检测性能，可参见文献[47]和[48]。

对韦布尔和对数正态杂波中目标 CFAR 检测研究，除本章介绍的方法和研究工作外，还有许多的其他的方法和研究。为克服对数 CFAR 检测器局限于瑞利分布的缺点，Sekine 等人提出了一种抑制韦布尔杂波的单元平均对数 CFAR 检测器[49]，并在地杂波和气象杂波应用方面开展了深入研究[50]。杂波背景根据分布信息的已知程度可细分为三种情况[51]：杂波分布完全未知(分布自由)、杂波分布完全已知和杂波分布部分已知。第一种情况适用于非参量方法。第三种情况可具体分为四种环境、未知水平环境、未知斜度参数环境、几何对称环境和几何平均环境，前三者依次为后者的子集。文献[52]用多项式拟合了非平稳高斯杂波功率水平对数值随距离的变化，分析了使用对数放大器的加权 CFAR 检测器的性能。结果表明，当实际环境与设计环境相符合时，虚警概率不依赖于多项式系数。

雷达系统中频率捷变技术的应用已很普遍，雷达工程师利用脉冲积累来提高信噪比，相应地也提高了系统的检测性能。脉冲积累可以是相参的[53]，也可以是非相参的[23]。关于非高斯背景中采用多脉冲进行目标检测，文献[53]研究了相参韦布尔背景中的目标检测问题。对于均值类 CFAR 检测器在韦布尔背景中的性能分析，由于很难找到其进行性能分析的关于检测概率和虚警概率的数学解析表达式，往往需要借助于计算机仿真方法进行分析。文献[54]基于拖尾外推理论研究了韦布尔背景下 CFAR 检测问题。文献[55]给出了 CA-CFAR 在韦布尔背景中的一种关于虚警概率的近似公式，而文献[56]则推导了韦布尔背景下的虚警概率和检测概率闭型表达式。另外，文献[57]和[58]讨论了韦布尔背景下基于 Bayesian 方法的 CFAR 处理及干扰控制问题；文献[59]对不变性理论在韦布尔背景下的目标 CFAR 检测推广方面进行了有益探索。关于多脉冲情况下非高斯背景中经典 CFAR 检测器的非相参积累性能评估，仍然有许多工作值得去做[60-61]。

随着 CFAR 技术的广泛应用，非高斯杂波下目标 CFAR 检测已拓展至 UHF(Ultra High Frequency)被动雷达[2]、MIMO 雷达[26]、声呐系统[62]等相关应用领域，多干扰目标自适应抑制[29]、杂波边缘自适应定位[27]等实际问题引起了重视；CFAR 处理手段日益丰富，涉及神经网络[10]、高阶统计量[7]、非整数阶矩估计[63]等新的信号处理技术。另外，非高斯杂波建模更为精细，韦布尔杂波和对数正态杂波共存的情况也引起了重视[46]，二者的有效鉴别[64]值得积极探讨；α 稳定分布杂波下目标CFAR 检测理论尚有待进一步完善[6]。

参考文献

[1] 何友，黄勇，关键，等. 海杂波中的雷达目标检测技术综述[J]. 现代雷达，2014，36(12)：1-9.

[2] Maestre N D R, Amores J M P, Moya M D, et al. Machine learning techniques for coherent CFAR detection based on statistical modeling of UHF passive ground clutter[J]. IEEE Journal of Selected Topics in Signal Processing, 2018, 12(1): 104-118.

[3] Sayama S. CFAR processing by converting Weibull to Rayleigh distributions[J]. Electronics and Communications in Japan, 2020, 103: 3-10.

[4] Sayama S. Detection of target and suppression of sea and weather clutter in stormy weather by Weibull/CFAR[J]. IEEE Transactions on Electrical and Electronic Engineering, 2021, 16(2): 180-187.

[5] 周亮，匡华星，张玉涛，等. 基于对数正态分布的海杂波修正概率密度分布函数[J]. 雷达与对抗，2021，41(1)：18-22.

[6] Aalo V A, Peppas K P, Efthymoglou G. Performance of CA-CFAR detectors in nonhomogeneous positive alpha-stable clutter[J]. IEEE Transactions on AES, 2015, 51(3): 2027-2038.

[7] Zebiri K, Mezache A. Radar CFAR detection for multiple-targets situations for Weibull and log-normal distributed clutter[J]. Signal Image and Video Processing, 2021, 15: 1671-1678.

[8] Schleher D C. Radar detection in Weibull clutter[J]. IEEE Transactions on AES, 1976, 12(6): 736-734.

[9] Trunk G V, George S. Detection of targets in non-Gaussian sea clutter[J]. IEEE Transactions on AES, 1970, 6(5): 620-628.

[10] Saeed T R, Hatem G M, Sadah J W A. Classification of radar non-homogenous clutter based on statistical features using neural network[J]. International Journal of Reasoning-based Intelligent Systems, 2020, 12(2): 138-148.

[11] Xin Z, Zhang R, Sheng W, et al. Intelligent CFAR detector for non-homogeneous Weibull clutter environment based on skewness[C]. Oklahoma: IEEE Radar Conference, 2018: 322-326.

[12] Goldstein G B. False alarm regulation in log-normal and Weibull clutter[J]. IEEE Transactions on AES, 1973, 9(1): 84-92.

[13] Bentoumi A, Mezache A, Kerba H T. Performance of non-parametric CFAR detectors in log-normal and K radar clutter[C]. Algiers: Proceedings of the 2018 International Conference on Electrical Sciences and Technologies in Maghreb, 2018: 1-4.

[14] Weber P, Haykin S. Ordered statistic CFAR processing for two-parameter distributions with variable skewness[J]. IEEE Transactions on AES, 1985, 21(6): 819-821.

[15] Levanon N, Shor M. Order statistics CFAR for Weibull background[J]. IEEE Proceedings Radar Sonar Navigation, 1990, 137(3): 157-162.

[16] Ravid R, Levanon N. Maximum-likelihood CFAR for Weibull background[J]. IEEE Proceedings Radar Sonar and Navigation, 1992, 139(3): 256-264.

[17] Guida M, Longo M, Lops M. Biparametric linear estimation for CFAR against Weibull clutter[J]. IEEE Transactions on AES, 1992, 28(1): 138-151.

[18] Shor M, Levanon N. Performance of order statistics CFAR[J]. IEEE Transactions on AES, 1991, 27(2): 214-224.

[19] Hansen V G. Constant false alarm rate processing in search radars[C]. London: IEEE International Radar Conference, 1973: 325-332.

[20] 何友，Rohling H. 有序统计恒虚警(OS-CFAR)检测器在 Weibull 干扰背景中的性能[J]. 电子学报，1995，23(1)：88-94.

[21] He You. Performance of the order statistics constant false alarm rate in Weibull clutter background[J]. Ortung und Navigation，1993，2：133-155.

[22] Levanon N. Detection loss due to interfering targets in ordered statistics CFAR[J]. IEEE Transactions on AES，1988，24(6)：678-681.

[23] El Mashade M B. Binary integration performance analysis of CA family of CFAR strategies in homogeneous Weibull clutter[J]. Radioelectronics and Communications Systems，2020，63(1)：24-41.

[24] Souad C，Laroussi T，Mezache A. Automatic WH-based edge detector in Weibull clutter[C]. Lisbon：Proceedings of the Signal Processing Conference，2014：1706-1710.

[25] Meng Xiangwei. Performance analysis of ordered-statistic greatest of constant false alarm rate with binary integration for M-sweeps[J]. IET Radar，Sonar and Navigation，2010，4(1)：37-48.

[26] Baadeche M，Soltani F，Gini F. Performance comparison of mean-level CFAR detectors in homogeneous and non-homogeneous Weibull clutter for MIMO radars[J]. Signal Image and Video Processing，2019，13(1)：1677-1684.

[27] Pourmottaghi A. A CFAR detector in a nonhomogenous Weibull clutter[J]. IEEE Transactions on AES，2012，48(2)：1747-1758.

[28] Meng Xiangwei. Performance analysis of OS-CFAR with binary integration for Weibull background[J]. IEEE Transactions on AES，2013，49(2)：1357-1366.

[29] Chabbi S，Laroussi T，Barkat M. Performance analysis of dual automatic censoring and detection in heterogeneous Weibull clutter：A comparison through extensive simulations[J]. Signal Processing，2013，93(11)：2879-2893.

[30] Zhang Baiqiang，Zhou Jie，Xie Junhao，et al. Weighted likelihood CFAR detection for Weibull background[J]. Digital Signal Processing，2021，115：103079.

[31] Cohen A C，Whitten B J. Parameter estimation in reliability and life span models[M]. New York：Marcel Dekler，1988.

[32] Finn H M，Johnson R S. Adaptive detection mode with threshold control as a function of spatially sampled clutter-level estimates[J]. RCA Review，1968，29：414-464.

[33] Harter H L，Moore A H. Maximum-likelihood estimation of the parameter of Gamma and Weibull populations from complete and from censored samples[J]. Technometerics，1965，7(4)：639-643.

[34] Epstein B，Sobel M. Life testing[J]. Journal of America Statistical Association，1953，48(263)：486-502.

[35] Guida M，Longo M，Lops M. Biparametric CFAR procedure for log-normal clutter[J]. IEEE Transactions On AES，1993，29(3)：798-809.

[36] Feller W. An introduction to probability theory and its applications[M]. New York：Wiley，1971.

[37] Mann N R，Schafer R E，Singpurwalla N D. Methods for statistical analysis of reliability and life data[M]. New York：Wiley，1974.

[38] Sarhan A E，Greenberg B G. Contributions to order statistics[M]. New York：Wiley，1962.

[39] Gonzalez J G，Paredes J L，Arce G R. Zero-order statistics a mathematical framework for the processing and characterization of very impulsive signals[J]. IEEE Transactions on Signal Processing，2006，54(10)：3839-3851.

[40] Tsihrintzis G A，Nikias C L. Evaluation of fractional，lower-order statistics-based detection algorithms on real radar sea-clutter data[J]. IEE Proc. Radar，Sonar and Navigation，1997，144(1)：29-37.

[41] Pierce R D. Application of the positive alpha-stable distribution[C]. Banff：Proceedings of the IEEE Signal Processing Workshop on Higher-Order statistics，1997，420-424.

[42] Tsakalides P，Trinic F，Nikias C L. Performance assessment of CFAR processors in Pearson-distributed clutter[J]. IEEE Transactions on AES，2000，36(4)：1377-1386.

[43] Samorodnitsky G，Taqqu M S. Stable non-Gaussian random processes：stochastic models with infinite variance[M]. New York：Chapman and Hall，1994.

[44] Achim A, Kuruoglu E E, Zerubia J. SAR image filtering based on the heavy-tailed Rayleigh model[J]. IEEE Transactions on Image Processing, 2006, 15(9): 2686-2693.

[45] Xu Xiaolan, Zheng Rosa, Chen Genshe, et al. Performance analysis of order statistic constant false alarm rate (CFAR) detectors in generalized Rayleigh environment[C]. San Diego CA: Proceedings of SPIE, 2007: 1-10.

[46] Tang Shensheng, Zhu Jinwen, Xie Yi. Radar CFAR processing and design for hybrid Weibull and lognormal clutters[C]. Kansas: Proceedings of the 2011 IEEE Radar Conf, 2011: 639-642.

[47] Meziani H A, Soltani F. Performance analysis of some CFAR detectors in homogeneous and non-homogeneous Pearson-distributed clutter[J]. Signal Processing, 2006, 86(8): 2115-2122.

[48] Nadarajah S. Comments on "Performance analysis of some CFAR detectors in homogeneous and non-homogeneous Pearson-distributed clutter"[J]. Signal Processing, 2007, 87(5): 1169-1170.

[49] Sekine M, Musha T, Tomita Y, et al. Suppression of Weibull-distributed clutters using a cell-averaging LOG/CFAR receiver[J]. IEEE Transactions On AES, 1978, 14(5): 823-826.

[50] Sekine M, Ohtani S, Musha T, et al. Suppression of ground and weather clutter [J]. IEE Proceedings Communications, Radar and Signal Processing, 1981, 128(3): 175-178.

[51] Nitzberg R. Constant-false-alarm-rate signal processors for several types of interference[J]. IEEE Transactions on AES, 1972, 8(1): 27-34.

[52] Nitzber R. Constant false alarm rate processors for locally nonstationary clutter[J]. IEEE Transactions on AES, 1973, 9(3): 399-405.

[53] Farina A, Russo A, Scannapieco F. Radar detection in coherent Weibull clutter[J]. IEEE Transactions on Signal Processing, 1987, 35(6): 893-895.

[54] De Miguel G, Casar J R. CFAR detection for Weibull and other log-log-linear tail clutter distributions[J]. IEE Proc. Radar Sonar Navigation, 1997, 144(2): 64-70.

[55] Vela G D M, Portas J A B, Corredera J R C. Probability of false alarm of CA-CFAR detector in Weibull clutter[J]. Electronics Letters, 1998, 34(8): 806-807.

[56] Fernando D A G, Andrea C F R, Gustavo F, et al. CA-CFAR detection performance in homogeneous Weibull clutter[J]. IEEE Geoscience and Remote Sensing Letters, 2019, 16(6): 887-891.

[57] Stephen D H, Graham V W. Optimal predictive inference and noncoherent CFAR detectors[J]. IEEE Transactions on AES, 2020, 56(4): 2603-2615.

[58] Zhang Baiqiang, Xie Junhao, Zhou Wei. A Bayesian CFAR detector for interference control in Weibull clutter[J]. Digtal Signal Processing, 2020, 104: 102781.

[59] Weinberg G V, Bateman L, Hayden P. Development of non-coherent CFAR detection processes in Weibull background[J]. Digtal Signal Processing, 2018, 75: 96-106.

[60] Meng Xiangwei. Performance evaluation of OSSO-CFAR with binary integration in Weibull background[J]. Journal of Electronics, 2013, 30(1): 83-90.

[61] Baadeche M, Soltani F. Performance analysis of mean level constant false alarm rate detectors with binary integration in Weibull background[J]. IET Radar, Sonar and Navigation, 2015, 9(3): 233-240.

[62] Gao Jue, Li Haisen, Chen Baowei, et al. Fast two-dimensional subset censored CFAR method for multiple objects detection from acoustic image[J]. IET Radar, Sonar and Navigation, 2017, 11(3): 505-512.

[63] Gouri A, Mezache A, Oudira H. Radar CFAR detection in Weibull clutter based on zlog(z) estimator[J]. Remote Sensing Letters, 2020, 11(6): 581-589.

[64] Sazajnowski W J. Discrimination between log-normal and Weibull clutter[J]. IEEE Transactions on AES, 1977, 13(5): 480-485.

第8章 复合高斯杂波中的CFAR处理

CHAPTER 8

8.1 引言

相关非高斯杂波中目标的自动检测是现代雷达信号处理的一个主要发展方向。就杂波的包络分布来说,常用的是瑞利分布,这种包络分布意味着接收回波的同相正交分量是联合高斯过程。相应的物理解释是,杂波回波是由大量相同的、相互统计独立的散射体的回波矢量叠加合成的,因此根据中心极限定理,得到的随机过程就是高斯的。然而,在低入射角或高分辨率雷达等情况下,杂波回波包络往往不再服从瑞利分布,其经验分布通常呈现出长拖尾和大的标准差-均值比,此时单个包络的 PDF 不能适应各种不同场景,必须研究一种涵盖瑞利分布的包络分布族。另外还需考虑杂波的相关性,即杂波包络在各个距离单元之间的起伏性。幅度分布和相关性是杂波统计建模中必须考虑的两个约束条件,因此,文献[1]提出了复合高斯模型,该模型将杂波回波建模为快变复高斯过程与独立的慢变调制过程的乘积。复合高斯模型是一个分布族,广义 Laplace、广义 Cauchy、广义高斯、学生氏 t、Weibull、Rician、Rayleigh、K 分布及高斯分布都属于这个模型[1]。本章主要讨论典型的 K 分布,并基于杂波包络的 K 分布形式讨论 CFAR 检测问题,同时简单阐述杂波复幅度的复合高斯形式。

K 分布包络的复合模型表示为瑞利分布快变过程与 Gamma 分布慢变过程的乘积。对海杂波回波的包络模拟研究表明[2-4],K 分布这种复合形式可很好地与观测数据包络匹配。这个模型不仅在很宽的条件范围内与杂波包络分布匹配,且可正确地模拟杂波回波脉冲间的相关性,这一特点对于精确预测脉冲间积累后的目标检测性能十分重要。该模型得到了实验海杂波数据的支持[5-6],不同于以往的概率模型,K 分布在杂波散射机理上可得到很好的解释[4]。用 K 分布这种复合形式表示海杂波是基于如下的假设：在每个给定的距离-方位单元中的海杂波包络服从瑞利起伏分布(称为散斑),其方差在时间和空间上服从 Gamma 分布。这相当于用 Gamma 分布的随机变量在时间和空间上调制散斑功率,故称之为"调制过程"。也就是用空间上服从 Gamma 分布的强度来乘以一个独立的瑞利散斑分量[2]。这种乘积形式的杂波模型与文献[4]从二维随机游动出发建立的 K 分布模型是一致的,然而该 K 分布模型[4]没有考虑杂波样本间的相关性。文献[7]提出了相关 K 分布模型,而文献[8]表明,对于高分辨率雷达,当分辨单元内散射点个数起伏时,其杂波服从相关 K 分布。

对于杂波背景下的相参检测问题,一般不是利用杂波包络的复合模型,而是需要利用杂波复幅度的复合高斯模型。复合高斯模型为描述杂波复幅度的多维 PDF 提供了方便,在合理描述复幅度统计特性的同时,也具有较好的数学可操作性。基于复合高斯模型,许多学者研究了非高斯相关杂波背景下的相参雷达检测问题。Sangston[8]研究了杂波背景中的 Swerling Ⅰ型目标信号的相参检测问题,其中杂波包络服从 K 分布,杂波的相关性由一般的相关矩阵描述。文献[7]和[9]~[12]研究了复合高斯杂波中的信号最优和次优检测问题,其中信号只有随机或确定性的复尺度因子是未知的。文献[13]则研究了复合高斯杂波中的部分相关目标的建模。研究表明,悬停直升机的后向散射信号可以建模为子空间信号,而对于相关复合高斯杂波中具有非满秩协方差矩阵的高斯随机信号检测问题,可利用匹配子空间检测器推导其 CFAR 算法。文献[14]在杂波纹理完全相关的复合高斯杂波背景下,研究了子空间随机信号的相参检测问题。在上述这些研究中,均假设杂波纹理分量在雷达系统的 CPI 中是完全相关的。这个假设对于不太长的 CPI 来说是合理的,然而随着 CPI 的增加,杂波纹理将随时间缓慢变化。部分相关杂波纹理的情况更为复杂[15-16],Gini 和 Greco 等利用海杂波的循环平稳特性对此进行了探讨[17]。

本章在 8.2 节中介绍了复合高斯复幅度模型及相应的 K 分布包络模型,并考虑了相关 K 分布情况;8.3 节重点分析了 K 分布杂波加热噪声中的 CFAR 检测方法;8.4 节针对 K 分布中调制过程的 3 种相关程度,结合 CA、GO 和 OS 等典型的 CFAR 检测器,评估了相应的检测性能;8.5 节侧重于讨论复合高斯杂波下最优 CFAR 检测问题;8.6 节围绕杂波协方差矩阵结构估计,讨论球不变随机杂波下相参 CFAR 检测问题;8.7 节主要讨论先验分布下贝叶斯 CFAR 检测问题。

8.2 复合高斯分布

8.2.1 复合高斯复幅度模型

高分辨率雷达杂波往往建模为非均匀高斯过程,其中杂波功率水平是一个随机变量,其取决于雷达后向散射过程的空间和时间变化程度。这个模型被称为乘积模型[1],由 Sangston 和 Gerlach 等根据散射过程的现象图得到[8],其结果是以球不变随机向量(SIRV)形式表示的杂波多维统计模型。这里将杂波 $\boldsymbol{x}$ 建模为 SIRV,根据 Yao 的表示定理及其在复矢量中的延伸[1],$\boldsymbol{x}$ 可表示为两个独立随机变量的积,即

$$\boldsymbol{x}=\sqrt{\tau}\boldsymbol{s} \tag{8.1}$$

其中,$\boldsymbol{s}$ 是 m 维零均值复圆高斯矢量,其协方差矩阵为 $\mathrm{E}\{\boldsymbol{s}\boldsymbol{s}^{\mathrm{H}}\}=\boldsymbol{M}_s$,且 $\boldsymbol{M}_s$ 满足 $\boldsymbol{M}_{s,ii}=1, 1\leqslant i\leqslant m$;$\boldsymbol{M}_s$ 是 Hermitian 矩阵,即满足 $\boldsymbol{M}_s=\boldsymbol{M}_s^{\mathrm{H}}$,上标 H 表示复共轭转置。为了简化书写,记 $\boldsymbol{s}\sim\mathrm{CN}(\boldsymbol{0},\boldsymbol{M}_s)$,$\boldsymbol{s}$ 通常称为散斑。正随机变量 τ 通常称为纹理,表示距离单元中杂波的局部功率,它描述了观测场景的特征。平均杂波功率由 $\mu=\mathrm{E}\{\tau\}$ 给出。

在零假设下,对于给定 τ 值,$\boldsymbol{x}\sim\mathrm{CN}(\boldsymbol{0},\tau\boldsymbol{M}_s)$,因此其条件 PDF 表示为

$$f(\boldsymbol{x}\mid\tau)=\frac{1}{(\pi\tau)^m\left|\boldsymbol{M}_s\right|}\exp\left(-\frac{\boldsymbol{x}^{\mathrm{H}}\boldsymbol{M}_s^{-1}\boldsymbol{x}}{\tau}\right) \tag{8.2}$$

$f(\boldsymbol{x}|\tau)$对 τ 进行统计平均可得 $\boldsymbol{x}$ 的 PDF,即

$$f(\boldsymbol{x})=\mathrm{E}_\tau\{f(\boldsymbol{x}\mid\tau)\}=\int_0^\infty\frac{1}{(\pi\tau)^m\left|\boldsymbol{M}_s\right|}\exp\left(-\frac{\boldsymbol{x}^{\mathrm{H}}\boldsymbol{M}_s^{-1}\boldsymbol{x}}{\tau}\right)f(\tau)\,\mathrm{d}\tau \tag{8.3}$$

其中,$f(\tau)$表示 τ 的 PDF。上述乘积模型精确描述了在 CPI 级别观察时间间隔内的散射机制[1,8]。

高斯杂波是复合高斯杂波的一种特殊情况,事实上,令式(8.3)中的 $f(\tau)=\delta(\tau-\sigma^2)$即可获得高斯杂波的 PDF,其中 σ^2 为杂波功率。在高斯杂波中,局部功率 τ 不再是随机的,而是依概率 1 等于 σ^2,此时 $\boldsymbol{x}\sim\mathrm{CN}(\boldsymbol{0},\sigma^2\boldsymbol{M}_s)$,其 PDF 按照式(8.2)的形式表示为 $f(\boldsymbol{x})=f(\boldsymbol{x}|\sigma^2)$。因此,相关高斯杂波与白高斯热噪声之和构成的干扰属于复合高斯系列,但复合高斯杂波与白高斯热噪声之和不再是复合高斯过

程，因此该问题不在本书讨论范围之内。

需要注意的是，式(8.1)～式(8.3)中的矢量及其 PDF 都是复数域形式，而本章也将涉及实数域矢量及其 PDF，可通过将 m 维复矢量的实部与虚部串联构成 $2m$ 维的实矢量，以实现两者之间的转换。

8.2.2 K 分布杂波包络模型

复合形式的 K 分布包络是一个功率受随机过程调制的瑞利随机变量，其中功率调制过程是 Gamma 分布的。因此也可以用两个独立随机变量(y 与 s)的乘积形式描述 K 分布杂波包络 x 的统计特性[2]，这两个分量具有不同的时间相关性。其中，散斑分量 s 是具有短相关时间的快起伏分量，用瑞利分布描述，其由每一脉冲样本得到的，相关时间的典型值为 10ms，并可采用频率捷变使 s 在脉冲间完全不相关，通常认为散斑分量在每个距离-方位单元中是 IID 的。功率调制分量 y 表征了杂波包络局部均值水平，y^2 的 PDF 可用 Gamma 分布来描述，它是一个有很长相关时间的慢起伏分量，且不受频率捷变的影响。由实测数据发现，一般情况下，当方位分辨率不小于 120m，距离分辨率不小于 4.2m 时，分辨单元的海杂波回波总是高斯的，这表明它是由大量散射体合成的一个回波，与中心极限定理相吻合。但是这个大量散射体的假设在分辨单元小于某个极限值(尚未知)时就不成立了。散斑分量具有由功率调制分量确定的均值水平[18]。

于是，K 分布杂波包络的 PDF 为

$$f(x)=\int_0^{\infty} f(x \mid y)f(y)\mathrm{d}y,\quad 0\leqslant x\leqslant\infty \tag{8.4}$$

其中

$$f(x \mid y)=\frac{\pi x}{2y^2}\exp\left(-\frac{\pi x^2}{4y^2}\right) \tag{8.5}$$

$$f(y)=\frac{2b^{2\gamma}y^{2\gamma-1}\exp(-b^2y^2)}{\Gamma(\gamma)},\quad 0\leqslant y\leqslant\infty \tag{8.6}$$

因此，K 分布的 PDF 为

$$f(x)=\frac{4c}{\Gamma(\gamma)}(cx)^{\gamma}\mathrm{K}_{\gamma-1}(2cx) \tag{8.7}$$

其中，$\mathrm{K}_{\gamma}(x)$是 γ 阶第二类修正贝塞尔函数，$c=b\sqrt{\pi/4}$，b 是式(8.6)中 Gamma 分布的尺度参数。杂波平均功率为 γ/c^2。γ 是形状参数，取决于杂波尖峰程度。对于高分辨率海杂波，$0.1<\gamma\leqslant\infty$。其中，$\gamma=0.1$ 代表杂波尖峰十分严重，$\gamma\to\infty$时，$f(x)$趋于瑞利分布，代表热噪声包络的 PDF。

在 K 分布杂波加噪声背景下，回波包络不再服从 K 分布，但可采用提高回波的散斑分量的平均功率做有效的修正。把杂波加噪声回波的包络称为 a，那么新的散斑分量的 PDF 可以写成[19]

$$f(a \mid y)=\frac{a}{\sigma^2+2y^2/\pi}\exp\left(-\frac{-a^2}{2\sigma^2+4y^2/\pi}\right) \tag{8.8}$$

其中，σ^2 是噪声功率水平。

杂波加噪声的包络 CDF 可以写成

$$\begin{aligned}F(t)&=\int_0^t\int_0^{\infty} f(a \mid y)f(y)\mathrm{d}y\mathrm{d}a\\&=1-\int_0^{\infty}\exp\left(\frac{-t^2}{2\sigma^2+4y^2/\pi}\right)\frac{2b^{2\gamma}}{\Gamma(\gamma)}y^{2\gamma-1}\exp(-b^2y^2)\mathrm{d}y\end{aligned} \tag{8.9}$$

杂波加噪声包络分布的 n 阶矩为

$$m_n=\int_0^{\infty}\left(2\sigma^2+\frac{4y^2}{\pi}\right)^{\frac{n}{2}}\Gamma\left(\frac{n+2}{2}\right)\frac{2b^{2\gamma}}{\Gamma(\gamma)}y^{2\gamma-1}\exp(-b^2y^2)\,\mathrm{d}y$$

对于偶数 n,可以解得

$$m_2=2\sigma^2+\frac{4\gamma}{\pi b^2} \tag{8.10}$$

$$m_4=8\sigma^4+\frac{32\sigma^2}{\pi}\frac{\gamma}{b^2}+\frac{32}{\pi^2}\frac{\gamma(\gamma+1)}{b^4} \tag{8.11}$$

$$m_6=48\sigma^6+\frac{288\sigma^4}{\pi}\frac{\gamma}{b^2}+\frac{576\sigma^2}{\pi^2}\frac{\gamma(\gamma+1)}{b^4}+\frac{384}{\pi^3b^6}\gamma(\gamma+1)(\gamma+2) \tag{8.12}$$

8.2.3 相关K分布杂波幅度模型

8.2.2 节中的 K 分布包络模型未体现杂波相关性,然而表示杂波包络空间变化特性的分量在空间相邻样本之间可能是相关的。这个作为杂波调制过程的 Gamma 分布分量与海面外形有关,其已在机理上得到了解释[4-5]。因此,分辨单元之间的调制过程相关程度取决于海浪的空间相关特性,调制过程的相关距离与海浪相关距离具有相同数量级。在很好的波浪条件下,周期分量也将出现在调制过程的自相关函数(Auto-Correlation Function,ACF)中[20]。另外,可将调制过程的相关程度与 CFAR 滑窗内调制分量局部变化程度相对应。因为大多数 CFAR 方法试图估计背景信号功率水平的均值,而 CFAR 滑窗中功率水平均值的变化将明显影响 K 分布杂波下目标检测性能。正确建立相关 K 分布模型对于精确预测 K 分布杂波下检测性能十分必要。

假设杂波局部功率水平在一组 m 个脉冲间是完全相关的,即杂波功率水平在 m 个脉冲间为常数,则建立相关 K 分布杂波模型的关键在于建立多维相关高斯 PDF。令 $s_k=s_{\mathrm{I}k}+js_{\mathrm{Q}k}(k=1,2,\cdots,m)$ 表示 K 分布乘积模型中散斑分量构成的 m 个随机变量,$\boldsymbol{s}_{\mathrm{I}},\boldsymbol{s}_{\mathrm{Q}}$ 分别表示以 $s_{\mathrm{I}k},s_{\mathrm{Q}k}$ 构成的列矢量;令 $\boldsymbol{s}=[\boldsymbol{s}_{\mathrm{I}}^{\mathrm{T}},\boldsymbol{s}_{\mathrm{Q}}^{\mathrm{T}}]^{\mathrm{T}}$ 表示 m 个同相分量 $s_{\mathrm{I}k}$ 和 m 个正交分量 $s_{\mathrm{Q}k}$ 构成的 $2m$ 维列矢量。$\boldsymbol{s}$ 的 $2m$ 个分量的联合 PDF 是均值为零、方差为 1 且 $2m$ 维协方差矩阵为 $\boldsymbol{M}_s=\mathrm{E}\{\boldsymbol{s}\boldsymbol{s}^{\mathrm{T}}\}$ 的高斯 PDF。

设 y 是服从式(8.6)分布的与 $\boldsymbol{s}$ 独立的随机变量,根据 K 分布乘积模型,杂波回波幅度 $\boldsymbol{x}=y\boldsymbol{s}$。因此,$\mathrm{E}\{\boldsymbol{x}\}=\boldsymbol{0},\boldsymbol{M}_x=\mathrm{E}\{\boldsymbol{x}\boldsymbol{x}^{\mathrm{T}}\}=\mathrm{E}\{y^2\}\boldsymbol{M}_s=\gamma/b^2\boldsymbol{M}_s$。以 $y=\sqrt{\tau}$ 为条件的 $\boldsymbol{x}$ 的条件 PDF 为

$$f(\boldsymbol{x}\mid y)=\frac{1}{(2\pi)^m|\boldsymbol{M}_s|^{1/2}y^{2m}}\exp\left(-\frac{\boldsymbol{x}^{\mathrm{H}}\boldsymbol{M}_s^{-1}\boldsymbol{x}}{2y^2}\right) \tag{8.13}$$

其中,$|\boldsymbol{M}_s|$ 是 $\boldsymbol{M}_s$ 的行列式,因此可以认为杂波幅度服从零均值和随机参数 y 的多维高斯分布,$\boldsymbol{x}$ 的 PDF 为[7]

$$f(\boldsymbol{x})=\int_0^{\infty}f(\boldsymbol{x}\mid y)f(y)\mathrm{d}y=\frac{1}{(2\pi)^m|\boldsymbol{M}_s|^{1/2}\Gamma(\gamma)2^{\gamma-1}}[aR(\boldsymbol{x})]^{\gamma-m}\mathrm{K}_{\gamma-m}[aR(\boldsymbol{x})] \tag{8.14}$$

其中,$\mathrm{K}_{\gamma-m}$ 是第二类 $\gamma-m$ 阶修正贝塞尔函数,$a=\sqrt{2b}$,$R(\boldsymbol{x})=(\boldsymbol{x}^{\mathrm{H}}\boldsymbol{M}_s^{-1}\boldsymbol{x})^{1/2}$。$m=1$ 时,就是与式(8.7)相似的 K 分布形式,不同的是式(8.7)表示包络的 PDF,而式(8.14)表示幅度的 PDF。这一相关模型中假设杂波结构分量 y 在脉冲间是完全相关的,因此杂波功率水平在 m 个脉冲间是常数。文献[20]建立了杂波结构分量 y 在一组 m 个脉冲间完全不相关情况的 K 分布杂波模型。上述两种是极端情况。对于更实际的部分相关情况,文献[15]考虑了散斑分量和结构分量均有各自的可定义相关特

性,并建立了相关 K 分布杂波模型。文献[1]从 SIRP(Spherically Invariant Random Processes)角度剖析了 K 分布杂波,利用此模型可以引入杂波的脉冲间相关性。

8.2.4 K 分布杂波的仿真

在没有明晰的性能解析表达式可用时,对相关杂波的准确仿真是精确预测性能的基础[21]。如果可以仿真出服从假定分布的杂波数据,就可以避免在各种环境中烦琐的杂波测量。相关 K 分布是相关 Gamma 分布过程对乘性散斑分量作用的结果,因此在仿真相关 K 分布过程中,仿真相关 Gamma 分布是重要环节。Oliver[22]提出了对 Gamma 分布白噪声进行线性滤波产生空间相关 Gamma 过程的近似方法,但该方法只能获得精确的一阶和二阶矩,所得分布并非精确的相关 K 分布。

Armstrong 等[16]提出了一种精确的方法,它可以产生具有任意空间相关程度的 K 分布样本。相关 Gamma 模型的构成机理是:在形状参数为半整数时,Gamma 分布就变成了 Chi 方分布。Chi 方分布可以由独立零均值高斯子过程的平方和构成,然后在高斯子过程中引入相关性。散斑分量的相关性通过协方差矩阵引入。在 Gamma 分布的形状参数为任意正数时,需要采用无记忆非线性变换,这种方法的主要缺点是 ACF 的形式不能任意选择。而对于非线性变换,输入和输出的协方差矩阵间的关系很复杂,根据输出的需要控制输入是很困难的,因此 Rangaswamy 等[9]提出了基于 SIRP 描述的相关非高斯杂波的仿真方法;因为 SIRP 模型属于外生模型,所以允许对边缘 PDF 和相关函数进行独立控制,从而弥补了前述仿真方法的不足。

Blacknell[23]提出的仿真方法是用一组滑动平均滤波器对 IID 的 Gamma 分布随机变量序列滤波,然后对输出序列求和,通过调整滑动平均滤波器的参数,可以使 Gamma 分布随机变量序列具有所希望的相关性;进而用所产生的相关 Gamma 分布调制散斑分量的功率来获得相关 K 分布仿真数据。此外,文献[24]也研究了非高斯杂波的建模与仿真问题,国内学者也对该问题进行了积极探讨[25-26]。

8.3 K 分布杂波加热噪声中的检测性能

文献[27]讨论了在 K 分布海杂波环境下的目标检测性能,然而实际应用中当杂波和噪声功率相当时,则有必要分析 K 分布杂波加热噪声下的目标检测性能[28]。对于这个问题,文献[19]做了进一步分析,并扩展到了多脉冲情况,同时考虑了杂波在脉冲间相关性的影响。

8.3.1 K 分布与记录数据的匹配

在不存在热噪声的环境中,已经证实 K 分布能很好地与海杂波的观测数据匹配。基于 X 波段高分辨率机载雷达,文献[3]在很宽的条件范围内给出了大量杂波数据。其中,雷达脉冲宽度为 30ns,天线波束宽度为 1.2°,并且使用脉冲间频率捷变,重点针对长宽各大约 800m 区域进行雷达观测,在一段时间内记录了大量的距离单元雷达回波数据。图 8.1 给出了两组杂波数据的概率曲线(分别用○和×表示),它表示杂波包络的平方超过给定值的概率。另外,图中给出了相应的 K 分布拟合曲线,拟合曲线由上至下依次为 $\gamma=\infty$(瑞利分布,代表热噪声),$\gamma=1.5$(曲线 B),$\gamma=0.26$(曲线 A)。结果表明,分布 A 与 $\gamma=0.26$ 的 K 分布能较好地匹配,特别是在曲线尾部。图 8.1 中所示数据的杂噪比较大,如果记录数据的杂噪比较低,则热噪声影响增加,包络分布将严重偏离标准 K 分布。

在没有热噪声的 K 分布杂波环境和 K 分布杂波加噪声($\gamma=0.6$,CNR=0dB)两种情况下,图 8.2 给出了相应的概率曲线。由图 8.2 可见,加性噪声对分布的低包络值具有较大的影响。有多种方法可以用来估计与记录数据相匹配的合适 γ 值,如果杂噪比已知,记录数据的概率分布可以与各种 γ 值对应的分布进行比较,以确定最合适的 γ 值。一种简便方法是利用矩估计来确定最合适的 γ 值,进而用来匹配数据。如果记录的数据质量很高,且有足够多的独立样本用于精确估计各高阶矩,则由式(8.10)至式(8.12)可解得

$$\gamma=\frac{18(m_4-2m_2^2)^3}{(12m_2^3-9m_2m_4+m_6)^2}$$

$$2\sigma^2=m_2-\left[\frac{\gamma}{2}(m_4-2m_2^2)\right]^{1/2}$$

$$b^2=\frac{4\gamma}{\pi(m_2-2\sigma^2)}$$

于是,由 $\mathrm{CNR}=4\gamma/\pi b^2\sigma^2$ 可计算杂噪比。在没有噪声时有

$$\gamma=\left(\frac{m_4}{2m_2^2}-1\right)^{-1} \tag{8.15}$$

如果热噪声相对较小可以忽略,则可以考虑由式(8.15)估计 γ。在热噪声不可忽略时,可先通过式(8.15)估计 γ,再用式(8.16)进行修正得到有效值[5],即

$$\gamma_{\mathrm{eff}}=\gamma\left(1+\frac{1}{\mathrm{CNR}}\right) \tag{8.16}$$

在图 8.2 中,$\gamma=0.6$ 且 CNR=0dB 时对应的 $\gamma_{\mathrm{eff}}=2.4$。由图 8.2 可见,$\gamma_{\mathrm{eff}}=2.4$ 对应的曲线与包含加性噪声的曲线尾部能很好匹配,但是在低包络水平段匹配得较差。

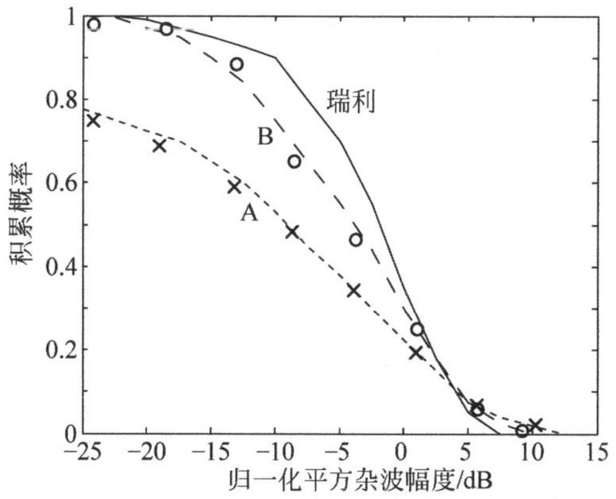

图 8.1 典型实际数据与最适合的 K 分布的比较

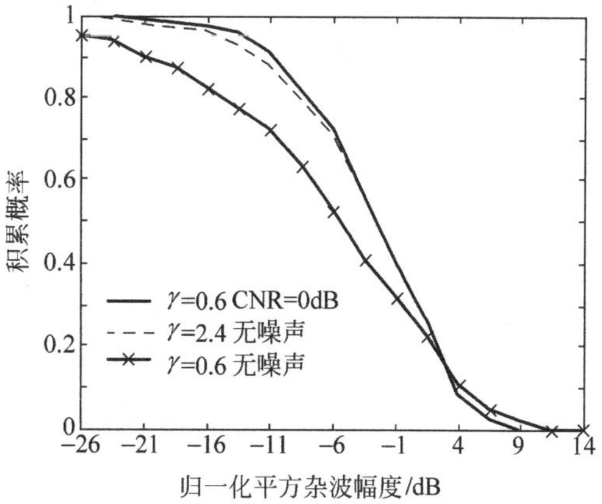

图 8.2 加性噪声对 K 分布杂波的概率分布的影响

8.3.2 杂波加噪声中目标检测的计算

本节和 8.3.3 节将分析 K 分布杂波加热噪声背景下的目标检测性能。对于机载对海警戒雷达中普遍采用的脉冲重复间隔,除瑞利分布这种极限情况外,在海杂波中由脉冲到脉冲间难以得到 K 分布杂波的独立样本。下面假设雷达脉冲积累时间比局部均值水平分量的相关期短,并且假设杂波散斑分量在积累期间上是完全相关或者不相关的,而脉冲间热噪声样本是独立的。下面主要研究两种情况,情况 1 假设杂波散斑分量和噪声在脉冲间是不相关的,情况 2 假设杂波散斑分量在脉冲间是完全相关的。

1. 目标回波、噪声与脉冲间不相关的杂波散斑分量之和

式(8.8)给出了热噪声加K分布杂波的包络PDF。对于一个非起伏目标，在给定杂波局部均值水平 y 的条件下，杂波加噪声环境下目标单脉冲回波的包络具有莱斯分布形式，即

$$f(z \mid y)=\frac{z}{\sigma^2+(2y^2/\pi)}\exp\left[-\frac{z^2+A^2}{2\sigma^2+(4y^2/\pi)}\right]I_0\left[\frac{Az}{\sigma^2+(2y^2/\pi)}\right]$$

其中，z 是目标回波加杂波与噪声的合成包络，A 是非起伏目标的包络。

对于一个起伏目标(Swerling Ⅰ或 Swerling Ⅱ型)，单脉冲包络为

$$f(z \mid y)=\frac{z}{\sigma^2+(2y^2/\pi)+(\overline{A^2}/2)}\exp\left[-\frac{z^2}{2\sigma^2+(4y^2/\pi)+\overline{A^2}}\right]$$

$\overline{A^2}$ 是瑞利分布起伏目标回波包络的均方值。

对于 N 个脉冲积累，在脉冲间得到的杂波加噪声是服从式(8.8)和式(8.9)分布的独立样本。为得到在 N 个脉冲积累后的 P_{fa} 和 P_d 值，要用式(8.6)中 y 的PDF对所有可能的 y 值进行统计平均。

2. 脉冲间完全相关的杂波散斑分量

如果散斑分量是完全相关的，并且处于没有严重热噪声的环境中，检测性能的计算就变成单脉冲检测的情况(至少对于非起伏或慢起伏目标是这样的)。然而，如果存在着显著的噪声，则必须考虑脉冲间噪声独立、杂波散斑完全相关的情况。

在这种情况下可以把杂波加噪声的PDF重新写成

$$f(a)=\int_0^\infty\int_0^\infty f(a \mid x)f(x \mid y)f(y)\,\mathrm{d}x\,\mathrm{d}y \tag{8.17}$$

其中

$$f(a \mid x)=\frac{a}{\sigma^2}\exp\left(-\frac{a^2+x^2}{2\sigma^2}\right)I_0\left(\frac{ax}{\sigma^2}\right) \tag{8.18}$$

$f(x|y)$ 和 $f(y)$ 分别见式(8.5)和式(8.6)。$f(a|x)$ 是海面上点回波的条件PDF，这个回波由一个固定的杂波回波 x 和功率为 σ^2 的噪声组成，且服从莱斯分布。

3. 脉冲间完全相关杂波中的虚警和检测概率

利用式(8.17)可得到杂波散斑完全相关的 N 个脉冲积累后的虚警概率 P_{fa}，即

$$P_{fa}=\int_0^\infty\int_0^\infty P_{fa}(x,N)f(x \mid y)f(y)\,\mathrm{d}x\,\mathrm{d}y \tag{8.19}$$

其中，$P_{fa}(x,N)$ 是在 N 个服从式(8.18)所示莱斯分布 $f(a|x)$ 的独立样本积累后得到的虚警概率。

当存在非起伏目标时，式(8.18)的 x 应在加入固定目标回波包络 A_t 后予以修正。当杂波回波相关时，一块海区回波的PDF可近似为只有杂波和噪声时的莱斯分布。若把这个目标回波加杂波分量称为 A_{tc}，则莱斯分布可以写成

$$f(z \mid A_{tc})=\frac{z}{\sigma^2}\exp\left(-\frac{z^2+A_{tc}^2}{2\sigma^2}\right)I_0\left(\frac{zA_{tc}}{\sigma^2}\right) \tag{8.20}$$

其中，A_{tc} 包含了目标包络 A_t 和固定杂波包络 x。在 x 给定时，A_{tc} 包络PDF为

$$f(A_{tc} \mid x)=\frac{A_{tc}}{\pi A_t x}\left[1-\left(\frac{A_t^2+x^2-A_{tc}^2}{2A_t x}\right)^2\right]^{-1/2},\quad |A_t-x| \leqslant A_{tc} \leqslant |A_t+x|$$

总的检测概率可写成

$$P_d = \int_0^{\infty}\int_0^{\infty}\int_{|A_t - x|}^{|A_t + x|} P_d(A_{tc}, N) f(A_{tc} \mid x) f(x \mid y) f(y) \mathrm{d}A_{tc} \mathrm{d}x \mathrm{d}y$$

其中，$P_d(A_{tc}, N)$是对莱斯分布（见式(8.20)）的 N 个样本积累之后的检测概率。

对于 Swerling Ⅱ型快起伏目标，假设目标幅度样本在脉冲间是独立的，则有

$$f(z \mid x) = \frac{2z}{2\sigma^2 + \overline{A_t^2}} \exp\left(-\frac{z^2 + x^2}{2\sigma^2 + \overline{A_t^2}}\right) I_0\left(\frac{2zx}{2\sigma^2 + \overline{A_t^2}}\right) \tag{8.21}$$

其中，$\overline{A_t^2}$ 是目标包络的均方值。

总的检测概率为

$$P_d = \int_0^{\infty}\int_0^{\infty} P_d(x, N) f(x \mid y) f(y) \mathrm{d}x \mathrm{d}y \tag{8.22}$$

其中，$P_d(x,N)$是由服从式(8.21)所示莱斯分布的 N 个样本积累后的检测概率。

8.3.3 性能分析

一般来说，在海杂波中的检测性能会变化很大，这取决于检测阈值设置的方式。一种极端方式是固定阈值。在这种情况下，P_{fa} 和 P_d 将随杂波局部均值水平 y 在时间上变化，并且也在空间上变化。然而在某些情况下，阈值可以自适应于杂波局部均值水平（或杂波加噪声的平均水平），从而提供一个不在空间上变化的 P_{fa} 值。当很好地获得了这个适应性时，称检测性能为"理想 CFAR"。

图 8.3 在固定阈值和理想 CFAR 条件下，给出了 Swerling Ⅱ型目标单脉冲检测时杂噪比随形状参数 γ 的变化曲线，其中 $P_{fa}=10^{-4}$，$P_d=0.5$。由图 8.3 可知，随着杂噪比下降，理想 CFAR 的检测性能对杂噪比的变化更敏感。也就是说，在杂波强度不变时，对热噪声的变化更敏感，并且更快地趋近于 $\gamma=\infty$ 时的检测性能。对于固定阈值检测，在低 γ 值对应的杂波尖峰情况下，其性能在很大程度上取决于杂波的尖峰程度，甚至在较低的杂噪比值时也是如此。

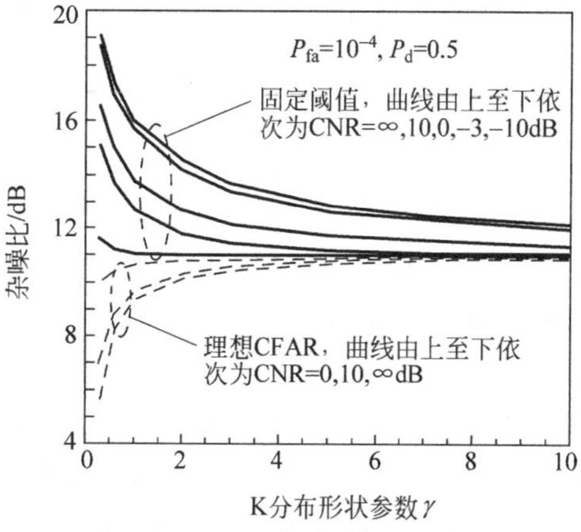

图 8.3 杂波加噪声中的单脉冲检测曲线

由于其阈值可随杂波局部均值水平变化而自适应变化，理想 CFAR 检测器优势很明显，实际应用中理想 CFAR 情况难以达到，但优于固定阈值的性能则可以得到。在某些情况下，如杂波尖峰十分严重时，阈值也许不可能跟上杂波局部均值水平的变化，此时固定阈值的性能代表了可以得到的最好性能。

在某些情况下可用简化分析获得近似结果。对于单脉冲检测，基于杂噪比对式(8.16)中的形状参数 γ_{eff} 进行合理修正，可对检测性能进行合理近似分析。对于杂噪比 CNR≤0dB 时的单脉冲理想 CFAR 检测，假设为瑞利统计量通常可以得到对性能的偏好估计。

文献[19]研究了两种极端情况下 K 分布杂波中的理想检测，即相邻距离单元间调制过程完全不相关（$\rho=0$）和完全相关（$\rho=1$）情况，但介于二者之间的部分相关情况更为常见。对于部分相关情况，图 8.4 给出了 $P_{fa}=10^{-6}$，$P_d=0.5$ 时的检测损失与相关系数 ρ 的关系曲线，由图 8.4 可知，相比于不相关调制过程，适当提高 ρ 值可获得几分贝的增益。

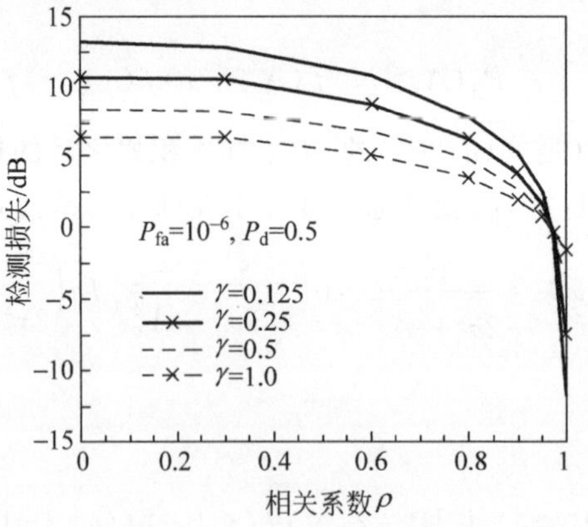

图 8.4　在空间相关 K 分布杂波中的检测损失与相关系数的关系

8.4　经典 CFAR 检测器在 K 分布杂波中的性能分析

Watts 等[5]用记录的杂波数据和一种估计性能的半解析方法,分析了 K 分布杂波中一种特殊形式的多脉冲 CA-CFAR 检测器的性能。在各种杂波空间相关条件下和一定的 CFAR 检测器参数范围内,Armstrong 等分析了 3 种典型的 CFAR 检测器在 K 分布杂波中的性能[29],其研究主要集中于两个方面,即 3 种 CFAR 检测器在 K 分布杂波中的检测损失,以及调制过程的空间相关条件对检测损失的影响。

下面的分析局限于对瑞利起伏目标的单脉冲检测。Watts 和 Ward[3,27]已经分析了多脉冲检波后积累的影响,相应结果可与下面给出的单脉冲检测结果一起使用,以评估检波后积累的 CFAR 检测器性能。在下面的分析中假设目标模型为 Swerling Ⅱ型,故相邻两次检测之间目标回波是不相关的。在对 K 分布杂波中的 CFAR 检测性能分析中,首先讨论两种极端的情况。

(1) 调制过程在空间上不相关,CFAR 滑窗中的参考样本是独立的,并且不随 CFAR 滑窗位置变化。

(2) 调制过程在空间上完全相关,因此 CFAR 滑窗中的调制过程是常数,但是调制过程按照 Gamma 分布随着 CFAR 滑窗位置在很宽范围内变化。

下面再讨论中间情况,调制过程既在 CFAR 滑窗内变化,又随着 CFAR 滑窗位置变化。为直接分析具有任意 ACF 的相关 Gamma 分布调制过程对 CFAR 性能的影响,要引入 N 阶多变量 Gamma 分布。这样一个多变量 Gamma 分布虽没有唯一的表达形式,但是有一些可供选择的表达方法。广义多变量 Gamma 分布(Generalized Multivariate Gamma Distribution,GMGD)表达形式可作为调制过程的模型,但 $N>2$ 时的数学公式即使在没有结合散斑分量之前就已异常复杂,因此需另寻他法。把调制过程作为非平稳过程处理,即在 CFAR 窗内把调制样本视为来自一个分布母体,该母体比全局 Gamma 调制分布更窄小,且具有与 CFAR 窗位置相关的一个或多个条件参数,即产生一个所有边缘 PDF 均为 Gamma 分布的多变量 Gamma 分布,从而使 CFAR 滑窗中调制过程样本间具有相关性。

8.4.1　调制过程不相关的 K 分布杂波下 CFAR 检测

本节假设噪声电平与杂波信号相比可以忽略,在均匀 K 分布杂波环境中分析几种经典 CFAR 检测器的检测性能，并且假设调制过程在距离单元之间是完全不相关的，检测损失是相对瑞利噪声中的理想检测而言，阈值因子基于对杂波形状参数 γ 的正确估计。本节最后还分析了偏离上述假设的情况。

假设 $f_Z(z)$ 为杂波包络均值水平估计 Z 的 PDF，则虚警概率为[29]

$$P_{\mathrm{fa}}=\int_0^{\infty}\left[\frac{2c^{\gamma}}{\Gamma(\gamma)}(TZ)^{\gamma}\mathrm{K}_{\gamma}(2cTZ)\right]f_Z(z)\mathrm{d}z \tag{8.23}$$

其中，{ · }内给出的是 K 分布杂波在 TZ 处 CDF 的补(Complementary CDF，CCDF)，相关参数含义参见式(8.7)。用数值方法解此式，可得满足特定 P_{fa} 的阈值因子 T。注意到，$\gamma=0.5$ 时可得较简单的解析式，可用于验证数值方法的精度。

对杂噪比为 SCR 的目标的检测概率为

$$P_{\mathrm{d}}=\int_0^{\infty}\Pr\{D>TZ\}f_Z(z)\mathrm{d}z \tag{8.24}$$

其中，目标加杂波的 CCDF 可表示为

$$\begin{aligned}\Pr\{D>TZ\}&=\int_0^{\infty}\left[\int_{TZ}^{\infty}\frac{2x}{\overline{A_{\mathrm{t}}^2}+\dfrac{4y^2}{\pi}}\exp\left(-\frac{x^2}{\overline{A_{\mathrm{t}}^2}+\dfrac{4y^2}{\pi}}\right)\mathrm{d}x\right]f(y)\mathrm{d}y\\&=\int_0^{\infty}\frac{2b^{2\gamma}}{\Gamma(y)}y^{2\gamma-1}\exp\left[-\frac{(TZ)^2}{\overline{A_{\mathrm{t}}^2}+\dfrac{4y^2}{\pi}}-b^2y^2\right]\mathrm{d}y\end{aligned} \tag{8.25}$$

其中，$\overline{A_{\mathrm{t}}^2}$ 是目标回波功率，即目标幅度的均方值，等于 $\gamma\mathrm{SCR}/c^2$；$f(y)$ 由式(8.6)给出。式(8.25)没有闭型表达式，然而对于杂噪比 CNR>10dB 的情况，可把此式简化成一个瑞利目标的 CCDF 表达式，其中瑞利目标的信号强度等于杂波加目标回波功率。这种近似是可接受的，因为一旦阈值因子 T 被确定后，分布的尾部对于检测概率只有很轻微的影响。只有目标加杂波 PDF 的中心“铃形”区域是重要的，并且低于目标信号 10dB 的小杂波分量的影响，主要体现在功率归一化的目标加杂波 PDF 的远处拖尾部分。这个近似的目标加杂波的 CCDF 为

$$\Pr\{D>TZ\}\approx\exp\left[-\frac{(TZ)^2}{(1+\mathrm{SCR})\gamma/c^2}\right] \tag{8.26}$$

1. CA-CFAR 检测器

假设每个参考单元中均匀杂波是独立的，CA 中杂波包络均值水平估计 Z 的 PDF 可以由杂波 PDF 与其本身的 R 重卷积得到。其中，R 是估计背景杂波或噪声水平所使用的参考单元数。在杂波服从 K 分布的情况下，Z 的 PDF 通常没有闭型表达式。对于杂波形状参数 $\gamma=m+1/2(m=0,1)$ 的特殊情况，可得到闭型表达式[29]。其中，$\gamma=0.5$ 时，有

$$f_Z(x)=\frac{2c}{\Gamma(R)}(2cx)^{R-1}\mathrm{e}^{-2cx}$$

$\gamma=3/2$ 时，有

$$f_Z(x)=\frac{(2c)^{2R}}{\Gamma(2R)}x^{2R-1}\mathrm{e}^{-2cx}$$

$\gamma>m+3/2$ 时，令

$$t=2R+j_0+j_1+\cdots+j_{m-1},\quad y=2cx,\quad \tau=c\sqrt{\pi}/[2^m\Gamma(\gamma)]$$

则

$$f_z(y)=\tau^R \mathrm{e}^{-y}\sum_{j_0=0}^{R}\sum_{j_1=0}^{j_0}\sum_{j_2=0}^{j_1}\cdots\sum_{j_{m-1}=0}^{j_{m-2}}\binom{R}{j_0}\binom{j_0}{j_1}\cdots\binom{j_{m-2}}{j_{m-1}}\frac{\beta_m^{R-j_0}\beta_{m-1}^{j_0-j_1}\cdots\beta_0^{j_{m-1}}}{\Gamma(t)}y^{m-1} \tag{8.27}$$

其中，$\beta_i=[(m+i)!(m+1-i)]/(2^i i!)$，$i=0,1,\cdots,m$。

值得一提的是，$\gamma=0.5$ 的简单形式对于核对数值计算精度很有帮助。当 $\gamma<0.5$ 时，杂波 PDF 在 x 趋近于 0 时趋近于无穷。这使得对 CA 的检测统计量的数值分析变得复杂，因为对包含奇点数据的数值卷积是不可靠的。但可以通过引入对杂波 PDF 的量化，使之变成对于所有 x 均为有限的离散 PDF；只要量化间隔设置得足够小，例如设置为杂波方差的 1/200，精度下降不会特别严重。

在几种 γ 取值、不同虚警率和参考单元数条件下，图 8.5 给出了 $P_{\mathrm{d}}=0.5$ 时 CA 的检测损失与形状参数 γ 的关系曲线。很明显，检测损失严重地依赖于 γ，并且与参考单元数和虚警概率设计值密切相关。对于 $\gamma=0.5$ 的杂波尖峰情况，检测损失达到 20dB 左右。参考单元数越小，且 P_{fa} 设计值越低，则检测器对杂波尖峰的增加就越敏感。

2. GO-CFAR 检测器

设 GO-CFAR 检测器的前沿和后沿滑窗长度均为 $R/2$，这些参考样本之和的 PDF 可由杂波 PDF 与其本身的 $R/2$ 重卷积得到，用与 CA 中描述的相同方法计算，区别在于用 $R/2$ 代替 R，即得到两个局部估计的概率密度函数 $f(z)$ 和概率分布函数 $F(z)$。杂波包络均值水平估计 Z 的 PDF 为[29]

$$f_Z(z)=2f(z)F(z) \tag{8.28}$$

把式(8.28)代入式(8.23)和式(8.24)，用与 CA 相同的方法可以计算检测性能。

图 8.6 给出了在 $P_{\mathrm{d}}=0.5$ 时 GO 的检测损失与 γ 的关系曲线，检测损失对各参数的依赖性与 CA 类似。比较图 8.5 和图 8.6 可知，GO 遭到的检测损失要比 CA 稍大一些。这是预料之中的结果，因为尽管杂波条件是非瑞利的，但仍然假设为均匀的，CA 在均匀背景中的优越性能是众所周知的，另外 GO 对 R 下降的变化也比 CA 敏感一些。

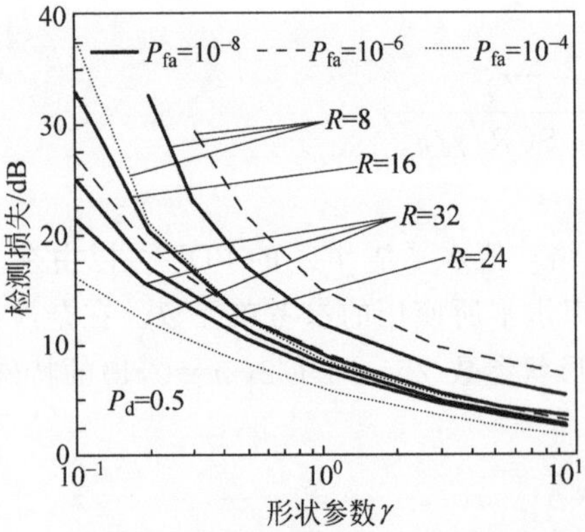

图 8.5 CA-CFAR 检测器的检测损失与 γ 的关系

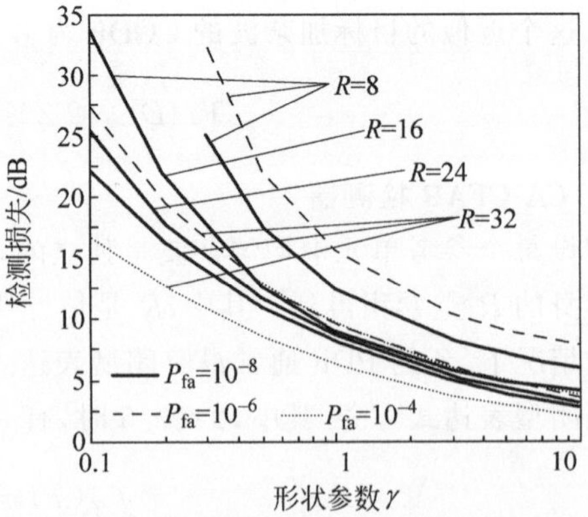

图 8.6 GO-CFAR 检测器的检测损失与 γ 的关系

3. OS-CFAR 检测器

OS-CFAR 的杂波包络均值水平估计 Z 是有序样本中的特定值，即从小到大排序后选取第 k 个样本，那么 Z 的 PDF 为[29]

$$f_Z(x)=k\binom{R}{k}f(x)F(x)^{k-1}[1-F(x)]^{R-k}$$

其中，$f(x)$和$F(x)$分别是均匀K分布杂波包络样本的PDF和CDF，进一步得到Z的PDF为

$$f_Z(x)=2ck\binom{R}{k}\left[\frac{2(cx)^{\gamma}}{\Gamma(\gamma)}\right]^{R-k+1}K_{\gamma-1}(2cx)K_{\gamma}^{R-k}(2cx)\left[1-\frac{2(cx)^{\gamma}}{\Gamma(\gamma)}K_{\gamma}(2cx)\right]^{k-1} \tag{8.29}$$

当$\gamma=m+1/2(m=0,1)$时，式(8.29)可进一步简化。在$\gamma=0.5$和1.5时，Z的PDF分别为

$$f_Z(x)=k\binom{R}{k}2c\mathrm{e}^{-2c(R-k+1)}(1-\mathrm{e}^{-2cx})^{k-1},\quad \gamma=0.5$$

$$f_Z(x)=k\binom{R}{k}4c^2x\mathrm{e}^{-2c(R-k+1)x}(1+2cx)^{R-k}[1-(1+2cx)\mathrm{e}^{-2cx}]^{k-1},\quad \gamma=1.5$$

把这些表达式代入式(8.23)和式(8.24)，用与前面相同的方法可计算检测性能。

在与前述CA和GO相同条件下，图8.7给出了$P_{\mathrm{d}}=0.5$时OS的检测损失与γ的关系曲线。其中，k选择在$3R/4$处，这是权衡了检测性能、抗干扰目标及抗边缘杂波能力间的一种选择。因为假设杂波是均匀的，所以选择k的值在$R/2\sim7R/8$时不会使OS性能起太大变化。由图8.7可见，OS对各种条件的依赖关系与CA和GO相似。对图8.5～图8.7的比较结果表明，OS遭到的损失明显比CA和GO大，特别是对于小γ值。OS对R下降的敏感性也比CA强。

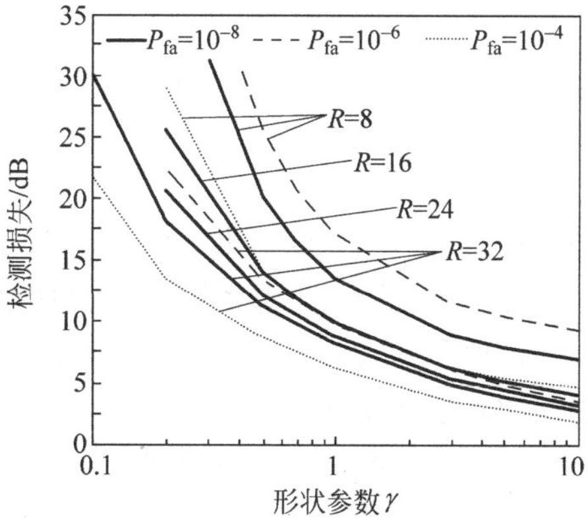

图8.7 OS-CFAR检测器的检测损失与γ的关系

4. CFAR阈值与检测损失的关系

前面三个曲线图中的检测损失均基于同一原因：进行目标检测所面临的是K分布杂波，而不再是瑞利杂波或噪声。为了单独确定由CFAR阈值选用造成的检测损失，要排除杂波尖峰的直接影响，有必要确定一个理想线性检测器在相同尖峰杂波下遭到的损失，并且从总损失中将其扣除。理想检测器损失是关于P_{fa}和γ的函数，可表示为$L_{\mathrm{i}}(\gamma,P_{\mathrm{fa}})$，该函数的确定可参考不相关K分布杂波下关于理想固定阈值的检测方法[27]。这样可得纯粹与CFAR阈值相关的CFAR损失L_{c}为

$$L_{\mathrm{c}}(\gamma,P_{\mathrm{fa}},R)=L_{\mathrm{t}}(\gamma,P_{\mathrm{fa}},R)-L_{\mathrm{i}}(\gamma,P_{\mathrm{fa}}) \tag{8.30}$$

其中，$L_{\mathrm{t}}(\gamma,P_{\mathrm{fa}},R)$是图8.5～图8.7中给出的总损失。CFAR损失$L_{\mathrm{c}}$按照使用的CFAR检测器的类型而有所不同。严格地说，所要求的检测概率也会影响式(8.30)中的所有三项。然而，P_{d}在0.3～0.9范围内变化所产生的影响实际上是可以忽略的，并且在分析中已忽略了P_{d}变化的影响。

在$P_{\mathrm{fa}}=10^{-6}$，$P_{\mathrm{d}}=0.5$条件下，表8.1提供了几种γ取值时，$R=16$和$R=32$的CA、GO和OS检测器的CFAR损失。第二列给出了理想固定阈值检测(理想线性检测器)的损失$L_{\mathrm{i}}(\gamma,10^{-6})$。由表8.1可见，大于2dB的CFAR损失在尖峰杂波中是常见的。在R较小时和杂波尖峰十分严重的极端情况下，超过10dB的较大CFAR损失也是可能的。CA的优越性能又一次表现得比较明显。使用大量参考单元的优势在γ较低时更显著，在$\gamma=0.1$时，R由32变为16将使损失增加10～17dB。

表 8.1 几种 γ 值时的 CA、GO 和 OS 的 CFAR 损失

γ	$L_i(\gamma, 10^{-6})$	CA		GO		OS	
		$R=16$	$R=32$	$R=16$	$R=32$	$R=16$	$R=32$
0.10	13.95	19.49	8.20	19.73	10.00	34.78	17.33
0.25	10.83	7.43	3.52	7.73	3.87	11.68	5.48
0.50	8.59	4.24	2.07	4.67	2.38	6.10	3.02
1.5	5.42	2.50	1.24	2.80	1.45	3.50	1.77
9.5	1.83	2.13	1.05	2.37	1.22	2.95	1.45
∞	0	2.12	1.04	2.35	1.20	2.93	1.44

5. 形状参数估计误差的影响

前面的讨论假设形状参数 γ 是确切已知的，γ 值改变主要影响阈值因子改变。因此，γ 估计的误差将导致不正确的 T 值，引起检测器性能下降。如果阈值设置太高（γ 值估计得太低），检测损失将上升；如果阈值设置得太低（γ 值估计得太高），则虚警概率将上升。由 γ 估计误差导致的损失可由前面的检测损失曲线近似得到。γ 的估计值与真值之间的损失差别，就是由 γ 估计误差引起的近似附加损失。在不同的 P'_{fa} 值下，图 8.8 给出了 $R=32$ 时 CA、OS 和 GO 的阈值因子 T 和 γ 的关系曲线。为确定由 γ 的估计误差引起的 P_{fa} 上升，首先要找出对应于 γ 估计值和虚警概率设计值的 T，然后与这个 T 对应的 γ 的估计值和真实值之间 P'_{fa} 的差别就是 P_{fa} 的上升。各曲线之间的水平差距，可作为虚警概率对 γ 估计误差的敏感性度量。

图 8.8 虽然只给了 $R=32$ 的曲线，但是由此可以得到一些总的变化趋势。

(1) 这三种检测器对 γ 估计误差的敏感性相似。

(2) 这三种 CFAR 检测器在低 $P'_{fa}(10^{-8})$ 时均比高 $P'_{fa}(10^{-4})$ 时对 γ 估计误差更敏感。

(3) 与 γ 值较大的情况相比，γ 值较小时，这三种 CFAR 检测器对 γ 估计误差的敏感性均明显增强。因此，杂波尖峰增强的影响掩盖了检测过程中可能存在的其他问题。

最后，应该说明的是，参考单元数没有严重影响 CFAR 检测器对 γ 估计误差的敏感性。

表 8.2 给出了 P_{fa} 的增加与 γ 的估计值和真实值之间的函数关系[29]。表 8.2 中的数据针对的是 $P'_{fa}=10^{-6}$，$R=32$ 时的 CA 检测器。对于其他使用不同 R 值的检测器，结果不会有太大的不同。

表 8.2 由 γ 估计误差引起的 CA 检测器 P_{fa} 的增加情况

γ_{est}	γ_{true}		
	$0.75\gamma_{est}$	$0.5\gamma_{est}$	$0.25\gamma_{est}$
10	$10^{-5.5}$	$10^{-5.0}$	$10^{-4.2}$
2	$10^{-5.2}$	$10^{-4.5}$	$10^{-3.3}$
0.5	$10^{-5.0}$	$10^{-4.0}$	$10^{-2.5}$

在 K 分布杂波尖峰环境下，若检测阈值是针对瑞利环境所设计的，虚警概率将随 γ 的改变而变化。由图 8.8 可见，在瑞利噪声条件下的虚警概率设计值为 10^{-6} 时，在适度的尖峰杂波（$\gamma \leqslant 2$）中可能产生大于 10^{-2} 的 P_{fa}。这说明在 CFAR 检测器设计中，构造一个正确的杂波包络统计量的重要性。

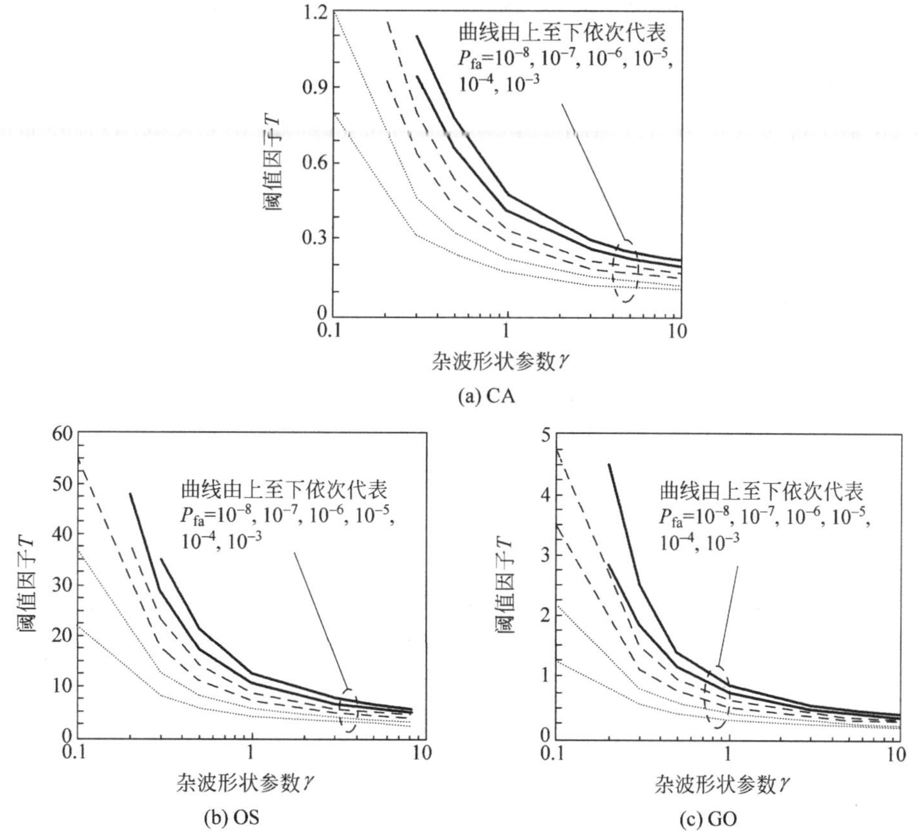

图 8.8　T 与形状参数 γ 的关系曲线($R=32$)

8.4.2　调制过程完全相关的 K 分布杂波下 CFAR 检测

如果调制过程在 CFAR 滑窗内是完全相关的,即调制过程在 CFAR 滑窗内为常数,则杂波包络在 CFAR 滑窗内为均匀瑞利分布。这意味着功率调制过程的 PDF 是在平均功率这一点上的一个 δ 函数。仍然假设杂波是 K 分布的,其局部瑞利杂波的平均功率随 CFAR 滑窗位置而变化,服从 Gamma 分布。因为 CFAR 滑窗中的杂波现在是瑞利分布的,所以除由杂波功率调制过程在信杂比中引入了一个附加起伏源外,CFAR 检测器将针对瑞利杂波工作。

对于给定的 CFAR 检测器和特定的 P_{fa},在瑞利杂波中的检测概率用 $P_d(\text{SCR}|\mu)$表示,其为杂波功率 μ 条件下局部信杂比的函数。杂波功率 μ 的 PDF 用 $f_\mu(x)$表示,它是一个形状参数等于 γ 的 Gamma 分布,其 PDF 为

$$f_\mu(x)=\frac{\beta^\gamma x^{\gamma-1}}{\Gamma(\gamma)}\mathrm{e}^{-\beta x} \tag{8.31}$$

其中,β 是尺度参数。总的检测概率为

$$P_d=\int_0^\infty P_d(\text{SCR}\mid\mu)f_\mu(x)\mathrm{d}x \tag{8.32}$$

可见,总的检测概率取决于 $P_d(\text{SCR}|\mu)$。$P_d(\text{SCR}|\mu)$在理论上取决于 CFAR 检测器的类型和使用的参考单元数。然而,对于瑞利杂波条件,各种情况下 $P_d(\text{SCR}|\mu)$的形式非常相似。用瑞利杂波中对

Swerling Ⅱ型目标进行理想检测的表达式 $P_d(\mathrm{SCR}|\mu)$，可给出一个很好的近似，即

$$P_d(\mathrm{SCR} \mid \mu) = \exp\left[-\frac{\ln P_{fa}^{-1}}{1+(\mathrm{SCR} \mid \mu)}\right] \tag{8.33}$$

其条件是 CFAR 损失被包含在 $\mathrm{SCR}|\mu$ 值中。把式(8.31)和式(8.33)代入式(8.32)，经变量置换，可得 P_d 的表达式为

$$P_d = \frac{\left(\beta\overline{A_t^2}\right)^{\gamma}}{\Gamma(\gamma)}\int_0^{\infty}\frac{1}{x^{\gamma+1}}\exp\left[-\frac{\beta\overline{A_t^2}}{x}-\frac{\ln P_{fa}^{-1}}{1+x}\right]\mathrm{d}x \tag{8.34}$$

其中，$\overline{A_t^2}$ 是目标包络的均方值，$\mathrm{SCR}|\mu=\overline{A_t^2}/\mu$。图 8.9 给出了不同检测概率下形状参数 γ 与检测损失的关系曲线，检测损失为调制过程中完全相关的 K 分布杂波相对于瑞利杂波的损失。

由图 8.9 可见，对于给定的 CFAR 检测器结构和参数，在尖峰杂波中可以获得相对于瑞利杂波的明显增益，特别是对于中等偏低的 P_d 值。但是，应该强调的是，杂波调制在 CFAR 检测器参考滑窗中完全相关的情况，在实际应用中出现较少。

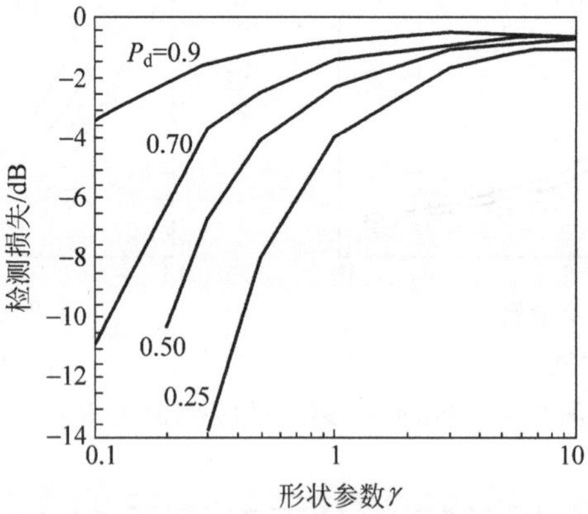

图 8.9　调制过程完全相关的 K 分布杂波的检测损失与 γ 的关系

8.4.3　调制过程部分相关时 K 分布杂波下 CFAR 检测

比较图 8.9 和图 8.5～图 8.7 可以看出，检测损失依赖于杂波调制过程中的空间相关程度。对于中等 γ 值，检测损失(增益)可以变化 8～12dB，对于更小的 $\gamma(\gamma\leqslant 0.5)$ 可以变化 20～30dB 或更多。上面讨论的两种空间相关情况代表了实际中不常遇到的极端情况，表明了检测损失可能出现的范围。如此宽的变化范围使实际的性能预测十分困难，只能给出性能可能变化十几或几十分贝的结论。为此，还需要研究如何把 CFAR 滑窗覆盖区域的杂波调制过程的空间相关程度进一步定量化。

1. 调制过程的复合非中心 Chi 方模型

这里不讨论调制过程的相关机理问题，只假设杂波功率调制过程在每个距离单元中的 PDF 是 Gamma 分布的。因为调制过程在距离单元间相关，给定一个距离单元功率时，其相邻距离单元中调制过程的条件 PDF 将比母体为 Gamma 分布 PDF 的分散度更窄小，而且不一定是 Gamma 分布的。

对于任何给定的调制过程自相关函数 ACF，在检测单元中的 PDF 已知时，不同参考单元中调制过程的条件 PDF 将具有不同的方差，具体取决于它们与检测单元的相关函数，且参考单元之间也不是独立的。因此，如果要确定检测统计量的 PDF，必须确立调制过程的多变量 PDF。前面提到 GMGD 模型的分析很

复杂，为此，基于非中心 Chi 方(Non-Central Chi-Square，NCCS)分布，建立另外一种近似模型。

对于形状参数 $\gamma=m+1/2(m=0,1)$，Gamma 分布变成了自由度为 2γ 的 Chi 方分布。众所周知，它可由 $n=2\gamma$ 个 IID 的零均值高斯过程的平方和产生。如果子过程是非平稳的，它们的方差相等，均值非零且可变化，那么 n 个这样过程的平方和具有非中心参数为 λ 的 n 阶非中心 Chi 方分布，如果子高斯过程的可变均值也是零均值高斯过程，那么非中心参数 λ 服从 Chi 方分布。这样可以得到如下结论：如果一个 NCCS 分布的非中心参数服从 Chi 方分布，则它是一个 Gamma 分布。把它与调制过程联系起来，令子高斯过程的均值随 CFAR 检测器位置变化，但假设它们在 CFAR 滑窗中保持为常数。那么，CFAR 滑窗中的调制过程就由一组独立 NCCS 分布样本组成。此 NCCS 分布的非中心参数 λ 在 CFAR 滑窗中为常数，但随着滑窗位置按 Chi 方分布而随机变化。尽管调制过程样本具有独立性，但由于其具有较窄的分布(只定义在大于零的范围内)，因此具有较高的归一化相关性。以上是对调制过程部分相关的 K 分布杂波的近似描述，在上述模型中，调制样本间的这种独立性大大简化了检测统计量的计算。

下面用数学形式复述上述模型。为不失一般性，假设总体复合杂波的平均功率为 1(这样，功率调制过程的均值就为 $\pi/4$)。功率调制过程 U 的 PDF 可以写成

$$f_U(u)=\left(\frac{4\gamma}{\pi}\right)^{\gamma}\frac{u^{\gamma-1}}{\Gamma(\gamma)}\mathrm{e}^{-4\gamma u/\pi},\quad u\geqslant 0 \tag{8.35}$$

其中，$\gamma=m+1/2$，$m=0,1$。这个调制过程被认为是 2γ 个独立同分布高斯过程 G_i 的平方和，即

$$U=\sum_{i=1}^{2\gamma}G_i^2 \tag{8.36}$$

若式(8.35)成立，则每个 G_i 的方差为 $\pi/8\gamma$，均值为零。式(8.35)对所有距离上的总体 PDF 都是适用的，但如果 G_i 的均值随 CFAR 滑窗位置变化，则可以写成

$$G_i=G_i'+\alpha_i \tag{8.37}$$

其中，G_i'服从 $\mathrm{N}(0,\sigma^2)$，表示子高斯过程的局部随机分量；α_i 服从 $\mathrm{N}(0,\alpha^2)$，表示子高斯过程的变化均值，且

$$\sigma^2+\alpha^2=\frac{\pi}{8\gamma}$$

把式(8.37)代入式(8.36)得到 U 的条件 PDF 为

$$f_U(u\mid A)=\frac{1}{2\sigma^2}\left(\frac{u}{A^2}\right)^{(\gamma-1)/2}\mathrm{e}^{-(u^2+A^2)/(2\sigma^2)}\mathrm{I}_{r-1}\left(\frac{A\sqrt{u}}{\sigma^2}\right),\quad u\geqslant 0 \tag{8.38}$$

其中

$$A^2=\sum_{i=1}^{2\gamma}\alpha_i^2$$

由式(8.38)定义的 PDF 对应非中心 Chi 方分布，其以 A^2 为条件。A^2 被称为非中心参数，它的 PDF 是形状参数为 γ、尺度参数为 $2\alpha^2$ 的 Gamma 分布(因为 α_i 之间独立)。根据假设，A^2 在 CFAR 滑窗内为常数，但随 CFAR 滑窗移动而变换的调制过程，具有以 A 为条件的概率密度函数 $f_U(u|A)$，即式(8.38)。那么，整个功率调制过程具有如下 PDF：

$$f_U(u)=\int_0^{\infty}f_U(u\mid A)f_A(A)\,\mathrm{d}A \tag{8.39}$$

其中，$f_U(u)$是形状参数为 γ、尺度参数为 $\pi/4\gamma$ 的 Gamma 分布(见式(8.35))。由式(8.38)可得电平调

制过程 $V=\sqrt{U}$ 的条件 PDF 为

$$f_V(v\mid A)=\frac{v^{\gamma}}{\sigma^2}\left(\frac{1}{A^2}\right)^{(\gamma-1)/2}\mathrm{e}^{-(v^2+A^2)/(2\sigma^2)}\,\mathrm{I}_{r-1}\left(\frac{Av}{\sigma^2}\right),\quad v\geqslant 0 \tag{8.40}$$

现在要知道的是,该 NCCS 模型中两个样本间的相关程度。根据条件分布的条件方差为 $\sigma^2=(1-\rho^2)E_0$,其中 E_0 为各距离单元上的无条件方差,$E_0=\sigma^2+\alpha^2$,ρ 为相关系数;因而用 $r=\alpha^2/(\alpha^2+\sigma^2)$ 描述两样本间的相关程度,它是局部对全局的方差系数。

如果由波浪结构引入调制相关性,当 CFAR 滑窗覆盖一个或更多的波浪波长时,这意味着 r 将非常低。例如,如果使用 32 个参考单元和两个保护单元,当雷达距离分辨率为 20m 时,CFAR 滑窗将覆盖一个 700m 的距离区域。只有在极少数情况下,CFAR 滑窗才覆盖少于一个波浪波长的区域。因此,十分低的 r 值是常见的,除非雷达分辨单元十分小或者参考单元数减少,但参考单元数的减少将导致附加损失。

2. CFAR 检测性能分析

下面将在基于复合 NCCS 模型调制过程的部分相关 K 分布杂波下,分析 CFAR 检测性能。CFAR 滑窗中复合杂波由电平调制过程 V 调制瑞利散斑分量得到。瑞利散斑的条件 PDF 为 $f(x|v)$,方差为 $4v^2/\pi$,电平调制过程 V 的条件 PDF 为 $f_V(v|A)$,其由式(8.40)给出。因此复合杂波的条件 PDF 为

$$\begin{aligned}f(x\mid A)&=\int_0^{\infty}f(x\mid v)f_V(v\mid A)\,\mathrm{d}v\\&=\frac{x\pi\mathrm{e}^{-A^2/(2\sigma^2)}}{2\sigma^2A^{r-1}}\int_0^{\infty}v^{r-2}\exp\left(-\frac{\pi x^2}{4v^2}-\frac{v^2}{2\sigma^2}\right)\mathrm{I}_{r-1}\left(\frac{Av}{\sigma^2}\right)\mathrm{d}v,\quad x\geqslant 0\end{aligned} \tag{8.41}$$

式(8.41)难以获得闭型表达式。$f(x|A)$ 可用来确定杂波包络均值水平估计 Z 的 PDF。令 $\lambda=A^2$,当给定 λ 时,虚警概率为 $P_{\mathrm{fa}}(\lambda,T)$,总的虚警概率为

$$P_{\mathrm{fa}}=\int_0^{\infty}P_{\mathrm{fa}}(\lambda,T)f_{\lambda}(\lambda)\,\mathrm{d}\lambda \tag{8.42}$$

由式(8.42)可以解得阈值因子 T。类似地,检测概率 $P_{\mathrm{d}}(\lambda,\mathrm{SCR})$ 是杂噪比和 λ 的函数,总的检测概率为

$$P_{\mathrm{d}}(\mathrm{SCR})=\int_0^{\infty}P_{\mathrm{d}}(\lambda,\mathrm{SCR})f_{\lambda}(\lambda)\,\mathrm{d}\lambda \tag{8.43}$$

对于给定的 r,也就是调制过程的相关程度,由式(8.43)可以确定检测损失。

图 8.10 给出了 CA 和 OS 的相对损失与方差系数 r 的关系曲线。由图 8.10 可知,完全相关的相对

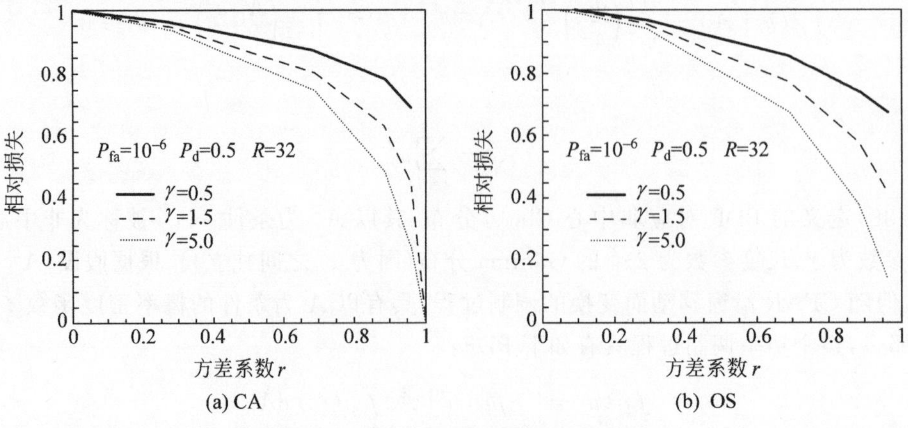

图 8.10 相对损失与方差系数 r 的关系

损失为 0，完全不相关时的损失为 100%，对多数实际的 r 值，相对损失将大于 50%；对于小 r（大约为 0.5），相对损失大于 75%。另外，随着杂波相关性增加，CA 和 OS 中相对损失的变化方式相近，这是因为它们都没有利用参考单元样本间的空间相关性，且 CFAR 滑窗中杂波 ACF 是相同的。

对于 CA 检测方法，图 8.11 给出了 P_{fa} 和 P_d 设计值对相对损失的影响，仍采用相对损失与方差系数的关系曲线。由图 8.11 可知，对于小 γ，P_d 设计值对相对损失有严重影响，随着 P_d 设计值增加，这种影响增强，并且影响程度也随 γ 下降而增强。P_{fa} 设计值对相对损失影响不如 P_d 那么显著，并且只有 $\gamma \leqslant 2$ 时才变得较为严重。虽然这里没有提供图形说明，但也可以注意到，尽管检测损失变化很大，但使用的参考单元数不会严重影响相对损失。对于比图 8.10 和图 8.11 中更大的 γ 值，图 8.11 中的相对损失将随相关性增加而更快下降；然而对于较低的 γ 值，曲线将不会下降那么快。对于 OS 检测方法，也有与上述相似的结果。

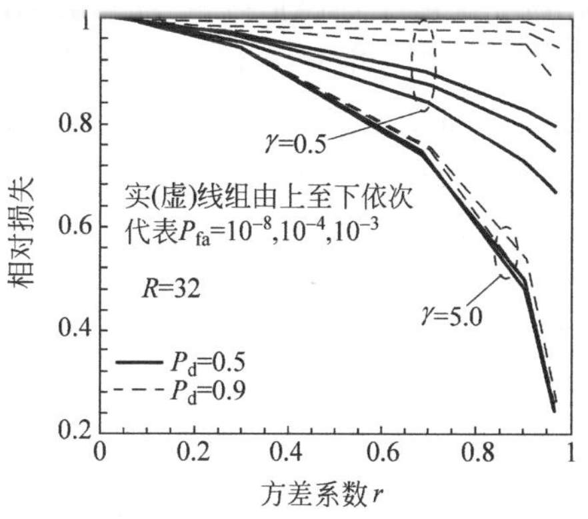

图 8.11 CA-CFAR 检测器的相对损失与方差系数 r 的关系

在杂波空间相关性对 CA-CFAR 性能影响的分析中，Watts[30] 采用空间相关长度（相关样本数）表示相关程度。杂波空间相关性越强，相关样本数越大。分析过程采用了计算机仿真和数学分析结合的方法，得到了与文献[5]中基于实际杂波获得的相同结论。结果表明，如果 CA 的滑窗长度与杂波的空间相关长度匹配得很好，则可以获得明显的性能优势。在杂波空间相关性很弱时，CA 的性能比固定阈值检测稍差，而且滑窗越短性能越差。随着杂波相关性增强，CA 的性能提高，而且滑窗越短，性能越好。当相关参考样本数大于某一值时，CA 的性能优于固定阈值检测的性能，并且差距随相关样本数的增加而增加；但是 CA 的性能始终比理想 CFAR 的性能差很多。对于尖峰杂波，短滑窗可以使阈值跟随杂波的空间变化而获得良好性能，但由于散斑分量引入的阈值起伏，短滑窗将使 CFAR 损失增加。当相关性较弱且形状参数 γ 较大时，长滑窗将获得较好的性能。由虚警率的空间分布来看，高虚警率区不是在强杂波区而是在快变杂波包络更加随机分布的区域。只有在极少数情况下，例如雷达距离分辨率十分高或者参考单元数减少时，CFAR 滑窗才覆盖少于一个海浪波长的区域；然而参考单元数减少会使检测损失上升。此时，CA 将提供总的杂波包络均值水平的估计，阈值因子将依赖于形状参数。如果形状参数在雷达工作区域中是变化的，那么可以利用反馈机制提供适当的阈值因子以保持 CFAR。

8.5 复合高斯杂波中的最优 CFAR 检测器

8.5.1 复合高斯杂波包络中的最优 CFAR 检测

CA、GO 和 OS 等典型的 CFAR 检测器都是在瑞利杂波背景下设计的。在 K 分布杂波背景中，上述检测器对杂波功率水平的估计，尤其是在相关 K 分布杂波背景中不一定是最优的。如果利用相关 K 分布模型参数的最优估计设计 CFAR 检测阈值，则为最优 CFAR 检测器。在相关 K 分布杂波背景中，这种最优 CFAR 检测器性能将比在瑞利背景下设计的 CFAR 检测器具有更好的性能。在相关 K 分布杂波模型中，调制分量在时间和空间上都将具有相关性，这时调制分量的联合分布包含了杂波结构信息,因此也称为结构分量。相关 K 分布杂波的相关性主要是由结构分量在时间和空间上的相关性引入的。

在复合高斯杂波(包括 K 分布和复合 Weibull 分布杂波)的结构分量部分相关条件下,Bucciarelli 等[31]研究了最优 CFAR 检测,推导出了四个基于局部结构分量估计的 CFAR 检测器。这四个检测器的区别在于利用功率值还是功率的对数值,以及是否利用先验信息。CA 和最优线性滤波(Optimum Linear Filter,OLF)利用功率值,而对数单元平均(Cell Averaging Logarithm,CAL)和对数最大后验(Logarithm Maximum a Posteriori,LMAP)利用功率的对数值。CA 和 CAL 不需要先验信息,在各种相关条件下都易于实现,LMAP 和 OLF 则需要由模型参数的精确先验信息构成的杂波图。随着形状参数(结构分量的归一化方差的倒数)的增加,检测概率随相关系数的变化范围明显减小。因此,当形状参数较大时,用结构分量的估计值代替真实值设定阈值,不会使检测性能受到太大的影响。

考虑利用杂波回波功率值估计结构分量的方法。在结构分量完全相关条件下,结构分量的最大似然估计与 CA 形式相同,在结构分量部分相关情况下则采用 MAP 估计。MAP 估计需要解大量的非线性方程组,其计算复杂性的解决方法是 OLF[32],OLF 利用参考样本的线性函数作估计,根据最小均方误差准则确定系数。在结构分量部分相关时,CA 和 OLF 的性能相差无几。在完全相关和完全不相关时,CA 和 OLF 完全重合。由此可见,OLF 的性能相对于 CA 没有多少提高,并且需要杂波分布的先验信息,而 CA 不需要任何附加的先验信息。于是,文献[31]得到了与文献[32]相同的结论,即估计器的线性限制不能获得相对 CA 的明显改善。因此,需要采用局部结构分量的非线性估计改善检测器性能。

对数转换就是典型的非线性运算,乘积模型经过对数转换变成了和的形式,两个分量的对数值的分布可以用具有相同一阶和二阶统计特性的高斯分布近似。首先,在结构分量完全相关条件下,由对数域的 ML 估计得到了与 CA 相似的形式,称之为 CAL。在部分相关条件下,采用对数域的 MAP 估计得到 LMAP。随着相关性减弱,CAL 的性能急速下降,在大信噪比和高检测概率区甚至比固定阈值性能还差。LMAP 的性能介于理想 CFAR 和固定阈值检测之间,对相关性的依赖较弱。在形状参数变化时,LMAP 的性能始终明显优于 CAL,与理想 CFAR 接近。因此,在对数域中利用结构分量的先验信息是值得的。相比之下,OLF 和 CAL 是应该放弃的方法。LMAP 有相对于 CA 的优势,它们之间的差距取决于结构分量的方差。在具有相同一阶和二阶矩的 K 分布和复合 Weibull 分布杂波中,CA 在两种环境中性能非常相似,LMAP 也是这样。因此,它们对统计特性服从不同乘积模型的杂波具有内在的适应性。估计器对参数的先验信息不是很敏感,因此足够精确的杂波图可以使 LMAP 得以实现。当先验信息的误差很大时,基于先验信息的检测性能严重下降,这种影响还会被对数运算放大。在实际应用中,可以采取某些措施克服这种影响[31]。文献[31]的扩展研究主要体现在,用结构分量的估计值代替其准确值,并把方法推广到所有乘积模型。

8.5.2 复合高斯杂波中的最优相参子空间 CFAR 检测

对于复合高斯杂波中的最优相参子空间 CFAR 检测问题[33],已有许多学者进行了研究。文献[7]～[12]研究了复合高斯杂波背景中的 Swerling Ⅰ型目标的相参检测问题,文献[13]则研究了复合高斯杂波中子空间目标的检测问题。在这些研究中,杂波的纹理分量都假设在雷达系统的 CPI 中是完全相关的,这个假设对于不太长的雷达处理时间来说是合理的。文献[14]推导了最优 Neyman-Pearson(NP)检测器、广义似然比检验(GLRT)检测器及 CFAR 检测器,并讨论了它们之间的区别。它们之间的区别在于所需的目标信号先验知识的程度不一样。文中特别强调了三者在性能和计算复杂度方面的比较。

为了对每个检测器是如何考虑杂波的非高斯特性有更深入的理解，该文还分析了与高斯情况之间的联系，给出了估计器-相关器、门限依赖数据的匹配滤波器这两种解释。基于这一点，文献[14]是文献[11]中结论在子空间目标随机模型中的推广，并指出了与秩 1 情况[11]的类似与区别之处。文献[13]分析了文献[14]中检测算法的性能，讨论 NP、GLRT 和 CFAR 检测器的性能对各种杂波和信号参数的依赖性。文中采用两种方法给出了数值结果，一种是采用 Monte-Carlo 方法进行计算机仿真；另一种则基于实测高分辨海杂波数据，即 McMaster 大学的 X 波段 IPIX 雷达数据。接着还分析了两种场景：空时自适应处理(STAP)场景和地基监视系统场景。

式(8.44)～式(8.46)分别给出了 NP、GLRT 和 CFAR 三种检测器的判决形式[14]，即

$$\Lambda_{\mathrm{NP}}(\boldsymbol{z})=\frac{\mathrm{E}_{\tau}\mathrm{E}_{\boldsymbol{\beta}}\{f_{z|\tau,\mathrm{H}_0}(\boldsymbol{z}-\boldsymbol{U}_{\mathrm{t}}\boldsymbol{\beta}\mid\tau,\mathrm{H}_0)\}}{\mathrm{E}_{\tau}\{f_{z|\tau,\mathrm{H}_0}(\boldsymbol{z}\mid\tau,\mathrm{H}_0)\}}\underset{\mathrm{H}_0}{\overset{\mathrm{H}_1}{\gtrless}}\eta \tag{8.44}$$

其中，$\boldsymbol{z}$ 是待检测距离单元的观测矢量；目标观测矢量 $\boldsymbol{t}$ 建模为高斯线性模型 $\boldsymbol{t}=\boldsymbol{U}_{\mathrm{t}}\boldsymbol{\beta}$，$\boldsymbol{U}_{\mathrm{t}}$ 和$\boldsymbol{\beta}$ 完全已知；$\boldsymbol{\beta}$ 服从均值为零矢量、协方差矩阵为 $\sigma_{\mathrm{t}}^2\boldsymbol{\Lambda}_{\mathrm{t}}$的复高斯分布。$\mathrm{E}_{\boldsymbol{\beta}}\{\cdot\}$表示关于$\boldsymbol{\beta}$ 的统计均值。从式(8.44)可知，要想执行最优 NP 检测策略，必须预先已知目标模式权矢量$\boldsymbol{\beta}$ 的 PDF(用 $f_{\boldsymbol{\beta}}(\boldsymbol{\beta})$表示)和杂波纹理 τ 的 PDF(用 $f_{\tau}(\tau)$表示)。

如果 $f_{\boldsymbol{\beta}}(\boldsymbol{\beta})$未知，就意味着不知道 σ_{t}^2 和/或$\boldsymbol{\Lambda}_{\mathrm{t}}$(或者甚至不能假设$\boldsymbol{\beta}$ 是高斯分布的)，那么必须将$\boldsymbol{\beta}$建模为一个未知的确定性的矢量，并采取另外一种方法。在这种情况下，一致最优势(UMP)检验就不存在了。一种合理的方法就是所谓的 GLRT 方法，该准则中的未知参数被替换为它的最大似然估计。具体来说，未知矢量$\boldsymbol{\beta}$ 被替换为其最大似然估计$\hat{\boldsymbol{\beta}}_{\mathrm{ML}}$，然后用这个似然比与门限 η 进行比较决策：

$$\Lambda_{\mathrm{GLRT}}(\boldsymbol{z})=\frac{\mathrm{E}_{\tau}\{f_{z|\tau,\mathrm{H}_0}(\boldsymbol{z}-\boldsymbol{U}_{\mathrm{t}}\hat{\boldsymbol{\beta}}_{\mathrm{ML}}\mid\tau,\mathrm{H}_0)\}}{\mathrm{E}_{\tau}\{f_{z|\tau,\mathrm{H}_0}(\boldsymbol{z}\mid\tau,\mathrm{H}_0)\}}\underset{\mathrm{H}_0}{\overset{\mathrm{H}_1}{\gtrless}}\eta \tag{8.45}$$

换言之，由于不能通过统计平均将未知参量$\boldsymbol{\beta}$ 从 $\Lambda_{\mathrm{NP}}(\boldsymbol{z};\boldsymbol{\beta})$中剔除，而 $f_{\boldsymbol{\beta}}(\boldsymbol{\beta})$是未知的(或者其计算量太大)，所以就用相应的最大似然估计来代替未知的$\boldsymbol{\beta}$。这两种不同的方法(最优 NP 方法与 GLRT 方法)反映了对观测信号统计模型的不同了解程度。

如果 $f_{\tau}(\tau)$也未知，则需采用类似的方法[34]：在两个假设条件下，分别将 τ 替换为相应的最大似然估计 $\hat{\tau}_{\mathrm{ML},0}$ 和 $\hat{\tau}_{\mathrm{ML},1}$，得到的结果为

$$\Lambda_{\mathrm{CFAR}}(\boldsymbol{z})=\frac{p_{z|\tau,\mathrm{H}_0}(\boldsymbol{z}-\boldsymbol{U}_t\hat{\boldsymbol{\beta}}_{\mathrm{ML}}\mid\hat{\tau}_{\mathrm{ML},1},\mathrm{H}_0)}{p_{z|\tau,\mathrm{H}_0}(\boldsymbol{z}\mid\hat{\tau}_{\mathrm{ML},0},\mathrm{H}_0)}\underset{\mathrm{H}_0}{\overset{\mathrm{H}_1}{\gtrless}}\eta \tag{8.46}$$

因此，就从两个条件 PDF 之比中剔除了对未知参数$\boldsymbol{\beta}$ 和 τ 的依赖性。值得强调的是，上述两个参数是未知且随机的；另外，它们的先验 PDF($f_{\boldsymbol{\beta}}(\boldsymbol{\beta})$和 $f_{\tau}(\tau)$)是未知的。

针对复合高斯分布情况，由式(8.44)～式(8.46)，上述三个检测器的检验统计量可具体表示为

$$\Lambda(\boldsymbol{z})=\frac{\displaystyle\int_0^{\infty}\frac{1}{\tau^m\left(\dfrac{\sigma_t^2}{\tau}\boldsymbol{p}\boldsymbol{M}^{-1}\boldsymbol{p}^{\mathrm{H}}+1\right)}\exp\left[-\frac{1}{\tau}\left(\boldsymbol{z}^{\mathrm{H}}\boldsymbol{M}^{-1}\boldsymbol{z}-\frac{|\boldsymbol{p}^{\mathrm{H}}\boldsymbol{M}^{-1}\boldsymbol{z}|^2}{\boldsymbol{p}\boldsymbol{M}^{-1}\boldsymbol{p}^{\mathrm{H}}+\dfrac{\tau}{\sigma_t^2}}\right)\right]f_{\tau}(\tau)\mathrm{d}\tau}{\displaystyle\int_0^{\infty}\frac{1}{\tau^m}\exp\left(-\frac{\boldsymbol{z}^{\mathrm{H}}\boldsymbol{M}^{-1}\boldsymbol{z}}{\tau}\right)f_{\tau}(\tau)\mathrm{d}\tau}\underset{\mathrm{H}_0}{\overset{\mathrm{H}_1}{\gtrless}}\eta \tag{8.47}$$

$$\Lambda_{\mathrm{GLRT}}(\boldsymbol{z})=\frac{\int_0^{\infty}\frac{1}{\tau^m}\exp\left[-\frac{\boldsymbol{z}^{\mathrm{H}}(\boldsymbol{M}^{-1}-\boldsymbol{Q}_2)\boldsymbol{z}}{\tau}\right]f_\tau(\tau)\mathrm{d}\tau}{\int_0^{\infty}\frac{1}{\tau^m}\exp\left(-\frac{\boldsymbol{z}^{\mathrm{H}}\boldsymbol{M}^{-1}\boldsymbol{z}}{\tau}\right)f_\tau(\tau)\mathrm{d}\tau}\underset{\mathrm{H}_0}{\overset{\mathrm{H}_1}{\gtrless}}\eta \tag{8.48}$$

$$\boldsymbol{z}^{\mathrm{H}}\boldsymbol{Q}_2\boldsymbol{z}\underset{\mathrm{H}_0}{\overset{\mathrm{H}_1}{\gtrless}}\eta\boldsymbol{z}^{\mathrm{H}}\boldsymbol{M}^{-1}\boldsymbol{z} \tag{8.49}$$

其中,$\boldsymbol{p}$ 是 Swerling Ⅰ型目标条件下的 $\boldsymbol{U}_{\mathrm{t}}$,$\boldsymbol{Q}_2=\boldsymbol{M}^{-1}\boldsymbol{p}(\boldsymbol{p}^{\mathrm{H}}\boldsymbol{M}^{-1}\boldsymbol{p})^{-1}\boldsymbol{p}^{\mathrm{H}}\boldsymbol{M}^{-1}$,$\boldsymbol{M}$ 表示杂波散斑分量的协方差矩阵。上述三式中,式(8.47)是最优 NP 检测器,式(8.48)是 GLRT 检测器,式(8.49)是 CFAR 检测器。

8.6 球不变随机杂波下相参 CFAR 检测

8.5 节重点阐述了复合高斯杂波下最优 CFAR 检测问题,本节将讨论复合高斯杂波下一般意义上的点目标 CFAR 检测问题。估计未知杂波的协方差矩阵(或杂波谱属性)是雷达目标自动检测中需要解决的重要问题[35-36]。实际中常常利用与被检测单元邻近的辅助数据,对未知杂波协方差矩阵进行估计。实测数据表明[2],在高分辨率或低掠地角情况下,雷达会接收到类似于目标的尖峰,这种含尖峰的杂波往往服从复合高斯分布,而 SIRV 则为多维复合高斯建模提供了有力的数学工具[1]。根据中心极限定理的局部有效性,SIRV 是一个时间和空间"慢变化"的纹理分量(反映了受照块的反射率),与一个变化"更快"的"散斑"高斯向量的乘积。在实际的雷达工作环境中,由于杂波的统计特性往往是未知的,需要从观测值中估计得到。

在假设 SIRV 杂波协方差矩阵结构已知的情况下,文献[10]和文献[12]分别得到了归一化匹配滤波器(NMF),文献[12]还利用不含目标的辅助数据估计 SIRV 杂波协方差矩阵结构,并将估计矩阵代入 NMF,获得了自适应归一化匹配滤波器(ANMF)。文献[37]研究了 SIRV 协方差矩阵结构的 ML 估计问题,并指出 ML 估计具有加权样本协方差矩阵的形式,但没有闭型表达式,且涉及超越方程的求解。本节将重点围绕协方差矩阵结构估计[38],讨论球不变随机杂波下相参 CFAR 检测问题。

8.6.1 最大似然估计问题

为了简要说明问题,以纯杂波数据情况为例。由于纹理分量 τ_0 的 PDF 未知,SIRV 杂波服从具有未知方差 τ_0 的条件高斯分布。参照式(8.13)的复合高斯模型,在散斑分量$\boldsymbol{\eta}_0$ 的协方差矩阵结构$\boldsymbol{\Sigma}$ 未知的条件下,纯杂波距离单元对应的同相与正交分量构成的复观测值 $\boldsymbol{z}_0$ 的 PDF 可表示为

$$f(\boldsymbol{z}_0\mid\tau_0,\boldsymbol{\Sigma})=\frac{1}{\pi^N\tau_0^N\det(\boldsymbol{\Sigma})}\times\exp\left(-\frac{1}{\tau_0}\boldsymbol{z}_0^{\mathrm{H}}\boldsymbol{\Sigma}^{-1}\boldsymbol{z}_0\right) \tag{8.50}$$

其中,上标 H 表示共轭转置,det(·)表示取矩阵行列式,N 为相参积累数。

由于存在未知参数,可以利用 GLRT 准则设计检测器。根据 GLRT 理论[39],未知参数可用相应的 ML 估计来代替。式(8.50)中,未知参数 τ_0 和$\boldsymbol{\Sigma}$ 的 ML 估计分别可以表示为[37]

$$\hat{\tau}_0=\frac{1}{N}\boldsymbol{z}_0^{\mathrm{H}}\hat{\boldsymbol{\Sigma}}^{-1}\boldsymbol{z}_0 \tag{8.51}$$

和

$$\hat{\boldsymbol{\Sigma}} = \boldsymbol{z}_0 \boldsymbol{z}_0^{\mathrm{H}} / \hat{\tau}_0 \tag{8.52}$$

由式(8.51)和式(8.52)可知,参数 τ_0 和$\boldsymbol{\Sigma}$ 的 ML 估计均涉及超越方程的求解,而协方差矩阵结构$\boldsymbol{\Sigma}$的 ML 估计可以进一步表示为

$$\hat{\boldsymbol{\Sigma}} = \frac{N \boldsymbol{z}_0 \boldsymbol{z}_0^{\mathrm{H}}}{\boldsymbol{z}_0^{\mathrm{H}} \hat{\boldsymbol{\Sigma}}^{-1} \boldsymbol{z}_0} \tag{8.53}$$

需要说明的是,式(8.53)关于$\boldsymbol{\Sigma}$ 的解无明确的解析表达式,这就使得未知$\boldsymbol{\Sigma}$ 条件下基于 GLRT 的检测器难以实现。文献[40]在已知纹理分量分布的假设下,利用期望最大化算法求解 ML 估计的超越方程,但是该算法的收敛速率较慢,且计算复杂度较高。

8.6.2 CFAR 检测问题

由于未知$\boldsymbol{\Sigma}$ 条件下基于 GLRT 的检测器难以实现,文献[10]在已知杂波协方差矩阵结构$\boldsymbol{\Sigma}$ 的条件下,将被检测单元的纹理分量 τ_0 看作未知确定量,利用 GLRT 理论,获得了 NMF 检测器。为了实现自适应检测,文献[12]基于两步法 GLRT,首先在已知$\boldsymbol{\Sigma}$ 假设下建立 NMF 检测器,然后利用纯杂波辅助数据获得杂波协方差矩阵结构的合适估计$\hat{\boldsymbol{\Sigma}}$,并用之代替未知的$\boldsymbol{\Sigma}$,最终获得自适应归一化匹配滤波器(ANMF),待检测单元对应的复观测值 $\boldsymbol{z}$ 的检测统计量可表示为[12]

$$\lambda_{\mathrm{ANMF}} = \frac{|\boldsymbol{p}^{\mathrm{H}} \hat{\boldsymbol{\Sigma}}^{-1} \boldsymbol{z}|^2}{(\boldsymbol{p}^{\mathrm{H}} \hat{\boldsymbol{\Sigma}}^{-1} \boldsymbol{p})(\boldsymbol{z}^{\mathrm{H}} \hat{\boldsymbol{\Sigma}}^{-1} \boldsymbol{z})} \tag{8.54}$$

其中,$\boldsymbol{p}$ 表示已知的目标导向矢量。

如何选择杂波协方差矩阵结构的合适估计$\hat{\boldsymbol{\Sigma}}$,成为 ANMF 检测器在 SIRV 杂波下有效实现目标自适应检测的关键[41]。

在高斯杂波背景下,经典的样本协方差矩阵 SCM 是相应的 ML 估计,并有着广泛的应用。为了便于比较,这里直接给出基于辅助数据的样本协方差矩阵 SCM[35]

$$\hat{\boldsymbol{\Sigma}}_{\mathrm{SCM}} = \frac{1}{R} \sum_{t=1}^{R} \boldsymbol{z}_t \boldsymbol{z}_t^{\mathrm{H}} \tag{8.55}$$

其中,$\boldsymbol{z}_t (t=1,2,\cdots,R)$表示辅助数据集,每个这样的辅助数据都不包含有用的目标回波,且具有与待检测距离单元同分布的 SIRV 杂波。一般令 $R \geqslant N$,由此确保由辅助数据获得的协方差矩阵是非奇异的[37,42]。

另外,在 SIRV 杂波背景下,考虑到不同距离单元间杂波功率水平的起伏,往往先对辅助数据进行归一化,去除杂波纹理分量的影响,然后利用归一化后的杂波辅助数据进行矩阵估计,即可获得基于归一化辅助数据的归一化样本协方差矩阵(NSCM)[43]

$$\hat{\boldsymbol{\Sigma}}_{\mathrm{NSCM}} = \frac{1}{R} \sum_{t=1}^{R} \left(\frac{N}{\boldsymbol{z}_t^{\mathrm{H}} \boldsymbol{z}_t} \right) \boldsymbol{z}_t \boldsymbol{z}_t^{\mathrm{H}} \tag{8.56}$$

需要指出的是,经典的 SCM 是高斯背景下的 ML 估计,但不是 SIRV 杂波背景下的 ML 估计[40];而 NSCM 虽然消除了纹理分量的影响,也同样不是 SIRV 杂波背景下的 ML 估计。将 SCM 和 NSCM 估计矩阵分别代入式(8.54),可分别得到基于 SCM 的 ANMF(简称 ANMF-SCM)和基于 NSCM 的 ANMF(简称 ANMF-NSCM)检测器。

8.6.3 性能分析

为便于仿真分析,假设纹理分量服从形状参数为 L、均值为 1 的 Gamma 分布[38];散斑分量对应的

协方差矩阵结构$\boldsymbol{\Sigma}$用一阶相关系数为γ的指数相关结构建模。在参考单元数$R=24$和相参积累数$N=8$的条件下,图8.12和图8.13分别给出了$L=0.1,1$和$\gamma=0,0.9$时ANMF-SCM和ANMF-NSCM的检测阈值与虚警概率的关系曲线。从图8.12中可以看出,在L相同的条件下,不同的杂波一阶相关系数γ对应的曲线完全重合,而不同的杂波尖峰L对应的曲线相差甚远,这说明ANMF-SCM检测器对杂波协方差矩阵结构具有CFAR特性,但对杂波功率水平起伏不具有自适应特性。而从图8.13中可以看出,在相同的条件下,不同的杂波尖峰L对应的曲线完全重合,但不同的杂波一阶相关系数γ对应的曲线存在起伏,这说明ANMF-NSCM检测器对杂波功率水平起伏具有CFAR特性,但对杂波协方差矩阵结构不具有自适应特性。

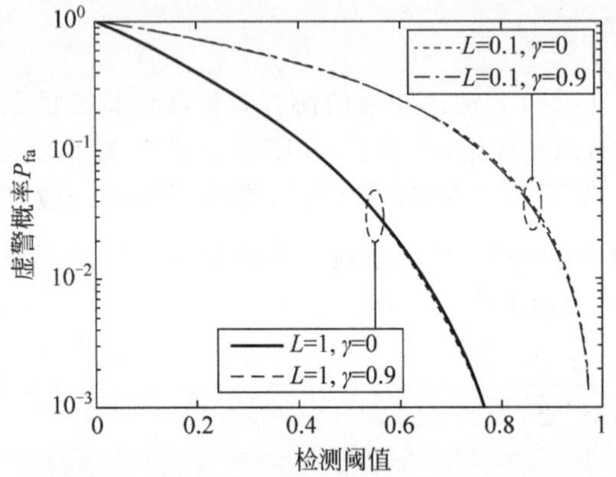

图8.12 $R=24,N=8,L=0.1,1,\gamma=0,0.9$时ANMF-SCM的检测阈值与虚警概率的关系曲线

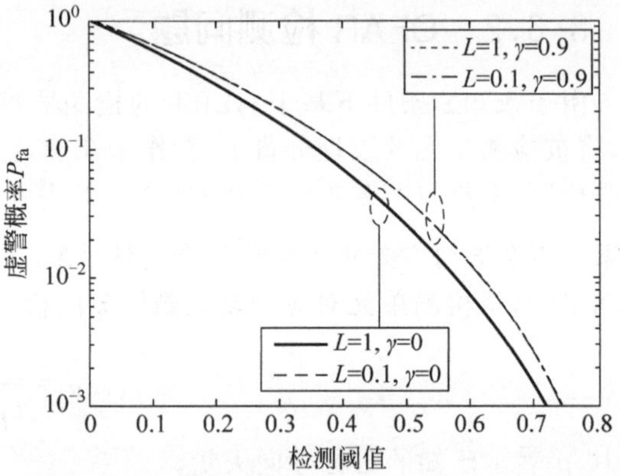

图8.13 $R=24,N=8,L=0.1,1,\gamma=0,0.9$时ANMF-NSCM的检测阈值与虚警概率的关系曲线

综合来看,虽然在高斯背景下,ANMF-SCM对协方差矩阵具有自适应特性[43],但其在SIRV非高斯背景下只具有部分自适应能力,即只对杂波协方差矩阵结构是CFAR的,但对不同距离单元的杂波功率水平不是CFAR的,除非不同距离单元间的纹理分量完全相关[38]。另外,在SIRV背景下,ANMF-NSCM也只具有部分自适应能力,即对不同距离单元的纹理分量是CFAR的,但对杂波协方差矩阵结构不是CFAR的[43]。由于SIRV背景下协方差矩阵结构$\boldsymbol{\Sigma}$的ML无明确的解析表达式,为实现CFAR检测或近似CFAR检测,诸多学者从近似最大似然估计[44-45]和杂波按功率分组估计[46-47]等方面进行了有益探索。

8.7 复合高斯杂波中的贝叶斯自适应检测器

在非均匀复合高斯杂波环境下,辅助数据的选取规模往往很有限。为弥补辅助数据不足造成的信息缺失,引入先验知识十分必要[48]。与传统的非贝叶斯方法相比,贝叶斯方法可以更加有效地引入先验信息,降低对辅助数据的需求,进而提升雷达目标检测性能[36]。在8.5节的自适应检测器设计过程中,若纹理分量的PDF已知,可得到最优NP检测器或GLRT检测器;若纹理分量的PDF未知,将其视为未知确定量,并利用不同假设下纹理分量的ML估计进行替代,则可得到前述式(8.49)的CFAR检测器[14]。值得注意的是,在复合高斯杂波模型中,纹理分量被建模为正随机变量,在其实际统计分布未知的情况下,若能引入合适的先验分布对其建模,有利于构造具有广泛适用性的自适应检测器。

本节在复合高斯杂波环境下，依据贝叶斯方法，将均匀分布作为纹理分量的先验分布，利用辅助数据估计纹理分量的后验 PDF，进而给出相应的贝叶斯自适应检测器，并在 K 分布和 t 分布两种典型复合高斯杂波下进行检测器仿真分析。

8.7.1 问题描述

复合高斯杂波下的雷达目标检测问题可由如下二元假设检验表示：

$$\begin{aligned} &\mathrm{H}_0 : \boldsymbol{z} = \boldsymbol{c} \\ &\mathrm{H}_1 : \boldsymbol{z} = \alpha \boldsymbol{p} + \boldsymbol{c} \end{aligned} \tag{8.57}$$

其中，$\boldsymbol{z}$、$\boldsymbol{c}$ 和 $\boldsymbol{p}$ 分别表示 $N \times 1$ 维的观测向量、杂波分量和已知导向矢量；α 为未知的目标复幅度。

在 H_0 假设下，复合高斯杂波分量 $\boldsymbol{c}$ 服从零均值协方差矩阵为 $\tau\boldsymbol{\Sigma}$ 的条件复高斯分布。假定纹理分量 $\boldsymbol{\tau}$ 的 PDF 为 $f_\tau(x)$，则 H_0 假设下 $\boldsymbol{z}$ 的 PDF 可以表示为

$$f_z(\boldsymbol{z}) = \int_0^{+\infty} \frac{1}{(x\pi)^N |\boldsymbol{\Sigma}|} \exp\left(-\frac{\boldsymbol{z}^{\mathrm{H}} \boldsymbol{\Sigma}^{-1} \boldsymbol{z}}{x}\right) f_\tau(x) \mathrm{d}x \tag{8.58}$$

显然，τ 的分布特性决定了杂波分量 $\boldsymbol{c}$ 的分布特性，τ 选取不同概率分布即可获得不同的复合高斯杂波。例如，典型的 K 分布[49]对应于 Gamma 分布的纹理分量，而 t 分布[50]则对应于逆 Gamma 分布的纹理分量。当 τ 的分布已知时，可直接依据检验准则求取相应的最优检测器[14]，但此处 τ 的确切分布是无法获得的，可对其 PDF 进行适当估计，进而进行检测器设计。

8.7.2 贝叶斯自适应检测器设计

1. 纹理分量 PDF 的估计

为估计待检测观测数据中杂波分量的统计特性，假设 R 个辅助数据只包含与待检测距离单元 IID 的纯杂波，记为 $\boldsymbol{z}_t, t=1,2,\cdots,R$。根据贝叶斯公式，基于辅助数据的纹理分量 PDF 估计可表示为

$$\hat{f}_\tau(x) = \frac{1}{R}\sum_{t=1}^{R} f_\tau(x \mid \boldsymbol{z}_t) = \frac{1}{R}\sum_{t=1}^{R} \frac{f_z(\boldsymbol{z}_t \mid x) f_{\tau_0}(x)}{f_{z_0}(\boldsymbol{z}_t)} \tag{8.59}$$

其中，$f_\tau(x|\boldsymbol{z}_t)$表示基于辅助数据 $\boldsymbol{z}_t$ 的纹理分量后验 PDF，$f_z(\boldsymbol{z}_t|x)$表示基于纹理分量的辅助数据 $\boldsymbol{z}_t$ 的条件 PDF，$f_{\tau_0}(x)$表示纹理分量的先验 PDF，$f_{z_0}(\boldsymbol{z}_t)$表示基于纹理分量先验分布的辅助数据 PDF，可由式(8.60)求得

$$f_{z_0}(\boldsymbol{z}_t) = \int_0^{+\infty} f_z(\boldsymbol{z}_t \mid x) f_{\tau_0}(x) \mathrm{d}x \tag{8.60}$$

根据前述分析，辅助数据 $\boldsymbol{z}_t$ 的条件分布 $f_z(\boldsymbol{z}_t|x)$为复高斯分布，由式(8.59)和式(8.60)可知，给定纹理分量的先验 PDF，即可估计纹理分量 PDF。

关于纹理分量先验分布的选取，存在诸多不同选择。最常用的为逆 Gamma 分布[51]，另外逆 Gaussian 分布[52]也可作为相应的纹理分量先验分布，这些分布的共同特点是都属于共轭分布[50]，在刻画纹理分量的分布特性的同时，易于推导运算。但特定分布的先验信息往往需要满足特定应用条件，若关于纹理分量无明确的统计先验信息，则特定先验信息的引入可能导致与实际不符，进而影响检测器性能。从复合高斯模型中，可以明确的先验信息是纹理分量为正随机变量，考虑到随机变量取值的普适性，此处选取均匀分布作为纹理分量的非信息先验分布[53]。

根据纹理分量的取值范围，$f_{\tau_0}(x)$可表示为

$$f_{\tau_0}(x)=\frac{1}{\theta},\quad x\in(0,\theta),\quad \theta\to+\infty \tag{8.61}$$

将式(8.61)代入式(8.60)和式(8.59)中,经化简计算可得

$$\hat{f}_\tau(x)=\frac{1}{R}\sum_{t=1}^{R}\frac{Z_t^{N-1}}{x^N\Gamma(N-1)}\exp\left(-\frac{Z_t}{x}\right),\quad x>0 \tag{8.62}$$

式中,$Z_t=\boldsymbol{z}_t^{\mathrm{H}}\boldsymbol{\Sigma}^{-1}\boldsymbol{z}_t,t=1,2,\cdots,R$。

2. 检测统计量的构建

在前述假设检验问题的基础上,此处采用两步法 GLRT 准则构建 $\boldsymbol{z}$ 的检测统计量 λ_{B},即首先假设协方差矩阵结构$\boldsymbol{\Sigma}$ 已知,按成检测统计量构建后再由其估计值将其替换。注意到在两种假设下,待检测距离单元的观测向量 $\boldsymbol{z}$ 的条件 PDF 均为复高斯分布,因此检测统计量可初步表示为

$$\lambda_{\mathrm{B}}=\frac{\displaystyle\int_0^{+\infty}\frac{1}{x^N}\exp\left[-\frac{A_1(\boldsymbol{y})}{x}\right]\hat{f}_\tau(x)\mathrm{d}x}{\displaystyle\int_0^{+\infty}\frac{1}{x^N}\exp\left[-\frac{A_0(\boldsymbol{y})}{x}\right]\hat{f}_\tau(x)\mathrm{d}x} \tag{8.63}$$

式中,$A_0(\boldsymbol{z})=\boldsymbol{z}^{\mathrm{H}}\boldsymbol{\Sigma}^{-1}\boldsymbol{z}$,$A_1(\boldsymbol{z})=(\boldsymbol{z}-\alpha_{\mathrm{ML}}\boldsymbol{p})^{\mathrm{H}}\boldsymbol{\Sigma}^{-1}(\boldsymbol{z}-\alpha_{\mathrm{ML}}\boldsymbol{p})$。注意到 H_1 假设下未知目标幅度 α 的 ML 估计为

$$\hat{\alpha}_{\mathrm{ML}}=\frac{\boldsymbol{p}^{\mathrm{H}}\boldsymbol{\Sigma}^{-1}\boldsymbol{z}}{\boldsymbol{p}^{\mathrm{H}}\boldsymbol{\Sigma}^{-1}\boldsymbol{p}} \tag{8.64}$$

将式(8.62)和式(8.64)代入式(8.63)中,经运算化简后,可得检测统计量为[54]

$$\lambda_{\mathrm{B}}=\frac{\displaystyle\sum_{t=1}^{R}Z_t^{N-1}[A_1(\boldsymbol{z})+Z_t]^{1-2N}}{\displaystyle\sum_{t=1}^{R}Z_t^{N-1}[A_0(\boldsymbol{z})+Z_t]^{1-2N}} \tag{8.65}$$

式中,$A_1(\boldsymbol{z})=\boldsymbol{z}^{\mathrm{H}}\boldsymbol{\Sigma}^{-1}\boldsymbol{z}-|\boldsymbol{p}^{\mathrm{H}}\boldsymbol{\Sigma}^{-1}\boldsymbol{z}|^2/(\boldsymbol{p}^{\mathrm{H}}\boldsymbol{\Sigma}^{-1}\boldsymbol{p})$。

根据两步法 GLRT,将式(8.56)基于辅助数据的归一化样本协方差矩阵$\hat{\boldsymbol{\Sigma}}_{\mathrm{NSCM}}$ 代入式(8.65),替换其中的未知$\boldsymbol{\Sigma}$,即可得贝叶斯自适应检测器(Bayesian Adaptive Detector,BAD)。

8.7.3 性能分析

为便于对比分析,主要考虑 K 分布和 t 分布两种典型的复合高斯分布情形,其中 K 分布模型中纹理分量服从形状参数为 v、尺度参数为 b 的 Gamma 分布[55],t 分布模型中纹理分量服从形状参数为 w、尺度参数为 d 的逆 Gamma 分布[51]。对比检测器主要包括:式(8.54)的自适应归一化匹配滤波器(ANMF),其中假定纹理分量为未知确定量;针对 t 分布纹理分量所设计的最优 t 检测器(Optimum Student-t Detector,OtD)[51],其中的分布参数(w,d)假定已知。为不失一般性,设定虚警概率 $P_{\mathrm{fa}}=10^{-2}$,$N=5$,$R=10$。

图 8.14 给出了形状参数 $v=0.5$ 的 K 分布杂波环境下,BAD 与 ANMF 的检测性能对比结果。由图 8.14 可知,二者在典型的 K 分布复合高斯杂波下均表现出较好的检测性能,其中 BAD 在低 SCR 条件下的检测性能更好,而 ANMF 在高 SCR 条件下略有优势。

图 8.15 在形状参数 $v=20$ 的 K 分布杂波下分析了二者的检测性能。由图 8.15 可知,随着形状参

数 v 的增大，复合高斯分布逐渐逼近高斯分布，BAD 的检测性能优于 ANMF，且在低 SCR 条件下更为突出。

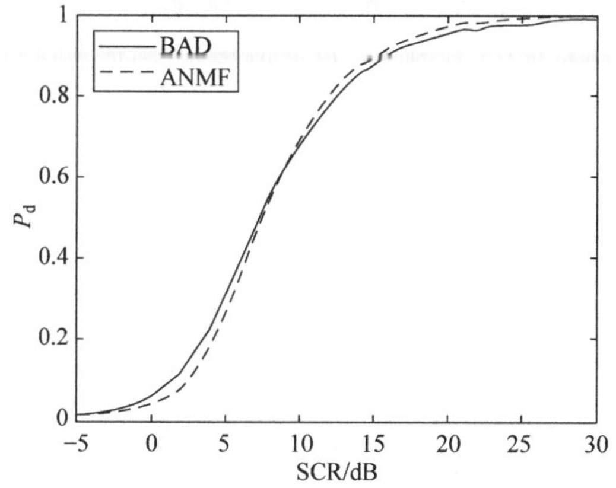

图 8.14　$v=0.5$ 的 K 分布杂波下 BAD 与 ANMF 的检测性能

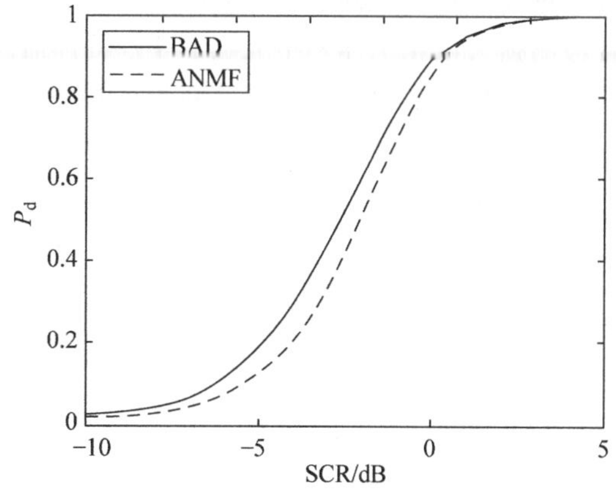

图 8.15　$v=20$ 的 K 分布杂波下 BAD 与 ANMF 的检测性能

图 8.16 进一步在典型 t 分布复合高斯杂波环境下，比较了 BAD、ANMF 及 OtD 的检测性能。由图 8.16 可知，在 t 分布杂波环境下，无杂波分布先验信息的 BAD 与已知杂波分布参数的最优检测器 OtD 相当，二者均明显优于 ANMF，这一性能优势在低 SCR 环境下尤为明显。

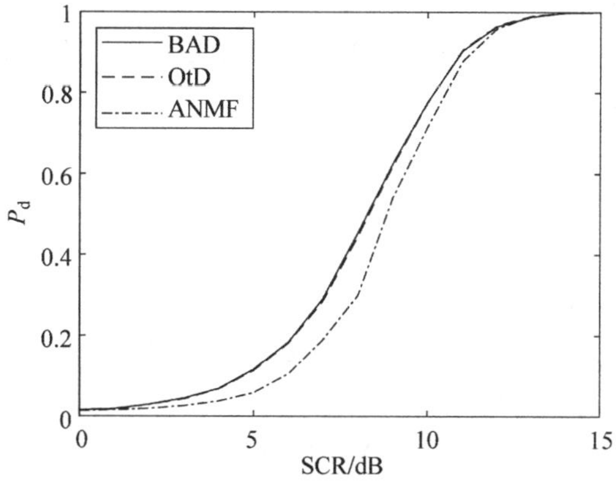

图 8.16　$w=5$ 和 $d=1$ 的 t 分布杂波下 BAD、ANMF 与 OtD 的检测性能

8.8　小结

本章讨论了复合高斯杂波复幅度模型和 K 分布包络模型，并介绍了一种相关 K 分布模型[7]，重点分析了不同复合高斯杂波下的目标 CFAR 检测问题。

8.3 节分析了 K 分布海杂波加热噪声中的检测性能，并且给出了预测目标检测性能(包括脉冲积累的影响)的方法。分析结果表明检测性能在很宽的范围内的变化不仅取决于杂波条件，还取决于产生阈值的方法和杂噪比。8.3 节还分析了瑞利背景中三种典型 CFAR 检测器(CA、GO 和 OS)在具有两种极端相关情况(调制过程在空间上完全不相关和调制过程在空间上完全相关这两种情况)的 K 分布杂波中的性能。

8.4.1 节在调制过程不相关的 K 分布杂波条件下分析了 CA、GO 和 OS 的性能。在相邻单元的杂波调制过程不相关条件下，它们在杂波环境中(图 8.5～图 8.7)表现出了相当大的检测损失(约 10dB)，并且检测损失依赖于参考单元数和 P_{fa} 设计值。参考单元数越小，并且 P_{fa} 设计值越低，则检测器对杂波尖峰的增加越敏感。检测概率对检测损失没有太明显的影响。OS 的损失明显比 CA 和 GO 的高，特别是对于小 γ，比 CA 高几分贝。对于 CA、GO 和 OS，参考单元数的减小也使检测损失上升，其作用同降低 P_{fa} 设计值一样。OS 对 R 下降的敏感性也比 CA 强。GO 在尖峰杂波中的性能介于 OS 和 CA 之间。γ 的估计误差对 P_{fa} 的影响分析表明，上述三种检测器对估计误差的敏感性相似，并且参考单元数对这种敏感性没有很强的影响。P_{fa} 设计值和 γ 值越低，CFAR 检测器对误差越敏感。为瑞利环境设置的 CFAR 阈值用在 γ 值约为 0.5 的 K 分布杂波环境中可能导致 P_{fa} 有 1000～10 000 倍的增加。

8.4.2 节在调制过程完全相关的条件下分析了 CFAR 检测器的性能。结果发现，在调制过程完全相关的尖峰杂波中可以获得相对于瑞利杂波中更强的检测增益，特别是对于中等偏低的 P_d 值(见图 8.9)。

8.4.3 节建立了更实际的调制过程部分相关的 K 分布杂波模型。对于一个给定的相关程度 r，杂波越尖，P_{fa} 设计值越低和 P_d 设计值越高，损失就越趋近于完全不相关情况。在多数情况下，方差系数(相关程度)r 是十分小的，这也说明了损失将趋近于完全不相关情况。因此，在多数调制过程相关条件下，CA、GO 和 OS 的检测损失将趋近于最差情形的损失，也就是完全不相关条件下的损失。在杂波调制过程部分相关的某些情况下，检测性能可以得到明显的改善。在 CA、GO 和 OS 中，OS 最容易受尖峰杂波的直接或间接影响，它的损失随杂波尖峰的增加而增加得更快；并且对于相同的损失，OS 需要更多的参考单元，这意味着 CFAR 滑窗中杂波的相关程度更低。然而在杂波边缘和干扰目标情况下，OS 具有明显优势。

CA、GO 和 OS 等典型的 CFAR 检测器都是在瑞利杂波背景条件下设计的，在复合高斯杂波背景条件下的工作性能不一定是最优的。如果利用复合高斯杂波参数的最优估计设置检测阈值，可以获得复合高斯杂波背景中的最优 CFAR 检测，8.5 节讨论了这个问题。针对复合高斯杂波包络中的目标最优 CFAR 检测问题，8.5.1 节讨论了文献[31]提出的几种方案，结果表明，在对数域中利用结构分量的先验信息是值得的。相比之下，OLF 和 CAL 是应该放弃的方法；LMAP 有相对于 CA 的优势，它们之间的差距取决于结构分量的方差。而针对复合高斯杂波背景中的相参 CFAR 检测问题，8.5.2 节简单讨论了子空间模型下的最优 NP、GLRT 和 CFAR 检测器[14]，三者的区别在于所需的目标信号先验知识的程度不一样。

SIRV 杂波协方差矩阵结构的 ML 估计无闭型解，导致自适应 GLRT 检测器难以实现。8.6 节重点分析了 SIRV 杂波下一般意义上的点目标 CFAR 检测问题，从 SIRV 协方差矩阵结构的 ML 估计和 ANMF 的 CFAR 实现两个角度，讨论了 SCM 和 NSCM 的对检测性能的影响，二者对应的 ANMF 分别只对杂波协方差矩阵结构和杂波功率水平具有 CFAR 性能，相关的研究还可参考文献[49]和[56]。

为降低辅助数据需求，提升雷达目标检测性能[36]，引入先验知识十分必要，而贝叶斯方法在引入先验分布方面具有独特优势。8.7 节在复合高斯杂波环境下，讨论了贝叶斯自适应检测方法，将无先验信息的均匀分布作为纹理分量的先验分布，基于辅助数据估计纹理分量后验 PDF，构建了相应的贝叶斯自适应检测器，在 K 分布和 t 分布两种典型复合高斯杂波下均展现了良好的检测性能和适应性。

复合高斯杂波纹理分量的经典模型为 Gamma 分布，相应的杂波包络服从 K 分布[50]。随着杂波建模的精细化，纹理分量模型逐渐拓展到逆 Gamma 分布[50,57-59]、逆高斯分布[60]等；另外，杂波散斑分量协方差矩阵也可采用随机矩阵建模[36,61]。面向实际应用，目标信号失配[62-63]、杂波参数失配[64]等失配情况下CFAR检测问题值得深入研究。关于检验准则的选择方面，除常用的GLRT准则[65]，Rao检验准则[66]、Wald检验准则[67]、Bayes准则[54]等有待进一步分析。随着CFAR处理技术的广泛应用，复合高斯背景下的CFAR处理已延伸到极化处理[68]、MIMO雷达[48]、天基雷达[69]、雷电信号处理[70]、多基地雷达[6]、模糊软阈值判决[71]等信号处理领域。总之，对复合高斯杂波中的自动检测和CFAR处理方法研究还有许多问题需要解决，并且越来越引起相关研究者的兴趣。

参考文献

[1] Conte E, Longo M. Characterization of radar clutter as a spherically invariant random process[J]. IEE Proc.-F, 1987, 134(2): 191-197.

[2] Ward K D. Compound representation of high resolution sea clutter[J]. Electronics Letters, 1981, 17(16): 561-563.

[3] Ward K D, Watts S. Radar sea clutter[J]. Microwave Journal, 1985, 28(6): 109-121.

[4] Jakeman E, Pusey P N. A model for non-Rayleigh sea echo[J]. IEEE Transactions on AP, 1976, 24(6): 806-814.

[5] Ward K D, Baker C J, Watts S. Maritime surveillance radar[J]. IEE Proc.-F, 1990, 137(2): 51-72.

[6] Palamà R, Greco M, Gini F. Multistatic adaptive CFAR detection in non-Gaussian clutter[J]. EURASIP Journal on ASP, 2016(1): 107.

[7] Pentini F A, Farina A, Zirilli F. Radar detection of targets located in a coherent K-distributed clutter background[J]. IEE Proc.-F, 1992, 139(3): 239-245.

[8] Sangston K J, Gerlach K R. Coherent detection of radar targets in a non-Gaussian background[J]. IEEE Transactions on AES, 1994, 30(2): 330-340.

[9] Rangaswamy M, Weiner D, Ozturk A. Computer generation of correlated non-Gaussian radar clutter[J]. IEEE Transactions on AES, 1995, 31(1): 106-115.

[10] Gini F. Suboptimum coherent radar detection in a mixture of K-distributed and Gaussian clutter[J]. IEE Proc.-F, 1997, 144(1): 39-48.

[11] Sangston K J, Gini F, Greco M V, et al. Structures for radar detection in compound-Gaussian clutter[J]. IEEE Transactions on AES, 1999, 35(2): 445-458.

[12] Conte E, Lops M, Ricci G. Asymptotically optimum radar in compound-Gaussian clutter[J]. IEEE Transactions on AES, 1995, 31(2): 617-625.

[13] Gini F, Farina A. Matched subspace CFAR detection of hovering helicopters[J]. IEEE Transactions on AES, 1999, 35(4): 1293-1305.

[14] Gini F, Farina A. Vector subspace detection in compound-Gaussian clutter part I: survey and new results[J]. IEEE Transactions on AES, 2002, 38(4): 1295-1311.

[15] Lombardo P, Farina A. Coherent radar detection against K-distributed clutter with partially correlated texture[J]. Signal Processing, 1996, 48(1): 1-15.

[16] Armstrong B C, Griffiths H D. Modeling spatially correlated K-distributed clutter[J]. Electronics Letters, 1991, 27(15): 1355-1356.

[17] Gini F, Giannakis G B, Greco M, et al. Time-averaged subspace methods for radar clutter texture retrieval[J]. IEEE Transactions on Signal Process., 2001, 49(9): 1886-1898.

[18] Guan J, He Y, Peng Y N. CFAR detection in K-distributed clutter[C]. Beijing: Proceedings of 4th International Conference on Signal Processing, 1998: 1513-1516.

[19] Watts S. Radar detection prediction in K-distributed sea clutter and thermal noise[J]. IEEE Transactions on AES, 1987, 23(1): 40-45.

[20] Long M W. Radar reflectivity of land and sea[M]. London: Artech House, 1983.

[21] Yi L, Yan L, Han N. Simulation of inverse Gaussian compound Gaussian distribution sea clutter based on SIRP[C]. Ottawa: IEEE Workshop on Advanced Research and Technology in Industry Applications, 2014: 1026-1029.

[22] Oliver C J, Tough R J A. On simulation of correlated K-distributed random clutter[J]. Optica Acta, 1986, 33(3): 223-250.

[23] Blacknell D. New method for the simulation of correlated K-distributed clutter[J]. IEE Proc.-F, 1994, 141(1): 53-58.

[24] Conte E, Longo M, Lops M. Modeling and simulation of non-Rayleigh radar clutter[J]. IEE Proc.-F, 1991, 138(2): 121-130.

[25] 欧阳文，何友，靳煜. 基于统计模型的时-空相关海杂波仿真[J]. 系统仿真学报，2006，18(2): 467-471.

[26] 张彦飞，关键，周伟，等. 基于球不变随机过程的高分辨率雷达杂波仿真[J]. 现代雷达，2006，28(2): 46-49.

[27] Watts S. Radar detection prediction in sea clutter using the compound K-distributed model[J]. IEE Proc.-F, 1985, 132(7): 613-620.

[28] Guidoum N, Soltani F, Zebiri K, et al. Robust non parametric CFAR detector in compound Gaussian clutter in the presence of thermal noise and interfering targets[C]. Berlin: International Conference on Image and Signal Processing, 2018: 186-193.

[29] Armstrong B C, Griffiths H D. CFAR detection of fluctuating targets in spatially correlated K-distributed clutter[J]. IEE Proc.-F, 1991, 138(2): 139-152.

[30] Watts S. Cell-averaging CFAR gain in spatially correlated K-distributed clutter[J]. IEE Proc.-F, 1996, 143(5): 321-327.

[31] Bucciarelli T, Lombardo P, Tamburrini S. Optimum CFAR detection against compound Gaussian clutter with partially correlated texture[J]. IEE Proc.-F, 1996, 143(2): 95-104.

[32] Armstrong B C. Processing techniques for improved radar detection in spiky clutter[D]. London: University College of London, 1992.

[33] 邓晓波，施长海，高超. 复合高斯杂波中子空间信号检测[J]. 系统工程与电子技术，2013，35(9): 1836-1840.

[34] Zhang Yangzhong, Zhang Yu, Tang Bo. Persymmetric adaptive detection of subspace signals in compound-Gaussian clutter[C]. Guangzhou: CIE International Conference on Radar, 2016.

[35] 王智，简涛，何友，等. 杂波协方差矩阵结构的融合估计方法[J]. 控制与决策，2019，34(9): 2010-2014.

[36] 苗旭炳，简涛，丁彪. 非均匀杂波协方差矩阵的知识辅助估计方法[J]. 电光与控制，2016，23(10): 45-48.

[37] Gini F, Greco M. Covariance matrix estimation for CFAR detection in correlated heavy tailed clutter[J]. Signal Process., 2002, 82(12): 1847-1859.

[38] He Y, Jian T, Su F, et al. CFAR assessment of covariance matrix estimators for non-Gaussian clutter[J]. Science China: Information Sciences, 2010, 53(11): 2343-2351.

[39] Jian T, He Y, Wang H P, et al. Persymmetric generalized adaptive matched filter for range-spread targets in homogeneous environment[J]. IET Radar, Sonar and Navigation, 2019, 13(8): 1234-1241.

[40] Rangaswamy M. Statistical analysis of the nonhomogeneity detector for non-Gaussian interference backgrounds[J]. IEEE Transactions on Signal Process., 2005, 53(6): 2101-2111.

[41] 王智，简涛，何友. 复合高斯背景下基于最优控制参数的自适应检测器[J]. 控制与决策，2018，33(8): 1532-1536.

[42] 简涛，苏峰，何友，等. 复合高斯杂波下距离扩展目标的自适应检测[J]. 电子学报，2012，40(5): 990-994.

[43] Gini F, Michels J H. Performance analysis of two covariance matrix estimators in compound Gaussian clutter[J]. IEE Proc.-F, 1999, 146(3): 133-140.

[44] Pascal F, Chitour Y, Ovarlez J P, et al. Covariance structure maximum-likelihood estimates in compound Gaussian noise: existence and algorithm analysis[J]. IEEE Transactions on Signal Process., 2008, 56(1): 34-48.

[45] 顾新锋，简涛，何友，等. 协方差矩阵结构的广义近似最大似然估计[J]. 应用科学学报，2013，31(6): 585-592.

[46] Conte E, De Maio A, Ricci G. CFAR detection of distributed targets in non-Gaussian disturbance[J]. IEEE

Transactions on AES，2002，38(2)：612-621.
[47] 顾新锋，简涛，何友，等. 协方差矩阵结构的广义杂波分组估计方法[J]. 宇航学报，2012，33(12)：1794-1800.
[48] Zhang T X，Cui G L，Kong L J，et al. Adaptive Bayesian detection using MIMO radar in spatially heterogeneous clutter[J]. IEEE Signal Process. Letters，2013，20(6)：547-550.
[49] He Y，Jian T，Su F，et al. Adaptive detection application of covariance matrix estimation for correlated non-Gaussian clutter[J]. IEEE Transactions on AES，2010，46(4)：2108-2117.
[50] 高永婵. 复杂场景下多通道阵列自适应目标检测算法研究[D]. 西安：西安电子科技大学，2015.
[51] Jay E，Ovarlez J P，Declercq D，et al. BORD：Bayesian optimum radar detector[J]. Signal Processing，2003，83(6)：1151-1162.
[52] Ollila E，Tyler D E，Koivunen V，et al. Compound-Gaussian clutter modeling with an inverse Gaussian texture distribution[J]. IEEE Signal Process. Letters，2012. 19(12)：876-879.
[53] 苗旭炳. 异质环境下基于知识的雷达目标自适应检测方法研究[D]. 烟台：海军航空工程学院，2015.
[54] Miao X B，Jian T，He Y. A novel Bayesian adaptive detector against non-Gaussian clutter[C]. Hangzhou：IET International Radar Conference 2015，2015.
[55] Zhang T X，Cui G L，Kong L J，et al. Phase-modulated waveform evaluation and selection strategy in compound-Gaussian clutter[J]. IEEE Transactions on Signal Process.，2013，61(5)：1143-1148.
[56] Zhang Y F，Guan J. Adaptive subspace detection of range distributed targets in compound-Gaussian clutter[C]. Rome：Proceedings of IEEE Radar Conference，2008.
[57] Sangston K J，Gini F，Greco M S. Coherent radar target detection in heavy-tailed compound-Gaussian clutter[J]. IEEE Transactions on AES，2012，48(1)：64-77.
[58] Weinberg G V. Coherent CFAR detection in compound Gaussian clutter with inverse Gamma texture[J]. EURASIP Journal on ASP，2013(1)：105.
[59] 刘明. 海杂波中微弱运动目标自适应检测方法研究[D]. 西安：西安电子科技大学，2016.
[60] 薛健. 复合高斯海杂波背景雷达目标检测算法[D]. 西安：西安电子科技大学，2020.
[61] De Maio A. Generalized CFAR property and UMP invariance for adaptive signal detection[J]. IEEE Transactions on Signal Processing，2013，61(8)：2104-2115.
[62] 王泽玉. 雷达目标自适应检测算法研究[D]. 西安：西安电子科技大学，2018.
[63] 许述文，石星宇，水鹏朗. 复合高斯杂波下抑制失配信号的自适应检测器[J]. 雷达学报，2019，8(3)：326-334.
[64] 简涛，何友，苏峰. 复合高斯杂波协方差矩阵估计的失配性能分析[J]. 电子学报，2011，39(4)：963-966.
[65] Shang X，Song H. Radar detection based on compound-Gaussian model with inverse Gamma texture[J]. IET Radar Sonar and Navigation，2011，5(3)：315-321.
[66] Jian T，He Y，Su F，et al. Adaptive detection of range-spread targets without secondary data in multichannel autoregressive process[J]. Digital Signal Processing，2013，23(5)：1686-1694.
[67] Maio D A，Han S，Orlando D. Adaptive radar detectors based on the observed FIM[J]. IEEE Transactions on Signal Processing，2018，66(14)：3838-3847.
[68] 庞晓宇. 分布式 MIMO 雷达的自适应检测研究[D]. 哈尔滨：哈尔滨工业大学，2014.
[69] 王海涛，叶琦，刘爱芳. 基于自适应波形设计的天基雷达目标检测方法[J]. 宇航学报，2013,34(8)：1130-1136.
[70] 杜海明. 雷电信号检测方法及相关问题研究[D]. 武汉：华中科技大学，2012.
[71] Xu Y W，Yan S F，Ma X C，et al. Fuzzy soft decision CFAR detector for the K distribution data[J]. IEEE Transactions on AES，2015，51(4)：3001-3013.

第 9 章 非参量 CFAR 处理

CHAPTER 9

9.1 引言

前面讨论的方法都假设杂波包络的分布类型已知，只需要估计一些未知参数，使之在该假设下具有 CFAR，这类方法称为参量 CFAR 方法。在实际中，雷达环境的杂波类型往往是未知且常是时变的，这时采用非参量（Nonparametric）检测或自由分布（Distribution Free）检测就具有一定的优势。当参量 CFAR 方法中所假设的杂波分布与实际杂波环境不一致时，它就失去了恒虚警能力[1]，这种情况下虚警概率与杂波分布类型无关的非参量 CFAR 方法就显出了优势。非参量或者自由分布检测器的目的是，在分布不确定或变化的背景噪声或杂波包络统计量中提供虚警控制能力[2]。

在前面几章中讨论的检测策略都是假设在一个观测间隔上进行检测时，除有限个参数之外，杂波包络的分布是已知的，即除一个或更多的分布参数可能变化之外，分布类型是固定的。然而，非参量检测器提供了不需要关于背景噪声或杂波分布的先验假设的检测。严格地说，统计学的非参量指这样一个概念：可能的噪声或杂波分布构成了一个不能用有限个数的实参数标志的大集合。在雷达检测理论文献中常采用一个不太精确的定义[3]，即非参量检测器是在关于背景噪声或杂波统计特性的弱假设下具有固定（常数）虚警概率的检测器。这种弱假设描述的是，在单脉冲匹配滤波器输出处所得到的乃是一类（如中值已知的独立分布）杂波或纯噪声样本数据的集合，它是关于杂波或纯噪声随机过程的唯一信息，没有其他可以利用的杂波或纯噪声分布的函数形式信息。

非参量检测器的基本结构是把杂波或纯噪声输入数据集转换成检测统计量。这个检测统计量与一个固定检测阈值进行比较，以获得关于背景噪声或杂波环境统计特性弱假设下的恒虚警率。在给定的假设集下，完成这一功能的转换不一定是唯一的。但是，通常在更严格的条件下不成立时，能完成这一功能的检测策略优于采用更严格统计条件推导出的基于 CFAR 的最优检测策略。

非参量处理方面的文献很多，但将其应用于雷达等方面的文献却较有限。非参量检测有单样本和双样本非参量检测之分。双样本非参量检测是指在检测器输入端有两个样本集合可供利用，一个样本集合可能包含目标信号，是需要统计判决的集合；而另一个集合是作为参考样本的观测噪声样本。本章先讨论对非参量检测器进行衡量的传统准则，即渐近相对效率（Asymptotic Relative Efficiency，ARE），然后介绍单样本非参量检测器的典型代表：符号检测器和 Wilcoxon 检测器。符号检测器和 Wilcoxon 检测器均要求杂波中值水平已知，同时，它们的检测性能损失对于有限的样本数目是很大的，但这些缺点可以通过两样本检测策略克服。本章将在 9.4 节讨论两样本非参量检测器，并研究它们的检测性能，而 9.5 节将研究次优秩检验非参量检测器，广义符号检测器和 Savage 检测器可看成它的特例。9.6 节在韦布尔分布的典型非高斯背景下，重点对量化秩与广义符号两种非参量检测器进行虚警和检测性能分析。9.7 节讨论利用逆正态得分函数对秩检验进行改进的非参量检测方法，并给出仿真分析结果。

9.2 非参量检测器的渐近相对效率

在实际雷达系统中选择一个非参量检测策略通常要考虑其在工作环境中的检测性能，以及与雷达系统中其他处理方法的兼容能力(如模拟与数字滤波器、相关波形处理能力、存储容量等)。但是，一个非参量检测器的性能究竟如何？是否可取？前面曾指出，当参数 CFAR 方法中所假设的杂波分布与实际杂波环境不一致时，它就失去了恒虚警能力，这种情况下非参量 CFAR 方法就显出了优势。但若干扰的统计特性为已知，则参量型检测器往往是更优的。在这种情况下，由于非参量检测器的针对性差，没有充分利用干扰的统计知识，其性能一般不如参量型检测。这是一种定性的说明，实际上到底差到什么程度，这是人们所关心的，必须要有一个参数能定量地比较两个检测器的性能。虽然可以定义出许多相对效率来比较一种检测器相对于另一种检测器的相对效率，但最常使用的来自 Pitman。Pitman 首先提出了用 ARE 来衡量两个不同检测器的性能。应该指出，渐近相对效率是在弱信号假设下导出的，原因在于该假设下导出的表达式较简单。若没有这个假设，许多计算从解析的观点将会变得不可能，实际上即使在弱信号的假设下，渐近相对效率表达式的推导过程也是很繁杂的。

与前几章讨论的参量型检测器一样，非参量型检测器也是对二元假设检验依据统计独立的观测样本 $w_1, w_2, \cdots, w_n$(其概率分布记为 P)做出选择。令 $N_1(\alpha,\beta,P)$表示检测器 D_1 在零假设 H 下虚警概率为 α 时为了在备择假设 K 下达到检测概率 β 所需要的最少观测样本数目，而 $N_2(\alpha,\beta,P)$表示检测器 D_2 在同样的条件下虚警概率为 α 时为了达到检测概率 β 所需要的最少观测样本数目。则检测器 D_1 相对于检测器 D_2 的相对效率定义为

$$e_{1,2}=\frac{N_1(\alpha,\beta,P)}{N_2(\alpha,\beta,P)} \tag{9.1}$$

显然，相对效率是关于 K，α，N_1 和 N_2 的函数。但是对于任意的备择假设 K，α，N_1 和 N_2 来说，相对效率的计算是非常困难的。一种简单的方法是令 N_1 和 N_2 趋于无穷大，这时对于固定的备择假设 K 来说，检测器 D_1 和 D_2(假定检测器 D_1 和 D_2 为相合检验)的检测概率逼近于 1，但是令备择假设 K 逼近于零假设 H，这样仍然会让 N_1 和 N_2 趋于无穷大时检测概率 β(在统计学中也称为功效函数)保持不变。这样，就有了渐近相对效率 ARE(D_1,D_2)的概念，即

$$\mathrm{ARE}(D_1,D_2)=\lim_{\substack{\mathrm{K}\to\mathrm{H}\\ N_1,N_2\to\infty}}\frac{N_1(\alpha,\beta,\mathrm{K})}{N_2(\alpha,\beta,\mathrm{K})} \tag{9.2}$$

备择假设 K→H，也就是信噪比逼近于零，这是一种弱信号情况下检测器检测性能的衡量。渐近相对效率将极限情况下一种检测器在满足虚警概率约束时为达到给定检测概率所需要的样本数目与另一种检测器所需要的数目联系起来。显然，检测器 D_1 相对于二元假设检验对(H，K)下的最优检测器的渐近相对效率会小于 1，另外，检测器 D_1 相比于对背离统计假设不敏感的次优检测器的渐近相对效率

会大于 1。

这里考虑的二元假设检验仍然是一个门限检测问题，而门限需要预先从给定的虚警概率中确定。在许多的具体计算中，用来计算渐近相对效率(ARE)的检测统计量和门限的函数形式是非常烦琐的。可考虑另外一种形式的渐近相对效率的计算公式，即采用两种检测器效力的比值来计算渐近相对效率，即

$$\mathrm{ARE}(D_1, D_2)=\frac{\varepsilon_1}{\varepsilon_2} \tag{9.3}$$

其中，ε_i 为第 i 个检测器的效力。采用效力的优点在于对于复杂的检验问题，其计算易于进行，而效力定义为

$$\varepsilon=\lim_{N\to\infty}\frac{\left.\left[\dfrac{\partial}{\partial \bar{s}}\mathrm{E}(T\mid\bar{s})\right]^2\right|_{\bar{s}=0}}{N\sigma_0^2(T)} \tag{9.4}$$

其中，$\bar{s}$ 表示信号噪声平均功率比；N 表示观测次数或脉冲探测次数；$\sigma_0^2(T)$表示在零假设下只有噪声出现时检测统计量 T 的方差。$\mathrm{E}(T\mid\bar{s})$表示信号噪声平均功率比为 $\bar{s}$ 时检测统计量 T 的数学期望。可以证明，式(9.4)和式(9.2)给出的渐近相对效率是相同的。

9.3 单样本非参量检测器

9.3.1 符号检测器

现在考虑比($\mathrm{H}_0, \mathrm{K}_0$)和($\mathrm{H}_1, \mathrm{K}_1$)假设描述的检测问题更为广泛的情况，即将高斯分布中的均值(Mean)推广为统计分布的中位数(Median)(累积分布函数 $F(x)$的中位数 m 定义为$F(m)=1/2$)。假定输入序列 $\boldsymbol{w}=(w_1, w_2, \cdots, w_n)$统计独立且同分布，它们共同的累积分布函数为 $F(x)=P\{w_i\leqslant x\}$。定义概率 p 为

$$p=\Pr\{w_i>0\}=1-F(0) \tag{9.5}$$

考虑如下的非参量零假设 H_2 和非参量备择假设 K_2：

$$\mathrm{H}_2: p=1/2 \tag{9.6}$$

$$\mathrm{K}_2: p>1/2 \tag{9.7}$$

这等效于 $\boldsymbol{w}=(w_1, w_2, \cdots, w_n)$是中位数为零的统计独立噪声样本与观测样本中出现加性正值信号时中位数大于零的假设检验判决。在 K_2 假设下，观测样本中信号的存在会引起观测样本中位数的一个正偏移，而符号检测器(Sign Detector)提供了对这个偏移的检验。符号检测器定义为

$$d_{\mathrm{s}}=\sum_{i=1}^{n}u(w_i)\begin{cases}<C_2\Rightarrow D_2(\boldsymbol{w})=0\\ \geqslant C_2\Rightarrow D_2(\boldsymbol{w})=1\end{cases} \tag{9.8}$$

其中，$u(w_i)$表示单位阶跃函数，即

$$u(w_i)=\begin{cases}1, & w_i>0\\ 0, & w_i<0\end{cases}$$

可以证明，符号检测器 $D_2(\boldsymbol{w})$是关于假设检验($\mathrm{H}_2, \mathrm{K}_2$)的一个 Neyman-Pearson 准则下的最优检测器[4]。图 9.1 给出了符号检测器的原理框图。

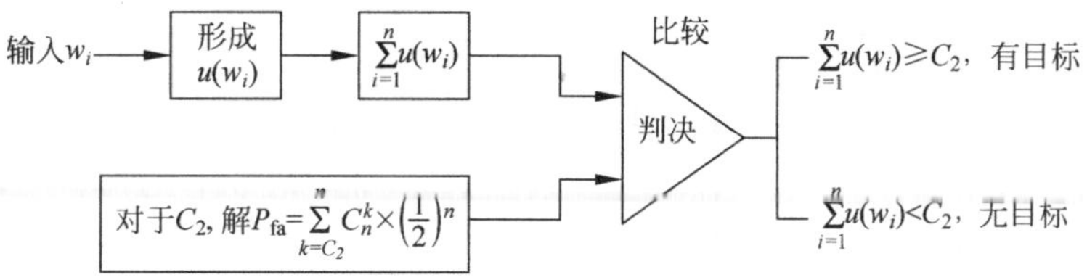

图 9.1 符号检测器的原理框图

检测统计量 $d_s=\sum_{i=1}^{n}u(w_i)$ 是正的观测样本的数目,并且是一个服从参数为 p 的二项式分布的随机变量,即

$$\Pr\{d_s=k\}=\frac{n!}{k!(n-k)!}p^k(1-p)^{n-k}=C_n^k p^k(1-p)^{N-k},\quad k=0,1,\cdots,n \tag{9.9}$$

在备择假设 K_2 下,符号检测器 $D_2(\boldsymbol{w})$的检测概率为

$$P_{\mathrm{d}}=\Pr\{d_s\geqslant C_2\mid K_2\}=\sum_{k=C_2}^{n}C_n^k p^k(1-p)^{N-k} \tag{9.10}$$

在零假设 H_2 下,$p=1/2$,符号检测器 $D_2(\boldsymbol{w})$的虚警概率为

$$P_{\mathrm{fa}}=\Pr\{d_s\geqslant C_2\mid H_2\}=\sum_{k=C_2}^{n}C_n^k\times\left(\frac{1}{2}\right)^n \tag{9.11}$$

符号检测器 $D_2(\boldsymbol{w})$的检测门限 C_2 由给定的虚警概率通过式(9.11)来确定。由此可见,符号检测器除了要求样本分布的中位数为零,对统计分布函数形式没有其他的要求,门限值 C_2 对于所有零假设 H_2 下的分布来说是一样的。因此,符号检测器对 H_2 来说为一个非参量检测器。显然,符号检测器对 H_2 的一个子集来说也是非参量的,如对于假设 H_0 和 H_1。但是,符号检测器对 H_2 的一个子集来说只能是次优检测器。符号检测器的优势在于易于工程实现,并且在一般情况下它的渐近性能(观测样本个数很多时)相比最优检测器并不太差。

文献[4]给出了符号检测器在高斯噪声中检测一个直流信号时相对于最优的线性检测器的 ARE 为 $2/\pi$。也就是说,符号检测器具有线性检测器大约 64%的效率。另外,对其他的非高斯杂波(如具有零均值的对称双边指数分布 $f(x)=\frac{\alpha}{2}\mathrm{e}^{-\alpha|x|},-\infty<x<\infty$)环境,符号检测器能够提供比线性检测器大得多的检测效率。

9.3.2 Wilcoxon 检测器

符号检测器在 Neyman-Pearson 准则下对假设检验(H_2,K_2)是一个最优检测器,对于检测中位数的移动是一个自然的选择。但是,它利用了相对少的输入样本信息。若将输入样本中相对于原点的距离信息考虑进去,用检测统计量 $\sum_{i=1}^{n}\lambda_i u(w_i)$ 来代替符号检测器中的检测统计量 $\sum_{i=1}^{n}u(w_i)$,就得到了 Wilcoxon 检测器。它在非参量统计学中称为 Wilcoxon 符号秩检验。图 9.2 给出了 Wilcoxon 检测器的检测原理框图。

Wilcoxon 符号秩检测器对输入样本序列 $\boldsymbol{w}=(w_1,w_2,\cdots,w_n)$按绝对值大小进行排序,得

$$|w_{(1)}|<|w_{(2)}|<\cdots<|w_{(n)}| \tag{9.12}$$

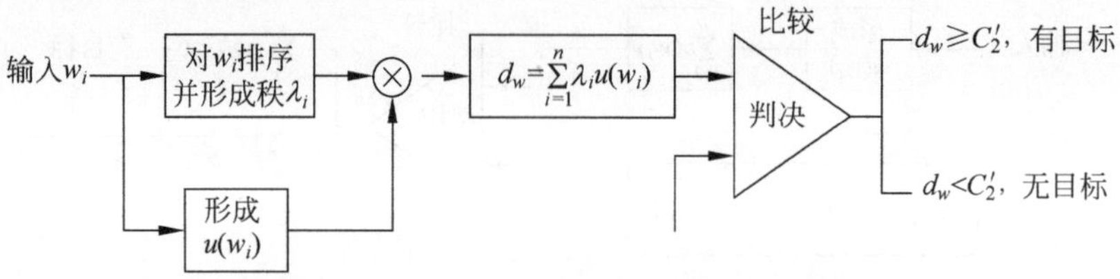

图 9.2 Wilcoxon 检测器的检测原理框图

令 λ_i 表示 w_i 的秩，则

$$\lambda_i = \begin{cases} j, & w_i = w_{(j)} > 0 \\ 0, & w_i = w_{(j)} \leqslant 0 \end{cases} \tag{9.13}$$

Wilcoxon 检测器的检测统计量定义为

$$d_w = \sum_{i=1}^{n} \lambda_i u(w_i) \tag{9.14}$$

它的判决准则为

$$d_w = \sum_{i=1}^{n} \lambda_i u(w_i) \begin{cases} < C'_2 \Rightarrow D'_2(\boldsymbol{w}) = 0 \\ \geqslant C'_2 \Rightarrow D'_2(\boldsymbol{w}) = 1 \end{cases} \tag{9.15}$$

在零假设 H_2 下，Wilcoxon 检测器的检测统计量 $d_w = \sum_{i=1}^{n} \lambda_i u(w_i) = k$ 时的概率分布为

$$\Pr\{d_w = k\} = \begin{cases} c_n(k)/2^n, & k = 0, 1, \cdots, \dfrac{n(n+1)}{2} \\ 0, & \text{其他} \end{cases} \tag{9.16}$$

其中，$c_n(k)$ 是和数恰为 k 的 $\{1, 2, \cdots, n\}$ 的子集的个数。这样，在零假设 H_2 下，Wilcoxon 检测器 $D'_2(\boldsymbol{w})$ 的虚警概率为

$$P_{\text{fa}} = \Pr\{d_w \geqslant C'_2 \mid H_2\} = \sum_{k=C_2}^{n} \Pr\{d_w = k\} \tag{9.17}$$

显然，Wilcoxon 检测统计量的检测门限不依赖于杂波噪声分布形式，因此 Wilcoxon 检测器对 H_2 来说也是一个非参量检测器。

文献[4]利用 ARE 统计方法，评价了 Wilcoxon 检测器在高斯杂波噪声环境相对于最优参量线性检测器的检测效率。在高斯杂波噪声环境中，Wilcoxon 检测器在大样本数量时的效率大约是线性检测器的 $3/\pi$，也就是大约 95%，远高于符号检测器的 64%。另外，对于一些非高斯环境，文献[4]的分析表明，Wilcoxon 检测器像符号检测器一样能提供大于线性检测器的检测效率。

9.4 两样本非参量检测器

前面介绍了单样本的符号检测器和 Wilcoxon 非参量检测器，通过前面的分析可以看出，它们要求检测样本统计分布的中位数为零或是已知的。这个条件通常把它们限制在背景噪声中而不是杂波环境中。在数字系统中可以利用两样本非参量检测器克服这种限制。与单样本非参量检测器不同，两样本非参量检测器是指在检测器的输入端存在两个可供利用的观测样本集合[5]，一个集合由 M 个样本组成，其中可能包含目标信号；而另外一个集合由 MN 个参考样本构成。它们可由雷达在连续 M 个重复周期的视频信号输出中得到。

在天线波束范围内(或电扫描雷达天线波束的某一指定方向上)发射 M 个检测脉冲，则在 M 个重复周期内，接收机的视频输出如图 9.3 所示。图 9.3 中假定 t_0 处的信号对应距离 R_0 处的目标且在所有的 M 个探测周期内信噪比是相同的。

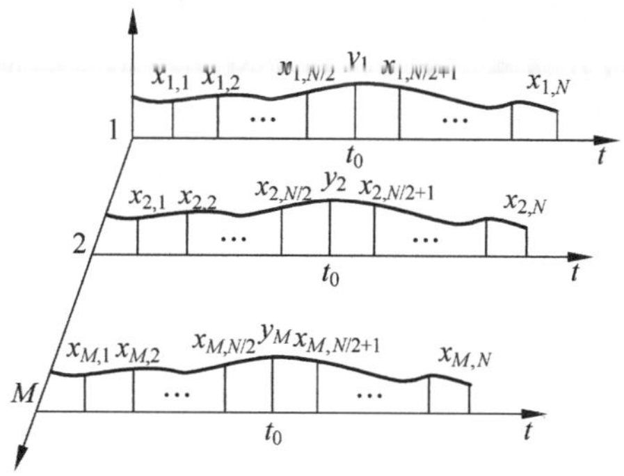

图 9.3 M 个连续探测周期内接收机的视频输出

在 M 个检测周期中，检测单元的样本(t_0 时刻的采样)用 y_j 表示($j=1,2,\cdots,M$)，参考单元的样本用 x_{ki} 表示($i=1,2,\cdots,N$；$k=1,2,\cdots,M$)(有时考虑到目标回波延伸而不仅仅占据一个分辨单元，可以在检测单元两边空开一个或几个保护单元再取参考单元)。如果把所有这些样本的结果保存下来，可表示为 $M\times N$ 样本存储矩阵：

$$\begin{matrix} x_{1,1} & x_{1,2} & \cdots & x_{1,N/2} & y_1 & x_{1,[(N/2)+1]} & \cdots & x_{1,(N-1)} & x_{1,N} \\ x_{2,1} & x_{2,2} & \cdots & x_{2,N/2} & y_2 & x_{2,[(N/2)+1]} & \cdots & x_{2,(N-1)} & x_{2,N} \\ \vdots & \vdots & & \vdots & \vdots & \vdots & & \vdots & \vdots \\ x_{M,1} & x_{M,2} & \cdots & x_{M,N/2} & y_M & x_{M,[(N/2)+1]} & \cdots & x_{M,(N-1)} & x_{M,N} \end{matrix} \tag{9.18}$$

这些观测样本成为构造两样本非参量检验统计量的基础。本节将对广义符号(Generalized Sign，GS)检测器、Savage(S)检测器、修正的 Savage(Modified Sign，MS)检测器、Mann-Whitney(MW)检测器、秩方(Rank Square，RS)检测器、修正的秩方(Modified Rank Square，MRS)检测器进行介绍，并对其检测性能进行比较和分析。

9.4.1 广义符号检测器

GS 检测器的检测统计量为[6]

$$T_{\mathrm{GS}}=\sum_{j=1}^{M} r_j=\sum_{j=1}^{M}\sum_{i=1}^{N} u(y_j-x_{ji}) \tag{9.19}$$

其中

$$r_j=\sum_{i=1}^{N} u(y_j-x_{ji}) \tag{9.20}$$

且

$$u(y_j-x_{ji})=\begin{cases}1, & y_j>x_{ji}\\ 0, & y_j<x_{ji}\end{cases} \tag{9.21}$$

这里把 y_j 量化成 0 和 1 两种值，但量化的比较标准为 x_{ji}，而不是真正按 x_{ji} 的符号，因此称为 GS 检测器。当 $y_j=x_{ji}$ 时，$u(y_j-x_{ji})$ 的取值可以为零，也可以为 1。有时为了使结果更准确，规定当 $y_j=x_{ji}$ 时有

$$u(y_j-x_{ji})=\begin{cases}1, & i-j \text{ 为奇数}\\ 0, & i-j \text{ 为偶数}\end{cases} \tag{9.22}$$

这样做是为了使 $u(t)$ 在 $t=0$ 时为 1 和 0 的机会均等。r_j 是检测单元的秩值，即对第 j 个单元进行检测，其样本值为 y_j，将它与参考单元的值 $x_{j1},x_{j2},\cdots,x_{jN}$ 相比较，比较的结果按 $u(y_j-x_{ji})$ 函数处理之后相加即得 r_j。这正是检测单元样本值 y_j 与诸参考单元的值 $x_{ji}(i=1,2,\cdots,N)$ 按从小到大的顺序排列时，y_j 处的序号，所以称 r_j 为检测单元的秩值。也正是由于这一点，也把 GS 称秩和非参量检测器。

检测统计量 T_{GS} 与检测阈值进行比较，若大于检测阈值则判定信号存在，反之认为仅有噪声存在。

9.4.2 Mann-Whitney 检测器

MW 检测器的检测统计量为[7]

$$T_{\mathrm{MW}}=\sum_{j=1}^{M}r_j=\sum_{j=1}^{M}\sum_{k=1}^{M}\sum_{i=1}^{N}u(y_j-x_{ki}) \tag{9.23}$$

这种检测统计量与广义符号检验统计量的差别在于 T_{GS} 中 y_j 只与它所在检测周期的参考单元的样本 $x_{ji}(i=1,2,\cdots,N)$ 比较，而 T_{MW} 是 y_j 与 M 个探测周期中的所有参考单元的样本 $x_{ki}(k=1,2,\cdots,M;i=1,2,\cdots,N)$ 作比较，即

$$r_j=\sum_{k=1}^{M}\sum_{i=1}^{N}u(y_j-x_{ki}) \tag{9.24}$$

这两种检验统计量相比，显然 T_{MW} 检验统计量对参考数据利用更充分，相应的运算量更大，相应设备量也大，但可以预期其检测性能比 GS 检测器要好一些。

9.4.3 Savage 检测器与修正的 Savage 检测器

Savage(S)检测器的检测统计量为[8]

$$T_{\mathrm{S}}=\sum_{j=1}^{M}a_j(r_j) \tag{9.25}$$

其中

$$r_j=\sum_{k=1}^{M}\sum_{i=1}^{N}u(y_j-x_{ki}) \tag{9.26}$$

$$a_j(r_j)=\sum_{\ell_j=NM+1-r_j}^{NM+1}(\ell_j)^{-1},\quad 0\leqslant r_j\leqslant NM \tag{9.27}$$

如果对于所有的 j，都有 $a_j(r_j)=r_j$，则 S 检测器退化为熟知的 MW 检测器，即

$$T_{\mathrm{MW}}=\sum_{j=1}^{M}\sum_{k=1}^{M}\sum_{i=1}^{N}u(y_j-x_{ki}) \tag{9.28}$$

在 MS 检测器中，对应于 y_j 的修正 Savage 统计量为[9]

$$T_{\mathrm{MS}}=\sum_{j=1}^{M}a_j(r_j) \tag{9.29}$$

与 S 检测器不同的是

$$r_j=\sum_{i=1}^{N}u(y_j-x_{ji}) \tag{9.30}$$

$$a_j(r_j) = \sum_{\ell_j = N+1-r_j}^{N+1} (\ell_j)^{-1}, \quad 0 \leqslant r_j \leqslant N \tag{9.31}$$

T_S、T_{MS} 分别与各自的检测阈值进行比较，如果大于检测阈值则判定目标存在，反之，判定目标不存在。

9.4.4 秩方检测器与修正的秩方检测器

如果在 Savage 的检验统计量中令 $a_j(r_j)=r_j^2$，则可得到 RS 检测器的检测统计量为

$$T_{RS} = \sum_{j=1}^{M} r_j^2 \tag{9.32}$$

其中，r_j 的定义与式(9.24)相同，即

$$r_j = \sum_{j=1}^{M}\sum_{i=1}^{N} u(y_j - x_{ji})$$

MRS 检测器的检测统计量为[9]

$$T_{MRS} = \sum_{j=1}^{M} r_j^2 \tag{9.33}$$

其中，r_j 的定义与式(9.20)相同，即

$$r_j = \sum_{i=1}^{N} u(y_j - x_{ji})$$

由此可见，MRS 检测器与 GS 检测器类似，不同的是 GS 检测器采用检测单元在参考单元中秩的和作为检测统计量，而 MRS 检测器采用检测单元在参考单元中秩平方的和作为检测统计量。

9.4.5 几种非参量检测器的渐近相对效率

在传统的非参量检测器的研究文献中，关于检测器检测性能的评估通常采用渐近相对效率来进行衡量，它是非参量 CFAR 检测器相对于固定门限最优检测器在极限情况下的性能比较。假定在雷达系统中采用平方率检波，GS、MW、S、MS、RS、MRS 检测器在高斯背景中相对于最优线性检测器的渐近相对效率分别给出如下[9]：

$$\text{ARE}_{GS} = \frac{0.75}{1+\dfrac{2}{N}} \tag{9.34}$$

$$\text{ARE}_{MW} = \frac{0.75}{1+\dfrac{1}{N}} \tag{9.35}$$

$$\text{ARE}_{S} = \frac{1}{1+\dfrac{1}{N}} \tag{9.36}$$

$$\text{ARE}_{MS} = 1 - \frac{1}{N+1}\sum_{i=1}^{N+1}\frac{1}{i} \tag{9.37}$$

$$\text{ARE}_{RS} = \frac{0.868}{1+\dfrac{1}{N}} \tag{9.38}$$

$$\mathrm{ARE}_{\mathrm{MRS}}=\frac{5N(10N-1)^2}{36(2N+1)(8N^2+13N-6)} \tag{9.39}$$

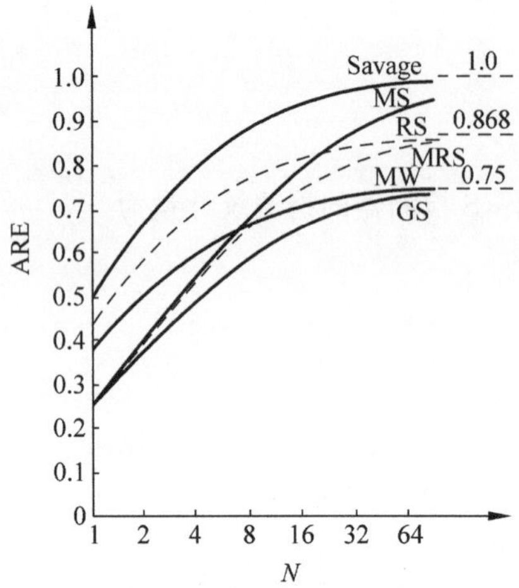

图 9.4 相对于最优线性检测器的渐近相对效率

图 9.4 给出了上述几种检测器的渐近相对效率随参考单元数目 N 的变化曲线。可以看出，随着 N 的增大(大于 64)，MS 检测器的渐近性能逼近于 S 检测器的渐近性能，且它们的值趋向于 1；MRS 检测器的渐近性能逼近于 RS 检测器的渐近性能，接近于 0.868；GS 检测器的渐近性能逼近于 MW 检测器的渐近性能，接近于 0.75。对于所有的 N 值，MS、MRS 检测器的性能处在 GS 和 S 检测器之间。对于 $N>8$，MRS 和 MS 检测器的渐近性能均比 MW 检测器的要好。S 检测器的渐近性能对于所有的 N 均比 MW 检测器的要好。MS 和 MRS 检测器均比 S 和 MW 检测器容易实现，但是比 GS 检测器复杂。

9.4.6 非参量检测器采用有限样本时的检测性能

对非参量检测器性能的传统评估采用渐近相对效率来衡量，其主要比较不同检测器在目标信噪比及脉冲积累数在极限情况下的性能，采用这种分析方法对实际雷达目标的检测来说是不充分的，因为实际雷达总是采用有限的样本在一定的信噪比情况下对目标进行检测[10]。下面将分析 GS、MW、MS、MRS 在有限脉冲数和有限参考噪声样本情况下的检测性能[9]。假定雷达杂波噪声服从高斯分布并采用平方率检波，目标模型为参数为 K 的 χ^2 分布，其中 $K=1$ 对应 Swerling Ⅱ 型目标，$K=2$ 对应 Swerling Ⅳ 型目标，$K=\infty$ 对应非起伏目标。

图 9.5 和图 9.6 分别给出了 GS、MW、MS、MRS 在脉冲积累个数 $M=12$ 和 $M=16$ 时的检测性能曲线，而参考样本个数均为 $N=4$，设定的虚警概率为 $P_{\mathrm{fa}}=10^{-6}$。在图 9.5 中，MS、MRS 检测器对于起伏目标和非起伏目标均有相同的检测性能；对于起伏目标来说，MS、MRS 检测器都比 GS 检测器的性能好。但在图 9.6 中，当脉冲积累数变多，MS 检测器的检测性能比 MRS 检测器要好，但后者对于起伏目标和非起伏目标的检测性能均比 GS 检测器的性能好。另外，从图 9.6 中可以看出，MW、MS、MRS 检测器的检测性能都比较接近。

图 9.7 和图 9.8 分别给出了 GS、MW、MS、MRS 检测器在参考样本个数 $N=12$ 和 $N=16$ 时的检测性能曲线，而脉冲积累个数均为 $M=8$，设定的虚警概率为 $P_{\mathrm{fa}}=10^{-6}$。当参考样本个数变多，对于 $K=1$ 的 Swerling Ⅱ 型目标，MS、MRS 检测器比 GS 检测器的检测性能优势变得更明显。另外，可以注意到 MS 检测器的检测性能比 MRS 检测器要好。当 K 变大即目标起伏变小，MS、MRS 检测器相比 MW 和 GS 检测器的性能差别变小。通过观察比较图 9.5 和图 9.6 及图 9.7 和图 9.8，可以看出，脉冲积累个数对检测器性能的影响比参考样本个数对检测器性能的影响要大，因为在检测中采用较多的脉冲个数会导致更好的检测性能。

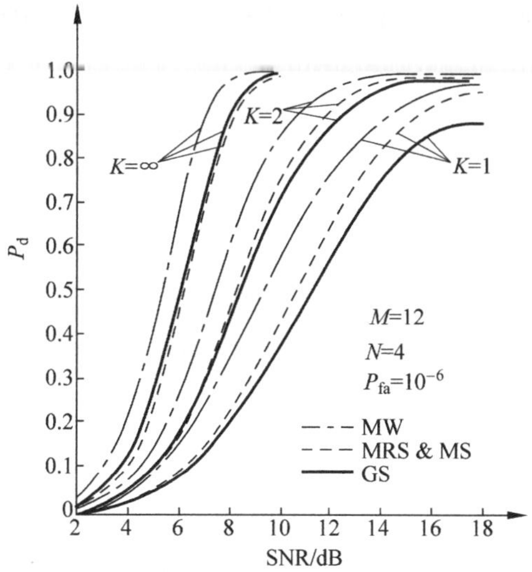

图 9.5　非参量检测器的检测概率随信噪比 SNR 的变化曲线($M=12$)

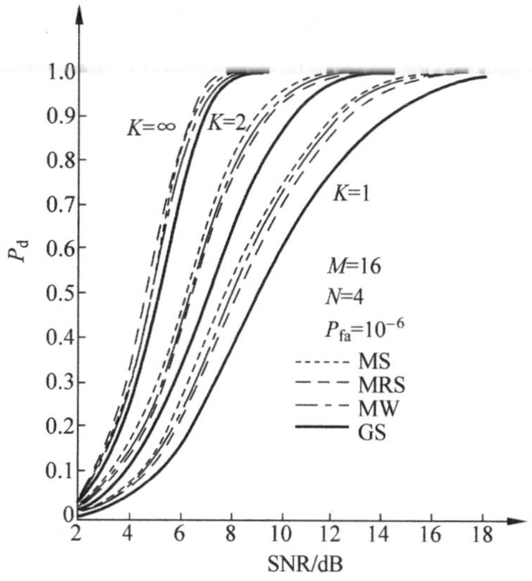

图 9.6　非参量检测器的检测概率随信噪比 SNR 的变化曲线($M=16$)

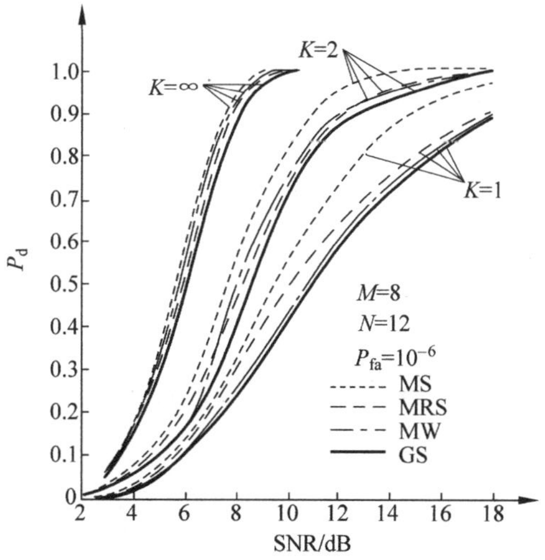

图 9.7　非参量检测器的检测概率随信噪比 SNR 的变化曲线($N=12$)

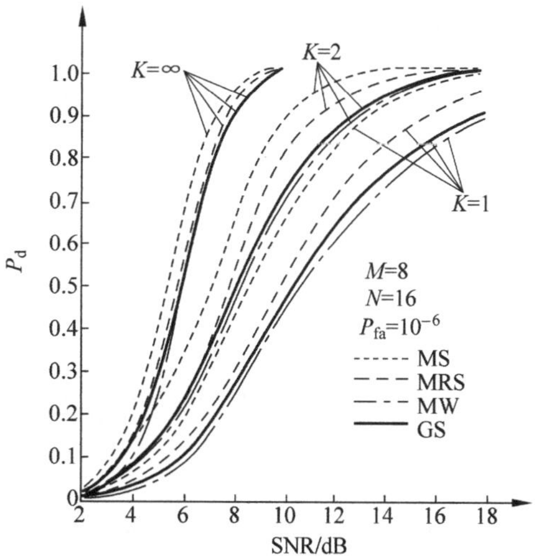

图 9.8　非参量检测器的检测概率随信噪比 SNR 的变化曲线($N=16$)

9.5　次优秩非参量检测器

在这里仍然考虑图 9.3 所示的二维脉冲雷达检测问题，即假设 M 个检测脉冲和 N 个参考单元。参考式(9.18)，考虑如下向量：

$$\boldsymbol{w}_i=(x_{i1},x_{i2},\cdots,x_{iN},y_i),\quad i=1,2,\cdots,M \tag{9.40}$$

在假设 H_0 下，有

$$H_0:F(\boldsymbol{w}_1,\boldsymbol{w}_2,\cdots,\boldsymbol{w}_N\mid H_0)=\prod_{i=1}^{M}F_{0i}(y_i)\prod_{j=1}^{N}F_{0i}(x_{ij}) \tag{9.41}$$

$F_{0i}(x)$对应于第 i 个脉冲的噪声样本的 CDF。注意到,在假设 H_0 下检测单元和参考单元的噪声样本是统计独立且同分布的。在 H_1 假设下,有

$$H_1: F(\boldsymbol{w}_1,\boldsymbol{w}_2,\cdots,\boldsymbol{w}_N \mid H_1)=\int F(\boldsymbol{w}_1,\boldsymbol{w}_2,\cdots,\boldsymbol{w}_N \mid \boldsymbol{s},H_1)\cdot p_{S_0}(\boldsymbol{s})\mathrm{d}\boldsymbol{s} \tag{9.42}$$

其中

$$F(\boldsymbol{w}_1,\boldsymbol{w}_2,\cdots,\boldsymbol{w}_N \mid H_1)=\prod_{i=1}^{M} F_{y_i}(y_i \mid s_i,H_1)\prod_{j=1}^{N} F_{0i}(x_{ij}) \tag{9.43}$$

其中,$\boldsymbol{s}=(s_1,s_2,\cdots,s_M)$是信噪比向量,$s_i(i=1,2,\cdots,M)$是检测单元对应于第 i 个脉冲的信噪比。$F_{y_i}(y_i \mid s_i,H_1)$是在 s_i 和 H_1 条件下检测单元变量 y_i 的 CDF。$p_{S_0}(\boldsymbol{s})$是目标的 PDF,$\boldsymbol{s}_0=(s_{01},s_{02},\cdots,s_{0M})$是分布的参数。当 $\boldsymbol{s}_0\to 0$ 时,意味着 $H_1\to H_0$。

这里定义检测单元样本的秩 r_i 为

$$r_i=\sum_{j=1}^{N} u(y_i-x_{ij}),\quad 0\leqslant r_i\leqslant N,\quad i=1,2,\cdots,M \tag{9.44}$$

可以证明由检测单元的秩 r_i 构成的秩向量 $\boldsymbol{r}=(r_1,r_2,\cdots,r_M)$是检验假设 H_0 和 H_1 的充分统计量。

在假设 H_0 下,对检测单元的秩向量 $\boldsymbol{r}=(r_1,r_2,\cdots,r_M)$,有

$$H_0:\Pr\{\boldsymbol{R}=\boldsymbol{r}\mid H_0\}=\prod_{i=1}^{M}\Pr\{R_i=r_i \mid H_0\}=\left(\frac{1}{N+1}\right)^M \tag{9.45}$$

在 H_1 假设下,又有

$$H_1:\Pr\{\boldsymbol{R}=\boldsymbol{r}\mid H_1\}=\int\prod_{i=1}^{M}\Pr\{R_i=r_i \mid s_i,H_1\}\cdot p_{S_0}(\boldsymbol{s})\mathrm{d}\boldsymbol{s} \tag{9.46}$$

其中

$$\Pr\{R_i=r_i \mid s_i,H_1\}=\binom{N}{r_i}\int_0^{\infty}[F_{0i}(x)]^{r_i}\times[1-F_{0i}(x)]^{N-r_i} f_{y_i}(x \mid s_i)\mathrm{d}x \tag{9.47}$$

$$0\leqslant r_i\leqslant N$$

其中,$f_{y_i}(x \mid s_i)$是信噪比为 s_i 时检测单元 y_i 的 PDF。

式(9.45)和式(9.46)建立了 H_0 和 H_1 假设下关于秩检验的数学模型,下一步需找到 Neyman-Pearson 准则下的最佳秩检测器。

9.5.1 局部最优秩检测器

针对式(9.45)和式(9.46)应用 Neyman-Pearson 准则,得

$$\frac{\Pr\{\boldsymbol{R}=\boldsymbol{r}\mid H_1\}}{\Pr\{\boldsymbol{R}=\boldsymbol{r}\mid H_0\}}=(N+1)^M\times\int\prod_{i=1}^{M}\Pr\{R_i=r_i \mid s_i,H_1\}\cdot p_{S_0}(\boldsymbol{s})\mathrm{d}\boldsymbol{s}\ \underset{H_0}{\overset{H_1}{\gtrless}}\ \lambda \tag{9.48}$$

式(9.48)依赖于目标模型和平均信噪比 $\boldsymbol{s}_0$,因此通常来说,一致最优检测器(Uniformly Most Powerful)是不存在的。但当 $\boldsymbol{s}_0\to 0$ 时,可以找到局部最优秩检测器(Locally Most Powerful Rank Detector,LORD)[10]。该检测器是独立于目标模型的,可表示为

$$\sum_{i=1}^{M} G_i^2 \left(\frac{\partial \Pr\{R_i = r_i \mid s_i\}}{\partial s_i}\right)_{s_i=0} \underset{H_0}{\overset{H_1}{\gtrless}} T_0' \tag{9.49}$$

其中，G_i 是天线第 i 次扫描的功率增益。

假定背景为高斯白噪声，回波包络的 PDF 在 H_0 和 H_1 假设下分别为

$$\frac{\mathrm{d}F_{0i}(x)}{\mathrm{d}x} = f_{0i}(x) = x\exp(-x^2/2),\quad x \geqslant 0 \tag{9.50}$$

$$f_{y_i}(x/s_i) = x\mathrm{I}_0(x\sqrt{2s_i})\exp\left(-\frac{x^2+2s_i}{2}\right)\quad x \geqslant 0,\quad i=1,2,\cdots,M \tag{9.51}$$

$\mathrm{I}_0(\cdot)$是第一类零阶修正的贝塞尔函数，$s_i = E_i/N_0$ 是信噪比。E_i 是目标反射信号的能量，$N_0/2$ 是噪声功率谱密度。

由式(9.47)、式(9.50)和式(9.51)可得

$$\Pr\{R_i = r_i \mid s_i, H_1\} = \binom{N}{r_i}\sum_{k=0}^{r_i}(-1)^k\binom{r_i}{k}\frac{1}{N-r_i+k+1}\times\exp\left(-s_i\frac{N-r_i+k}{N-r_i+k+1}\right) \tag{9.52}$$

式中，$0 \leqslant r_i \leqslant N$。由式(9.52)可得

$$\left[\frac{\partial \Pr\{R_i = r_i \mid s_i, H_1\}}{\partial s_i}\right]_{s_i=0} = \frac{1}{N+1}\left(\sum_{k=0}^{r_i}\frac{1}{N-k+1} - 1\right) \tag{9.53}$$

最后，由式(9.49)和式(9.53)可得

$$T_{\mathrm{L}} = \sum_{i=1}^{M} a_i(r_i) \underset{H_0}{\overset{H_1}{\gtrless}} T_{0\mathrm{L}} \tag{9.54}$$

其中

$$a_i(r_i) = G_i^2\sum_{k=0}^{r_i}\frac{1}{N-k+1},\quad 0 \leqslant r_i \leqslant N,\quad i=1,2,\cdots,M \tag{9.55}$$

式(9.49)和式(9.54)构成了 LOAD 检测器的实现结构，其中假定噪声样本属于高斯白噪声随机过程。

LOAD 检测器的 ARE 由式(9.56)给出：

$$\mathrm{ARE}(T_{\mathrm{L}}) = 1 - \frac{1}{N+1}\sum_{k=1}^{N}\frac{1}{k+1} \tag{9.56}$$

事实上，如果天线增益 $G_i = 1, i = 1, 2, \cdots, M$，则 LOAD 检测器就是 MS 检测器。

9.5.2 次优秩检测器

在高斯白噪声条件下，非起伏目标和 Swerling Ⅱ 目标模型，由式(9.48)得到的最优秩检测器结构可近似表示为[11]

$$T' = \sum_{i=1}^{M} a_i(r_i) \underset{H_0}{\overset{H_1}{\gtrless}} T_0' \tag{9.57}$$

其中

$$a_i(r_i)=A_i\ln\left(\frac{N+1+\theta_i}{N-r_i+\theta_i}\right),\quad 0\leqslant r_i\leqslant N,\theta_i>0,A_i>0,\quad i=1,2,\cdots,M \tag{9.58}$$

检测器参数 A_i 和 $\theta_i(i=1,2,\cdots,M)$ 取决于目标模型和信噪比 s_i。

如果取 $A_i=G_i^2$ 和 $\theta_i=0.56$，则可以证明式(9.57)和式(9.58)对应的检测器与式(9.54)和式(9.55)对应的 LORD 检测器是近似相同的[11]。当不考虑目标模型时，将式(9.57)和式(9.58)对应的检测器称为次优秩检测器。

如果 $A_i=\theta_i\to\infty$，式(9.58)变成 $a_i(r_i)=r_i+1$，这样，GS 检测器[6]变为次优秩检测器的特例。如果 $A_i=-1/\ln\theta_i,\theta_i\to 0$，并且 $a_i(r_i)=u(r_i-N)$，MS 检测器[9]也变为次优秩检测器的特例。

假定天线增益为矩形窗，则 $a_i(r_i)$ 仅依赖于 r_i，式(9.57)和式(9.58)对应的检测器变为

$$T=\sum_{i=1}^{M}a(r_i)\underset{\mathrm{H}_0}{\overset{\mathrm{H}_1}{\gtrless}}T_0 \tag{9.59}$$

$$a(r_i)=A\ln\left(\frac{N+1+\theta}{N-r_i+\theta}\right) \tag{9.60}$$

式中，A 可作归一化处理，θ 是检测器参数。它的 ARE 由式(9.61)给出：

$$\mathrm{ARE}(T)=\frac{\left[\sum_{r_i=0}^{N}\left(1-\sum_{k=0}^{r_i}\frac{1}{N-k+1}\right)\ln(N-r_i+\theta)\right]^2}{(N+1)\sum_{r_i=0}^{N}[\ln(N-r_i+\theta)]^2-\left[\sum_{r_i=0}^{N}\ln(N-r_i+\theta)\right]^2} \tag{9.61}$$

经分析可知，对一个给定的 N，ARE 在 $\theta=0.56$ 时取最大值，与 LOAD 检测器取相同的 ARE 值；这从另一个角度说明 LOAD 检测器是次优秩检测器的一个特例。同时，ARE 对 θ 的变化不是很敏感。

9.5.3 性能分析

下面通过 Monte-Carlo 仿真方法，分析当参数取不同值时由式(9.59)确定的次优秩检测器的检测性能[11]。在仿真中，考虑了虚警概率 P_{fa} 为 10^{-6} 和 10^{-8} 两种设定值，脉冲积累个数 M 为 8、16、30 和参考单元数 N 为 7、15。检测器参数取 $\theta=10^3$（几乎与 GS 检测器等价），$\theta=0.56$（LOAD 检测器），$\theta=\theta_{\mathrm{opt}}$（使 P_{d} 在区间 $0.2\leqslant P_{\mathrm{d}}\leqslant 0.95$ 达到最大值）。其中，为便于仿真分析，设定 $A=10(1+\theta/10)$。

图 9.9 和图 9.10 分别给出了 $M=16$、$N=7$ 和 $M=30$、$N=7$ 时次优秩检测器的检测概率 P_{d} 相对于信噪比的变化曲线，目标模型为 Swerling Ⅱ型。可以看出，最优值 θ_{opt} 随着脉冲积累个数 M 的增加而增加。当 $M=30$ 时，次优秩检测器 $\theta_{\mathrm{opt}}=0.3$ 对应的检测性能曲线与局部最优秩检测器 $\theta=0.56$ 的检测性能曲线非常接近；故当 $M\geqslant 30$ 时，可以将 LOAD 检测器作为最优秩检测器。

在 H_0 假设下，在参考单元中存在干扰目标意味着 P_{fa} 将低于没有干扰目标时的虚警概率设定值；而在 H_1 假设下，在参考单元中存在干扰目标将引起检测器检测概率 P_{d} 的恶化。图 9.11 给出了在参考单元中存在干扰目标时次优秩检测器的检测概率 P_{d} 相对于信噪比的变化曲线，主目标和干扰目标同为 Swerling Ⅱ型，干扰目标数分别为 0、1、2，脉冲积累个数 $M=8$，参考单元数 $N=15$。图 9.12 则给出了干扰目标数分别为 0、1、2、3 时 $M=30$、$N=15$ 的情况。图 9.11 和图 9.12 中“1”表示参考单元中存在 1 个具有与主目标相同功率的干扰目标。由图 9.11 和图 9.12 可知，随着干扰目标数的增加，检测概率 P_{d} 逐步恶化；增大脉冲积累个数和参考单元个数，次优秩检测器抗击干扰目标的能力得到加强。下

面给出一些定量的结果，假设虚警概率 $P_{fa}=10^{-6}$，P_d 在合理范围内($0.2\leqslant P_d\leqslant 0.95$)：①当参考单元数 $N=7$，对抗一个干扰目标时的脉冲积累个数 M 须大于16，对抗两个干扰目标时的脉冲积累个数 M 须大于30；②对参考单元数 $N=15$，对抗两个干扰目标时的脉冲积累个数 M 须大于16，对抗三个干扰目标时的脉冲积累个数 M 须大于30。

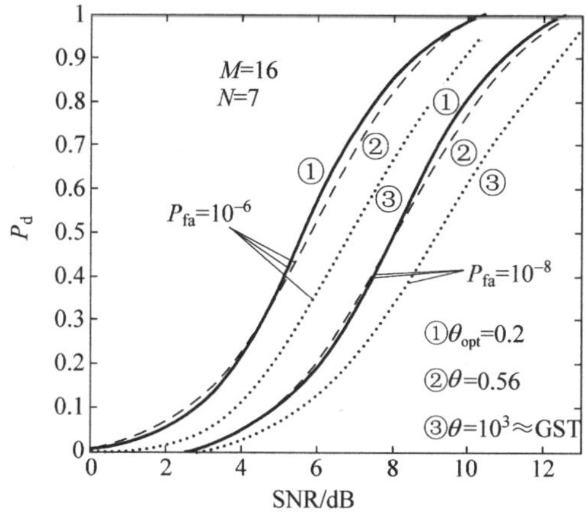

图9.9　对 Swerling Ⅱ型目标的检测概率 P_d 相对于信噪比的变化曲线

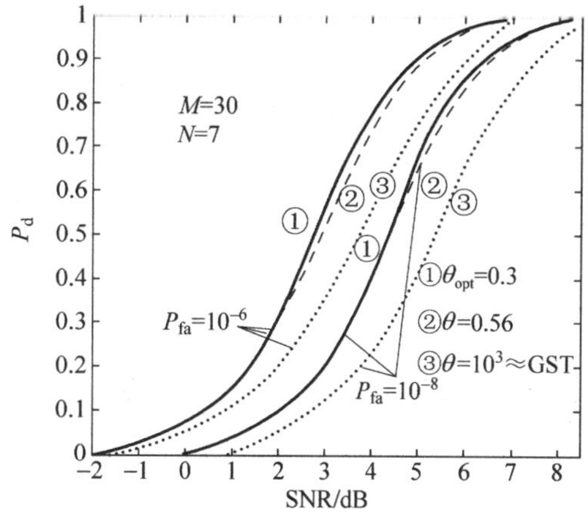

图9.10　对 Swerling Ⅱ型目标的检测概率 P_d 相对于信噪比的变化曲线

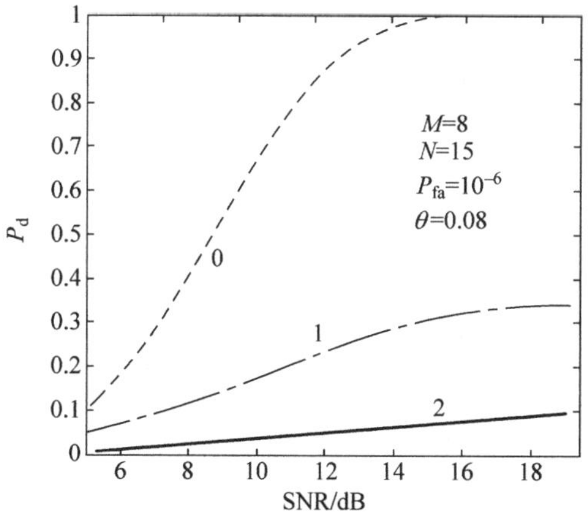

图9.11　多目标情况下次优秩检测器的检测概率 P_d 相对于信噪比的变化曲线

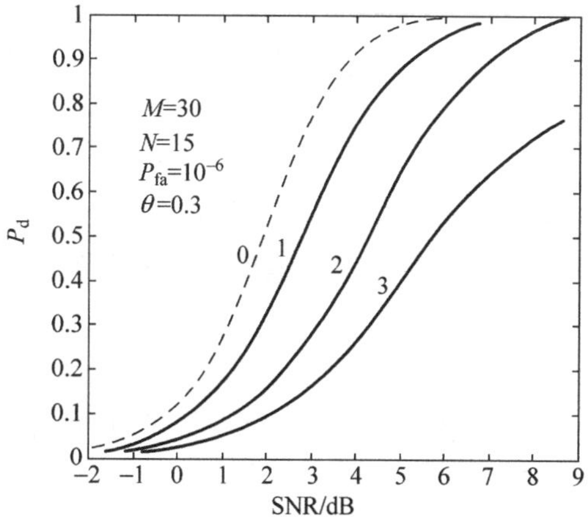

图9.12　多目标情况下次优秩检测器的检测概率 P_d 相对于信噪比的变化曲线

9.6　韦布尔杂波下非参量检测器的性能分析

非参量检测器若要保持CFAR特性，杂波统计特性只需满足IID的条件。经典非参量检测器一般假定背景为高斯分布，而实际应用中往往面临非高斯时变背景。因此，有必要在非高斯背景中分析非参量检测器的性能。在诸多非参量检测器中，较为典型的是秩和检测器和量化秩(RQ)检测器[12]。秩和检测器即为9.4.1节中描述的广义符号(GS)检测器[6]，而量化秩检测器也被称为修正的符号检测器。

本节将在典型的韦布尔分布非高斯背景下，对 RQ 检测器和 GS 检测器进行虚警和检测概率的分析，并简要讨论其在均匀、多目标和杂波边缘情形下的检测性能。

9.6.1 韦布尔背景下量化秩非参量检测器

本节将采用解析方法对量化秩检测器在韦布尔分布中的检测和虚警性能进行分析，并考虑均匀、多目标和杂波边缘情形。

1. 均匀杂波背景下 RQ 的虚警与检测概率

雷达接收信号经平方律检波后，按照距离单元进行统计独立样本[5]。在待检测距离单元附近选择 N 个参考单元，假定脉冲串含 M 个脉冲。第 i 个脉冲对应的样本矢量为

$$\boldsymbol{x}_i=(x_{i1},x_{i2},\cdots,x_{iN},x_i),\quad i=1,2,\cdots,M \tag{9.62}$$

其中，标量 x_i 对应待检测距离单元的回波样本。在 H_0 假设下，因无目标存在，可假定矢量 $\boldsymbol{x}_i$ 的 $N+1$ 个分量是 IID 的；在 H_1 假设下，除 x_i 外，$\boldsymbol{x}_i$ 的剩余 N 个分量是 IID 的[13]。

对第 i 次扫描来说，待检测距离单元回波 x_i 的秩 r_i 为

$$r_i=\sum_{j=1}^{N}u(x_i-x_{ij}),\quad 0\leqslant r_i\leqslant N,\quad i=1,2,\cdots,M \tag{9.63}$$

将相应的秩 r_i 与秩量化门限 K 进行比较，可得二进制秩值

$$B(r_i)=\begin{cases}1, & r_i\geqslant K\\ 0, & r_i<K\end{cases} \tag{9.64}$$

接着进行二进制积累，则 RQ 检测器的检测统计量为

$$Z=\sum_{i=1}^{M}B(r_i) \tag{9.65}$$

由于参考样本 $x_{i1},x_{i2},\cdots,x_{iN}\,(i=1,2,\cdots,M)$ 是 IID 的，假定其 PDF 为 $f(x)$，CDF 为 $F(x)$，$B(r_i)=1$ 的概率即 H_0 假设下单次检测的虚警概率

$$P_{\mathrm{fa1}}(K)=\int_{-\infty}^{\infty}\sum_{i=K}^{N}\binom{N}{i}[F(x)]^{i}[1-F(x)]^{N-i}f(x)\mathrm{d}x \tag{9.66}$$

经变量替换和数学运算可将式(9.66)化简为

$$P_{\mathrm{fa1}}(K)=\frac{N+1-K}{N+1} \tag{9.67}$$

基于单次检测的虚警概率结果，则二进制积累后 RQ 的虚警概率为

$$P_{\mathrm{fa}}(T,K)=\frac{1}{(N+1)^{M}}\sum_{r=T}^{M}\binom{M}{r}(N+1-K)^{r}K^{M-r} \tag{9.68}$$

从式(9.68)可知，RQ 的虚警概率与噪声的统计特性无关，体现了相应的自由分布特征和 CFAR 特性。

在 H_1 假设下，$B(r_i)=1$ 的概率即为雷达目标在单次检测中的检测概率

$$P_{\mathrm{d1}}(K)=\int_{-\infty}^{\infty}\sum_{i=K}^{N}\binom{N}{i}[F(x)]^{i}[1-F(x)]^{N-i}g(x)\mathrm{d}x \tag{9.69}$$

式(9.69)中 $g(x)$ 为在 H_1 假设下 x_i 的 PDF。假定平方律检波后杂波服从功率尺度参数为 ρ、形状参数为 c 的韦布尔分布，记为 $W(\rho,c)$，$F(x)$ 为相应的 CDF，可参见式(7.12)，注意与式(7.1)相比有 $\rho=$

b^2。假定为 Swerling Ⅱ型目标，因韦布尔分布杂波下尚难以获得目标加杂波的 PDF 闭型表达式，可用瑞利分布来近似检测单元杂波[14]，其平均功率为

$$\rho_c=\rho\Gamma\left(1+\frac{2}{c}\right) \tag{9.70}$$

即近似后的杂波平均功率与参考单元中韦布尔杂波相同。该近似方法信杂比较大时可较准确地计算检测概率，经过近似处理后，目标加杂波的 PDF 可表示为[15]

$$g(x)=\frac{1}{\rho_c(1+\lambda)}\exp\left[-\frac{x}{\rho_c(1+\lambda)}\right],\quad x\geqslant 0 \tag{9.71}$$

其中，λ 表示信杂比。

将式(9.71)代入式(9.69)中，经变量替换可得

$$P_{\mathrm{d1}}(K)=\frac{1}{\rho_c(1+\lambda)}\Phi_1(s)\bigg|_{s=\frac{1}{\rho_c(1+\lambda)}} \tag{9.72}$$

其中，$\Phi_1(s)$表示拉普拉斯变换，即

$$\Phi_1(s)=\rho\sum_{i=K}^{N}\binom{N}{i}\sum_{\ell=0}^{i}(-1)^{\ell}\binom{i}{\ell}\sum_{f=0}^{\infty}(-1)^{f}\frac{(N-i+\ell)^{f}}{f!}\frac{\Gamma\left(\frac{fc}{2}+1\right)}{(\rho s)^{fc/2+1}} \tag{9.73}$$

当形状参数 $c=2$ 时，$\Phi_1(s)$可简化为

$$\Phi_1(s)=\sum_{i=K}^{N}\binom{N}{i}\sum_{\ell=0}^{i}(-1)^{\ell}\binom{i}{\ell}\frac{1}{s+(N-i+\ell)/\rho} \tag{9.74}$$

二进制积累后 RQ 的检测概率为

$$P_{\mathrm{d}}(T,K)=\sum_{r=T}^{M}\binom{M}{r}\left[P_{\mathrm{d1}}(K)\right]^{r}\left[1-P_{\mathrm{d1}}(K)\right]^{M-r} \tag{9.75}$$

2. 杂波边缘情况下 RQ 的虚警概率

杂波边缘的功率建模为一个阶跃函数[16]，假定 IID 的强杂波样本有 L 个，服从 $\mathrm{W}(\gamma\rho,c)$分布；IID 的弱杂波样本有 $N-L$ 个，服从 $\mathrm{W}(\rho,c)$分布；其中 γ 表示强、弱杂波的功率比。若待检测单元处于强杂波区，则 $B(r_i)=1$ 的虚警概率为[17]

$$\begin{aligned}P_{\mathrm{fa2}}(K)=\int_0^{\infty}\sum_{i=K}^{N}\sum_{j=\max(0,i-L)}^{\min(i,N-L)}\binom{N-L}{j}\binom{L}{i-j}F^{j}(x)\\ \left[1-F(x)\right]^{N-L-j}\bar{F}^{i-j}(x)\left[1-\bar{F}(x)\right]^{L-i+j}\mathrm{d}\bar{F}(x)\end{aligned} \tag{9.76}$$

式中，$F(x)$和 $\bar{F}(x)$分别表示强、弱杂波的 CDF。

利用韦布尔分布的 CDF，经过化简可得[17]

$$\begin{aligned}P_{\mathrm{fa2}}(K)=\sum_{i=K}^{N}\sum_{j=\max(0,i-L)}^{\min(i,N-L)}\binom{N-L}{j}\binom{L}{i-j}\sum_{i_1=0}^{j}\binom{j}{i_1}\frac{(-1)^{i_1}}{\gamma^{c/2}}\\ \sum_{j_1=0}^{i-j}\frac{\binom{i-j}{j_1}(-1)^{j_1}}{i_1+N-L-j+(j_1+L-i+j+1)/\gamma^{c/2}}\end{aligned} \tag{9.77}$$

假定杂波功率在各次扫描间保持不变，则在杂波边缘环境下二进制积累后 RQ 检测器的虚警概率为

$$P_{\mathrm{fa}}(T,K)=\sum_{r=T}^{M}\binom{M}{r}\left[P_{\mathrm{fa2}}(K)\right]^{r}\left[1-P_{\mathrm{fa2}}(K)\right]^{M-r} \tag{9.78}$$

3. 仿真分析

(1) 分析均匀背景下 RQ 检测器的性能。

图 9.13 在不同的二进制积累门限 T 条件下，给出了 RQ 的虚警概率与秩量化门限 K 的关系曲线，其中 $M=8,N=32$。可以看出，对于给定的虚警概率，RQ 只能成对选择 K 和 T 的取值，即整数 K 和 T 相互关联。因此，对于给定的虚警概率 α，为获得较准确的虚警概率数值，可采用随机化处理技术，即找出同时满足 $P_{\mathrm{fa}}(T,K+1)<\alpha$ 和 $P_{\mathrm{fa}}(T,K)\geqslant\alpha$ 的门限 K，同时找出满足 $P_{\mathrm{fa}}(T,K+1)+P(A)[P_{\mathrm{fa}}(T,K)-P_{\mathrm{fa}}(T,K+1)]=\alpha$ 的概率 $P(A)$。当 $N=32$、$M=12$ 和 $\alpha=10^{-5}$ 时，RQ 需要的 $P(A)$ 和 (T,K) 的数值如表 9.1 所示。

表 9.1 RQ 检测器需使用的 $P(A)$ 和 (T,K) 的数值（$N=32,M=12$）

(T,K)	(6,31)	(7,30)	(8,29)	(9,27)	(10,25)	(11,23)	(12,20)
$P(A)$	0.2874	0.3266	0.6430	0.1832	0.1022	0.3817	0.5398

图 9.14 给出了不同形状参数 c 下 RQ 的检测性能曲线。可以看出，当 c 从 2 变为 0.8 时，RQ 的检测性能逐步下降。可能的原因是，随着形状参数的减少，杂波尖峰增多，目标被背景遮蔽的概率增加。值得注意的是，虽然 RQ 的虚警概率与杂波分布形式无关，但其检测性能仍受杂波环境影响。因此，为了使 RQ 对形状参数变化具有一定的检测鲁棒性，不建议采用具有较高 K 值的门限对(T,K)。

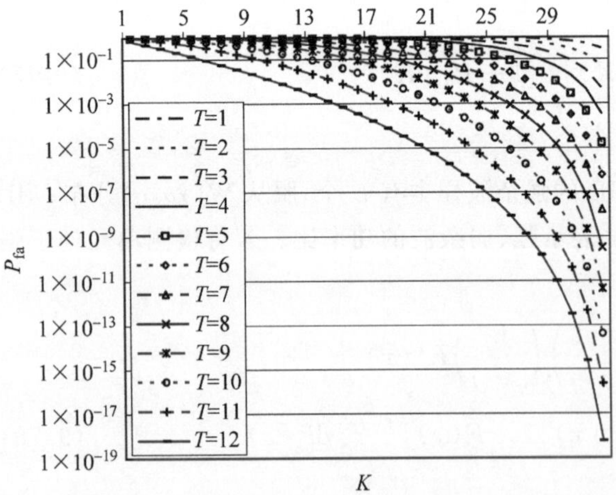

图 9.13 RQ 检测器的虚警概率 $P_{\mathrm{fa}}(T,K)$ 随 K 和 T 的变化关系（$N=32,M=8$）

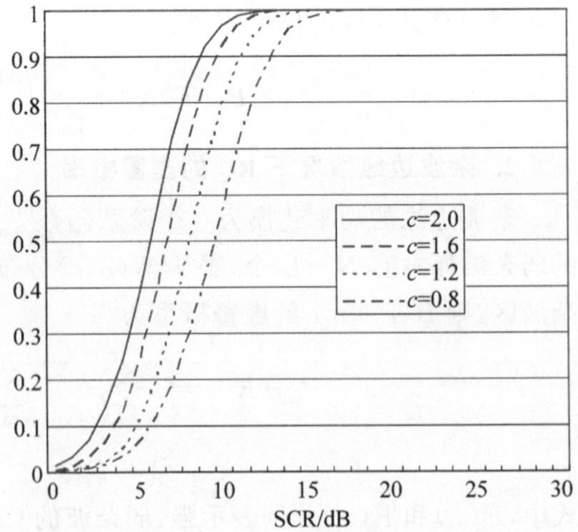

图 9.14 $c=2.0,1.6,1.2,0.8$ 时 $P_{\mathrm{d}}(6,31)$ 对应的 RQ 检测器的检测性能曲线（$N=32,M=12$）

(2) 分析多目标环境下 RQ 的检测性能。

考虑干扰功率远大于杂波的强干扰目标情况，假定干扰目标数为 r。此时可通过将式(9.75)中的 N 替换成$(N-r)$，用以计算 RQ 的检测概率。图 9.15 给出了不同 r 值下 RQ 的检测性能曲线。可以看出，RQ 的检测性能随着干扰目标数的增加而下降。事实上，RQ 只能对抗$(N-K-1)$个干扰目标，若 $r>(N-K-1)$，RQ 的检测性能将急剧恶化。例如，图 9.15 中当 $r=5>(N-K-1)=4$ 时，检测性能陡然下降，RQ 已无法正常工作。

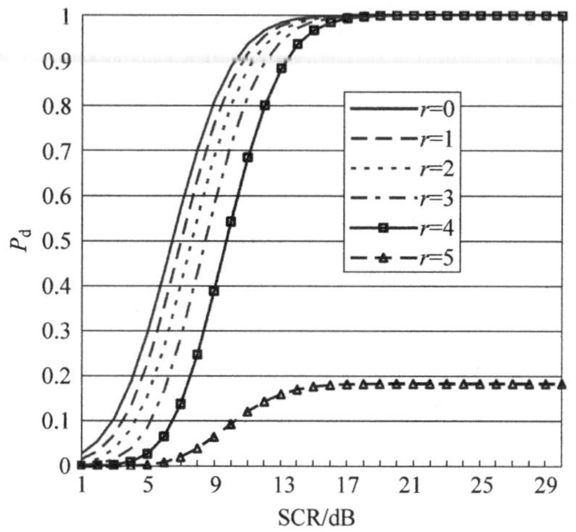

图 9.15 RQ 检测器在有干扰目标时 $P_d(9,27)$随信杂比的变化曲线($N=32,M=12,\alpha=10^{-5}$)

另外,针对杂波边缘环境下 RQ 的仿真分析表明[13]:采用较高秩量化门限 K 时杂波边缘的虚警性能要好于较低 K 值的情况;随着形状参数 c 的减小,RQ 在杂波边缘仍有较强的虚警控制能力,体现出一定的鲁棒性。

综上分析,RQ 采用高秩量化门限时具有较好的杂波边缘虚警控制能力,但较低秩量化门限会提高 RQ 对多干扰目标的对抗能力,因此秩量化门限 K 的取值,需兼顾多目标干扰、重拖尾杂波和杂波边缘多种情况的检测需求。

9.6.2 韦布尔背景下广义符号非参量检测器

本节将采用解析方法对广义符号检测器在韦布尔分布中的检测和虚警性能进行分析,并考虑均匀、多目标和杂波边缘情形。

1. 均匀杂波背景下 GS 的虚警和检测概率

采用 9.6.1 节中平方律检波的目标检测模型,基于检测单元 x_i 在样本矢量 $\boldsymbol{x}_i=(x_{i1},x_{i2},\cdots,x_{iN},x_i)$中的秩 $r_i(i=1,2,\cdots,M)$,对 M 次脉冲扫描中检测单元的秩进行求和,可得 GS 的检测统计量为[6]

$$r=\sum_{i=1}^{M} r_i \tag{9.79}$$

在 H_0 假设下,假定 IID 的参考样本 $x_{i1},x_{i2},\cdots,x_{iN}(i=1,2,\cdots,M)$服从韦布尔分布 $W(\rho,c)$[18],且 PDF 为 $f(x)$,CDF 为 $F(x)$,因此 $r_i=k$ 的概率为

$$P_0(r_i=k)=\int_{-\infty}^{\infty}\binom{N}{k}[F(x)]^k[1-F(x)]^{N-k}f(x)\mathrm{d}x,\quad i=1,2,\cdots,M;\ k=0,1,\cdots,N \tag{9.80}$$

经过变量代换和化简,可得

$$P_0(r_i=k)=\frac{1}{N+1} \tag{9.81}$$

在假定扫描间、参考单元间杂波样本为 IID 的,考虑到统计独立随机变量之和的 PDF 等于各自 PDF 卷积,利用 Z 变换中的卷积定理,GS 的检测统计量 r 的 PDF 可表示为

$$P'_0(r=k)=Z^{-1}\left\{\prod_{i=1}^{M}Z\{P_0(r_i=k)\}\right\}=Z^{-1}\left\{\frac{1}{(N+1)^M}\frac{(1-z^{-N-1})^M}{(1-z^{-1})^M}\right\} \tag{9.82}$$

式中 $Z\{\}$和 $Z^{-1}\{\}$表示取 Z 变换和 Z 反变换。

因此,GS 的虚警概率为[19]

$$P_{\text{fa}}=\sum_{k=T}^{\infty}P'_0(r=k),\quad i=1,2,\cdots,M;\ k=0,1,\cdots,N \tag{9.83}$$

由式(9.83)可知,GS 的虚警概率与噪声的统计特性无关,体现了相应的自由分布特征和 CFAR 特性[20]。

H_1 假设下,秩 $r_i=k$ 的概率为

$$P_1(r_i=k)=\int_{-\infty}^{\infty}\binom{N}{k}[F(x)]^k[1-F(x)]^{N-k}g(x)\mathrm{d}x \tag{9.84}$$

式中,$g(x)$是 H_1 假设下检测单元回波 x_i 的 PDF。

采用 9.6.1 节中的杂波和目标模型进行检测性能分析,将式(9.71)代入式(9.84)中,经变量替换可得

$$P_1(r_i=k)=\frac{1}{\rho_c(1+\lambda)}\Phi_1(s)\bigg|_{s=\frac{1}{\rho_c(1+\lambda)}} \tag{9.85}$$

式中,$\Phi_1(s)$对应拉普拉斯变换:

$$\Phi_1(s)=\rho\binom{N}{k}\sum_{\ell=0}^{k}(-1)^{\ell}\binom{k}{\ell}\sum_{f=0}^{\infty}(-1)^f\frac{(N-k+\ell)^f}{f!}\frac{\Gamma\left[\frac{fc}{2}+1\right]}{(\rho s)^{fc/2+1}} \tag{9.86}$$

在形状参数 $c=2$ 时,$\Phi_1(s)$可简化为

$$\Phi_1(s)=\binom{N}{k}\sum_{\ell=0}^{k}(-1)^{\ell}\frac{1}{s+(N-k+\ell)/\rho} \tag{9.87}$$

假定检测单元样本扫描间是 IID 的,利用 Z 变换的卷积定理,可得 GS 检测统计量 r 的 PDF 为

$$P'_1(r=k)=Z^{-1}\{[Z\{P_1(r_i=k)\}]^M\} \tag{9.88}$$

因此,GS 的检测概率可表示为

$$P_{\text{d}}=\sum_{k=T}^{\infty}P'_1(r=k) \tag{9.89}$$

其中,T 表示 GS 的检测阈值。

2. 杂波边缘情况下 GS 的虚警概率

仍采用 9.6.1 节中的杂波边缘模型[21],如果检测单元处于强杂波区,则秩 $r_i=k$ 的概率为

$$P_2(r_i=k)=\int_0^{\infty}\sum_{j=\max(0,k-L)}^{\min(k,N-L)}\binom{N-L}{j}\binom{L}{k-j}F^j(x)[1-F(x)]^{N-L-j}\bar{F}^{k-j}(x)[1-\bar{F}(x)]^{L-k+j}\mathrm{d}\bar{F}(x) \tag{9.90}$$

式中,$F(x)$和 $\bar{F}(x)$参见式(9.79)。

利用韦布尔分布的 CDF,经适当化简可得[22]

$$P_2(r_i=k)=\sum_{j=\max(0,k-L)}^{\min(k,N-L)}\binom{N-L}{j}\binom{L}{k-j}\sum_{i_1=0}^{j}\binom{j}{i_1}\frac{\Gamma(m_\gamma+1+L-k+j)\Gamma(k-j+1)}{(-1)^{i_1}\Gamma(m_\gamma+1+L+1)} \tag{9.91}$$

式中，$m_{\gamma}=\gamma^{c/2}(i_1+N-L-j)$。

在杂波边缘情况下，假定扫描间样本是IID的，GS检测统计量r的PDF可表示为

$$P_2'(r=k)=Z^{-1}\{[Z\{P_2(r_i=k)\}]^M\} \tag{9.92}$$

因此，GS在杂波边缘的虚警概率为

$$P_{\mathrm{fa}}=\sum_{k=T}^{\infty}P_2'(r=k) \tag{9.93}$$

3. 仿真分析

(1) 均匀背景下GS检测器的性能。

与RQ检测器类似，对于GS的检测门限来说，虽然其对应的虚警概率也不能准确地达到设定值，但其与设定值相差较小，可直接采用。当$N=32$时，在不同的积累门限M的情况下，表9.2给出了达到设定概率$\alpha=10^{-6}$的最接近门限值T及其虚警概率值P_{fa}。

表9.2 $N=32$和不同M情况下最接近$\alpha=10^{-6}$的门限值T及对应的虚警概率值

M	4	6	8	10	12
T	128	185	238	287	335
P_{fa}	8.43×10^{-7}	1.33×10^{-6}	1.11×10^{-6}	1.25×10^{-6}	1.04×10^{-6}

当参考单元数$N=32$时，图9.16给出了积累脉冲数$M=6$、8、10、12对应的GS检测性能曲线。可以看出，随着参考单元数和积累脉冲数的增大，检测性能有所改善。图9.17进一步给出了不同形状参数$c=0.6,0.8,1.2,1.6,2.0$对应的GS检测性能曲线。可以看出，GS的检测概率在形状参数为[1.2, 2.0]的区间变化不大，但当形状参数为0.8和0.6时，GS检测性能有所改善。可能的原因在于，在信杂比不变的情况下，随着形状参数的减小，相应尺度参数的杂波强度也会减小，从而导致目标易于被检测出来[2]。

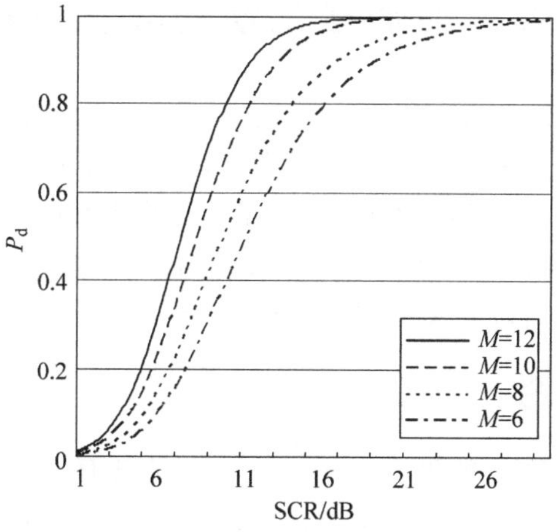

图9.16 当$c=2.0,\alpha=10^{-6},N=32,M=6$、8、10、12时GS的检测性能

(2) 多目标环境下GS检测器的检测性能。

图9.18给出了干扰目标数$r=0,1,2,3,4$对应的GS检测性能曲线，其中$M=12,N=32,c=2.0$，检测门限$T=238$。由图9.18可知，此时GS能对抗的干扰目标数最多为4，相应数值可对$(M\times N-T)/N$进行取整获得。另外，针对杂波边缘环境下GS检测器的仿真结果表明[22]，GS在杂波边缘的虚

警性能并未随形状参数的减小而变差。可能的原因在于,虚警是否上升主要取决于杂波强度,而杂波强度与尺度参数有关,在信杂比给定时,形状参数与尺度参数正相关变化,因此 GS 检测器在形状参数减小时虚警性能并未恶化。

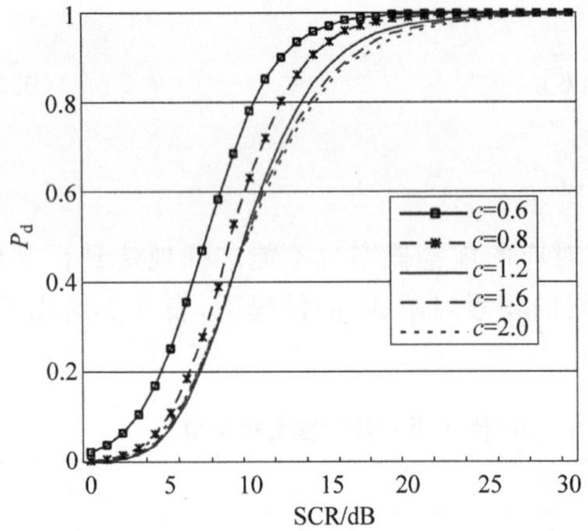

图 9.17 当 $N=32,M=8,T=179$, $c=0.6,0.8,1.2,1.6,2.0$ 时 GS 的检测性能

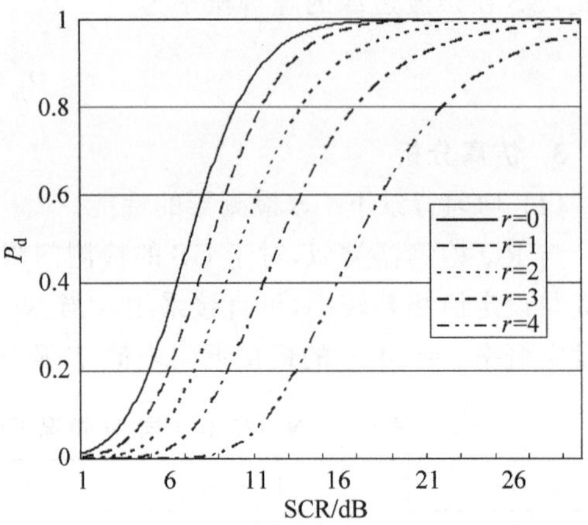

图 9.18 当 $N=32,M=12,T=238$, $\alpha=1.04\times10^{-6}$ 时 GS 多目标情况下的检测性能

实际检测中为达到设定的虚警概率,GS 检测器需要较多的参考单元数和积累脉冲数。当参考单元数 $N=24$ 时,积累脉冲数至少为 5,才能达到 $\alpha=10^{-6}$ 的虚警概率,而此时其抗干扰目标能力又极为有限[19]。总体来说,为了改善 GS 在均匀背景中的检测性能和多目标分辨能力,需采用较多的参考单元数和积累脉冲数;但与 RQ 检测器相比,GS 检测器中可采用相对较少的参考单元数和积累脉冲数。

9.7 利用逆正态得分函数修正秩的非参量检测器

非参量检测器只需假设杂波数据的 PDF 关于原点对称,且不同观测间的样本是 IID 的,相比于参量检测器,其假设的约束条件要更为宽松[23-24]。事实上,观测数据的整数秩本身对原始数据的代表性仍较为有限,如何从未知统计分布的观测数据中提取出更具代表性的检测统计量[25],是一个值得深入探讨的问题。本节将利用逆正态得分函数(Inverse Normal Score Function,INSF)对秩检验进行改进,进而设计非参量检测方法,并在高斯和 K 分布杂波下分别给出仿真分析结果。

9.7.1 基本设计思路

由概率积分变换定理可知[26],如果 $X_1,X_2,\cdots,X_n$ 是具有连续 CDF 的任意总体中的随机取样,那么其相应的分布函数 $F_X(X_1),F_X(X_2),\cdots,F_X(X_n)$ 就是服从(0,1)区间上连续均匀分布的一个随机取样。如果 $X_{(1)},X_{(2)},\cdots,X_{(n)}$ 是前述原始样本的顺序统计量,且累积分布函数 F_X 连续,则

$$F_X(X_{(1)})<F_X(X_{(2)})<\cdots<F_X(X_{(n)}) \tag{9.94}$$

为服从均匀分布的顺序统计量。依据变换

$$U_{(r)}=F_X(X_{(r)}) \tag{9.95}$$

可得一个随机变量 $U_{(r)}$，其中 $U_{(r)}$ 表示服从(0,1)区间均匀分布的第 r 个顺序统计量，而与 $F_X(x)$ 究竟具有何种分布形式无关，即 $U_{(r)}$ 与 $F_X(x)$ 具体形式无关，为分布自由随机变量。由式(9.95)可得

$$X_{(r)}=F_X^{-1}(U_{(r)})=Q_X(U_{(r)}) \tag{9.96}$$

其中，$Q_X(\cdot)$ 是累积分布函数的反函数，亦称作随机变量 X 的分位数函数。

本节将利用概率积分变换的非线性变换特性，构造分布自由检验统计量，进而设计非参量检测器。

当样本数 n 趋于无穷大，且 $r/n\to p$，$0<p<1$ 时，存在极限[26]

$$\lim_{n\to\infty}\Pr\{U_{(r)}\leqslant t\}=\Phi\left(\frac{t-\mu}{\sigma}\right) \tag{9.97}$$

即当样本数 n 很大时，顺序统计量 $U_{(r)}$ 的分布可由均值为 μ、方差为 σ^2 的正态分布逼近。

对于取自任意连续累积分布函数 $F_X(x)$ 的第 r 个顺序统计量 $X_{(r)}$，由式(9.96)可知，只要代入恰当的均值和方差，$X_{(r)}$ 的渐近分布也逼近正态分布。假设 $X_1,X_2,\cdots,X_n$ 和 $Y_1,Y_2,\cdots,Y_m$ 是分别取自连续累积分布函数 $F_X(x)$ 和 $F_Y(y)$ 的独立随机样本，定义原假设

$$F_X(x)=F_Y(x)=F(x) \tag{9.98}$$

对于任意 x 成立且 $F(x)$ 的具体分布未知。将两组样本合并得到数量为 $N=m+n$ 且取自同一未知分布 $F(x)$ 的随机样本组合，可定义元素的秩为 $1,2,\cdots,N$ 之间的整数。

为方便表示联合有序样本，定义指示随机变量 $Z_i(i=1,2,\cdots,N)$，若联合有序样本中第 i 个随机变量是 X 就令 $Z_i=1$，反之若为 Y 则令 $Z_i=0$。此时 Vander Waerden 检验统计量定义为

$$X_N=\sum_{i=1}^{N}\Phi^{-1}\left(\frac{i}{N+1}\right)Z_i \tag{9.99}$$

其中，$\Phi^{-1}(x)$ 是标准正态分布的 CDF 的反函数，称为 INSF。为探讨逆正态得分变换后，原始数据或随机变量值的代表性，下面将考虑利用逆正态得分变换构造顺序统计量[27]。

9.7.2　检测器设计

利用 INSF，对 9.4.1 节 GS 检验统计量 T_{GS} 中的 r_j 进行修正，可得广义符号修正秩(GSMR)检测器，其检验统计量可表示为

$$T_{\mathrm{GSMR}}=\sum_{j=1}^{N}\Phi^{-1}\left(\frac{r_j}{M+1}\right),\quad j=1,2,\cdots,N \tag{9.100}$$

式中，r_j 参见式(9.20)。值得注意的是，r_j 代表第 j 个脉冲回波中检测单元的顺序统计量，当 M 很大时 $r_j/(M+1)$ 服从(0,1)区间上的均匀分布；由 9.7.1 节的分析可知，当 $M\to\infty$ 时，$r_j/(M+1)$ 逼近标准正态分布。因为 $\Phi^{-1}[r_j/(M+1)]$ 表示标准正态分布的第 $r_j/(M+1)$ 个分位点，因此将检测单元的顺序统计量用标准正态分布的第 $r_j/(M+1)$ 个分位点代替，这表明该检验统计量与背景分布无关，是一种非参量秩检测器。在 CFAR 特性方面，检验统计量 T_{GSMR} 在无目标的假设下只与 M、N 和 R_j 有关，而与杂波具体分布类型无关，即检测阈值不依赖于杂波具体统计分布形式，表明该非参量检测器具有 CFAR 特性。后续几种修正秩检测器的非参量特性和 CFAR 特性与之类似。

与 GSMR 检验统计量的构造方式相同，将 9.4.2 节中 MW 检验统计量 T_{MW} 中的 r_j 用其 INSF 进行修正，可得 Mann-Whitney 修正秩(MWMR)检测器，其检验统计量可表示为

$$T_{\mathrm{MWMR}}=\sum_{j=1}^{N}\Phi^{-1}\left(\frac{r_j}{NM+1}\right),\quad j=1,2,\cdots,N \tag{9.101}$$

式中,r_j 参见式(9.24)。

类似地,将 9.4.4 节中 RS 和 MRS 检测器中的秩方统计量分别用其 INSF 进行修正,可分别得到逆正态秩方(Inverse Normal Rank Square,INRS)检测器和逆正态修正秩方(Inverse Normal Modified Rank Square,INMRS)检测器,两种非参量检测器的检验统计量分别如下:

$$T_{\mathrm{INRS}}=\sum_{j=1}^{N}\Phi^{-1}\left(\frac{r_j^2}{N^2M^2+1}\right),\quad j=1,2,\cdots,N \tag{9.102}$$

式中,r_j 参见式(9.24)。

$$T_{\mathrm{INMRS}}=\sum_{j=1}^{N}\Phi^{-1}\left(\frac{r_j^2}{M^2+1}\right),\quad j=1,2,\cdots,N \tag{9.103}$$

式中,r_j 参见式(9.20)。

9.7.3 性能分析

本节将分别在高斯杂波和 K 分布杂波背景下,分析对 Swerling Ⅱ型起伏目标的检测性能。

1. 高斯背景下的检测性能分析

图 9.19～图 9.22 分别给出了 GSMR、MWMR、INMRS、INRS 四种基于 INSF 的检测器与相应非参量检测器的检测性能对比。由图 9.19～图 9.22 可知,在高斯杂波背景下,基于 INSF 的修正秩检测器均不同程度地优于相应经典秩检测器,性能改善程度依次为 INRS＞MWMR＞GSMR＞INMRS,表明整数秩的 INSF 比整数秩本身对原始高斯分布数据更具有代表性。其中,在检测概率为 0.5 条件下,GSMR 相比 GS 的性能改善约为 0.6dB,MWMR 相比 MW 的性能改善约为 1.6dB,INMRS 相比 MRS 的性能改善约为 0.2dB,INRS 相比 RS 的性能改善约为 2.4dB。

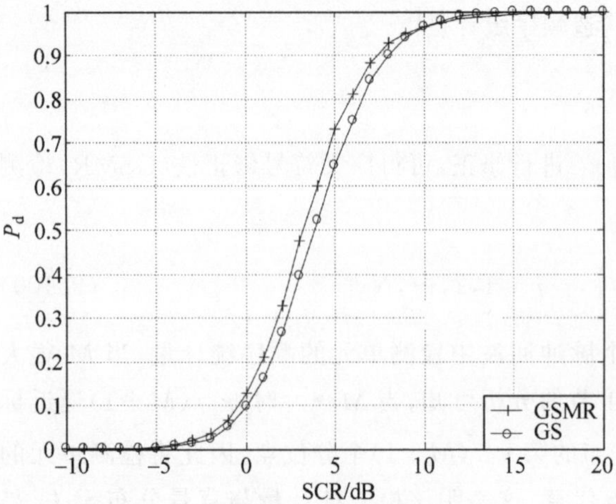

图 9.19　GSMR 与 GS 检测器在高斯背景下的检测性能($N=10,M=16$)

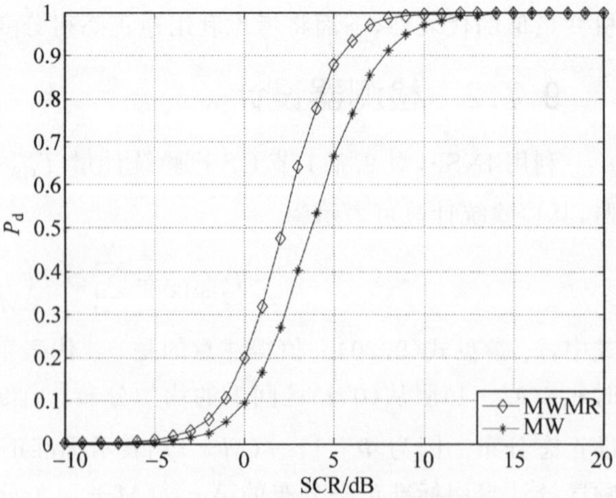

图 9.20　MWMR 与 MW 检测器在高斯背景下的检测性能($N=10,M=16$)

图 9.23 对四种基于 INSF 的检测器进行了检测性能对比。由图 9.23 可知,INRS 的检测性能最好,GSMR 和 INMRS 的检测性能较差,且二者的检测性能曲线几乎完全重合,而 MWMR 的检测性能则处于三者中间。

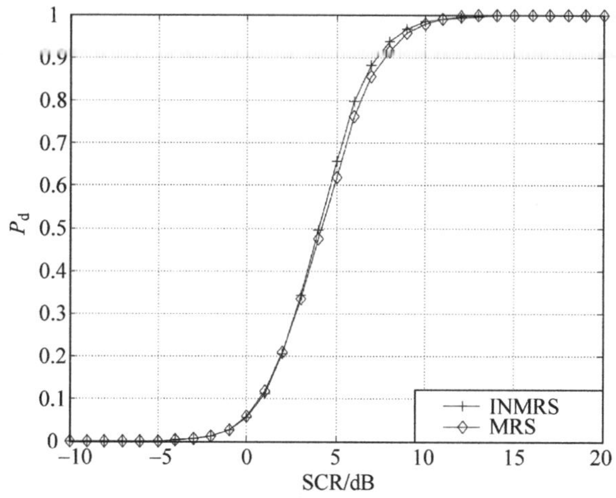

图 9.21　INMRS 与 MRS 检测器在高斯背景下的检测性能($N=10,M=16$)

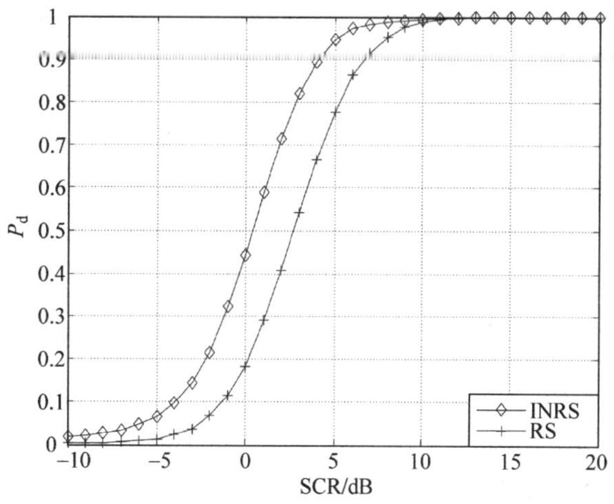

图 9.22　INRS 与 RS 检测器在高斯背景下的检测性能($N=10,M=16$)

2. K 分布背景下的检测性能分析

图 9.24 给出了 GSMR 和 MWMR 与相应非参量检测器的检测性能对比,其中,K 分布杂波尺度参数 $b=1$,形状参数 $c=0.1$。由图 9.24 可知,在杂波尖峰较严重的情况下,MWMR 要明显优于另外三种非参量检测器,经典的 GS 检测器性能优于 GSMR 检测器,而 MW 的性能最差。图 9.25 在 $b=1$ 和 $c=0.1$ 的典型尖峰 K 杂波情况下,给出了 INRS 和 INMRS 与相应非参量检测器的检测性能对比。由图 9.25 可知,检测性能优劣依次为: MRS>INRS>RS>INMRS。从图 9.24 和图 9.25 所示结果综合来看,除 MWMR 优于相应经典非参量检测器外,利用 INSF 修正后的非参量检测器在对抗尖峰杂波方面均不同程度劣于相应的经典非参量检测器。

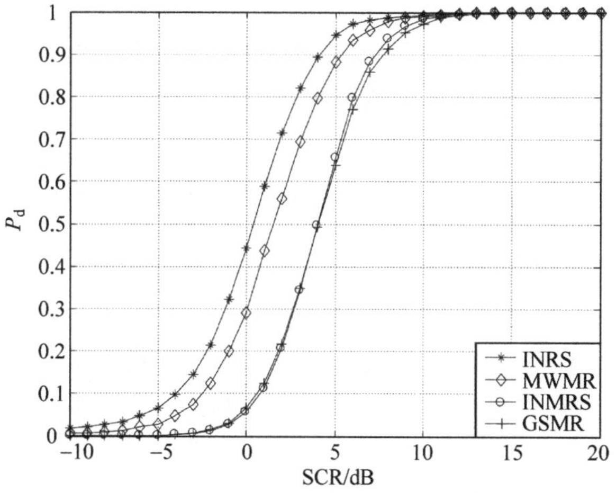

图 9.23　GSMR、INMRS、MWMR、INRS 检测器在高斯背景下的检测性能($N=10,M=16$)

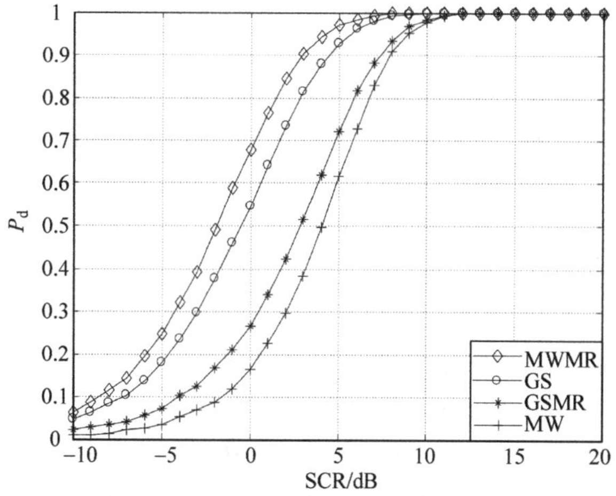

图 9.24　GSMR 和 MWMR 在 K 分布杂波下的检测性能($N=10,M=16,b=1,c=0.1$)

对比高斯杂波和 K 分布杂波背景下的检测性能可知,基于 INSF 的秩检验方法更适合于尖峰较少、起伏较小的杂波背景。考虑到实际应用中可能面临的不同杂波分布[28-29],建议采用鲁棒性较强的 MWMR 检测器。

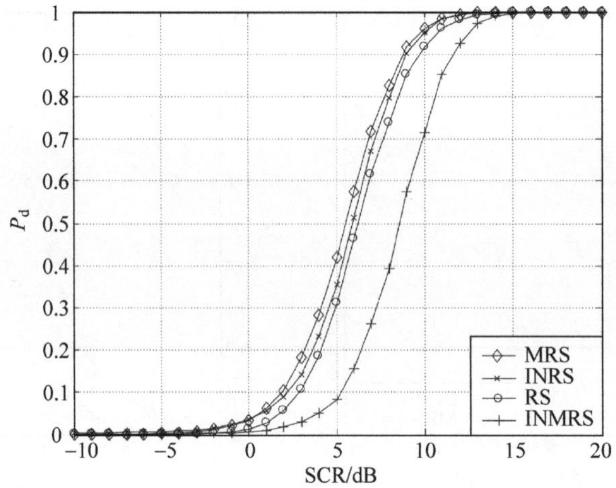

图 9.25 INRS 和 INMRS 在 K 分布杂波下的检测性能($N=10,M=16,b=1,c=0.1$)

9.8 比较与总结

常见的非参量检测器有单样本和两样本之分。9.3 节介绍了单样本非参量检测器,其中符号检测器和 Wilcoxon 检测器要求杂波样本分布的中位数为零或已知,这制约了其在雷达杂波背景中的应用。9.4 节介绍的样本非参量检测方案,在实际雷达系统中应用更为广泛,其虚警概率不依赖于杂波分布类型,只要参考滑窗内杂波样本的分布满足 IID 条件即可。相比参量检测器,非参量检测器在杂波类型已知的场合下,其检测性能要略弱一些,但在其他的场合下其检测性能和虚警性能要优于前者。较常见的非参量检测器包括 GS、MW、S、MS、RS、MRS 等。总体来说,S 检测器的性能优于 RS 检测器,RS 检测器的性能优于 MW 检测器,MW 检测器的性能优于 GS 检测器。而 GS 检测器相对于 MW 检测器、MS 检测器相对于 S 检测器、MRS 检测器相对于 RS 检测器在结构实现方面要简单一些,前者采用检测单元信号在本次脉冲扫描中与参考噪声样本的比较结果实施检测,而后者则采用检测单元信号与所有脉冲扫描中参考噪声样本的比较结果实施检测。

9.5 节讨论的次优秩检测器是在 Neyman-Pearson 准则下基于秩的似然比近似得到的,其最优检测参数 θ_{opt} 随着脉冲积累数的增加而增大。另外,次优秩检测器抗干扰目标能力有赖于脉冲积累数和参考单元数的增加。值得一提的是,MS、GS 和 MS 等检测器均可视为次优秩检测器的特例。

9.6 节在韦布尔分布的典型非高斯背景下,重点讨论了 RQ 和 GS 两种非参量检测器。在韦布尔杂波下,采用高秩量化门限时,RQ 具有较好的杂波边缘虚警控制能力,但较低秩量化门限能提高 RQ 对多干扰目标的对抗能力,因此秩量化门限 K 的取值,需兼顾多目标干扰、重拖尾杂波和杂波边缘多种情况的检测需求。为了改善 GS 在均匀韦布尔杂波中的检测性能和多目标分辨能力,参考单元数和积累脉冲数均需较多;但与 RQ 相比,GS 对参考单元数和积累脉冲数的要求稍少。

9.7 节分析了基于逆正态得分函数的改进秩检验方法。研究结果表明,基于 INSF 的秩检验方法更适合尖峰较少、起伏较小的杂波背景,考虑到实际应用中可能面临的不同杂波分布,建议采用鲁棒性较强的 MWMR 检测器。

非参量检测器因具有与背景杂噪统计分布无关的特性,正得到越来越多的重视,除了非参量方法研究有所发展,其在各种新体制雷达中的应用研究也成为当前的研究热点。将参量 CFAR 检测方法中的处理思路引入非参量检测器设计，可对非参量方法进行改进。例如，文献[24]利用参量 CFAR 检测中的选大选小处理，对广义符号检测器进行改进；文献[10]基于排序数据方差(ODV)对干扰目标进行自动删除后，再进行广义符号检测；而文献[21]和[30]基于可变指标(VI)的阈值选择思想，分别尝试对多种广义符号检测器或秩方检测器进行整合切换。置换检验也是一种非参量检验[31]，秩检验和条件检验都是它的一个子集，前面介绍的非参量检测器都属于秩检验，置换检验的最大缺点是实现复杂度大，但随着数字信号处理芯片速度的提高，该方法在非参量检测具有较好的潜力。另外，文献[32]对检测统计量的经验 CDF 的高值区域进行简单多项式逼近，避开了杂波统计建模过程，但该方法只适用于杂波样本充足的情形；文献[25]利用核密度估计处理参考单元数据，通过估计杂波的 PDF 进行检测阈值设置。将分数阶矩估计[33]、分形[29]、经验模态分解[34]等信号处理新技术引入非参量检测中，亦可为非参量检测方法研究提供拓展。虽然非参量方法种类繁多，但非参量 CFAR 检测的研究还有许多问题需要解决，其在实际目标检测过程中同样也会遇到多目标、杂波边缘等复杂情况，对非参量检测器在传统雷达中的应用实现[28,35]，在多输入多输出(MIMO)雷达[36]、太赫兹雷达[37]等新体制雷达中的应用等方面，仍有大量的研究工作有待开展。

参考文献

[1] 徐从安. 雷达目标自适应 CFAR 检测方法研究[D]. 烟台：海军航空工程学院，2012.

[2] 张林. 海杂波中目标非参量广义符号恒虚警检测算法研究[D]. 烟台：海军航空工程学院，2012.

[3] 孟祥伟. 雷达目标恒虚警率检测方法研究[D]. 烟台：海军航空工程学院，2008.

[4] Thomas，J B. Nonparametric detection[J]. IEEE Proceedings，1970，58(5)：623-631.

[5] 赵志坚. 海杂波中目标的非参量恒虚警检测算法研究[D]. 烟台：海军航空工程学院，2010.

[6] Hansen V G，Olsen B A. Nonparametric radar extraction using a generalized sign test[J]. IEEE Trans. on AES，1971，7(5)：942-950.

[7] Zeoli W G. Fong S T. Performance of a two sample Mann-Whitney nonparametric detector in a radar application[J]. IEEE Trans. on AES，1971，7(5)：951-959.

[8] Savage I R，Contributions to the theory of rank order statistics-The two-sample case[J]. Ann. Math. Statistics，1956，27(3)：590-615.

[9] Hussaini E K，Badran F M，Turner L F. Modified savage and modified rank squared nonparametric detectors[J]. IEEE Trans. on AES，1978，14(2)：242-250.

[10] 张林，黄勇，关键，等. 基于排序方差的非参量自动删除检测算法[J]. 雷达科学与技术，2012，10(3)：281-285.

[11] Sanz G J L，Figueiras V. A suboptimum rank test for nonparametric radar detection[J]. IEEE Trans. on AES，1986，22(6)：670-680.

[12] 金伟，刘向阳，许稼. K 分布雷达杂波中两种非参量检测器性能分析[J]. 雷达科学与技术，2010，8(4)：357-361.

[13] Meng Xiangwei. Performance evaluation of RQ non-parametric CFAR detector in multiple target and nonuniform clutter[J]. IET Radar，Sonar and Navigation，2020，14(3)：415-424.

[14] Ravid R，Levanon N. Maximum-likelihood CFAR for Weibull background[J]. IEE Proc. Radar Sonar and Navigation，1992，139(3)：256-264.

[15] 孟祥伟. 韦布尔杂波下非参量量化秩检测器的性能[J]. 电子学报，2009，37(9)：2030-2034.

[16] 徐从安，简涛，何友，等. 一种改进的 VI-CFAR 检测器[J]. 信号处理，2011，27(6)：926-931.

[17] 孟祥伟. 量化秩非参数 CFAR 检测器在杂波边缘中的性能分析[J]. 电子学报，2020，48(2)：384-389.

[18] Weinberg V G，Bateman L，Hayden P. Development of non-coherent CFAR detection processes in Weibull

background[J]. Digital Signal Processing，2018，75：96-106.
[19] 孟祥伟. 非参数秩和检测器的性能分析[J]. 电子与信息学报，2013，35(8)：2029-2032.
[20] Meng Xiangwei. Rank Sum Nonparametric CFAR Detector in Nonhomogeneous Background[J]. IEEE Trans. on AES，2021，57(1)：397-403.
[21] 张林，赵志坚，关键，等. 基于自适应阈值选择的非参量 GS 检测算法[J]. 雷达学报，2012，1(4)：387-392.
[22] 孟祥伟. 秩和非参数检测器在杂波边缘中的性能[J]. 电子与信息学报，2019，41(12)：2859-2864.
[23] 赵志坚，关键. 海杂波中非参量恒虚警检测器性能分析[J]. 雷达科学与技术，2010，8(1)：65-68.
[24] 张林，黄勇，关键，等. 基于广义符号最大或最小选择检测器[J]. 雷达科学与技术，2011，9(3)：259-263.
[25] 郝凯利，易伟，董天发，等. 未知杂波背景下恒虚警检测门限获取方法[J]. 雷达科学与技术，2015，13(2)，183-189.
[26] Gibbons J D，Chakraborti S. Nonparametric Statistical Inference[M]. Berlin：Springer，2014.
[27] Zhao Z J，Xu R L，Huang Y，et al. New nonparametric detectors under K-distributed sea clutter in radar applications. Proceedings of 2011 IEEE CIE International Conference on Radar[C]. Chengdu：2010.
[28] 刘彬. 船载导航雷达目标检测技术研究[D]. 哈尔滨：哈尔滨工业大学，2016.
[29] 张林，李秀友，刘宁波，等. 基于分形特性改进的 EMD 目标检测算法[J]. 电子与信息学报，2016，5：1041-1046.
[30] Li X D，Zhang Y H，Pei B N. An improved nonparametric detector for nonhomogeneous[C]. Weihai：IEEE Symposium on Computer Applications and Communications，2014.
[31] Gonzalez G J E，Sanz G J L，Alvarez V F. Nonparameter permutation tests versus parametric tests in radar detection under K-distributed clutter[C]. USA：2005 IEEE International Radar Conference，2005.
[32] 冉世领，赵宏钟，付强. 基于局部累积概率密度函数估计的 CFAR 检测门限获取新方法[J]. 信号处理，2012，28(12)：1692-1699.
[33] 李军，王雪松，王涛. 基于分数阶矩估计的非参量 CFAR 检测. 电子与信息学报[J]. 2011，33(3)：642-645.
[34] 张林，黄勇，薛永华，等. 基于 IMF 能量分布重构的目标检测技术[J]. 海军航空工程学院学报，2019，34(5)：401-406.
[35] Prokopenko I，Prokopenko K，Vovk V. CFAR Rank detection and estimation of Doppler radar signals[C]. Dresden：16th International Radar Symposium，2015.
[36] Sinitsyn R B，Yanovsky F J. MIMO radar copula ambiguity function[C]. Amsterdam：Proceedings of the 9th European Radar Conference，2012.
[37] Liu T，Min R，Pi Y，et al. Binary integration nonparametric detection for range-spread targets in distributed terahertz radar network under unknown clutter[J]. EURASIP Journal on Advances in Signal Processing，2016.

第10章 杂波图CFAR处理

CHAPTER 10

10.1 引言

根据对参数的估计方法可以将参量CFAR方法分为两大类：一类是空域CFAR处理方法，它根据检测单元的邻近单元(距离、角度或多普勒频率单元)样本值来形成所需参数的估计，这类方法适用于在空域上平稳的杂波背景中；另一类是杂波图(Clutter Map，CM)CFAR处理方法(也称时序CFAR方法)，它根据检测单元以往多次扫描测量值形成检测单元的杂波背景强度估计，该类方法更适合于时域上平稳的杂波环境。传统杂波图一般存储每个方位-距离单元背景电平的估计值，每个值依靠新的和以前若干次的扫描测量值来进行迭代更新，并把它作为当前的杂波背景强度估计值。当空域杂波强度变化剧烈时，若采用传统的均值类恒虚警方案只能采用很少的参考单元，因而恒虚警损失很大，且虚警不易保持恒定。一般情况下，杂波虽然在距离和方位上的变化十分剧烈，但若同一距离单元的杂波强度随时间变化是较缓慢的，则可以采用"时间单元"恒虚警处理方法，在时间上对以往各次雷达回波的测量值进行迭代处理，也就是所谓的杂波图CFAR方法[1]。近年来随着杂波图技术在硬件实现和工程应用中的日益成熟[2-5]，其在地面侦察雷达[6]、警戒雷达[7]、毫米波雷达[8]、线性调频航管雷达[9]、超宽带雷达[10]、船舶交通管理系统导航雷达[11]、分布式传感器网络[12]等领域得到广泛应用，相关概念已拓展到多普勒域的速度杂波图[13]、距离-方位-仰角三维空间的立体杂波图[14-15]、杂波区域划分的轮廓杂波图[16]、杂波图更新方式中的静态/动态杂波图[2]等。

Nitzberg等[1]早期提出了杂波图CFAR方法，在文献中通常称为Nitzberg杂波图方法，这一处理过程利用了雷达杂波环境时域的相对平稳性。因此，杂波图处理不受诸如地杂波或海杂波在空域非平稳性的影响[15]。Levanon等给出了另外一种计算Nitzberg杂波图技术虚警概率的公式[17]，它可以较快收敛。为了增强杂波图的鲁棒性，文献[18]提出了混合杂波图技术(CM/L-CFAR)，即将空域处理与时域处理结合起来，将几个雷达分辨单元的回波信号组合成一个杂波图单元，对杂波图单元内的回波样本基于L滤波器进行空域处理作为当前杂波图单元的输入信号；再对每个杂波图单元的输入信号对以往各次的结果进行迭代，得到检测位置处杂波强度的估计以检测目标，其中考虑的是高斯背景。文献[19]基于位置-尺度分布分析了混合杂波图技术在非高斯背景中的目标检测应用潜力。文献[20]提出了一种适合对海浪杂波进行处理的杂波图单元平均CFAR平面技术，其先对杂波图单元中各个分辨单元幅值进行单元平均处理后再进行迭代处理。文献[21]和[22]进一步对杂波图处理的点技术和平面技术进行了拓展。此外，还有学者讨论了杂波图CFAR技术在韦布尔背景中的检测性能[23-25]。事实上，对于对数正态分布等包含形状参数和尺度参数的杂波情况，前述仅基于杂波幅度的单参数杂波图方法会存在一定的性能损失,文献[26]提出了双参数杂波图设计思路,通过样本均值和方差估计值共同设置检测门限[27],获得了更好的虚警控制性能。

本章首先在10.2节中介绍经典的Nitzberg杂波图技术,并进行性能分析;接着在10.3节中讨论杂波图单元平均CFAR平面检测技术,并与Nitzberg杂波图所代表的点技术进行对比分析;在10.4节中,重点讨论空域处理与时域处理结合的混合CM/L-CFAR技术;10.5节中重点介绍双参数杂波图CFAR处理技术。

10.2 Nitzberg杂波图技术

10.2.1 Nitzberg杂波图检测的原理

最经典的杂波图检测技术是Nitzberg等提出的[1],该方法把雷达空间分成杂波图单元进行处理,根据检测单元多次扫描测量值形成检测单元杂波背景的强度估计。杂波图存储每个方位-距离单元杂波强度的幅值,每个值依靠新的和以前若干次的扫描测量值的迭代来更新,并把它作为当前杂波背景的强度估计。

杂波图处理主要包括两个步骤:杂波图更新和杂波图检测[12]。Nitzberg杂波图处理方法利用式(10.1)进行迭代更新,从而得出每个杂波图单元的背景功率估计。事实上,式(10.1)所表示的系统为一阶自回归模型,其检测原理框图见图10.1。

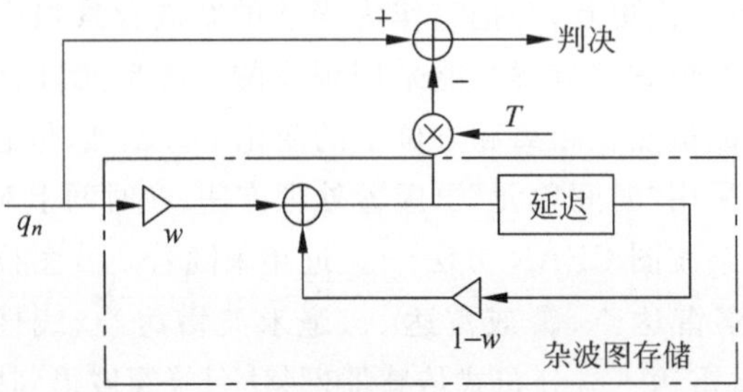

图10.1 Nitzberg杂波图技术的检测原理框图

$$\hat{p}_n(k)=(1-w)\hat{p}_{n-1}(k)+wq_n(k) \tag{10.1}$$

将式(10.1)展开可以看出,上述迭代过程的实质是对先前所有回波信息进行指数加权平均:

$$\hat{p}_n(k)=w\sum_{l=0}^{\infty}(1-w)^l q_{n-l}(k) \tag{10.2}$$

式中,$\hat{p}_n$是对第k个空间单元的第$n+1$次扫描的估值;q_n是第k个空间单元的第n次回波输出的样本值,由于空间单元在方位上为半个波瓣宽度,因此q_n可考虑为对应的多次雷达扫描回波的样本平均值;w是衰减因子。若现时刻为第n次扫描,则可利用上一次回波样本的杂波强度更新值$\hat{p}_{n-1}$来设置检测阈值。杂波图实际所存储的是每个空间单元的检测阈值($\hat{p}_{n-1}$),且这个值会随着新回波的到来不断更新。

假定热噪声加杂波服从瑞利分布(当杂波包含许多小的散射体回波时,这样的假设是有效的),目标起伏模型为Swerling Ⅰ型,即各次扫描间的回波输出相互独立且服从同一瑞利分布。那么对于平方律检波,每次扫描输出q的PDF为

$$f(q)=\begin{cases}\dfrac{1}{\mu}\exp\left(-\dfrac{q}{\mu}\right), & \mathrm{H}_0\\ \dfrac{1}{\mu(1+\lambda)}\exp\left[-\dfrac{q}{\mu(1+\lambda)}\right], & \mathrm{H}_1\end{cases} \tag{10.3}$$

式中，μ 代表热噪声加杂波的功率强度；λ 是信号对整个干扰（热噪声加杂波）的平均信杂比；H_0 表示单元中不存在目标的假设；H_1 表示检测单元中存在目标的假设。

自适应判决准则为：

$$\frac{q_n}{\hat{p}_{n-1}} \underset{\mathrm{H}_0}{\overset{\mathrm{H}_1}{\gtrless}} T \tag{10.4}$$

式中，T 指阈值因子。由于 $\hat{p}_{n-1}$ 是随机变化的，那么检测概率 P_{d} 便需要利用 $\hat{p}_{n-1}$（简写为 $\hat{p}$）的 PDF 进行统计表征，即

$$P_{\mathrm{d}}=\int_0^{\infty}\exp\left[-\frac{T\hat{p}}{\mu(1+\lambda)}\right]f(\hat{p})\,\mathrm{d}\hat{p} \tag{10.5}$$

其中，$f(\hat{p})$ 为杂波强度估计 $\hat{p}$ 的 PDF。若用 $\hat{p}$ 的 MGF $M_{\hat{p}}(u)$ 表示式(10.5)，则有

$$P_{\mathrm{d}}=M_{\hat{p}}(u)\big|_{u=T/\mu(1+\lambda)} \tag{10.6}$$

由式(10.3)知，在 H_0 假设条件下有 $q\propto\Gamma(1,\mu)$ 和 $M_q(u)=(1+\mu u)^{-1}$。那么由 $wq\propto\Gamma(1,w\mu)$ 可得 $M_{wq}(u)=(1+w\mu u)^{-1}$。进一步地，再利用式(10.2)得

$$\hat{p}_n=w\sum_{l=0}^{\infty}(1-w)^l q_{n-l}=wq_n+w(1-w)q_{n-1}+w(1-w)^2q_{n-2}+\cdots \tag{10.7}$$

因此

$$\begin{aligned}M_{\hat{p}}(u)&=(1+w\mu u)^{-1}[1+w(1-w)\mu u]^{-1}[1+w(1-w)^2\mu u]^{-1}\cdots\\&=\prod_{l=0}^{\infty}[1+w(1-w)^l\mu u]^{-1}\end{aligned} \tag{10.8}$$

因此

$$P_{\mathrm{d}}=M_{\hat{p}}(u)\big|_{u=T/\mu(1+\lambda)}=\prod_{l=0}^{\infty}\left[1+\frac{Tw(1-w)^l}{1+\lambda}\right]^{-1} \tag{10.9}$$

这里的 w 被作为一个衰减因子。令式(10.9)中的 $\lambda=0$ 便可以得到 Nitzberg 杂波图处理方法的虚警概率 P_{fa} 的解析表达式，根据指定的 P_{fa} 值即可预先确定阈值因子 T，然后再将其代入式(10.9)来计算检测概率 P_{d}[17]。

Nitzberg 杂波图处理方法中 ADT 的解析表达式为

$$\mathrm{ADT}=\frac{\mathrm{E}\{T\hat{p}_{n-1}\}}{\mu}=-\frac{T}{\mu}\frac{\mathrm{d}M_{\hat{p}}(u)}{\mathrm{d}u}\bigg|_{u=0}=T\lim_{L\to\infty}\sum_{i=1}^{L}w(1-w)^{i-1} \tag{10.10}$$

ADT 值越小，检测性能越好。

根据式(10.9)，对于不同的衰减因子 w（$w=0/8,1/8,\cdots,8/8$），图 10.2 给出了虚警概率为 10^{-6} 时 Nitzberg 杂波图的检测概率随信杂比的变化曲线。可以看出，随着衰减因子 w 的增大，CFAR 损失也增加。

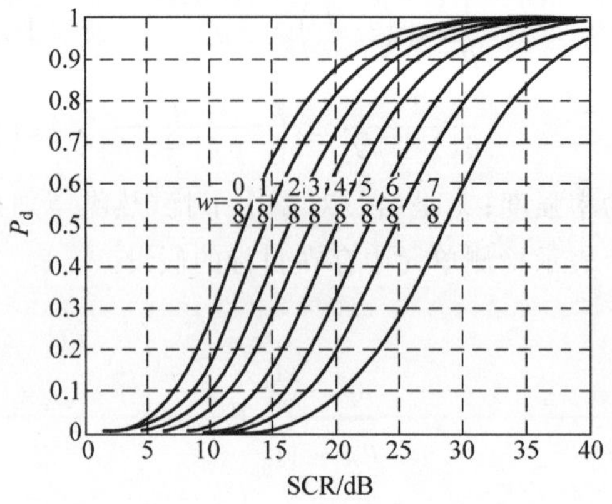

图 10.2 Nitzberg 杂波图处理方法的检测性能曲线

10.2.2 Nitzberg 杂波图 ADT 值和虚警指标对 w 取值的约束

在前面假设模型下，这里计算了一组随 w 和 L 变化的 T 和 ADT 值，虚警概率设定值 $P_{fa}=10^{-6}$。表 10.1 中每组的 T 和 ADT 值都是在给定 w 和 L 时为达到虚警指标计算出的结果。可以看到，在 P_{fa} 和 L 一定的情况下，w 越小则 ADT 值越小。在 P_{fa} 和 w 确定时，阈值因子 T 值随着 L 的增加而逐渐减小至某一个值，这个值也是最小的 ADT 值。即在虚警指标给定的情况下，一个 w 值将对应着一个最低的 T 和 ADT 值。因此，阈值因子 T 值其实也是表征检测性能的一个度量。

表 10.1 一组随 w 和 L 变化的 T 和 ADT 值

L	$w=0.5$		$w=0.25$		$w=0.125$	
	T	ADT	T	ADT	T	ADT
15	76.6532	76.6508	33.2621	32.8176	28.8182	24.9297
50	76.6222	76.6222	31.5428	31.5427	21.0063	20.9799
100	76.6222	76.6222	31.5427	31.5427	20.9509	20.9509
200	76.6222	76.6222	31.5427	31.5427	20.9509	20.9509
400	76.6222	76.6222	31.5427	31.5427	20.9509	20.9509
L	$w=0.0625$		$w=0.03125$		$w=0.001$	
	T	ADT	T	ADT	T	ADT
15	37.4207	23.2078	60.1950	22.8066	1522.506	22.6784
50	18.1255	17.4063	20.4946	16.3045	326.1285	15.9132
100	17.1014	17.0745	16.2000	15.5229	155.6191	14.8162
200	17.0630	17.0620	15.5321	15.3994	103.8553	14.4729
400	17.0615	17.0615	15.4022	15.3753	48.8826	14.3055

当选择 $L=50$ 且虚警概率满足 10^{-6} 的 T 值作为每个 w 取值的统一 T 值时，相应的 ADT 值如表 10.2 所示，其中 ADT_i 表示扫描 i 次后的 ADT 值。

表 10.2 一组随 w 和 L 变化的 ADT 值

w	0.5	0.25	0.125	0.0625	0.03125	0.001
T	76.6222	31.5428	21.0063	18.1255	20.4946	326.1285
ADT_{50}	76.6222	31.5427	20.9799	17.4063	16.3045	15.9132
ADT_{200}	76.6222	31.5428	21.0063	18.1255	20.4588	59.1440
ADT_{400}	76.6222	31.5428	21.0063	18.1255	20.4945	107.5615
ADT_{800}	76.6222	31.5428	21.0063	18.1255	20.4946	170.6480

由表 10.2 可以看到，w 值越大，其达到稳态的速度就越快，反之则越慢。这里的稳态应理解为满足虚警指标，且检测性能也能基本稳定。仅就表 10.2 中的 ADT 值来看，w 值的选取范围是有考虑的。$w=0.0625$ 时所获得的检测性能要好于其他取值情形。可以看出，当 w 取值在 0.06 左右时，ADT 值最小。从降低 ADT 值（提高检测性能）的角度考虑，在扫描 $L=50$ 次后虚警概率满足 10^{-6} 的要求下，w 取值在 0.06 左右是值得推荐的。因此，衰减因子 w 的选择是有一定范围的，而不是越小越好。

w 值不能取得过小的另一个原因与虚警概率的控制有关。前面所得的 w 值是在 L 超过 50 次时满足虚警 10^{-6} 的情况下得到的。对于其他 $L=10,20,30,40$ 次时满足虚警概率 $P_{fa}=10^{-6}$ 的情况，可以利用性能公式计算出其阈值因子 T 随衰减因子 w 的变化曲线，如图 10.3 所示。可以看到，使 T 值达到最小的最佳 w 值将随着虚警指标的变化而不断地变化。具体来说，$L=10$，20，30，40，50 的虚警指标情况对应的最佳 w 值分别为 0.22，0.12，0.10，0.08，0.06。虚警指标要求越严（达到同样 P_{fa} 时的 L 值越小），其最佳的 w 值越大。对于不同的虚警指标情况都存在不同的可选择的 w 取值范围。

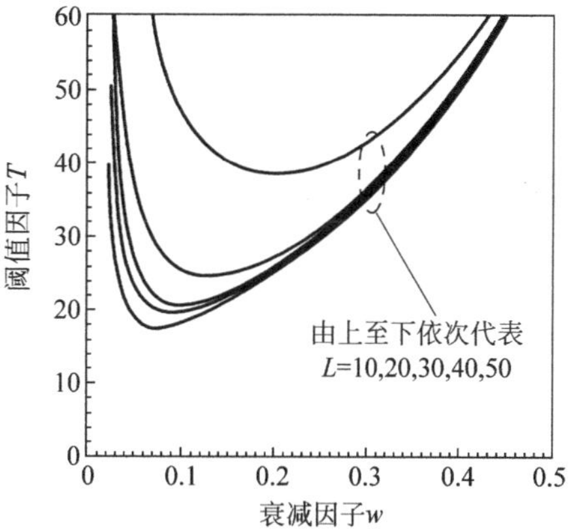

图 10.3 不同虚警指标情况下 w 随 T 的变化曲线

综上所述，Nitzberg 杂波图中衰减因子 w 的取值受两方面因素的制约：即尽量提高检测性能（小的 ADT 值）和必须满足的虚警指标（在给定的扫描次数下达到要求的虚警指标）。一般对于相同的虚警指标情况，都有一个可选的最佳 w 的取值，但这不是绝对的。比如，若要求检测性能较高，那么可将 w 值取得稍小些，即牺牲一点虚警控制性能以换取好的检测概率。同样，若对 P_{fa} 的控制要求比较严格，那么只能将适当提高 w 值；此时的检测损失也就会大一些。总体来说，在满足虚警指标要求的情况下，尽量选择稍小的 w 值以提高检测性能。

10.2.3 Nitzberg 杂波图在韦布尔分布中的性能

实际应用中雷达杂波统计特性常常呈现非高斯性[27]，评估 Nitzberg 杂波图方法在常用非高斯韦布尔分布下的性能，具有较重要的应用价值[25]。文献[23]对距离单元回波数据进行指数变换，使得变换后的杂波数据从韦布尔分布变成了负指数分布，为分析韦布尔杂波下的 Nitzberg 杂波图方法提供了便利条件。本节在韦布尔杂波背景下，重点分析 Nitzberg 杂波图方法的扫描次数影响、虚警和检测性能[24]。

在韦布尔杂波形状参数 $c=2$，衰减因子 $w=1/8,2/8,\cdots,8/8$ 条件下，图 10.4 分析了 Nitzberg 杂波图的平均判决门限（ADT）随扫描次数 L 的变化。由图 10.4 可知，ADT 随着衰减因子 w 的减小而降低，意味着杂波图趋于稳定所需时间越长。当扫描次数 L 大于 30 时，不同衰减因子 w 对应的 ADT 值基本不变，即杂波图已处于稳定状态。

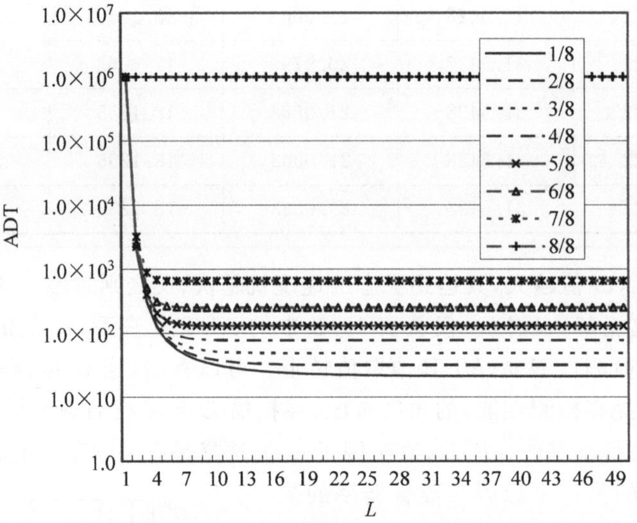

图 10.4　Nitzberg 杂波图的 ADT 随扫描次数 L 的变化曲线($P_{fa}=10^{-6}$, $c=2$)

在衰减因子 $w=1/8, 2/8, \cdots, 8/8$ 条件下，通过采用重采样技术，图 10.5 和图 10.6 分别给出了 $c=1.4$ 和 $c=0.8$ 对应的 Nitzberg 杂波图虚警概率曲线。由图 10.5 和图 10.6 可知，对于给定的虚警概率，相应的门限因子 T 随着衰减因子 w 的减小而降低；另外，门限因子 T 随着形状参数 c 的减小而增大，即杂波尖峰增多导致门限因子提高。

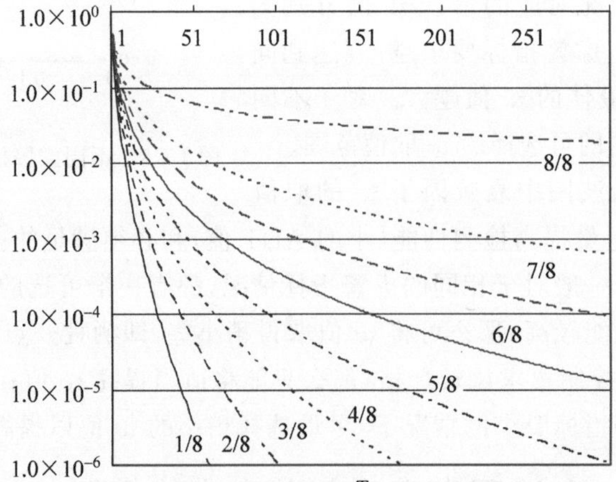

图 10.5　Nitzberg 杂波图的虚警概率与门限因子 T 的关系曲线($c=1.4$, $L=100$)

图 10.7 和图 10.8 分别分析了 $c=1.4$ 和 $c=0.8$ 时 Nitzberg 杂波图的检测性能，杂波功率水平已知的理想最优检测器作为比较基准给出，其与 Nitzberg 杂波图间的性能差别体现了 CFAR 损失。对比两图可知，随着形状参数 c 的减小，杂波尖峰增多，Nitzberg 杂波图的检测性能有所下降，即在杂波尖峰增多情况下，达到相同检测概率所需信杂比更高。对于形状参数 $c=0.8$ 的杂波尖峰较严重的情况，检测概率 0.5 所对应的 SCR 约为 30dB，即常规 Nitzberg 杂波图方法在这种杂波尖峰环境下难以有效发挥作用。

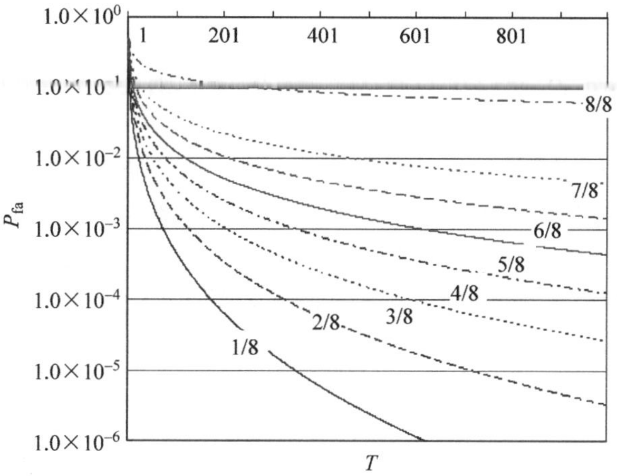

图 10.6　Nitzberg 杂波图的虚警概率与门限因子 T 的关系曲线($c=0.8, L=100$)

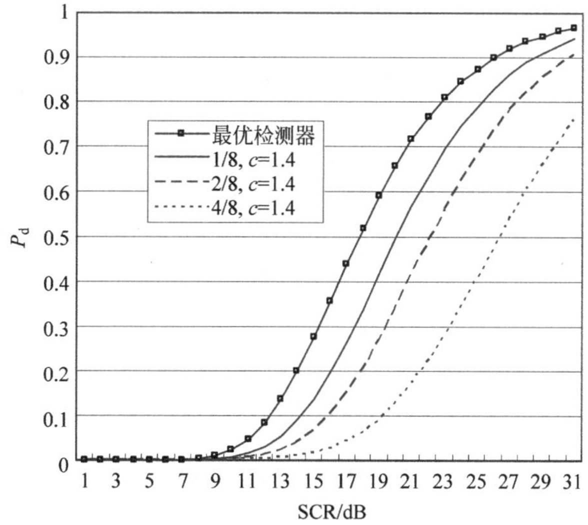

图 10.7　Nitzberg 杂波图的检测性能($c=1.4, L=100, P_{fa}=10^{-6}$)

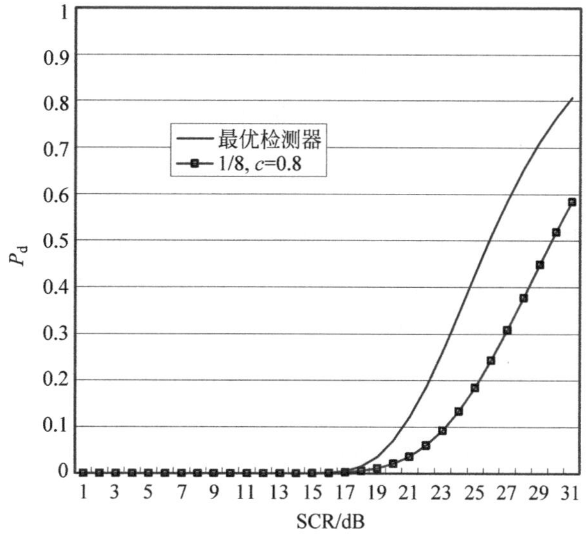

图 10.8　Nitzberg 杂波图的检测性能($c=0.8, L=100, P_{fa}=10^{-6}$)

10.3 杂波图单元平均 CFAR 平面检测技术

雷达工作时，将其周围环境的杂波回波幅度按距离和方位二维平面有序地存储下来，从而建立杂波图。将整个检测空域划分成若干个杂波图单元，而每个杂波图单元对应于在方位和距离二维平面上连续的几个分辨单元的结合。与 Nitzberg 杂波图处理的点技术相比，该方法可以利用方位和距离二维平面数据来估计杂波图检测门限，将其称为杂波图面技术的处理[14,16]。这里先介绍杂波图单元平均 CFAR 处理方式[20]，10.4 节将讨论混合杂波图(CM/L-CFAR)处理技术[18]。

10.3.1 基本模型描述

这里通过一个简单示意图来说明杂波图 CFAR 检测面技术的基本原理，如图 10.9 所示。以检测单元 A 为中心，取单元数为 MN 的一个空间区域，M、N 分别代表方位和距离上的杂波图单元数。其中 pq 个单元(图 10.9 中阴影部分)作为检测窗，其余空间单元用来估计检测单元 A 的杂波功率水平，称为参考或杂波背景单元。检测窗不参与杂波功率估计，同普通 CFAR 处理相似，这也是为了防止检测单元的信号泄漏到背景单元中而影响到对背景杂波的功率估计。

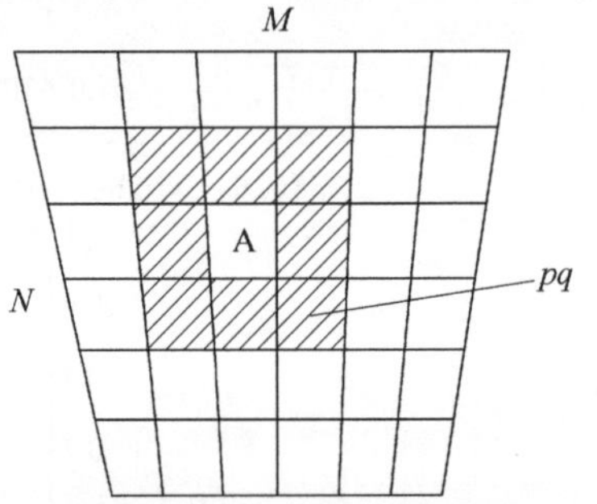

图 10.9 杂波图面技术的示意图

杂波图单元平均 CFAR 平面检测原理为：在每个扫描周期先对杂波图上分辨单元幅值进行单元平均处理，然后对各个杂波图单元平均值进行迭代得到检测单元的背景估计值，所以这种方法称为杂波图单元平均 CFAR 平面检测技术[20]。

一般地，检测判决应在建立了比较稳定的杂波图后进行。即前 L 个(L 常取 10～20)扫描周期进行迭代估计，以建立起较平稳的杂波图。假设目标起伏模型服从 Swerling Ⅰ型，即各次扫描间的回波输出相互独立且服从同一瑞利分布，经平方律检波后每个杂波图单元输出服从指数分布。重写式(10.1)和式(10.2)如下：

$$\hat{p}_n(k)=(1-w)\hat{p}_{n-1}(k)+wq_n(k) \tag{10.11}$$

$$\hat{p}_n(k)=w\sum_{l=0}^{\infty}(1-w)^l q_{n-l}(k) \tag{10.12}$$

在本节中，q_n 的含义有所改变，它表示对参考单元上进行单元平均处理后的值，由式(10.13)表示：

$$q_n=\sum_{i=1}^{K}q_n^i \tag{10.13}$$

其中，$K=MN-pq$，q_n^i 代表第 n 次扫描中第 i 个杂波图分辨单元的输出。在假设 H_0 条件下，有 $q_n^i\propto\Gamma(1,\mu)$，$q_n\propto\Gamma(K,\mu)$ 和 $M_{q_n}(u)=(1+\mu u)^{-K}$。那么由 $wq_n\propto\Gamma(K,w\mu)$ 可得 $M_{wq_n}(u)=(1+w\mu u)^{-K}$，因此有

$$\begin{aligned}M_{\hat{p}}(u)&=(1+w\mu u)^{-K}[1+w(1-w)\mu u]^{-K}[1+w(1-w)^2\mu u]^{-K}\cdots\\&=\prod_{l=0}^{\infty}[1+w(1-w)^l\mu u]^{-K}\end{aligned} \tag{10.14}$$

可得检测概率为

$$P_{\mathrm{d}}=M_{\hat{\mu}}(u)\big|_{u=T/\mu(1+\lambda)}=\prod_{l=0}^{\infty}\left[1+\frac{Tw(1-w)^{l}}{1+\lambda}\right]^{-K} \tag{10.15}$$

这里的 w 为衰减因子。令式(10.15)中的 $\lambda=0$ 便可以得到虚警概率 P_{fa} 的解析表达式。这样根据指定的虚警概率 P_{fa} 设计值即可预先确定阈值因子 T，然后再将其代入式(10.15)可计算出杂波图单元平均CFAR平面处理技术的检测概率 P_{d}。

根据平均判决门限(ADT)的定义式，可以得到杂波图单元平均CFAR平面处理技术的ADT，即

$$\mathrm{ADT}=TK\lim_{L\to\infty}\sum_{i=1}^{L}w(1-w)^{i-1} \tag{10.16}$$

10.3.2 均匀背景中的性能分析

利用式(10.15)和式(10.16)所得性能数据如表10.3所示。这里取 L 大于50次后满足虚警指标的 T 值作为每种 w 取值情况的统一 T 值，其中，$P_{\mathrm{fa}}=10^{-6}$，$K=MN-pq=7\times7-3\times3$。

表10.3 杂波图平面处理技术的性能数据

w		0.5		0.125		0.0625		0.001	
T		0.3656		0.3498		0.3618		7.1029	
L		50	800	50	800	50	800	50	800
ADT		14.624	14.624	13.975	13.991	13.899	14.473	13.863	93.704
P_{d}	$\lambda=5$dB	0.0313	0.0313	0.0352	0.0350	0.0356	0.0311	0.0359	0.0000
	$\lambda=10$dB	0.2666	0.2666	0.2811	0.2807	0.2829	0.2685	0.2837	0.0000
	$\lambda=15$dB	0.6393	0.6393	0.6517	0.6513	0.6537	0.6418	0.6538	0.0083
	$\lambda=20$dB	0.8653	0.8653	0.8708	0.8707	0.8714	0.8665	0.8717	0.2124

由表10.3可知，在 T 可变的情况下，与杂波图点技术类似，平面技术采用CA方法，其检测性能也将随 w 的减小而提高。若 T 为定值，w 的选择同样也有一个折中考虑的范围。过小的 w 会导致过大的 T 值，从而会出现检测性能随着 L 的增加而不断下降的现象[21]。取 L 大于50次后满足虚警指标的 T 值作为每种 w 取值情况的统一 T 值，列出几种 w 取值情况的ADT数据如表10.4所示。其中，ADT_i 表示经 i 次扫描后的平均判决门限值。由表10.4可知，当 w 取0.125左右时其ADT值最为理想。进一步计算发现，当 w 取0.11～0.13范围内的值时，其ADT值都低于14。

表10.4 几种 w 取值情况下杂波图平面处理技术的ADT值

w	0.5	0.25	0.125	0.0625	0.03125	0.001
T	0.3656	0.3540	0.3498	0.3618	0.4360	7.1029
ADT_{50}	14.6240	14.1583	13.9754	13.8990	13.8730	13.8633
ADT_{200}	14.6240	14.1583	13.9910	14.4726	17.4117	51.5242
ADT_{400}	14.6240	14.1583	13.9910	14.4726	17.4118	93.7038

10.3.3 面技术与点技术的性能比较

表10.5列出一组点技术与面技术的检测概率 P_{d} 的对比数据，由表10.5可以看出，面技术的检测性能相对点技术有较大的提高。原因在于，面技术参与杂波强度估计的参考单元数目是点技术的 K 倍(此处为40倍)。表10.5中点、面技术都采用 L 大于50时满足虚警指标的阈值因子 T，表10.5中数据

是在 $L=800$ 时计算得出的。

表 10.5 点、面技术的检测概率 P_d 的比较($P_{fa}=10^{-6}$)

采用技术	$\lambda=5$dB	$\lambda=10$dB	$\lambda=15$dB	$\lambda=20$dB	$\lambda=25$dB
点技术	0.0128	0.1653	0.5323	0.8134	0.9361
面技术	0.0350	0.2806	0.6513	0.8706	0.9569

图 10.10(a)和图 10.10(b)分别给出了 L 为 20 和 50 时的点技术阈值因子随衰减因子 w 变化的曲线,图 10.11(a)和图 10.11(b)分别给出了 L 为 20 和 50 时的面技术阈值因子随衰减因子 w 变化的曲线。每个图中的曲线由上至下分别代表 $P_{fa}=10^{-8},10^{-7},10^{-6},10^{-5},10^{-4},10^{-3},10^{-2}$ 的情形;每条曲线上的阈值因子应能使在扫描了 L 次后满足相应的虚警指标。由图 10.10 和图 10.11 可以看出,面技术同点技术类似,随着 L 的增加,对虚警控制时间的要求变弱,阈值因子 T 普遍下降,检测性能变好。在 L 和 w 一定的情况下,P_{fa} 值越低,则阈值因子 T 值越大,反之阈值因子 T 值越小。在每种 L 情况下,对于每个虚警指标值都有一个最佳的衰减因子 w 取值,并且这个衰减因子 w 将随着 L 和虚警指标值的变化作相应的变化。值得注意的是,面技术的 T 值对 w 取值变化不敏感。

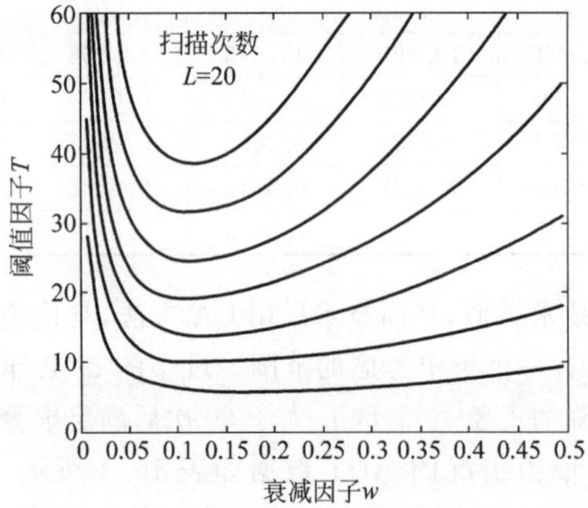

(a) $L=20$

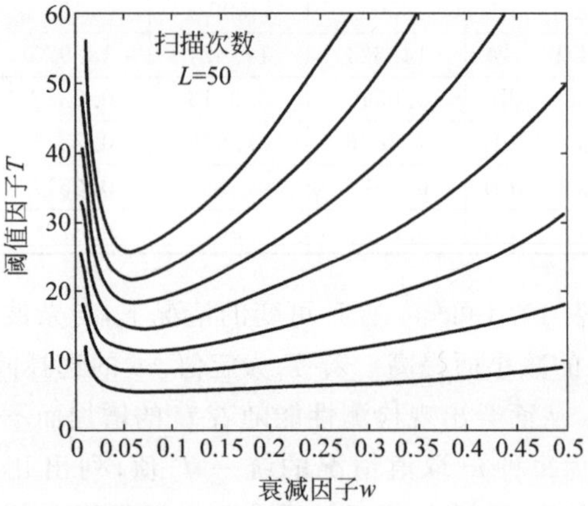

(b) $L=50$

图 10.10 杂波图点技术的阈值因子 T 随衰减因子 w 的变化曲线

在 $L=20, P_{fa}=10^{-2},10^{-5},10^{-8}$ 的情况下,图 10.12 比较了点技术与面技术的阈值因子 T 随 w 的变化曲线。图 10.12 中实线组和虚线组由上至下分别代表 $P_{fa}=10^{-8},10^{-5},10^{-2}$ 的情况。从图 10.12 中可以看到,面技术的 T 值对 w 取值变化的敏感性明显比点技术要弱得多。即点技术为取得较好的检测性能对 w 取值的依赖很强,而面技术则不是这样。由此不难看出,面技术的检测性能对 w 取值的依赖性较点技术要弱得多。

下面对点技术和面技术的最佳 w 值和对应的最低 T 值(ADT 值)进行比较和讨论。图 10.13 和图 10.14 分别给出了点、面技术在不同虚警控制要求下的最佳 w 值和最低 T 值的曲线。图 10.13 中的实线组和虚线组由上至下分别代表 $L=10,20,30,40,50$ 的情况,而在图 10.14 中相反。由图 10.13 可以看出,虚警要求越高则最佳的 w 值越低;不过相同条件下,面技术选取的 w 值要大于点技术的 w 值。图 10.14 表明了点技术对 L 的变化(虚警控制时间的要求)比较敏感,而面技术则几乎没有影响。如虚警概率设定值 $P_{fa}=10^{-8}$,那么对于点技术,$L=10$ 比 $L=50$ 时的 T 值(ADT 值)要超出 44.0502。这

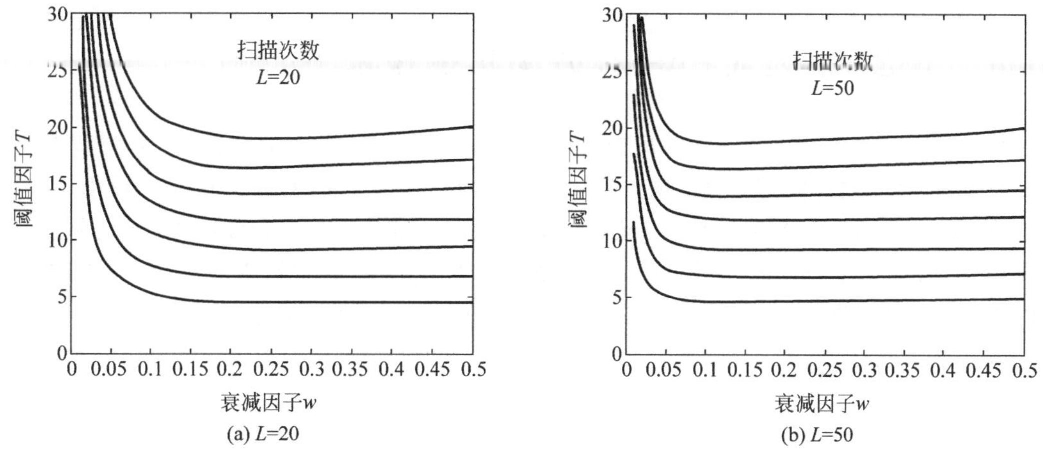

图 10.11 杂波图面技术的阈值因子 T 随衰减因子 w 的变化曲线

样，相对严格的虚警控制时间要求会导致 ADT 值出现很大增长，从而也导致点技术检测性能的严重下降。相比而言，面技术对虚警控制时间的要求则不敏感。如虚警概率设定值 $P_{fa}=10^{-8}$ 时，$L=10$ 比 $L=50$ 时的 T 值（ADT 值）仅高出 0.8855。

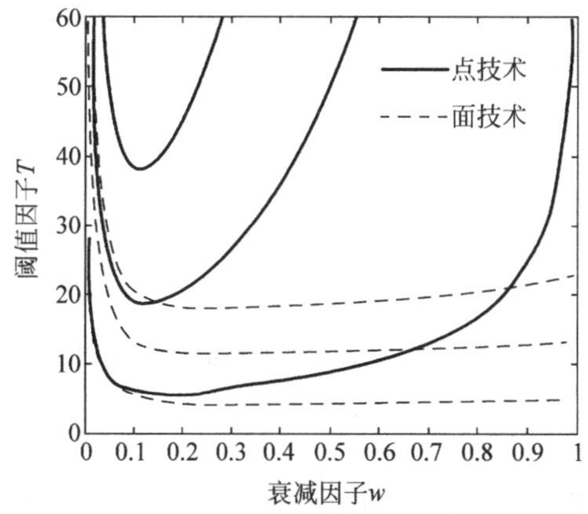

图 10.12 $L=20$ 时点技术与面技术的阈值因子 T 随 w 的变化曲线

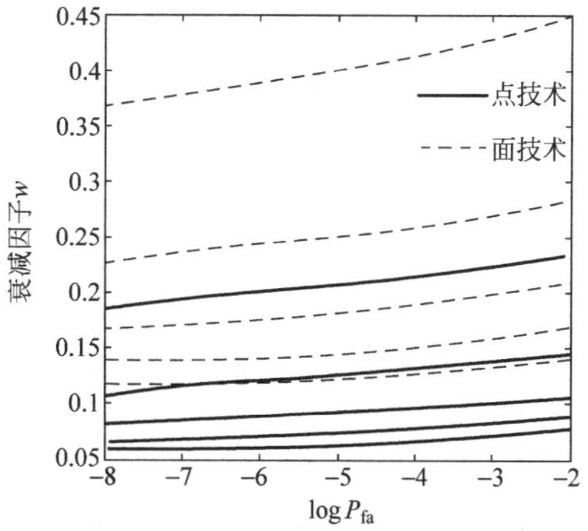

图 10.13 点技术和面技术的最佳 w 值曲线

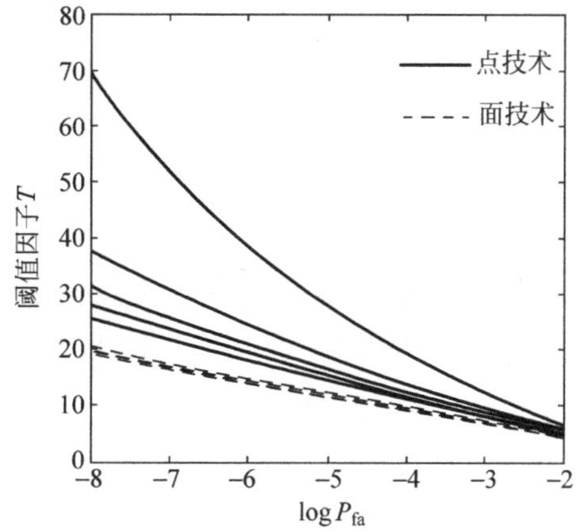

图 10.14 点技术和面技术的最佳 T 值曲线

10.4 混合 CM/L-CFAR 杂波图检测技术

经典的 Nitzberg 杂波图处理方法在出现慢速移动目标时，目标信号在几个雷达扫描周期内会占据同一杂波图单元，会导致目标“自遮蔽”现象[10]。文献[18]提出了混合 CM/L-CFAR 技术，即将空域处理与时域处理结合起来。它将几个雷达分辨单元的回波信号组合成一个杂波图单元，对杂波图单元内的回波样本基于 L 滤波器进行空域处理作为当前杂波图单元的输入信号，再对每个杂波图单元的输入信号对以往各次的结果进行迭代，得到杂波图对检测位置处杂波强度的估计。

10.4.1 基本模型

混合 CM/L-CFAR 检测器的检测原理如图 10.15 所示。

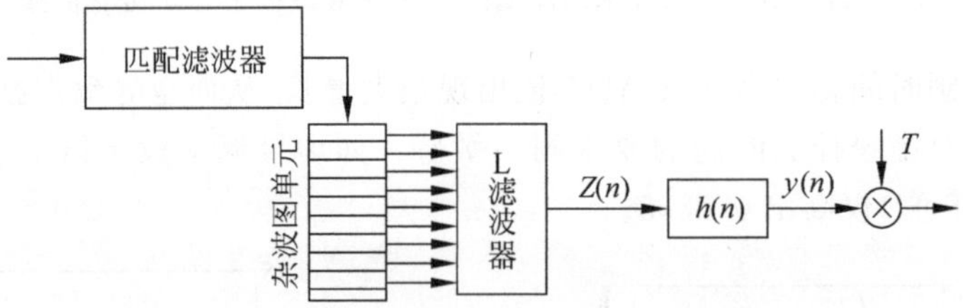

图 10.15 混合 CM/L-CFAR 检测器的检测原理框图

用 M 维向量 $\boldsymbol{x}(n)=[x_1(n),x_2(n),\cdots,x_M(n)]^{\mathrm{T}}$ 表示第 n 次扫描得到的杂波回波，M 表示邻近距离分辨单元的个数，M 个雷达分辨单元形成一个杂波图单元。对 $\boldsymbol{x}(n)$ 中元素按升序排列，得到向量 $\boldsymbol{x}_r(n)$，即

$$\boldsymbol{x}_r(n)=[x_{(1)}(n),x_{(2)}(n),\cdots,x_{(M)}(n)]^{\mathrm{T}} \tag{10.17}$$

其中，$x_{(j)}(n)$ 表示 $\boldsymbol{x}_r(n)$ 中的第 j 个有序统计量，对向量 $\boldsymbol{x}_r(n)$ 中元素基于 L 滤波器进行线性组合，得到统计量

$$z(n)=\boldsymbol{c}^{\mathrm{T}}\boldsymbol{x}_r(n) \tag{10.18}$$

其中，$\boldsymbol{c}$ 表示 M 维权值向量。对向量进行排序和线性组合运算两步操作即为 L 滤波器的处理方式。再将序列 $z(n)$ 输入脉冲响应为 $h(n)=w(1-w)^n u(n)$ 的线性离散系统中，输出 $y(n)$ 即为杂波功率水平的估计。也就是将每次扫描的处理值 $z(n)$ 作为式(10.1)描述的一阶离散系统的输入，而 $h(n)=w(1-w)^n u(n)$ 则为该系统的脉冲响应。

自适应判决准则为

$$x_i(n+1)\underset{\mathrm{H}_0}{\overset{\mathrm{H}_1}{\gtrless}} Th(n)*z(n)=Ty(n),\quad i=1,2,\cdots,M \tag{10.19}$$

其中，$*$ 表示卷积，T 为阈值因子，根据设定的虚警概率来确定。注意到，若 $M=1$，则混合 CM/L-CFAR 处理系统就是经典的 Nitzberg 杂波图技术。下面的讨论中，设 r 为将向量 $\boldsymbol{x}_r(n)$ 中从高端起删除的样本个数，则权向量 $\boldsymbol{c}$ 为 $\boldsymbol{c}=[c_1,c_2,\cdots,c_{M-r},0,0,\cdots,0]^{\mathrm{T}}$。

10.4.2 均匀背景中的性能分析

假定杂波包络服从瑞利分布，目标起伏模型为 Swerling Ⅱ 型，各分辨单元间和各次扫描间的雷达回波信号相互独立。对于平方律检波，每 n 次扫描中第 i 个距离分辨单元的输出 $q_i(n)$ 的PDF为

$$f_{q_i(n)}(t)=\begin{cases}\dfrac{1}{\mu}\exp\left(-\dfrac{t}{\mu}\right), & H_0\\[2ex] \dfrac{1}{\mu(1+\lambda)}\exp\left[-\dfrac{t}{\mu(1+\lambda)}\right], & H_1\end{cases} \tag{10.20}$$

式中，μ 代表热噪声加杂波的功率水平；λ 是信号对整个干扰(热噪声加杂波)的平均信杂比；H_0 表示单元中不存在目标的假设；H_1 表示检测单元中存在目标的假设。

根据自适应判决准则式(10.19)，在第 $n+1$ 次扫描中对第 k 个分辨单元的检测概率为

$$\begin{aligned}P_{\mathrm{d}}&=\Pr[x_k(n+1)\geqslant Ty(n)]\\&=\mathrm{E}\left\{\exp\left[-\frac{T}{\mu(1+\lambda)}\sum_{j=-\infty}^{n}\boldsymbol{c}^{\mathrm{T}}\boldsymbol{x}_r(n)h(n-j)\right]\right\}\end{aligned} \tag{10.21}$$

其中，E代表统计平均，虚警概率 P_{fa} 可令式(10.21)中信杂比 $\lambda=0$ 得到。由于每次扫描间的回波是统计独立的，式(10.21)又可写成

$$P_{\mathrm{d}}=\prod_{j=-\infty}^{n}\mathrm{E}\left\{\exp\left[-\frac{T}{\mu(1+\lambda)}\boldsymbol{c}^{\mathrm{T}}\boldsymbol{x}_r(n)h(n-j)\right]\right\} \tag{10.22}$$

进一步化简式(10.22)的关键在于对排序统计量的线性组合 $\boldsymbol{c}^{\mathrm{T}}\boldsymbol{x}_r(n)$ 的处理。作如下的变量转换，有

$$\boldsymbol{c}^{\mathrm{T}}\boldsymbol{x}_r(j)=\boldsymbol{v}^{\mathrm{T}}\boldsymbol{z}(j) \tag{10.23}$$

其中，向量 $\boldsymbol{z}(j)$ 中的第 m 个分量定义为

$$z_m(j)=\begin{cases}Mx_{(1)}(j), & m=1\\ (M+1-m)[x_{(m)}(j)-x_{(m-1)}(j)], & 2\leqslant m\leqslant M\end{cases} \tag{10.24}$$

向量 $\boldsymbol{v}$ 取决于向量 $\boldsymbol{c}$，向量 $\boldsymbol{v}$ 中的第 m 个分量为

$$v_m=\frac{1}{M+1-m}\sum_{k=m}^{M}c_k \tag{10.25}$$

由文献[28]可知，向量 $\boldsymbol{z}(j)$ 中的每个分量 $z_m(j)$ 均具有相同的分布 $f(t)=\dfrac{1}{\mu}\exp\left(-\dfrac{t}{\mu}\right)u(t)$。将式(10.23)代入式(10.22)，再取统计平均可得混合CM/L-CFAR检测器的检测概率

$$P_{\mathrm{d}}=\prod_{j=-\infty}^{n}\prod_{m=1}^{M}\frac{1}{1+\dfrac{T}{1+\lambda}v_m h(n-j)} \tag{10.26}$$

令式(10.26)中信杂比 $\lambda=0$，可得到混合CM/L-CFAR检测器的虚警概率为

$$P_{\mathrm{fa}}=\prod_{j=-\infty}^{n}\prod_{m=1}^{M}\frac{1}{1+Tv_m h(n-j)} \tag{10.27}$$

注意到，若令 $h(n)=\delta(n)$，式(10.26)和式(10.27)将分别变为

$$P_{\mathrm{d}}=\prod_{m=1}^{M}\frac{1}{1+\dfrac{T}{1+\lambda}v_m} \tag{10.28}$$

$$P_{\mathrm{fa}}=\prod_{m=1}^{M}\frac{1}{1+Tv_m} \tag{10.29}$$

在这种情况下,杂波图技术仅仅利用空域临近参考单元的样本进行杂波强度估计,而没有进行时域的迭代处理,就是基于L滤波器处理的情形。若令 $M=1$,式(10.26)和式(10.27)将变为经典的Nitzberg杂波图处理方法[1]。

混合CM/L-CFAR检测器先对空域 M 个临近分辨单元采用L滤波器进行空域处理,再对所得结果在时域进行迭代处理。采用L滤波器进行空域处理,首先要对 M 个分辨单元进行排序,再对排序样本进行线性加权组合,其中存在着如何确定权向量 $\boldsymbol{c}$ 的问题。文献[18]利用拉格朗日乘子法,确定出权向量 $\boldsymbol{c}$ 每个分量 c_m 的最优加权值为

$$c_m=\begin{cases}\dfrac{1}{M-r}, & m\neq M-r\\[2mm] \dfrac{r+1}{M-r}, & m=M-r\end{cases} \tag{10.30}$$

假定 $P_{\mathrm{fa}}=10^{-4}$,图10.16～图10.19分别给出了不同参数条件下,混合CM/L-CFAR检测器的检测性能曲线。图10.16对应没有筛除临近单元样本的情况($r=0$),且不同临近分辨单元数目 $M=1,2,4,8,16$,衰减因子 $w=0.2$,图10.16中最右边一条曲线对应 $M=1$。可以看出,临近分辨单元个数 M 越多,检测器的性能越好。图10.17对应衰减因子 $w=0.2$,临近分辨单元数目 $M=8$,筛除不同数目临近单元样本 $r=0,1,2,3,4,5,6,7$ 的情况,图10.17中最右边一条曲线对应 $r=7$。可以看出,筛除掉临近单元样本的个数 r 越多,检测器的损失越大。图10.18对应 $M=8$、$w=1$(没有时域迭代处理时),筛除不同数目临近单元样本 $r=0,1,2,3,4,5,6,7$ 的情况,图10.18中最右边一条曲线对应 $r=7$。可以看出,筛除掉临近单元样本的个数 r 越多,检测器的损失越大;当没有时域迭代处理,筛除样本个数多(r 大)时,检测器的检测损失比采用时域迭代处理的混合CM/L-CFAR检测器要大得多。因而,采用混合CM/L-CFAR检测器可以对临近参考单元样本进行较多的筛除而不致使检测性能过分下降。图10.19对应 $M=8$、$r=0$,采用不同的衰减因子 $w=0.1,0,3,0.5,0.7,0.9,1$ 的情况,图10.19中最右边一条曲线对应 $w=1$。可以看出,衰减因子越小,检测性能越好,衰减因子越小意味着时域积累的时间越长。

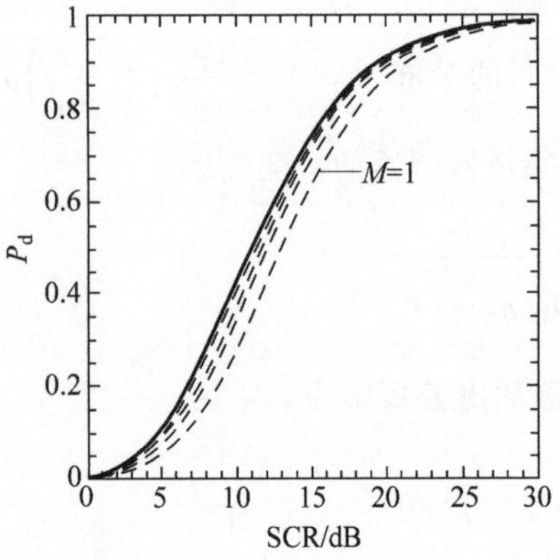

图10.16 混合CM/L-CFAR检测器在均匀背景中的检测性能曲线
($M=1,2,4,8,16,w=0.2,r=0,P_{\mathrm{fa}}=10^{-4}$)

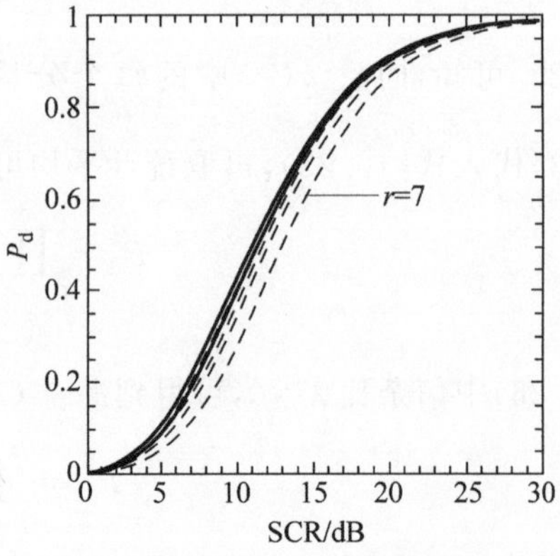

图10.17 混合CM/L-CFAR检测器在均匀背景中的检测性能曲线
($r=0,1,2,3,4,5,6,7,w=0.2,M=8,P_{\mathrm{fa}}=10^{-4}$)

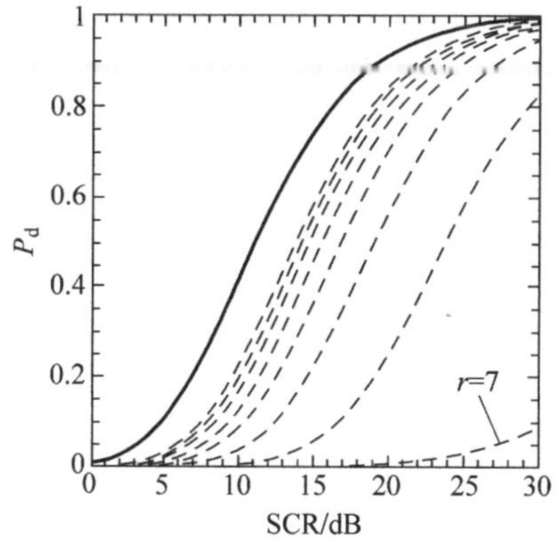

图 10.18 混合 CM/L-CFAR 检测器在均匀背景中的检测性能曲线

($r=0,1,2,3,4,5,6,7,w=1,M=8,P_{\mathrm{fa}}=10^{-4}$)

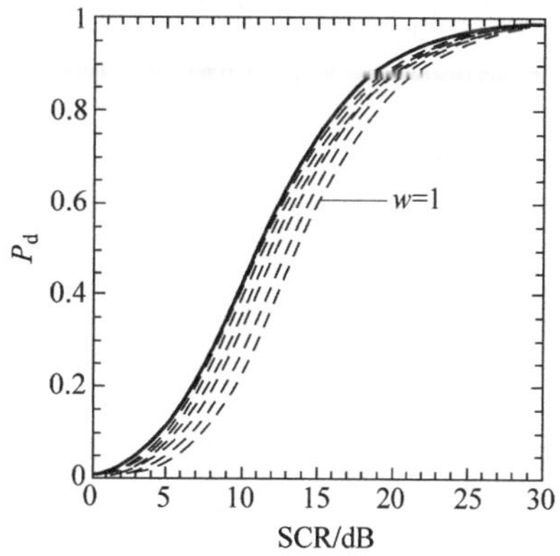

图 10.19 混合 CM/L-CFAR 检测器在均匀背景中的检测性能曲线

($w=0.1,0.3,0.5,0.7,0.9,1,r=0,M=8,P_{\mathrm{fa}}=10^{-4}$)

10.4.3 存在干扰目标时的性能分析

Nitzberg 杂波图处理方法在出现慢速移动目标时，目标信号在几个雷达扫描周期内会占据同一杂波图单元，会导致目标“自遮蔽”现象。而混合 CM/L-CFAR 技术是几个雷达分辨单元的回波信号组合成的一个杂波图单元，对杂波图单元内的回波样本基于 L 滤波器处理后再进行时域迭代处理，这样就可以通过 L 滤波器的功能将进入的杂波图单元内的干扰目标信号剔除掉。

假定从第 $n+1$ 次扫描周期起，有 N_i 个目标进入杂波图单元中并持续 L 个扫描周期，且各个目标的回波在各扫描周期是不相关的且具有相同的功率强度。第 k 个距离单元在第$(n+L+1)$次扫描的检测概率为

$$\begin{aligned}\Pr[q_k(n+L+1)\geqslant Ty(n+L)\mid \mathrm{H}_1]&=\mathrm{E}\left\{\exp\left[-\frac{T}{\mu(1+\lambda)}\sum_{j=-\infty}^{n+L}\boldsymbol{c}^{\mathrm{T}}\boldsymbol{x}_r(j)h(n+L-j)\right]\right\}\\&=\prod_{j=-\infty}^{n}\mathrm{E}\left\{\exp\left[-\frac{T\boldsymbol{c}^{\mathrm{T}}\boldsymbol{x}_r(j)h(n+L-j)}{\mu(1+\lambda)}\right]\right\}\times\\&\quad\prod_{j=n+1}^{n+L}\mathrm{E}\left\{\exp\left[-\frac{T\boldsymbol{c}^{\mathrm{T}}\boldsymbol{x}_r(j)h(n+L-j)}{\mu(1+\lambda)}\right]\right\}\end{aligned}\tag{10.31}$$

注意到，当 $j\leqslant n$ 时，$\boldsymbol{c}^{\mathrm{T}}\boldsymbol{x}_r(j)$表示均匀背景中有序统计量的线性组合，因而有

$$\prod_{j=-\infty}^{n}\mathrm{E}\left[\exp\left(-\frac{T}{\mu(1+\lambda)}\boldsymbol{c}^{\mathrm{T}}\boldsymbol{x}_r(j)h(n+L-j)\right)\right]=\prod_{j=-\infty}^{n}\prod_{m=1}^{M}\frac{1}{1+\dfrac{T}{1+\lambda}v_m h(n+L-j)}\tag{10.32}$$

但是，式(10.31)右边第二项中的有序统计量 $\boldsymbol{x}_r(j)$来自于干扰目标和噪声样本的两种母体分布。在这种情况下，计算式(10.31)右边第二项中的统计平均是非常麻烦的，这时可以考虑一种极限情况来进行衡量。即假定干扰目标的强度足够大，因此 N_i 个干扰目标总是占据有序统计量的前 N_i 个最大样本位置。基于以上假设，将 M 维有序统计量 $\boldsymbol{x}_r(j)$分为两组，其中一组为只包含杂波样本的 $M-N_i$ 个有序统计量，另一组为包含杂波和干扰信号的 N_i 个有序统计量。当 $j\geqslant n+1$ 时，有

$$\boldsymbol{c}^{\mathrm{T}}\boldsymbol{x}_r(j)=\sum_{k=1}^{M-N_i}c_k x_{(k)}(j)+\sum_{k=M-N_i+1}^{M}c_k x_{(k)}(j)=\sum_{k=1}^{M-N_i}\zeta_k\eta_k+\sum_{k=1}^{N_i}\beta_k\theta_k \tag{10.33}$$

其中

$$\zeta_k=\frac{1}{M-N_i-k+1}\sum_{l=k}^{M-N_i}c_l,\quad k=1,2,\cdots,M-N_i$$

$$\beta_k=\frac{1}{N_i-k+1}\sum_{l=M-N_i+k}^{M}c_l,\quad k=1,2,\cdots,N_i \tag{10.34}$$

式中,η_k 和 θ_k 分别为相互统计独立的服从指数分布的随机变量,且平均功率分别为 μ 和 $\mu(1+\lambda)$(假定干扰目标和主目标信号具有相同的功率强度)。根据式(10.31)~式(10.34),混合 CM/L-CFAR 检测器在干扰目标环境中的检测概率变为

$$P_{\mathrm{d}}=\prod_{j=-\infty}^{n}\prod_{m=1}^{M}\frac{1}{1+\dfrac{T}{1+\lambda}v_m h(n+L-j)}\times\prod_{j=n+1}^{n+L}\prod_{m=1}^{M-N_i}\frac{1}{1+\dfrac{T}{1+\lambda}\zeta_m h(n+L-j)}\times$$
$$\prod_{j=n+1}^{n+L}\prod_{m=1}^{N_i}\frac{1}{1+T\beta_m h(n+L-j)} \tag{10.35}$$

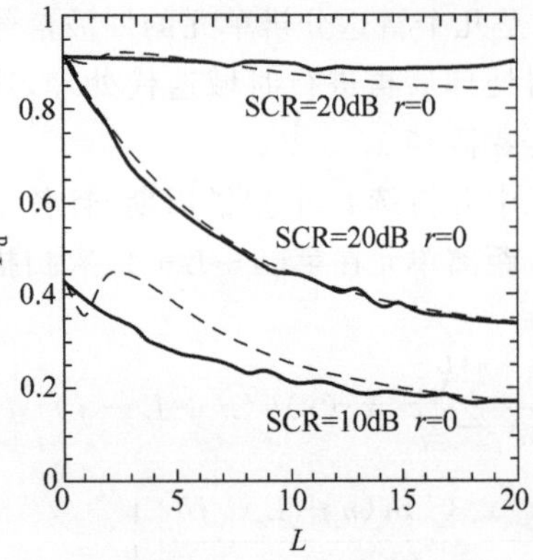

图 10.20 检测概率与扫描周期数 L 的关系

当 $L\to\infty$ 时,滤波器脉冲响应趋于零,式(10.35)变为

$$P_{\mathrm{d}}=\prod_{j=0}^{\infty}\prod_{m=1}^{M-N_i}\frac{1}{1+\dfrac{T}{1+\lambda}\zeta_m h(j)}\prod_{j=0}^{\infty}\prod_{m=1}^{N_i}\frac{1}{1+T\beta_m h(j)} \tag{10.36}$$

假定只有一个目标在 L 次扫描周期一直停留在杂波图单元中,图 10.20 给出了式(10.35)的计算结果和计算机仿真的曲线,图 10.20 中虚线为仿真结果,实线为式(10.35)的计算结果。可以看出,当 SCR=20dB 时,式(10.35)的计算结果和仿真曲线吻合得很好;当 SCR=10dB 时,二者间的差别较大。从图 10.20 中可以看出,当只有一个目标停留在杂波图单元中时,只要筛选掉 $r=1$ 个样本,就可使检测器的性能几乎不受影响。如果将混合 CM/L-CFAR 技术与参考单元筛选处理相结合[29],可进一步提高在多目标环境下的检测性能。

10.5 双参数杂波图检测技术

对于对数正态分布等包含形状和尺度双参数的杂波情况,前述单参数杂波图方法会存在一定的性能损失[30-31],为此文献[26]提出了双参数杂波图设计思路,通过样本均值和方差估计值共同设置检测门限,以提高虚警控制性能。本节将以对数放大预处理为例介绍双参数杂波图检测技术。

10.5.1 双参数杂波图基本模型

为了对比分析,首先简要介绍单参数杂波图方法。为了获得较大的动态范围,假定输入信号 Z_n 经

对数放大器后为 $X_n=\ln(Z_n)$。

在杂波为主的环境下，一般假定无目标情况下 X_n 服从正态分布，其 PDF 可表示为

$$f_X(x)=\frac{1}{\sqrt{\pi\ln k}}\exp\left[-\frac{(x-\ln\sqrt{P_c/k})^2}{\ln k}\right] \tag{10.37}$$

其 PDF 右侧的拖尾扩展程度形状因子 k 可描述为

$$k=\exp\left(\frac{\sigma^2}{2}\right) \tag{10.38}$$

其中，σ^2 对应于正态分布的方差，P_c 表示对数放大之前的杂波功率。

在噪声为主的环境下，一般假定无目标情况下 X_n 服从 Gumbel 分布，其 PDF 可表示为

$$f_X(x)=\frac{1}{b}\exp\left(\frac{x-a}{b}\right)\exp\left[-\exp\left(\frac{x-a}{b}\right)\right] \tag{10.39}$$

其中

$$a=\frac{\ln(2P_n)}{\ln 100} \tag{10.40}$$

$$b=\frac{1}{\ln 100} \tag{10.41}$$

其中，P_n 表示对数放大之前的噪声功率。

上述两种情况下，相应的 PDF 均为位置-尺度类型。但当杂波和噪声共存且均不占优的杂噪混合背景下，其 PDF 将不具有位置-尺度结构。单参数杂波图方法的典型结构如图 10.21 所示[26]。其中的虚框中采用的是单极点滤波器，可实现对输入信号 $X_n(k)$ 均值的渐进无偏估计，本质上与图 10.1 中的 Nitzberg 杂波图处理方式一致。其中，α 可被认为是杂波图累积的等效脉冲数，即表示杂波图的存储长度。该方法只涉及 PDF 中单个参数的估计，且在单参数的瑞利杂波背景下具有 CFAR 特性，因此常常被称为单参数杂波图方法。在图 10.21 中，通过对单极点滤波器的输出进行 μ 的偏移获得检测阈值。当形状参数 k 已知时，在杂波为主的杂噪混合环境下，杂噪比 CNR＞15dB 时单参数杂波图可保持 CFAR 特性，但当杂噪比进一步减小时，将会出现一定的检测损失。

为了应对单参数杂波图面临的上述难题，文献[26]给出了一种双参数杂波图的设计思路，其典型结构如图 10.22 所示。其中用了两个单极点滤波器，这两个滤波器分别用来估计输入信号 X 的均值和均方值，进而估计相应的标准差，而检测阈值 T 则设定为均值估计 $\hat{m}$ 和标准差估计 $\hat{\sigma}$ 的线性组合，即

$$T=\hat{m}+H\hat{\sigma} \tag{10.42}$$

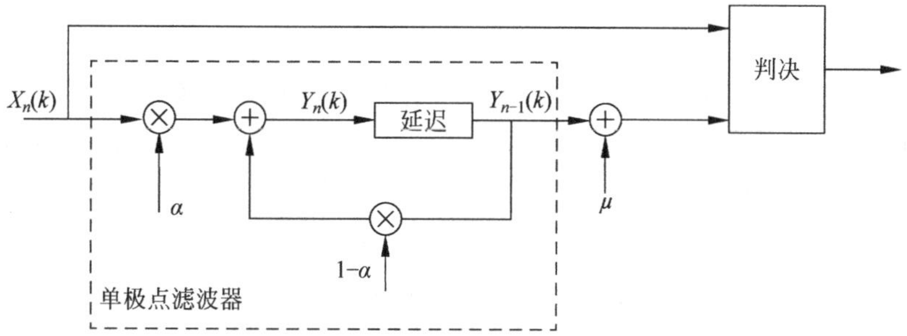

图 10.21　单参数杂波图典型结构示意图

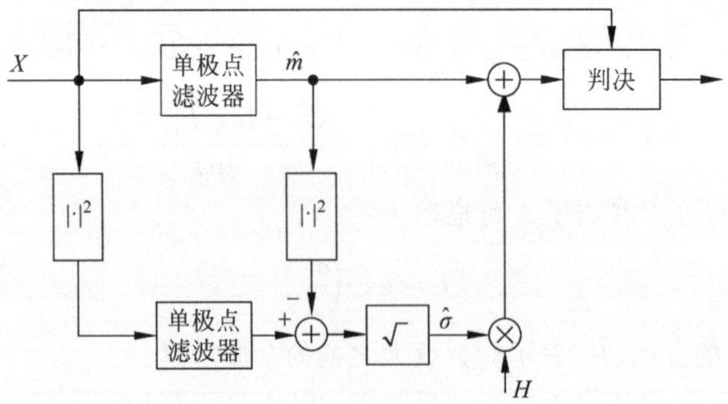

图 10.22 双参数杂波图典型结构示意

对于具有位置-尺度型 PDF 的杂波或噪声分布来说,其检测阈值可表示为位置参数估计值 $\hat{\theta}_L$ 和尺度参数估计值 $\hat{\theta}_S$ 的线性组合,即

$$T=\hat{\theta}_L+\alpha\hat{\theta}_S \tag{10.43}$$

当上述两个估计值具有等变性时,相应的检测阈值可保持 CFAR 特性[32]。估计值 $\hat{\theta}_L$ 和 $\hat{\theta}_S$ 的等变性具体表现为:对于任意样本序列 $x(n)$、正常数 c 和常数 d,基于样本序列 $x(n)$的估计值 $\hat{\theta}_L$ 和 $\hat{\theta}_S$ 均满足

$$\begin{aligned}\hat{\theta}_L\{c[x(n)]+d\}&=c\hat{\theta}_L\{[x(n)]\}+d\\ \hat{\theta}_S\{c[x(n)]+d\}&=c\hat{\theta}_S\{[x(n)]\}\end{aligned} \tag{10.44}$$

例如,对于对数正态杂波来说,经过对数变换后,其相应的分布变为正态分布,位置参数和尺度参数分别对应均值和标准差,此时式(10.42)和式(10.43)是等价的。而对于噪声为主的环境来说,对数变换后相应的分布变为 Gumbel 分布,其均值和标准差与位置参数 θ_L 和尺度参数 θ_S 的关系可表示为

$$m=\theta_L+\gamma\theta_S \tag{10.45}$$

$$\sigma=\frac{\pi}{\sqrt{6}}\theta_S \tag{10.46}$$

其中,γ 表示欧拉常数。此时式(10.43)对应的检测阈值可表示为

$$T=\hat{m}+\frac{\sqrt{6}}{\pi}(\alpha-\gamma)\hat{\sigma} \tag{10.47}$$

式(10.47)相当于式(10.42)中 $H=\sqrt{6}(\alpha-\gamma)/\pi$ 时的情形,双参数杂波图方法与位置-尺度型阈值是兼容的。需要注意的是,虽然式(10.42)的双参数杂波图阈值能同时适应对数正态杂波和噪声为主的环境,但不同情况下具体的 H 值是不同的。

相关研究表明[26],在杂波和噪声共存且均不占优的杂噪混合背景下,其混合后的 PDF 将不具有位置-尺度结构,式(10.42)的阈值将失去理论 CFAR 特性,但若能根据整体干扰功率估计值自适应调整 H 值,则双参数杂波图仍能获得较好的虚警控制性能。另外,在杂噪比 CNR>20dB 时双参数杂波图可保持 CFAR 特性;与单参数杂波图不同,在噪声为主的低杂噪比 CNR<−12dB 环境下,双参数杂波图

仍能获得 CFAR 能力。在形状参数 k 在[1.3,2.0]区间变化时,相比于单参数杂波图,双参数杂波图的虚警概率控制误差从 4 个数量级减小为 2 个数量级。

10.5.2　对目标自遮蔽的处理

虽然双参数杂波图能快速适应环境的变化,但过快的自适应变化也将导致对慢速目标的自遮蔽现象。为了改善对慢速目标自遮蔽效应,文献[27]对双参数杂波图进行了进一步改进,改进后的双参数杂波图典型结构如图 10.23 所示。为了对抗自遮蔽,杂波图结构主要增加了两个处理模块:移动窗口限制器和阈值锁定器。

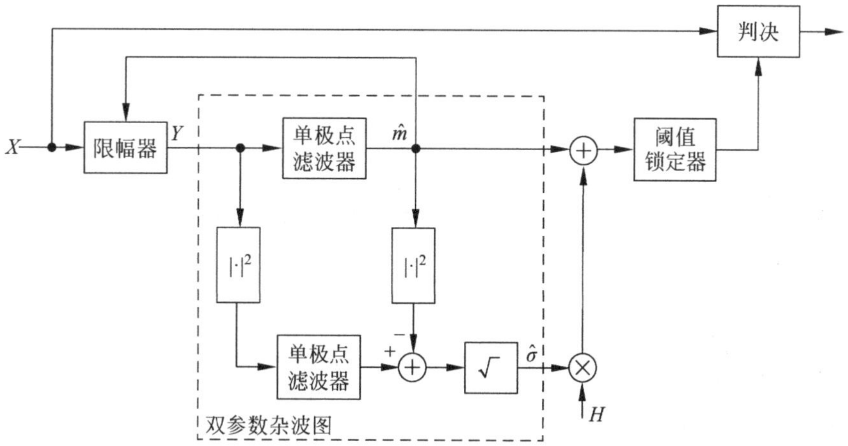

图 10.23　基于自遮蔽处理的双参数杂波图典型结构示意

移动窗口限制器位于杂波图的输入端,其本质为限幅器,它可限制单次扫描中检测阈值的最大变化量,其基于这样一种假定:过强输入信号的出现可能与目标有关,因此不应按原规则提高检测阈值。经过对数放大后获得幅值 X 作为限制器的输入,其输出 Y 按如下规则设定:

$$Y=\begin{cases}\hat{m}-u, & X<\hat{m}-u\\ X, & \hat{m}-u\leqslant X\leqslant\hat{m}+u\\ \hat{m}+u, & X>\hat{m}+u\end{cases} \tag{10.48}$$

其中,选择以 $\hat{m}$ 为中心的移动窗口,相应宽度为 $2u$,如此可使杂波超过窗口限制的概率 P 保持在较低水平。如果 P 过低,则移动窗口限制器难以有效发挥作用;但若输入过频地超过窗口限制,相应参数的估计值则发生相应改变;窗口限制的参数设置需重点考虑标准差估计值 $\hat{\sigma}$ 的因素。

阈值锁定器位于杂波图的输出端,它可抑制连续 r 次检测出目标后检测阈值的上升。因为连续检测出信号的情况极有可能由慢速目标引起,为了防止自遮蔽,不应提高阈值;而在连续多次未检测到目标信号之后,阈值锁定应被移除。需要注意的是,杂波激活阈值锁定器的概率应设置得很小。

相关研究表明[27],在对抗目标自遮蔽效应方面,虽然阈值锁定器可以单独使用,但它与移动窗口限制器相结合更为有效。在连续三次检测到目标后锁定阈值,在杂波为主的环境下,将由杂波引起激活输入端限制的概率设置为 10^{-4},即采用不太紧的移动窗口限制器,则由自遮蔽引起的检测损失可降低到 4dB 以下,而在噪声为主的环境下,由自遮蔽引起的检测损失可降低到 2dB 以下。

10.6 比较和总结

本章研究了杂波图 CFAR 检测技术，这一处理技术适用于在空域上变化激烈而时域上变化平稳的杂波环境。经典的杂波图处理技术是 Nitzberg 杂波图处理方法，这一方法在时间上对以往各次雷达回波的测量值进行迭代处理，对杂波强度进行估计并设置门限。Nitzberg 杂波图处理方法利用单个雷达分辨单元回波信号进行处理，属于杂波图处理的点技术。一般来说，在满足虚警指标要求的情况下，尽量选择稍小的衰减因子以提高检测性能；在杂波尖峰较多的非高斯环境下点技术存在较大的性能损失。为改善和提高杂波图技术的检测性能和克服慢速移动目标引起的“自遮蔽”现象，可采用杂波图面技术处理方式，该技术还有助于检测切向运动目标。10.3 节对杂波图单元平均 CFAR 面技术和点技术进行了详细的性能分析，可以看到，杂波图单元平均 CFAR 面技术比点技术可以带来检测性能方面的改善；另外，对于衰减因子的选择应该在检测性能和适应杂波环境之间做折中考虑。而杂波图单元平均 CFAR 面技术也增加了干扰目标进入杂波强度估计的概率。10.4 节讨论的混合 CM/L-CFAR 检测器则可以在一定程度上克服干扰目标引起的“自遮蔽”现象，只要将进入杂波图单元中的干扰目标信号剔除，混合 CM/L-CFAR 检测器对主目标的检测就几乎不受影响。对于对数正态分布等包含形状和尺度双参数的杂波情况，10.5 节讨论的双参数杂波图技术能获得更好的虚警控制性能，但其对慢速目标自遮蔽效应缺乏鲁棒性，在杂波图处理的前端和后端分别设置移动窗口限制器和阈值锁定器，为对抗慢速目标自遮蔽提供了处理思路。

在许多实际情况中，雷达杂波的统计分布会背离高斯分布而呈现非高斯特征[15]。文献[19]将混合 CM/L-CFAR 检测器的处理方法推广到了非高斯位置-尺度分布中。事实上，其对位置和尺度两参数的估计采用的是最优线性无偏估计，估计过程与两参数 CFAR 技术中类似，故不再赘述。关于 Nitzberg 杂波图技术在韦布尔背景下的性能分析表明[23-24]，Nitzberg 杂波图技术在拖尾严重的非高斯杂波中检测性能恶化。双参数杂波图技术为对数正态[27]和韦布尔[25]等双参数分布下的 CFAR 检测提供了可能，并在算法硬件实现[2,14]、机场跑道异物检测[25]、超宽带雷达人体检测[10]等方面得到了一定的应用。基于多脉冲非相参处理方式，将杂波图技术与二进制积累相结合[33-35]，一定程度上可改善杂波图处理的检测性能。

随着脉冲多普勒模式的广泛应用，利用不同多普勒通道进行多通道杂波图处理研究[5-6,13]，已得到越来越多的重视。在实际工程应用中，杂波图技术常常作为复杂杂波环境中的预处理手段[8,10,36-37]，其中衰减因子的自适应选择[15]仍是值得重视的问题。面对多目标干扰和慢速移动目标“自遮蔽”等难题，如何将参考单元筛选技术与杂波图处理进行有效结合[29,38]，是值得积极探索的研究方向。另外，分布式检测中杂波图 CFAR 处理方面[3,12]仍有许多细致的工作有待开展。

参考文献

[1] Nitzberg R. Clutter map CFAR analysis[J]. IEEE Trans. on AES，1986，22(4)：419-421.

[2] 林彦，彭应宁，王秀坛，等. 一种双模式双参数杂波图恒虚警检测器的实现[J]. 系统工程与电子技术，2004，25(2)：133-136.

[3] Benseddik H，Cherki B，Hamadouche M，et al. FPGA-based real-time implementation of distributed system CA-

CFAR and clutter MAP-CFAR with noncoherent integration for radar detection[C]. Tangiers: International Conference on Multimedia Computing and Systems, 2012.

[4] 徐丽. 脉冲多普勒雷达信号处理并行技术研究[D]. 成都：电子科技大学，2019.

[5] 丁施健."低小慢"目标探测雷达信号处理机的设计[D]. 南京：南京理工大学，2019.

[6] 曾志云. 地面侦察雷达杂波抑制技术研究[D]. 南京：南京理工大学，2012.

[7] 王艳丽. 警戒雷达的杂波图组合实现方法[J]. 信息化研究，2019，45(03)：8-10,30.

[8] 王宝帅，兰竹，李正杰，等. 毫米波雷达机场跑道异物分层检测算法[J]. 电子与信息学报，2018，40(11)：2676-2683.

[9] 张鑫. 线性调频航管雷达风电场杂波抑制及目标检测[D]. 天津：中国民航大学，2019.

[10] Lee B H, Lee S, Yoon Y J, et al. Adaptive clutter suppression algorithm for human detection using IR-UWB radar[C]. Glasgow: 2017 IEEE Sensors, 2017.

[11] 练学辉，丁春. 一种用于VTS导航雷达的杂波图恒虚警率处理技术[J]. 雷达与对抗，2013，33(3)：35-37，49.

[12] Bouchlaghem H E, Hamadouche M, Soltani F, et al. Adaptive clutter-map CFAR detection in distributed sensor networks[J]. AEU-Int. J. Electron. Commun., 2016, 70(9): 1288-1294.

[13] 剡熠琛. 复杂背景下的小目标检测技术[D]. 西安：西安电子科技大学，2020.

[14] 毛云. 雷达杂波图CFAR检测算法研究及实现[D]. 西安：西安电子科技大学，2018.

[15] 张海龙，李赛辉，张宁，等. 某雷达杂波数据分析及杂波图技术研究[J]. 雷达与对抗，2020，40(1)：22-26.

[16] 刘宁. 地杂波背景下雷达目标检测方法的研究[D]. 西安：西安电子科技大学，2018.

[17] Levanon N. Numerically efficient calculations of clutter map CFAR performance[J]. IEEE Trans. on AES, 1987, 23(6): 813-814.

[18] Lops M. Hybrid clutter-map/L-CFAR procedure for clutter rejection in nonhomogeneous environment[J]. IEE Proc.-Radar, Sonar and Navigation, 1996, 143(4): 239-245.

[19] Conte E, Lops M, Tulino A M. Hybrid procedure for CFAR in non-Gaussian clutter[J]. IEE Proc.-Radar, Sonar and Navigation, 1997, 144(6): 361-369.

[20] 沈福民，刘峥. 杂波图CFAR平面检测技术[J]. 系统工程与电子技术，1996，18(7)：9-14.

[21] 何友，刘永，孟祥伟. 杂波图CFAR平面技术在均匀背景中的性能[J]. 电子学报，1999，27(3)：119-120.

[22] 刘永，何友，孟祥伟. 幅度杂波图恒虚警处理中的点技术研究[J]. 系统工程与电子技术，1998，20(9)：7-10.

[23] Hamadouche M, Barakat M, Khodja M. Analysis of the clutter CFAR map in Weibull clutter[J]. Signal Process., 2000, 80: 117-123.

[24] Meng X W. Performance analysis of Nitzberg's clutter map for Weibull distribution[J]. Digital Signal Process., 2010, 20(3): 916-922.

[25] 吴静. 机场跑道异物检测技术研究与实现[D]. 成都：电子科技大学，2014.

[26] Naldi M, Beccarini A. Threshold control for a millimetre-wave miniradar: biparametric vs. monoparametric clutter maps[C]. Edinburgh: Proc. of the 1997 International Radar Conference, 1997.

[27] Naldi M. False alarm control and self-masking avoidance by a biparametric clutter map in a mixed interference environment[J]. IEE Proc.-Radar, Sonar and Navigation, 1999, 146(4): 195-200.

[28] David H A. Order Statistics[M]. New York: Wiley, 1981.

[29] Zhang R L, Sheng W X, Ma X F, et al. Clutter map CFAR detector based on maximal resolution cell[J]. Signal Image Video Process., 2015, 9(5): 1151-1162.

[30] Wang C Y, Pan R Y. Clutter suppression and target detection based on biparametric clutter map CFAR[C]. Hangzhou: IET International Radar Conference, 2015.

[31] 王辉辉，袁子乔. 一种自适应双参数杂波图检测方法[J]. 火控雷达技术，2020，49(2)：54-59，65.

[32] Conte E, M. Bisceglie D, Lops M. A clutter-map procedure for CFAR in Weibull environment[C]. Paris: International Conference on Radar, 1994.

[33] Meng X W. Performance of clutter map with binary integration against Weibull background[J]. AEU-International

Journal of Electronics and Communications，2013，67：611-615.

[34] 安文，孟祥伟．一种改进的杂波图检测方法[J].武汉大学学报：信息科学版，2015，40(9)：1176-1179，1208.

[35] 王蓓．基于杂波图的恒虚警处理技术研究[D].西安：西安电子科技大学，2018.

[36] Lai Y K. Foreign object debris detection method based on fractional Fourier transform for millimeter-wave radar[J]. Journal of Applied Remote Sensing，2020，14(1)：1-15.

[37] 张林．海杂波中目标非参量广义符号恒虚警检测算法研究[D].烟台：海军航空工程学院，2012.

[38] Meng X W，Qu F Y. Adaptive clutter map detector in nonhomogeneous environment[C]. Beijing：Proc. of 2010 IEEE International Conference on Signal Processing，2010.

第 11 章 变换域 CFAR 处理

CHAPTER 11

11.1 引言

在雷达、声呐等探测系统中对微弱目标的检测一直是比较困难的，主要原因是这类目标的回波强度小，信号微弱，幅度被噪声或杂波所淹没，经典的时域 CFAR 检测难以有效检测出目标。为了检测强背景中的运动目标，除了常规的杂波抑制、抗干扰和降低系统噪声等措施，一种比较有效的方法是将时域信号变换到频域、时频域及小波域等其他变换域进行 CFAR 处理，也称作变换域 CFAR 检测[1]。该方法的本质是，利用背景噪声或杂波与目标信号在变换域的能量分布不同或特征差异来提高检测性能。例如，相参积累增强目标回波，用时间换取能量，改善积累增益后再检测。本章将重点讨论频域、小波域、分数阶傅里叶变换域、Hilbert-Huang 变换域和稀疏表示域的目标检测器设计及相应的检测方法。

11.2 频域 CFAR 检测

频域 CFAR 处理研究较多，其背景杂波包括接收机热噪声、旁瓣杂波、主瓣杂波剩余。对瑞利分布的地物、气象和海浪杂波进行中频 CFAR 处理已有许多方法，如对数中放、时间增益控制、瞬时自动增益控制、近程自动增益控制等。频域邻近单元样本的“宽带放大-限幅-窄带放大”电路是利用硬限幅对瑞利分布的地物、气象和海洋杂波进行中频 CFAR 处理的最早方法。11.2.1 节讨论信号和杂波噪声等干扰在频域中的不同特性。机载脉冲多普勒(Pulse Doppler，PD)体制雷达的频域 CFAR 检测器普遍先以多普勒滤波器组对输入信号进行相参积累(DFT 变换)，然后再作相应的 CFAR 处理[1-3]。美国 F-15、F-16 机载雷达已经在多普勒滤波器组后采用了频域 CFAR 技术[4-5]，MTI-FFT-CFAR 是这种雷达信号处理机的典型结构[6-7]，11.2.2 节和 11.2.3 节对该处理器进行较详细的讨论。11.2.5 节分析采用奇偶处理对频域 CFAR 检测器性能的影响。

11.2.1 信号和杂波噪声的离散傅里叶变换处理

在采样周期为 T 的均匀采样下，N 点序列 $x(n)(n=0,1,\cdots,N-1)$ 的离散傅里叶变换(Discret Fourier Transform，DFT)用 N 点序列表示为

$$X(k)=\sum_{n=0}^{N-1}x(n)\mathrm{e}^{-\mathrm{j}2\pi nk/N},\quad k=0,1,\cdots,N-1 \tag{11.1}$$

如果 $x(n)$ 是输入信号复包络样本值，则 $x(n)\mathrm{e}^{-\mathrm{j}2\pi nk/N}$ 可看作输入信号频率降低 k/NT 后的复包

络样本值，$X(k)$则看作移频后的输入信号复包络样本序列的相参积累。因此，N 点 DFT 相当于 N 路相参积累器，信号在各支路里频移量不同，在 k 支路里的频移量为 k/NT。

如果信号振幅为 A，多普勒频率为 f_d，初相角为 θ，则其复包络样本序列为 $x_S(n)=A\mathrm{e}^{\mathrm{j}(2\pi f_d nT+\theta)}$ $(n=0,1,\cdots,N-1)$。当 $f_d=k/NT$ 时，其 DFT 系数为

$$X_S(k)=NA\mathrm{e}^{\mathrm{j}\theta} \tag{11.2}$$

$|X_S(k)|$在采用 N 个值同相叠加时达到最大，这是信号与相参积累器匹配的情况。当 $f_d\neq k/NT$ 时，$|X_S(k)|$减小，这就是 DFT 的频率选择性。由于 DFT 相当于多路相参积累器，因此在分析检测性能时可把 DFT 当成相参积累来考虑。

杂波复包络样本序列为 $x_C(n)=A_n\mathrm{e}^{\mathrm{j}\varphi_n}$ $(n=0,1,\cdots,N-1)$，其中 A_n 和 φ_n 分别表示振幅和相位。假设杂波样本是 IID 的，杂波的振幅和相位相互独立，杂波相位在$[0,2\pi]$上均匀分布。杂波复包络样本序列的 DFT 为

$$X_C(k)=\sum_{n=0}^{N-1}A_n\mathrm{e}^{\mathrm{j}(\varphi_n-2\pi nk/N)} \tag{11.3}$$

令 $\varphi_n=\varphi_n-2\pi nk/N$，则

$$X_C(k)=\sum_{n=0}^{N-1}A_n\mathrm{e}^{\mathrm{j}\varphi_n} \tag{11.4}$$

由于 φ_n 也在$[0,2\pi]$上均匀分布，并且 φ_n 与 A_n 也互相独立，各 $A_n\mathrm{e}^{\mathrm{j}\varphi_n}$ 也是 IID 的，利用中心极限定理可以证明 $X_C(k)$是渐近复高斯的，模$|X_C(k)|$是渐近瑞利分布的。

当 N 有限时，若输入杂波是高斯的，即其包络是瑞利分布时，根据高斯过程的线性变换仍是高斯过程这个已知结论，可知$|X_C(k)|$服从瑞利分布。当 N 不大时，如果输入杂波是非高斯的，其包络不是瑞利分布的，$|X_C(k)|$便偏离瑞利分布。针对某种输入杂波的概率分布函数和具体的有限 N 值，可以通过 Monte Carlo 仿真求得经 DFT 后概率分布的近似形式。$|X_C(0)|$近似服从韦布尔分布。$|X_C(0)|$的概率分布函数为

$$F(x)=1-\exp\left[-(\ln 2)\left(\frac{x}{x_m}\right)^c\right] \tag{11.5}$$

其中，x_m 是中位数，c 是形状参数。式(11.5)可表示为如下的对数形式

$$\lg\left(\ln\left(\frac{1}{1-F(x)}\right)\right)=\frac{c}{10}\left[10\lg x+\frac{10}{c}\lg(\ln 2)-10\lg x_m\right] \tag{11.6}$$

式中，x 代表$|X_C(0)|$。在 $N=30$ 时，只要输入杂波包络概率密度函数的尾部衰减不太慢，$|X_C(0)|$都可以近似看作瑞利分布。但当输入杂波包络的概率密度函数尾部衰减很慢时，$|X_C(0)|$不服从瑞利分布。根据式(11.4)可知，对于$|X_C(k)|(k\neq 0)$的情况也有相同的结论。

在随后的讨论中均假定杂波背景为下列三种类型：①输入杂波包络是瑞利分布的，杂波 DFT 的模也是瑞利分布的；②输入杂波包络是非瑞利分布的，杂波 DFT 的模是瑞利分布的；③输入杂波包络是非瑞利分布的，杂波 DFT 的模也是非瑞利分布的。暂不考虑天线方向图对回波信号幅度的调制，认为脉冲多普勒雷达检测的目标信号是具有某一多普勒频率且初相未知的等幅相参脉冲串。考虑到 DFT 与相参积累的上述对应关系，在一般的讨论中，由于每种方案中都用 DFT 来处理未知 f_d 的信号，并且只考虑一路相参积累器，故为不失一般性，可把待检测信号看成 $f_d=0$ 的初相未知的等幅相参脉冲串。至于信号的 f_d 与 DFT 失配时由 DFT 频率选择性引起的信噪比损失，将在 11.4 节中进行讨论。

在比较各种 CFAR 处理方案的信噪比损失时，统一采用在已知平均功率的瑞利杂波中检测初相未知、等幅相参脉冲串信号的最佳参量检测器作为比较基准。这种最佳参量检测器在没有帧间积累时，是由相参积累、包络检波和门限比较组成的，如图 11.1 所示，简称为线性检测器或相参检测器。如果脉冲多普勒雷达要求帧间积累，则上述最佳参量检测器在包络检波之后还要进行某种形式的视频积累。

图 11.1　线性检测器框图

11.2.2　频域 CA-CFAR 检测器

在 DFT 之后采用频域 CA-CFAR 处理方案时，检测器的结构如图 11.2 所示。如前面约定的，把 DFT 看成相参积累，图 11.2 中的平方律检波器实际上是取 DFT 系数的模的平方。频域 CA-CFAR 处理器是一种参量型的 CFAR 方案，它能在瑞利包络杂波条件下保持 CFAR，当杂波包络偏离瑞利分布时，便失去 CFAR 能力。因此，这种方案在 11.2.1 节中讨论的①和②型杂波背景中可以采用，而在③型杂波背景中不能采用。设 M 是视频积累脉冲个数，在脉冲多普勒雷达里是积累帧数，R 表示参考单元数。

图 11.2　频域 CA-CFAR 处理器

设 $P_{fa}=10^{-6}$，$P_d=0.5$。在 $M=1$（对应无帧间积累情况），参考单元数 $R=10$ 时，信噪比损失 $L_{SNR}\approx 3.3\text{dB}$；$R=20$ 时，$L_{SNR}\approx 1.6\text{dB}$。在 $M=2$ 时，$R=10$ 的 $L_{SNR}\approx 1.9\text{dB}$，$R=20$ 的 $L_{SNR}\approx 1.0\text{dB}$。不论 DFT 输入杂波包络是否服从瑞利分布，相参积累器对信噪比的改善值在图 11.1 的线性检波器和图 11.2 的 CA-CFAR 检测器两种情况下是一样的。

11.2.3　MTI-FFT-频域 CA-CFAR 方案

文献[6-8]对机载 PD 雷达的 MTI-FFT-CFAR 进行了分析，下面讨论系统的检测性能及计算机仿真结果。信号是在 MTI 之前加入的，主要目的是比较不同的 MTI 滤波器及加权对检测性能的影响。

1. 系统结构

MTI-FFT-CFAR 系统框图如图 11.3 所示。对于机载脉冲多普勒雷达，由于其载机的高度、速度、方位扫描角及地形等因素的变化，其主旁瓣杂波强度和主瓣宽度有较大的变化。对于中等 PRF 工作的下视雷达，其输入的杂波功率谱如图 11.4 所示。主瓣旁瓣杂波比（Main Side Lobe Ratio，MSLR）高达 40～50dB，图中谱宽 B 为 PRF，d 为主瓣杂波半宽度。主瓣杂波宽度可按式（11.7）估算[3]

$$2d=\theta_1\frac{2v}{\lambda}\sin\theta \tag{11.7}$$

其中，v 为飞机速度，λ 为波长，θ 为波束指向与速度方向夹角，θ_1 为天线波束两零点之间的宽度。

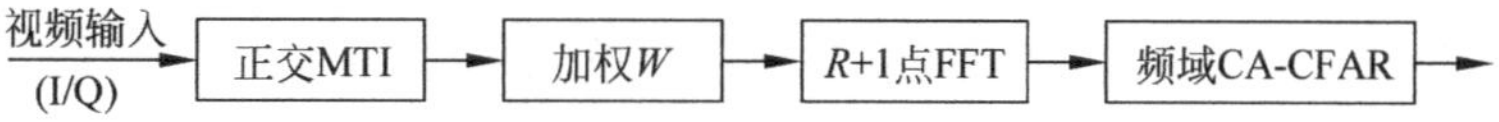

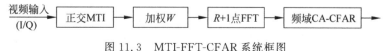

图 11.3　MTI-FFT-CFAR 系统框图

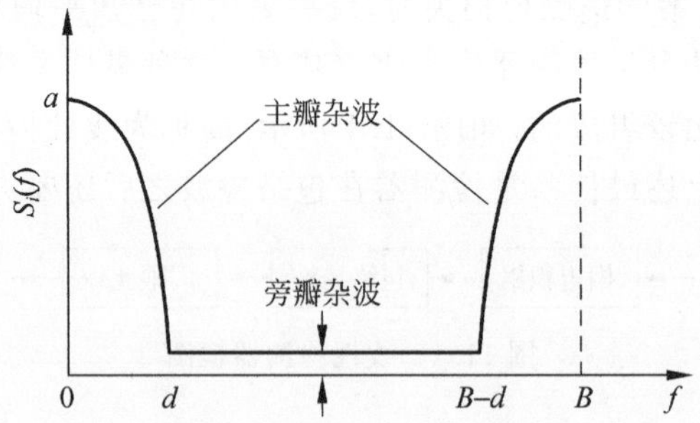

图 11.4　输入杂波功率谱

输入杂波谱可表示为

$$S_i(f)=\begin{cases}a\cos(bf), & 0\leqslant f<d \\ 1, & d\leqslant f<B-d \\ a\cos[b(B-f)], & B-d\leqslant f<B\end{cases} \tag{11.8}$$

选定 MSLR 以后,即可确定输入谱的参数 a 和 b。杂波背景除上述主瓣杂波剩余和旁瓣杂波外,还有接收机热噪声,这里把热噪声和旁瓣杂波一起加以考虑。为对消主瓣杂波,设置正交两路 MTI 对消器。MTI 结构形式可采用非递归和递归滤波器,分别为 MTI-0 型和 MTI-1 型,如图 11.5 所示。

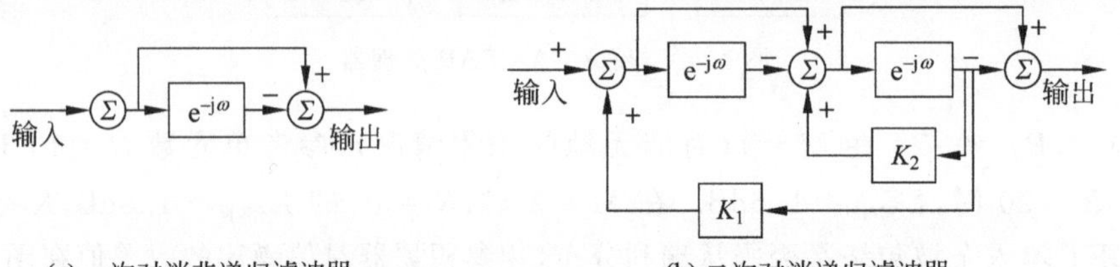

图 11.5　MTI 对消器的两种结构

MTI-0 型,即一次对消非递归滤波器的传递函数为

$$H(e^{j\omega})=1-e^{j\omega} \tag{11.9}$$

MTI-1 型,即二次对消递归滤波器的传递函数为

$$H(e^{j\omega})=\frac{(e^{j\omega}-1)^2}{e^{j2\omega}-(K_1+K_2)e^{j\omega}+K_1} \tag{11.10}$$

文献[6]中选取 $K_1=0.5$,$K_2=0.25$。

加权是为了减小主瓣杂波的泄漏,使主瓣杂波影响减小,从而改善检测性能。文献[6]采用海明加权函数

$$\omega(n)=0.54-0.46\cos\left(\frac{2\pi n}{R+1}\right),\quad n=0,1,\cdots,R \tag{11.11}$$

$R+1$ 为相参脉冲数。若 FFT 输入序列为 $x(n)(n=0,1,\cdots,R)$,则加权 FFT 的第 k 个输出为

$$X(k)=\sum_{n=0}^{R}x(n)\omega(n)e^{-j2\pi nk/(R+1)},\quad k=0,1,\cdots,R \tag{11.12}$$

它相当于 $R+1$ 个多普勒滤波器输出,于是可在频域进行单元平均 CFAR 处理。

图 11.6 给出了频域 CA-CFAR 具体实施流程,包括两种实施方式,其中图 11.6(a)为在噪声背景下的频域 CA-CFAR 实施流程,由于噪声频谱的均匀性,因此可在一个距离单元内、沿多普勒轴进行 CA-

CFAR 处理，滤波器 F_l 为检测单元，其余为参考单元。对参考单元中的噪声功率进行平均之后，能自适应地调节门限，以保持 CFAR 能力。图 11.6(b)为频域 CA-CFAR 的通用处理流程，待检测距离单元编号为 n，即，对第 n 个距离单元进行检测，具体做法是，首先将每个距离单元 $R+1$ 个相参脉冲的回波变换到频域后形成多普勒-距离二维数据空间；然后依次对每个多普勒通道、沿距离维进行 CA-CFAR 检测，如图 11.6(b)所示，在第 l 个多普勒通道中，对第 n 个距离单元进行检测，选取第 $n-1\sim n-N/2$、$n+1\sim n+N/2$ 个距离单元中第 l 个多普勒通道的数据作为参考样本，进行检测单元背景功率水平估计；最后对第 n 个距离单元中的 $R+1$ 个多普勒通道的检测结果进行融合处理，在一定的融合规则下给出第 n 个距离单元中是否存在目标的判决结果。

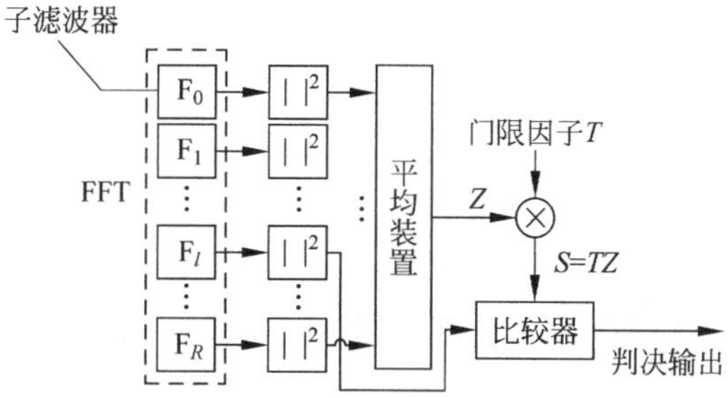

(a) 噪声背景下频域CA-CFAR实施流程

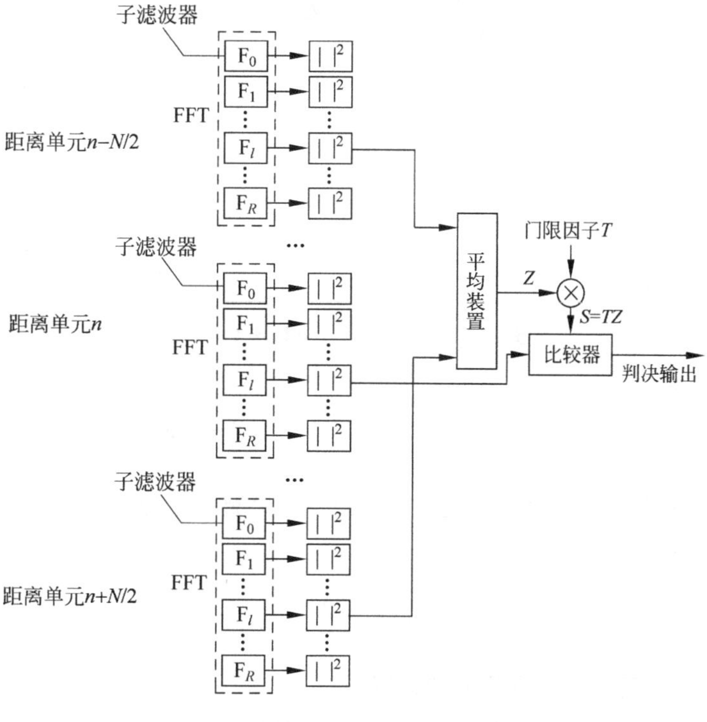

(b) 多个距离单元条件下频域CA-CFAR实施流程

图 11.6　频域 CA-CFAR 具体实施流程

2. FFT 输出的概率分布

CA-CFAR 只有在样本统计独立，服从瑞利分布的情况下有 CFAR 能力。图 11.6 是在 FFT 后经平方律检波器再作 CFAR 处理，这就需要了解 FFT 输出的分布情况。输入 MTI 的是相关复高斯序列，

经 MTI 线性变化以后仍为相关复高斯序列,即 FFT 的输入序列为

$$x(n)=x_1(n)+\mathrm{j}x_2(n),\quad \mathrm{E}\{x\}=0 \tag{11.13}$$

且 $x_1(n)$ 与 $x_2(n)$ 是相互独立的,因而它们的分布为

$$f(x)=\frac{1}{2\pi\sigma^2}\exp\left[-\frac{x_1^2(n)+x_2^2(n)}{2\sigma^2}\right] \tag{11.14}$$

FFT 的输入矢量为 $\boldsymbol{Z}_{\mathrm{I}}=[x(0),x(1),\cdots,x(R)]^{\mathrm{T}}$,矢量 $\boldsymbol{Z}_{\mathrm{I}}$ 的联合分布为多维复高斯分布

$$f(\boldsymbol{Z}_{\mathrm{I}})=\frac{1}{\pi^{R+1}|\boldsymbol{B}_{\mathrm{I}}|}\exp(-\boldsymbol{Z}_{\mathrm{I}}^{\mathrm{H}}\boldsymbol{B}_{\mathrm{I}}^{-1}\boldsymbol{Z}_{\mathrm{I}}) \tag{11.15}$$

其中,$(\cdot)^{\mathrm{H}}$ 表示共轭转置,$\boldsymbol{B}_{\mathrm{I}}$ 为 MTI 输出矢量的协方差矩阵,有

$$\boldsymbol{B}_{\mathrm{I}}=\mathrm{E}\{\boldsymbol{Z}_{\mathrm{I}}\boldsymbol{Z}_{\mathrm{I}}^{\mathrm{H}}\}=\begin{vmatrix} r(0) & r(1) & \cdots & r(R)\\ r^*(1) & \cdots & \cdots & r(R-1)\\ \vdots & \vdots & & \vdots\\ r^*(R) & \cdots & \cdots & r(0)\end{vmatrix} \tag{11.16}$$

由于输入功率谱密度是偶对称的,所以 MTI 的输出,即 FFT 的输入的协方差矩阵是实对称阵。设 FFT 的输出矢量为

$$\boldsymbol{Z}_{\mathrm{O}}=[X(0),X(1),\cdots,X(R)]^{\mathrm{T}} \tag{11.17}$$

DFT 变换对应的 $(R+1)\times(R+1)$ 阶矩阵为

$$\boldsymbol{F}=(\mathrm{e}^{-\mathrm{j}2\pi nk/(R+1)})_{(R+1)\times(R+1)} \tag{11.18}$$

加权对角阵为

$$\boldsymbol{W}=\mathrm{diag}[\omega(0),\omega(1),\cdots,\omega(R)] \tag{11.19}$$

则 FFT 的输出矢量为

$$\boldsymbol{Z}_{\mathrm{O}}=\boldsymbol{FWZ}_{\mathrm{I}} \tag{11.20}$$

所以,FFT 的输出协方差矩阵为

$$\boldsymbol{B}_{\mathrm{O}}=\mathrm{E}[\boldsymbol{Z}_{\mathrm{O}}\boldsymbol{Z}_{\mathrm{O}}^{\mathrm{H}}]=\boldsymbol{FWB}_{\mathrm{I}}\boldsymbol{WF}^{\mathrm{H}} \tag{11.21}$$

则复矢量 $\boldsymbol{Z}_{\mathrm{O}}$ 的概率密度为

$$f(\boldsymbol{Z}_{\mathrm{O}})=\frac{1}{\pi^{R+1}|\boldsymbol{B}_{\mathrm{O}}|}\exp(-\boldsymbol{Z}_{\mathrm{O}}^{\mathrm{H}}\boldsymbol{B}_{\mathrm{O}}^{-1}\boldsymbol{Z}_{\mathrm{O}}) \tag{11.22}$$

图 11.7 中各单元 FFT 输出的模平方记为 $|X(k)|^2$。检测单元的杂波功率估值为 Z,它由 R 个频域邻近参考单元的平均求得,即

$$Z=\frac{1}{R}\sum_{\substack{k=0\\k\neq l}}^{R}|X(k)|^2 \tag{11.23}$$

3. 参考样本相关条件下标称化因子 T 的选取

参考样本独立时,门限 $S(S=TZ)$ 服从 χ^2 分布。虚警概率为[9]

$$P_{\mathrm{fa}}=\left(\frac{1}{1+T/R}\right)^R \tag{11.24}$$

由此可确定门限因子 T。

当参考样本相关时,门限 S 的分布不是 χ^2 分布。一般来说,输入是相关的,那么检测单元样本与参考样本也是相关的,因而与门限 S 也是相关的。这样反而会使虚警概率有所减小,所以待选标称化

因子应略小于理论(样本独立时)的标称化因子。

4. 系统性能的分析结果

由于 MTI 并不能实现杂波的理想白化,加上加权的影响,FFT 输出杂波在频域上不再是 IID 的。对于中等 PRF 机载 PD 雷达,计算表明 FFT 输出及各频率单元的杂波不均匀,且相互之间有很强的相关性。因此系统的检测性能单靠解析方法不容易求得,故采用解析公式和 Monte Carlo 仿真结合的分析方法。图 11.7 所示为 P_{fa} 与标称化因子 T 的关系曲线,并同独立时的理论值作了比较。相关时的 P_{fa} 低于独立时的值,即在相同的 P_{fa} 情况下,应选较小的标称化因子。图 11.8 比较了加权的 MTI-1 型滤波器和 MTI-0 型滤波器的检测性能。图 11.9 比较了加权与不加权对检测性能的影响。由图 11.7～图 11.9 可以看出,加权时的检测性能优于不加权的情况,采用递归滤波器的检测性能优于非递归的情况。

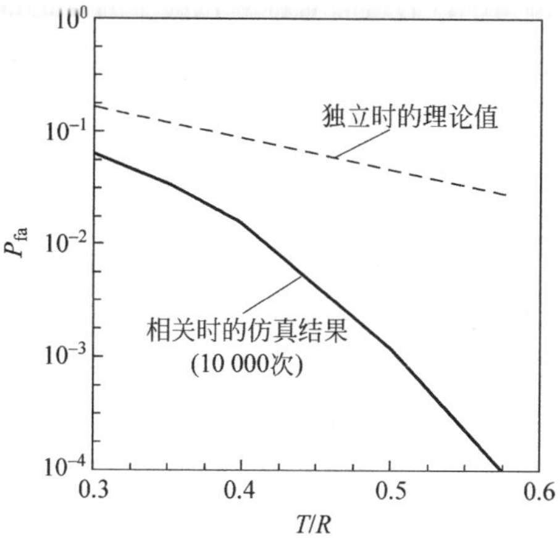

图 11.7　P_{fa} 与 T/R 的关系曲线

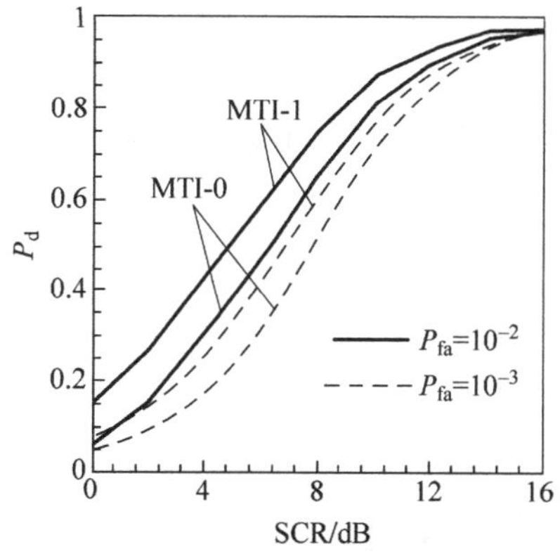

图 11.8　在加权情况下采用不同滤波器的性能比较

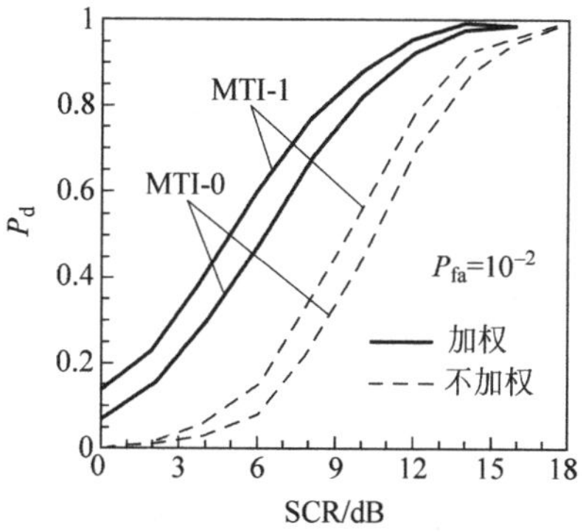

图 11.9　加权与不加权时的性能比较

对这里所讨论的两种 MTI 滤波器和海明加权的结果表明,当 $P_{fa}=10^{-2}$ 和 10^{-3},$P_d=0.5$ 时,经加权处理,可将信号旁瓣杂波比损失减小 1～4dB;加权后,若采用递归滤波器还能将信杂比损失减小约 1dB。由此可见,FFT 加权的形式(窗函数)和 MTI 结构对改善检测性能有较大的影响。因此,在雷达系统设计时必须精心选择 FFT 的窗函数和 MTI 滤波器结构,并且可以用文献[6]中给出的方法进行检测性能的计算机仿真分析。为便于比较,表 11.1 给出了 Monte Carlo 仿真结果,SCR 为所需信号旁瓣杂波比。

表 11.1　Monte Carlo 仿真结果(SCR/dB)比较($P_d=0.5$,$R=31$)

滤波器类型	检测单元序号	P_{fa}		
		10^{-2}		10^{-3}
		加权	不加权	加权
MTI-1	17	5.1	9.2	7
MTI-0	17	6.2	10.4	7.9

在中等 PRF 机载 PD 雷达的 MTI-FFT-频域 CA-CFAR 处理中，文献[7]分析了频域各杂波单元明显偏离 IID 假设时对标称化因子的修正，以及这种修正对系统性能的影响。

(1) 由于各频率单元杂波之间的相关性，如按 IID 条件选取标称化因子，则 P_{fa} 大为下降，为保持给定的虚警概率，标称化因子应比 IID 条件下的低。

(2) 在检测区内，当各频率单元的杂波功率大致均匀时，频域单元平均检测器大体上具有 CFAR 能力。

(3) 如按 IID 选取标称化因子，则 P_d 下降；而按给定 P_{fa} 修正标称化因子，则 P_d 提高而且可以高于 IID 条件下的值。

(4) 主瓣杂波谱宽度越窄，检测概率越高，而谱太宽时检测性能变坏。

总之，如按 IID 条件选取标称化因子，P_{fa} 大为下降，如果适当修正标称化因子，检测性能可得到改善。

11.2.4 频域奇偶处理检测器

1. 奇偶处理检测器的系统结构

MTD 雷达主要通过对目标回波的相参脉冲串作准匹配处理来检测目标。MTD 的核心为多普勒滤波器组，采用离散傅里叶变换 DFT 算法，实际应用中往往利用快速傅里叶变换实现。在相参处理时间内，对每个距离单元的 N 个脉冲样本进行多普勒滤波，可以获得频域数据。当目标信号的多普勒频率不能和多普勒滤波器组的中心频率完全匹配时，由于在其他滤波器中也存在信号分量，采用均值类估计而引起的“信号污染”畸变不仅导致信噪比损失，也会引起检测概率的降低。文献[10]通过传统的填零处理(Zero Padding，ZP)来避免畸变，并提出一种奇偶处理(Odd-Even Processing，OEP)技术来阻止检测概率的降低。OEP 检测器的原理框图如图 11.10 所示。首先对来自 I/Q 通道的复数据进行填零处理，增加输入数据序列的长度，然后进行 FFT 获得频域数据，最后采用 OEP 方法对频域数据进行目标检测。

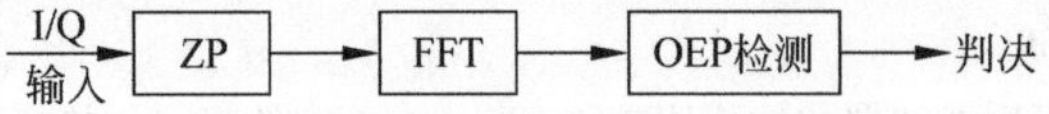

图 11.10 OEP 检测器的原理框图

2. ZP 的频域特性

多普勒滤波器组通过 DFT 来实现，如果 $u_0, u_1, \cdots, u_{N-1}$ 是信号的同相分量和正交分量的复样本值，不进行填零或加窗处理时的 DFT 定义为

$$U_n = \sum_{k=0}^{N-1} u_k \exp(-\mathrm{j}2\pi nk/N) \tag{11.25}$$

其中，$n=0,1,\cdots,N-1$。

由 Parseval 能量守恒定律有如下关系

$$\sum_{k=0}^{N-1} |u_k|^2 = \frac{1}{N}\sum_{n=0}^{N-1} |U_n|^2 \tag{11.26}$$

考虑在原始数据 $u_0, u_1, \cdots, u_{N-1}$ 后添加 N 个零，原则上可以添加任意数目的零，但这里只考虑所加零数目是 N 的奇数倍的情况。填零后新序列的 DFT 的长度是 $2N$

$$V_n=\sum_{k=0}^{N-1}u_k\exp(-\mathrm{j}\pi nk/N) \tag{11.27}$$

其中，$n=0,1,\cdots,2N-1$。因为新序列后面的 N 个样本均为零，式(11.27)中的求和运算在 $N-1$ 时就终止了。由式(11.25)和式(11.27)可知 $V_{2n}=U_n$，而根据式(11.26)，由 Parseval 公式可得

$$\sum_{k=0}^{N-1}|u_k|^2=\frac{1}{2N}\sum_{n=0}^{2N-1}|V_n|^2=\frac{1}{2N}\left(\sum_{n=0}^{N-1}|V_{2n}|^2+\sum_{n=0}^{N-1}|V_{2n+1}|^2\right) \tag{11.28}$$

因为 $V_{2n}=U_n$，根据式(11.26)和式(11.28)可知，式(11.28)括号中的两项是相等的。已经证明：如果 DFT 的输入数据是 IID 的复圆高斯过程的样本，那么 DFT 的输出数据在没有填零的情况下也是 IID 的；然而，当输入序列添加 N 个零后，DFT 输出数据的奇数项是 IID 的，其偶数项也是 IID 的，但它们之间未必是独立的。

3. OEP 检测器的虚警概率

假设 $x_0,x_1,\cdots,x_{N-1}$ 是参数为 σ^2 的循环高斯过程的 IID 样本值。那么未加权且未填零条件下的 DFT 系数 $X_0,X_1,\cdots,X_{N-1}$ 是参数为 $N\sigma^2$ 的循环高斯分布[11]。因此，输出幅度的平方值 $G_0,G_1,\cdots,G_{N-1}$ 是 IID 的指数分布，概率密度函数为

$$f(x)=\frac{1}{2N\sigma^2}\exp\left(-\frac{x}{2N\sigma^2}\right) \tag{11.29}$$

定义 S_k 为

$$S_k=\sum_{\substack{n=0\\n\neq k}}^{N-1}G_n \tag{11.30}$$

则每个 DFT 输出端的虚警概率为

$$P_{\mathrm{fa}}=\Pr\{G_k>TS_k\}=(1+T)^{-(N-1)} \tag{11.31}$$

其中，$k=0,1,\cdots,N-1$。在每个滤波器输出端利用式(11.31)，则在给定 P_{fa} 的条件下，可以得到标称化因子 T。为了进行 CFAR 检测，总的虚警概率 P_{FA} 与 P_{fa} 的关系为

$$P_{\mathrm{fa}}=P_{\mathrm{FA}}/N \tag{11.32}$$

如果在 DFT 输入数据之后添上 N 个零，可以利用所有 $2N$ 个 DFT 输出数据进行均值水平估计，则式(11.30)变为

$$S_h=\sum_{\substack{n=0\\n\neq h}}^{2N-1}G_n \tag{11.33}$$

但当目标信号存在时，式(11.33)的估计值将会遭到严重的“信号污染”，最终会降低高信噪比条件下的检测概率。从式(11.28)可以看出，信号能量完全包含在奇数项系数之和与偶数项系数之和中。因此，可以采用 OEP 方法，它可以有效避免严重的“信号污染”现象。如果下标 h 是奇数，可用式(11.34)计算均值水平

$$S_h=\sum_{n=0}^{N-1}G_{2n+1}-G_h \tag{11.34}$$

如果 h 是偶数，则用式(11.35)进行计算

$$S_h=\sum_{n=0}^{N-1}G_{2n}-G_h \tag{11.35}$$

如上所述,OEP 需要判断 h 是奇数还是偶数,从而决定用式(11.34)还是式(11.35)进行均值水平估计。根据式(11.28)的结论,可以对 S_h 的计算进行简化

$$S_h = \frac{1}{2}\sum_{n=0}^{2N-1} G_n - G_h \tag{11.36}$$

式(11.36)的计算结果与分别利用式(11.34)和式(11.35)时的计算结果是相同的,但式(11.36)无须考虑 h 的奇偶性,具有更强的通用性。

如前所述,DFT 输出数据的偶数项(或奇数项)系数是 IID 的随机变量,因此式(11.34)和式(11.35)(等同于式(11.36))涉及 IID 随机变量的求和。这意味着可以利用式(11.31)来计算标称化因子 T,并将其应用到每个多普勒单元中。为了保持 FAR 恒定,式(11.31)应该进行如下修正

$$P_{\text{fa}} = P_{\text{FA}}/(2N) \tag{11.37}$$

4. OEP 检测器的分析结果

这里对 OEP 处理和不采用零填充处理(NZP)的性能进行了比较,同时也给出了已知信号多普勒频率和噪声功率水平下的最优处理(OP)与只知道多普勒频率情况下均值类检测器(MLD)的性能分析结果。OEP 和 NZP 的结果通过仿真得到,针对每个值进行 2000 次仿真,归一化多普勒频率均匀分布在区间$(0.5/N, 1/N)$上。MLD 的结果通过文献[12]中的递推公式获得,而 OP 的结果利用正态分布概率密度函数进行精确的计算获得。所有的结果均采用典型值 $N=32$,这样所涉及的运算量较小,能及时进行仿真。与 $N=32$ 相比,N 取较大值时的结果可能有所不同,但对于不同的处理方法来说,所得的检测性能曲线的相对关系是相同的。

图 11.11 给出了 $P_{\text{FA}}=10^{-5}$,$N=32$ 的情况下检测概率 P_{d} 随信噪比变化的函数曲线。OP 与 MLD 所需的信噪比差别大约只有 2dB,这与文献[12]中所给的结果相符。在多普勒频率未知情况下的 OEP 所需的信噪比,只比在多普勒频率已知情况下的 MLD 所需的信噪比多出不到 1dB;而 NZP 所需信噪比明显高于 OEP 或 MLD,特别是在 $P_{\text{d}}>0.5$ 时,在较大信噪比上检测概率的下降是很明显的。图 11.12 进一步减小虚警概率($P_{\text{FA}}=10^{-10}$)进行分析,可以看出,此时 NZP 的检测性能几乎无法接受,特别是在 $P_{\text{d}}>0.5$ 的情况下。

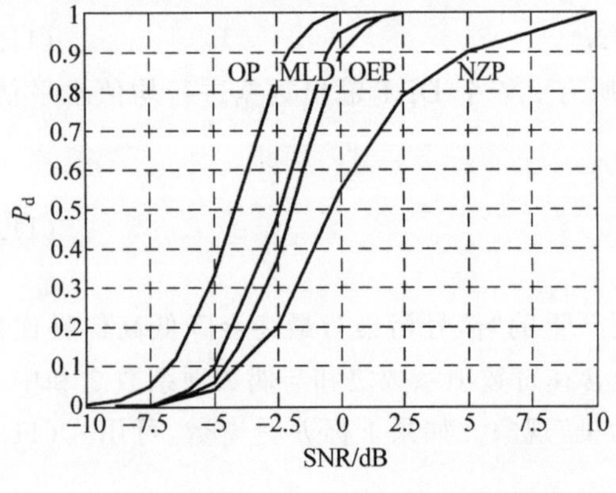

图 11.11　$P_{\text{FA}}=10^{-5}$、$N=32$ 时不同检测器的检测性能

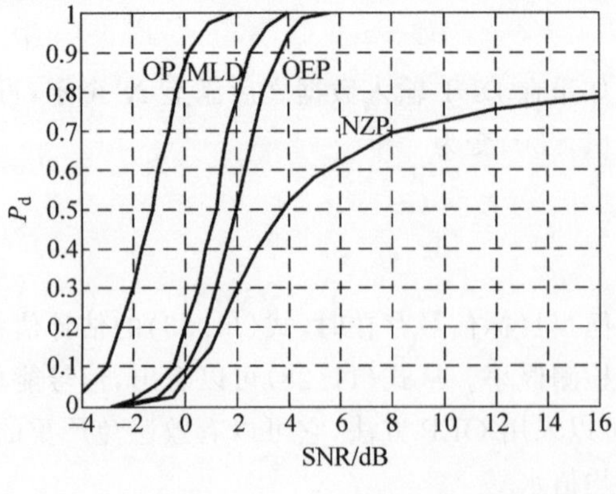

图 11.12　$P_{\text{FA}}=10^{-10}$、$N=32$ 时不同检测器的检测性能

11.3　小波域 CFAR 检测

在某些情况下,单纯依靠时域或频域检测方法并不能有效检测目标,关于其他变换域如小波域的雷

达目标 CFAR 检测也引起了广泛的关注，并取得了初步的研究进展。小波变换（Wavelet Transform，WT）既保持了傅里叶方法的优点，又具有短时傅里叶变换的良好局域特性，已成为应用数学和信号处理等领域的研究热点[13]。随着雷达信号处理技术的发展和深入，对突变信号和非平稳信号的处理已成为雷达信号处理中的关键问题，雷达信号的非平稳性使小波变换的应用进一步得到了推广。小波变换适合对信号的局部现象进行分析，在处理非平稳信号和微弱信号方面有着独特的优势[14-15]。本节主要讨论两种小波域 CFAR 处理方法。首先利用删除技术（Censored Method，CM）准确估计小波域中的噪声功率水平，介绍一种基于离散小波变换的 CM-CFAR 检测方法。通过分析高斯白噪声在正交小波域中的特性，进一步研究基于正交小波变换的 CA-CFAR 方法。通过仿真实验比较小波软阈值处理和硬阈值处理的检测效果，分别得到适应两种小波域 CFAR 处理方法的阈值处理方法。

11.3.1　基于离散小波变换的 CM-CFAR 检测方法

文献[13]首先对含杂波信号进行离散小波变换，利用删除技术估计小波域中的噪声功率水平，提出一种基于离散小波变换的 CM-CFAR 检测方法。在雷达一类的信号检测中，很难知道先验概率和代价函数，所以一般采用 Neyman-Pearson 准则进行判决。假设在一定小波分解尺度 j 下，信号的小波系数表示为

$$\mathrm{H}_1: d_{\mathrm{X}}^j = d_{\mathrm{S}}^j + d_{\mathrm{C}}^j \tag{11.38}$$

$$\mathrm{H}_0: d_{\mathrm{X}}^j = d_{\mathrm{C}}^j \tag{11.39}$$

式中，H_1 表示有目标存在的假设，H_0 表示无目标存在的假设，d_{S}^j 是尺度 j 下信号的小波系数，d_{C}^j 是尺度 j 下杂波的小波系数。由小波变换的理论可知，高斯噪声经过小波变换之后依然是高斯噪声，在杂波背景服从高斯分布的前提下，可以假设噪声的小波系数 d_{C}^j 服从均值为 μ 方差为 σ^2 的高斯分布。

由于小波变换是线性变换，可得

$$f(d_{\mathrm{X}}^j \mid \mathrm{H}_0) = \frac{1}{\sqrt{2\pi}\sigma}\exp\left[-\frac{(d_{\mathrm{X}}^j - \mu)^2}{2\sigma^2}\right] \tag{11.40}$$

$$f(d_{\mathrm{X}}^j \mid \mathrm{H}_1) = \frac{1}{\sqrt{2\pi}\sigma}\exp\left[-\frac{(d_{\mathrm{X}}^j - A_j - \mu)^2}{2\sigma^2}\right] \tag{11.41}$$

式中，A_j 是目标信号小波系数均值的估计值。

由 Neyman-Pearson 准则获得的似然比检验统计量可以表示为

$$\lambda = \frac{d_{\mathrm{X}}^j - \mu}{\sigma} \underset{\mathrm{H}_0}{\overset{\mathrm{H}_1}{\gtrless}} \gamma \tag{11.42}$$

式中，γ 为阈值，则虚警概率为[13]

$$P_{\mathrm{FA}} = \int_{\gamma}^{+\infty} \frac{1}{\sqrt{2\pi}}\exp\left(-\frac{z^2}{2}\right)\mathrm{d}z = \alpha \tag{11.43}$$

由此可以得到 $P_{\mathrm{FA}} = \alpha$ 时的阈值 γ，表 11.2 是不同虚警概率下的阈值 γ。

表 11.2　γ 与 P_{FA} 的关系

P_{FA}	10^{-7}	10^{-6}	10^{-5}	10^{-4}	10^{-3}	10^{-2}	10^{-1}
γ	5.199	4.753	4.265	3.719	3.090	2.326	1.282

当 $P_{\mathrm{FA}}=\alpha$ 时,可得检测概率 P_{d} 的值为

$$P_{\mathrm{d}}=\int_{\gamma}^{+\infty}\frac{1}{\sqrt{2\pi}}\exp\left[-\frac{(z-\widetilde{A}_j)^2}{2}\right]\mathrm{d}z \tag{11.44}$$

式中,$\widetilde{A}_j^2=(A_j/\sigma)^2$ 是信杂比。

在小波系数服从均值为 μ、方差为 σ^2 的正态分布的前提下,可以得到尺度 j 下的小波系数的阈值为

$$Th=\gamma\sigma+\mu \tag{11.45}$$

然后使用软阈值或硬阈值方法对小波系数进行处理。设对小波系数 W_j 处理后的小波系数为 $\hat{W}_j$,软阈值方法原理如下

$$\hat{W}_j=\begin{cases}\mathrm{sgn}(W_j)(|W_j|-Th), & |W_j|>Th\\ 0, & \text{其他}\end{cases} \tag{11.46}$$

而硬阈值方法原理如下

$$\hat{W}_j=\begin{cases}W_j, & |W_j|>Th\\ 0, & \text{其他}\end{cases} \tag{11.47}$$

由式(11.42)知统计量 $\lambda=(d_{\mathrm{X}}^j-\mu)/\sigma$ 是小波系数 d_{X}^j 对噪声小波系数的均值和标准差归一化的结果。由 $\lambda\sim\mathrm{N}(0,1)$ 知,λ 与噪声功率水平 σ 无关,而判决值 γ 完全由虚警率 $P_{\mathrm{FA}}=\alpha$ 确定。在 γ 不变的情况下对 λ 进行检测即可保持恒定的虚警率。以上分析表明,在高斯噪声背景下,该信号检测方法具有 CFAR 特性。

因为噪声背景的均值 μ 和方差 σ^2 均是未知,阈值 Th 中还有未知参数需要估计。本节通过分析噪声和信号小波系数不同的变化特点,采用删除均值 CFAR 检测器中的删除技术 CM,对噪声小波系数的参量进行准确估计来保证检测方法的 CFAR 特性。删除技术先将尺度 j 下的小波系数按绝对值大小进行排序,然后删除从最大值起始的一部分小波系数,认为这些被删除的小波系数是目标信号的小波系数,取剩余的小波系数作为噪声小波系数的估计 $\hat{d}_i^j(i=1,2,\cdots,N_j^{\mathrm{C}})$。均值 μ 和标准差 σ 的无偏估计量分别为

$$\hat{\mu}=\frac{1}{N_j^{\mathrm{C}}}\sum_{i=1}^{N_j^{\mathrm{C}}}\hat{d}_i^j \tag{11.48}$$

$$\hat{\sigma}=\frac{1}{N_j^{\mathrm{C}}-1}\sum_{i=1}^{N_j^{\mathrm{C}}}(\hat{d}_i^j-\hat{\mu}) \tag{11.49}$$

对于小波系数删除多少才是最合适的,这里引入噪声置信度 $\beta(0<\beta\leqslant 1)$ 来量化表示。尺度 j 下的 β_j 的含义是该尺度下噪声小波系数所占小波系数总数 N_j 的百分比,即表示绝对值最大的 $N_j(1-\beta)$ 个小波系数将被删除。β_j 的具体取值与目标信号总能量、噪声与目标信号的小波系数特点有关。

Donoho 阈值消噪方法是工程实践中最常用的方法,其所得最优阈值水平如下

$$\sigma_j=\frac{\mathrm{median}(|d_{\mathrm{X}}^j|)}{0.6745} \tag{11.50}$$

$$Th_j=\sigma_j\sqrt{2\lg n} \tag{11.51}$$

式中,σ_j、n 和 Th_j 分别表示尺度 j 下的小波系数的标准差、原始信号长度和阈值水平,median 表示求中位数。通过该阈值对小波系数进行处理,可以得出尺度 j 下低于该阈值的小波系数的个数,从而得出

尺度 j 下的噪声置信度 β_j。然后利用参数估计式(11.48)、式(11.49)和阈值公式(11.45)对含噪信号进行 CFAR 处理。

由于噪声具有随机性，因此不能采用固定的 β_j 值，由 Donoho 所提的阈值计算公式可以自适应地确定 β_j，从而保证对噪声水平的准确估计。基于离散小波变换的 CM-CFAR 检测方法的原理框图如图 11.13 所示，其中 J 是小波分解的最大尺度水平，$S'(n)$是检测出来的信号。

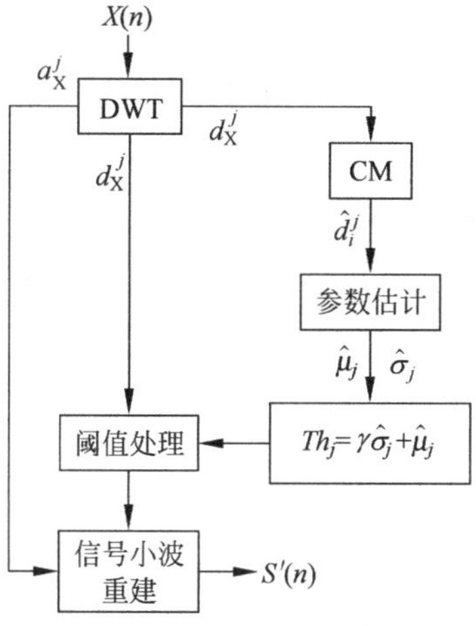

图 11.13　基于离散小波变换的 CM-CFAR 检测方法的原理框图

利用基于离散小波变换的 CM-CFAR 检测方法对不同杂波背景下的信号进行处理。原始目标信号采用正弦、方波和三角波混合信号，在高斯杂波、瑞利杂波、韦布尔杂波和对数正态杂波背景情况下的检测信号。从对高斯杂波、瑞利杂波和韦布尔杂波的检测结果来看，在相同的杂波背景条件下，硬阈值处理方法在不同的虚警概率下的检测信号相差无几，可见其对虚警概率的变化不敏感，检测过程中存在过平滑的问题；从软阈值处理方法的检测结果来看，虚警概率越高，检测出信号所含噪声越多，而虚警概率过低会导致信号的细节部分有所损失。而对于含异常值较多的对数正态杂波，硬阈值方法有比软阈值方法更好的检测效果，在虚警概率达到 10^{-6} 时，硬阈值方法基本去除了峰值较大的杂波信号，相对于其他三种杂波环境，由于对数正态杂波含有较多的异常值，基于高斯分布的小波 CM-CFAR 检测方法效果并不是很理想。总体来说，软阈值方法比硬阈值方法更适合于 CM-CFAR 检测方法，虽然 CM-CFAR 方法建立在高斯模型的基础上，但对不同杂波背景下的检测结果显示，该方法在不同杂波背景下仍可以很好地检测出信号，具有一定的鲁棒性和自适应特性。

11.3.2　基于正交小波变换的 CA-CFAR 检测方法

首先对含高斯噪声信号进行正交小波多尺度分析，基于高斯噪声在小波域中的特性，通过对小波系数进行平方律处理，建立基于正交小波变换的 CFAR 检测器模型，并给出相应的虚警概率和检测概率公式，分析信号小波系数序列长度对检测性能的影响，通过仿真实验得到适合于小波域 CA-CFAR 的阈值处理方法。

基于正交小波变换的 CA-CFAR 检测器的原理框图如图 11.14 所示。首先将雷达回波信号经过正交小波变换，对尺度 j 下的小波系数进行平方律处理，然后将所得结果进行 CA-CFAR 检测，超过检测门限的小波系数被认为是目标信号的小波系数，经过处理后加以保留，否则被认为是噪声的小波系数而被剔除，最后对处理后的小波系数进行重构得到检测出的信号。

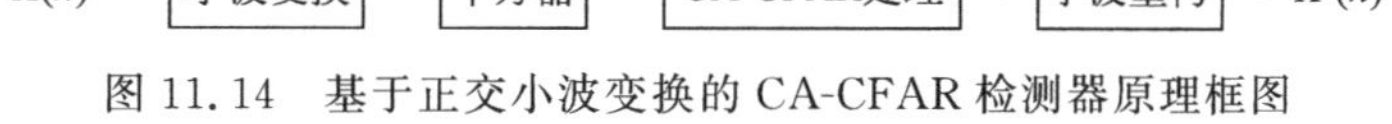

图 11.14　基于正交小波变换的 CA-CFAR 检测器原理框图

雷达回波信号 $X(n)$一般由两个可能的分量组成：一个来自目标反射信号 $S(n)$，另一个来自噪声环境 $N(n)$。由于独立高斯噪声经过离散正交小波变换后保持了原有的方差和独立性，在一定小波分解尺度 j 下，回波信号的小波系数表示为

$$\mathrm{H}_0: d_j^{\mathrm{X}} = w_j^{\mathrm{N}} \tag{11.52}$$

$$\mathrm{H}_1: d_j^{\mathrm{X}} = d_j^{\mathrm{S}} + d_j^{\mathrm{N}} \tag{11.53}$$

式中，d_j^{S} 和 d_j^{N} 分别表示目标信号和噪声在尺度 j 下的小波系数。在独立噪声 $N(n) \sim \mathrm{N}(0,\sigma_0^2)$ 的假设下，可知 $d_j^{\mathrm{N}} \sim \mathrm{N}(0,\sigma_0^2)$，且仍保持相互独立。若用 $G_{j,k}^{\mathrm{X}}$ 表示小波系数通过平方律检测的结果，则每个参考单元样本 $G_{j,k}^{\mathrm{X}}$ 服从指数分布，其概率密度函数为

$$f(x) = \mathrm{e}^{-x/\lambda'}/\lambda', \quad x \geqslant 0 \tag{11.54}$$

在 H_0 的假设下，λ' 是背景杂波和热噪声总的平均功率水平，用 μ 表示；在 H_1 的假设下，λ' 是 $\mu(1+\lambda)$。其中，λ 是目标信号的平均功率与杂噪功率比。于是有

$$\lambda' = \begin{cases} \mu, & \mathrm{H}_0 \\ \mu(1+\lambda), & \mathrm{H}_1 \end{cases} \tag{11.55}$$

在均匀杂波背景中，参考单元样本是 IID 的，并且它们的 λ' 都是 μ。为不失一般性，用 G_k 来表示 $G_{j,k}^{\mathrm{X}}$，假设在尺度 j 下小波系数序列长度为 M，定义 S_q 为

$$S_q = \sum_{\substack{k=0 \\ k \neq q}}^{M-1} G_k \quad q = 0,1,\cdots,M-1 \tag{11.56}$$

若在背景杂波功率水平 μ 确知的假设下进行最优检测，则只需要一个固定阈值 Z_0 来判定目标是否存在，这时的虚警概率为

$$P_{\mathrm{fa}} = \frac{1}{\mu}\int_{Z_0}^{\infty} \mathrm{e}^{-x/\mu}\,\mathrm{d}x = \mathrm{e}^{-Z_0/\mu} \tag{11.57}$$

式中，Z_0 是固定的最优阈值，最优检测的检测概率 P_{d_1} 为

$$P_{\mathrm{d}_1} = \frac{1}{\mu(1+\lambda)}\int_{Z_0}^{\infty} \mathrm{e}^{-x/\mu(1+\lambda)}\,\mathrm{d}x = \mathrm{e}^{-Z_0/[\mu(1+\lambda)]} \tag{11.58}$$

在背景杂波功率水平 μ 未知的情况下，由于阈值 $Z_q = TS_q$ 是一个随机变量，因而可以用 Z_q 的统计特征将虚警概率表示为

$$p_0 = \int_0^{\infty} f_{S_q}(s)\int_{Ts}^{\infty} \mathrm{e}^{-x/\mu}/\mu\,\mathrm{d}x\,\mathrm{d}s = (1+T)^{-(M-1)} \tag{11.59}$$

在每个滤波器的输出端利用阈值 Z_q 进行判决。在给定 p_0 的条件下，可以得到归一化阈值 T。为了方便比较，将保持虚警率恒定，即

$$p_0 = P_{\mathrm{FA}}/M \tag{11.60}$$

将式(11.60)代入式(11.59)可以确定 T。同理可得未知噪声功率水平下的检测概率为

$$P_{\mathrm{d}_2} = [1 + T/(1+\lambda)]^{-(M-1)} \tag{11.61}$$

在噪声功率水平已知和未知两种情况下，比较不同 M 值时二者的检测性能差异，并假设 $P_{\mathrm{fa}} = p_0 = P_{\mathrm{FA}}/M$。为方便比较，定义检测概率差值 P_{Diff}，它表征了在相同的虚警概率和输入信噪比条件下，噪声功率水平已知和未知时检测概率的差异大小，具体如下

$$P_{\mathrm{Diff}} = P_{\mathrm{d}_1} - P_{\mathrm{d}_2} \tag{11.62}$$

假设 $P_{\mathrm{FA}} = 10^{-4}$，$P_{\mathrm{Diff}}$ 随输入信噪比变化情况如图 11.15 所示。随着 M 的增大，噪声功率水平未知时的检测概率与已知时的检测概率越来越接近。在 $M=64$ 时，两种情况的检测概率已极为相近，而当$M \geqslant 256$ 时，两种情况的检测概率几乎相同，因为在大样本数量下，估计值十分接近真实值。在与图 11.15 相同条件下，未知噪声水平时的 P_{d_2} 随输入信噪比变化情况如图 11.16 所示，M 取值与曲线形状

的对应情况与图 11.15 相同。由图 11.16 可知，在 $M<64$ 时，检测概率随着 M 的增大而增加，而当 $M\geqslant 64$ 时，检测概率随着 M 的增大反而有所减小，但总的来说，$M\geqslant 64$ 时的检测性能相差不大。由于实际应用中，噪声功率水平一般是未知的，所以 $M=64$ 是较好的选择。

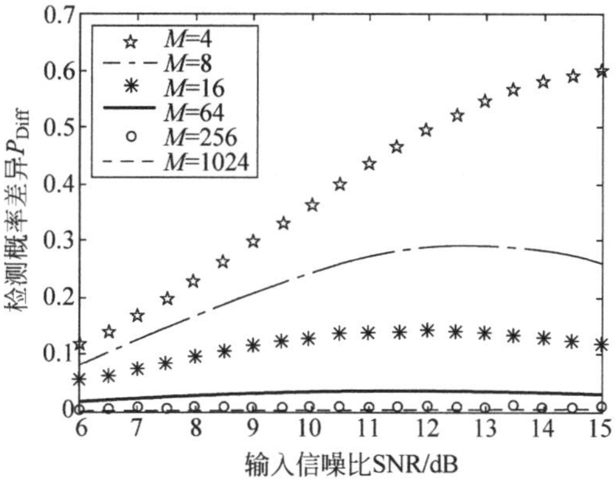

图 11.15　噪声功率水平已知和未知情况下的检测性能差异 P_{Diff} 曲线

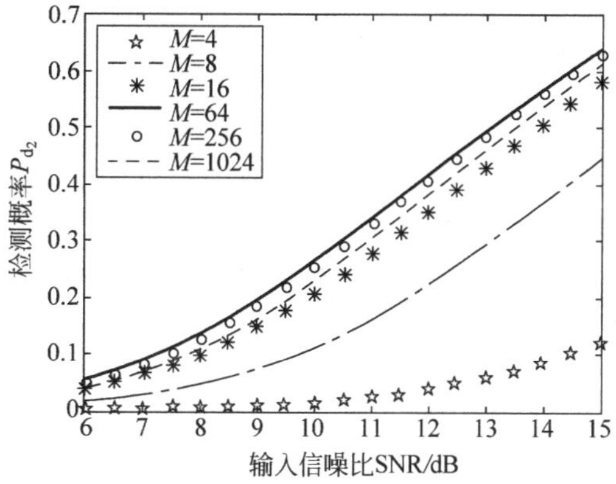

图 11.16　噪声功率水平未知情况下的检测性能比较

在尺度 j 下，由式(11.56)和式(11.59)可确定阈值 $Z_q=TS_q$，而小波阈值可确定为

$$T_{j,q}=\sqrt{TS_q} \tag{11.63}$$

根据式(11.63)所得阈值，可以利用 Donoho 的小波阈值处理方法对回波信号进行处理，最终利用处理后的小波系数重构得到检测信号。

利用基于正交小波变换的 CA-CFAR 检测方法对不同杂波背景下的目标信号进行处理。实验中原始目标信号采用线性调频信号，虚警率为 $P_{FA}=10^{-2}$、$P_{FA}=10^{-4}$ 和 $P_{FA}=10^{-6}$，在高斯杂波、瑞利杂波、韦布尔杂波和对数正态杂波背景下检测目标信号。根据检测信号的特征，实验中小波函数采用具有正交性的 db4 小波，最大的分解尺度为 6。与所得的检测结果相对应，不同处理方法检测后的信噪比提高值如表 11.3 所示。检测结果表明，虚警概率越高，检测出信号所含噪声越多，而虚警概率过低会导致信号的细节部分有所损失，这一点在软阈值处理方法中尤为突出。在 $P_{FA}=10^{-6}$ 时，四种杂波背景下软阈值方法检测的信号几乎都丢失了部分高频信息，而硬阈值方法在 $P_{FA}=10^{-6}$ 时基本能保留较好的高频信息。在 $P_{FA}=10^{-2}$ 时，两种阈值方法检测出来的信号均含有较多的噪声分量，但软阈值方法相对硬阈值方法较好些，表 11.3 中的数据很好地说明了这一点，从信噪比提高值来看，软阈值方法比硬阈值方法高出约 3dB。在 $P_{FA}=10^{-4}$ 时，虽然硬阈值方法检测的信号比软阈值方法含较多的噪声分量，但其较好地保留了目标信号的高频信息。从表 11.3 来看，对于高斯杂波、瑞利杂波和韦布尔杂波来说，在虚警概率较高时，软阈值处理优于硬阈值处理，但随着虚警概率的降低，硬阈值处理要好于软阈值处理，在 $P_{FA}=10^{-6}$ 时，硬阈值处理的信噪比提高值比软阈值处理至少大 2dB。从对数正态杂波的处理效果来看，软硬阈值的处理效果均不理想，在 $P_{FA}=10^{-6}$ 时，检测出的信号仍含有较多噪声。表 11.3 中的数据进一步表明，对数正态杂波在原始输入信噪比较大(0.3390dB)时，处理后的信噪比提高值仍然远小于其他三种杂波。

总的来说，与软阈值方法相比，硬阈值方法在有效检测目标信号的同时，较好地保留了目标高频细节信息，更适合基于正交小波变换的 CA-CFAR 检测方法。虽然该方法建立在高斯模型的基础上，但通过对不同杂波背景下信号的检测结果来看，其在瑞利杂波和韦布尔杂波背景下仍可以较好地检测出信号，具有一定的鲁棒性和自适应特性。

表 11.3 不同杂波背景下软硬阈值方法处理后的信噪比提高值

杂波类型	输入信噪比	软阈值处理方法			硬阈值处理方法		
		$P_{FA}=10^{-2}$	$P_{FA}=10^{-4}$	$P_{FA}=10^{-6}$	$P_{FA}=10^{-2}$	$P_{FA}=10^{-4}$	$P_{FA}=10^{-6}$
高斯杂波	−0.1086	7.7209dB	6.9240dB	6.2057dB	4.9940dB	7.9671dB	8.2293dB
瑞利杂波	−0.2419	8.2251dB	7.1813dB	6.4626dB	5.3403dB	8.3579dB	8.6189dB
韦布尔杂波	0.1609	7.3383dB	6.2573dB	5.4055dB	4.8274dB	7.3163dB	7.6947dB
对数正态杂波	0.3390	5.8151dB	5.2255dB	4.5636dB	2.4455dB	3.5570dB	4.8986dB

11.4 分数阶傅里叶变换域目标检测

对于运动目标的检测，通常通过傅里叶变换转变到频域，能够提高目标能量，前提是目标近似为匀速运动。然而，很多应用场景下动目标具有机动特性，如探测机动的飞机、低空导弹、海面起伏舰船等目标，这时回波近似建模为 LFM 信号，具有时变特性，其频谱发散，能量难以积累，从而不利于目标的检测。为此，本节讨论基于分数阶傅里叶变换(Fractional Fourier Transform，FRFT)的动目标检测方法，通过旋转一定的角度，在 FRFT 最佳变换域使信号能量得到最大程度的积累，进而检测出目标信号，提高雷达对于机动目标的检测能力。

FRFT 实质上是一种统一的时频变换，同时反映了信号在时域和频域的信息[16-18]。FRFT 将信号分解在 FRFT 域的一组正交的 chirp 基上，因而更适于用来分析或处理某些时变的非平稳信号，特别是 LFM 信号[19-20]。现代雷达系统中，目标多普勒频率与目标速度近似成正比，在较短的观测时间范围内，可用 LFM 信号作为运动目标回波的一阶近似模型[21]。传统的傅里叶变换不能对 LFM 信号进行有效的能量积累，而与常用二次型时频分布不同的是，FRFT 采用单一变量表示时频信息，没有交叉项干扰，又是一种线性变换，从而在加性噪声的干扰情况下更具有优势，并且具有比较成熟的快速离散算法，使得 FRFT 具有较好的工程实用基础。对于 LFM 信号，当旋转角度与信号相匹配时，可得到冲激信号，其能量聚集性最强；当旋转角度与信号不匹配时，仍然变换为广义的 LFM 信号。当信号分量之间和信号与噪声之间在时域或频域存在较强的耦合时，经典的时频分析方法和滤波方法难以实现有效的信号分离和信噪分离。而通过旋转一定的角度，FRFT 能够很容易实现有效的信号分离和滤波。因此，FRFT 扩展了传统 MTD 方法，是一种雷达动目标检测的有效工具[22-23]。本节从基于 FRFT 的 LFM 信号检测与估计基本原理出发，介绍 FRFT 域雷达动目标检测的主要技术途径。

11.4.1 基于 FRFT 的 LFM 信号检测与参数估计

1. LFM 信号的 FRFT

FRFT 为线性算子，信号的 FRFT 可解释为信号的表示轴在时频平面的旋转，定义为[16]

$$F_\alpha[x](u)=\int_{-\infty}^{+\infty}x(t)K_\alpha(t,u)\mathrm{d}t \tag{11.64}$$

$K_\alpha(t,u)$为核函数

$$K_\alpha(t,u)=\begin{cases}A_\alpha \mathrm{e}^{\mathrm{j}\left(\frac{1}{2}t^2\cot\alpha-ut\csc\alpha+\frac{1}{2}u^2\cot\alpha\right)}, & \alpha\neq n\pi\\ \delta(t-u), & \alpha=2n\pi\\ \delta(t+u), & \alpha=(2n+1)\pi\end{cases} \tag{11.65}$$

式中，$A_\alpha=\sqrt{\dfrac{1-\mathrm{j}\cot\alpha}{2\pi}}$，$n$ 取整数，α 为变换角度，与变换阶数 p 的关系为 $\alpha=p\pi/2, p\in(-2,2]$。式(11.64)说明信号 $x(t)$可被分解为 u 域上一组 LFM 基的线性组合。

噪声背景下的 LFM 信号模型可表示为

$$x(t)=s(t)+w(t)=A(t)\exp(\mathrm{j}2\pi f_0 t+\mathrm{j}\pi\mu t^2)+w(t) \tag{11.66}$$

式中，$A(t)$是信号幅度时间的函数，f_0 和 μ 分别为 LFM 信号的中心频率和调频率，$w(t)$为加性高斯白噪声。则 $x(t)$的 FRFT 为

$$\begin{aligned}F_\alpha[x(t)]&=A_\alpha \mathrm{e}^{\frac{\mathrm{j}u^2\cot\alpha}{2}}\int_{-\infty}^{\infty}[s(t)+w(t)]\,\mathrm{e}^{\mathrm{j}\left(\frac{1}{2}t^2\cot\alpha-ut\csc\alpha\right)}\,\mathrm{d}t\\&=A(t)A_\alpha \mathrm{e}^{\frac{\mathrm{j}u^2\cot\alpha}{2}}\int_{-\infty}^{\infty}\mathrm{e}^{\mathrm{j}\frac{(\cot\alpha+2\pi\mu)}{2}t^2+\mathrm{j}(2\pi f_0-u\csc\alpha)t}\,\mathrm{d}t+F_\alpha[w(t)]\end{aligned} \tag{11.67}$$

当变换角度与 LFM 信号调频率相匹配时，即 $\alpha_0=\arctan\left(-\dfrac{1}{2\pi\mu}\right)$时，则

$$|F_{\alpha_0}[x(t)]|=|A(t)A_{\alpha_0}\delta(2\pi f_0-u\csc\alpha_0)|+|F_{\alpha_0}[w(t)]| \tag{11.68}$$

由文献[17]可知，高斯函数的 FRFT 是具有复变量的高斯函数。由式(11.68)可知，LFM 信号在 FRFT 域呈现冲激函数，而噪声不会呈现明显的能量聚集，利用这一特性可实现噪声背景下的 LFM 信号检测。LFM 信号在时频平面的谱分布如图 11.17 所示。在频域，LFM 信号的能量分布于很宽的频谱范围内。旋转时频轴，使得 LFM 信号与某组基的调频率相匹配，在该组基上形成峰值，信号能量得到最大限度积累，说明 LFM 信号在 FRFT 域上具有良好的时频聚集性，u_0 域称为最佳 FRFT 域。

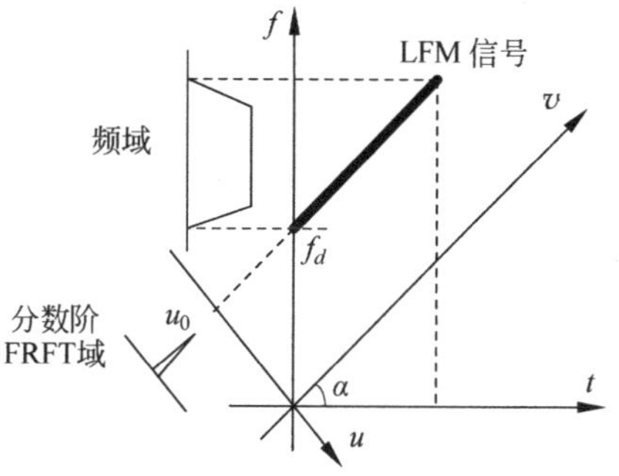

图 11.17　LFM 信号在时频平面的谱分布

对回波信号进行 FRFT，信号能量在参数平面(α,u)上形成二维分布，通过阈值搜索此二维平面的峰值点，可确定信号的最佳变换角度，即

$$(\alpha_0,u_0)=\underset{\alpha,u}{\arg\max}|F_\alpha(u)| \tag{11.69}$$

上述方法存在以下两方面不足：一是参数估计精度由扫描步长决定，当精度要求比较高时，就需要采用很小的搜索步长，使得计算量增大；二是仅适用于高信噪(杂)比环境，变换域信号峰值易受到噪声或杂波干扰，不能有效积累信号能量。因此，人们分别从降低运算量和提高变换域信噪(杂)比的角度对传统二维峰值搜索方法进行改进，主要方法有步进式粗搜和拟 Newton 法精搜方法、分级迭代峰值搜索法、FRFT 极值混合优化算法、最大分数阶时宽带宽比值法和最小基带带宽法、黄金分割的优化峰值搜索方法等[24-25]，这些方法在一定程度上提高了算法的运算效率。文献[26]利用高阶统计量抑制噪声的特性，通过计算雷达回波信号在 FRFT 域的峰度值，采用分级迭代的搜索方法，可有效确定低信噪(杂)比下 LFM 信号的最佳变换角度。

2. FRFT 域 LFM 信号的参数估计

根据 FRFT 域峰值点坐标(α_0,u_0)，对于式(11.66)中的 LFM 信号模型，量纲归一化处理后的参数估计方法为

$$
\begin{cases}
\hat{\mu} = -\cot\alpha_0 / S^2 \\
\hat{f}_0 = u_0 \csc\alpha_0 / S \\
\hat{A}(t) = \mathrm{Re}\left[x(t)\exp(-2\mathrm{j}\pi\hat{f}_0 t + \mathrm{j}\pi\hat{\mu}t^2)\right]
\end{cases}
\tag{11.70}
$$

式中,$\hat{\mu}$、$\hat{f}_0$ 和 $\hat{A}(t)$分别为 LFM 信号的调频率、中心频率和幅度的估计,尺度因子 $S=\sqrt{T/f_s}$。

11.4.2 FRFT 域动目标检测器设计

1. FRFT 域动目标检测原理

对动目标回波模型的讨论得知,目标平动的建模包括匀速运动和匀加速运动(有限高阶运动和无穷高阶运动)。采用 FRFT 检测动目标的基本原理是,在雷达发射单频信号或 LFM 信号的前提下,目标多普勒频率与目标速度近似成正比,目标的运动状态不同,参数估计方法略有不同,根据 Weierstrass 近似原理,其回波信号可由足够阶次的多项式相位信号近似表示,在一段短的观测时间范围内,可采用二次相位信号,即 LFM 信号作为动目标回波模型。通过选择合适的变换阶数,将最佳 FRFT 域的信号幅值作为检测统计量,与门限进行比较后判断目标的有无。因此,采用 FRFT 检测杂波背景下的动目标具有很大的优势[27]。

目标的高速运动和较远的雷达观察距离导致多普勒模糊、距离模糊和低 SNR 等问题,使得高速微弱动目标检测一直以来是弹道目标、空间目标的预警和探测以及外辐射源雷达信号处理领域的难题。利用长时间相参处理进行信号积累是提高微弱运动目标检测能力的一种有效方法,可改善 SCR,即利用时间换取能量[28-29]。由于传统采用傅里叶变换的相参积累技术仅是对含线性相位信号的最佳匹配滤波器,而目标径向加速度产生二次相位调制,因此会导致积累增益下降。此时,可借鉴 MTD 多普勒滤波器组的思想,将 p 阶 FRFT 看成一组扫频滤波器组,采用 FRFT 同时对中心频率和调频率补偿,然后通过构建 FRFT 域(p,u)检测单元图对动目标进行 CFAR 检测,如图 11.18 所示,使得 FRFT 适于实现二次相位补偿的长时间相参处理,有效增强雷达在强杂波背景下对微弱运动目标的检测能力。

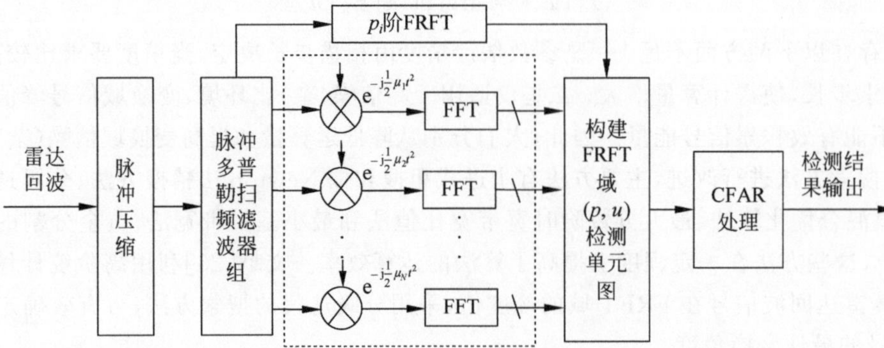

图 11.18　基于 FRFT 的加速度补偿相参积累检测原理框图

遍历所有距离单元,形成不同距离单元回波信号的 $M\times N$ 维最佳 FRFT 谱,其中 M 为距离单元数,N 为获取的回波采样点数

$$
\boldsymbol{S}_{\mathrm{FRFT}}^{\alpha_{i_0}}(u)=\begin{bmatrix} S_{\mathrm{FRFT}}^{\alpha_{1_0}}(1)\mid_{r_1}, & S_{\mathrm{FRFT}}^{\alpha_{1_0}}(2)\mid_{r_1}, & \cdots & ,S_{\mathrm{FRFT}}^{\alpha_{1_0}}(N)\mid_{r_1} \\ S_{\mathrm{FRFT}}^{\alpha_{2_0}}(1)\mid_{r_2}, & S_{\mathrm{FRFT}}^{\alpha_{2_0}}(2)\mid_{r_2}, & \cdots & ,S_{\mathrm{FRFT}}^{\alpha_{2_0}}(N)\mid_{r_2} \\ \vdots & \vdots & & \vdots \\ S_{\mathrm{FRFT}}^{\alpha_{M_{1_0}}}(1)\mid_{r_M}, & S_{\mathrm{FRFT}}^{\alpha_{M_{1_0}}}(2)\mid_{r_M}, & \cdots & ,S_{\mathrm{FRFT}}^{\alpha_{M_{1_0}}}(N)\mid_{r_M} \end{bmatrix}_{M\times N} \tag{11.71}
$$

式中，$\boldsymbol{S}_{\mathrm{FRFT}}^{\alpha}(u)\mid_{r_i}$ 表示距离 r_i 处回波的 FRFT 域表示，则最佳 FRFT 谱为 $\boldsymbol{S}_{\mathrm{FRFT}}^{\alpha_{i_0}}(u)\mid_{r_i}$。

将 $\boldsymbol{S}_{\mathrm{FRFT}}^{\alpha_{i_0}}(u)$幅值作为检测统计量，与给定虚警率条件下的检测门限进行比较判决

$$
|\boldsymbol{S}_{\mathrm{FRFT}}^{\alpha_{i_0}}(u)| \underset{\mathrm{H}_0}{\overset{\mathrm{H}_1}{\gtrless}} \eta \tag{11.72}
$$

式中，η_2 为检测门限，由虚警概率确定，如果检测统计量低于检测门限，判决为该距离单元没有机动目标，若检测统计量高于检测门限，则判决为该距离单元存在机动目标。

2. FRFT 域动目标检测器设计过程

由于 LFM 信号在不同的 FRFT 域上呈现出不同的能量聚集性，检测含有未知参数的 LFM 信号的基本思路是以旋转角为变量进行扫描，求观测信号的 FRFT，从而形成信号能量在参数(α,u)平面上的二维分布，在此平面上按阈值进行峰值点的二维搜索即可检测单个动目标回波信号并估计其运动参数。建立动目标检测模型[30]

$$
\begin{cases} \mathrm{H}_1: F_p(u)=S_p(u)+N_p(u) \\ \mathrm{H}_0: F_p(u)=N_p(u) \end{cases} \tag{11.73}
$$

式中，$S_p(u)$、$N_p(u)$分别为目标 $s(t)$和高斯杂波 $n(t)$的 p 阶 FRFT。

$F_p(u)$为复信号，其实部和虚部分别为 $\mathrm{Re}[F_p]$和 $\mathrm{Im}[F_p]$。将雷达回波信号量纲归一化处理，进行 FRFT，经过包络检波后输出结果为

$$
M=\sqrt{\mathrm{Re}^2[F_p]+\mathrm{Im}^2[F_p]}=|F_p(u)| \tag{11.74}
$$

设随机过程 $n(t)$的协方差为

$$
C_n(\tau)=\mathrm{E}\{n(t)n^*(t+\tau)\}=N_0\delta(\tau) \tag{11.75}
$$

则 $n(t)$的 FRFT 域的协方差为

$$
C_N(\tau)=\mathrm{E}\{N_p(t,u)N_p^*(t+\tau,u)\}=\int_0^T n(t)n^*(t+\tau)K_p(t,u)K_p^*(t,u)\mathrm{d}t=A_\alpha^2N_0T \tag{11.76}
$$

则随机向量 $\mathrm{Re}[F_p]\mid\mathrm{H}_0$、$\mathrm{Im}[F_p]\mid\mathrm{H}_0$ 服从零均值、方差为 $A_\alpha^2N_0T/2$ 的高斯分布，其概率密度函数为

$$
f(\mathrm{Re}[F_p]\mid\mathrm{H}_0)=f(\mathrm{Im}[F_p]\mid\mathrm{H}_0)=\frac{1}{\pi A_\alpha^2N_0T}\exp\left[-\frac{|\mathrm{Re}([F_p])|^2}{A_\alpha^2N_0T}\right] \tag{11.77}
$$

令 $A_\alpha^2N_0T/2=\sigma^2$，则包络峰值 $M\mid\mathrm{H}_0$ 服从 Rayleigh 分布，其 PDF 为

$$
f(M\mid\mathrm{H}_0)=\frac{M}{\sigma^2}\exp\left(-\frac{M^2}{2\sigma^2}\right),\quad M\geqslant 0 \tag{11.78}
$$

当有目标存在时，考虑到 $n(t)$ 的均值为零，则向量 $\mathrm{Re}[F_p]|\mathrm{H}_1$、$\mathrm{Im}[F_p]|\mathrm{H}_1$ 同样服从高斯分布，数学期望为

$$\mathrm{E}\{\mathrm{Re}[F_p] \mid \mathrm{H}_1\}=\mathrm{E}\{\mathrm{Im}[F_p] \mid \mathrm{H}_1\}=\frac{1}{2}\mathrm{E}\{S_p(u)\}=\frac{1}{2}AA_\alpha T\mathrm{e}^{\mathrm{j}\frac{1}{2}u^2\cot\alpha} \tag{11.79}$$

$$f(\mathrm{Re}[F_p] \mid \mathrm{H}_1)=f(\mathrm{Im}[F_p] \mid \mathrm{H}_1)=\frac{1}{2\pi\sigma^2}\exp\left(-\frac{|\mathrm{Re}[F_p]-\mathrm{E}[\mathrm{Re}[F_p] \mid \mathrm{H}_1]|^2}{2\sigma^2}\right) \tag{11.80}$$

那么包络峰值 $M|\mathrm{H}_1$ 服从莱斯分布，其 PDF 为

$$f(M \mid \mathrm{H}_1)=\frac{M}{\sigma^2}\mathrm{I}_0\left(\frac{AA_\alpha T}{2\sigma^2}M\right)\exp\left[-\frac{M^2+(AA_\alpha T/2)^2}{2\sigma^2}\right] \tag{11.81}$$

$I_0()$是第一类零阶贝塞尔函数。

由式(11.78)和式(11.81)得似然比为

$$\Lambda=\frac{f(M \mid \mathrm{H}_1)}{f(M \mid \mathrm{H}_0)}=\exp\left(\frac{A^2T}{4N_0}\right)I_0\left(\frac{AA_\alpha T}{2\sigma^2}M\right) \tag{11.82}$$

由于函数 I_0 为单调上升函数，故以 $I_0\left(\frac{AA_\alpha T}{2\sigma^2}M\right)$ 进行判决完全等效于以统计量 M 进行判决，于是判决规则可以写为

$$M=|F_p(u)|\underset{\mathrm{H}_0}{\overset{\mathrm{H}_1}{\gtrless}}\beta \tag{11.83}$$

式中，β 是判决门限，其值由给定的虚警概率和杂波功率水平确定。因此，回波经 FRFT，包络检波后，与门限比较，如果大于门限，则认为有目标，相反则认为不存在目标。当检测到目标时，可估计信号参数。

讨论高斯杂波背景下 FRFT 域动目标检测系统的检测性能。检测器变量 $M|\mathrm{H}_0$ 服从 Rayleigh 分布，所以虚警概率等于

$$P_{\mathrm{fa}}=\int_\beta^{+\infty}f(M \mid \mathrm{H}_0)\mathrm{d}M=\exp\left(-\frac{\beta^2}{2\sigma^2}\right) \tag{11.84}$$

检测概率等于

$$P_{\mathrm{d}}=\int_\beta^{+\infty}f(M \mid \mathrm{H}_1)\mathrm{d}M=\int_\beta^{+\infty}\frac{M}{\sigma^2}\exp\left[-\frac{M^2+(AA_\alpha T/2)^2}{2\sigma^2}\right]I_0\left(\frac{AA_\alpha T}{2\sigma^2}M\right)\mathrm{d}M \tag{11.85}$$

3. FRFT 检测器输出信噪比分析

由于 FRFT 具有在二维 FRFT 域聚集信号而分离噪声的性质，传统的信噪比定义，即平均信号功率与平均噪声功率的比值，已不再适用。文献[31]提出把信号在 FRFT 域的峰值平方作为信号功率，该处的噪声方差作为噪声功率。采用 FRFT 检测动目标的检测统计量为

$$|X_\alpha(u)|^2=|S_\alpha(u)+W_\alpha(u)|^2 \tag{11.86}$$

其中，$X_\alpha(u)$、$S_\alpha(u)$、$W_\alpha(u)$分别为 $x(t)$、$s(t)$和 $w(t)$的 α 旋转角度的 FRFT。则检测器的输出信噪比为

$$\mathrm{SNR}_{\mathrm{out}}=\frac{|S_\alpha(u)|^4}{\mathrm{var}[|X_\alpha(u)|^2]}=\frac{[(2N+1)A^2/\sigma_n^2]^2}{2[(2N+1)A^2/\sigma_n^2+1]}=\frac{(Tf_s\mathrm{SNR}_{\mathrm{in}})^2}{2(Tf_s\mathrm{SNR}_{\mathrm{in}}+1)} \tag{11.87}$$

其中，σ_n^2 为噪声方差，输入信噪比 $\mathrm{SNR_{in}}$ 定义为 A^2/σ_n^2。显然增加数据长度 N，即增加观测时长 T 能够改善输出信噪比，提高参数估计精度。

11.4.3 FRFT 域长时间相参积累检测方法

常规雷达采用机械扫描，每个指向的波束驻留时间较短，因而可利用的脉冲积累回波数量较少[29,32]。通常，长时间积累可分为非相参积累和相参积累两种方法，二者的不同之处在于是否利用了信号的相位信息[33-34]。非相参积累对系统没有严格的相参要求，在工程实现上比较简单，但其积累增益和对 SNR 的改善能力较差，对于复杂背景下微弱动目标检测难以取得理想效果。相参积累方法利用了目标的运动特性和多普勒信息，与非相参积累方法相比，得到的积累增益更高。目前，相参积累处理方法在实施过程中主要存在两个方面的困难：其一，由于目标运动速度快、雷达距离分辨力不断提高，目标回波会产生距离徙动效应(Across Range Unit，ARU)[35]，这种效应是包络在不同脉冲周期之间弯曲和走动所产生的，会导致距离向目标能量分散，传统的基于距离单元的 MTD 方法已不能有效适应该类目标的检测；其二，目标运动表现为高阶变速运动、转动等复杂形式，造成回波相位变化，使得雷达回波信号具有非平稳时变特性，并且其相位为高阶形式，导致目标回波产生多普勒频率徙动(Doppler Frequency Modulation，DFM)，目标能量在频域分散，进而导致相参积累增益降低[36]。

包络相关法是一种常用的 ARU 补偿方法，但其在 SNR/SCR 较低的情形下，由于相邻回波不具有良好的相关性，导致包络对齐效果较差；Keystone 变换(Keystone Transform，KT)方法适用于多目标、低 SNR 环境[37]，但不能正确校正由目标径向加速度引起的距离弯曲。长时间相参积累(Long-Time Coherent Integration，LTCI)技术同时利用了目标回波的幅度和相位信息，具有积累增益高，抗杂波性能好等优点，能对具有威胁的微弱动目标进行早期预警，非常适于复杂环境下微弱动目标的检测[38-45]。Radon-傅里叶变换法(Radon-Fourier Transform，RFT)通过联合搜索参数空间中目标参数的方式解决了距离徙动与相位调制耦合的问题[44]，通过长时间的能量积累，提高雷达的探测性能。机动目标回波信号具有时变特性，因此采用高阶信号如 LFM 信号或二次调频(QFM)信号能够很好地表征目标的加速度和急动度信息。目前的多普勒徙动补偿方法主要是以时频分析技术和高阶信号处理技术为主，包括高阶模糊函数(HAF)[46]、多项式傅里叶变换(PFT)[47]及多项式相位变换(PPT)[48]等方法。上述算法计算复杂，高阶次的非线性变换会产生交叉项，影响参数估计和信号检测，而且可利用信号脉冲数的多少会对其补偿性能产生严重影响。综上所述，如何有效、同步地完成 ARU 和 DFM 效应的补偿是决定微弱运动目标长时间相参积累性能的关键性问题。

1. Radon-FRFT 长时间相参积累

针对具有加速度的机动目标，假设 $x(t_\mathrm{m},r_\mathrm{s})\in C$ 是 $(t_\mathrm{m},r_\mathrm{s})$ 平面的二维复函数，根据目标初始运动参数，确定参数化曲线 $r_\mathrm{s}=r_0+vt_\mathrm{m}+at_\mathrm{m}^2/2$。将连续 RFRFT 定义为[35]

$$G_r(\alpha,u)=\mathcal{F}^\alpha[x(t_\mathrm{m},r_\mathrm{s})](u)=\int_{-\infty}^{\infty}x(t_\mathrm{m},r_0+vt_\mathrm{m}+at_\mathrm{m}^2/2)K_\alpha(t_\mathrm{m},u)\mathrm{d}t \tag{11.88}$$

式中，$\mathcal{F}^\alpha$ 表示变换角 $\alpha\in(0,\pi]$ 的 RFRFT 算子，参数 r、v 和 a 均有明确的物理含义，核函数为

$$K_\alpha(t_\mathrm{m},u)=\begin{cases}A_\alpha\exp\left[\mathrm{j}\left(\dfrac{1}{2}t_\mathrm{m}^2\cot\alpha-ut_\mathrm{m}\csc\alpha+\dfrac{1}{2}u^2\cot\alpha\right)\right], & \alpha\neq n\pi\\ \delta\left[u-(-1)^n t_\mathrm{m}\right], & \alpha=n\pi\end{cases} \tag{11.89}$$

则具有加速度的机动目标回波的 RFRFT 表达式为

$$S_{\mathrm{RFRFT}}=\int s_{\mathrm{PC}}\left[2(r_0+v_0t_{\mathrm{m}}+a_{\mathrm{s}}t_{\mathrm{m}}^2/2)/c,t_{\mathrm{m}}\right]K_{\alpha}(t_{\mathrm{m}},u)\mathrm{d}t_{\mathrm{m}} \tag{11.90}$$

量纲归一化后,RFRFT 所需的变换角 α 与变换阶数 p 的关系为

$$p_i=-\frac{2\operatorname{arccot}(\mu_i S^2)}{\pi}+2=-\frac{2\operatorname{arccot}(2a_i S^2/\lambda)}{\pi}+2 \tag{11.91}$$

式中,$S=\sqrt{T_{\mathrm{n}}/f_{\mathrm{r}}}$ 为量纲归一化的尺度因子。当 $p=1$ 时,估计的加速度等于零,则 RFRFT 为 RFT 的广义形式;当距离走动量不超过一个距离单元,即 $\Delta r_{\mathrm{s}}\leqslant\rho_{\mathrm{r}}$ 时,FRFT 又是 RFRFT 的特例。

RFRFT 是线性变换,不受交叉项干扰。图 11.19 对几种常见的运动目标相参积累方法和 RFRFT 的积累时间进行了比较,容易得出以下几个结论。

(1) RFRFT 既能够获得 RFT 的长相参积累增益,又如同 FRFT 一样适用于分析时变、非平稳信号,该方法结合了二者的优势。

(2) RFRFT 的核函数 $K_{\alpha}(t_{\mathrm{m}},u)$ 可以补偿由于目标变速运动导致的回波脉冲间起伏和走动,得到明显峰值。

(3) 可将 RFRFT 看作一种由变换阶数决定的广义多普勒滤波器组,能够同时匹配目标运动速度和加速度。

(4) RFRFT 的相参积累时间明显长于 MTD 和 FRFT 方法,因此该方法具有优秀的杂波抑制性能,能够进一步提高雷达对低可观测、远距离、高速高机动目标的检测能力。

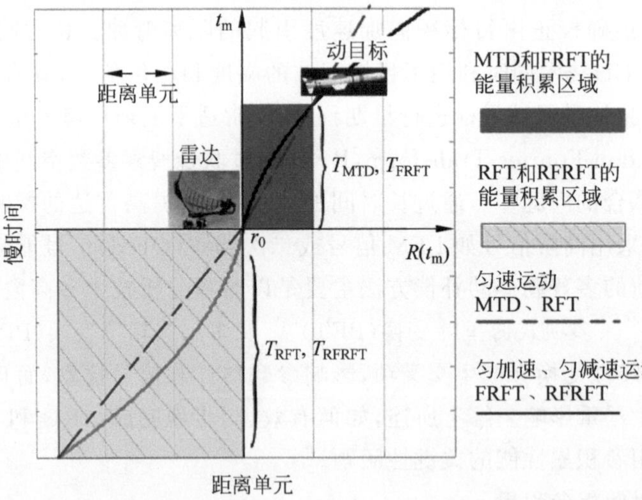

图 11.19　相参积累处理方法示意图(MTD,FRFT,RFT 和 RFRFT 的相参积累时间比较)

2. Radon-分数阶模糊函数(RFRAF)长时间相参积累

针对具有急动度的高机动目标,首先定义长积累时间下的瞬时自相关函数(Long-time Instantaneous ACF,LIACF):假设 $f(t_{\mathrm{m}},r_{\mathrm{s}})\in C$ 是定义在距离-慢时间平面 $(t_{\mathrm{m}},r_{\mathrm{s}})$ 的二维复函数,$r_{\mathrm{s}}=r_0+v_0t_{\mathrm{m}}+a_{\mathrm{s}}t_{\mathrm{m}}^2/2+g_{\mathrm{s}}t_{\mathrm{m}}^3/6$ 表示此平面内的任意一条曲线,代表目标复杂的高阶运动,其中 v_0、a_{s} 和 g_{s} 表示目标的搜索运动参数,则 $f(t_{\mathrm{m}},r_{\mathrm{s}})$ 的 LIACF 定义为[38]

$$R_f(t_{\mathrm{m}},\tau)=f\left(t_{\mathrm{m}}+\frac{\tau}{2},r_{\mathrm{s}}\right)f^*\left(t_{\mathrm{m}}-\frac{\tau}{2},r_{\mathrm{s}}\right) \tag{11.92}$$

表示沿曲线 r_{s} 提取位于 $(t_{\mathrm{m}},r_{\mathrm{s}})$ 二维平面中的目标观测值 $f(t_{\mathrm{m}},r_{\mathrm{s}})$,并对其进行自相关运算,$f(t_{\mathrm{m}},$

r_s)的连续 RFRAF 定义为

$$\text{RFRAF}[f(t_m,r_s)](\tau,u)=\mathcal{R}_f^\alpha(\tau,u)=\int_{-\infty}^{\infty}R_f(t_m,\tau)K_\alpha(t_m,u)\mathrm{d}t_m \tag{11.93}$$

式中，$\mathcal{R}^\alpha()$表示 RFRAF 算子，$\alpha\in(0,\pi]$为旋转角度。逆 RFRAF(Inverse RFRAF，IRFRAF)由$-\alpha$参数的 RFRAF 确定，即$\mathcal{R}^\alpha(\mathcal{R}^{-\alpha})=\mathcal{R}^{\alpha-\alpha}=\mathcal{R}^0=I$。

由 RFRAF 的定义及其物理含义可知，RFRAF 表示瞬时自相关函数$R_f(t_m,\tau,r_s)$在(t_m,r_s)平面内的一种仿射变换。图 11.20 为基于 RFRAF 的长时间相参积累原理图，RFRAF 能够很好地匹配和积累建模为 QFM 信号的机动目标或微动目标回波信号，并通过 LIACF 和对时频平面的旋转在 RFRAF 形成峰值，峰值坐标为(p,u)或(τ,u)。RFRAF 根据目标的运动参数提取位于距离-慢时间(方位向)二维平面中的目标观测值，进行瞬时自相关的降阶运算，然后通过 FRFT 对该观测值进行相参积累，达到匹配动目标信号，改善 SCR/SNR 的目的。

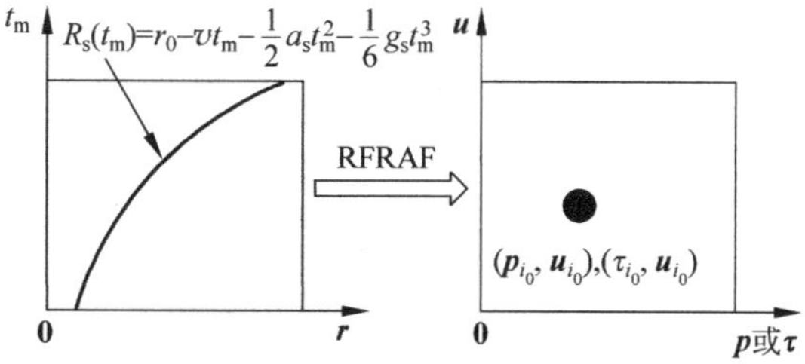

图 11.20　基于 RFRAF 的长时间相参积累原理图

3. Radon-FRFT 机动目标检测方法

图 11.21 给出了利用高阶相位信息的雷达动目标检测方法流程图。该方法主要解决非匀速运动和高速机动目标的检测问题，根据预先设定的目标运动搜索参数范围，提取位于距离-慢时间二维平面中的运动目标观测值，然后在 RFRFT 和 RFRAF 域选择一系列的变换角度对该观测值进行匹配和积累，从而实现微弱机动目标能量的长时间相参积累。

主要分为以下几个步骤。

(1) 雷达回波距离向解调、脉压，完成脉内积累。

(2) 根据波束驻留时间和雷达系统参数，对脉间相参积累时间T_n、距离搜索范围、间隔等进行设定，根据待测目标的运动状态、类型，确定需要补偿的初速度、加速度以及急动度的间隔和搜索范围，实现长时间相参积累参数的初始化。

(3) RFRFT、RFRAF 补偿距离和多普勒徙动，完成长时间脉间相参积累，RFRFT 和 RFRAF 根据目标的运动参数提取位于距离-慢时间二维平面中的目标观测值，然后通过不同变换阶数下的广义多普勒滤波器对该观测值进行长时间相参积累，因此目标的运动参数分别对应 RFRFT、RFRAF 域中的坐标。

(4) 构建距离-RFRFT、RFRAF 域检测单元图，并进行 CFAR 检测，通过对所有参数搜索范围进行遍历，得到某距离单元内的最佳 RFRFT、RFRAF 域幅值，继续计算其他距离单元，将距离单元-最佳 RFRFT、RFRAF 域检测统计量与给定虚警概率下的自适应检测门限进行比较，判断目标的有无。

需要说明的是，最佳变换域的确定仍是关键技术问题。在实际工程应用中，可根据待检测目标的特性及应用场景，设定一定数量的变换参数，分别对应匀速运动、机动、高机动等运动状态，依次对不同距离单元内的脉冲进行多个变换参数条件下的 RFRFT 或 RFRAF 运算，实现动目标的快速有效

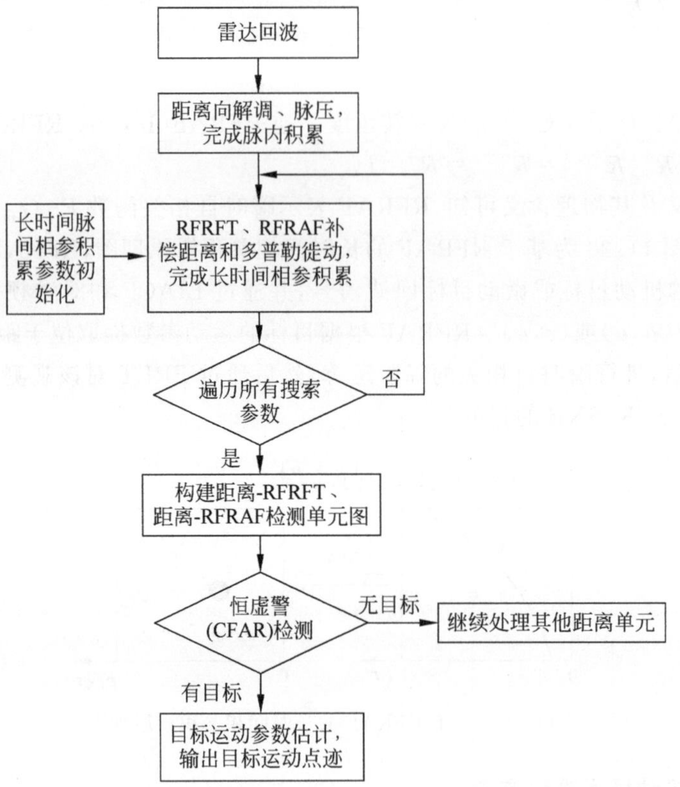

图 11.21　基于 RFRFT 和 RFRAF 的机动目标检测方法流程图

检测。

4. 海上机动目标检测结果与分析

受复杂海洋环境和目标特性的影响，海杂波背景下的微弱运动目标检测一直以来是雷达信号处理领域的难题。这里采用南非科学与工业研究理事会(Council for Scientific and Industrial Research, CSIR)采集的对海雷达数据验证算法性能。分别选取两组数据，其中一组数据包含加速运动的海上合作目标(CSIR-TFC15-038)，另一组数据目标做复杂机动(TFC17-006)。

CSIR-TFC15-038 数据的目标雷达回波描述及特性分析如图 11.22 所示，从图 11.22(a)可明显看出，由风速产生的周期性海面起伏，通过显著波高分析可知试验时的海况等级较高，而且其频谱具有非平稳时变特性，多普勒分布在 50～150Hz，谱宽较宽，表明观测方向为逆风向。图 11.22(a)中一合作海面目标从第 17 个距离单元远离雷达运动，白色曲线为目标的实际 GPS 距离，该目标雷达回波极其微弱，被强海杂波所覆盖。由图 11.22 可知，目标在较长的观测时间内，跨越了约 30 个距离单元，产生了距离走动和距离弯曲，目标回波存在二次和高阶相位，需要进行距离和多普勒徙动补偿以积累目标能量。

图 11.23 对基于 MTD、FRFT 和 RFRFT 方法的处理结果进行了比较，其中受限于 ARU 效应，MTD 的相参积累时间设为 2.25 s，目标的多普勒谱被海杂波和噪声所覆盖，难以利用多普勒滤波器组最大输出检测和估计目标。通过比较 MTD 和 FRFT 的检测结果可知，在最佳 FRFT 域($p_{\text{opt}}=0.970$)目标能量有所集中，估计目标的运动参数为 $v_0=-4.335$ 海里/小时，$a_s=-0.276\text{m/s}^2$，然而由于积累脉冲数量有限，目标和杂波峰值比仍较低，导致存在部分虚警。采用长时间相参积累的方法处理机动目

标回波，初始搜索时间和搜索距离为 50s 和第 27 个距离单元，相参积累时间设为 9s，得到图 11.23(c)，可知 RFRFT 在获得高积累增益的同时能够抑制海杂波。

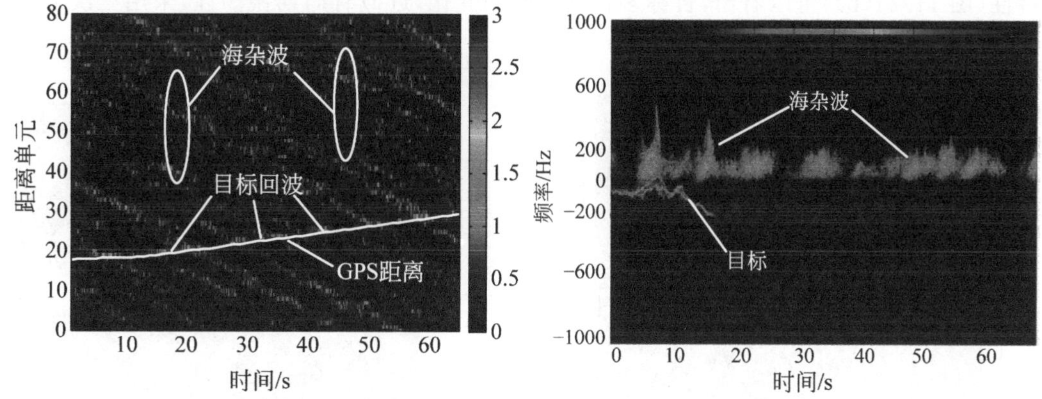

(a) 目标雷达回波距离-时间图　　(b) 目标回波时频分析(距离单元20)

图 11.22 CSIR-TFC15-038 数据特性分析

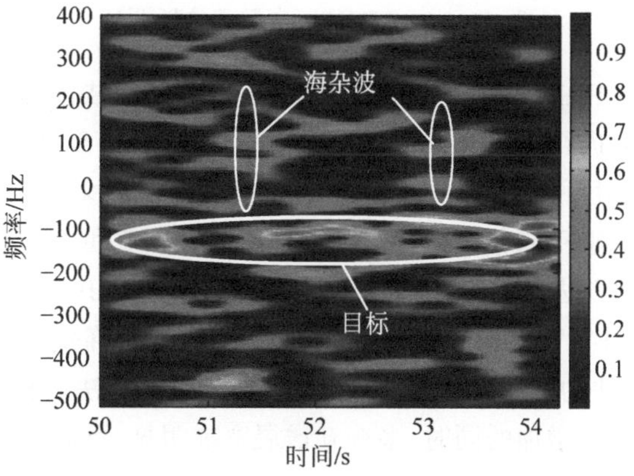

(a) 目标STFT谱(T_n=2.25 s)

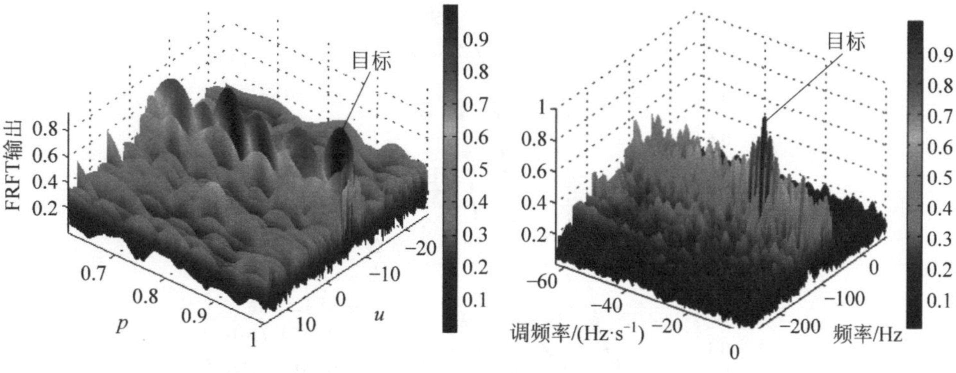

(b) 目标FRFT谱(T_n=2.25 s)　　(c) 目标RFRFT谱(T_n=9 s)

图 11.23 不同方法的相参积累结果比较

采用 CSIR 数据库中的 TFC17-006 数据验证所提算法，在高海况条件下由 Fynmeet 雷达采集合作目标乘浪者号充气橡皮艇的回波数据。图 11.24(a)的雷达回波距离-时间图表明雷达观测范围约为 45 个距离单元,观测时间为 100s,仅通过幅度信息很难从强海杂波背景中检测出目标。进一步分析回波的时频特性(图 11.24(b))可以看出,目标多普勒随时间变化,近似有周期振荡性,采用三次多项式函数能够很好地拟合目标运动轨迹。

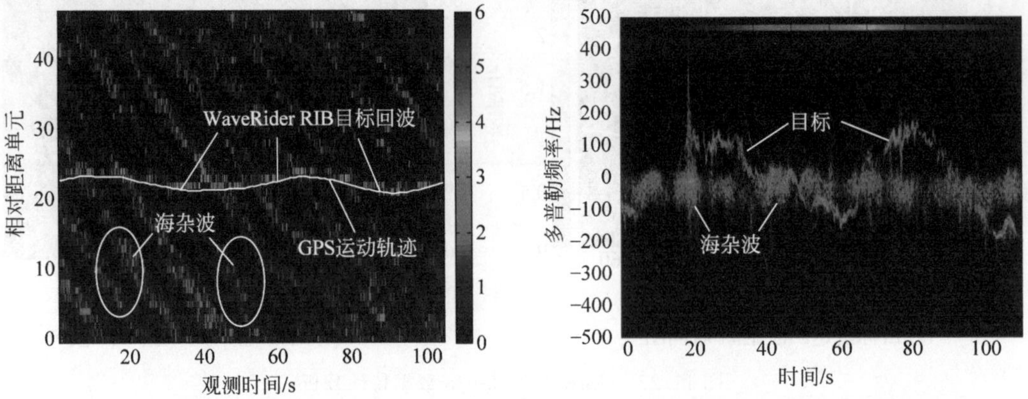

(a) 目标雷达回波距离-时间图　　(b) 目标回波时频分析(距离单元20)

图 11.24　TFC17-006 数据特性分析

将 MTD、FRFT、RFT 和 RFRFT 方法与双参数 CFAR 检测器相结合,形成相应的检测方法,并分别在噪声背景和实测海杂波背景下进行 Monte Carlo 仿真分析与验证,得到图 11.25 所示的目标发现概率与 SCR 的关系曲线,$P_{fa}=10^{-4}$。由图 11.25 可以看出,MTD 和 FRFT 检测方法的积累脉冲数有限,因此当 SNR<−10dB 时,检测概率急剧下降,信号幅值被噪声淹没,而 RFT 和 RFRFT 利用长时间相参积累,有效积累脉冲数远远大于其余两种方法,进一步改善了 SNR,使得在−10dB 也能达到较好的检测性能;达到相同的检测概率($P_d=0.8$),RFRFT 相对于 RFT 算法,对 SNR 的需求降低 3dB 左右;随着 SNR 的提高,尤其是 SNR>−8dB 时,RFT 算法的检测性能却增加缓慢,与 FRFT 算法的检测性能曲线存在交叉,这是因为 RFT 未补偿多普勒徙动,SNR 增加的同时,多普勒谱展宽越明显,目标能量发散。

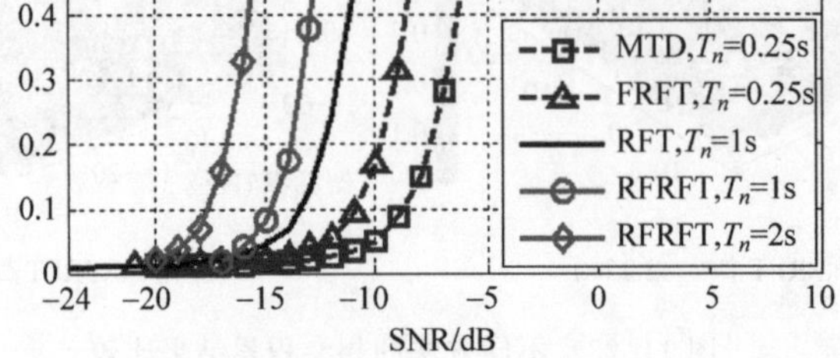

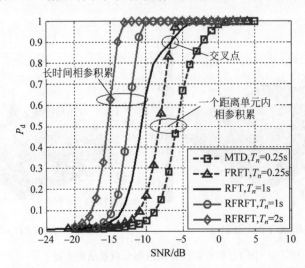

图 11.25　不同积累时间下的 MTD、RFT、FRFT 和 RFRFT 检测器检测概率($P_{fa}=10^{-4}$)

11.5　Hilbert-Huang 变换域目标检测

Hilbert-Huang 变换(Hilbert-Huang Transformation,HHT)从根本上摆脱了傅里叶分析的限制,提出了固有模态函数(Intrinsic Mode Function,IMF)和经验模态分解(Empirical Mode Decomposition,EMD)的概念,使得信号的瞬时频率具有了物理意义,从而能得到非平稳信号完整的时频谱,具有优秀的时频分析能力和自适应分解信号能力等。本节介绍 HHT 的原理,并以海杂波中目标检测为例,从海杂波的 IMF 特性方面出发,讨论基于固有模态奇异值熵的微弱目标检测方法。

11.5.1　HHT 基本原理

Hilbert-Huang 变换(HHT)[49]是由黄锷博士等提出的一种信号处理技术,被认为是近年来对以傅里叶变换为基础的线性或平稳信号分析的一大突破,自提出以来,就受到了众多学者的青睐。HHT 的主要创新在于从根本上摆脱了傅里叶分析的限制,提出了固有模态函数(IMF)和经验模态分解(EMD)的概念,使得信号的瞬时频率具有了物理意义,从而能得到非平稳信号完整的时频谱。HHT 方法具有优秀的时频分析能力和自适应分解信号能力,非常适合处理具有非平稳、时变和非线性特性的雷达信号,在噪声抑制[50]、目标检测[51-55]、特征提取[56-57]等方面有广泛应用。文献[52-55]将 HHT 和 EMD 用于海杂波抑制和目标检测,通过构建 Hilbert 谱脊线、EMD 能量熵、Hilbert 边际谱、Hilbert 谱脊线盒维数等特征,用于区分海杂波和目标,提高了海杂波中微弱目标检测性能。研究结果表明,HHT 方法是一种有效的时频分析方法,能较好地描述和提取信号时频特征。

HHT 的基本原理为:通过 EMD,提取其自身固有的一族模态函数,即 IMF,然后对各个 IMF 进行 Hilbert 变换构造解析信号,进而得到其瞬时角频率和振幅,最后得到完整的 Hilbert 谱及其边际谱[54]。HHT 是一种具有自适应能力的时频分析方法,它可根据信号的局部时变特征进行自适应分解,无须预先设定基函数,其基函数在分解过程中自适应产生,消除了人为因素的影响,克服了传统方法中用无意义的谐波分量表示非平稳、非线性信号的缺陷,并可得到极高的时频分辨率,具有良好的时频聚集性,非常适合对非平稳、非线性信号进行分析。

1. 特征尺度参数

描述信号特征的基本参数包括时间和频率,频率虽能反映信号的本质特征,但不直观。有时直接从时域观察信号的变化过程可以直观得到类似频率的信号特征,这就是特征尺度。通过观察,很容易获得信号特定点之间的时间跨度,称之为时间尺度参数,它和频率一样,都能够描述信号的本质。在傅里叶变换中,基函数的时间尺度参数与频率具有定量的关系,表明了谐波函数的周期长度。而对于非平稳信号,时间尺度参数是基于信号特征点的特征参数,虽然与傅里叶频谱没有定量的关系,但更能够反映非平稳信号的特征。

时间尺度参数定义为信号在特定点之间的时间跨度,数学上对于任意信号 $s(t)$的时间尺度参数在数学上可由信号的过零点获得,信号的过零点为式(11.94)的 t 值,即

$$s(t)=0 \tag{11.94}$$

在两个相邻的过零点之间的时间跨度就是过零尺度参数。

如果通过信号的极值点定义,可得极值尺度参数。信号的极值点为满足式(11.95)的 t 值,即

$$\frac{\mathrm{d}s(t)}{\mathrm{d}t}=0 \tag{11.95}$$

在两个相邻的极值点之间的时间跨度就是极值尺度参数。在 HHT 的信号分解方法中,采用了基于极值点的特征尺度参数。

2. IMF

为获得信号的瞬时频谱,必须计算其瞬时频率。对于单分量信号 $s(t)=a(t)\cos(\varphi(t))$,称 $a(t)$ 为瞬时振幅,$\varphi(t)$ 为瞬时相角,$\mathrm{d}\varphi(t)/\mathrm{d}t$ 为其瞬时频率。由于瞬时频率仅对单分量信号才有意义,因此为计算瞬时频率,必须将多分量信号分解为单分量信号的线性组合。IMF 就是为计算信号的瞬时频率而定义的,它是满足单分量信号解释的一类信号,从而使得瞬时频率具有了物理意义。IMF 应满足以下两个条件。

(1) 在信号长度内,极值点数目和过零点数目相等或最多相差为 1。

(2) 在任意时刻,上包络线和下包络线均值为零,即两者相对时间轴局部对称。其中,上包络线由信号的局部极大值点构成,下包络线由信号的局部极小值点构成。

第(1)个条件类似于高斯平稳过程的传统窄带要求,而第(2)个条件则能保证由 IMF 计算的瞬时频率有意义。从这两个条件可以看出,IMF 反映了信号内部固有的波动性,在它的每个周期上,仅包含一个波动模态,不存在多个波动模态混叠的现象。

3. EMD

对于 IMF,可由 Hilbert 变换得到其解析信号,然后计算瞬时频率。而当信号不满足 IMF 的条件时,应首先采用 EMD 方法对其进行筛选,获得其 IMF。

假设任一复杂信号都由一些不同的 IMF 组成,每一 IMF 的极值点和过零点数目相同,在相邻的两个过零点之间只有一个极值点,且上、下包络线关于时间轴对称,任意两个模态之间相互独立,则 EMD 方法可将任一复杂信号 $x(t)$ 分解成若干个 IMF 的和,分解步骤如下。

(1) 确定信号 $x(t)$ 所有的局部极值点,然后用三次样条线将所有的局部极大值点连接起来形成上包络线,再用三次样条线将所有的局部极小值点连接起来形成下包络线,上下包络线应包络所有的数据点。

(2) 上下包络线的平均值记为 $m_1(t)$,求出

$$x(t)-m_1(t)=h_1(t) \tag{11.96}$$

如果 $h_1(t)$ 是一个 IMF,那么 $h_1(t)$ 就是 $x(t)$ 的第 1 个 IMF 分量。

(3) 如果 $h_1(t)$ 不满足 IMF 的条件,将 $h_1(t)$ 作为 $x(t)$,重复步骤(1)~(2),得到上下包络线的平均值 $m_{11}(t)$,再判断

$$h_{11}(t)=h_1(t)-m_{11}(t) \tag{11.97}$$

是否满足 IMF 的条件,如不满足则重复循环 k 次,直到 $h_{1k}(t)$ 满足 IMF 条件

$$h_{1k}(t)=h_{1(k-1)}(t)-m_{1k}(t) \tag{11.98}$$

记 $c_1(t)=h_{1k}(t)$,则 $c_1(t)$ 为信号 $x(t)$ 的第 1 个满足 IMF 条件的分量。

(4) 将 $c_1(t)$ 从 $x(t)$ 中分离出来,得到

$$r_1(t)=x(t)-c_1(t) \tag{11.99}$$

将 $r_1(t)$ 代替原始信号 $x(t)$ 重复步骤(1)~(3),得到 $x(t)$ 的第 2 个满足 IMF 条件的分量 $c_2(t)$,重复循环 n 次,得到 n 个满足 IMF 条件的分量,即

$$\begin{cases} r_1(t)-c_2(t)=r_2(t) \\ r_2(t)-c_3(t)=r_3(t) \\ \quad\vdots \\ r_{n-1}(t)-c_n(t)=r_n(t) \end{cases} \tag{11.100}$$

当 $r_n(t)$ 为一单调函数不能再提取满足 IMF 条件的分量时，循环结束，$r_n(t)$ 称为余项。这样由式(11.99)和式(11.100)可得

$$x(t)=\sum_{i=1}^{n}c_i(t)+r_n(t) \tag{11.101}$$

式中，$c_i(t)(i=1,2,\cdots,n)$ 为 IMF，$r_n(t)$ 为余项，代表信号变化的平均趋势。

EMD 分解信号的过程其实是一个“筛分”过程，在该过程中，一方面消除了模态波形叠加，另一方面使波形轮廓更加对称。从上面的介绍可知，EMD 方法从特征时间尺度出发，首先分离信号中特征时间尺度最小的 IMF，然后分离特征时间尺度较大的 IMF，最后分离特征时间尺度最大的 IMF。因此，可将 EMD 方法看成一组具备不同通频带的滤波器。

4. Hilbert 谱与 Hilbert 边际谱

对式(11.101)中的每个固有模态函数 $c_i(t)$ 做 Hilbert 变换得

$$\hat{c}_i(t)=\frac{1}{\pi}\int_{-\infty}^{\infty}\frac{c_i(\tau)}{t-\tau}\mathrm{d}\tau \tag{11.102}$$

构造解析信号

$$z_i(t)=c_i(t)+\mathrm{j}\hat{c}_i(t) \tag{11.103}$$

得到幅值函数和相位函数

$$a_i(t)=\sqrt{c_i^2(t)+\hat{c}_i^2(t)} \tag{11.104}$$

$$\varphi_i(t)=\arctan\frac{\hat{c}_i(t)}{c_i(t)} \tag{11.105}$$

进一步可求出瞬时频率

$$f_i(t)=\frac{1}{2\pi}\omega_i(t)=\frac{1}{2\pi}\times\frac{\mathrm{d}\varphi_i(t)}{\mathrm{d}t} \tag{11.106}$$

然后可得 Hilbert 谱，记作

$$H(f,t)=\mathrm{Re}\left\{\sum_{i=1}^{n}a_i(t)\mathrm{e}^{\mathrm{j}\int 2\pi f_i(t)\mathrm{d}t}\right\} \tag{11.107}$$

式中，Re{·}为取实部。

对式(11.107)进行积分，可得 Hilbert 边际谱，即

$$H(f)=\int_0^T H(f,t)\,\mathrm{d}t \tag{11.108}$$

式中，T 为信号的总长度。

Hilbert 谱 $H(f,t)$ 精确描述了信号的幅值在整个频段上随时间和频率的变化规律，而 Hilbert 边际谱 $H(f)$ 反映了信号幅值在整个频段上随频率的变化情况。

11.5.2　基于 IMF 特性的微弱目标检测方法

在 HHT 中，IMF 是在 EMD 过程中自适应产生的，它体现了信号本身固有的本质的物理特性，分析杂波的 IMF 可以获得对杂波信号更多更详尽的信息。以海杂波为例，海杂波频率成分较多，目标频率成分较少，当海杂波中出现目标时，海杂波能量在某些频带内会增加，而在其他频带内会减少，与纯海杂波的能量分布相比，两者差异较大。而 IMF 是信号频带的一种自动划分，不同的 IMF 包含的不同的

频率成分,故海杂波与目标能量在频段间的分布差异必将反映到海杂波与目标的 IMF 特性差异中。本节从海杂波的 IMF 特性方面出发,讨论一种基于固有模态奇异值熵的微弱目标检测算法。

1. 固有模态奇异值定义

设实矩阵 $\boldsymbol{A}$ 行数为 N,列数为 M,按式(11.109)对其进行分解,称为奇异值分解,即

$$\boldsymbol{A}=\boldsymbol{U\Lambda V}^{\mathrm{T}} \tag{11.109}$$

式中,$\boldsymbol{U}=[u_1,u_2,\cdots,u_N]\in\boldsymbol{R}^{N\times N}$,$\boldsymbol{U}^{\mathrm{T}}\boldsymbol{U}=\boldsymbol{I}$;$V=[v_1,v_2,\cdots,v_M]\in\boldsymbol{R}^{M\times M}$,$\boldsymbol{V}^{\mathrm{T}}\boldsymbol{V}=\boldsymbol{I}$,$\boldsymbol{\Lambda}\in\boldsymbol{R}^{N\times M}$ 为矩阵$[\mathrm{diag}\{\sigma_1,\sigma_2,\cdots,\sigma_p\}:0]$或其转置形式,与 $N<M$ 或 $N<M$ 有关,$p=\min(N,M)$,$\sigma_1\geqslant\sigma_2\geqslant\cdots\geqslant\sigma_p\geqslant 0$,$\sigma_1,\sigma_2,\cdots,\sigma_p$ 为矩阵 $\boldsymbol{A}$ 的奇异值。当矩阵 $\boldsymbol{A}$ 的各行向量相等或成比例时,仅 $\sigma_1\neq 0$,$\sigma_2=\cdots=\sigma_p=0$;当矩阵 $\boldsymbol{A}$ 的各行向量不相等或不成比例时,矩阵 $\boldsymbol{A}$ 有多个不为 0 的奇异值。

为了得到时间序列的奇异值,需构建其特征矩阵 $\boldsymbol{A}$。目前在构建时间序列的特征矩阵时常采用延时嵌入法,该方法重构特征矩阵的过程如下:设时间序列为 $X=\{x(i)\}$,$i=1,2,\cdots,N$ 为重构特征矩阵,需建立 m 维嵌入空间,这时可对时间序列 X 进行延时采样,得到$[N-(m-1)J]$个时间序列,即

$$a_i=x(i),x(i+J),x(i+2J),\cdots,x(i+(m-1)J)\quad i=1,2,\cdots,N-(m-1)J \tag{11.110}$$

式中,$J=\tau/\Delta t$,τ 为延迟时间,且为 Δt 的整数倍,即 J 为整数。a_i 对应于一个行向量,$N-(m-1)J$ 个向量就可以重构特征矩阵 $\boldsymbol{A}$。

$$\boldsymbol{A}=\begin{bmatrix}a_1\\a_2\\\vdots\\a_n\end{bmatrix}=\begin{bmatrix}x(1) & x(1+J) & \cdots & x[1+(m-1)J]\\x(2) & x(2+J) & \cdots & x[2+(m-1)J]\\\vdots & \vdots & & \vdots\\x[N-(m-1)J] & x[N-(m-2)J] & \cdots & x(N)\end{bmatrix} \tag{11.111}$$

在重构特征矩阵时,嵌入维数 m 和延迟时间 τ 是需要重点考虑的问题。这里将 EMD 方法引入奇异值分解的特征矩阵构建环节,以解决嵌入维数 m 和延迟时间 τ 难以确定的问题。EMD 方法是 Hilbert-Huang 变换的核心,通过 EMD 方法可将信号分解成一系列的 IMF,IMF 反映了信号本身的固有属性,不同的 IMF 具备不同的时间尺度,而不同的时间尺度对应不同的频率,故不同的 IMF 包含了信号不同的频率成分。如果认为系统是由不同频率的子系统构成的,就可以采用 IMF 构建信号的特征矩阵。采用 IMF 构建特征矩阵的过程如下:假设通过 EMD 方法对信号 $s(k)$,$k=1,2,\cdots,K$,进行分解后,得到 n 个 IMF $c_1(k),c_2(k),\cdots,c_n(k)$,将这 n 个 IMF 组成初始特征矩阵 $\boldsymbol{A}$,则 $\boldsymbol{A}$ 可以表示为

$$\boldsymbol{A}=\begin{bmatrix}c_1(1) & c_1(2) & \cdots & c_1(K)\\c_2(1) & c_2(2) & \cdots & c_2(K)\\\vdots & \vdots & & \vdots\\c_n(1) & c_n(2) & \cdots & c_n(K)\end{bmatrix} \tag{11.112}$$

式中,n 为信号 IMF 的数目,对于海杂波信号,K 可取为 $T/\Delta t$,T 为雷达相参积累时间,Δt 为采样时间。对矩阵 $\boldsymbol{A}$ 进行奇异值分解,就可以得到矩阵 $\boldsymbol{A}$ 的奇异值 $\sigma_A=[\sigma_A^1,\sigma_A^2,\cdots,\sigma_A^n]$,称为固有模态奇异值。

2. 目标对海杂波固有模态奇异值的影响

采用 IMF 建立特征矩阵 $\boldsymbol{A}$,然后进行奇异值分解,可获得海杂波的固有模态奇异值。IPIX 雷达的 280#和 310#数据(VV 极化)的固有模态奇异值如图 11.26 和图 11.27 所示。从图 11.26 和图 11.27 中可以看出,对于不含目标的纯海杂波信号,前 3 个固有模态奇异值相对较大,而后 6 个相对较小;而

当目标出现时，含目标海杂波的前 3 个固有模态奇异值变化较小，而后 6 个固有模态奇异值明显增大，各奇异值间数值差异减小。这主要是因为：与海杂波相比，目标的频谱较窄，频率成分较少，而 IMF 是按照频率大小对信号进行的一种自动划分，故目标不能覆盖海杂波所有的 IMF，仅可能出现在海杂波的一个或几个 IMF 中。固有模态奇异值反映了特征矩阵 $\boldsymbol{A}$ 本身的固有属性，由固有模态奇异值的计算过程可知，特征矩阵 $\boldsymbol{A}$ 是由 IMF 建立的，不同的固有模态奇异值提取了特征矩阵中不同频段 IMF 的特征，即固有模态奇异值能够反映特征矩阵 $\boldsymbol{A}$ 中的各特征向量 IMF 在目标出现时的变化。因此，当目标出现时会引起海杂波相应的固有模态奇异值增大。在 280＃和 310＃数据中，目标为静止的慢起伏目标，多普勒频率较低，其成分仅出现在后分解出的海杂波低频 IMF 中，故当目标出现时，海杂波的后 6 个固有模态奇异值明显增大。

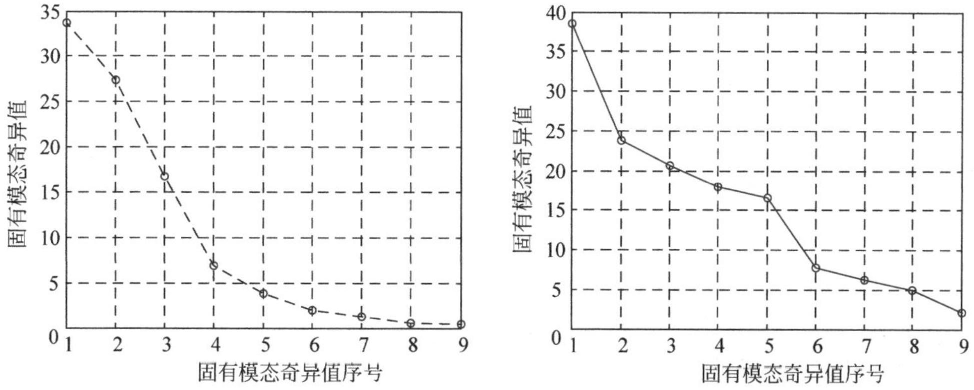

(a) 纯海杂波信号的固有模态奇异值　　(b) 含目标海杂波信号的固有模态奇异值

图 11.26　280＃数据海杂波信号的固有模态奇异值(VV 极化)

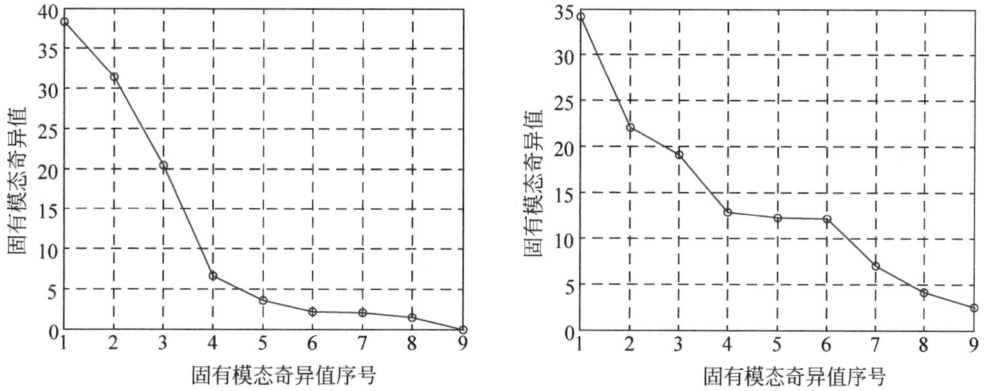

(a) 纯海杂波信号的固有模态奇异值　　(b) 含目标海杂波信号的固有模态奇异值

图 11.27　310＃数据海杂波信号的固有模态奇异值(VV 极化)

这两组数据距离单元的固有模态奇异值熵及其均值分别如图 11.28 和图 11.29 所示。从图 11.28 和图 11.29 中可以看出，当无目标时，纯海杂波信号的固有模态奇异值熵较小，而当目标出现时，海杂波信号的固有模态奇异值熵明显增大。由此可知，固有模态奇异值熵可以准确地描述目标对海杂波固有模态奇异值的影响，采用固有模态奇异值熵检测海杂波中微弱目标是可行的。

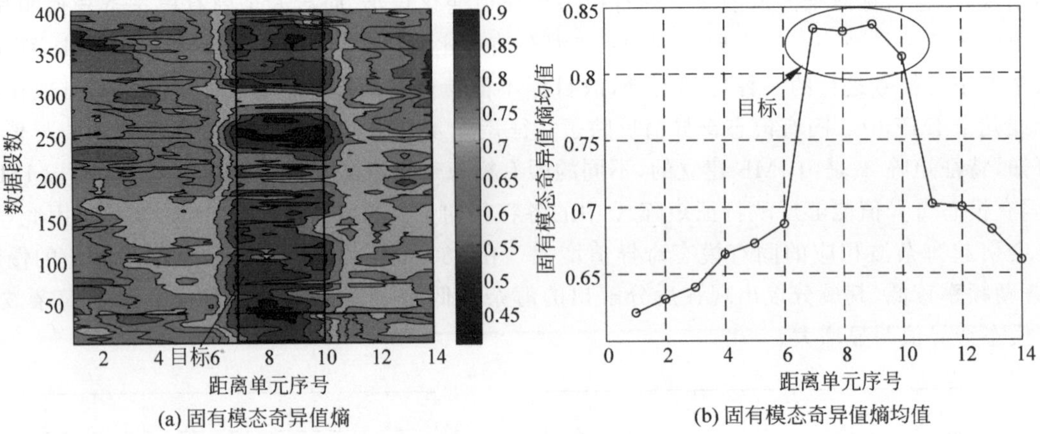

图 11.28　280＃数据海杂波的固有模态奇异值熵及其均值(VV 极化)

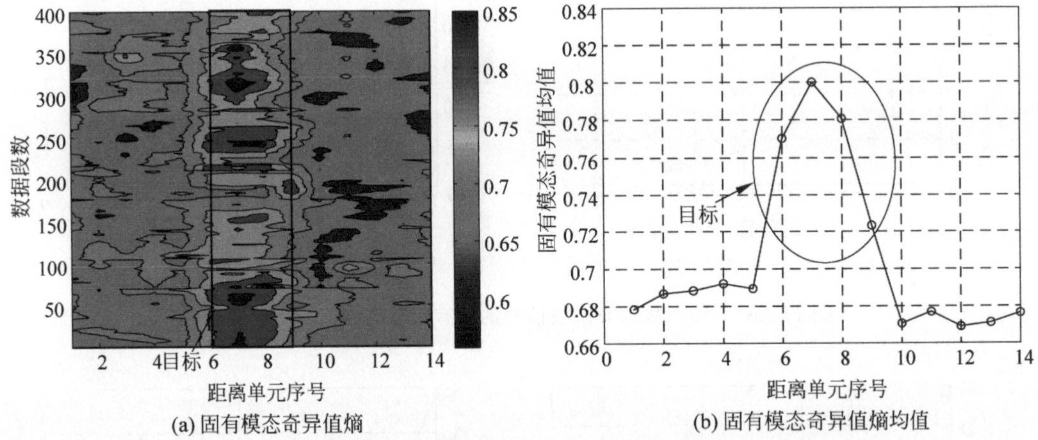

图 11.29　310＃数据海杂波的固有模态奇异值熵及其均值(VV 极化)

3. 检测器设计

通过上述分析可知，当目标出现时，海杂波的固有模态奇异值将产生明显变化。为刻画这种变化，这里将信息熵引入奇异值分解中，定义固有模态奇异值熵。假设海杂波信号的固有模态奇异值为 $\sigma_{\mathbf{A}}^1, \sigma_{\mathbf{A}}^2, \cdots, \sigma_{\mathbf{A}}^n$，根据信息熵的定义，相应的海杂波信号的固有模态奇异值熵定义为

$$H = -\sum_{i=1}^{n} p_i \log p_i \tag{11.113}$$

式中，$p_i = \sigma_{\mathbf{A}}^i / \sigma$，为第 i 个固有模态奇异值所占比重，σ 为所有固有模态奇异值的和，$\sigma = \sum_{i=1}^{n} \sigma_{\mathbf{A}}^i$。

由信息熵性质可知，p_i 分布越均匀，熵 H 越大，且当各 p_i 相等时，达到最大值 $\log n$；反之，熵 H 越小。从上面分析可知，当无目标时，各固有模态奇异值差异较大，故固有模态奇异值熵较小；而当存在目标时，各固有模态奇异值差异减小，故固有模态奇异值熵增大。因此，可以考虑采用固有模态奇异值熵对海杂波中的微弱目标进行检测，其原理框图如图 11.30 所示。

该算法主要流程如下。

(1) 对海杂波数据进行 EMD 处理，获得其 n 个 IMF $c_1(k), c_2(k), \cdots, c_n(k), k=1,2,\cdots,K$。

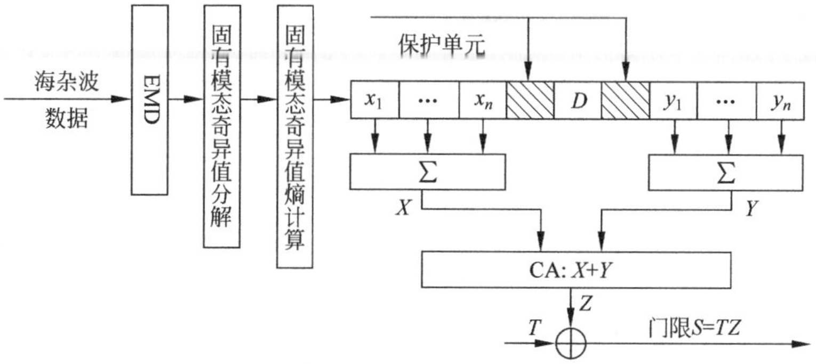

图 11.30　基于固有模态奇异值熵的微弱目标检测算法原理框图

(2) 采用 n 个 IMF 构建特征矩阵 $\boldsymbol{A}$，并进行奇异值分解，获得海杂波的 n 个固有模态奇异值 $\sigma_{\boldsymbol{A}}^1$，$\sigma_{\boldsymbol{A}}^2$，…，$\sigma_{\boldsymbol{A}}^n$，并计算固有模态奇异值熵。

(3) 将检测单元 D 的固有模态奇异值熵与门限 S 比较，如检测单元 D 的熵值大于门限 S，则判为有目标，反之则判为无目标。门限可采用 CA-CFAR 门限计算方法获得，即各参考单元的熵值取平均，再乘以阈值因子，得到门限。

采用该算法对微弱目标检测，需满足两个假设条件：①在相邻的距离单元中，海杂波具有相同的相关性；②目标为低速慢起伏目标。

4. 检测性能分析

1) 实测目标数据的检测性能

采用 IPIX 雷达的 280＃和 310＃数据分析基于固有模态奇异值熵的微弱目标检测算法对实测目标的检测性能。在这两组数据中，目标信号较弱，SCR 较低，但具体数值未知。因此，为分析检测算法的性能，将其与另一种基于盒维数的典型微弱目标检测算法的检测性能进行了对比，检测性能对比曲线($P_{\mathrm{d}} \sim P_{\mathrm{fa}}$)如图 11.31 所示。对于这两组数据，在虚警概率 $P_{\mathrm{fa}}=10^{-3}$ 时，基于固有模态奇异值熵检测算法的检测概率 P_{d} 能达到 70%以上，与基于盒维数的微弱目标检测算法的检测概率相比，平均高约 25%以上。因此，对于实测目标数据，基于固有模态奇异值熵检测算法的检测性能优于基于盒维数的检测算法的检测性能。

2) SCR 对检测性能的影响

采用 Monte Carlo 仿真实验的方法分析 SCR 对检测性能的影响。海杂波数据为 ISAR 雷达的 I1＃数据，目标采用仿真方法产生，为 Swerling Ⅰ型的固定目标，SCR 在－15dB～0dB 变化，幅度服从瑞利分布，在 100 个脉冲中保持不变。受数据量的限制，表 11.4 仅给出了 $P_{\mathrm{fa}}=0.01$ 时，所述检测算法与基于盒维数的检测算法和多脉冲 CA-CFAR 检测算法的检测概率。

表 11.4　3 种方法的检测概率($P_{\mathrm{fa}}=0.01$)

检测方法	SCR(dB)			
	－15	－10	－5	0
基于固有模态奇异值熵的微弱目标算法	85%	99%	100%	100%
基于盒维数的微弱目标检测算法	2%	4%	19%	72%
多脉冲 CA-CFAR 算法	—	19%	57%	82%

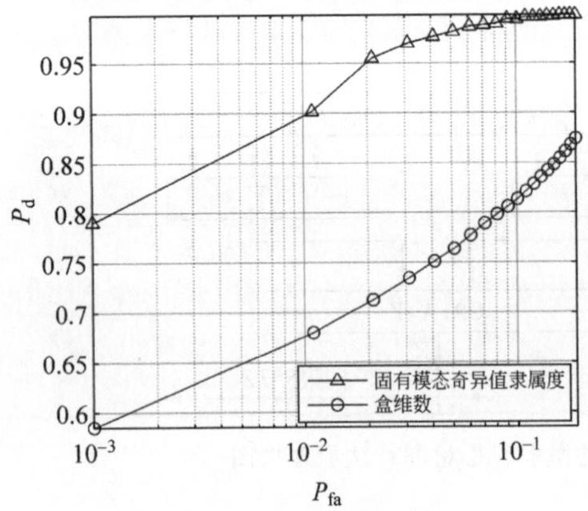

(a) 280#数据

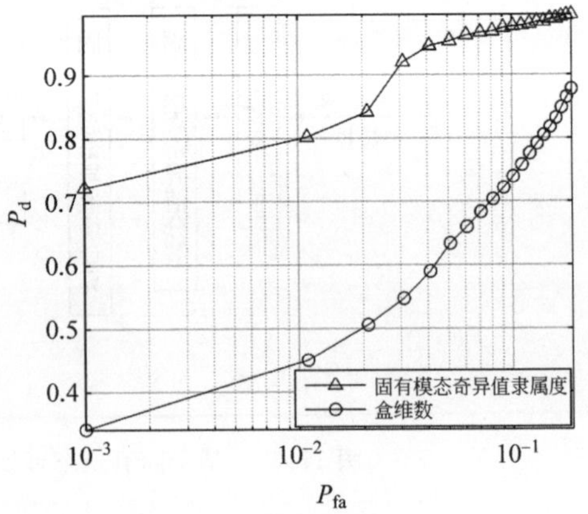

(b) 310#数据

图 11.31 对 IPIX 雷达不同数据的虚警概率与检测概率关系曲线(VV 极化)

从表 11.4 可知：

(1) 当 SCR≥−5dB 时，基于固有模态奇异值熵的微弱目标检测算法的检测概率能够达到 100%，对目标的检测性能较好，优于其他两种算法的检测性能；基于盒维数算法对目标的检测概率较低，检测性能弱于多脉冲 CA-CFAR(100 个脉冲)方法的检测性能。

(2) 当 SCR 从 0dB 下降到 −15dB 时，基于固有模态奇异值熵的微弱目标检测算法的检测性能下降较慢，而其他两种算法的检测性能下降较快。

因此，通过上面对实测目标数据检测性能的分析以及 SCR 对检测性能影响的分析可知，基于固有模态奇异值熵微弱目标检测算法适用于低 SCR 条件下的微弱目标检测。

11.6 稀疏表示域目标检测

目标雷达回波可视为少数强散射中心回波的叠加，回波具有稀疏特性。利用动目标回波信号具有稀疏性的特点，将稀疏分解的局部优化思想和稀疏变换方法引入时频分析或变换域处理，在稀疏表示域设计目标检测器，能够有效提高算法运算效率、时频分辨率和参数估计性能，从而更有利于目标检测。本节主要讨论基于稀疏优化求解的稀疏表示域目标检测方法，分别介绍信号稀疏表示模型及求解方法、信号稀疏特性及杂波背景下的动目标检测方法。

11.6.1 信号稀疏表示模型及求解方法

在雷达对目标的探测和定位过程中，目标相对于雷达观测场景通常是非常稀疏的，如广阔的大气、空中、海上只有少数的飞机、舰船等目标，这就造成雷达系统的复杂性、海量观测数据与目标的稀疏性之间的不平衡。对于采样率问题，最理想的办法是降低采样率；对于存储问题，最直接的办法是压缩数据；对于系统时间资源问题，最简单的方法是寻找一种解决低脉冲重复条件下工作存在问题的信号处理技术。压缩感知(Compressed Sensing,CS)理论正是用于解决信号处理中采样率高、数据量大、实时处理困难等问题的[58]。CS 理论主要涉及以下 3 方面：信号的稀疏变换、观测矩阵的设计及稀疏信号的重构算法。对于雷达目标检测，主要集中在 CS 的第一个阶段，即稀疏表示阶段，利用信号在某个域中的稀疏特性，采用少量的观测样本，通过求解最优化问题，在稀疏域中实现对该信号的高分辨率表示。

实质上,信号的时频表示及变换处理,也可以看成稀疏表示的特例,在信号去噪、雷达成像、目标检测和识别等方面有广泛的应用[59-61]。文献[62-64]基于雷达海面回波中固有的稀疏性信息,利用海杂波和目标信号组成成分的差异性,对这两种信号分别进行稀疏表示,准确地提取海杂波信号并将其抑制,从而达到提高微弱动目标检测性能的目的。下面简单介绍稀疏表示数学模型及其求解方法。

给定一个信号 $\boldsymbol{x}\in\boldsymbol{R}^{n\times 1}$,$n$ 为其长度,稀疏理论试图寻找一种对该信号的线性描述方式,以使该描述方式中的大部分元素都为0。为此,定义一个子空间 $\boldsymbol{D}\in\boldsymbol{R}^{n\times m}$,其中,$m\gg n$,即 $\boldsymbol{D}$ 是冗余的,则信号 $\boldsymbol{x}$ 可以表示为[58]

$$\boldsymbol{x}=\boldsymbol{D}\boldsymbol{z} \tag{11.114}$$

其中,$\boldsymbol{z}\in\boldsymbol{R}^{m\times 1}$ 为信号 $\boldsymbol{x}$ 在 $\boldsymbol{D}$ 中的表达,m 为表达的长度。在稀疏表达研究领域中,通常称 $\boldsymbol{D}$ 为字典,称其各列向量为基元。如图11.32所示,如果稀疏表示向量 $\boldsymbol{z}$ 中只有 k 个非零元素,则称 $\boldsymbol{z}$ 为信号的 k 稀疏表示。$\boldsymbol{z}$ 中白色方块代表对应分量为零,其余方块代表分量不为零。

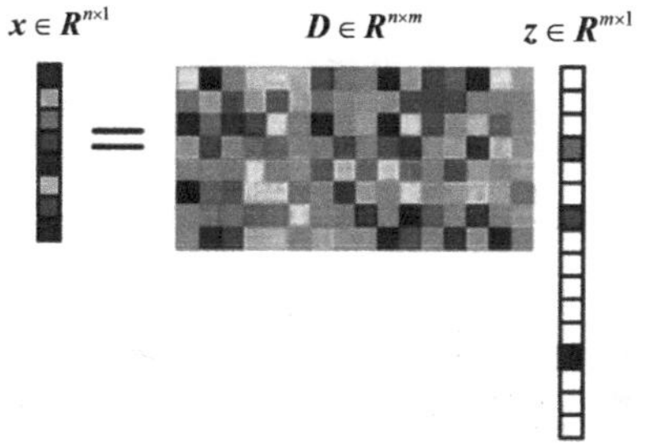

图11.32　稀疏表示模型示意图

显然,式(11.114)为一个欠定线性方程组,其解向量 $\boldsymbol{z}$ 有无穷多个。为此,需要在式(11.114)中加入其他的约束项,来保证求得的解 $\boldsymbol{z}$ 是足够稀疏的。为了衡量一个解向量是否稀疏,通常使用 l_0"范数"来度量一个向量的稀疏性,记为 $\|\boldsymbol{z}\|_0$,定义 $\|\boldsymbol{z}\|_0=\sum_i|z_i|^0$,即求向量 $\boldsymbol{z}$ 中非0元素的数量。$\|\boldsymbol{z}\|_0$ 越小,$\boldsymbol{z}$ 越稀疏(非0元素越少)。基于 l_0"范数"的概念,可将要解决的问题改写为

$$\min_{\boldsymbol{z}}\|\boldsymbol{z}\|_0,\ \text{s.t.}\ \boldsymbol{x}=\boldsymbol{D}\boldsymbol{z} \tag{11.115}$$

即寻找可以准确描述信号 $\boldsymbol{x}$ 的表达 $\boldsymbol{z}$,且 $\boldsymbol{z}$ 是所有可行的表达中最稀疏的。

当 $n>m$ 时,待处理信号 $\boldsymbol{x}$ 中某些向量不能完整地用基向量字典 $\boldsymbol{D}$ 来线性表示,称基向量字典 $\boldsymbol{D}$ 为非完备的;当 $m=n$ 时,待处理信号 $\boldsymbol{x}$ 能用基向量字典 $\boldsymbol{D}$ 唯一地表示,称基向量字典 $\boldsymbol{D}$ 为完备的;当 $n<m$ 时,待处理信号 $\boldsymbol{x}$ 在基向量字典 $\boldsymbol{D}$ 分解得到无穷多组解,此时称基向量字典 $\boldsymbol{D}$ 为超完备的。

稀疏表示就是信号基于超完备字典,求得其最佳的稀疏表示或者稀疏逼近的过程。但是,在超完备字典下,由于 $n<m$,式(11.115)是一个NP-hard问题,是无法直接求解的[60]。由于其中涉及了 l_0"范数",而 l_0"范数"是非凸的,这导致直接求解的计算复杂度高。近年来,人们提出许多求解式(11.114)的高效数值算法:①贪心算法,如匹配追踪、正交匹配追踪、弱匹配追踪,阈值算法等;②凸松弛方法,如基追踪、迭代重加权最小二乘等;③快速近似算法,如LASSO(Least Absolute Shrinkage and Selection Operator)、迭代阈值收缩算法等。其中,贪心算法将式(11.115)中的问题分解为若干个子问题,在求解的每步中都试图用一个最优匹配来求解一个子问题,最后的解由所有子问题的解综合而成。凸松弛方法将要求解的非凸问题转化成一个凸优化问题,从而可以快速高效地求解。快速近似算法则是允许式(11.115)中所陈述的问题中存在噪声的干扰,即将原问题转化成以下的新问题

$$\min_{\boldsymbol{z}}\|\boldsymbol{z}\|_0,\ \text{s.t.}\ \|\boldsymbol{x}-\boldsymbol{D}\boldsymbol{z}\|_2\leqslant\varepsilon \tag{11.116}$$

其中,ε 为加性噪声的功率,式(11.116)中的问题通常也会应用凸松弛技术将其转化成一个凸优化问题,进而快速高效地求解。

11.6.2　基于稀疏时频分布的雷达目标检测方法

尽管稀疏表示分析方法突破采样定理的限制,具有分辨率高、对噪声不敏感、稳健性高等优点,但仍

有很多问题亟待解决和研究。信号时频处理方法，在分析非平稳信号时具有一定的优势，但估计性能受时频分辨率的限制，而且若目标特性与变换方法不相匹配，则对信杂比的改善不明显。而非平稳时变信号在时频域往往具有较好的稀疏性，受稀疏表示技术的启发，国内外一些学者将稀疏分解的局部优化思想引入时频分析，即采用稀疏时频分布(Sparse Time-Frequency Distribution，STFD)的方法对目标特性进行研究，能够有效提高信号时频聚集性和参数估计性能[65-68]。美国麻省理工学院(MIT)提出了一种稀疏快速傅里叶变换方法(Sparse FT，SFT)[67]，通过分筐操作将 N 点长序列转化为 B 点短序列再作离散傅里叶变换(Discrete FT，DFT)，比经典 FFT 更为高效，被 MIT《技术评论》评选为当年十大颠覆性技术之一。相比经典 FFT，SFT 能够较大地降低运算量，提高运算效率。文献[69]结合 SFT 的优势对 Pei 的 FRFT 方法进行重新设计，研究了一种新的快速算法，即稀疏 FRFT(Sparse FRFT，SFRFT)，进一步降低了离散 FRFT 的复杂度。文献[70-72]利用动目标回波信号具有稀疏性的特点，将稀疏分解的局部优化思想引入时频分析，提出了多种杂波背景下基于 STFD 的目标检测方法。这里首先介绍信号稀疏表示模型及相应的求解方法，在此基础上，结合时频分析方法分析雷达回波信号稀疏特性，给出基于稀疏表示的雷达动目标检测方法，最后以海上目标检测为例，给出具体的分析结果。

1. STFD 原理

信号 $x(t)$ 的 TFD 是稀疏表示的特例，其字典可由信号在 TFD 域中的频率估计组成

$$\rho_x(t,f)=\sum_{i=1}^{I}\beta_i(t)h(t)g_i(t,f) \tag{11.117}$$

式中，$\rho_x(t,f)$ 为信号 $x(t)$ 的稀疏时频分布，$h(t)$ 为窗函数，$g_i(t,f)$ 为稀疏表示的字典。

求解式(11.117)的信号稀疏表示问题可转化为最优化问题，可近似采用 l_1 范数的最小化求解[63]

$$\rho_x=\arg\min_{\rho_x}\|\rho_x(t,f)\|_1,\ \text{s.t.}\ o\{\rho_x(t,f)\}=b \tag{11.118}$$

式中，$\rho_x\in R^N$，$b\in R^K$，b 为实数，o 为 $K\times N$ 的稀疏算子。式(11.118)可松弛为不等约束，即

$$\rho_x=\arg\min_{\rho_x}\|\rho_x(t,f)\|_1,\ \text{s.t.}\ \|o\{\rho_x(t,f)\}-b\|_2\leqslant\varepsilon \tag{11.119}$$

当 $\varepsilon=0$ 时，式(11.118)和式(11.119)具有相同的形式。当 o 为 FRFT 时，b 为 FRFT 域幅值，则短时稀疏 FRFT(ST-SFRFT)可表示为

$$\mathcal{F}^\alpha=\arg\min_{\mathcal{F}^\alpha}\|\mathcal{F}^\alpha(t,f)\|_1,\ \text{s.t.}\ \|o\{\mathcal{F}^\alpha(t,f)\}-f(\alpha,u)\|_2\leqslant\varepsilon \tag{11.120}$$

式中，$\mathcal{F}^\alpha(t,f)$ 为 ST-SFRFT 时频分布，α 为旋转角，u 为 ST-SFRFT 域。

2. 算法流程

以典型的海上目标检测为例，基于 ST-STFD 的检测方法流程图如图 11.33 所示，具体实现步骤如下。

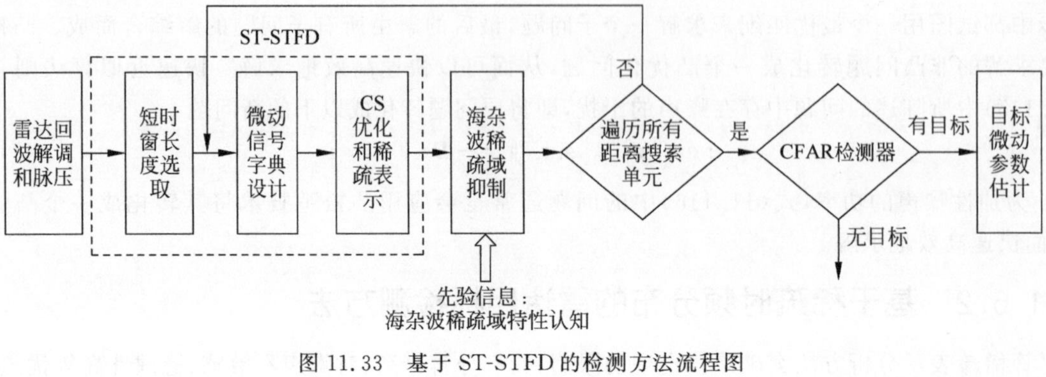

图 11.33　基于 ST-STFD 的检测方法流程图

(1) 雷达回波解调和脉冲压缩,实现距离高分辨,并选取待检测距离单元。

(2) 雷达信号稀疏表示运算,主要包括短时窗长度选取、动目标信号字典设计、CS优化和稀疏表示等。动目标信号稀疏表示的字典可根据待观测目标类型和海况高低等先验信息确定。对于以非匀速平动为主要运动方式的目标,如低空掠海飞行器、快艇等,稀疏表示字典可采用 chirp 字典

$$\boldsymbol{G}_x=\begin{bmatrix} g_x(f_1,\mu_1) & g_x(f_1,\mu_2) & \cdots & g_x(f_1,\mu_M) \\ g_x(f_2,\mu_1) & g_x(f_2,\mu_2) & \cdots & g_x(f_2,\mu_M) \\ \vdots & \vdots & & \vdots \\ g_x(f_L,\mu_1) & g_x(f_L,\mu_2) & \cdots & g_x(f_L,\mu_M) \end{bmatrix} \tag{11.121}$$

$$g_x(f_l,\mu_m)=\exp(\mathrm{j}2\pi f_l t+\mathrm{j}\pi\mu_m t^2),\quad l=1,2,\cdots,L;\quad m=1,2,\cdots,M \tag{11.122}$$

式中,μ 为调频率,L 和 M 代表参数个数。而对于以转动为主要运动方式或者高机动的海面目标,其回波具有周期调频性,稀疏表示字典可采用 QFM 或周期调频函数作为稀疏分解字典,即

$$g_x(\mu_m,k_n)=\exp\left(\mathrm{j}\pi\mu_m t^2+\frac{1}{3}\pi k_n t^3\right),\quad m=1,2,\cdots,M;\ n=1,2,\cdots,N \tag{11.123}$$

$$g_x(\omega_l)=\exp(\mathrm{j}\omega_l t+\varphi_r),\quad l=1,2,\cdots,L \tag{11.124}$$

式中,k 代表由急动度产生的调频率,ω 为转动角速度。

(3) 海杂波稀疏域抑制。通过对纯海杂波单元的稀疏域特性认知,在海杂波稀疏域设计杂波抑制方法,达到改善 SCR 的目的。

(4) 遍历所有距离搜索单元,进行微动目标信号稀疏域 CFAR 检测。与给定虚警概率下的 CFAR 检测门限进行比较,如果检测统计量低于检测门限,判决为没有动目标信号,继续处理后续的检测单元,若检测统计量高于检测门限,则判决为存在动目标信号。

(5) 目标微动参数估计。将 STFD 域检测目标后的剩余峰值点坐标对应的频率、调频率、急动度或转动角速度等,作为特征的参数估计值。

11.6.3　雷达目标检测结果与分析

采用南非 CSIR 对海雷达数据库中的 TFA17-014 数据分析海上目标微动特征在变换域中的稀疏特性。在高海况条件下由 Fynmeet 雷达采集合作目标乘浪者号充气橡皮艇的回波数据。图 11.34 为 CSIR 对海雷达动目标探测试验数据描述(TFA17-014),其中图 11.34(a)的距离-时间图表明雷达观测时间为 100 s,机动目标处于 17～22 距离波门内,黑色曲线为 GPS 记录的运动轨迹。可知,目标在观测时间内跨越了多个距离单元,具有高机动特性,其回波微弱,仅通过幅度难以从强海杂波中发现目标。进一步分析目标单元的时频谱图(图 11.34(b))可以看出,目标多普勒随时间变化,近似有周期振荡性,具有微动特性,海杂波频谱较宽,覆盖了大部分目标频谱。

分别采用以傅里叶信号和 chirp 信号为字典的稀疏表示方法,即短时-稀疏傅里叶变换(ST-SFT)和短时-稀疏 FRFT(ST-SFRFT)对海上微动目标检测,分析比较结果,如图 11.36 和图 11.37 所示。图 11.36(a)和图 11.36(b)为 ST-SFT 及其 20.2s 处的切面图,通过对比图 11.35 的 STFT 分析结果可知:

(1) ST-SFT 具有较高的时频分辨率,目标微动信号特征较为明显,原因在于利用了稀疏分解的优点,仅保留最为稀疏的信号成分,因此既保证了信号的能量聚集,又实现了微动信号的时频表示。

(2) 目标运动参数估计精度得到提升,根据 STFD 的稀疏分解系数可估计出目标的运动瞬时速度为 2.03m/s。

(3) 采用 ST-SFRFT 方法能够更好地表征微动信号的时变特征(见图 11.37),目标峰值明显,经过检测后,剩余虚警明显降低,表明海杂波在该 STFD 域中无稀疏峰值,从而进一步提高了雷达对微动信号的检测和参数估计能力。

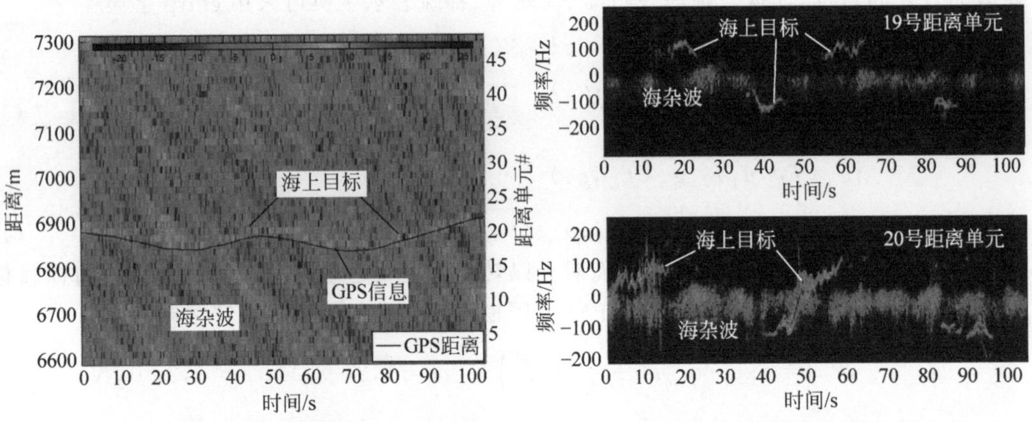

(a) 距离-时间图　(b) 目标单元时频分布(距离单元19和20)

图 11.34 CSIR 对海雷达动目标探测试验数据描述(TFA17-014)

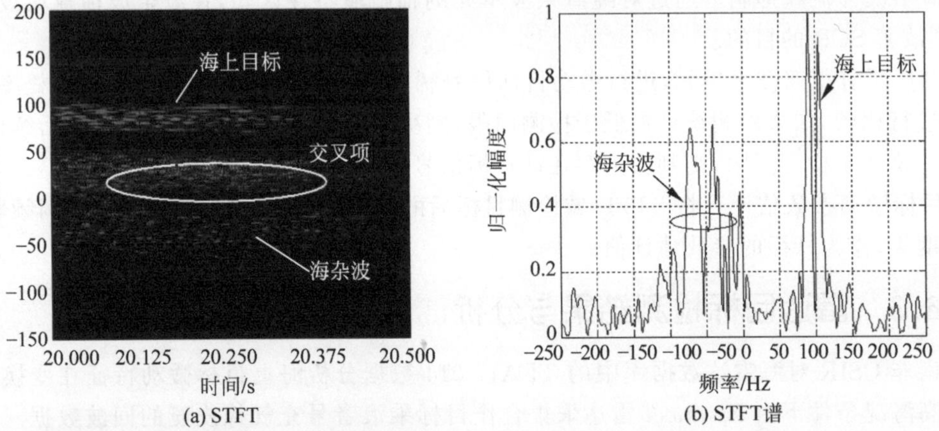

(a) STFT　(b) STFT谱

图 11.35 基于 STFT 的海上微动目标检测结果(起始时间=20s)

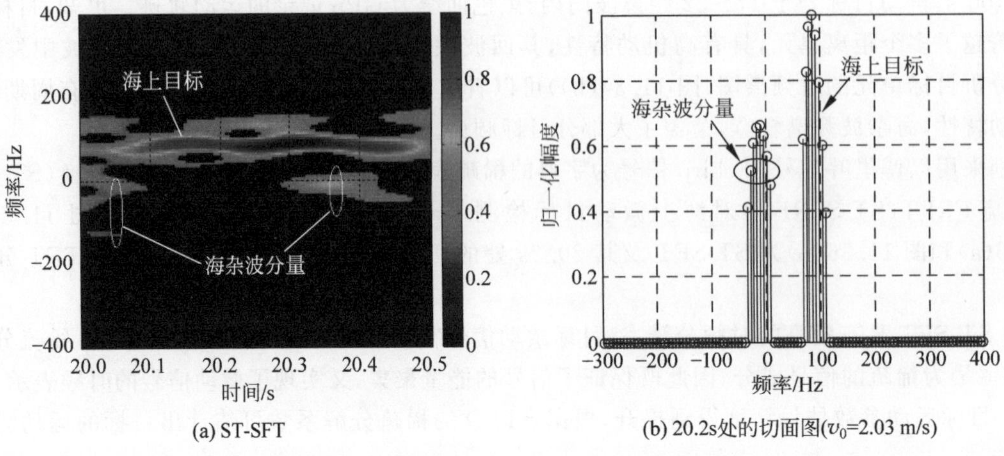

(a) ST-SFT　(b) 20.2s处的切面图(v_0=2.03 m/s)

图 11.36 基于 ST-SFT 的海上微动目标检测结果(起始时间=20s)

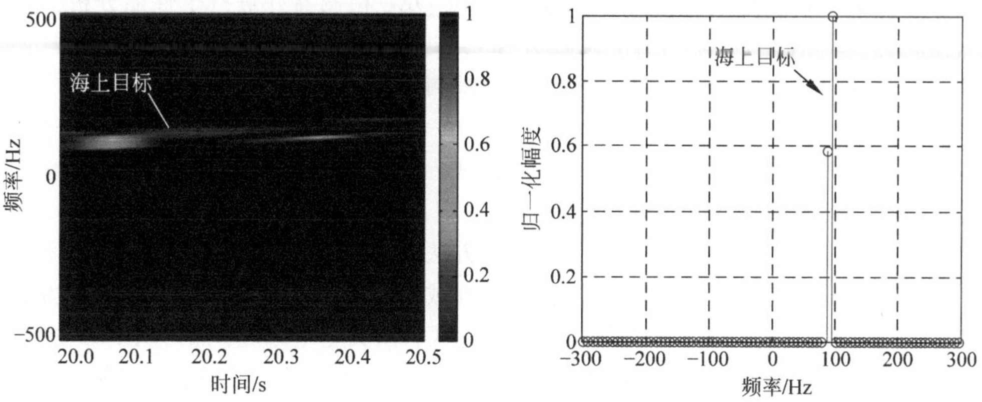

(a) ST-SFRFT

(b) 20.2s处的切面图(v_0=2.04 m/s, a_s=0.45 m/s^2)

图 11.37　基于 ST-SFRFT 的海上微动目标检测结果(起始时间＝20s)

进一步定量分析算法的检测性能。表 11.5 对传统 TFD 类动目标检测方法(MTD、FRFT、WVD 和 SPWVD)和 STFD 方法进行对比,数据采用仿真微动信号＋TFA17-014 海杂波,并通过 Monte Carlo 仿真计算不同 SCR 条件下的算法检测性能及计算时间。通过对比可知:

(1) 传统 TFD 类方法中,FRFT 的检测概率最高,但其运算量也相应增加。

(2) 传统 TFD 和所提方法对比,后者的检测性能有明显提高,原因在于 STFD 方法可通过稀疏优化算法,在微动信号最优稀疏域设计检测方法,提升能量聚集程度。

(3) 分析比较算法的运算时间可知,传统 TFD 方法运算量较大,基于 CS 优化理论的稀疏分析方法需要求解目标函数,其运算效率仍有待进一步提高,可基于 SFT 算法开发相应的快速稀疏变换方法,提高运算效率。

表 11.5　不同方法海上微动目标检测性能比较(仿真微动信号＋TFA17-014 海杂波,$P_{fa}=10^{-4}$)

参　数	MTD	FRFT	WVD	SPWVD	ST-SFT	**ST-SFRFT**
P_d(%)(SCR＝－5dB)	39.26	57.26	35.68	55.24	49.21	**71.35**
P_d(%)(SCR＝0dB)	52.84	76.84	62.27	72.58	63.28	**85.69**
计算时间*(ms)	10.54	17.57	15.21	19.66	12.14(1) 8.12(2)	**17.46(1)** **11.68(2)**

* 计算机配置:Intel Core i7-4790 3.6GHz CPU; 16G RAM; Matlab R2014a。

(1) 基于 CS 优化理论计算方法,如 BPDN,(2) 基于 SFFT(code version 2)的快速算法[67]。

11.7　小结

变换域目标检测的方法是对雷达回波信号进行一定的变换(FFT、DWT、时频变换、FRFT、HHT、稀疏表示等),以有效区分目标加杂波及纯杂波。一般来说,变换域 CFAR 处理就是根据有用信号和杂波的各种变换域特性的差别,利用回波信号的变换谱检测有用信号。形成检测统计量过程中的重要环节是回波信号的谱估计及变换域特征提取问题。11.2 节讨论了频域 CA-CFAR 检测器。11.2.2 和 11.2.3 分析表明,FFT 加权形式和 MTI 滤波器结构对改善检测器性能都有较大的影响。11.2.4 节针对均值类估计引起的“信号污染”畸变问题,研究了频域 OEP 检测器并进行了性能分析。11.3 节主要在高斯背景下讨论了两种小波域 CFAR 处理方法,并进行了理论分析,仿真结果表明,两种小波域 CFAR 检测方法均能较好的适应不同杂波环境,有效检测目标信号,具有较好的鲁棒性。

11.4 节～11.6 节分别介绍了 FRFT、HHT 和稀疏表示域雷达目标检测方法。11.4.1 节讨论了 FRFT 理论在雷达非平稳信号处理和检测中的应用,总结了基于 FRFT 的 LFM 信号检测与估计方法。11.4.2 节重点介绍了 FRFT 动目标检测器设计,对输出信噪比进行了定量分析。11.4.3 节给出了雷达在长时间观测时,解决跨距离和多普勒单元问题的方法,并采用海上机动目标雷达数据进行了验证比较。FRFT 在目标检测领域中发挥着重要作用,是继频域检测后又一重要的变换域检测方法。11.5 节讨论了 HHT 域目标检测,首先简要介绍了 HHT 研究现状及其基本原理;然后从 IMF 的角度出发,研究了基于固有模态奇异值熵的微弱目标检测算法,结合雷达实测数据对不同海况和极化条件下的数据进行了分析和验证,可知与盒维数算法和多脉冲 CA-CFAR 算法相比,有明显的性能提升。稀疏表示方法突破采样定理的限制,具有时频分辨率高、抗杂波及适合多分量信号分析等优点,为雷达杂波抑制和动目标检测提供了新的思路和方向。11.6 节结合传统时频变换动目标检测和稀疏表示方法的优势,提出稀疏时频分布目标检测方法,在稀疏域实现时变信号的高分辨表示和检测。实测雷达数据分析了海杂波和海上目标的稀疏特性,并验证了方法的有效性。

从本章的讨论和分析可以看出,变换域 CFAR 是变换方法及 CFAR 检测器的有机结合,变换域 CFAR 研究与时域 CFAR 相比,在相关技术和工程应用等方面还需进一步探索和研究。随着现代雷达技术的快速发展及应用,变换域 CFAR 处理有着很好的应用前景。

参考文献

[1] Aubry A, Maio A De, Huang Y, et al. Robust design of radar Doppler filters[J]. IEEE Transactions on Signal Processing, 2016, 64(22): 5848-5860.

[2] 贲德, 韦传安, 林幼权. 机载雷达技术[M]. 北京: 电子工业出版社, 2006.

[3] 陈小龙, 薛永华, 张林, 等. 机载雷达系统与信息处理[M]. 北京: 电子工业出版社, 2021.

[4] Brennan L, Mallett J, Reed I. Adaptive arrays in airborne MTI radar[J]. IEEE Transactions on Antennas and Propagation, 1976, 24(5): 607-615.

[5] Ringel M B, Mooney D H, Long W H. F-16 Pulse Doppler radar(AN/APG-16) performance[J]. IEEE Trans. on AES, 1983, 19(1): 147-158.

[6] 朱兆达. 机载 PD 雷达中 MTI-FFT-CFAR 系统的性能分析[R]. 南航科技资料, 1985: 1-16.

[7] 朱兆达, 叶蔡如, 黄新平. 机载 PD 雷达频域 CFAR 处理性能分析[C]. 北京: 第四届全国雷达会议文集, 1987.

[8] Yoganandam Y, Reddy V U. Analysis of a CFAR scheme at the output of MTI-FFT processor[C]. Proc. of IRSI, 1983: 339-334.

[9] Finn H M, Johnson R S. Adaptive detection mode with threshold control as a function of spatially sampled clutter-level estimates[J]. RCA Review, 1968, 29: 414-464.

[10] Dillard G M, Summers B F. Mean-level detection in the frequency domain[J]. IEE Proc. -Radar, Sonar and Navigation, 1996, 143(5): 307-312.

[11] 简涛, 何友, 苏峰, 等. 基于 ZP-FFT 提高雷达检测性能的仿真分析[J]. 信号处理, 2007, 23(3): 370-373.

[12] Dillard G M. Mean-level detection of nonfluctuating signals[J]. IEEE Trans. on AES, 1974, 10(6): 795-799.

[13] 曲长文, 何友, 刘卫华, 等. 框架理论及应用[M]. 北京: 国防工业出版社, 2009.

[14] 简涛, 何友, 苏峰, 等. 一种基于小波变换的信号恒虚警率检测方法[J]. 信号处理, 2006, 22(3): 430-433.

[15] Jian T, He Y, Su F, et al. A novel DWT-CFAR linear detector[C]. International Conference on Sensing

Computing and Automation, 2006, 7: 3440-3444.

[16] Namias V. The fractional order Fourier transform and its application to quantum mechanics[J]. Journal of Institute Applied Math, 1980, 25(3): 241-265.

[17] 陶然，邓兵，王越. 分数阶傅里叶变换及其应用[M]. 北京：清华大学出版社，2009.

[18] 陈小龙，刘宁波，黄勇，等. 雷达目标检测分数域理论及应用[M]. 北京：科学出版社，2022.

[19] Chen X L, Guan J, Wang G Q, et al. Fast and refined processing of radar maneuvering target based on hierarchical detection via sparse fractional representation[J]. IEEE Access, 2019, 7: 149878-149889.

[20] Guo Y X. Joint modulation format identification and frequency offset estimation based on superimposed lfm signal and FrFT[J]. IEEE Photonics Journal, 2019, 11(5): 1-12.

[21] 关键，李宝，刘加能，等. 两种海杂波背景下的微弱匀加速运动目标检测方法[J]. 电子与信息学报，2009，31(8)：1898-1902.

[22] Zhang Y D, Wang S H, Yang J F, et al. A Comprehensive survey on fractional Fourier transform[J]. Fundamental Informaticae, 2017, 151(1-4): 1-48.

[23] Chen X L, Guan J, Li X Y, et al. Effective coherent integration method for marine target with micromotion via phase differentiation and radon-Lv's distribution[J]. IET Radar Sonar & Navigation, 2015, 9(9): 1284-1295.

[24] 马金铭，苗红霞，苏新华，等. 分数傅里叶变换理论及其应用研究进展[J]. 光电工程，2018，45(6)：170747.

[25] 陈小龙，关键，黄勇，等. 分数阶 Fourier 变换在动目标检测和识别中的应用：回顾和展望[J]. 信号处理，2013，29(01)：85-97.

[26] Guan J, Chen X L, Huang Y, et al. Adaptive fractional fourier transform-based detection algorithm for moving target in heavy sea clutter[J]. IET Radar Sonar & Navigation, 2012, 6(5): 389-401.

[27] Chen X L, Guan J, Huang Y, et al. Radar signal processing for low-observable marine target-challenges and solutions[C]. Chongqing: 2019 IEEE International Conference on Signal, Information and Data Processing (ICSIDP), 2019.

[28] Jin K, Li G Q, Lai T, et al. A novel long-time coherent integration algorithm for doppler-ambiguous radar maneuvering target detection[J]. IEEE Sensors Journal, 2020, 20(16): 9394-9407.

[29] 陈小龙，黄勇，关键，等. MIMO 雷达微弱目标长时积累技术综述[J]. 信号处理，2020，36(12)：1947-1964.

[30] 陈小龙，关键，郭海燕，等. 基于 WPT-FRFT 的微弱动目标检测及性能分析[J]. 雷达科学与技术，2010，8(2)：139-145.

[31] Barbarossa S. Analysis of multicomponent LFM signals by a combined Wigner-Hough transform[J]. IEEE Transactions on Signal Processing, 1995, 43(6): 1511-1515.

[32] 孙艳丽，陈小龙，柳叶. 雷达动目标变换域相参积累检测及性能分析[J]. 太赫兹科学与电子信息学报，2019，17(3)：457-461.

[33] 张月，邹江威，陈曾平. 泛探雷达长时间相参积累目标检测方法研究[J]. 国防科技大学学报，2010，32(6)：15-20.

[34] 关键，陈小龙，于晓涵. 雷达高速高机动目标长时间相参积累检测方法[J]. 信号处理，2017，33(S1)：1-8.

[35] Chen X L, Guan J, Liu N B, et al. Maneuvering target detection via Radon-fractional Fourier transform-based long-time coherent integration[J]. IEEE Transactions on Signal Processing, 2014, 62(4): 939-953.

[36] Chen X L, Guan J, Liu N B, et al. Detection of a low observable sea-surface target with micromotion via the Radon-linear canonical transform[J]. IEEE Geoscience and Remote Sensing Letters, 2014, 11(7): 1225-1229.

[37] Li D, Zhan M Y, Liu H Q, et al. A robust translational motion compensation method for isar imaging based on keystone transform and fractional fourier transform under low SNR environment[J]. IEEE Transactions on Aerospace and Electronic Systems, 2017, 53(5): 2140-2156.

[38] Chen X L, Huang Y, Liu N B, et al. Radon-fractional ambiguity function-based detection method of low-observable maneuvering target[J]. IEEE Transactions on Aerospace and Electronic Systems, 2015, 51(2): 815-833.

[39] Chen X L, Guan J, Huang Y, et al. Radon-linear canonical ambiguity function-based detection and estimation method for marine target with micromotion[J]. IEEE Transactions on Geoscience and Remote Sensing, 2015, 53(4): 2225-2240.

[40] Chen X L, Yu X H, Huang Y, et al. Adaptive Clutter Suppression and Detection Algorithm for Radar Maneuvering Target with High-order Motions via Sparse Fractional Ambiguity Function[J]. IEEE Journal of Selected Topics in Applied Earth Observations and Remote Sensing, 2020, 13: 1515-1526.

[41] Li X L, Cui G L, Yi W, et al. Fast coherent integration for maneuvering target with high-order range migration via TRT-SKT-LVD[J]. IEEE Transactions on Aerospace and Electronic Systems, 2016, 52(6): 2803-2814.

[42] Li X L, Cui G L, Yi W, et al. Sequence-reversing transform-based coherent integration for high-speed target detection[J]. IEEE Transactions on Aerospace and Electronic Systems, 2017, 53(3): 1573-1580.

[43] Li X L, Sun Z, Yeo T S. Computational efficient refocusing and estimation method for radar moving target with unknown time information[J]. IEEE Transactions on Computational Imaging, 2020, 6: 544-557.

[44] Xu J, Xia X G, Peng S B, et al. Radar maneuvering target motion estimation based on generalized Radon-Fourier transform[J]. IEEE Transactions on Signal Processing, 2012, 60(12): 6190-6201.

[45] Niu Z Y, Zheng J B, Su T, et al. Radar high speed target detection based on improved minimalized windowed RFT[J]. IEEE Journal of Selected Topics in Applied Earth Observations and Remote Sensing, 2021, 14: 870-886.

[46] Wang Y, Kang J, Jiang Y C. ISAR imaging of maneuvering target based on the local polynomial wigner distribution and integrated high-order ambiguity function for cubic phase signal model[J]. IEEE Journal of Selected Topics in Applied Earth Observations and Remote Sensing, 2014, 7(7): 2971-2991.

[47] Wu W, Wang G B, Sun J P. Polynomial radon-polynomial Fourier transform for near space hypersonic maneuvering target detection[J]. IEEE Transactions on Aerospace and Electronic Systems, 2018, 54(3): 1306-1322.

[48] Wang Y, Huang X, Cao R. Novel approach for ISAR cross-range scaling based on the multidelay discrete polynomial-phase transform combined with keystone transform[J]. IEEE Transactions on Geoscience and Remote Sensing, 2020, 58(2): 1221-1231.

[49] Huang N E, Shen Z, Long S R, et al. The empirical mode decomposition and the Hilbert spectrum for nonlinear and non-stationary time series analysis[J]. Proceedings Mathematical Physical & Engineering Science, 1998, 454 (1971): 903-995.

[50] Elgamel S A, Soraghan J J. Using EMD-FrFT Filtering to mitigate very high power interference in chirp tracking radars[J]. IEEE Signal Processing Letters, 2011, 18(4): 263-266.

[51] Li Y, Yang Y H, Zhu X Y. Target detection in sea clutter based on multifractal characteristics after empirical mode decomposition[J]. IEEE Geoscience and Remote Sensing Letters, 2017, 14(9): 1547-1551.

[52] 时艳玲,刘子鹏,张学良,等. 基于EMD能量占比的海面漂浮小目标特征检测[J]. 系统工程与电子技术, 2021, 43(2): 300-310.

[53] 张林,李秀友,刘宁波,等. 基于分形特性改进的EMD目标检测算法[J]. 电子与信息学报, 2016, 38(5): 1041-1046.

[54] 张建,关键,董云龙,等. 基于局部Hilbert谱平均带宽的微弱目标检测算法[J]. 电子与信息学报, 2012, 34(1): 121-127.

[55] 张建,关键,黄勇,等. 基于Hilbert谱脊线盒维数的微弱目标检测算法[J]. 电子学报, 2012, 40(12): 2404-2409.

[56] Bai X R, Xing M D, Zhou F, et al. Imaging of micromotion targets with rotating parts based on empirical-mode decomposition[J]. IEEE Transactions on Geoscience and Remote Sensing, 2008, 46(11): 3514-3523.

[57] Zhao Y C, Su Y. The extraction of micro-Doppler signal with emd algorithm for radar-based small uavs' detection[J]. IEEE Transactions on Instrumentation and Measurement, 2020, 69(3): 929-940.

[58] Donoho David L. Compressed sensing[J]. IEEE Transactions on Information Theory, 2006, 52(4): 1289-1306.

[59] 李刚，夏向根. 参数化稀疏表征在雷达探测中的应用[J]. 雷达学报，2016,5(1)：1-7.

[60] 焦李成，杨淑媛，刘芳，等. 压缩感知回顾与展望[J]. 电子学报，2011，39(7)：1651-1662.

[61] Duarte M F，Eldar Y C. Structured compressed sensing：from theory to applications[J]. IEEE Transactions on Signal Processing，2011，59(9)：4053-4085.

[62] 陈小龙，关键，何友，等. 高分辨稀疏表示及其在雷达动目标检测中的应用[J]. 雷达学报，2017，6(3)：239-251.

[63] 陈小龙，关键，于晓涵，等. 雷达动目标短时稀疏分数阶傅里叶变换域检测方法[J]. 电子学报，2017，45(12)：3030-3036.

[64] 陈小龙，关键，董云龙，等. 稀疏域海杂波抑制与微动目标检测方法[J]. 电子学报，2016，44(4)：860-867.

[65] Zhang Z，Xu Y，Yang J，et al. A survey of sparse representation：algorithms and applications[J]. IEEE Access，2015：3，490-530.

[66] Ali Gholami. Sparse time-frequency decomposition and some applications[J]. IEEE Transactions on Geoscience and Remote Sensing，2013，51(6)：3598-3604.

[67] Gilbert A，Indyk P，Iwen M，et al. Recent developments in the sparse Fourier transform：A compressed Fourier transform for big data[J]. IEEE Signal Processing Magazine，2014，31(5)：91-100.

[68] Zhang H，Shan T，Liu S，et al. Optimized sparse fractional Fourier transform：Principle and performance analysis[J]. Signal Processing，2020，174：107646.

[69] Liu S，Shan T，Tao R，et al. Sparse discrete fractional Fourier transform and its applications[J]. IEEE Trans. on Signal Processing，2014，62(24)：6582-6595.

[70] Chen X L，Guan J，Chen W S，et al. Sparse long-time coherent integration-based detection method for radar low-observable maneuvering target[J]. IET Radar Sonar and Navigation，2019，14(4)：538 -546.

[71] Yu X H，Chen X L，Huang Y，et al. Fast detection method for low-observable maneuvering target via robust sparse fractional fourier transform[J]. IEEE Geoscience & Remote Sensing Letters. 2020，17(6)：978-982.

[72] Yu X H，Chen X L，Huang Y，et al. Radar moving target detection in clutter background via adaptive dual-threshold sparse Fourier transform[J]. IEEE Access，2019，7：58200-58211.

第 12 章 高分辨率雷达目标检测

CHAPTER 12

12.1 引言

随着雷达技术的不断进步，人们对雷达探测目标提出了更高的要求，不仅希望可以探测到感兴趣目标的有无，而且还希望能够对目标进行成像进而判别目标的类别属性[1]。高距离分辨率(High Range Resolution，HRR)雷达采用了脉冲压缩等技术，使得雷达发射信号具有很大的时宽带宽积，从而获得了距离高分辨的能力。HRR 雷达距离分辨率可达亚米级，一般目标的回波分布在不同的径向距离单元中，形成"距离扩展目标"[2]。

前面几章主要研究了点目标的 CFAR 检测方法，对于距离扩展目标，这些方法的检测性能可能会下降，甚至无法检测目标[3-4]。对距离扩展目标检测问题的理论阐述最早可以追溯到 Van Trees 教授的专著 *Detection，Estimation and Modulation Theory*，其中首先提出了这一问题，并初步论述了 HRR 雷达距离扩展目标的回波统计特性[5]。几乎同时，荷兰学者 Gerard 在高斯白噪声背景下研究了 Swerling Ⅱ型距离扩展目标的单脉冲检测问题[6]。随后美国海军研究实验室的 Hughes 尝试沿目标所占据的距离单元进行非相参积累检测[7]。自 20 世纪 90 年代中期以来，对距离扩展目标的检测引起雷达界越来越多的关注，出现了大量的研究文献。以意大利的 Gini[8]、Conte[9]、De Maio[10] 和美国海军研究实验室的 Gerlach[11] 等为代表，学者们对距离扩展目标检测问题进行了深入的研究，从高斯背景[12-17]入手，随着对高分辨率雷达杂波的深入认识[18]，逐渐扩展到部分均匀[19-21]和非高斯背景[22-24]，目标信号模型也从秩 1 模型向多秩模型发展[17,25-26]。

将高距离分辨率与高方位分辨率相结合，还可获得高分辨的雷达二维图像。其中合成孔径雷达(Synthetic Aperture Radar，SAR)基于天线沿轨迹运动形成的虚拟孔径，改善了实孔径雷达的方位分辨率，获得了更加精确的目标方位信息，已广泛应用于监视监测等军民领域[1]。近几十年来，世界各国相继研制基于多极化、多波段、多分辨率的 SAR 系统，基于匹配滤波[27]、稀疏优化[28]、深度学习[29]等的 SAR 成像算法大量涌现，有力推动了 SAR 技术领域的发展，显著提升了 SAR 图像的质量。

SAR 图像解译是指从复杂成像背景中提取有价值的目标信息。SAR 图像自动目标识别(Automatic Target Recognition，ATR)是 SAR 图像解译领域的一个引人注目的重要分支。美国麻省理工学院林肯实验室最早开展了针对 SAR ATR 的研究，并将 SAR ATR 分为检测、鉴别和分类/识别三个处理层次[30]，已成为 SAR ATR 算法设计的常规流程[31]。目标检测旨在从宽幅大场景的 SAR 图像中去除明显不含感兴趣目标的区域，生成目标有无的判决结果，但是这个阶段会产生许多虚警；鉴别阶段作为检测阶段的后处理，在保留真实目标信息的前提下进一步减少虚警；目标分类/识别阶段通过对小范围感兴趣目标区域的特征提取与分类，获得目标类别等精细信息。其中，目标检测是SAR ATR处理的首要步骤，该阶段常面临SAR图像中的复杂非均匀杂波、斑点噪声、十字斑、目标散焦、目标密集、目标尺寸大差异多等导致的大量虚警，因此如何实现精准高效的SAR图像目标检测，对提升后续ATR流程的效率和准确性至关重要。CFAR处理是目前应用最广泛的SAR图像目标检测方法之一[31-41]，其具备恒定虚警率的自动目标检测能力。

针对高分辨率雷达目标CFAR检测问题，本章主要从一维像距离扩展目标检测和二维SAR图像目标检测两个方面进行分析。

在距离扩展目标CFAR检测方面，针对秩1和多秩信号模型，讨论非高斯杂波背景下的距离扩展目标检测方法。12.2节对距离扩展目标的信号模型进行了简要的分析；12.3节采用球不变随机过程对复合高斯杂波进行建模，描述复合高斯杂波中多秩距离扩展目标的子空间检测方法；12.4节针对接收机热噪声不能忽略的情况，对热噪声进行等效处理，研究复合高斯杂波加热噪声中的距离扩展目标检测方法；12.5节针对特定实测杂波用复合高斯杂波模拟效果不佳的问题，采用对称alpha稳定(Symmetric alpha Stable，SαS)分布对杂波建模，并探讨了SαS分布杂波中的距离扩展目标检测方法。

在SAR图像目标CFAR检测方面，重点梳理了杂波样本选取、杂波统计建模、知识辅助检测和高效费比处理等四个主要方面的研究进展。12.7节从自动审查和超像素边界两个角度，分析了SAR图像CFAR检测中杂波单元的选取方法；12.8节在广义Gamma分布杂波背景下，侧重于SAR图像CFAR检测的虚警控制性能分析；12.9节讨论了基于语义知识辅助的SAR图像CFAR处理方法；12.10节通过挖掘海上舰船目标的空域稀疏性，研究了基于密度特征的SAR图像CFAR检测快速实现方法。

12.2 距离扩展目标的信号模型

12.2.1 秩1信号模型

假设一个具有N_a个阵元的均匀线阵(Uniformly Linear Array，ULA)在一个CPI中共发射了N_p个相参脉冲，由于距离扩展目标信号不只局限于一个距离单元，而是随机分布在多个距离单元中。当假设目标相对于雷达视角只有平动而忽略转动时，则距离扩展目标在每个距离单元中的有用信号可以表示为[26]

$$\boldsymbol{s}_t=\alpha_t\boldsymbol{p}(\theta,f_d),\quad t=1,2,\cdots,H \tag{12.1}$$

$\boldsymbol{p}(\theta,f_d)$是信号的空时导向矢量，并且有

$$\boldsymbol{p}(\theta,f_d)=\boldsymbol{b}(\theta)\otimes\boldsymbol{a}(f_d) \tag{12.2}$$

符号⊗表示Kronecker乘积，其中空域导向矢量为

$$\boldsymbol{b}(\theta)=\left[1,\exp\left(-\mathrm{j}\frac{2\pi}{\lambda}d\sin\theta\right),\cdots,\exp\left(-\mathrm{j}\frac{2\pi}{\lambda}d(N_a-1)\sin\theta\right)\right]^{\mathrm{T}} \tag{12.3}$$

而时域导向矢量为

$$\boldsymbol{a}(f_d)=\left[1,\exp\left(-\mathrm{j}2\pi\frac{f_d}{f_r}\right),\cdots,\exp\left(-\mathrm{j}2\pi(N_p-1)\frac{f_d}{f_r}\right)\right]^{\mathrm{T}} \tag{12.4}$$

其中，θ是信号的到达角，f_d是目标信号的多普勒频率，假设θ和f_d都是已知的或可从实测数据中估计得到[26]。λ是雷达工作波长，f_r是脉冲重复频率，d是天线阵元间隔距离。当$d=\lambda/2$时，式(12.3)简

化为

$$\boldsymbol{b}(\theta)=[1,\exp(-\mathrm{j}\pi\sin\theta),\cdots,\exp(-\mathrm{j}\pi(N_a-1)\sin\theta)]^{\mathrm{T}} \tag{12.5}$$

由前述假设可知,有用信号的空时导向矢量 $\boldsymbol{p}(\theta,f_{\mathrm{d}})$ 是 $N\times1$ 维的列矢量,其中 $N=N_{\mathrm{a}}N_{\mathrm{p}}$。特别地,对于普通的非阵列雷达天线,阵元数 $N_{\mathrm{a}}=1$,则相应的导向矢量变为

$$\boldsymbol{p}(f_{\mathrm{d}})=\boldsymbol{a}(f_{\mathrm{d}}) \tag{12.6}$$

用 $\alpha_t(t=1,2,\cdots,H)$ 表示距离扩展目标在不同距离单元中的复幅度,其相位均匀分布在$[0,2\pi]$上,幅度 $|\alpha_t|$ 用相关 χ^2 分布建模[9],即 $|\alpha_t|$ 的 PDF 为

$$f_{|\alpha_t|}(x)=\frac{2m^m x^{2m-1}}{\Gamma(m)(\varepsilon_t^2)^m}\exp\left(-m\frac{x^2}{\varepsilon_t^2}\right)u(x) \tag{12.7}$$

其中,$u(x)$ 是单位阶跃函数;自然数 m 是 χ^2 分布的自由度,用来表示 $|\alpha_t|$ 的起伏深度,m 越小,目标起伏越剧烈。$m=1$ 表示瑞利分布(Swerling Ⅰ)目标;$m=2$ 表示瑞利主加分布(Swerling Ⅲ)目标。$m=\infty$ 表示非起伏的 Swerling 0 目标。ε_t^2 是信号幅度 $|\alpha_t|$ 的均方值,$\Gamma(\cdot)$ 表示 Gamma 函数。假定 $|\alpha_t|$ 在各个距离单元之间是部分相关的,其协方差矩阵的元素为

$$\mathrm{cov}(|\alpha_h|^2,|\alpha_k|^2)=\frac{\varepsilon_h^2\varepsilon_k^2}{m}\rho^{|h-k|},\quad h,k\in\{1,2,\cdots,H\},\rho\in(0,1] \tag{12.8}$$

12.2.2 多秩子空间信号模型

秩1信号模型假设待检测的距离扩展目标信号处在观测空间的一维线性子空间上[42]。实际上,更为一般的情形是,距离扩展目标信号是一个多秩的子空间随机信号,即目标信号处在观测空间的有限维的线性子空间上,许多情况都会引起目标信号多秩。例如,当目标相对于雷达视角除平动之外,还有相对转动时,则每个距离单元中目标信号的多普勒频率彼此是各不相等的,需要用多秩子空间信号[8]来建立统计模型。

一个随机信号的秩被定义为其协方差矩阵的秩。根据距离扩展目标的多主散射点(Multiple Dominant Scattering,MDS)模型[43],距离扩展目标在每个距离单元内的回波,等价于该单元内有限个孤立强散射点回波的矢量和,则距离扩展目标在第 t 个距离单元中复回波的第 n 次采样可表示为

$$s_t(n)=\sum_{k=1}^{N_t}a_{t,k}\exp[\mathrm{j}2\pi(n-1)f_{t,k}],\quad t=1,2,\cdots,H;\quad n=1,2,\cdots,N \tag{12.9}$$

式中,N_t 是第 t 个距离单元内目标散射点总数目;$a_{t,k}$ 是第 t 个距离单元内第 k 个散射点的幅度;归一化频率 $f_{t,k}=f_{\mathrm{d}}(t,k)/f_{\mathrm{r}}$;$f_{\mathrm{d}}(t,k)$ 表示第 t 个距离单元内第 k 个散射点的 Doppler 频率;N_t 取决于 HRR 雷达所观测到的具体目标[1],$a_{t,k}$ 是慢变的,N_t 和 $a_{t,k}$ 都与采样数 n 无关。将式(12.9)写成矩阵形式为[26]

$$\boldsymbol{s}_t=\boldsymbol{E}_t\boldsymbol{a}_t,\quad t=1,2,\cdots,H \tag{12.10}$$

其中,$\boldsymbol{s}_t=[s_t(1),s_t(2),\cdots,s_t(N)]^{\mathrm{T}}$ 是 $N\times1$ 维的列矢量;$\boldsymbol{a}_t=[a_{t,1},a_{t,2},\cdots,a_{t,N_t}]^{\mathrm{T}}$ 是 $N_t\times1$ 维的列矢量,$N\times N_t$ 维的矩阵 $\boldsymbol{E}_t$ 为

$$\boldsymbol{E}_t=\begin{pmatrix}1 & 1 & \cdots & 1\\ \exp(\mathrm{j}2\pi f_{t,1}) & \exp(\mathrm{j}2\pi f_{t,2}) & \cdots & \exp(\mathrm{j}2\pi f_{t,N_t})\\ \vdots & \vdots & & \vdots\\ \exp(\mathrm{j}2\pi(N-1)f_{t,1}) & \exp(\mathrm{j}2\pi(N-1)f_{t,2}) & \cdots & \exp(\mathrm{j}2\pi(N-1)f_{t,N_t})\end{pmatrix} \tag{12.11}$$

对 $\boldsymbol{E}_t$ 进行奇异值分解，得 $\boldsymbol{E}_t=\boldsymbol{U}_t\boldsymbol{\Lambda}_t\boldsymbol{V}_t^{\mathrm{H}}$，$(\cdot)^{\mathrm{H}}$ 表示矩阵的共轭转置。其中，$\boldsymbol{U}_t$ 是由左奇异矢量构成的维数为 $N\times N_t$ 维的酉矩阵，$\boldsymbol{\Lambda}_t$ 是奇异值为对角元素的 $N_t\times N_t$ 维对角阵，$\boldsymbol{V}_t$ 是右奇异矢量构成的酉矩阵，则 $\boldsymbol{s}_t$ 可进一步表示为

$$\boldsymbol{s}_t=\boldsymbol{U}_t\boldsymbol{b}_t \tag{12.12}$$

注意到 $\boldsymbol{b}_t=\boldsymbol{\Lambda}_t\boldsymbol{V}_t^{\mathrm{H}}\boldsymbol{a}_t$。式(12.12)说明，距离扩展目标的回波信号可以用线性子空间模型来建模，即距离扩展目标回波处在信号子空间 $\langle\boldsymbol{U}_t\rangle$ 上，该子空间由酉矩阵 $\boldsymbol{U}_t$ 的列矢量张成；$N_t\times 1$ 维的列矢量 $\boldsymbol{b}_t$ 被称为位置矢量。距离扩展目标的回波信号 $\boldsymbol{s}_t$ 是子空间信号，但其在子空间中的位置矢量 $\boldsymbol{b}_t$ 却是未知的。模式矩阵 $\boldsymbol{U}_t$ 的秩确定了信号子空间 $\langle\boldsymbol{U}_t\rangle$ 的维数：$\dim(\langle\boldsymbol{U}_t\rangle)=\mathrm{rank}(\boldsymbol{U}_t)=\mathrm{rank}(\boldsymbol{E}_t)=N_t$，即在第 t 个距离单元内，距离扩展目标回波所在的信号子空间 $\langle\boldsymbol{U}_t\rangle$ 的维数，等于该距离单元内目标主散射点的数目。

本章假设信号子空间 $\langle\boldsymbol{U}_t\rangle$ 及其维数 N_t 是已知的，在实际应用中，维数 N_t 可以从实测数据中估计得到，而信号子空间 $\langle\boldsymbol{U}_t\rangle$ 可以采用超分辨谱估计等算法求解[26]。

12.3　复合高斯杂波中多秩距离扩展目标的子空间检测器

12.3.1　问题描述

高分辨率雷达接收到的基带复数据观测矢量可表示为

$$\boldsymbol{z}_t=\boldsymbol{s}_t+\boldsymbol{c}_t+\boldsymbol{n}_t,\quad t=1,2,\cdots,H \tag{12.13}$$

首先考虑一种简单情况，当匹配滤波器输出的杂波与噪声的功率比很大时，可考虑忽略内部热噪声 $\boldsymbol{n}_t$ 的影响[44]。上述观测模型简化为

$$\boldsymbol{z}_t=\boldsymbol{s}_t+\boldsymbol{c}_t \tag{12.14}$$

根据式(12.14)可将距离扩展目标的检测问题归结为如下的二元假设检验

$$\begin{cases}\mathrm{H}_0:\ \boldsymbol{z}_t=\boldsymbol{c}_t\\ \mathrm{H}_1:\ \boldsymbol{z}_t=\boldsymbol{s}_t+\boldsymbol{c}_t\end{cases} \tag{12.15}$$

$\boldsymbol{z}_t$ 是基带复数据观测矢量，$\boldsymbol{c}_t$ 是杂波向量，t 是距离单元编号，H 是待检测目标所占据的距离单元个数。$\boldsymbol{s}_t$ 是距离扩展目标的信号矢量，采用式(12.12)的多秩子空间随机信号建模。杂波向量 $\boldsymbol{c}_t$ 采用球不变随机过程(Spherically Invariant Random Processes，SIRP)建模[11,18]，即

$$\boldsymbol{c}_t=\sqrt{\tau_t}\,\boldsymbol{x}_t \tag{12.16}$$

其中，快起伏的散斑分量 $\boldsymbol{x}_t$ 是归一化协方差矩阵为 $\boldsymbol{M}_x$ 的零均值复高斯随机矢量，即 $\boldsymbol{x}_t\sim\mathrm{CN}(\boldsymbol{0},\boldsymbol{M}_x)$；纹理分量 τ_t 是一个正的随机变量，代表杂波功率水平。在上述观测模型中，已知的数据有 $N\times 1$ 维的复基带数据观测矢量 $\boldsymbol{z}_t$，维数为 $N\times N_t$ 维的模式矩阵 $\boldsymbol{U}_t$，即为距离扩展目标信号的多秩子空间[45]。接下来需针对二元假设检验，构造出相应检验统计量进行判决。

12.3.2　广义匹配子空间检测器的设计

根据 Neyman-Pearson 准则，上述假设检验问题可以用 GLRT 来求解。为简化分析，可根据“两步法”GLRT 检验理论设计检测器[9]：先假设散斑分量的归一化协方差矩阵 $\boldsymbol{M}_x$ 是已知的，将其他未知参数用其最大似然(Maximum Likelihood，ML)估计代替。

在 H_0 假设下,待检测单元内的复观测矢量 $z_t(t=1,2,\cdots,H)$ 的条件 PDF 为

$$f_{z_t}(z_t \mid \tau_{t0},H_0)=\frac{1}{(\pi\tau_{t0})^N\,|M_x|}\exp\left(-\frac{z_t^H M_x^{-1} z_t}{\tau_{t0}}\right) \tag{12.17}$$

z_t 在 H_1 假设下的条件 PDF 为

$$f_{z_t}(z_t \mid \tau_{t1},b_t;\,H_0)=\frac{1}{(\pi\tau_{t1})^N\,|M_x|}\exp\left[-\frac{(z_t-U_t b_t)^H M_x^{-1}(z_t-U_t b_t)}{\tau_{t1}}\right] \tag{12.18}$$

其中,$|M_x|$表示矩阵 M_x 的行列式。

根据"两步法"GLRT 检验理论,可得检测统计量为

$$\Lambda_1(z_1,z_2,\cdots,z_H)=\frac{\max\limits_{\{\tau_{11},\cdots,\tau_{H1}\}}\max\limits_{\{b_1,\cdots,b_H\}}\prod\limits_{t=1}^{H} f_{z_t}(z_t \mid \tau_{t1},b_t;H_1)}{\max\limits_{\{\tau_{10},\cdots,\tau_{H0}\}}\prod\limits_{t=1}^{H} f_{z_t}(z_t \mid \tau_{t0};H_0)}\ \underset{H_0}{\overset{H_1}{\gtrless}}\ G_1 \tag{12.19}$$

对其中的未知参数用 MLE 代替,并取自然对数可得检验统计量为

$$\ln\Lambda_1(z_1,z_2,\cdots,z_H)=-N\sum_{t=1}^{H}\ln\left(1-\frac{z_t^H Q_t z_t}{z_t^H M_x^{-1} z_t}\right) \tag{12.20}$$

式中

$$Q_t=M_x^{-1}U_t A_t U_t^H M_x^{-1} \tag{12.21}$$

$$A_t=(U_t^H M_x^{-1} U_t)^{-1} \tag{12.22}$$

令 $x_t=(z_t^H Q_t z_t)/(z_t^H M_x^{-1} z_t)$,注意到函数 $\ln(1-x_t)$是关于 x_t 的严格单调递减函数,因此前述对数似然比可进一步化简为

$$\Lambda(z_1,z_2,\cdots,z_H)=\sum_{t=1}^{H}\frac{z_t^H Q_t z_t}{z_t^H M_x^{-1} z_t}\ \underset{H_0}{\overset{H_1}{\gtrless}}\ G \tag{12.23}$$

其中,G 为等效的检测阈值。式(12.23)即复合高斯杂波中多秩距离扩展目标的子空间检测器,称为广义匹配子空间检测器(Generalized Matched Subspace Detector,GMSD)。

注意到"点目标"的匹配子空间检测器(Matched Subspace Detector,MSD)可表示为

$$T_{\mathrm{MSD}}(z_t)=\frac{z_t^H Q_t z_t}{z_t^H M_x^{-1} z_t} \tag{12.24}$$

则有

$$\Lambda(z_1,z_2,\cdots,z_H)=\sum_{t=1}^{H}T_{\mathrm{MSD}}(z_t)\ \underset{H_0}{\overset{H_1}{\gtrless}}\ G \tag{12.25}$$

由式(12.25)可知,在复合高斯杂波中检测多秩的距离扩展目标,可借用点目标的 MSD 来完成,即对各个距离单元分别采用 MSD 来检测,然后对各个距离单元输出的统计量做非相参积累[42],形成最终的检验统计量,进而与检测门限进行比较。GMSD 检测器的结构框图见图 12.1。

在 H_0 假设下有

$$T_{\mathrm{MSD}}(z_t \mid H_0)=T_{\mathrm{MSD}}(\sqrt{\tau_t}x_t)=\frac{(\sqrt{\tau_t}x_t)^H Q_t(\sqrt{\tau_t}x_t)}{(\sqrt{\tau_t}x_t)^H M_x^{-1}(\sqrt{\tau_t}x_t)}=T_{\mathrm{MSD}}(x_t) \tag{12.26}$$

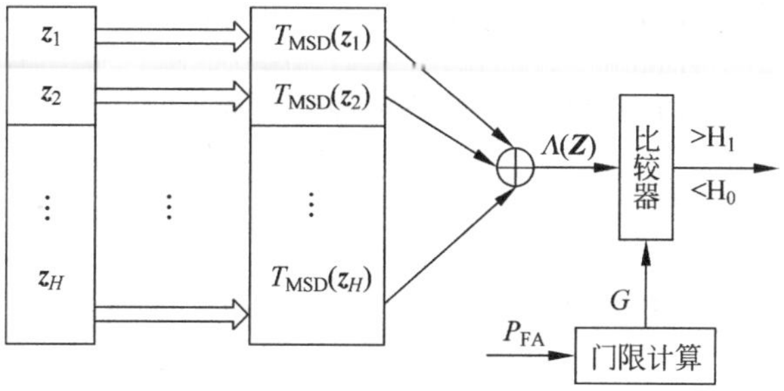

图 12.1　GMSD 检测器的结构框图

式(12.26)表明，组成 GMSD 检测器的每个 MSD 对杂波的纹理分量都是 CFAR 的，因此，由式(12.25)可知，GMSD 检测器对杂波的纹理分量也具有 CFAR 能力。

12.3.3　广义匹配子空间检测器虚警概率的计算

先计算式(12.25)中每个 MSD 的虚警概率[46]，然后再计算 GMSD 检测器的虚警概率。由于点目标 MSD 对复合高斯杂波的纹理分量是 CFAR 的，由式(12.26)得

$$P_{\mathrm{fa}} = \Pr\{T_{\mathrm{MSD}}(\boldsymbol{z}) > \lambda \mid \mathrm{H}_0\} = \Pr\left\{\frac{\boldsymbol{x}^{\mathrm{H}}\boldsymbol{Q}\boldsymbol{x}}{\boldsymbol{x}^{\mathrm{H}}\boldsymbol{M}_{\boldsymbol{x}}^{-1}\boldsymbol{x}} > \lambda\right\} \tag{12.27}$$

为便于书写，在不影响分析结果的基础上，式(12.27)忽略了距离单元下标 t。进一步对式(12.27)中的高斯散斑分量 $\boldsymbol{x}$ 做“白化”处理，即令 $\boldsymbol{x}=\boldsymbol{L}\boldsymbol{w}$ 或 $\boldsymbol{w}=\boldsymbol{L}^{-1}\boldsymbol{x}$，其中下三角矩阵 $\boldsymbol{L}$ 是杂波散斑分量 $\boldsymbol{x}$ 的归一化协方差矩阵 $\boldsymbol{M}_{\boldsymbol{x}}$ 的 Cholesky 分解因子，即有 $\boldsymbol{M}_{\boldsymbol{x}}=\boldsymbol{L}\boldsymbol{L}^{\mathrm{H}}$。将 $\boldsymbol{x}=\boldsymbol{L}\boldsymbol{w}$ 代入式(12.27)的右边，可得

$$P_{\mathrm{fa}} = \Pr\left\{\frac{\boldsymbol{w}^{\mathrm{H}}\boldsymbol{P}_q\boldsymbol{w}}{\boldsymbol{w}^{\mathrm{H}}\boldsymbol{w}} > \lambda\right\} \tag{12.28}$$

式中，$\boldsymbol{P}_q=\boldsymbol{L}^{\mathrm{H}}\boldsymbol{Q}\boldsymbol{L}$ 是信号在特定子空间 $\boldsymbol{U}_q$ 上的投影矩阵[43]，其中 $\boldsymbol{U}_q=\boldsymbol{L}^{-1}\boldsymbol{U}$。

利用矩阵的恒等变换 $(\boldsymbol{I}-\boldsymbol{P}_q)+\boldsymbol{P}_q=\boldsymbol{I}$，其中 $\boldsymbol{I}$ 是 $N\times N$ 维的单位阵，式(12.28)可变为

$$P_{\mathrm{fa}} = \Pr\left\{\frac{\boldsymbol{w}^{\mathrm{H}}\boldsymbol{P}_q\boldsymbol{w}}{\boldsymbol{w}^{\mathrm{H}}(\boldsymbol{I}-\boldsymbol{P}_q)\boldsymbol{w}} > \frac{\lambda}{1-\lambda}\right\} \tag{12.29}$$

在 H_0 假设下，二次型 $\boldsymbol{w}^{\mathrm{H}}\boldsymbol{P}_q\boldsymbol{w}$ 服从自由度为 $2r$ 的 χ^2 分布；$\boldsymbol{w}^{\mathrm{H}}(\boldsymbol{I}-\boldsymbol{P}_q)\boldsymbol{w}$ 服从自由度为 $2(N-r)$ 的 χ^2 分布；其中 r 是投影矩阵 $\boldsymbol{P}_q$ 的秩。注意到两个二次型是统计独立的，因此

$$F = \frac{\boldsymbol{w}^{\mathrm{H}}\boldsymbol{P}_q\boldsymbol{w}/2r}{\boldsymbol{w}^{\mathrm{H}}(\boldsymbol{I}-\boldsymbol{P}_q)\boldsymbol{w}/[2(N-r)]} \tag{12.30}$$

服从自由度为 $[2r,2(N-r)]$ 的 F 分布，即统计量 F 的 PDF 为

$$f_F(x) = \frac{1}{B(r,N-r)} r^r (N-r)^{N-r} x^{r-1} (rx+N-r)^{-N} u(x) \tag{12.31}$$

其中，$B(r,N-r)$ 是 Beta 函数，定义为

$$B(a,b) = \int_0^1 \zeta^{a-1}(1-\zeta)^{b-1}\mathrm{d}\zeta, \quad a>0, b>0 \tag{12.32}$$

根据式(12.30)，式(12.29)可进一步表示为

$$P_{\mathrm{fa}} = \Pr\left\{F > \frac{\lambda}{1-\lambda}\left(\frac{N}{r}-1\right)\right\} = \Pr\{F > F_0\} = \int_{F_0}^{\infty} f_F(x)\mathrm{d}x \tag{12.33}$$

其中,门限 $F_0=\frac{\lambda}{1-\lambda}\left(\frac{N}{r}-1\right)$ 可由式(12.33)反解。

由式(12.25)和式(12.28),可得 GMSD 检测器的虚警概率为

$$P_{\mathrm{FA}}=\Pr\{\Lambda(\boldsymbol{z}_1,\boldsymbol{z}_2,\cdots,\boldsymbol{z}_H)>G\mid \mathrm{H}_0\}=\Pr\left\{\sum_{t=1}^{H}\frac{\boldsymbol{w}_t^{\mathrm{H}}\boldsymbol{P}_{q_t}\boldsymbol{w}_t}{\boldsymbol{w}_t^{\mathrm{H}}\boldsymbol{w}_t}>G\mid \mathrm{H}_0\right\} \tag{12.34}$$

其中,$\boldsymbol{P}_{q_t}$ 是信号子空间 $\boldsymbol{U}_{q_t}=\boldsymbol{L}^{-1}\boldsymbol{U}_t$ 上的投影矩阵,令

$$\Gamma_1(t)=\frac{\boldsymbol{w}_t^{\mathrm{H}}\boldsymbol{P}_{q_t}\boldsymbol{w}_t}{\boldsymbol{w}_t^{\mathrm{H}}(\boldsymbol{I}-\boldsymbol{P}_{q_t})\boldsymbol{w}_t},\quad \Gamma_2(t)=\frac{\boldsymbol{w}_t^{\mathrm{H}}\boldsymbol{P}_{q_t}\boldsymbol{w}_t}{\boldsymbol{w}_t^{\mathrm{H}}\boldsymbol{w}_t} \tag{12.35}$$

则有

$$\Gamma_2(t)=\Gamma_1(t)/[\Gamma_1(t)+1] \tag{12.36}$$

由式(12.30)知,$\Gamma_1(t)=[r/(N-r)]F$,利用式(12.31)可得 $\Gamma_1(t)$ 的 PDF 为

$$f_{\Gamma_1(t)}(x)=\frac{1}{B(r,N-r)}x^{r-1}(x+1)^{-N}u(x) \tag{12.37}$$

同理,由式(12.36)和式(12.37),得 $\Gamma_2(t)(t=1,2,\cdots,H)$ 的 PDF 为

$$f_{\Gamma_2(t)}(y)=\frac{1}{B(r,N-r)}y^{r-1}(1-y)^{N-r-1},\quad 0<y\leqslant 1 \tag{12.38}$$

由式(12.34)和式(12.35),根据独立随机变量之和的 PDF 卷积公式,得到 GMSD 检验统计量在 H_0 假设下的 PDF 为

$$f_{\Lambda(\boldsymbol{z})}(y\mid \mathrm{H}_0)=f_{\Gamma_2(1)}(y)\otimes f_{\Gamma_2(2)}(y)\otimes\cdots\otimes f_{\Gamma_2(t)}(y)\otimes\cdots\otimes f_{\Gamma_2(H)}(y) \tag{12.39}$$

其中,$\otimes$表示卷积运算,$f_{\Gamma_2(t)}(y)$可由式(12.38)给出。计算式(12.39)中的卷积可用数值计算方法完成。则 GMSD 的虚警概率为

$$P_{\mathrm{FA}}=\Pr\{\Lambda(\boldsymbol{z})>G\mid \mathrm{H}_0\}=\int_G^H f_{\Lambda(\boldsymbol{z})}(y\mid \mathrm{H}_0)\mathrm{d}y \tag{12.40}$$

12.3.4 广义匹配子空间检测器的自适应实现

在距离扩展目标检测的实际应用中,GMSD 检测器自适应实现需估计的参数有:目标所占据的距离单元数目 H,杂波散斑分量的协方差矩阵 $\boldsymbol{M}_x$。下面分别给出估计方法。

1. 目标距离单元数目

在利用 GMSD 检测器进行有效检测前,需准确估计距离扩展目标所占据的距离单元数 H。低估该距离单元数可能会降低 GMSD 非相参积累效果,从而降低检测概率;而高估该距离单元数,则会增加检测器的虚警概率[47]。为达到较好的检测性能,必须根据观测数据估计出目标所占据的距离单元数,具体估计方法描述如下。

步骤一:计算平均一维距离幅度像

$$|z(r)|=\frac{1}{N}\sum_{n=1}^{N}|z_{nr}|,\quad r=1,2,\cdots,R \tag{12.41}$$

其中,N 是样本总数,R 是距离单元总数。

步骤二:设置最小滑窗。即在平均一维距离幅度像上任意选择一点 t,设置最小滑窗为

$$w_{\min}=[t-w_0,t+w_0],\quad w_0=\mathrm{int}(\Delta d_{\min}/2\Delta R) \tag{12.42}$$

其中,ΔR 是 HRR 雷达的距离分辨率,$\Delta d_{\min}$ 为预先设置的目标最小可能长度,int(·)表示对变元向上

取整数，ΔR 和 $\Delta d_{\min}$ 都是已知的。

步骤二：寻找平均一维距离幅度像的中心点。沿着平均一维距离幅度像滑动最小窗口 $w_{\min}$，对平均一维距离幅度像上的每个点 r $(r=1,2,\cdots,R)$，依次计算以该点为中心的最小窗口 $w_{\min}$ 内平均一维距离幅度像的能量之和，即

$$Y(r)=\sum_{r\in w_{\min}}|z(r)|^2 \tag{12.43}$$

得到序列 $Y(r)=\{Y(1),Y(2),\cdots,Y(R)\}$。定义平均一维距离幅度像的中心点

$$i_0=\underset{r}{\arg\max}\{Y(r),r=1,2,\cdots,R\} \tag{12.44}$$

其中，arg(・)表示取变元操作。

步骤四：计算平均一维距离幅度像的能量。在平均一维距离幅度像的中心点 $|z(i_0)|$ 处，逐渐增加最小窗口 $w_{\min}$ 的宽度，即令 $w_j=[i_0-w_0-j,i_0+w_0+j]$，$j=1,2,\cdots,L$，其中，$L=\mathrm{int}(\Delta d_{\max}/2\Delta R)$，$\Delta d_{\max}$ 为预先设置的目标最大可能长度；根据具体的应用情况，可预先对 $\Delta d_{\max}$ 进行粗略的保守估计。基于新的窗口序列 w_j $(j=1,2,\cdots,L)$，依次计算每个窗口序列 w_j 内一维距离幅度像的平均能量

$$\bar{E}(j)=\frac{1}{K_j}\sum_{r\in w_j}|z(r)|^2,\quad j=1,2,\cdots,L \tag{12.45}$$

以及窗口 w_j 之外的杂波的平均能量

$$\overline{E_c}(j)=\frac{1}{R-K_j}\sum_{r\notin w_j}|z(r)|^2,\quad j=1,2,\cdots,L \tag{12.46}$$

其中，$K_j=2(w_0+j)+1$ 是每个窗口 w_j 内包含的距离单元数。

步骤五：计算与每个窗口 w_j 相对应的平均信杂比

$$\overline{\mathrm{SCR}}(w_j)=\frac{\bar{E}(j)-\overline{E_c}(j)}{\overline{E_c}(j)} \tag{12.47}$$

则距离扩展目标所占据的距离单元数的估计值为

$$\hat{H}=2(w_0+\hat{j})+1 \tag{12.48}$$

其中，$\hat{j}=\underset{j}{\arg\max}(\overline{\mathrm{SCR}}(w_j))$是平均信杂比最大的窗口序列的编号。

2. 杂波散斑分量的协方差矩阵

假设可以获得与待检测单元邻近的 HK 个辅助单元观测数据 $\boldsymbol{z}_t$，$t=H+1,\cdots,H(K+1)$。该辅助数据不包含有用的目标信号，且和待检测单元杂波散斑分量有相同的协方差矩阵。则基于辅助数据的归一化样本协方差矩阵(NSCM)可表示为[48]

$$\boldsymbol{S}=\frac{N}{HK}\sum_{t=H+1}^{H(K+1)}\frac{\boldsymbol{z}_t\boldsymbol{z}_t^{\mathrm{H}}}{\boldsymbol{z}_t^{\mathrm{H}}\boldsymbol{z}_t} \tag{12.49}$$

按照"两步法"GLRT 设计方法，用 $\boldsymbol{S}$ 代替式(12.23)中的未知协方差矩阵 $\boldsymbol{M}_x$，可得距离扩展目标的自适应广义匹配子空间检测器(Adaptive GMSD，A-GMSD)为

$$\lambda=\sum_{t=1}^{H}\frac{\boldsymbol{z}_t^{\mathrm{H}}\hat{\boldsymbol{Q}}_t\boldsymbol{z}_t}{\boldsymbol{z}_t^{\mathrm{H}}\boldsymbol{S}^{-1}\boldsymbol{z}_t}\underset{\mathrm{H}_0}{\overset{\mathrm{H}_1}{\gtrless}}T \tag{12.50}$$

其中，T 表示等效阈值，$\hat{\boldsymbol{Q}}_t=\boldsymbol{S}^{-1}\boldsymbol{U}_t(\boldsymbol{U}_t^{\mathrm{H}}\boldsymbol{S}^{-1}\boldsymbol{U}_t)^{-1}\boldsymbol{U}_t^{\mathrm{H}}\boldsymbol{S}^{-1}$。

12.3.5 性能分析

由于子空间检测器 A-GMSD 的检测概率没有解析表达式，可采用 Monte Carlo 仿真进行性能分析。参数设置为：目标距离单元总数 $H=4$；虚警概率为 $P_{\mathrm{fa}}=10^{-4}$，各个散射点的幅度为服从自由度为 $2m$ 的 χ^2 起伏的 IID 随机矢量；各散射点的归一化频率 $f_{t,k}$ 的取值见表 12.1，表 12.1 中同时给出了距离单元内散射点数目 $N_t(t=1,2,3,4)$的值。

表 12.1 各散射点的归一化频率 $f_{t,k}$（距离单元数 $H=4$）

单元编号	1	2	3	4
归一化频率($f_{t,k}$)	{0.1,0.3}	{0.2}	{0.2,0.4}	{0.1,0.2,0.3,0.4}
散射点数(N_t)	2	1	3	4

复合高斯杂波建模为形状参数为 v、尺度参数为 b 的 K 分布，则杂波功率为 $\sigma^2=v/b^2$。信杂比定义为

$$\mathrm{SCR}=\frac{\sum_{t=1}^{H}(\boldsymbol{U}_t\boldsymbol{b}_t)^{\mathrm{H}}\boldsymbol{M}_x^{-1}(\boldsymbol{U}_t\boldsymbol{b}_t)}{N\sigma^2} \tag{12.51}$$

K 分布杂波散斑分量的功率谱密度选均值为 0.3、标准差为 0.2 的高斯谱，则对应的杂波散斑分量自相关函数为

$$R(\lambda)=\exp(-0.08\pi^2\lambda^2+\mathrm{j}0.6\pi\lambda) \tag{12.52}$$

由 $R(\lambda)$可以计算出 K 分布杂波的协方差矩阵 $\boldsymbol{M}_x$。在每个信杂比下进行 10^4 次 Monte Carlo 仿真来估计检测概率 P_{d}，接下来分析不同参数变化对检测性能影响。

(1) 目标参数对 A-GMSD 检测性能的影响。

由图 12.2 和图 12.3 可知，随着相参积累数 N 和距离单元数 H 的增加，A-GMSD 的检测性能均有不同程度的改善，其中图 12.3 中 $H=2$ 时各散射点的归一化频率的取值见表 12.2。

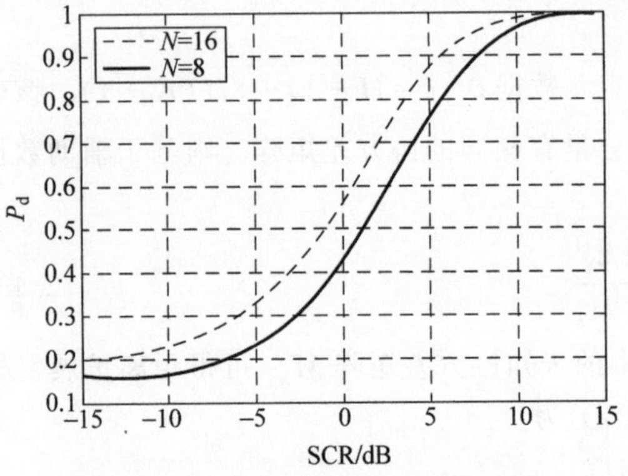

图 12.2 A-GMSD 的检测性能曲线
($H=4,K=8,v=0.5,m=1$)

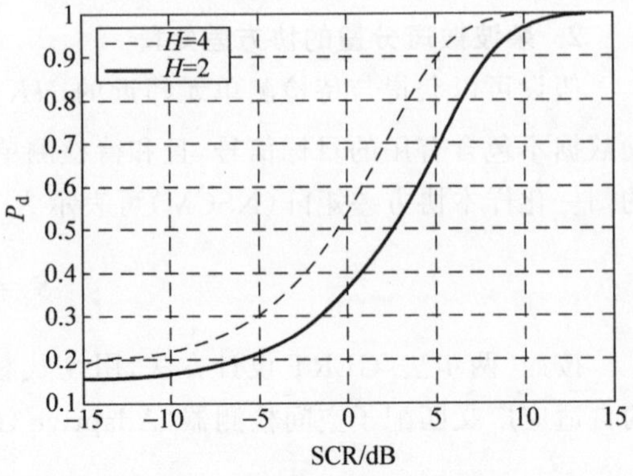

图 12.3 A-GMSD 的检测性能曲线
($N=16,K=8,v=0.5,m=1$)

表 12.2 各散射点的归一化频率 $f_{t,k}$（距离单元数 $H=2$）

单元编号	1	2
归一化频率($f_{t,k}$)	{0.1,0.3}	{0.1,0.2,0.4}
散射点数(N_t)	2	3

（2）目标信号子空间维数 N_t 对 A-GMSD 检测性能影响。

仿真参数取值见表 12.3，从图 12.4 的仿真结果可以看出，随着信号子空间维数的增大，A-GMSD 的检测概率有所下降。可能的原因在于，增加信号子空间维数，意味着信号子空间与满秩的“杂波子空间”越接近，导致从杂波中区分目标变得困难[49]。

表 12.3 不同目标模式下各散射点的归一化频率 $f_{t,k}$（距离单元数 $H=4$）

MSD 模式		距离单元编号			
		1	2	3	4
MSD 1	归一化频率	{0.1}	{0.2,0.3}	{0.2}	{0.1,0.2}
	散射点数	1	2	1	2
MSD 2	归一化频率	{0.1,0.3}	{0.2}	{0.2,0.4}	{0.1,0.2,0.3,0.4}
	散射点数	2	1	2	4

（3）杂波形状系数 v 对 A-GMSD 检测性能影响。

由图 12.5 可知，形状参数 v 越小，杂波尖峰越多，A-GMSD 的检测概率越高。可能的原因在于，形状参数 v 越小，杂波的非高斯特征越明显，针对非高斯环境所设计的 A-GMSD 检测器就越能充分发挥作用[26]。即 A-GMSD 检测器具有强非高斯杂波下有效检测目标的能力。

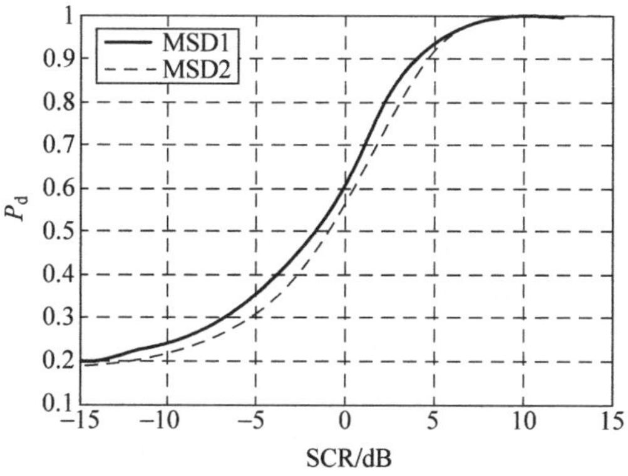

图 12.4 A-GMSD 的检测性能曲线

（$H=4, N=16, K=8, v=0.5, m=1$）

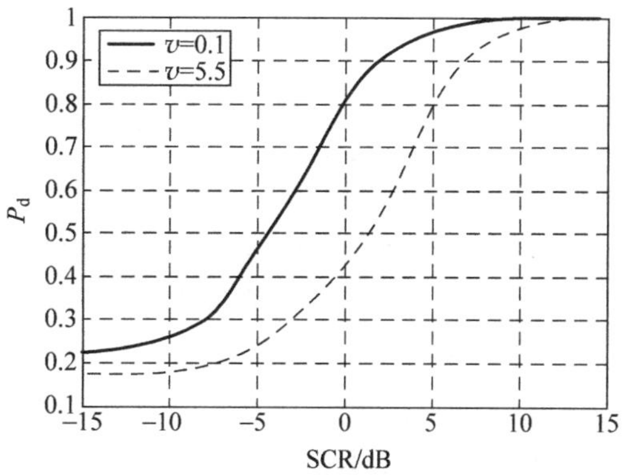

图 12.5 A-GMSD 的检测性能曲线

（$H=4, K=8, N=16, m=1$）

（4）A-GMSD 检测器对杂波的 CFAR 性能。

前述理论分析表明，在杂波散斑分量协方差矩阵 $\boldsymbol{M}_x$ 已知的条件下，非自适应的 GMSD 检测器对复合高斯杂波的纹理分量具有 CFAR 能力，即 GMSD 检测器的虚警概率不随杂波纹理分量的变化而改变，但 A-GMSD 检测器并不具备这种 CFAR 能力。

图 12.6 的仿真结果表明，随着杂波形状系数 v 的增加，A-GMSD 检测器的虚警概率逐渐接近理论设计值，这表明 A-GMSD 检测器在接近经典高斯杂波环境时才是渐进 CFAR 的。事实上，当杂波形状

系数 $v\to\infty$ 时，K 分布趋近于经典高斯分布，而此时的 A-GMSD 检测器实际的虚警概率等于理论设计值，相应的 A-GMSD 检测器也退化成高斯杂波下二阶高斯子空间信号的 CFAR 检测器[26]。

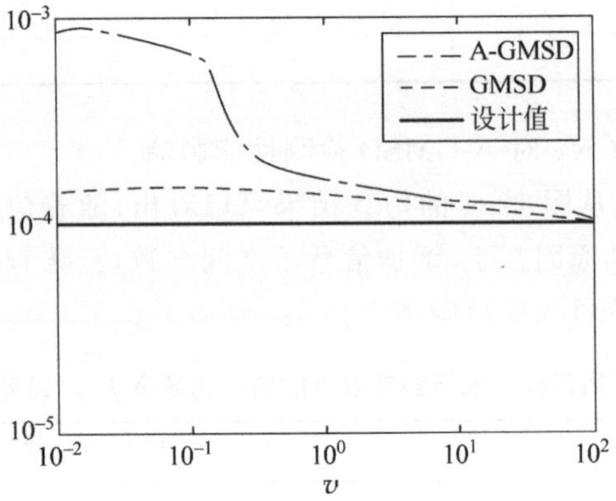

图 12.6 GMSD 和 A-GMSD 的虚警概率随形状系数 v 的变化

($H=4, N=16, K=8$，理论设计值 $P'_{fa}=10^{-4}$)

12.4 复合高斯杂波加热噪声中的距离扩展目标检测器

当匹配滤波器输出的杂噪比很大时，即雷达系统内部噪声影响可以忽略的情况下，可只考虑杂波的主要影响，进而研究距离扩展目标的检测问题。对于大多数的实际雷达系统，上述假设一般也都是可以满足的。但是当杂波抑制处理后或雷达所处环境的杂波强度较弱时，可能导致杂噪比在 $-3\sim10$dB[50]。此时检测器设计中若仍忽略噪声因素，观测数据模型将与实际情况出现严重失配，检测器的性能将退化甚至不能正常工作。本节主要针对热噪声不能忽略的情况下，给出复合高斯杂波背景下距离扩展目标的检测方法。

12.4.1 问题描述

考虑式(12.13)所示的距离扩展目标检测模型，即内部热噪声不可忽略的情形，此时干扰背景中，除 SIRP 杂波 $\boldsymbol{c}_t$ 外，还包含雷达系统内部噪声分量 $\boldsymbol{n}_t$。雷达系统的内部噪声 $\boldsymbol{n}_t$ 为零均值的白色复高斯分布，即有 $\boldsymbol{n}_t\sim \mathrm{CN}(\boldsymbol{0},\sigma^2\boldsymbol{I})$，$\sigma^2$ 是噪声功率水平，$\boldsymbol{I}$ 是 N 阶单位方阵，其中 $N=N_aN_p$，N_a 是天线阵元数或处理的通道数，N_p 是一个 CPI 内的相参脉冲数。$N\times1$ 维的 $\boldsymbol{s}_t$ 是距离扩展目标的复高斯信号矢量，仍采用多秩子空间信号模型，为简化问题，假设不同距离单元的模式矩阵相同，$\boldsymbol{s}_t$ 可建模为

$$\boldsymbol{s}_t=\boldsymbol{U}_s\boldsymbol{b}_t \tag{12.53}$$

其中，模式矩阵 $\boldsymbol{U}_s$ 是 $N\times r$ 维的酉矩阵，一般有 $1\leqslant r\leqslant N$。

12.4.2 热噪声的等效处理

假设外部杂波 $\boldsymbol{c}_t$ 和内部噪声 $\boldsymbol{n}_t$ 是统计独立的，当给定纹理分量 τ_t 时，包含杂波和噪声的总干扰基带复矢量 $\boldsymbol{d}_t$ 的条件协方差矩阵为

$$\boldsymbol{M}(\tau_t)=\sigma_d^2\boldsymbol{M}_d(\tau_t)=\mathrm{E}(\boldsymbol{d}_t\boldsymbol{d}_t^{\mathrm{H}}\mid\tau_t)=\tau_t\boldsymbol{M}_x+\sigma^2\boldsymbol{I} \tag{12.54}$$

其中，$\boldsymbol{M}_d(\tau_t),t=1,2,\cdots,H$ 是 $\boldsymbol{d}_t$ 的归一化条件协方差矩阵，σ_d^2 是总干扰的功率水平，$\boldsymbol{M}_x$ 是散斑分量

的归一化协方差矩阵。

通常相参雷达杂波的幅度按照式(12.55)来计算

$$|z_c|=\sqrt{I_c^2+Q_c^2} \tag{12.55}$$

其中,I_c 和 Q_c 分别是雷达杂波的同相和正交分量。当考虑内部噪声影响时,应将内部噪声的同相和正交分量分别添加到IQ通道中,得到总干扰的幅度为

$$|z_d|=\sqrt{(I_c+I_n)^2+(Q_c+Q_n)^2} \tag{12.56}$$

其中,I_n 和 Q_n 分别是内部噪声的同相和正交分量,均服从高斯分布。由于杂波的同相正交分量 I_c 和 Q_c 都不是高斯分布的,导致计算总干扰幅度的PDF非常复杂,难以获得解析表达式。为此,借鉴Watts的近似计算方法[50]:当考虑内部噪声 $\boldsymbol{n}_t$ 对外部杂波的影响时,可认为杂波散斑分量的平均功率在原来的 τ_t 基础上,增加了内部噪声 $\boldsymbol{n}_t$ 的功率水平 σ^2,变为 σ_{dt}^2,即

$$\sigma_{dt}^2=\tau_t+\sigma^2, \quad t=1,2,\cdots,H \tag{12.57}$$

由式(12.54)和式(12.57)可得总干扰复矢量 $\boldsymbol{d}_t$ 的归一化条件协方差矩阵为

$$\boldsymbol{M}_d(\tau_t)=\frac{\tau_t\boldsymbol{M}_x+\sigma^2\boldsymbol{I}}{\sigma_{dt}^2}=\frac{\mathrm{CNR}_t}{\mathrm{CNR}_t+1}\boldsymbol{M}_x+\frac{1}{\mathrm{CNR}_t+1}\boldsymbol{I} \tag{12.58}$$

其中,杂噪比定义为

$$\mathrm{CNR}_t=\tau_t/\sigma^2 \tag{12.59}$$

式(12.57)和式(12.58)表明,在SIRP杂波纹理分量给定的情况下,总干扰可以等效为条件复高斯矢量,即有 $\boldsymbol{d}_t\sim\mathrm{CN}[\boldsymbol{0},\sigma_{dt}^2\boldsymbol{M}_d(\tau_t)|\tau_t]$。

12.4.3 复合高斯杂波加热噪声中距离扩展目标检测器的设计

1. 多秩子空间信号情形

对内部热噪声影响做上述近似等效处理后,在外部SIRP杂波纹理分量给定的情况下,总干扰可以等效为条件复高斯矢量[51],其归一化条件协方差矩阵由式(12.58)给出。对比本节假设检验问题和式(12.15)的假设检验问题可知,二者的观测数据模型是类似的,式(12.58)中总干扰的归一化条件协方差矩阵 $\boldsymbol{M}_d(\tau_t)$,相当于式(12.16)的杂波散斑分量归一化协方差矩阵 $\boldsymbol{M}_x$。因此,可借鉴式(12.23)的多秩距离扩展目标子空间检测器,即将式(12.23)中的 $\boldsymbol{M}_x$ 用总干扰的归一化条件协方差矩阵 $\boldsymbol{M}_d(\tau_t)$ 代替,可得复合高斯杂波加热噪声下距离扩展目标的匹配子空间检测器(N-MSD)

$$\Lambda(\boldsymbol{z}_1,\boldsymbol{z}_2,\cdots,\boldsymbol{z}_H)=\sum_{t=1}^{H}\frac{\boldsymbol{z}_t^{\mathrm{H}}\boldsymbol{Q}(\tau_t)\boldsymbol{z}_t}{\boldsymbol{z}_t^{\mathrm{H}}[\boldsymbol{M}_d(\tau_t)]^{-1}\boldsymbol{z}_t}\underset{\mathrm{H}_0}{\overset{\mathrm{H}_1}{\gtrless}}G \tag{12.60}$$

式中,G 为等效阈值。

$$\boldsymbol{Q}(\tau_t)=[\boldsymbol{M}_d(\tau_t)]^{-1}\boldsymbol{U}_s\{\boldsymbol{U}_s^{\mathrm{H}}[\boldsymbol{M}_d(\tau_t)]^{-1}\boldsymbol{U}_s\}^{-1}\boldsymbol{U}_s^{\mathrm{H}}[\boldsymbol{M}_d(\tau_t)]^{-1} \tag{12.61}$$

要计算式(12.60)的检验统计量,需先获得式(12.58)的协方差矩阵 $\boldsymbol{M}_d(\tau_t)$,即需对杂波纹理分量 τ_t 作出估计。

可以将观测数据 $\boldsymbol{z}_t$ 向与信号子空间 $\boldsymbol{U}_s$(目标信号矢量 $\boldsymbol{s}_t$ 所隶属的线性子空间)正交的子空间上投影,得到不包含有用信号的总干扰数据为

$$\boldsymbol{z}_d(t)=(\boldsymbol{I}-\boldsymbol{P}_s)\boldsymbol{z}_t, \quad t=1,2,\cdots,H \tag{12.62}$$

其中投影矩阵 $\boldsymbol{P}_{\mathrm{s}}$ 可表示为

$$\boldsymbol{P}_{\mathrm{s}}=\boldsymbol{U}_{\mathrm{s}}(\boldsymbol{U}_{\mathrm{s}}^{\mathrm{H}}\boldsymbol{U}_{\mathrm{s}})^{-1}\boldsymbol{U}_{\mathrm{s}}^{\mathrm{H}}=\boldsymbol{U}_{\mathrm{s}}\boldsymbol{U}_{\mathrm{s}}^{\mathrm{H}} \tag{12.63}$$

注意到观测数据中目标、杂波和噪声3个分量相互统计独立，对 $\boldsymbol{z}_{\mathrm{d}}(t)$ 的能量求数学期望，可得

$$\begin{aligned}\mathrm{E}\{[\boldsymbol{z}_{\mathrm{d}}(t)]^{\mathrm{H}}[\boldsymbol{z}_{\mathrm{d}}(t)]\mid\tau_t\}&=\mathrm{E}\{[(\boldsymbol{I}-\boldsymbol{P}_{\mathrm{s}})\boldsymbol{z}_t]^{\mathrm{H}}[(\boldsymbol{I}-\boldsymbol{P}_{\mathrm{s}})\boldsymbol{z}_t]\mid\tau_t\}\\&=N\tau_t-\tau_t\,\mathrm{trace}(\boldsymbol{P}_{\mathrm{s}}\boldsymbol{M}_{\mathrm{x}})+\sigma^2N-\sigma^2r\end{aligned} \tag{12.64}$$

其中，trace(·)是矩阵的迹，$r=\mathrm{rank}(\boldsymbol{P}_{\mathrm{s}})$，式(12.64)的计算利用了投影矩阵$(\boldsymbol{I}-\boldsymbol{P}_{\mathrm{s}})$的幂等性质及矩阵迹等式。用样本值代替式(12.64)中的数学期望，得

$$\boldsymbol{z}_t^{\mathrm{H}}(\boldsymbol{I}-\boldsymbol{P}_{\mathrm{s}})\boldsymbol{z}_t=N\tau_t-\tau_t\,\mathrm{trace}(\boldsymbol{P}_{\mathrm{s}}\boldsymbol{M}_{\mathrm{x}})+\sigma^2N-\sigma^2r \tag{12.65}$$

反解式(12.65)，可得杂波纹理分量的功率水平 τ_t 的估计为

$$\hat{\tau}_t=\frac{\boldsymbol{z}_t^{\mathrm{H}}(\boldsymbol{I}-\boldsymbol{P}_{\mathrm{s}})\boldsymbol{z}_t-\sigma^2(N-r)}{N-\mathrm{trace}(\boldsymbol{P}_{\mathrm{s}}\boldsymbol{M}_{\mathrm{x}})},\quad t=1,2,\cdots,H \tag{12.66}$$

2. 秩1信号情形

对于秩1信号，有 $\boldsymbol{s}_t=\alpha\boldsymbol{p}(\theta,f_{\mathrm{d}})$，其中 $t=1,2,\cdots,H$，$N\times1$ 维的矢量 $\boldsymbol{p}(\theta,f_{\mathrm{d}})$ 是信号的空时导向矢量。由于此时的信号子空间是一维的，即有 $\boldsymbol{U}_{\mathrm{s}}=\boldsymbol{p}(\theta,f_{\mathrm{d}})$（为书写方便，简记为 $\boldsymbol{p}$），得到 $\boldsymbol{U}_{\mathrm{q}}=\boldsymbol{L}^{-1}\boldsymbol{p}$，其中下三角矩阵 $\boldsymbol{L}$ 是 $\boldsymbol{M}_{\mathrm{d}}(\tau_t)$ 的 Cholesky 分解因子，即 $\boldsymbol{M}_{\mathrm{d}}(\tau_t)=\boldsymbol{L}\boldsymbol{L}^{\mathrm{H}}$。经过白化处理后的投影矩阵为

$$\boldsymbol{P}_{\mathrm{q}}=\frac{\boldsymbol{L}^{-1}\boldsymbol{p}\boldsymbol{p}^{\mathrm{H}}(\boldsymbol{L}^{-1})^{\mathrm{H}}}{\boldsymbol{p}^{\mathrm{H}}[\boldsymbol{M}_{\mathrm{d}}(\tau_t)]^{-1}\boldsymbol{p}} \tag{12.67}$$

进一步可以得到

$$\boldsymbol{Q}(\tau_t)=\frac{[\boldsymbol{M}_{\mathrm{d}}(\tau_t)]^{-1}\boldsymbol{p}\boldsymbol{p}^{\mathrm{H}}[\boldsymbol{M}_{\mathrm{d}}(\tau_t)]^{-1}}{\boldsymbol{p}^{\mathrm{H}}[\boldsymbol{M}_{\mathrm{d}}(\tau_t)]^{-1}\boldsymbol{p}} \tag{12.68}$$

对秩1信号，式(12.60)的N-MSD检测器可简化为

$$\sum_{t=1}^{H}\frac{|\boldsymbol{p}^{\mathrm{H}}[\boldsymbol{M}_{\mathrm{d}}(\tau_t)]^{-1}\boldsymbol{z}_t|^2}{\{\boldsymbol{p}^{\mathrm{H}}[\boldsymbol{M}_{\mathrm{d}}(\tau_t)]^{-1}\boldsymbol{p}\}\{\boldsymbol{z}_t^{\mathrm{H}}[\boldsymbol{M}_{\mathrm{d}}(\tau_t)]^{-1}\boldsymbol{z}_t\}}\underset{\mathrm{H}_0}{\overset{\mathrm{H}_1}{\gtrless}}G \tag{12.69}$$

式中，G 为等效阈值。由式(12.69)可知，秩1信号的N-MSD恰好就是归一化匹配滤波器(Normalized Matched Filter，NMF)[48]在距离扩展目标情形下的推广形式。

若考虑阵列天线具有ULA结构的情况，在阵元间解耦的理想条件下，杂波散斑分量的归一化协方差矩阵 $\boldsymbol{M}_x$ 的秩由 Brennan 准则[52]给出

$$\mathrm{rank}(\boldsymbol{M}_x)=N_{\mathrm{a}}+\gamma(N_{\mathrm{p}}-1) \tag{12.70}$$

其中，N_{a} 是阵元数，N_{p} 是一个CPI内的相参处理脉冲数，$\gamma=(2vT)/d$ 是杂波环的斜率，v 是机载平台运动速度，T 是脉冲重复间隔，d 是阵元间距。一般机载相控阵雷达设计时将杂波环的斜率 γ 设计成比1稍微大的数值，此时有

$$\mathrm{rank}(\boldsymbol{M}_x)\approx N_{\mathrm{a}}+N_{\mathrm{p}}-1\ll N_{\mathrm{a}}N_{\mathrm{p}}=N \tag{12.71}$$

尤其是对很大的 N_{a} 和 N_{p}，可以确保 $\mathrm{rank}(\boldsymbol{M}_x)\ll N$ 成立，实际系统中这一条件完全可以满足，即对于具有ULA结构的相控阵雷达，与全空间维数 N 相比较，其杂波是低秩的。此时总干扰矢量 $\boldsymbol{d}_t$ 的协方差矩阵可以分解成[49]

$$\boldsymbol{M}_{\mathrm{d}}(\tau_t)=\sum_{i=1}^{\mathrm{rank}(\boldsymbol{M}_x)}(\tau_t\lambda_i+\sigma^2)\boldsymbol{u}_i\boldsymbol{u}_i^{\mathrm{H}}+\sum_{i=\mathrm{rank}(\boldsymbol{M}_x)+1}^{N}\sigma^2\boldsymbol{u}_i\boldsymbol{u}_i^{\mathrm{H}} \tag{12.72}$$

其中，λ_i 和 $\boldsymbol{u}_i(i=1,2,\cdots,N)$ 分别是 $\boldsymbol{M}_{\mathrm{d}}(\tau_t)$ 的特征值和相应的归一化特征向量。当条件 $(\tau_t\lambda_i)\gg\sigma^2$ 成立时，总干扰的协方差矩阵的逆矩阵可以采用如下的低秩形式近似

$$\boldsymbol{M}_{\mathrm{d}}(\tau_t)^{-1}\approx[\boldsymbol{I}-\boldsymbol{R}_{\mathrm{c}}(t)]/\sigma^2 \tag{12.73}$$

其中

$$\boldsymbol{R}_{\mathrm{c}}(t)=\sum_{i=1}^{\mathrm{rank}(\boldsymbol{M}_x)}\boldsymbol{u}_i\boldsymbol{u}_i^{\mathrm{H}} \tag{12.74}$$

将式(12.73)代入式(12.69)中，则 ULA 结构下秩 1 距离扩展目标的检测器为

$$\sum_{t=1}^{H}\frac{|\boldsymbol{p}^{\mathrm{H}}[\boldsymbol{I}-\boldsymbol{R}_{\mathrm{c}}(t)]\boldsymbol{z}_t|^2}{\{\boldsymbol{p}^{\mathrm{H}}[\boldsymbol{I}-\boldsymbol{R}_{\mathrm{c}}(t)]\boldsymbol{p}\}\{\boldsymbol{z}_t^{\mathrm{H}}[\boldsymbol{I}-\boldsymbol{R}_{\mathrm{c}}(t)]\boldsymbol{z}_t\}}\underset{\mathrm{H}_0}{\overset{\mathrm{H}_1}{\gtrless}}G \tag{12.75}$$

与式(12.69)相比，式(12.75)只需要计算前 rank($\boldsymbol{M}_x$)个大特征值和相应的归一化特征向量，达到了降秩的效果，便于算法的快速实现。

12.4.4 检测器的性能分析

由于 N-MSD 检测器的检测概率难以获得解析表达式，可采用 Monte Carlo 仿真进行性能分析，设定目标占据的距离单元数 $H=8$，虚警概率为 $P_{\mathrm{fa}}=10^{-3}$。

仿真中，设定距离扩展目标信号为服从复高斯分布的多秩子空间随机信号，即有 $\boldsymbol{s}_t\sim\mathrm{CN}(\boldsymbol{0},\sigma_{\mathrm{s}}^2\boldsymbol{M}_{\mathrm{s}})$ $(t=1,2,\cdots,H)$，其归一化协方差矩阵元素为

$$[\boldsymbol{M}_{\mathrm{s}}]_{k,l}=\frac{\sin^2[\pi B(k-l)]}{[\pi B(k-l)]^2}\exp[-2\pi\mathrm{j}f_{\mathrm{d}}(k-l)],\quad k,l\in\{1,2,\cdots,N\} \tag{12.76}$$

其中，B 是信号频谱的归一化带宽，f_{d} 是信号频谱的归一化多普勒中心频率。由于归一化协方差矩阵 $\boldsymbol{M}_{\mathrm{s}}$ 的特征值以概率 1 不等于 0，将 $\boldsymbol{M}_{\mathrm{s}}$ 的特征值降序排列可得 $\lambda_1\geqslant\lambda_2\geqslant\cdots\geqslant\lambda_N$，前面若干个大特征值之和首次超过 $0.99N$ 的最小特征值的序号，即为距离扩展目标信号的秩[8]。为不失一般性，设置高斯白噪声的功率水平 $\sigma^2=1$，相关 K 分布杂波散斑分量的归一化协方差矩阵的元素为

$$[\boldsymbol{M}_x]_{j,k}=\rho^{|j-k|},\quad j,k\in\{1,2,\cdots,N\},0<\rho\leqslant1 \tag{12.77}$$

其中，ρ 是一阶相关系数。

K 分布杂波的功率水平由信杂比 $\mathrm{CNR}=u/\sigma^2$ 确定，其中，$u=\mathrm{E}[\tau_t]=v/b^2$，$v$ 是 K 分布杂波形状参数，b 是尺度参数，定义信干比(Signal to Disturbance Ratio，SDR)为 $\mathrm{SDR}=\sigma_{\mathrm{s}}^2/(u+\sigma^2)$，信噪比为 $\mathrm{SNR}=\sigma_{\mathrm{s}}^2/\sigma^2$，易知杂噪比、信干比、信噪比三者的关系为 SDR=SNR/(CNR+1)。

以信噪比为参数(单位 dB)，信干比与杂噪比的关系曲线见图 12.7；以杂噪比为参数(单位 dB)，信干比与信噪比的关系曲线见图 12.8。从图 12.7 和图 12.8 中可以看出杂噪比和信噪比对信干比的影响：在给定杂噪比的情况下，信干比随信噪比的增加而增大，而在给定信噪比的情况下，信干比随杂噪比的增加而减小。

1. 热噪声对 N-MSD 检测性能的影响

图 12.9 是 N-MSD 检测器在不同杂噪比条件下检测概率随信干比的变化情况，并与式(12.23)中忽略噪声影响的 GMSD 检测器进行了对比。对比曲线 C 和 D 可知，N-MSD 在低杂噪比(CNR=0dB)时明显优于 GMSD 检测器，在高杂噪比(CNR=10dB)时两者的检测概率则十分接近(见曲线 A 和 B)，但是前者仍稍好于后者。可能的原因在于，当杂噪比很大时，与杂波功率相比，此时白噪声对信号功率

谱密度(Power Spectrum Density,PSD)的干扰作用变弱,起主要作用的是杂波的干扰。

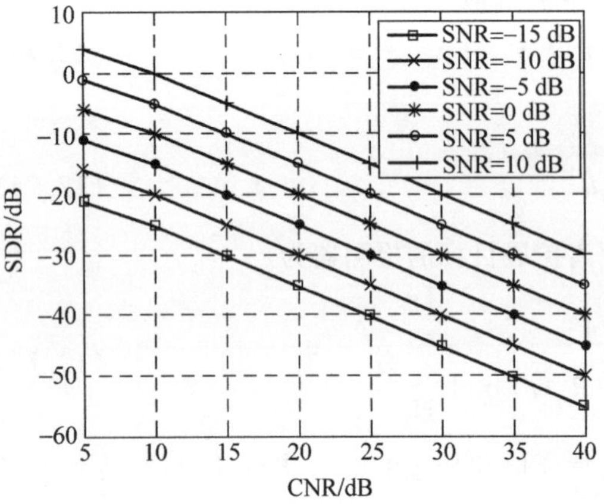

图 12.7 SDR 与 CNR 的曲线

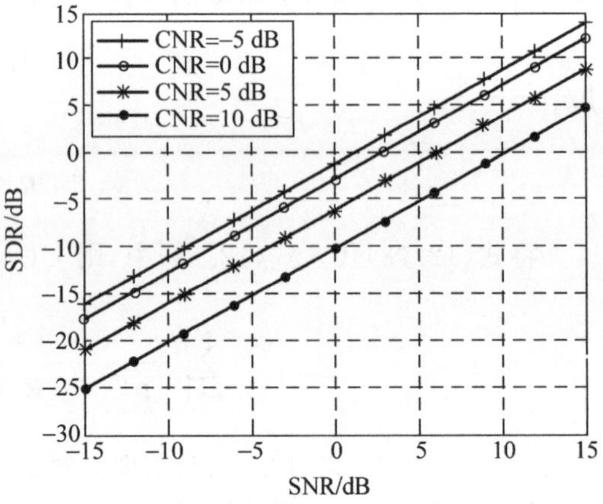

图 12.8 SDR 与 SNR 的曲线

2. 归一化多普勒中心频率 f_d 对 N-MSD 检测概率的影响

由图 12.10 可知,随着目标信号的归一化多普勒中心频率 f_d 增大,N-MSD 的检测概率也增大。由于 N-MSD 检测器利用信号 PSD 频段之外的干扰数据来对杂波纹理分量功率水平进行估计,因此当目标信号的 PSD 的中心频率 f_d 变小时,杂波和目标信号的 PSD 重叠部分变大,导致可用于估计纹理分量的干扰数据样本数量减小,估计误差增大,从而使检测概率下降。事实上,杂波和目标的频谱越分开,二者也更易区分。

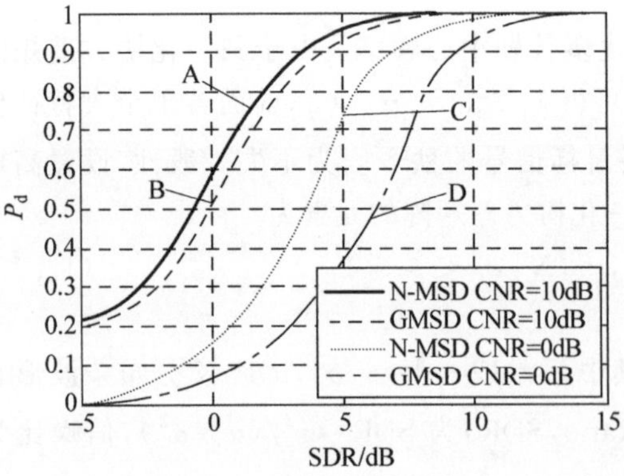

图 12.9 N-MSD 和 GMSD 在不同信干比下的检测概率

($P_{fa}=10^{-3}$,$\rho=0.8$,$\sigma^2=1$,

$H=8$,$N=16$,$v=0.5$,$B=f_d=0.2$)

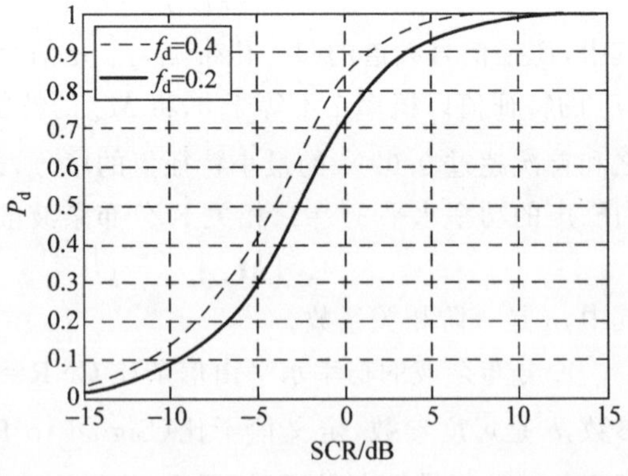

图 12.10 N-MSD 在不同中心频率 f_d 下的检测概率

($P_{fa}=10^{-3}$,$H=8$,$N=16$,$v=0.5$,

$\rho=0.8$,$\sigma^2=1$,$B=0.1$,CNR=0dB)

3. 目标信号频谱的归一化带宽 B 对 N-MSD 检测概率的影响

由图 12.11 可知,随着信号频谱归一化带宽 B 的增大,N-MSD 的检测概率降低。根据信号秩的经验公式,本例中信号秩近似为 $r=\min[\text{int}(2NB+1),N]$,$0<B\leqslant 0.5$。由此可见,随着归一化带宽 B 的增加,目标信号的秩也随之增加,即增加了信号子空间的维数,这意味着信号子空间与满秩的干扰子空间越来越靠近,加剧了目标信号和干扰的混叠程度,导致从干扰中区别目标变得越来越困难[49]。事实上,目标信号归一化带宽 B 的增大,将增加目标信号频谱与杂噪干扰频谱的重叠程度,进而加大了二者

的区分难度。

4. 白噪声条件下 N-MSD 检测器的 CFAR 能力

在只有 K 分布杂波而没有白噪声的环境中，设定虚警概率理论设计值为 $P'_{fa}=10^{-3}$。图 12.12 是 N-MSD 检测器在不同杂波形状参数($v=0.1,0.5,0.8,1.2$)下的实际虚警概率随杂噪比的变化情况。由图 12.12 可知，随着 v 的减小，杂波变得尖锐，实际虚警概率偏离理论设计值程度减弱，这说明当杂波形状参数 v 不是太大时，N-MSD 的实际虚警概率受杂噪比的影响较小，具有较好的虚警控制能力。另外，在大杂噪比(CNR>10dB)情况下，N-MSD 的实际虚警概率不随杂波形状参数的变化而改变，且与理论设计值基本保持一致，即 N-MSD 在大杂噪比(CNR>10dB)下具有近似 CFAR 能力。事实上，当杂噪比较大时，白噪声的影响很小，可以忽略不计，此时 N-MSD 检测器等价于 GMSD 检测器，因此保持了近似 CFAR 的能力。

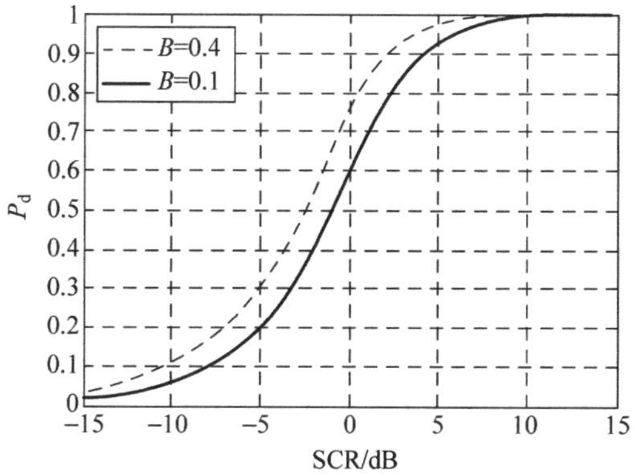

图 12.11 N-MSD 在不同归一化带宽 B 下的检测概率
($P_{fa}=10^{-3}, H=8, N=16, v=0.5, \rho=0.8, \sigma^2=1, f_d=0.2$, CNR=0dB)

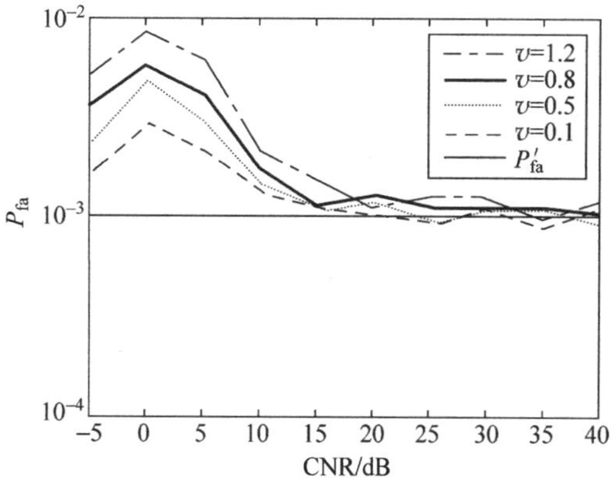

图 12.12 N-MSD 检测器在 $v=0.1,0.5,0.8,1.2$ 时的实际虚警概率与杂噪比的关系
(设计值 $P'_{fa}=10^{-3}, H=8, N=16, v=0.5, \rho=0.8, \sigma^2=1, B=0.1, f_d=0.4$)

12.5 SαS 分布杂波中的距离扩展目标检测器

实测杂波数据统计分析表明[53]，某些在特定的观测条件下，采用复合高斯统计模型描述实测高分辨率雷达杂波的效果欠佳，而对称 α 稳定分布统计模型能更好拟合实测杂波数据。本节针对 α 稳定分布杂波背景中距离扩展目标的检测问题，分析参量检测和非参量检测相结合的检测器。

12.5.1 SαS 分布及 PFLOM 变换

α 稳定分布主要用于描述显著冲击特性噪声，其模型极具吸引力的原因在于两个重要特点。

(1) α 稳定分布满足广义中心极限定理，即无穷多个方差不存在的 IID 的随机变量之和的分布就是稳定分布。

(2) 稳定性，IID 的 α 稳定分布变量的线性组合仍然是 α 稳定分布[54]。

由于 SαS 分布的方差不存在，为了能够有效抑制 SαS 分布样本的冲击特性，可采用保持相位的分数低阶矩(Phased Fractional Lower-Order Moment，PFLOM)变换[54]。对于服从参数为(α,γ)的 SαS

分布复数样本 z，其 a 阶 PFLOM 变换定义为

$$z^{\langle a\rangle}=\begin{cases}\dfrac{|z|^{a+1}}{z^*}, & z\neq 0\\ 0, & z=0\end{cases},\quad 0<a<\frac{\alpha}{2},\quad \alpha\in(0,2] \tag{12.78}$$

其中，$(\cdot)^*$ 是复共轭。如果将复数样本 z 用极坐标表示为 $z=|z|\mathrm{e}^{\mathrm{j}\phi(z)}$，其中 $\phi(z)$ 表示 z 的相角，则由式(12.78)可得 z 的 a 阶 PFLOM 变换为

$$z^{\langle a\rangle}=\begin{cases}|z|^{a}\mathrm{e}^{\mathrm{j}\phi(z)}, & z\neq 0\\ 0, & z=0\end{cases},\quad 0<a<\frac{\alpha}{2},\quad \alpha\in(0,2] \tag{12.79}$$

由式(12.79)可以看出，由于阶数 a 的作用，PFLOM 变换保持原来 SαS 分布样本的相位不变，只是将其幅度缩减了，可有效抑制 SαS 分布样本的冲击特性。

12.5.2 问题描述

若忽略内部热噪声影响，则距离扩展目标检测可归结为如下的二元假设检验

$$\begin{cases}\mathrm{H}_0: \boldsymbol{z}_t=\boldsymbol{c}_t, & t=1,2,\cdots,H\\ \mathrm{H}_1: \boldsymbol{z}_t=\boldsymbol{s}_t+\boldsymbol{c}_t, & t=1,2,\cdots,H\end{cases} \tag{12.80}$$

其中，$\boldsymbol{z}_t$ 是 $N\times 1$ 维的列矢量，t 是距离单元编号，H 是待检测的距离扩展目标所占据的距离单元总数，$\boldsymbol{c}_t$ 为杂波矢量，$\boldsymbol{s}_t$ 为目标信号矢量，为了简化问题，$\boldsymbol{s}_t$ 采用 12.2 节式(12.1)给出的秩 1 信号模型。

在 SαS 分布杂波背景中进行目标检测有两大困难，其一，SαS 分布的 PDF 没有闭型解析表达式，在基于 GLRT 理论来推导检验统计量时存在很大困难，因此只能考虑其他办法；其二，SαS 分布没有有限的二阶矩，服从 SαS 分布的样本具有强烈的脉冲特性，使得基于二阶矩的传统统计信号处理方法(如相关分析、功率谱、方差分析等)难以奏效。针对上述两个困难，接下来讨论如下两种检测器：基于 PFLOM 变换的距离扩展目标检测器和二元积累柯西检测器。

12.5.3 基于 PFLOM 变换的距离扩展目标检测器

如前所述，SαS 分布具有强烈的冲击特性，具有无穷方差，非二阶矩过程，使得基于二阶矩的处理方法难以适用。针对这一特性，可以对观测数据先进行 PFLOM 变换，在保持观测数据相位不变的前提下，尽量压缩观测数据的幅度，使经过这种 PFLOM 非线性变换的数据具有有限的二阶矩，然后再进行检测器设计。根据这一思路，基于 PFLOM 变换的距离扩展目标检测原理流程如图 12.13 所示。下面将详细分析该检测方案。

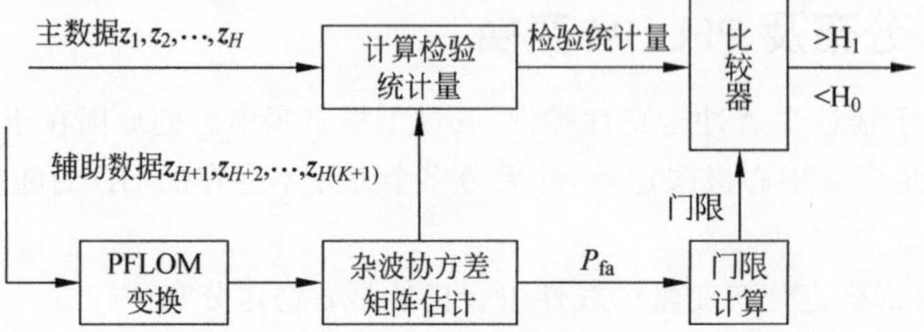

图 12.13 基于 PFLOM 变换的距离扩展目标检测原理流程图

假设可获得与待检测单元邻近的 HK 个辅助数据 $\boldsymbol{z}_t(t=H+1,H+2,\cdots,H(K+1))$，该辅助数据不包含有用的目标信号，仅含与待检测单元中杂波数据 IID 的 SαS 分布杂波数据。首先对该辅助数据的元素 z_{nt} 作式(12.78)的 PFLOM 变换，即

$$z_{nt}^{\langle a\rangle}=\begin{cases}\dfrac{|z_{nt}|^{a+1}}{z_{nt}^{*}}, & z_{nt}\neq 0\\ 0, & z_{nt}=0\end{cases},\quad n=1,2,\cdots,N;\quad t=H+1,H+2,\cdots,H(K+1)$$

(12.81)

经过上述的 PFLOM 变换，服从 SαS 分布的杂波数据幅度得到有效压缩，则变换后协方差矩阵 $\boldsymbol{R}_{\mathrm{PFLOM}}$ 的第 i 行第 j 列的元素可表示为

$$\boldsymbol{R}_{\mathrm{PFLOM}}^{ij}=\mathrm{E}[z_i^{\langle a\rangle}(z_j^{\langle a\rangle})^{*}],\quad i,j\in\{1,2,\cdots,N\} \tag{12.82}$$

为了符号简洁，式(12.82)中的观测值 z_i 省略了距离单元编号 t。相应的实部 $\mathrm{Re}\{\boldsymbol{R}_{\mathrm{PFLOM}}^{ij}\}$ 和虚部 $\mathrm{Im}\{\boldsymbol{R}_{\mathrm{PFLOM}}^{ij}\}$ 满足

$$\mathrm{Re}\{\boldsymbol{R}_{\mathrm{PFLOM}}^{ij}\}<\infty;\quad \mathrm{Im}\{\boldsymbol{R}_{\mathrm{PFLOM}}^{ij}\}<\infty \tag{12.83}$$

即经 PFLOM 变换后的杂波协方差矩阵是有界的，下面给出该结论的证明。事实上，只需证明实部是有界的，即 $\mathrm{Re}\{\boldsymbol{R}_{\mathrm{PFLOM}}^{ij}\}<\infty$，类似亦可以证明虚部是有界的。按照式(12.82)的定义有

$$\mathrm{Re}\{\boldsymbol{R}_{\mathrm{PFLOM}}^{ij}\}=\mathrm{Re}\{\mathrm{E}[z_i^{\langle a\rangle}(z_j^{\langle a\rangle})^{*}]\}=\mathrm{E}\{\mathrm{Re}[z_i^{\langle a\rangle}(z_j^{\langle a\rangle})^{*}]\} \tag{12.84}$$

注意到对任何复数 z，都有不等式 $\mathrm{Re}(z)\leqslant|z|$ 成立，则由式(12.84)可得

$$\mathrm{Re}\{\boldsymbol{R}_{\mathrm{PFLOM}}^{ij}\}=\mathrm{E}\{\mathrm{Re}[z_i^{\langle a\rangle}(z_j^{\langle a\rangle})^{*}]\}\leqslant\mathrm{E}[|z_i^{\langle a\rangle}(z_j^{\langle a\rangle})^{*}|]\leqslant\mathrm{E}(|z_i|^{a}|z_j|^{a}) \tag{12.85}$$

文献[54]已经证明：对于两个服从特征指数为 α 的 SαS 分布随机变量 X_1 和 X_2，以及两个正实数 p_1 和 p_2，当且仅当 $p_1+p_2<\alpha$ 时，有

$$\mathrm{E}(|X_1|^{p_1}|X_2|^{p_2})<\infty \tag{12.86}$$

令式(12.85)中 $X_1=z_i$，$X_2=z_j$ 及 $p_1=p_2=a$，并注意到限制条件 $0<a<\alpha/2,\alpha\in(0,2]$，可知 $p_1+p_2=2a<\alpha$，即式(12.85)满足式(12.86)成立的条件。因此有 $\mathrm{Re}[\boldsymbol{R}_{\mathrm{PFLOM}}^{ij}]<\infty$ 成立。采用类似的方法也可以证明虚部是有界的。因此，经特定 PFLOM 变换后的 SαS 分布杂波的协方差矩阵是有界的。

针对秩 1 的距离扩展目标，相应的广义自适应子空间检测器(Generalized Adaptive Subspace Detector，GASD)的检验统计量为[9]

$$\lambda_{\mathrm{GASD}}=\frac{\sum_{t=1}^{H}|\boldsymbol{p}(\theta,f_{\mathrm{d}})^{\mathrm{H}}\boldsymbol{S}^{-1}\boldsymbol{z}_t|^2}{[\boldsymbol{p}(\theta,f_{\mathrm{d}})^{\mathrm{H}}\boldsymbol{S}^{-1}\boldsymbol{p}(\theta,f_{\mathrm{d}})]\sum_{h=1}^{H}(\boldsymbol{z}_h^{\mathrm{H}}\boldsymbol{S}^{-1}\boldsymbol{z}_h)} \tag{12.87}$$

其中，$\boldsymbol{p}(\theta,f_{\mathrm{d}})$ 为空时导向矢量，$\boldsymbol{S}$ 为基于辅助数据的样本协方差矩阵(Sample Covariance Matrix，SCM)，即

$$\boldsymbol{S}=\sum_{t=H+1}^{H(K+1)}\boldsymbol{z}_t\boldsymbol{z}_t^{\mathrm{H}} \tag{12.88}$$

GASD 检测器要求 SCM 是有界的，若直接利用含 SαS 分布杂波的原始观测数据，因基于辅助数据的 SCM 是无界的，此时 GASD 检测器无法实现。可利用经 PFLOM 变换后的辅助数据来估计主数据中的杂波协方差矩阵，即

$$\boldsymbol{S}_{\text{PFLOM}}=\sum_{t=H+1}^{H(K+1)}\boldsymbol{z}_t^{\langle a\rangle}(\boldsymbol{z}_t^{\langle a\rangle})^{\text{H}} \tag{12.89}$$

由式(12.83)的结论可知,$\boldsymbol{S}_{\text{PFLOM}}$ 中的每个元素都是有界的。用 $\boldsymbol{S}_{\text{PFLOM}}$ 代替GASD检测器中的样本协方差矩阵 $\boldsymbol{S}$,则可得SαS分布杂波下基于PFLOM变换的距离扩展目标检测器(简记为GASD-PFLOM),其检验统计量为

$$\lambda_{\text{GASD-P}}=\frac{\sum_{t=1}^{H}\left|\boldsymbol{p}(\theta,f_{\text{d}})^{\text{H}}\boldsymbol{S}_{\text{PFLOM}}^{-1}\boldsymbol{z}_t\right|^2}{[\boldsymbol{p}(\theta,f_{\text{d}})^{\text{H}}\boldsymbol{S}_{\text{PFLOM}}^{-1}\boldsymbol{p}(\theta,f_{\text{d}})]\sum_{h=1}^{H}(\boldsymbol{z}_h^{\text{H}}\boldsymbol{S}_{\text{PFLOM}}^{-1}\boldsymbol{z}_h)} \tag{12.90}$$

为了检验GASD-PFLOM检测器在SαS分布杂波中的检测性能,仿真中采用ULA相控阵高分辨率雷达,相应参数设置见表12.4。

表12.4 仿真的高分辨率雷达的主要参数

参　数	数　值	参　数	数　值
阵元数(N_{a})	2	目标信号的到达角(θ)	60°
阵元间隔	雷达波长一半	目标信号的Doppler频率(f_{d})	80Hz
重复频率(f_{r})	800Hz	相参处理脉冲数(N_{p})	4

距离扩展目标信号建模为

$$s_{nt}=A_{nt}\exp[\text{j}\theta_{nt}+2\text{j}\pi(n-1)f_{\text{d}}],\quad n=1,2,\cdots,N;\quad t=1,2,\cdots,H \tag{12.91}$$

其中,幅度 A_{nt} 用Swerling Ⅱ模型建模,相角 θ_{nt} 均匀分布在$[0,2\pi]$区间上。

信杂比定义为

$$\text{SCR}=\sum_{t=1}^{H}\sum_{n=1}^{N}A_{nt}{}^2/N\gamma \tag{12.92}$$

其中,γ 是SαS分布的分散系数。

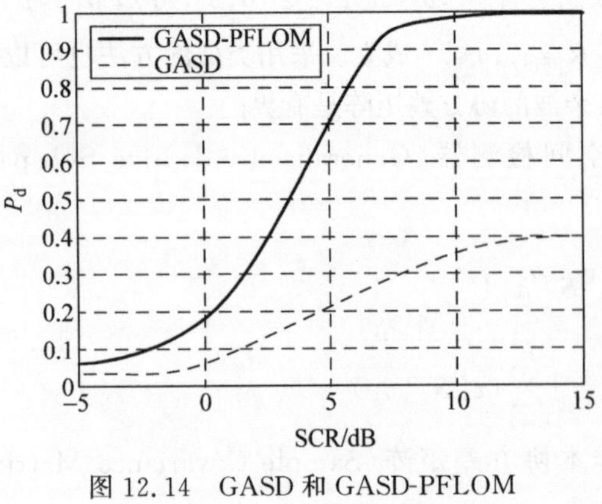

图12.14 GASD和GASD-PFLOM在SαS分布杂波中的检测概率($N=8,H=4,K=16,\alpha=1.5,\gamma=1,a=0.5,P_{\text{fa}}=10^{-4},f_{\text{d}}=0.1$)

在SαS分布杂波背景下,图12.14给出了GASD和GASD-PFLOM的检测性能对比曲线,其中虚警概率为10^{-4}。由图12.14可知,GASD-PFLOM明显优于普通的GASD检测器,GASD检测器未能对SαS分布杂波的强脉冲特性进行有效抑制,导致对主数据杂波协方差矩阵的估计出现较大误差,使得检测性能严重退化。图12.15是普通的GASD检测器和GASD-PFLOM检测器在高斯杂波($\alpha=2.0$)中的对比结果。由图12.15可见,在高斯杂波中,GASD-PFLOM检测器具有与GASD检测器相近的检测性能[49]。

图12.16是GASD-PFLOM的虚警概率随SαS杂波分散系数(相当于高斯杂波的方差)的变化曲线。其中,虚警概率的理论设计值 $P_{\text{fa}}'=10^{-4}$,检测门限对应固定分散系数 $\gamma=2$。由图12.16可知,若忽略仿真中的随机误差,实际的虚警概率与理论设计值基本相同,即GASD-PFLOM检测器在SαS分布杂波中具有很好的虚警控制能力。

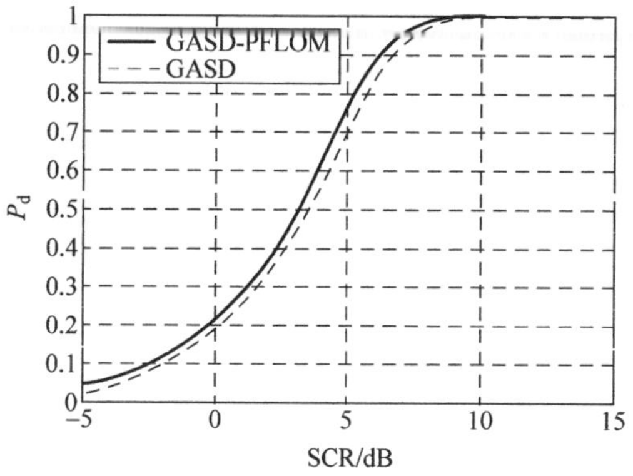

图 12.15 GASD 和 GASD-PFLOM 在高斯杂波中的检测概率（$N=8, H=4, K=16, \alpha=2.0, \gamma=1, a=0.5, P_{fa}=10^{-4}, f_d=0.1$）

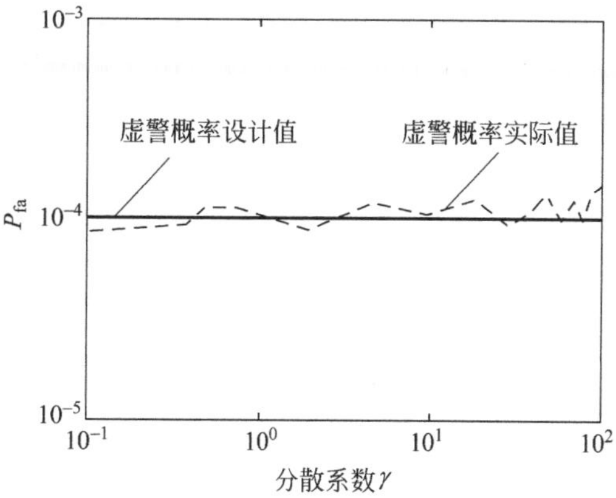

图 12.16 GASD-PFLOM 在 SαS 分布杂波中的虚警概率（$N=8, H=4, K=16, \alpha=1.5$，理论设计值 $P'_{fa}=10^{-4}, f_d=0.1, a=0.5$）

12.5.4 SαS 分布杂波中的二元积累柯西检测器

1. 二元积累检测器的设计

如前所述，在 SαS 分布杂波背景中进行目标检测的困难之一，在于 SαS 分布的 PDF 缺乏闭型解析表达式，因此无法采用 GLRT 理论来推导检验统计量。一种直观的解决思路是非参量检测方法，如秩检测器等，但是这类非参量检测器未能利用 SαS 分布的任何先验信息，导致检测性能较差，在实际应用中效果欠佳。幸运的是，柯西检测器在 SαS 分布杂波中表现出很好的鲁棒性，即与最佳检测器相比，柯西检测器在任意特征指数 α 下只有微小的检测性能损失。受此启发，对于在特征指数 α 未知或者时变的 SαS 分布杂波背景，可以利用柯西检测器的这种鲁棒性来构造距离扩展目标检测器。

另外，双门限检测作为准最佳检测方案，虽然比最佳检测器的检测性能稍差，但由于所需设备简单，便于工程实现，因而在雷达目标检测领域应用广泛。本节将点目标的柯西检测器和双门限检测相结合，在特征指数未知的 SαS 分布杂波背景下，给出了基于二元积累的距离扩展目标柯西检测器。

2. 二元积累柯西检测器

基于二元积累的距离扩展目标柯西检测器（简称为 BI-C）的原理框图见图 12.17。

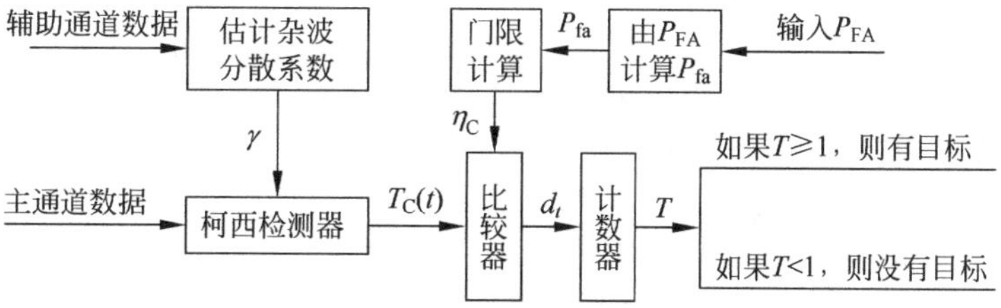

图 12.17 距离扩展目标的二元积累柯西检测器原理框图

首先，主数据中的每个距离单元观测数据逐一通过点目标柯西检测器，形成如下的检验统计量[54]

$$T_{\mathrm{C}}=\frac{1}{N}\sum_{n=1}^{N}\ln\left[\frac{2}{\pi}(\gamma^{2}+|z_{n}|^{2})^{3/2}I(x_{n})\frac{\sqrt{(\xi_{n})^{2}+\lambda_{n}}}{(\xi_{n})^{2}-\lambda_{n}}\right] \tag{12.93}$$

其中,$z_n(n=1,2,\cdots,N)$是给定某一距离单元的第 n 个观测复数据,γ 是柯西分布的分散系数,其他变量的含义如下

$$\begin{cases}\xi_{n}=z_{n}^{*}z_{n}+1+\gamma^{2};\quad \xi_{1n}=-2\mathrm{Re}(z_{n}p_{n}^{*});\quad \xi_{2n}=-2\mathrm{Im}(z_{n}p_{n}^{*})\\ \lambda_{n}=\sqrt{(\xi_{1n})^{2}+(\xi_{2n})^{2}};\quad x_{n}=\sqrt{\dfrac{2\lambda_{n}}{(\xi_{n})^{2}+\lambda_{n}}},\quad n=1,2,\cdots,N\end{cases} \tag{12.94}$$

其中,p_n 是式(12.2)中信号空时导向矢量的第 n 个元素,令 $y_n=\sqrt{1-x_n^2}$,则式(12.93)中的第二类完全椭圆积分为

$$I(x_{n})=\int_{0}^{\pi/2}\sqrt{1-x_{n}^{2}\sin^{2}u}\,\mathrm{d}u,\quad n=1,2,\cdots,N \tag{12.95}$$

可进一步用下面的多项式(误差小于 4×10^{-5})进行高精度近似

$$I(x_{n})\approx(1+0.46301y_{n}^{2}+0.10778y_{n}^{4})+(0.24527y_{n}^{2}+0.04124y_{n}^{4})\ln(1/y_{n}^{2}),\quad n=1,2,\cdots,N \tag{12.96}$$

其次,计算第 t 个距离单元的柯西检测器检验统计量 $T_{\mathrm{C}}(t)$,将 $T_{\mathrm{C}}(t)$与给定虚警概率下的检测门限 η_{C} 进行比较,将比较器的输出结果 d_t 进行二元量化

$$d_{t}=\begin{cases}1, & T_{\mathrm{C}}(t)\geqslant\eta_{\mathrm{C}}\\ 0, & T_{\mathrm{C}}(t)<\eta_{\mathrm{C}}\end{cases},\quad t=1,2,\cdots,H \tag{12.97}$$

将该二元量化的输出结果送计数器,对所有待检测的 H 个距离单元主数据均进行上述操作,计数器的输出结果为

$$T=\sum_{t=1}^{H}d_{t} \tag{12.98}$$

最后,依据如下准则做出判决

$$\text{BI-Cauchy:}\begin{cases}T\geqslant1:\text{有目标}\\ T<1:\text{没有目标}\end{cases} \tag{12.99}$$

为了最大限度地检测出上距离扩展目标,式(12.99)实际上将双门限检测方案中的二进制积累门限设置为1。

当BI-C检测器的总虚警概率 P_{FA} 给定后,易知 P_{FA} 与单个距离单元的点目标柯西检测器的虚警概率 P_{fa} 关系为

$$P_{\mathrm{FA}}=\sum_{k=1}^{H}\frac{H!}{k!\,(H-k)!}(P_{\mathrm{fa}})^{k}(1-P_{\mathrm{fa}})^{(H-k)} \tag{12.100}$$

类似地,BI-C检测器的总检测概率为

$$P_{\mathrm{D}}=\sum_{k=1}^{H}\frac{H!}{k!\,(H-k)!}(P_{\mathrm{d}})^{k}(1-P_{\mathrm{d}})^{(H-k)} \tag{12.101}$$

式中,P_{d} 是单个距离单元的点目标柯西检测器通过第一门限的检测概率。

3. 仿真结果及分析

由于难以获得柯西检测器检验统计量 T_{C} 在 H_0 和 H_1 假设下的PDF,因此相应的虚警和检测概率缺乏闭型解析表达式,为此采用Monte Carlo仿真来估计总的检测概率 P_{D}。仿真的雷达参数设置与

表 12.4 相同，距离扩展目标占据的距离单元数 $H=4$，单个距离单元幅度瑞利起伏($m=1$)，信号能量的分布模型采用简单的均匀分布模型。设置总的虚警概率 $P_{FA}=10^{-4}$，在四种典型的 SαS 分布($\alpha=1,1.5,1.8,2.0$)杂波背景中，二元积累柯西检测器对距离扩展目标的检测概率曲线见图 12.18。由图 12.18 可知，二元积累柯西检测器在不同特征指数下的检测性能变化不大，即其在 SαS 分布杂波下具有一定的检测鲁棒性。

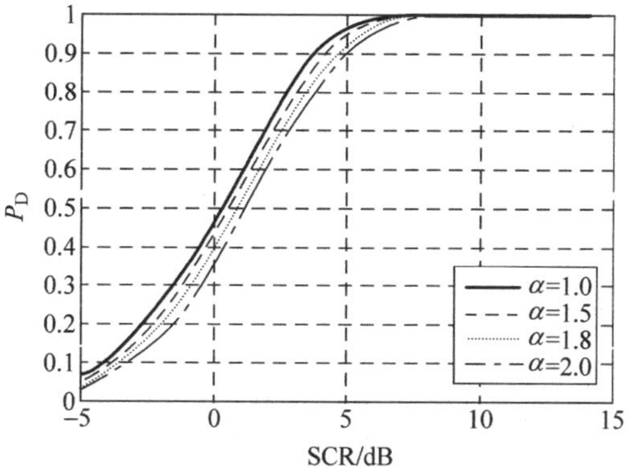

图 12.18 二元积累柯西检测器在 SαS 杂波中总的检测概率

($N=8,H=4,K=16,m=1,\gamma=2.0,P_{FA}=10^{-4}$)

12.6 SAR 图像 CFAR 检测研究的主要方面及杂波单元选取

12.6.1 SAR 图像 CFAR 检测研究的主要方面

在 SAR 图像目标检测中，CFAR 处理一般采用滑窗方式，基于距离-方位二维 SAR 图像中待检测像素或像素区域的灰度值，逐一与自适应判决阈值进行比较，进而输出二进制的目标检测结果，其中自适应判决阈值主要依据虚警率和 SAR 图像杂波模型进行设置。

根据 SAR 图像 CFAR 检测所涉及的多种因素，目前 SAR 图像 CFAR 检测的研究主要集中在以下四个方面。

1. CFAR 处理中杂波样本选取

CFAR 处理需要利用杂波样本来估计杂波的统计概率模型参数。杂波样本选取常用的方式是双窗法[55]，即以待检测单元为中心，由内向外依次是保护单元区域、杂波单元内窗、杂波单元外窗。现有的针对 CFAR 杂波样本的选取方法，旨在通过精炼双窗法确定的杂波样本，获得更加精确的杂波模型参数，有效提升 CFAR 检测方法在密集目标、非均匀杂波背景等复杂环境下的检测性能。

2. CFAR 处理中杂波统计建模

精确的杂波统计模型有利于更好地区分目标与杂波背景。常用的 SAR 图像参数化杂波统计模型包括高斯分布、瑞利分布、K 分布、Weibull 分布、Gamma 分布等，每种分布都有其具体的模型假设和应用范围，当打破其模型假设后，相应的杂波模型可能会失效，进而导致 CFAR 处理方法性能的退化。为了增强杂波统计模型的普适性，也有学者提出非参数化的杂波统计模型，采用核密度估计等策略实现对 SAR 图像杂波的自适应建模[34]。

3. 基于知识辅助的 CFAR 检测

传统的 CFAR 算法大多是基于目标与杂波背景的灰度特征差异进行方法设计。然而，在高分辨的

SAR 图像中，还存在其他多种可辅助目标检测的先验知识，对其进行充分利用可进一步增强目标与杂波的可区分度。例如，基于超像素的 CFAR 算法可以进一步开发目标的形状信息[56]，而基于目标候选框的 CFAR 算法可以充分利用 SAR 图像的梯度信息[39]，等等。

4. 高效费比 CFAR 处理

传统 CFAR 算法对 SAR 图像逐像素/逐区域精细滑窗处理，实际 SAR 图像常覆盖多达几十甚至几百平方千米的区域，但感兴趣的目标往往只占据少量的像素区域，如海域中的舰船目标、地面上的车辆目标等。考虑许多遥感任务的高时效需求，众多学者针对 SAR 图像 CFAR 处理，开展了保质低开销算法设计及实现研究，主要包括快速预减少 CFAR 测试单元数[57-58]、CFAR 并行化处理[59]等。

针对上述 SAR 图像 CFAR 检测研究的四个主要方面，后续章节将讨论相应的典型方法。

12.6.2 SAR 图像 CFAR 检测的杂波单元选取

对 SAR 图像杂波单元的选取策略可大致分为两种：对双窗法选取的杂波单元进行精炼[32,35,60-61]或者直接改变双窗法中"窗"的形状[56,58]来提升杂波样本的纯度。因此，本节将重点讨论两种选取策略：基于自动审查策略的 CFAR 杂波单元选取和基于超像素边界的 CFAR 杂波单元选取。

1. 基于自动审查策略的 CFAR 杂波单元选取

在海边的港口区域以及内陆的停车场等典型 SAR 观测场景中，舰船目标或者车辆目标往往呈现密集分布。在目标密集分布场景中，基于双窗法选取的 CFAR 杂波单元极易包含目标像素，进而使得杂波模型估计不够精确，降低 CFAR 算法的检测性能。为此，文献[32]提出了一种基于自动审查策略的 CFAR 杂波单元选取方法，其主要思路是：设计全局阈值，对目标和杂波像素做出预区分，在杂波模型估计时弃用已经被预判为目标的像素，进而提高密集目标场景下杂波模型精度。

基于自动审查策略的 CFAR 杂波单元选取方法的示意图如图 12.19 所示，该方法流程可分为全局阈值计算和杂波审查两个主要步骤[32]。

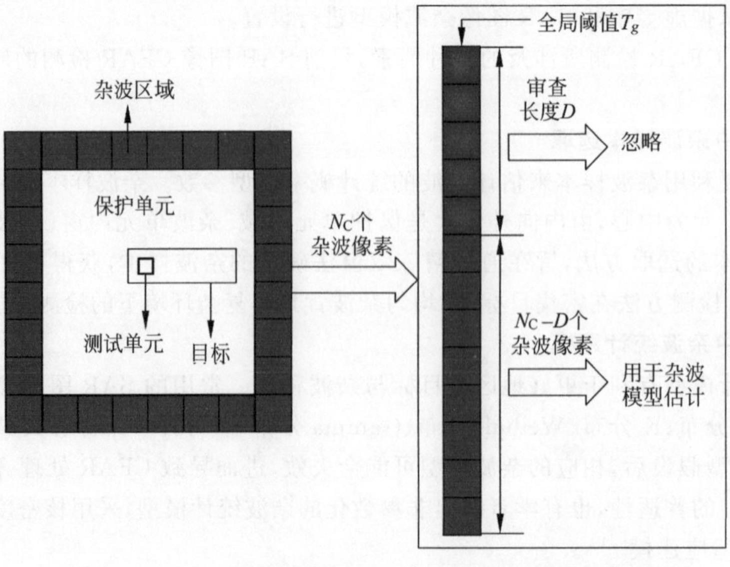

图 12.19 基于自动审查策略的 CFAR 杂波单元选取示意图

首先，全局阈值可根据待检测 SAR 图像的直方图来确定。全局阈值的作用在于，像素灰度值低于全局阈值的像素可视为杂波像素。一般来说，目标具有较强的后向散射系数，在 SAR 图像中比杂波背景呈现更高的灰度值，即目标样本通常出现在 SAR 图像直方图的“尾部”，而杂波样本往往出现在直方图的“前部”。令 $\varphi\in[0,1]$表示杂波像素占整个 SAR 图像像素的比例，可根据经验设置。令全局阈值为 T_g，则有

$$\Pr(x>T_g)=1-\varphi \tag{12.102}$$

令 $F(\cdot)$表示由 SAR 图像直方图获得的近似累积分布函数 CDF，则式(12.102)可进一步表示为

$$1-F(T_g)=1-\varphi \tag{12.103}$$

全局阈值 T_g 可通过求解式(12.103)得出。基于全局阈值，可获得一个和待检测 SAR 图像尺寸相同的索引二进制矩阵 $\mathbf{A}$，在该矩阵中，元素为 0 表示 SAR 图像相应的位置为杂波像素，元素为 1 表示 SAR 图像相应的位置为疑似目标像素。

其次，杂波审查过程将滤除由双窗法获得的杂波样本中的疑似目标像素。如图 12.19 所示，对于一个测试单元，首先根据双窗法获得初始的杂波样本。如果该杂波样本在索引二进制矩阵 $\mathbf{A}$ 中对应的元素是 1，则该样本不参与后续的杂波模型估计。在目标密集场景中，杂波审查可使 CFAR 滑窗选取的杂波样本中的目标像素数量显著降低。

图 12.20 给出了有/无杂波审查策略下，基于 X 波段下 HH 极化 SAR 图像的检测结果，其中，图像分辨率为 0.5m×0.5m，图像尺寸为 200×500，图像中包含 4 个距离较近的车辆目标，CFAR 处理的恒虚警率为 $P_{fa}=10^{-4}$，杂波模型为 G^0 分布。从图 12.20 可以看出[32]，无论是否包含杂波的自动审查，CFAR 均能成功检测到相对不拥挤的目标 1 和目标 4。而杂波自动审查的杂波选取方式的主要优势在于，对于比较拥挤的目标 2 和目标 3，可以更加精准地提纯杂波样本，进而提升目标检测性能。

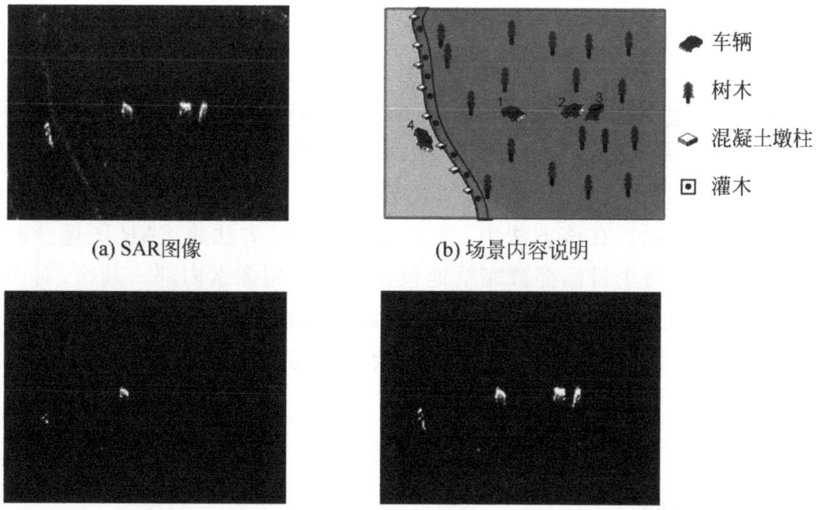

(a) SAR图像　(b) 场景内容说明

(c) 无杂波自动审查检测结果　(d) 有杂波自动审查的检测结果

图 12.20　基于杂波自动审查策略的 CFAR 检测结果($P_{fa}=10^{-4}$)

如图 12.19 所示，超参数 φ 决定了审查长度 D。因此，基于式(12.103)的自动审查策略的性能对超参数 φ 较为敏感。为此，可考虑基于迭代审查的杂波样本选取方法[35,62]，对目标检测结果和选取的杂波样本进行迭代更新，则无须人工设置超参数来决定审查长度 D，可一定程度提高 CFAR 处理方法的

自适应性。

2. 基于超像素边界的 CFAR 杂波单元选取

传统用于杂波单元选取的双窗法一般采用矩形窗，其主要缺点在于，对杂波样本中可能含有的目标像素，矩形窗自身并不具备相应的鉴别与去除能力，进而导致 CFAR 检测性能的损失。超像素的出现为解决这一问题提供了新的思路[56]，与基于自动审查策略的杂波样本选取方式不同，基于超像素的边界可自动鉴别并去除杂波样本中的目标像素，且无须加入额外的审查步骤。

超像素分割是一种性能优异的图像分割技术，旨在将图像中的像素进行局部聚类，进而将图像分割为多个内部均匀、具有视觉意义的不规则像素块，强化了图像的局部区域特征[63-65]。超像素分割已经在 SAR 图像处理领域受到了广泛的关注，主流的超像素分割算法包括基于图信号处理的方法、基于均值偏移的方法、基于局部 K-均值聚类的方法等[66-67]。其中，以简单线性迭代聚类(Simple Linear Iterative Clustering，SLIC)为代表的局部 K-均值聚类方法计算复杂度低，对存储空间的要求较低，成为最为广泛使用的超像素分割方法[67]。

假设预期的 SAR 图像中的每个超像素包含约 S^2 个像素，其中 S 被称为超像素的尺寸。SLIC 的主要流程如下。

步骤(1)：在 SAR 图像中，以间隔 S 均匀地设置超像素中心，再将每个超像素中心移动至其邻域内具有最小梯度值的位置，以防超像素中心落在图像内部的边缘上。

步骤(2)：对于每个超像素中心邻域内的 $2S\times2S$ 个像素，进行 K-均值聚类。

K-均值聚类中两像素的距离定义如下

$$D=\sqrt{(d^{\text{spat}})^2+\tau(d^{\text{inte}})^2} \tag{12.104}$$

其中，d^{spat} 表示欧式空间距离，d^{inte} 表示灰度距离，$\tau>0$ 表示权重参数。

步骤(3)：重复步骤(2)，直至超像素的中心位置趋于收敛。

步骤(4)：进行后处理，增强超像素内部的连通性。

基于超像素，文献[56]提出了一种代替双窗法中矩形窗的杂波单元选择方法。超像素最重要的特征是其内部像素的特征一致性，即超像素极少会“打断”图像中的边缘。超像素内部的一致性也意味着，如果设置的超像素尺寸显著大于待检测目标的尺寸，则目标自身及其周围的部分杂波区域会被完全包裹在一个超像素内部，如图 12.21 所示。鉴于此，文献[56]设计了基于超像素边界的 CFAR 杂波单元双窗选取方法，如图 12.22 所示。在该方法中，首先，利用 SLIC 方法将 SAR 图像分割为众多尺寸显著大于目标尺寸的超像素，使得每个目标都被完整地包含在一个超像素内部。其次，对于超像素内部的待检测像素，将其所处超像素的边界视为内窗，内窗以内的其他像素视为保护单元。最后，将与其所处超像素邻接的超像素内的像素视为杂波样本，用于估计杂波的统计模型。

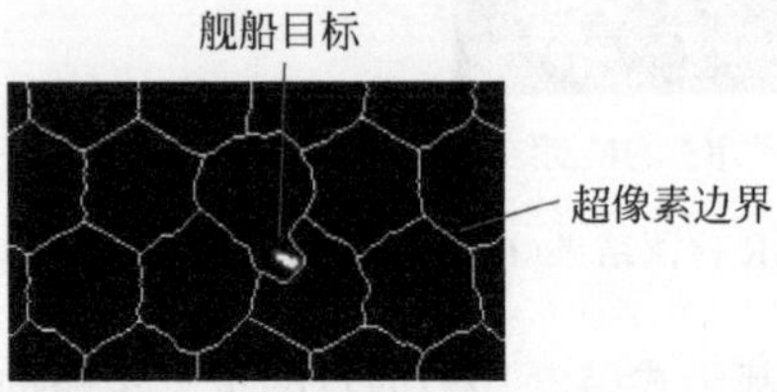

图 12.21 舰船目标被包围在超像素内部
(SAR 图像极化方式为 VH，来自哨兵-1 卫星[56])

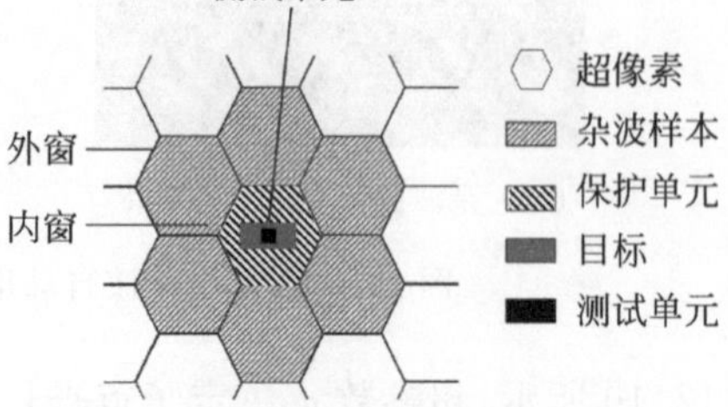

图 12.22 基于超像素边界的双窗法示意图

基于超像素边界的CFAR杂波单元选取方法具有两个主要优势。其一，当超像素的尺寸显著大于目标尺寸时，可以较好地将待检测单元之外的其他可能含目标像素的部分纳入保护单元中，避免污染杂波样本；其二，在多目标场景中，两个距离很近、尺寸相当的目标将很有可能被包含在同一个超像素内，确保了所选取的杂波样本的纯净性。

图12.23分别采用矩形窗和超像素边界两种杂波单元选取方式，给出了相应的CFAR检测对比结果。SAR图像来自TerraSAR-X卫星，分辨率为18.5m，图像尺寸为412×352。在图12.23中，CA-CFAR指经典的滑动平均CFAR处理，Weibull-CFAR指基于Weibull杂波统计模型的CFAR处理，SPCFAR(Superpixel-based CFAR)指采用超像素边界作为杂波单元选取方式的CFAR处理。由图12.23可知，基于超像素边界的杂波单元选取对CA-CFAR、Weibull-CFAR的改进效果十分突出，显著减少了CA-CFAR对目标的漏检和Weibull-CFAR的虚警数量。

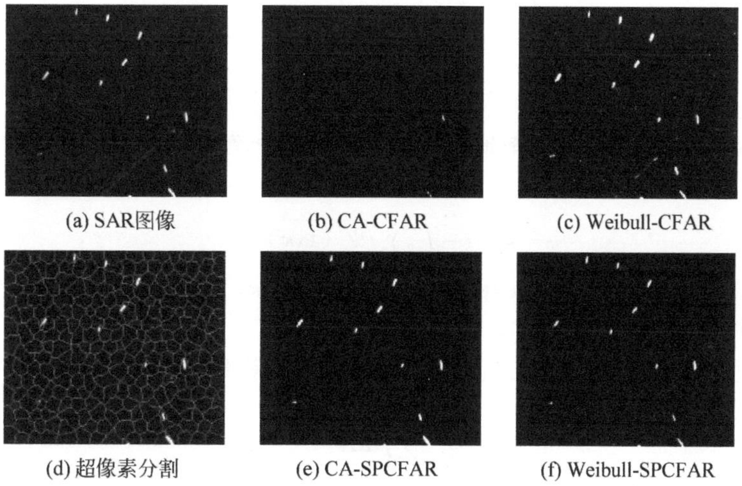

图12.23 基于矩形窗(b)(c)和超像素边界(e)(f)的杂波单元选取方式的CFAR检测结果($P_{fa}=10^{-3}$)[56]

基于超像素边界的CFAR杂波单元选取方法尚存在有待改进之处。当两个目标的距离恰好满足一个目标处于中心超像素内，而另一个目标处于与中心超像素邻接的超像素内时，估计得出的杂波模型会显著偏离真实情况。当然，这一问题在基于矩形窗的双窗法杂波单元选取中也同样存在。通过对所选取的杂波样本进行无监督式最近邻分类[55]，有望缓解上述基于超像素边界和矩形窗双窗法所面临的共性问题。

12.7 基于广义Gamma杂波模型的SAR图像CFAR检测

在低分辨SAR图像中，高斯分布是常用的杂波统计模型，其参数估计简单，易得出检测阈值的显式解[68]。近年来，随着SAR图像分辨率的不断提高，高斯分布难以精准刻画高分辨SAR图像的杂波特征，学术界提出了众多非高斯杂波模型，例如对数正态分布[33]、Wishart分布[69]、K分布[70]、G^0分布[70]、混合型分布[71]、截断型分布[38,41,72-73]等。相比于高斯杂波模型，基于非高斯杂波模型的CFAR算法在固定虚警率的情况下获得了更佳的检测性能。然而，诸多非高斯杂波模型存在杂波参数估计计算量大、无显式解的判决阈值数值求解时计算量大等问题。本节将讨论基于广义Gamma分布(Generalized Gamma Distribution，GΓD)的SAR图像CFAR检测方法[68]。

12.7.1 检测方法设计

GΓD 的概率密度函数 PDF 可表示为

$$f(x)=\frac{|v|\kappa^{\kappa}}{\sigma\Gamma(\kappa)}\left(\frac{x}{\sigma}\right)^{\kappa v-1}\exp\left[-\kappa\left(\frac{x}{\sigma}\right)^{v}\right],\quad \sigma,|v|,\kappa,x>0 \tag{12.105}$$

其中,x 表示图像的灰度信息,σ、v、κ 分别表示 GΓD 的尺度参数、功率参数和形状参数。GΓD 实际上是一簇分布,可退化为常见的瑞利分布、指数分布、Weibull 分布、Gamma 分布、逆 Gamma 分布等;对数正态分布也是 GΓD 的一种极限情况[74]。

基于对数累积量,可实现对 SAR 图像 GΓD 的参数估计,则 GΓD 的三个对数累积量的计算如下

$$L_1=\ln\sigma+[\Psi(\kappa)-\ln\kappa]/v \tag{12.106}$$

$$L_2=\Psi'(1,\kappa)/v^2 \tag{12.107}$$

$$L_3=\Psi'(2,\kappa)/v^3 \tag{12.108}$$

其中,$\Psi(x)=\mathrm{d}\ln\Gamma(x)/\mathrm{d}x$,$\Psi'(n,x)=\mathrm{d}^n\Psi(x)/\mathrm{d}x^n$。由式(12.107)和式(12.108)可知$(L_2)^3/(L_3)^2=[\Psi'(1,\kappa)]^3/[\Psi'(2,\kappa)]^2$,进而可通过数值算法快速求解出参数 κ 的值。式(12.106)~式(12.108)中的参数 v 和 σ 可通过以下方式获得

$$v=\mathrm{sgn}(-L_3)\sqrt{\Psi'(1,\kappa)/L_2} \tag{12.109}$$

$$\sigma=\exp\{L_1-[\Psi(\kappa)-\ln\kappa]/v\} \tag{12.110}$$

其中,sgn(·)表示符号函数。

对于 N_{C} 个杂波样本$\{X_1,X_2,\cdots,X_{N_{\mathrm{C}}}\}$,上述三个对数累积量可采用估计值[74]

$$\hat{L}_1=\frac{1}{N_{\mathrm{C}}}\sum_{i=1}^{N_{\mathrm{C}}}\ln X_i \tag{12.111}$$

$$\hat{L}_2=\frac{1}{N_{\mathrm{C}}}\sum_{i=1}^{N_{\mathrm{C}}}(\ln X_i-L_1)^2 \tag{12.112}$$

$$\hat{L}_3=\frac{1}{N_{\mathrm{C}}}\sum_{i=1}^{N_{\mathrm{C}}}(\ln X_i-L_1)^3 \tag{12.113}$$

根据上述 GΓD 参数估计结果,对于基于 GΓD 的 SAR 图像 CFAR 检测方法,相应的自适应判决阈值的计算采用如下步骤。首先,令 $y=\kappa(t/\sigma)^v$,则 GΓD 的 CDF 为

$$F(x)=\begin{cases}\dfrac{1}{\Gamma(\kappa)}\displaystyle\int_0^{\kappa(x/\sigma)^v}y^{\kappa-1}\exp(-y)\,\mathrm{d}y, & v>0\\ \dfrac{1}{\Gamma(\kappa)}\displaystyle\int_{\kappa(x/\sigma)^v}^{+\infty}y^{\kappa-1}\exp(-y)\,\mathrm{d}y, & v<0\end{cases} \tag{12.114}$$

根据不完备 Gamma 函数的定义,即

$$Q(x,a)=\frac{1}{\Gamma(a)}\int_0^x t^{a-1}\exp(-t)\,\mathrm{d}t,\quad a,x>0 \tag{12.115}$$

式(12.114)可改写为

$$F(x)=\begin{cases}Q[\kappa(x/\sigma)^v,\kappa], & v>0\\ 1-Q[\kappa(x/\sigma)^v,\kappa], & v<0\end{cases} \tag{12.116}$$

令 T 表示 CFAR 自适应判决阈值,根据式(12.116)和式(12.117)

$$P_{\mathrm{fa}}=1-F(T) \tag{12.117}$$

可得判决阈值为

$$T=\begin{cases}\sigma\left[\dfrac{1}{\kappa}Q_{\mathrm{inv}}(1-P_{\mathrm{fa}},\kappa)\right]^{1/v}, & v>0\\ \sigma\left[\dfrac{1}{\kappa}Q_{\mathrm{inv}}(P_{\mathrm{fa}},\kappa)\right]^{1/v}, & v<0\end{cases} \tag{12.118}$$

其中，$Q_{\mathrm{inv}}(\cdot,\cdot)$表示不完备 Gamma 函数的逆函数。

12.7.2　性能分析

图 12.24 给出了不同杂波分布下，设置的 CFAR 恒虚警率 P_{fa} 与真实虚警率的对比结果[68]，其中横坐标表示不同杂波样本组的序号，每组杂波样本为 200×200 的海杂波 Terra-X SAR 图像，极化方式为 VV，分辨率为 1.9m×3.3m，对比的 SAR 图像杂波模型包括高斯分布、Weibull 分布、K 分布、G^0 分布和 GΓD 分布等，具体 PDF 如表 12.5 所示[68]。可以看到，相比于其他杂波分布，GΓD 分布所生成的真实虚警率可以更好地贴近设定的 CFAR 恒虚警率，即该分布模型能更好地对 SAR 图像中的海杂波进行统计建模。

表 12.5　SAR 图像常用杂波模型

杂波模型	概率密度函数
高斯分布	$f(x)=\dfrac{1}{\sqrt{2\pi\sigma^2}}\exp\left[-\dfrac{(x-\mu)^2}{2\sigma^2}\right]$
Weibull 分布	$f(x)=\dfrac{c}{b}\left(\dfrac{x}{b}\right)^{c-1}\exp\left[-\left(\dfrac{x}{b}\right)^{c}\right]$
K 分布	$f(x)=\dfrac{4\lambda nx}{\Gamma(\alpha)\Gamma(n)}(\lambda nx^2)^{\frac{\alpha+n}{2}}K_{\alpha-n}(2x\sqrt{\lambda n})$，$K_\alpha(\cdot)$为二阶修正 Bessel 函数，$n$ 表示多视数
G^0 分布	$f(x)=\dfrac{2n^n\Gamma(n-\alpha)}{\gamma^\alpha\Gamma(n)\Gamma(-\alpha)}\dfrac{x^{2n-1}}{(\gamma+nx^2)^{n-\alpha}}$，$n$ 表示多视数
GΓD 分布	$f(x)=\dfrac{\lvert v\rvert\kappa^\kappa}{\sigma\Gamma(\kappa)}\left(\dfrac{x}{\sigma}\right)^{\kappa v-1}\exp\left[-\kappa\left(\dfrac{x}{\sigma}\right)^{v}\right]$

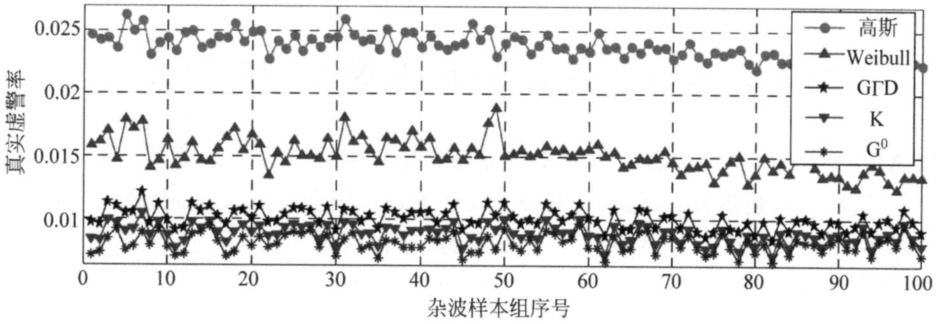

(a) 恒虚警率设定值P_{fa}=0.01

图 12.24　CFAR 恒虚警率设定值 P_{fa} 与真实虚警率的对比结果

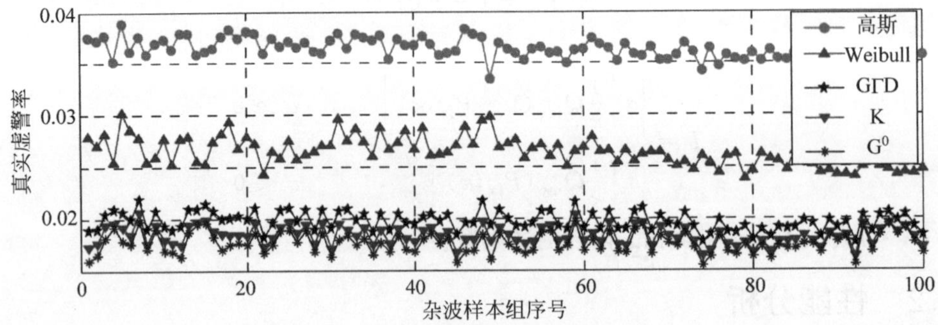

(b) 恒虚警率设定值P_{fa}=0.02

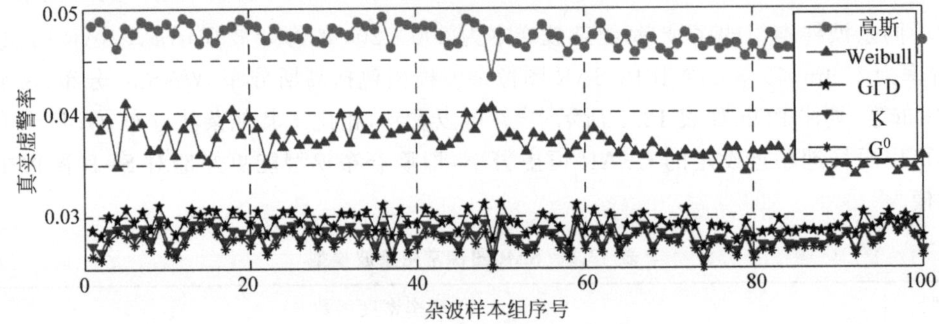

(c) 恒虚警率设定值P_{fa}=0.03

图 12.24 (续)

针对 200×200 尺寸的 SAR 图像,在特定计算环境下,表 12.6 进一步给出了三种较复杂的杂波分布 PDF 参数估计、判决阈值计算等所需时间的对比结果[68]。可以看到,GΓD 在降低杂波参数估计时间和判决阈值计算时间上具有明显优势,也体现了其低复杂度的优势。

表 12.6 杂波参数估计、判决阈值计算所需时间对比 (单位: ms)

杂波分布	杂波参数估计时间	判决阈值计算时间
K	10.93	2209.16
G^0	11.48	6.35
GΓD	7.00	0.10

12.8 基于语义知识辅助的 SAR 图像 CFAR 检测

传统的 SAR 图像 CFAR 检测算法依赖于对恒虚警率的精准设置。恒虚警率设计值过高,在检测到目标的同时会造成的大量虚警;相反,恒虚警率设计值过低则会造成大量的漏检。对 SAR 图像先验知识的引入,例如目标的形状、大小、结构[36,40,56]、目标/杂波区域的边缘特征[39]等的先验认知,或将为解决上述问题提供新的思路。为此,可采用基于语义辅助的 SAR 图像目标 CFAR 检测算法[37],通过使用高层次的语义关系,获得更加精确的认知结果。在基于语义辅助的 CFAR 检测算法中,可通过设置较高的恒虚警率来防止漏掉微弱目标,再结合目标的语义特征来抑制传统 CFAR 检测结果中的虚警[37]。

12.8.1 检测方法设计

以车辆目标的SAR图像为例[37]，车辆目标的关键特征是由SAR侧视特性引起的强散射和阴影效应，这构成了车辆目标的重要语义要素。图12.25显示了一个简化的车辆模型，雷达波束的入射角为θ；a表示车辆周围地面的散射系数区域；b和c分别表示来自车头和前挡风玻璃的强散射系数区域；d和f分别表示车顶和尾部的散射区域。车辆的底盘阻挡了部分车辆后部的雷达波束，进而形成阴影面e。车辆后部形成的强散射区域f的长度取决于车辆后部的长度。

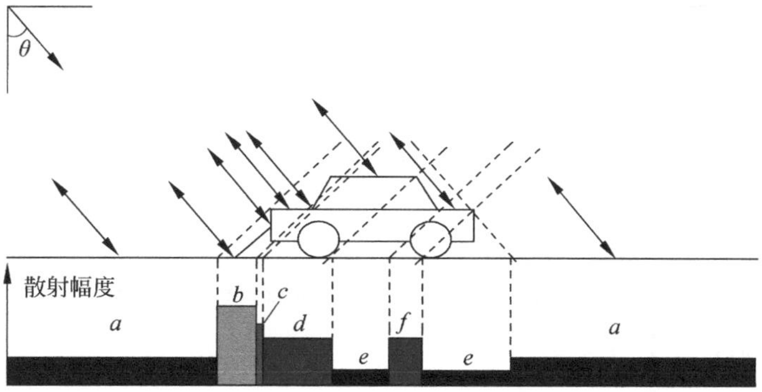

图12.25 SAR雷达处于左上方的车辆目标散射模型

在SAR图像中，车辆目标的亮像素区域由强散射面形成，暗像素区域由阴影面形成；可采用CFAR算法搜索亮像素，采用Otsu算法搜索暗像素，其中Otsu算法通过最大化类间方差和类内方差的比值来选择分割阈值[75]。在CFAR处理和Otsu算法之后，将彼此相邻的亮像素/暗像素分别合并到亮区域/暗区域中，并移除面积较小的亮区域/暗区域来降低图像噪声的影响。

造成强散射面和阴影面的车辆目标部件都具有结构化的特征，相应地，SAR图像中的亮区域和暗区域也满足结构化的特征，随着车辆方向、位置等方面的变化而呈现不同的形状。亮区域的面积和形状取决于车辆的表面，而暗区域的面积和形状取决于车辆遮挡雷达波束的底部。鉴于此，用于搜索满足车辆强散射面和阴影面的滑窗，其形状和大小如表12.7所示[37]。当雷达波入射方向和汽车主轴之间的角度接近0°、−45°、45°和90°时，分别使用1型、2型、3型和4型滑窗来进行搜索[37]。滑窗的尺寸可根据车辆的先验信息进行适当设置。

表12.7 滑窗形状与大小

角度	0°	−45°	45°	90°
强散射区域滑窗				
阴影区域滑窗				
尺寸	21×21	21×21	21×21	19×30 15×30
编号	1	2	3	4

值得注意的是，SAR 图像中的亮区域和暗区域也可能来自其他与车辆目标类似的目标。亮区域和暗区域是图像的灰度特征，而强散射面和阴影面的组合则是车辆目标的语义特征，图像的灰度特征距离目标的语义特征仍存在一定的差距。因此，可采用隶属度函数来描述亮区域/暗区域属于车辆的强散射面/阴影面的可能性[37]

$$\phi_b(x,y,i)=\exp\left[-(B(x,y,i)-\mu_1)\right]^2 \tag{12.119}$$

$$\phi_d(x,y,i)=\exp\left[-(D(x,y,i)-\mu_2)\right]^2 \tag{12.120}$$

其中，$B(x,y,i)$表示中心位于(x,y)的第 i 型滑窗中亮像素的比率，$D(x,y,i)$表示中心位于(x,y)的第 i 型滑窗中暗像素的比率，μ_1 和 μ_2 为常数。式(12.119)和式(12.120)利用隶属度函数将图像的灰度特征和目标的语义特征建立了联系，可设置阈值来对上述的隶属度进行处理，进而判断当前区域是否属于强散射面或阴影面。

相比灰度信息，目标的空间关系是更高层次的图像语义。强散射面位于车辆的前方，阴影面则位于车辆的后方；强散射面和阴影面沿入射角方向依次排列。针对同一个车辆目标，强散射面和阴影面之间存在一个约等于车辆长度的特定距离。如果发现亮像素区域和暗像素区域按照强散射面和阴影面的空间关系进行排布，则可将该区域检测为一个车辆目标。结合式(12.119)、式(12.120)及车辆目标的空间语义关系，可定义车辆目标隶属度为

$$\varphi(x_1,y_1,i)=\exp\left[-\left(\sqrt{(x_1-x_2)^2+(y_1-y_2)^2}-\mu_3\right)^2/\sigma^2\right]\times \phi_d(x_2,y_2,i)\times\phi_b(x_1,y_1,i) \tag{12.121}$$

其中，指数函数部分表示从强散射面到阴影面的距离对车辆身长的隶属度，μ_3 和 σ 为常数；(x_1,y_1)和(x_2,y_2)分别表示亮区域和暗区域的中心坐标。令 $\varphi(x_0,y_0,i_0)$表示 $\varphi(x,y,i)$的峰值点，则点(x_0,y_0)表示针对车辆目标具有最高隶属度的像素点位置。如果$\varphi(x_0,y_0,i_0)$高于设定的阈值，则表示一个车辆目标被检测到。车辆区域从强散射面开始，以阴影面结束。此外，可将匹配到同一阴影面的两个强散射面合并到同一个车辆目标中。如果两个车辆目标的区域相互重叠，可只保留具有较大隶属度的目标。上述基于语义知识辅助的 CFAR 算法处理流程如图 12.26 所示。

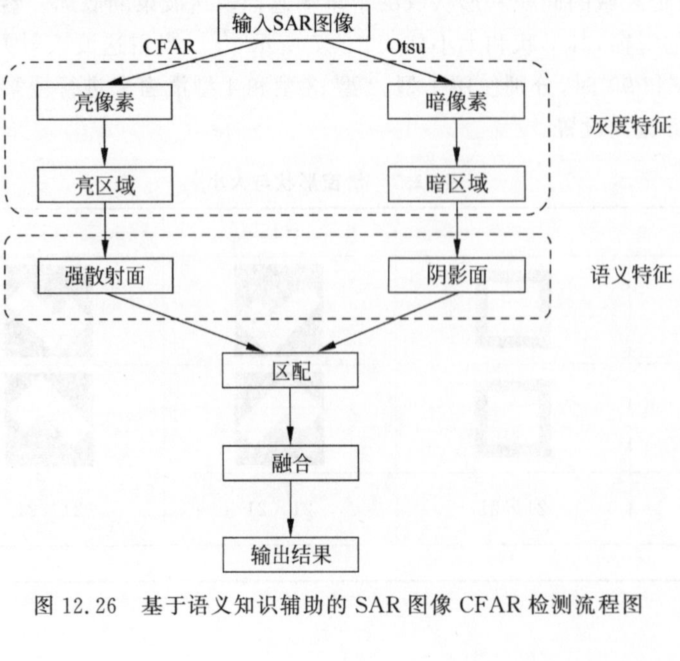

图 12.26 基于语义知识辅助的 SAR 图像 CFAR 检测流程图

12.8.2 性能分析

图 12.27 给出了两幅美国桑迪亚国家实验室公开的 MiniSAR 图像[37]，分辨率为 0.1m，图像中包含了房屋、树木、汽车等物体。基于语义知识辅助的 CFAR 算法的超参数值设置为：$\mu_1=0.36$，$\mu_2=1$，$\mu_3=48$，$\sigma=54$。

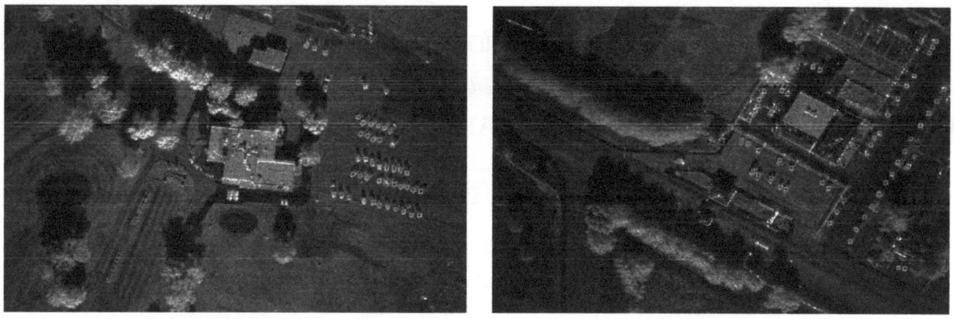

(a) SAR图像1　　(b) SAR图像2

图 12.27　两幅包含车辆目标的地面场景 SAR 图像

针对图 12.27 的检测场景，表 12.8 给出了基于语义知识辅助的 CFAR 算法和常规 CFAR 算法的检测结果[37]。对于图 12.27(a)的场景，语义知识对 CFAR 的辅助在保证检测率不降低的同时有效减少了虚警数；对于图 12.27(b)的场景，语义知识对 CFAR 的辅助则同步提升了目标检测率、降低了虚警率。上述结果表明，对语义知识的充分开发，可在一定程度上提升 SAR 目标 CFAR 检测性能。

表 12.8　基于语义知识辅助的 CFAR 算法和常规 CFAR 算法检测结果对比

检测结果	图 12.27(a)场景				图 12.27(b)场景			
	语义辅助 CFAR		常规 CFAR		语义辅助 CFAR		常规 CFAR	
	数量	比例	数量	比例	数量	比例	数量	比例
真实目标数	52	100%	52	100%	55	100%	55	100%
检测出目标数	47	90%	47	90%	45	82%	26	47%
虚警情况	28	37%	61	56%	53	54%	92	78%

12.9 基于密度特征的 SAR 图像 CFAR 检测快速实现

相比于红外、可见光、高光谱等传感器，SAR 成像受云雾、水汽等天气因素影响较小，具有全天时、全天候的工作能力，且扫描幅宽可达几十千米甚至上百千米，具有良好的大场景监视潜力。但在大场景监视的同时，存在对占比较少的重点目标检测效率偏低的问题，且在高时效的舰船目标检测场景中尤为突出。CFAR 舰船目标检测算法大多依赖于对 SAR 图像的逐像素/逐区域精细滑窗处理[55-56,64]，且每个 CFAR 滑窗操作均需要估计海杂波的模型，该估计过程往往包含烦琐的迭代、求逆等操作，计算量较大。然而，重点舰船目标在远海区域中极其稀疏，SAR 图像可覆盖几十甚至几百平方千米的海域，但其中所感兴趣的舰船目标只占据少量像素区域。

12.9.1 检测方法设计

相对于舰船目标的稀疏性,传统的全图像精细滑窗 CFAR 处理的检测效率较低,传统 SAR 图像舰船目标 CFAR 检测面临广阔海域稀疏舰船目标检测效率低的难题。为此,可考虑基于密度特征的 SAR 舰船目标 CFAR 快速检测方法[58],该方法基于如下事实:相对于海杂波,舰船目标在广阔海域中存在稀疏性,即低密度特征。

针对广阔海域的稀疏舰船目标,基于密度特征的 SAR 舰船目标 CFAR 快速检测算法,在不降低检测水平的前提下,可以将现有检测技术的检测时间减少约 1 个数量级。其处理流程如图 12.28 所示,主要分为三个步骤:超像素分割、密度筛选、局部 CFAR 检测。

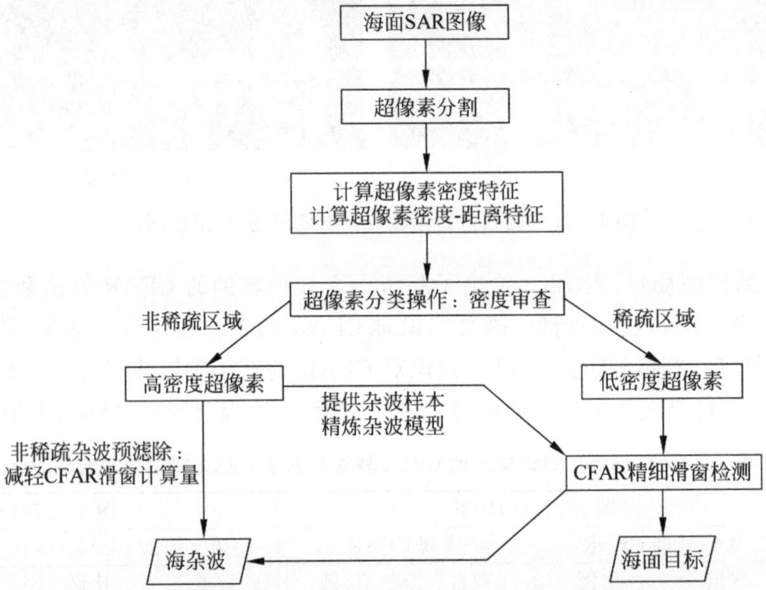

图 12.28 基于密度特征的 SAR 海面目标 CFAR 快速检测流程

超像素分割旨在将图像中像素进行局部聚类,进而将图像分割为多个具有视觉意义的局部像素块。与 12.7.2 节类似,此处仍采用经典的简单线性迭代聚类 SLIC 算法,对海面 SAR 图像进行超像素分割。需要指出的是,由于此处的超像素分割的目的在于将一个舰船目标分割为一个或者少数几个超像素,而非获取杂波单元,因此 SLIC 算法中的超像素大小需与舰船目标平均尺寸相匹配,SAR 图像及其超像素分割示意如图 12.29。

密度筛选是 CFAR 快速检测的核心步骤。首先利用两种基于密度的特征来对海杂波和舰船目标进行超像素鉴别。

(1) 密度特征:数据点的密度表示其与特征空间中其他点的相似程度[76],即该数据点在整个数据集形成的特征空间中的稀疏度。令 ω_i 表示 SAR 图像中第 i 个超像素内部像素灰度的平均值,其中 $i=1,2,\cdots,I$,I 表示 SAR 图像中超像素的数量。则第 i 个超像素的密度定义为

$$\rho_i = \sum_{j\in\{1,2,\cdots,I\}/i} \exp\left[-(D_{i,j}/D_c)^2\right] \tag{12.122}$$

其中,$D_{i,j}=|\omega_i-\omega_j|$ 表示第 i 个超像素平均灰度和第 j 个超像素平均灰度之间的距离,$D_c=\eta\times\max\limits_{\forall i,j}\{D_{i,j}\}$ 表示截断距离,$0<\eta<1$ 是一个常数。

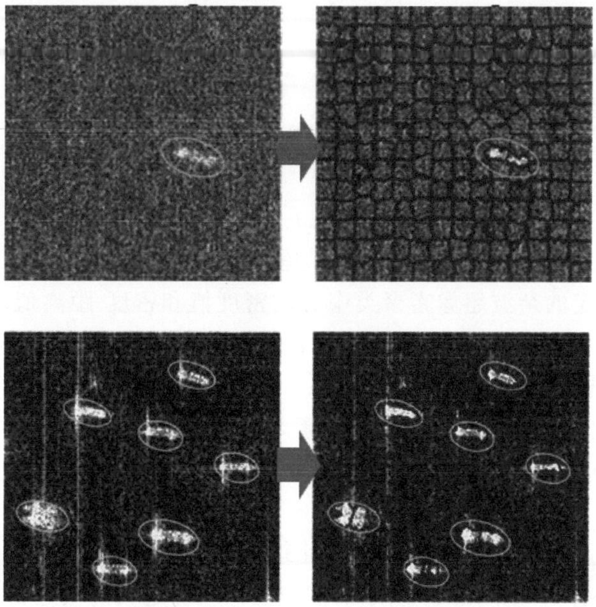

图 12.29　SAR 图像及其超像素分割示意

从密度特征的含义来看，在广域海面 SAR 图像中，对于一个非稀疏的杂波超像素，即使在复杂背景下，也有较多其他杂波超像素与之有相近的灰度距离，因此杂波超像素的密度值 ρ 较大；相反，由于海面舰船的稀疏性，在整个广域 SAR 图像中，与一个舰船超像素灰度距离接近的其他舰船超像素较少，因此舰船超像素的密度值 ρ 相对较小。

(2) 基于密度的距离特征：定义 $\Gamma_i \triangleq \{j \mid \rho_j < \rho_i, j=1,2,\cdots,I, j \neq i\}$，作为密度低于第 i 个超像素的其他超像素的索引集，则第 i 个超像素的密度-距离特征可定义为

$$\mathcal{D}_i = \begin{cases} \max_{j \in \Gamma_i} \{D_{i,j}\}, & \Gamma_i \neq \varnothing \\ \min_{j \in \{1,2,\cdots,I\}/i} \{D_{i,j}\}, & \Gamma_i = \varnothing \end{cases} \tag{12.123}$$

从密度-距离特征的含义来看，由于低密度的舰船超像素和高密度的海杂波超像素存在较大的灰度差异，因此高密度杂波超像素的密度-距离值 $\mathcal{D}_i$ 较大；相反，低密度舰船超像素的灰度整体上较为接近，并且当第 i 个超像素有最低的密度时($\Gamma_i = \varnothing$)，$\mathcal{D}_i$ 是 $\{D_{i,j}, \forall j, j \neq i\}$ 中的最小值，因此目标超像素的密度-距离值 $\mathcal{D}_i$ 相对较小。

基于式(12.122)中的密度 $\{\rho_i, \forall i\}$ 和式(12.123)中密度-距离 $\{\mathcal{D}_i, \forall i\}$，接下来通过无监督式的最近邻分类器，对 SAR 图像中的所有超像素进行密度筛选操作。首先使用 Min-Max 归一化操作分别将 $\{\rho_i, \forall i\}$ 和 $\{\mathcal{D}_i, \forall i\}$ 映射到区间[0,1]。由于目标超像素的 ρ、$\mathcal{D}$ 值较小，而杂波超像素的 ρ、$\mathcal{D}$ 值较大，因此目标和杂波的“聚类中心”分别由式(12.124)和式(12.125)获得

$$i_{\mathrm{T}} = \underset{\forall i}{\operatorname{argmin}} f(\rho_i, \mathcal{D}_i) \tag{12.124}$$

$$i_{\mathrm{C}} = \underset{\forall i}{\operatorname{argmax}} f(\rho_i, \mathcal{D}_i) \tag{12.125}$$

其中，$f(\cdot,\cdot)$ 表示融合策略，可采用加法、乘法或模糊融合等，此处采用简单的加法融合规则 $f(\rho_i, \mathcal{D}_i) = \rho_i + \mathcal{D}_i$。接下来，将最近邻分类器用于审查 SAR 图像中的每个超像素

$$\begin{cases}\delta_i = 1, & \theta_i^{\mathrm{C}} \geqslant \theta_i^{\mathrm{T}} \\ \delta_i = 0, & \theta_i^{\mathrm{C}} < \theta_i^{\mathrm{T}}\end{cases} \tag{12.126}$$

其中,杂波和目标的 θ_i 参数分别为

$$\theta_i^{\mathrm{C}} = \sqrt{(\rho_i - \rho_{\mathrm{C}i})^2 + (\mathcal{D}_i - \mathcal{D}_{\mathrm{C}i})^2} \tag{12.127}$$

$$\theta_i^{\mathrm{T}} = \sqrt{(\rho_i - \rho_{\mathrm{T}i})^2 + (\mathcal{D}_i - \mathcal{D}_{\mathrm{T}i})^2} \tag{12.128}$$

其中,$\rho_{\mathrm{T}i}$ 和$\mathcal{D}_{\mathrm{T}i}$分别表示由式(12.124)确定的目标超像素聚类中心的密度值和密度-距离值,$\rho_{\mathrm{C}i}$ 和$\mathcal{D}_{\mathrm{C}i}$分别表示由式(12.125)确定的杂波超像素聚类中心的密度值和密度-距离值。当 $\delta_i=1$ 时,表明第 i 个超像素属于杂波超像素区域;当 $\delta_i=0$ 时,表明第 i 个超像素是候选的目标超像素。在接下来的局部CFAR检测阶段,只对候选的目标超像素进行精细CFAR检测处理,其余已经被鉴别为杂波的超像素则不进行CFAR检测处理。密度筛选得出的超像素之所以被称为候选目标超像素,是因为其中除真实目标超像素外,还可能还包含SAR图像中少量稀疏的杂波超像素,例如舰船目标的旁瓣和异常的强杂波区域等。

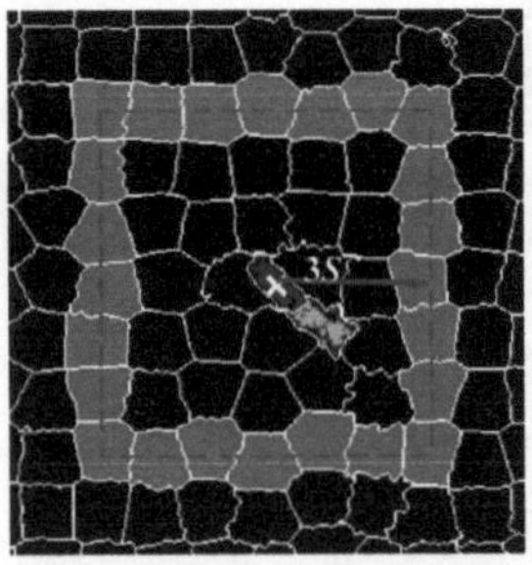

图 12.30 超像素双窗法示意

局部CFAR处理针对每个候选目标超像素进行。首先,使用如图12.30所示的双窗法[64],确定当前候选目标超像素的临近杂波超像素区域,其中灰色区域表示当前测试单元选中的临近杂波超像素区域。其次,使用式(12.126)筛选出的杂波超像素来精炼(当前候选目标超像素的)临近杂波超像素区域。最后,使用精炼后的临近杂波超像素区域估计杂波PDF中的参数,然后根据预设的恒虚警率计算CFAR检测的自适应判决阈值。在下文的试验验证环节,将选择经典高斯分布[32]作为杂波模型;需要指出的是,K分布、G^0 分布、Gamma分布[64,70]等其他分布也适用于这里的局部CFAR处理。将当前候选目标超像素中的像素灰度与CFAR自适应判决阈值进行比较,以产生最终的判决结果。计算候选目标超像素中检测出的目标像素数量与当前超像素中像素总数的比值,若该比值大于1个固定的常数(此处取20%),则该候选目标超像素被视为真正的目标超像素。

与现有SAR舰船目标CFAR检测算法相比[55-56,64],基于密度特征的SAR海面舰船目标CFAR快速检测算法具有三个优势。其一,在后续的局部CFAR检测阶段,基于密度筛选的操作有助于避免在背景杂波区域浪费大量计算时间。其二,用密度筛选操作鉴别得出的纯净杂波样本,可以获得更加精确的杂波PDF参数,以便更好确定CFAR自适应判决阈值,提升检测性能。其三,为了防止漏检重要目标,在局部CFAR检测阶段,即使提高CFAR恒虚警率,之前被滤除的非稀疏杂波区域也不会成为虚警,这使得最终检测结果的虚警数量可以保持在较低水平。

12.9.2 性能分析

图12.31给出了两个密度特征的示意图。图12.31(a)提供了具有异质杂波的SAR图像切片[77],其中椭圆表示舰船目标;图12.31(b)为相应的超像素分割图,其中包含两个目标超像素。图12.31(c)是图12.31(b)中所有超像素的 ρ 和$\mathcal{D}$在归一化后的结果,用于展示杂波超像素和目标超像素在密度特征上的显著区别。可以看到,利用密度特征,图12.31(b)中的目标超像素和杂波超像素得到了很好的区分。图12.31(d)为密度筛选后的超像素,可以看出,其中大部分非稀疏的杂波超像素已经被滤除。

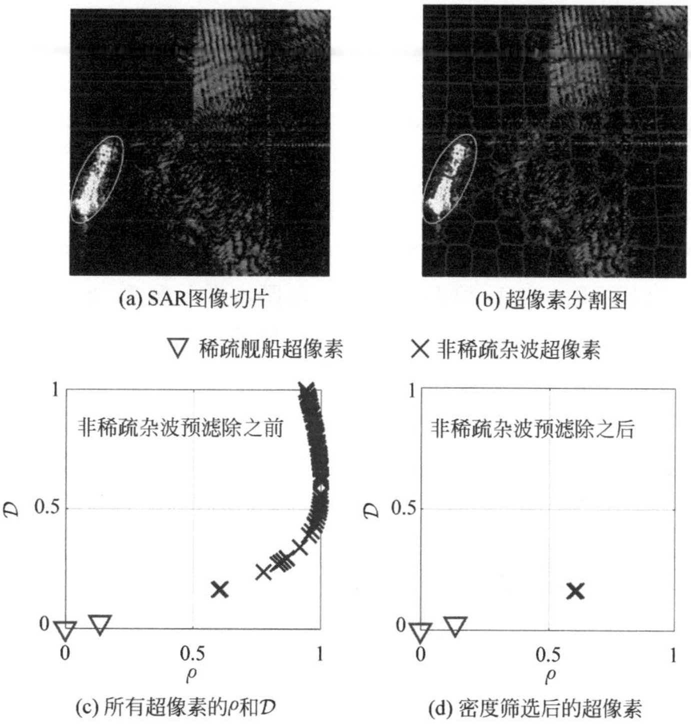

图 12.31　两个密度特征 ρ 和$\mathcal{D}$的示意图

基于 300 幅分辨率为 3m、大小为 256×256 的 SAR 图像切片[77]，图 12.32 对比了基于全图像精细 CFAR 滑窗的 SPCCFAR(Superpixel Clustering-based CFAR)[55]和基于密度特征的快速 CFAR 方法的 ROC 性能。由图 12.32 可知，基于密度特征的舰船目标 CFAR 快速检测方法具有更好的检测性能，原因在于，CFAR 杂波模型由密度筛选得出纯净杂波样本，进而使杂波模型参数估计更加精确，且在密度筛选操作中滤除的杂波超像素不会成为后续虚警。

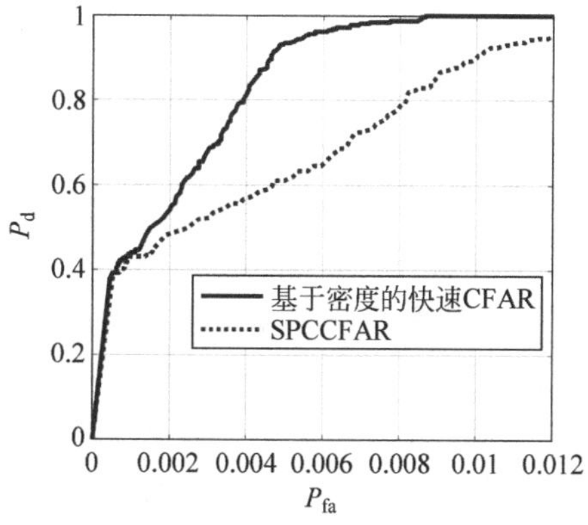

图 12.32　基于 300 幅 256×256 大小的 SAR 图像切片的 ROC 曲线对比

基于相同杂波模型[55]，利用 300 幅高分 3 号 SAR 图像切片，表 12.9 给出了特定计算环境下不同 CFAR 检测方法的平均运行时间对比。由表 12.9 可知，相比现有 SPCCFAR 检测算法，基于密度特征的舰船目标 CFAR 快速检测方法的计算效率提升了一个数量级以上。

表 12.9　两种 CFAR 处理方法的运行时间对比　　（单位：s）

检测算法	运行时间
SPCCFAR	3.45
基于密度的快速 CFAR	0.19

12.10 比较与小结

本章讨论了高分辨率雷达目标CFAR检测问题。从一维像的距离扩展目标和二维SAR图像两个方面进行CFAR检测分析。

在距离扩展目标CFAR检测方面。基于目标的秩1信号模型和多秩子空间信号模型,12.3节分析了复合高斯杂波下距离扩展目标的子空间检测。对于秩1模型和多秩信号模型的目标检测,当杂波背景为复合高斯杂波时,可以采用GMSD检测器;当目标距离单元数和杂波散斑分量协方差矩阵未知时,可采用A-GMSD检测器。对于A-GMSD检测器,信号子空间的维数增加将会导致检测性能的下降,而复合高斯杂波越尖锐,检测概率越高。

12.4节分析了复合高斯杂波加热噪声中的距离扩展目标检测问题。针对复合高斯杂波加热噪声的背景,通过对热噪声进行功率等效处理,采用N-MSD方案对距离扩展目标进行检测。在大杂噪比条件下,N-MSD检测器具有近似CFAR能力;当出现杂波尖峰时,N-MSD检测器的虚警概率受杂噪比的影响较小,有较好的虚警控制能力。

12.5节分析了SαS分布杂波下的距离扩展目标检测问题。由于SαS分布没有有限的二阶矩,其PDF没有闭型解析表达式,这些因素为检测器设计带来了一定的困难。一种解决思路是将PFLOM变换与GASD检测器相结合,构造GASD-PFLOM检测器,在保持对SαS分布杂波CFAR特性的同时,对杂波的强脉冲特性进行有效抑制,提高SαS分布杂波下的目标检测性能。另一种思路是将点目标的柯西检测器和双门限检测器结合起来,建立二元积累柯西检测器,这种检测器结构简单,便于实现,且在不同特征指数的SαS分布杂波下具有较好鲁棒性。

在SAR图像目标CFAR检测方面。12.6节梳理了SAR图像CFAR检测中杂波单元的主要选取方法;先讨论了基于自动审查的SAR图像杂波单元选取方法,该方法利用审查阈值对目标和杂波像素做出预区分,使得密集目标场景下用于杂波模型估计的杂波样本更纯;其次分析了基于超像素边界的杂波单元选取方法,该方法可自动改善传统基于矩形窗选取的杂波样本纯净程度。

12.7节基于广义Gamma分布杂波模型,分析了SAR图像CFAR检测的虚警控制性能。GΓD模型着重改善传统SAR图像杂波模型的参数估计计算量大、CFAR判决阈值缺乏显式解的问题,且能涵盖常见的瑞利分布、指数分布、Weibull分布、Gamma分布等。

12.8节讨论了基于语义知识辅助的SAR图像CFAR处理方法,该方法通过设置较高的恒虚警率来防止漏掉微弱目标,再结合目标的语义特征来抑制传统CFAR检测中的虚警;并以车辆目标的SAR图像为例,在CFAR检测之外融入目标不同部位的结构化语义特征,相比只利用灰度特征的CFAR处理方式,显著减少了SAR图像CFAR检测的虚警数量。

12.9节分析了基于密度特征的SAR图像CFAR检测快速实现方法,通过挖掘海上舰船目标的空域稀疏性,将稀疏的舰船目标归为低密度区域,并将CFAR计算资源集中在SAR图像中低密度区域;该方法通过超像素分割、密度筛选、局部CFAR检测三个主要步骤,在不降低检测性能的前提下实现了快速CFAR检测,在广阔海域重点舰船目标快速检测等方面具有较好应用潜力。

总的来看,距离扩展目标检测研究目前仍处于发展和完善阶段,尚存在诸多问题有待进一步研究[78],例如,检测器在失配环境中的鲁棒性分析[79]、基于目标多散射点特征的检测器设计[80-82]、基于杂波分组的自适应检测器设计[19,51]、利用杂波先验信息的检测器设计[25,83-85]、多秩距离扩展目标检测方法[25-26,45]、运动距离扩展目标检测[22-23,86]、导向矢量失配时的检测器设计[79,87]等。另外,MIMO雷达[84]、检测前跟踪[88]等新领域的距离扩展目标检测问题也逐渐引起了研究者的重视。

随着SAR在多领域的广泛应用，相应的SAR图像目标CFAR检测方法得到快速发展。针对不同SAR图像观测场景，相应的杂波自适应选取方式和杂波模型自适应选择问题仍有待进一步探索[31]。例如，可结合图像分类思想，在检测前先对当前SAR图像进行预分类，然后根据单目标、多目标、对陆观测、对海观测、杂波背景复杂程度、图像对比度等不同图像类别或场景，选取最适合当前SAR图像的杂波样本选取方法和杂波模型，有望进一步提升CFAR检测性能的稳健性。此外，用于辅助CFAR检测的先验知识仍有待进一步挖掘，例如，可尝试针对舰船、车辆、溢油、赤潮等SAR图像目标的形状[89]、多阶统计量[65]等特征，改善传统基于灰度特征的CFAR检测方法在复杂背景下虚警率高的问题。在对时敏目标的快速检测问题上，面向海面目标的CFAR快速检测算法已开展了部分研究[57-58]，但面向复杂背景中地物目标SAR图像，相应的CFAR快速检测算法仍有待进一步深入探讨。

参考文献

[1] 保铮，邢孟道，王彤. 雷达成像技术[M]. 北京：电子工业出版社，2005.

[2] 何友，顾新锋，简涛. 非高斯杂波背景中距离扩展目标的自适应积累检测器[J]. 中国科学，2013，43(4)：488-501.

[3] 简涛. 非高斯背景下雷达距离扩展目标检测方法研究[D]. 烟台：海军航空工程学院，2011.

[4] 顾新锋. 基于先验信息的距离扩展目标检测方法研究[D]. 烟台：海军航空工程学院，2013.

[5] Van Trees H L. Detection，Estimation and Modulation Theory. vol. III：Radar/Sonar Signal Processing and Gaussian Signals in Noise[M]，New York：John Wiley，1971.

[6] Gerard A，Van D S. Detection of a distributed target[J]. IEEE Trans. on AES，1971，7(5)：922-931.

[7] Hughes II P K. A high resolution radar detection strategy[J]. IEEE Trans. on AES，1983，19(5)：663-667.

[8] Gini F，Farina A. Matched subspace CFAR detection of hovering helicopters[J]. IEEE Trans. on AES，1999，35(4)：1293-1305.

[9] Conte E，De Maio A，Ricci. GLRT-based adaptive detection algorithms for range-spread targets[J]. IEEE Trans. on SP，2001，49(7)：1336-1348.

[10] De Maio A，Orlando D. Adaptive radar detection of a subspace signal embedded in subspace structured plus Gaussian interference via invariance[J]. IEEE Trans. on SP，2015，64(8)：2156-2167.

[11] Gerlach K. Spatially distributed targets detection in non-Gaussian Clutter[J]. IEEE Trans. on AES，1999，35(3)：926-934.

[12] 孙以平，陆林根. 距离扩展目标检测的研究[J]. 系统工程与电子技术，1994，8：36-45.

[13] 胡文明，关键，何友. 基于视频积累的距离扩展目标检测[J]. 火控雷达技术，2006，35(4)：16-18.

[14] 顾新锋，简涛，何友. 距离扩展目标的双门限恒虚警检测器及性能分析[J]. 电子与信息学报，2012，34(6)：1318-1323.

[15] Jian Tao，He You，Wang Haipeng，et al. Persymmetric generalized adaptive matched filter for range-spread targets in homogeneous environment[J]. IET Radar Sonar Nav.，2019，13(8)：1234-1241.

[16] Jian Tao，He You，Su Feng，et al. Adaptive detection of range-spread targets without secondary data in multichannel autoregressive process[J]. Digital Signal Process.，2013，23(5)：1686-1694.

[17] Liu Jun，Zhang Zijing，Cao Yunhe，et al. Distributed target detection in subspace interference plus Gaussian noise[J]. Signal Process.，2014，95：88-100.

[18] He You，Jian Tao，Su Feng，et al. Adaptive detection application of covariance matrix estimation for correlated non-Gaussian clutter[J]. IEEE Trans. on AES，2010，46(4)：2108-2117.

[19] 顾新锋，简涛，何友，等. 局部均匀背景中距离扩展目标的GLRT检测器及性能分析[J]. 电子学报，2013，41(12)：2367-2373.

[20] Hao Chengpeng, Orlando D, Foglia G, et al. Persymmetric adaptive detection of distributed targets in partially-homogeneous environment[J]. Digital Signal Process., 2014, 24: 42-51.

[21] Liu Weijian, Xie Wenchong, Liu Jun, et al. Adaptive double subspace signal detection in Gaussian background-Part II: Partially homogeneous environments[J]. IEEE Trans. on SP, 2014, 62(9): 2358-2369.

[22] Petrov N, Le Chevalier F, Yarovoy A G. Detection of range migrating targets in compound-Gaussian clutter[J]. IEEE Trans. on AES, 2017, 54(1): 37-50.

[23] Xu Shuwen, Shui Penglang, Cao Yunhe. Adaptive range-spread maneuvering target detection in compound-Gaussian clutter[J]. Digital Signal Process., 2015, 36: 46-56.

[24] He You, Jian Tao, Su Feng, et al. Novel range-spread target detectors in non-Gaussian clutter[J]. IEEE Trans. on AES, 2010, 46(3): 1312-1328.

[25] Linghong, Dai Fengzhou, Wang Xili. Knowledge-based wideband radar target detection in the heterogeneous environment[J]. Signal Process., 2018, 144: 169-179.

[26] Guan Jian, Zhang Yanfei, Huang Yong(黄勇). Adaptive subspace detection of range-distributed target in compound-Gaussian clutter[J]. Digital Signal Process., 2009, 19(1): 66-78.

[27] Wang Yan, Li Jingwen, Yang Jian. Wide nonlinear chirp scaling algorithm for spaceborne stripmap range sweep SAR imaging[J]. IEEE Trans. Geosci. Remote, 2017, 55(12): 6922-6936.

[28] Wang Xue Qian, Li Gang, Liu Yu, et al. Enhanced 1-bit radar imaging by exploiting two-level block sparsity[J]. IEEE Trans. Geosci. Remote, 2018, 57(2): 1131-1141.

[29] Xiong Kai, Zhao Guanghui, Wang Yingbin, et al. SPB-Net: A deep network for SAR imaging and despeckling with downsampled data[J]. IEEE Trans. Geosci. Remote, 2020, 59(11): 9238-9256.

[30] Novak L M, Owirka G J, Netishen C M. Performance of a high-resolution polarimetric SAR automatic target recognition system[J]. Lincoln Laboratory Journal, 1993, 6(1): 11-24.

[31] 杜兰,王兆成,王燕,等.复杂场景下单通道 SAR 目标检测及鉴别研究进展综述[J].雷达学报,2020,9(1):34-54.

[32] Gao Gui, Liu Li, Zhao Lingjun, et al. An adaptive and fast CFAR algorithm based on automatic censoring for target detection in high-resolution SAR images[J]. IEEE Trans. Geosci. Remote, 2008, 47(6): 1685-1697.

[33] Ai Jiaqiu, Qi Xiangyang, Yu Weidong, et al. A new CFAR ship detection algorithm based on 2-D joint log-normal distribution in SAR images[J]. IEEE Geosci. Remote S., 2010, 7(4): 806-810.

[34] Gao Gui. A parzen-window-kernel-based CFAR algorithm for ship detection in SAR images[J]. IEEE Geosci. Remote S., 2010, 8(3): 557-561.

[35] An Wentao T, Xie Chunhua, Yuan Xinzhe. An improved iterative censoring scheme for CFAR ship detection with SAR imagery[J]. IEEE Trans. Geosci. Remote, 2013, 52(8): 4585-4595.

[36] Hou Biao, Chen Xingzhong, Jiao Licheng. Multilayer CFAR detection of ship targets in very high resolution SAR images[J]. IEEE Geosci. Remote S., 2014, 12(4): 811-815.

[37] Huang Yong, Liu Fang. Detecting cars in VHR SAR images via semantic CFAR algorithm[J]. IEEE Geosci. Remote S., 2016, 13(6): 801-805.

[38] Tao Ding, Anfinsen S N, Brekke C. Robust CFAR detector based on truncated statistics in multiple-target situations[J]. IEEE Trans. Geosci. Remote, 2015, 54(1): 117-134.

[39] Dai Hua, Du Lan, Wang Yan, et al. A modified CFAR algorithm based on object proposals for ship target detection in SAR images[J]. IEEE Geosci. Remote S., 2016, 13(12): 1925-1929.

[40] Wang Chonglei, Bi Fukun, Zhang Weiping, et al. An intensity-space domain CFAR method for ship detection in HR SAR images[J]. IEEE Geosci. Remote S., 2017, 14(4): 529-533.

[41] Ai Jiaqiu, Yang Xuezhi, Song Jitao, et al. An adaptively truncated clutter-statistics-based two-parameter CFAR detector in SAR imagery[J]. IEEE Journal of Oceanic Engineering, 2017, 43(1): 267-279.

[42] Jian Tao, He You, Su Feng, et al. Cascaded detector for range-spread target in non-Gaussian clutter[J]. IEEE Trans. on AES, 2012, 48(2): 1713-1725.

[43] Jian Tao, He You, Su Feng, et al. Adaptive range-spread target detection based on modified generalised likelihood ratio test in non-Gaussian clutter[J]. IET Radar Sonar Nav., 2011, 5(9): 970-977.

[44] Jian Tao, He You, Su Feng, et al. Adaptive detection of sparsely distributed target in non-Gaussian clutter[J]. IET Radar Sonar Nav., 2011, 5(7): 780-787.

[45] 关键，张晓利，简涛，等. 分布式目标的子空间双门限 GLRT CFAR 检测[J]. 电子学报，2012，40(9)：1759-1764.

[46] Jian Tao, He You, Su Feng, et al. Robust detector for range-spread target in non-Gaussian background[J]. J. Syst. Eng. Electron., 2012, 23(3): 355-363.

[47] 顾新锋，简涛，何友，等. 非高斯背景下基于 ODV 的距离扩展目标检测器[J]. 电子学报，2012，40(3)：575-579.

[48] He You, Jian Tao, Su Feng, et al. CFAR assessment of covariance matrix estimators for non-Gaussian clutter [J]. Science China Information Sciences, 2010, 53(11): 2343-2351.

[49] 张彦飞. 非高斯杂波中距离扩展目标的相干检测[D]. 烟台：海军航空工程学院，2008.

[50] Watts S. Radar detection prediction in K-distributed sea clutter and thermal noise[J]. IEEE Trans. on AES, 1987, 23(1): 40-45.

[51] Jian Tao, He You, Su Feng, et al. High resolution radar target adaptive detector and performance assessment[J]. J. Syst. Eng. Electron., 2011, 22(2): 212-218.

[52] 王永良，彭应宁. 空时自适应信号处理[M]. 北京：清华大学出版社，2000.

[53] Tsakalides P, Nikias C L. Robust space-time adaptive processing(STAP) in non-Gaussian clutter environments[J]. IEE Proc.-F, 1999, 146(2): 84-93.

[54] Tsihrintzis G A, Nikias C L. Incoherent receivers in alpha stable impulsive noise[J]. IEEE Trans. on SP, 1995, 43(9): 2225-2229.

[55] Yu Wenyi, Wang Yinghua, Liu Hongwei, et al. Superpixel-based CFAR target detection for high-resolution SAR images[J]. IEEE Geosci. Remote S., 2016, 13(5): 730-734.

[56] Pappas O, Achim A, Bull D. Superpixel-level CFAR detectors for ship detection in SAR imagery[J]. IEEE Geosci. Remote S., 2018, 15(9): 1397-1401.

[57] Li T, Liu Z, Xie R, et al. An improved superpixel-level CFAR detection method for ship targets in high-resolution SAR images[J]. IEEE J-STARS, 2017, 11(1): 184-194.

[58] Wang X Q, Li G, Zhang X P, et al. A fast CFAR algorithm based on density-censoring operation for ship detection in SAR images[J]. IEEE Signal Process. Letters, 2021, 28: 1085-1089.

[59] Cui Zongyong, Quan Hongbin, Cao Zongjie, et al. SAR target CFAR detection via GPU parallel operation[J]. IEEE J-STARS, 2018, 11(12): 4884-4894.

[60] Zefreh R G, Taban M R, Naghsh M M, et al. Robust CFAR detector based on censored harmonic averaging in heterogeneous clutter[J]. IEEE Trans. on AES, 2020, 57(3): 1956-1963.

[61] Ai Jiaqiu, Mao Yuxiang, Luo Qiwu, et al. Robust CFAR ship detector based on bilateral-trimmed-statistics of complex ocean scenes in SAR imagery: a closed-form solution [J]. IEEE Trans. on AES, 2021, 57 (3): 1872-1890.

[62] Cui Yi, Zhou Guangyi, Yang Jian, et al. On the iterative censoring for target detection in SAR images[J]. IEEE Geosci. Remote S., 2011, 8(4): 641-645.

[63] Cui Zongyong, Hou Zesheng, Yang Hongzhi, et al. A CFAR target-detection method based on superpixel statistical modeling[J]. IEEE Geosci. Remote S., 2020, 18(9): 1605-1609.

[64] Li Tao, Peng Dongliang, Chen Zhikun, et al. Superpixel-level CFAR detector based on truncated gamma distribution for SAR images[J]. IEEE Geosci. Remote S., 2020, 18(8): 1421-1425.

[65] Wang Xueqian, He You(何友), Li Gang, et al. Adaptive superpixel segmentation of marine SAR images by aggregating fisher vectors[J]. IEEE J-STARS, 2021, 14: 2058-2069.

[66] Yin J J, Wang Tao, Du Yanlei, et al. SLIC superpixel segmentation for polarimetric SAR images[J]. IEEE Trans. Geosci. Remote, 2021, 60(1): 1-17.

[67] Achanta R, Shaji A, Smith K, et al. SLIC superpixels compared to state-of-the-art superpixel methods[J]. IEEE Trans. Pattern Anal., 2012, 34(11): 2274-2282.

[68] Qin Xianxiang, Zhou Shilin, Zou Huanxin, et al. A CFAR detection algorithm for generalized gamma distributed background in high-resolution SAR images[J]. IEEE Geosci. Remote S., 2012, 10(4): 806-810.

[69] Liu Tao, Yang Ziyuan, Yang Jian, et al. CFAR ship detection methods using compact polarimetric SAR in a K-Wishart distribution[J]. IEEE J-STARS, 2019, 12(10): 3737-3745.

[70] Frery A C, Muller H J, Yanasse C C F, et al. A model for extremely heterogeneous clutter[J]. IEEE Trans. Geosci. Remote, 1997, 35(3): 648-659.

[71] Tao D, Doulgeris A P, Brekke C. A segmentation-based CFAR detection algorithm using truncated statistics[J]. IEEE Trans. Geosci. Remote, 2016, 54(5): 2887-2898.

[72] Ai Jiaqiu, Luo Qiwu, Yang Xuezhi, et al. Outliers-robust CFAR detector of Gaussian clutter based on the truncated-maximum-likelihood-estimator in SAR imagery[J]. IEEE Trans. Intell. Transp., 2019, 21(5): 2039-2049.

[73] Liu T, Yang Z Y, Marino A, et al. Robust CFAR detector based on truncated statistics for polarimetric synthetic aperture radar[J]. IEEE Trans. Geosci. Remote, 2020, 58(9): 6731-6747.

[74] Li H C, Hong W, Wu Y R, et al. On the empirical-statistical modeling of SAR images with generalized gamma distribution[J]. IEEE J-STSP, 2011, 5(3): 386-397.

[75] Liao P S, Chen T S, Chung P C. A fast algorithm for multilevel thresholding[J]. Journal of Information Science and Engineering, 2001, 17(5): 713-727.

[76] Rodriguez A, Laio A. Clustering by fast search and find of density peaks[J]. Science, 2014, 344(6191): 1492-1496.

[77] Wang Y Y, Wang C, Zhang H, et al. A SAR dataset of ship detection for deep learning under complex backgrounds[J]. Remote Sensing, 2019, 11(7): 765.

[78] 简涛，何友，苏峰，等. 高距离分辨率雷达目标检测研究现状与进展[J]. 宇航学报，2010，31(12)：2623-2628.

[79] He Y, Jian T, Su F, et al. Two adaptive detectors for range-spread targets in non-Gaussian clutter[J]. Science China: Information Sciences, 2011, 54(2): 386-395.

[80] 顾新锋，简涛，何友，等. 非高斯杂波背景中的两个距离扩展目标检测器[J]. 宇航学报，2012，33(5)：648-654.

[81] 顾新锋，何友，简涛，等. 基于修正熵的距离扩展目标检测器[J]. 系统工程与电子技术，2012，34(6)：1136-1139.

[82] 胡文明，关键，何友. 基于二维积累的雷达分布式目标检测新方法[J]. 系统工程与电子技术，2006，28(9)：1335-1337.

[83] 顾新锋，简涛，何友，等. 复合高斯杂波中距离扩展目标的迭代近似GLRT检测器[J]. 航空学报，2013，34(5)：1140-1150.

[84] Gao Y C, Li H B, Himed B. Knowledge-aided range-spread target detection for distributed MIMO radar in nonhomogeneous environments[J]. IEEE Trans. on SP, 2016, 65(3): 617-627.

[85] Jian T, He Y, Liao G S, et al. Adaptive persymmetric detector of generalized likelihood ratio test in homogeneous environment[J]. IET Signal Process., 2016, 10(2): 91-99.

[86] Yang X L, Wen G J, Ma C H, et al. CFAR detection of moving range-spread target in white Gaussian noise using waveform contrast[J]. IEEE Geosci. Remote S., 2016, 13(2): 282-286.

[87] 顾新锋，简涛，何友，等. 非高斯杂波背景中距离扩展目标的盲积累检测器[J]. 航空学报，2012，33(12)：2261-2267.

[88] Jiang H C, Yi W, Cui G L, et al. Track-before-detect strategies for range distributed target detection in compound-Gaussian clutter[J]. Signal Process., 2016, 120: 462-467.

[89] An Q Z, Pan Z X, Liu L, et al. DRBox-v2: An improved detector with rotatable boxes for target detection in SAR images[J]. IEEE Trans. Geosci. Remote, 2019, 57(11): 8333-8349.

第13章 多传感器分布式CFAR处理

CHAPTER 13

13.1 引言

现代雷达面临着“四抗”(抗干扰、抗摧毁、抗隐身、抗低空突防)的问题。在空间上分布的多个传感器系统可以提高系统的生存能力和抗干扰能力(频率分集),增加覆盖区域和监视目标数,提供更高的总信杂噪比,并且可以提高系统的反应速度以及在单个传感器故障情况下的可靠性[1]。因此,利用多传感器的分布式信号处理格外受到重视。多传感器系统最初仅应用在军事指挥、控制和通信中,现在它的应用已经拓展到了众多领域中,如气象预报、医疗诊断、组织管理决策及雷达探测中[2-3]。

在经典的多传感器系统中,所有的局部传感器将原始观测数据全部直接传送给中心处理器,由中心处理器完成目标检测跟踪等任务。这种处理方式需要局部传感器和中心处理器间具有很大的通信带宽。在多传感器分布式处理系统中,每个传感器对观测先进行一定的预处理,然后将压缩的数据传送给其他的传感器,或者直接汇总到被称为融合中心的中心处理器。对数据的压缩性预处理降低了对通信带宽的要求。此外,多传感器分布式结构可以降低对单个传感器的性能要求。分散的信号处理方式可以增加计算容量,在利用高速通信网的条件下可以完成复杂的算法。

对分布式检测的理论研究兴起于20世纪80年代。1981年,Tenney和Sandell在文献[4]中将经典的贝叶斯理论拓展到并行结构的分布式传感器二元假设检验。文献[5-6]则分析了并行、串行等多种网络结构的分布式检测性能。1989年,Barkat和Varshney在文献[7]中将CA-CFAR引入基于局部二元判决的分布式检测中,并在文献[8]中拓展到多种网络结构中。文献[9-13]是基于局部二元判决,研究了CA、OS等局部CFAR处理,以及非均匀、非高斯环境等问题。13.2节将介绍基于局部二元判决的分布式CFAR检测的基本理论。

基于局部二元判决的分布式检测将局部观测信息压缩到了最低限度,这是由于通信带宽的限制。随着雷达和信息传输技术的快速进步,局部传感器可以传输更多的信息。文献[14]提出局部做OS-CFAR处理,向融合中心传送检测单元采样和参考单元采样的OS处理结果。关键于2000年完成的博士学位论文和有关论文[15-18]则重点研究了基于局部检测统计量的分布式CFAR检测,文献[19-25]进一步做了这方面的研究,13.3节介绍基于局部检测统计量的分布式CFAR检测。

随着通信带宽的进一步增加,向融合中心传输所有局部观测信息(包括局部观测的检测单元数据和参考单元数据)成为可能。13.4节针对多传感器分布式CFAR检测问题,以分布式MIMO雷达体制为研究对象,利用所有局部观测信息,从似然比检验出发导出了自适应匹配滤波器(Adaptive Matched Filter,AMF)检测器[26]。分析表明,该AMF检测器本质上仍属于13.3节中讨论的基于局部检测统计量的分布式 CFAR 检测的范畴。

13.2 基于局部二元判决的分布式 CFAR 检测

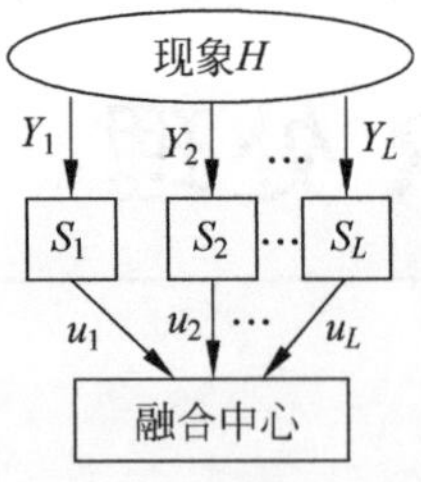

图 13.1 并行结构

本节介绍各局部传感器进行二元 CFAR 判决的分布式 CFAR 检测。考虑由一个融合中心和 L 个分布式 CFAR 检测器组成的系统，其结构图如图 13.1 所示。

假定第 i 个检测器的参考滑窗的长度为 $N_i(i=1,2,\cdots,L)$，设目标是慢起伏的 Swerling Ⅰ型目标，均匀背景噪声是具有未知功率水平的高斯噪声。各传感器虚警概率和检测概率分别为 $P_{\mathrm{fa},i}$ 和 $P_{\mathrm{d},i}(i=1,2,\cdots,L)$。假设在检测单元中有目标时，各局部检测器具有相同的目标信号与噪声功率比，即信噪比 $\lambda_1=\lambda_2=\lambda_L=\lambda$。对于各局部信噪比不相等的情况可以用类似的方法推广。对检测器 $i(i=1,2,\cdots,L)$，局部虚警概率表示为

$$P_{\mathrm{fa},i}=\int_0^{\infty}\Pr[y_i>T_iz_i\mid z_i,\mathrm{H}_0]f(z_i)\mathrm{d}z_i \tag{13.1}$$

式中，y_i 是检测单元采样，T_i 是标称化因子，z_i 为局部 CFAR 检测器 i 的噪声功率估计，$f(z_i)$ 表示 z_i 的概率密度函数。类似地，局部检测器 i 的检测概率可表示为

$$P_{\mathrm{d},i}=\int_0^{\infty}\Pr[y_i>T_iz_i\mid z_i,\mathrm{H}_1]f(z_i)\mathrm{d}z_i \tag{13.2}$$

每个局部 CFAR 检测器都把它的判决 $u_i(i=1,2,\cdots,L)$ 传送到信息融合中心。然后，中心基于这些局部判决做出全局判决 u_0。全局虚警概率和检测概率分别表示为 P_{FA} 和 P_{D}。对于固定融合规则的系统，其目标是融合中心在保持 P_{FA} 恒定条件下使 P_{D} 极大化，可应用拉格朗日乘子法来求解。下面研究采用 CA-CFAR 和 OS-CFAR 检测器作为局部检测器，“与”和“或”作为融合规则的分布式 CFAR 检测。简单起见，假设所有局部检测器都使用相同的 CFAR 处理。当然，应用不同局部 CFAR 处理的分布式 CFAR 检测也可用类似的方法设计。

13.2.1 分布式 CA-CFAR 检测

在均匀背景下，对具有 CA-CFAR 局部检测的分布式检测系统，当给定融合规则时，可通过最优地设置局部检测器的标称化因子 $T_i(i=1,2,\cdots,L)$ 来极大化全局检测概率。在所有局部节点具有相同信噪比的条件下，应用拉格朗日乘子方法得到的目标函数为[2]

$$J(T_1,T_2,\cdots,T_L)=P_{\mathrm{D}}(T_1,T_2,\cdots,T_L)+\gamma[P_{\mathrm{FA}}(T_1,T_2,\cdots,T_L)-\alpha] \tag{13.3}$$

式中，J 是需要极大化的目标函数，γ 为拉格朗日乘子，α 是融合中心的虚警概率设计值。为了确定 P_{D} 和 P_{FA}，需要给出局部检测器 $P_{\mathrm{d},i}$ 和 $P_{\mathrm{fa},i}$ 表达式。在均匀高斯噪声背景中，对 CA-CFAR

$$P_{\mathrm{d},i}=\frac{1}{[1+T_i/(1+\lambda)]^{N_i}},\quad P_{\mathrm{fa},i}=\frac{1}{(1+T_i)^{N_i}},\quad i=1,2,\cdots,L \tag{13.4}$$

在优化过程中，令 $J(T_1,T_2,\cdots,T_L)$ 关于 $T_i(i=1,2,\cdots,L)$ 的偏导数等于零，并根据约束方程 $P_{\mathrm{FA}}(T_1,T_2,\cdots,T_L)=\alpha$ 产生具有 $(L+1)$ 个未知量的 $(L+1)$ 个非线性方程，求解这些方程可得产生最高全局检测概率 P_{D} 的 $T_i(i=1,2,\cdots,L)$ 最优集，其中 $(L+1)$ 个未知量是 $T_i(i=1,2,\cdots,L)$ 和拉格朗日乘子 γ。下面给出采用“与”和“或”融合规则时的结果。

1. “与”融合规则

对“与”融合规则有

$$P_{\mathrm{D}}=\prod_{i=1}^{L} P_{\mathrm{d},i}, P_{\mathrm{FA}}=\prod_{i=1}^{L} P_{\mathrm{fa},i} \tag{13.5}$$

把式(13.4)代入式(13.5)，然后再把式(13.5)代入式(13.3)，则目标函数变为

$$J(T_1,T_2,\cdots,T_L)=\prod_{i=1}^{L}\frac{(1+\lambda)^{N_i}}{(1+\lambda+T_i)^{N_i}}+\gamma\left[\prod_{i=1}^{L}\frac{1}{(1+T_i)^{N_i}}-\alpha\right] \tag{13.6}$$

对 $J(T_1,T_2,\cdots,T_L)$ 求关于 $T_j(j=1,2,\cdots,L)$ 的偏导数，并令其为零，然后结合约束方程 $P_{\mathrm{FA}}=\prod_{i=1}^{L}\frac{1}{(1+T_i)^{N_i}}=\alpha$，即可求得 $T_i(i=1,2,\cdots,L)$。

在 $L=2$ 的特殊情况下，上面方程组的解为

$$T_1=T_2=-1+\alpha^{-\frac{1}{N_1+N_2}} \tag{13.7}$$

2. “或”融合规则

在“或”融合规则情况下，有

$$P_{\mathrm{M}}=\prod_{i=1}^{L} P_{\mathrm{m},i},\quad P_{\mathrm{FA}}=1-\prod_{i=1}^{L}(1-P_{\mathrm{fa},i}) \tag{13.8}$$

其中，$P_{\mathrm{m},i}$ 和 P_{M} 分别表示局部和全局的漏警概率。于是，目标函数可表示为

$$J(T_1,T_2,\cdots,T_N)=1-\prod_{i=1}^{L}\left[1-\frac{(1+\lambda)^{N_i}}{(1+\lambda+T_i)^{N_i}}\right]+\gamma\left[1-\prod_{i=1}^{L}\left(1-\frac{1}{(1+T_i)^{N_i}}\right)-\alpha\right] \tag{13.9}$$

对目标函数 $J(T_1,T_2,\cdots,T_L)$ 求关于 $T_j(j=1,2,\cdots,L)$ 的偏导数，并令其等于零，然后结合约束方程 $P_{\mathrm{FA}}=1-\prod_{i=1}^{L}\left[1-\frac{1}{(1+T_i)^{N_i}}\right]=\alpha$，即可求得 $T_j(j=1,2,\cdots,L)$。

对两个传感器的情况，不像“与”融合规则那样，得不到 T_1 和 T_2 的显式表达式，需要通过数值求解来确定。

13.2.2 分布式 OS-CFAR 检测

现在考虑局部为 OS-CFAR 检测器的分布式检测。在均匀背景环境下，对于给定的融合规则，全局检测概率的极大化是通过同时最优化各局部检测器的参数实现的。对局部检测器 i 来说，这些参数是指标称化因子 T_i 和有序值 k_i（即用第 k_i 个有序值作为检测统计量）。关于系统的最优化，仍然采用13.2.1节使用的拉格朗日乘子法求解。对每个局部传感器都假定有相同的信噪比 λ 和滑窗尺寸 $N_i(i=1,2,\cdots,L)$，于是，目标函数可表示为

$$J[(T_1,k_1),\cdots,(T_L,k_L)]=P_{\mathrm{D}}[(T_1,k_1),\cdots,(T_L,k_L)]+\gamma\{P_{\mathrm{FA}}[(T_1,k_1),\cdots,(T_L,k_L)]-\alpha\} \tag{13.10}$$

其中，α 是给定的融合中心虚警概率设计值，γ 是拉格朗日乘子。局部 OS-CFAR 检测器 $P_{\mathrm{fa},i}$ 和 $P_{\mathrm{d},i}$ 可分别表示为

$$P_{\mathrm{d},i}=\prod_{\ell=0}^{k_i-1}\frac{N_i-\ell}{(N_i-\ell)+\dfrac{T_i}{1+\lambda_i}},\quad P_{\mathrm{fa},i}=\prod_{\ell=0}^{k_i-1}\frac{N_i-\ell}{N_i-\ell+T_i} \tag{13.11}$$

在优化过程中，首先对 $k_i(i=1,2,\cdots,L)$ 选择一组特殊值，并令目标函数关于 $T_i(j=1,2,\cdots,L)$ 的偏导数等于零，然后结合约束方程 $P_{FA}[(T_1,k_1),\cdots,(T_L,k_L)]=\alpha$，产生具有 $(L+1)$ 个未知量的 $(L+1)$ 个非线性方程，从而，对于给定的一组 $k_i(i=1,2,\cdots,L)$ 值，求得最优 $T_i(i=1,2,\cdots,L)$ 集合，并同时计算对应的全局检测概率。对每组可能的 k_i 的取值重复这一过程，提供最大全局检测概率 P_D 的 k_i 取值集合和对应的最优 $T_i(i=1,2,\cdots,L)$ 值为局部检测器的最优解。

对"与"规则，目标函数 J 表示为

$$J=\prod_{i=1}^{L}\left[\prod_{\ell=0}^{k_i-1}\frac{N_i-\ell}{N_i-\ell+T_i(1+\lambda)}\right]+\gamma\left\{\prod_{i=1}^{L}\left[\prod_{\ell=0}^{k_i-1}\frac{N_i-\ell}{N_i-\ell+T_i}\right]-\alpha\right\} \tag{13.12}$$

对"或"规则，目标函数 J 表示为

$$J=1-\prod_{i=1}^{L}\left[1-\prod_{\ell=0}^{k_i-1}\frac{N_i-\ell}{N_i-\ell+T_i/(1+\lambda)}\right]+\gamma\left\{1-\prod_{i=1}^{L}\left[1-\prod_{\ell=0}^{k_i-1}\frac{N_i-\ell}{N_i-\ell+T_i}\right]-\alpha\right\} \tag{13.13}$$

令目标函数的偏导数等于零，在虚警概率的约束下，求解关于阈值标称化因子和拉格朗日乘子的方程，就可以获得分布 OS-CFAR 检测系统的参数。

13.2.3 分布式 CFAR 检测性能分析

考虑由两个局部检测器组成的分布式 CFAR 检测系统，系统的融合规则采用"或"或"与"规则。在均匀背景中，局部检测器的参考滑窗尺寸假定为 $N_1=24, N_2=16$。对 P_{FA} 的设计值是 10^{-4}，对"或"和"与"规则，目标是极大化 P_D。图 13.2 描述了 P_D 随信噪比 λ 变化的系统性能。像预料的那样，在均匀背景中，局部 CA-CFAR 的性能优于 OS-CFAR，"或"规则优于"与"规则。

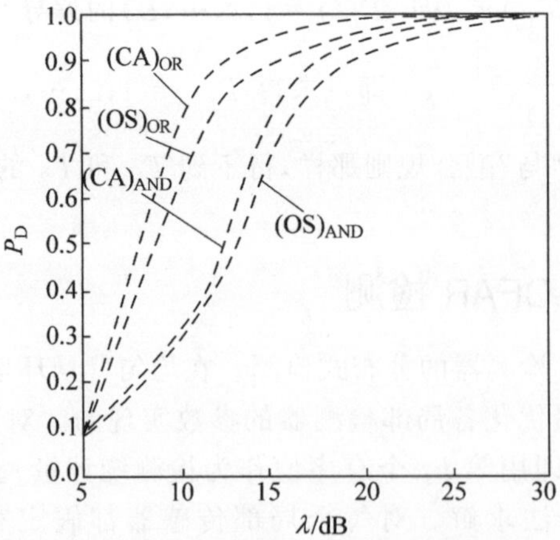

图 13.2 均匀背景噪声中分布式 CFAR 检测系统的性能($N_1=24, N_2=16$)

13.3 基于局部检测统计量的分布式 CFAR 检测

早期关于分布式 CFAR 检测的文献大多研究基于局部二元量化的分布式 CFAR 检测，如 13.2 节所述，还有文献研究了基于局部多位量化的分布式 CFAR 检测，这些都可以归结为基于局部量化的分布式 CFAR 检测。文献[15]研究了基于局部多位量化的分布式 CFAR 检测在 Neyman-Pearson 意义上的最优问题，证明上限性能是由局部检测统计量(Local Test Statistic, LTS)的似然比检验实现的。局部检测统计量是对局部原始观测数据的压缩处理结果，并且可以直接用于具有上限性能的似然比检验。这些因素促成了基于局部检测统计量的分布式CFAR检测的思路。

本节将在高斯杂波背景、平方律检波和Swerling Ⅰ型目标及各单元采样(检测单元和参考单元)相互统计独立的条件下，讨论文献[15]提出的两类局部检测统计量，即R类(Ratio)和S类(Substract)局部检测统计量，这两类局部检测统计量的形式简单，因此仍具有分布式处理的优势。

13.3.1 基于R类局部检测统计量的分布式CFAR检测

1. R类局部检测统计量的统计特性

R类局部检测统计量为

$$S_i=\frac{X_{i0}}{Z_i},\quad i=1,2,\cdots,L \tag{13.14}$$

这里只考虑图13.1所示的并行网络结构。每个局部处理器将局部预处理产生的局部检测统计量$S_i(i=1,2,\cdots,L)$传送给融合中心，融合形成全局检测统计量G，然后进行如下的全局判决，其中T是判决门限

$$H_1: G\geqslant T,\quad H_0: G<T \tag{13.15}$$

为使局部检测统计量是CFAR的，Z_i应由参考单元采样根据CFAR算法产生。这里讨论利用CA-CFAR和OS-CFAR算法产生Z_i的R类局部检测统计量，则相应的R类局部检测统计量分别用CA-R和OS-R表示。CA-R的Z_i为$Z_i=\sum_{j=1}^{N_i}X_{ij}$；OS-R的$Z_i$为$Z_i=X_{i(k_i)}$，$X_{i(k_i)}$是对$X_{ij}(j=1,2,\cdots,N_i)$进行由小到大排序得到的第$k_i$个采样。$k_i$是第$i$个局部处理器的OS算法序值，也被称为单站序值。

在非均匀杂波背景中，假设参考单元只包含两种强度的采样。也就是说，在多目标环境中，各干扰目标强度相同；在杂波边缘环境中，强杂波采样的强度也均相同。并且仍然假设检测单元与参考单元之间以及参考单元采样之间是统计独立的。

1) CA-R的统计特性

对于式(13.14)定义的R类局部检测统计量，其概率分布函数为

$$F_{S_i}(y_i)=\Pr[S_i<y_i]=1-\Pr\left[\frac{X_{i0}}{Z_i}\geqslant y_i\right]=1-\Pr[X_{i0}\geqslant y_iZ_i] \tag{13.16}$$

其中，$\Pr[X_{i0}\geqslant y_iZ_i]=\Phi_{Z_i}(u)\Big|_{u=\frac{y_i}{\mu_i(1+\lambda_i)}}$，$\Phi_{Z_i}(u)$是$Z_i$的特征函数。对于CA-R，杂波功率水平估计$Z_i$由CA方法产生。CA-R的$S_i$的概率密度函数为

$$f_{S_i}(y_i)=\frac{m_i(1+\beta_i)/(1+\lambda_i)}{\left(1+y_i\dfrac{1+\beta_i}{1+\lambda_i}\right)^{m_i+1}\left(1+\dfrac{y_i}{1+\lambda_i}\right)^{N_i-m_i}}+\frac{(N_i-m_i)/(1+\lambda_i)}{\left(1+y_i\dfrac{1+\beta_i}{1+\lambda_i}\right)^{m_i}\left(1+\dfrac{y_i}{1+\lambda_i}\right)^{N_i-m_i+1}} \tag{13.17}$$

在多目标环境中，m_i代表干扰目标数；在杂波边缘环境中，m_i代表强杂波单元数。β_i代表强干扰与噪声的平均功率之比。在均匀杂波背景中，S_i的概率密度函数可以通过令$\beta_i=0$或$m_i=0$得到。由式(13.17)可见，S_i的统计特性与μ_i无关，其中μ_i表示第i个局部处理器的背景噪声功率水平。因此，CA-R局部检测统计量是CFAR的。

2）OS-R 的统计特性

杂波功率水平估计 Z_i 由 OS 方法产生。经过推导，求得 Z_i 的特征函数为

$$\Phi_{Z_i}(u)=H_i\left[N_i,k_i;m_i,\beta_i;\frac{-1}{1+u\mu_i/a_i}\right] \tag{13.18}$$

其中

$$H_i[N_i,k_i;m_i,\beta_i;x_i]=\sum_{i_1=k_i}^{N_i}\sum_{i_2=\max(0,i_1-m_i)}^{\min(i_1,N_i-m_i)}\binom{N_i-m_i}{i_2}\binom{m_i}{i_1-i_2}\sum_{j_1=0}^{i_2}\binom{i_2}{j_1}(-1)^{j_1}\sum_{j_2=0}^{i_1-i_2}\binom{i_1-i_2}{j_2}(-1)^{j_2}x_i$$

$$a_i=j_1+N_i-m_i-i_2+(j_2+m_i-i_1+i_2)/(1+\beta_i)$$

于是，可得 S_i 的概率密度函数为

$$f_{S_i}(y_i)=H_i[N_i,k_i;m_i,\beta_i;q_i(y_i)] \tag{13.19}$$

其中，$q_i(y_i)=-\frac{a_i(1+\lambda_i)}{[y_i+a_i(1+\lambda_i)]^2}$。在多目标环境中，$m_i$ 代表干扰目标数；在杂波边缘环境中，m_i 代表强杂波单元数。在均匀杂波背景中，S_i 的概率密度函数可以通过令 $\beta_i=0$ 或 $m_i=0$ 得到。由式(13.19)可见 S_i 的统计特性与 μ_i 无关。因此，OS-R 是 CFAR 的。

2. R 类局部检测统计量在求和融合条件下的性能分析

1）检测性能的数学模型

下面研究求和融合(用 SUM 表示)，即全局检测统计量为 $G=\sum_{i=1}^{L}S_i$。若局部检测统计量为 CA-R，则用 CA-R-SUM 表示这种方案。同理，用 OS-R-SUM 表示对 OS-R 局部检测统计量进行求和融合的方案。

在假设 $S_i(i=1,2,\cdots,L)$ 之间统计独立的条件下，SUM 融合的全局检测统计量 G 大于 T 的概率为

$$\Pr[G>T]=\int\limits_{\sum_{i=1}^{L}y_i>T}\cdots\int\prod_{i=1}^{L}f_{S_i}(y_i)\mathrm{d}y_1\mathrm{d}y_2\cdots\mathrm{d}y_L \tag{13.20}$$

对于 CA-R-SUM，将式(13.17)代入式(13.20)可得全局检测统计量 G 大于 T 的概率。对于 OS-R-SUM，$f_{S_i}(y_i)$ 由式(13.19)给出。在 $L=2$ 时，OS-R-SUM 的全局检测统计量 G 大于 T 的概率为

$$\Pr[G>T]=H\left[\frac{b_1+b_2}{T+b_1+b_2}+\frac{b_1b_2\ln[(T+b_1)(T+b_2)/(b_1b_2)]}{(T+b_1+b_2)^2}\right] \tag{13.21}$$

其中，$H(x)=H_1[N_1,k_1;m_1,\beta_1;H_2(N_2,k_2;m_2,\beta_2;x)]$，$b_1=a_1(1+\lambda_1)$，$b_2=a_2(1+\lambda_2)$。

2）性能分析

下面将 CA-R-SUM 和 OS-R-SUM 与集中式 OS-CFAR 检测(用 COS 表示)在均匀背景、多目标和杂波边缘环境中进行对比。设传感器数为 2，局部处理器参数 $N_1=11$，$k_1=8$，$N_2=13$，$k_2=9$，融合中心的全局虚警概率设计值为 $P_{FA}=10^{-6}$。定义 $\rho=\mu_2/\mu_1$，μ_i 表示第 i 个局部处理器的背景杂波功率水平。

图 13.3 给出了 COS 在 ρ 值匹配情况下的检测概率曲线。由图 13.3 可见，基于 CA-R 和 OS-R 局

部检测统计量的分布式 CFAR 检测均具有不受 ρ 值影响的良好性质。CA-R-SUM 在多目标环境中检测概率的严重下降是由于局部 CA-CFAR 算法造成的，而不能归咎于 R 类局部检测统计量。

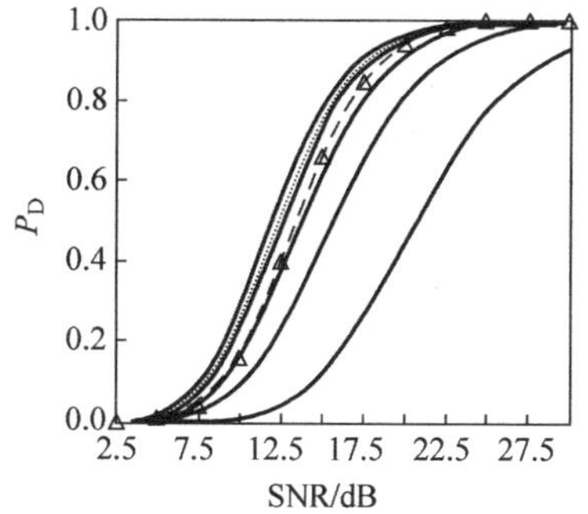

(a) 均匀背景

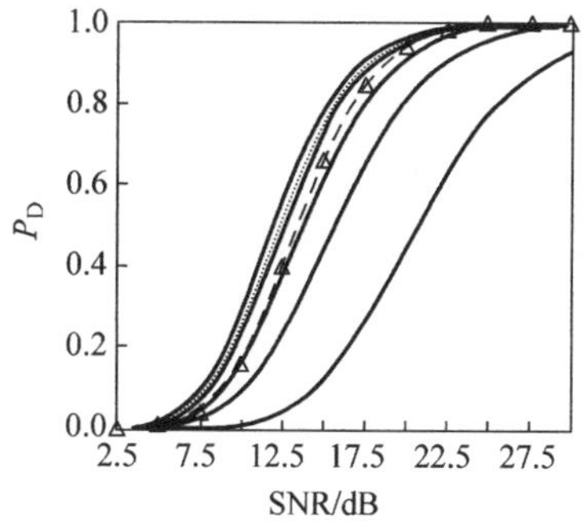

(b) 多目标环境

图 13.3　COS、CA-R-SUM 和 OS-R-SUM 的检测概率

($\lambda_1=\lambda_2=$SNR，$m_1=m_2=2$，$\beta_1=\beta_2=$SNR；△OS-R-SUM，----CA-R-SUM；其他为 COS，从左至右依次为 $\rho=1,0.5,0.25,10,0.01$)

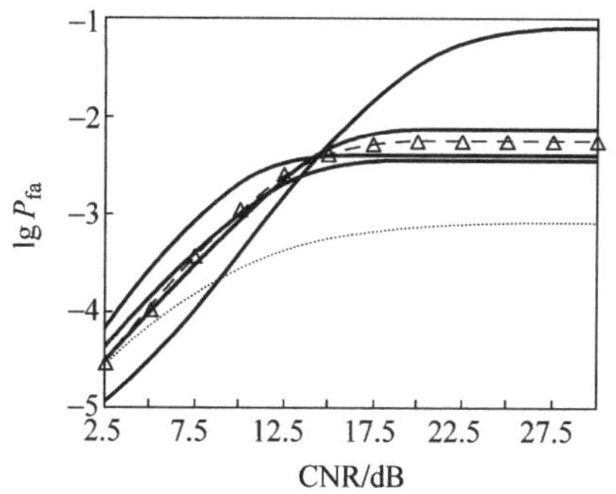

图 13.4　杂波边缘环境中的虚警概率

($m_1=6$，$m_2=7$，$\beta_1=\beta_2=$CNR；△OS-R-SUM，----CA-R-SUM；其他为 COS，在 CNR=22.5(dB) 处从上至下依次为 $\rho=0.01,10,1,0.25$)

在杂波边缘环境中，对于一个固定的杂噪比值，由图 13.4 可见 COS 的虚警概率随 ρ 的变化而明显地变化。总的看来，COS 只在 $|\log\rho|$ 较低时有效。然而，CA-R-SUM 和 OS-R-SUM 的虚警概率不受 ρ 值的任何影响。

综合图 13.3 和图 13.4 可知，在 $|\log\rho|$ 较低时，OS-R-SUM 的检测概率和虚警概率与 COS 均很接近。这说明 OS-R-SUM 的全局检测统计量与 COS 的全局检测统计量具有相近的统计特性。因此，OS-R 局部预处理和 SUM 融合在检测性能上是有效的。此外，相对于 COS，OS-R 的简单形式使局部处理器向融合中心传送的数据量大大减少，即降低了对通信带宽的要求。而且分布式 OS 处理也使 OS 处理的时间大大缩减。因此，无论从检测性能、对通信带宽的要求和算法所需时间上，OS-R-SUM 都是可取的。

3. R 类局部检测统计量在最优融合条件下的性能分析

求和融合方案尽管具有较好的性能，但是不具有某种意义上的最优。为了了解基于 R 类局部检测统计量的分布式检测的上限性能，下面介绍对 R 类局部检测统计量的最优融合和性能。

1) R 类局部检测统计量的最优融合

基于 R 类局部检测统计量的分布式 CFAR 检测的融合中心的最优全局检测统计量为如下的似然比(Likelihood Ratio，LR)

$$S=\frac{f_1(\boldsymbol{s})}{f_0(\boldsymbol{s})}=\frac{f(s_1,\cdots,s_{L-1},s_L\mid \mathrm{H}_1)}{f(s_1,\cdots,s_{L-1},s_L\mid \mathrm{H}_0)}\tag{13.22}$$

即对 R 类局部检测统计量的最优融合是似然比，称为似然比(LR)融合。在假设 $S_i(i=1,2,\cdots,L)$之间

统计独立的条件下，若采用对数似然比，则全局检测统计量为

$$S=\sum_{i=1}^{L}\Psi_i \tag{13.23}$$

其中，$\Psi_i=\log\left[\dfrac{f_{S_i}^{(\mathrm{h})}(y_i\mid \mathrm{H}_1)}{f_{S_i}^{(\mathrm{h})}(y_i\mid \mathrm{H}_0)}\right]$称为局部对数似然比，$f_{S_i}^{(\mathrm{h})}(y_i\mid \mathrm{H}_a)(a=0,1)$的上标“h”表示均匀背景，即$f_{S_i}^{(\mathrm{h})}(y_i\mid \mathrm{H}_a)$为均匀背景中的概率密度函数。

融合中心的全局判决为

$$\mathrm{H}_1: S\geqslant T,\quad \mathrm{H}_0: S<T \tag{13.24}$$

采用 Neyman-Pearson 准则，则 T 应满足下面的限制条件

$$\int_T^{\infty} f_S(y\mid \mathrm{H}_0)\mathrm{d}y=P_{\mathrm{FA}} \tag{13.25}$$

对于 OS-R，由式(13.19)可得

$$f_{S_i}^{(\mathrm{h})}(y_i\mid \mathrm{H}_1)=\sum_{i=k_i}^{N_i}\binom{N_i}{i}\sum_{j=0}^{i}\left\{\binom{i}{j}(-1)^{j+1}\frac{(j+N_i-i)(1+\lambda_i)}{[y_i+(j+N_i-i)(1+\lambda_i)]^2}\right\} \tag{13.26a}$$

$$f_{S_i}^{(\mathrm{h})}(y_i\mid \mathrm{H}_0)=\sum_{i=k_i}^{N_i}\binom{N_i}{i}\sum_{j=0}^{i}\left[\binom{i}{j}(-1)^{j+1}\frac{j+N_i-i}{(y_i+j+N_i-i)^2}\right] \tag{13.26b}$$

相应的检测方案用 OS-R-LR 表示。应该指出，总的检测阈值 T 不仅是局部处理器参数 N_i 和 $k_i(i=1,2,\cdots,L)$的函数，还取决于 LR 融合中实际的局部观测信噪比 $\lambda_i(i=1,2,\cdots,L)$。当实际的局部观测信噪比未知时，需用假设的局部观测信噪比 $\hat{\lambda}_i(i=1,2,\cdots,L)$代替实际值。

2）性能分析

下面分析 OS-R-LR 在均匀和非均匀杂波背景中的性能。考虑两传感器($L=2$)的并行结构，局部处理器参数为 $N_1=11,k_1=8,N_2=13,k_2=9$，融合中心总的虚警概率设计值为 $P_{\mathrm{FA}}=10^{-6}$。非均匀杂波背景包括多目标环境和杂波边缘环境。

对于 OS-R-LR，图 13.5～图 13.7 给出了检测性能曲线。对于检测概率曲线，横坐标 $\mathrm{SNR}=10\lg[(\lambda_1+\lambda_2)/2]$。在一个确定的信噪比值处，“最大”和“最小”分别是指这一组曲线在该信噪比处检测概率的最大和最小值。对于虚警概率曲线，横坐标 $\mathrm{CNR}=10\lg[(\beta_1+\beta_2)/2]$。设实际的局部观测信噪比之比 $\lambda_1/\lambda_2=r$，融合准则中假设的局部观测信噪比之比为 r_{d}。当 $r_{\mathrm{d}}=r$ 时称为“匹配”，否则称为“失配”。由于似然比函数与目标的统计特性有关，因此还要考虑目标模型是否匹配。这里假设最优融合准则与目标模型是匹配的。下面讨论中提到的匹配与否都是指 r_{d} 与 r 是否匹配。

首先由图 13.5 和图 13.6 可见，当 $r_{\mathrm{d}}=r$，即匹配时，“最大”和“最小”曲线的间距非常小，特别是在 $r=1$ 时几乎不能区分。当失配时，局部观测信噪比的假设值会对检测性能有较大影响。这说明在匹配条件下，局部观测信噪比的准确信息对均匀背景和多目标环境中检测性能的影响很微弱，检测性能主要取决于局部观测信噪比之间的相对比例关系 r 的先验信息的准确程度。

在杂波边缘环境中，定义 $r_{\mathrm{d}}=\hat{\beta}_1/\hat{\beta}_2$，$r=\beta_1/\beta_2$。当 $r_{\mathrm{d}}=r$ 时称为匹配。图 13.7 中 OS-R-LR 的各条曲线对应的 dB 值是 $10\lg[(\hat{\beta}_1+\hat{\beta}_2)/2]$。由图 13.7 可见，OS-R-LR 的虚警概率随该 dB 值变化而变化，在 $r_{\mathrm{d}}=1$ 时的变化范围比较小。

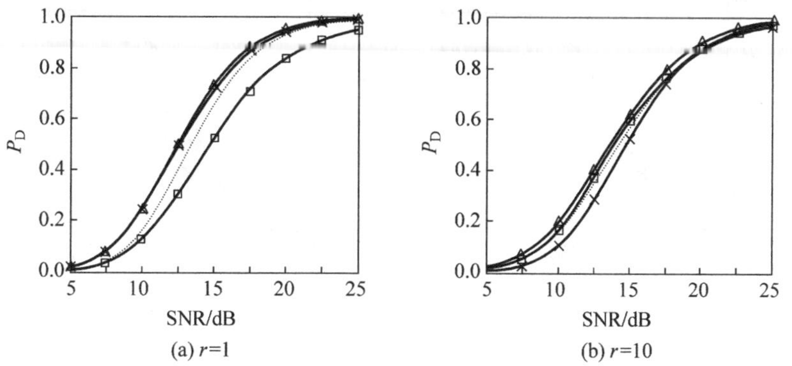

图 13.5 OS-R-LR 和 OS-R-SUM 在均匀杂波背景中的检测概率

（OS-R-LR：◇ 最大（$r_d=1$），× 最小（$r_d=1$），△ 最大（$r_d=10$），□ 最小（$r_d=10$）；…… OS-R-SUM）

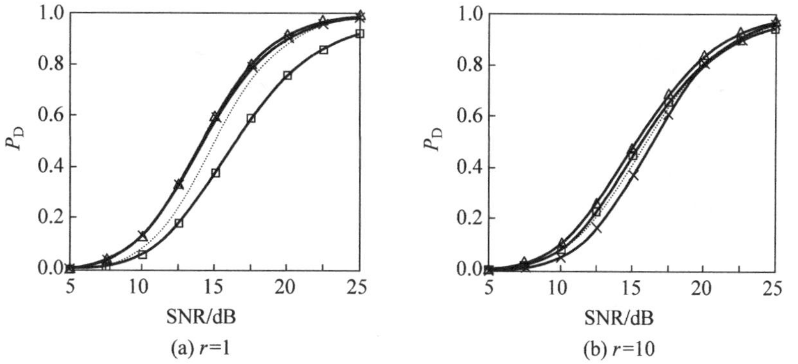

图 13.6 OS-R-LR 和 OS-R-SUM 在多目标环境中的检测概率

（$m_i=2, \beta_i=\lambda_i (i=1,2)$；OS-R-LR：◇ 最大（$r_d=1$），× 最小（$r_d=1$），

△ 最大（$r_d=10$），□ 最小（$r_d=10$）；…… OS-R-SUM）

图 13.5～图 13.7 中同时还给出了 OS-R-SUM 的性能曲线。首先，SUM 融合不需要假设局部信噪比和目标模型，因此不存在融合准则的匹配问题，融合准则失配对 LR 融合的所有影响对于 SUM 融合不存在。其次，由图 13.5 和图 13.6 可见，SUM 融合与相同条件下的 LR 融合的性能差距很小。在 $|\log r|$ 较高的情况下，OS-R-SUM 更接近于 OS-R-LR。在均匀杂波背景和多目标环境中，SUM 融合相对于 LR 融合的损失在 $r=1$ 时在 1dB 左右，在 $r=10$ 时在 0.5dB 以下，可以跟随 LR 融合的性能。在杂波边缘环境中，由图 13.7 可见，SUM 融合与 LR 融合的虚警概率在相同的数量级上。但是 SUM 融合不存在融合准则失配的问题，且 SUM 融合具有容易实现的优点。这些现象在一定程度上可以说明，融合中心的总信噪比是影响系统性能的主要因素之一。

4. R 类局部检测统计量在次优融合条件下的性能分析

在局部观测信噪比完全已知的条件下，基于 R 类局部检测统计量和 LR 融合的分布式 CFAR 检测在 NP 意义上是最优的。然而在实际应用中，局部观测信噪比往往是未知的，需要用估计值代替。对于 OS-R，用局部观测信噪比 λ_i 的 ML 估计代替式(13.26)中的 λ_i，这种方法被称为 GLRT。

1) 局部观测信噪比的 ML 估计

对于第 i 个局部处理器，其信噪比 λ_i 的 ML 估计可以通过解式(13.27)得到

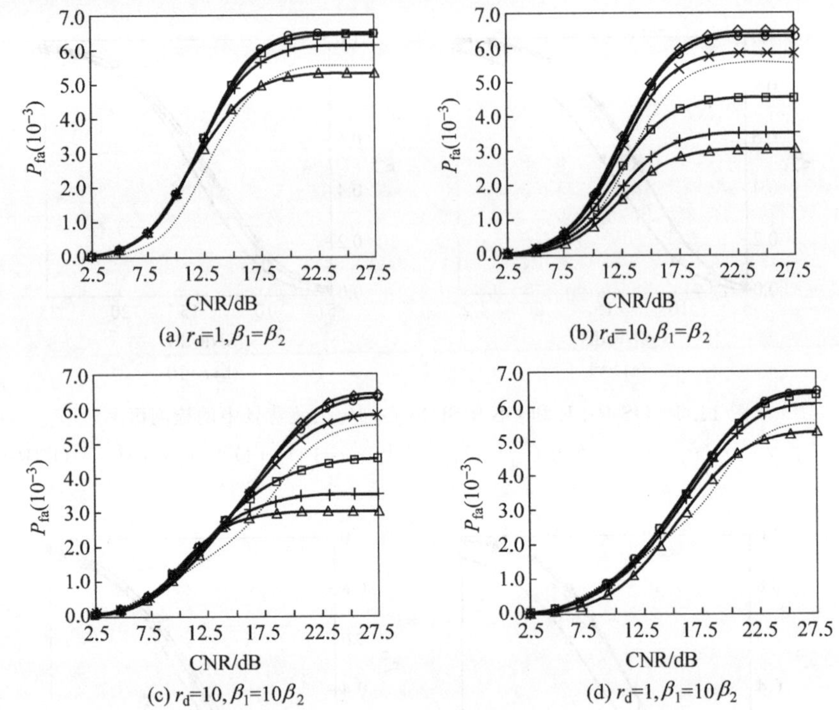

图 13.7 OS-R-LR 和 OS-R-SUM 在杂波边缘环境中的虚警概率

($m_1=6, m_2=7$；OS-R-LR：△ 5(dB)，+ 10(dB)，

□ 15(dB)，× 20(dB)，○ 25(dB)，◇ 30(dB)；…… OS-R-SUM)

$$\frac{\partial f_{S_i}^{(\mathrm{h})}(y_i \mid \mathrm{H}_1)}{\partial \lambda_i}=0 \tag{13.27}$$

于是，由 OS-R 局部检测统计量 S_i 的样本 y_i 得到的第 i 个局部处理器的局部观测信噪比 λ_i 的最大似然估计为

$$\hat{\lambda}_i=U[y_i-a_i]\left(\frac{y_i}{a_i}-1\right), \quad y_i>0 \tag{13.28}$$

其中，$U[\cdot]$代表阶跃函数，而 a_i 就是满足式$\left.\frac{\partial f_{S_i}^{(\mathrm{h})}(y_i \mid \mathrm{H}_1)}{\partial \lambda_i}\right|_{\lambda_i=0}=0$ 的 y_i。

2）OS-R-GLR 分布式 CFAR 检测

将式(13.28)代入局部对数似然比 Ψ_i 的表达式中可得如下的局部对数广义似然比

$$\tilde{\Psi}_i=U[y_i-a_i]\left\{\log\left[\frac{1}{y_i f_{S_i}^{(\mathrm{h})}(y_i \mid \mathrm{H}_0)}\right]+b_i\right\}, \quad y_i>0 \tag{13.29}$$

定义 $\tilde{\Psi}_i$ 为广义似然比(GLR)类局部检测统计量，其中 b_i 是与 N_i 和 k_i 对应的常数。融合中心的全局检测统计量为 $S=\sum_{i=1}^{L}\tilde{\Psi}_i$，全局判决仍然按照式(13.24)进行。但是这时的 T 值要做相应的改变。用 OS-R-GLR 表示这种基于 OS-R 局部检测统计量和广义似然比融合的分布式 CFAR 检测方案。

3）性能分析

OS-R-GLR 在均匀背景和多目标环境中的检测概率均与 OS-R-LR 比较接近，并且相对于 OS-R-SUM 有一定的性能改善。OS-R-GLR 的虚警概率比 OS-R-SUM 略有上升，但是仍然保持在与 OS-R-

SUM 同一数量级的水平上。与各种假设的杂噪比条件下最优融合的虚警尖峰相比，OS-R-GLR 的虚警也是适中的，并且不受局部观测信噪比假设失配的影响。

总体来说，OS-R-GLR 的检测性能介于 OS-R-SUM 和 OS-R-LR 之间，更趋近于 OS-R-LR。因此，OS-R-GLR 是一种折中的方案。相对于 OS-R-SUM，OS-R-GLR 以计算复杂性的增加换取性能的提高。相对于 OS-R-LR，OS-R-GLR 是次优的，但是 OS-R-GLR 不需要已知局部观测信噪比。三者在不同的条件下各有其优势。

13.3.2 基于 S 类局部检测统计量的分布式 CFAR 检测

1. S 类局部检测统计量的统计特性

S 类局部检测统计量如下所示，对于第 i 个局部处理器有

$$S_i = X_{i0} - T_i Z_i, \quad i=1,2,\cdots,L \tag{13.30}$$

X_{i0} 是检测单元采样，Z_i 由参考单元采样根据 CFAR 算法产生，T_i 是用于调节融合中心总的虚警概率的因子。在本节设 $T_1=T_2=\cdots=T_L=T$，后面的性能分析将表明这种选择也具有很好的性能。这里仍然讨论利用 CA-CFAR 和 OS-CFAR 算法产生 Z_i 的 S 类局部检测统计量，相应的 S 类局部检测统计量分别用 CA-S 和 OS-S 表示。

这里仍然考虑图 13.1 所示的并行网络结构。每个局部处理器将局部预处理产生的 S 类局部检测统计量 $S_i(i=1,2,\cdots,L)$ 传送给融合中心，融合形成全局检测统计量 G，然后进行如下的全局判决

$$\mathrm{H}_1: G \geqslant 0, \quad \mathrm{H}_0: G < 0 \tag{13.31}$$

在非均匀杂波背景中，假设参考单元只包含两种强度的采样。也就是说，在多目标环境中，各干扰目标强度相同；在杂波边缘环境中，强杂波采样的强度也均相同。并且仍然假设检测单元与参考单元之间及参考单元采样之间是统计独立的。

1) CA-S 的统计特性

对于 S 类方案，式(13.30)定义的 $S_i(i=1,2,\cdots,L)$ 的累积概率分布函数为

$$F_{S_i}(y_i) = \Pr[S_i < y_i] = 1 - \Pr[X_{i0} - TZ_i \geqslant y_i] \tag{13.32}$$

在以 CA 为局部处理器的方案中，杂波功率水平估计 Z_i 由 CA 方法产生。CA-S 在 $y_i>0$ 时的概率密度函数为

$$f_{S_i}(y_i) = \frac{g_i}{\mu_i(1+\lambda_i)} \exp\left[-\frac{y_i}{\mu_i(1+\lambda_i)}\right] \tag{13.33}$$

其中，$g_i = \sum_{n_1=1}^{m_i} \frac{A_{i,n_1}}{[1+T(1+\beta_i)/(1+\lambda_i)]^{n_1}} + \sum_{n_2=1}^{N_i-m_i} \frac{B_{i,n_2}}{[1+T/(1+\lambda_i)]^{n_2}}$。而当 $y_i<0$ 时，CA-S 的概率密度函数为

$$\begin{aligned} f_{S_i}(y_i) = {} & \exp\left[\frac{y_i}{T\mu_i(1+\beta_i)}\right] \sum_{n_1=1}^{m_i} \left\{A_{i,n_1} \sum_{j=0}^{n_1-1} \frac{(-1)^j}{\Gamma(j+1)} \left[1-\left(1+\frac{T(1+\beta_i)}{1+\lambda_i}\right)^{j-n_1}\right] \frac{y_i^j}{[T\mu_i(1+\beta_i)]^{j+1}}\right\} + \\ & \exp\left[\frac{y_i}{T\mu_i(1+\beta_i)}\right] \sum_{n_1=1}^{m_i} \left\{A_{i,n_1} \sum_{j=0}^{n_1-1} \frac{(-1)^j}{\Gamma(j)} \left[1-\left(1+\frac{T(1+\beta_i)}{1+\lambda_i}\right)^{j-n_1}\right] \frac{y_i^{j-1}}{[T\mu_i(1+\beta_i)]^j}\right\} + \\ & \exp\left(\frac{y_i}{T\mu_i}\right) \sum_{n_2=1}^{N_i-m_i} \left\{B_{i,n_2} \sum_{j=0}^{n_2-1} \frac{(-1)^j}{\Gamma(j+1)} \left[1-\left(1+\frac{T}{1+\lambda_i}\right)^{j-n_2}\right] \frac{y_i^j}{(T\mu_i)^{j+1}}\right\} + \end{aligned}$$

$$\exp\left(\frac{y_i}{T\mu_i}\right)\sum_{n_2=1}^{N_i-m_i}\left\{B_{i,n_2}\sum_{j=0}^{n_2-1}\frac{(-1)^j}{\Gamma(j)}\left[1-\left(1+\frac{T}{1+\lambda_i}\right)^{j-n_2}\right]\frac{y_i^{j-1}}{(T\mu_i)^j}\right\} \tag{13.34}$$

其中,m_i 在多目标环境中代表干扰目标数,在杂波边缘环境中代表强杂波单元数。在均匀背景中,S_i 的 PDF 可以通过令 $\beta_i=0$ 或 $m_i=0$ 得到。

2) OS-S 的统计特性

对于 OS-S,式(13.30)中的 Z_i 由 OS 方法产生。当 $y_i>0$ 时,OS-S 的概率密度函数为

$$f_{S_i}(y_i)=\frac{H_i[N_i,k_i;m_i,\beta_i;b_i]}{\mu_i(1+\lambda_i)}\exp\left[-\frac{y_i}{\mu_i(1+\lambda_i)}\right] \tag{13.35}$$

其中,$b_i=-\dfrac{a_i}{a_i+Q_i}$,$Q_i=\dfrac{T}{1+\lambda_i}$,$a_i$ 和 m_i 在前面已有定义,$H[\cdot]$见式(13.18)。

当 $y_i<0$ 时,OS-S 的概率密度函数为

$$f_{S_i}(y_i)=H_i\left[N_i,k_i;m_i,\beta_i;\frac{a_ic_i}{T\mu_i}\exp\left(\frac{a_iy_i}{T\mu_i}\right)\right] \tag{13.36}$$

其中,$c_i=-\dfrac{Q_i}{a_i+Q_i}$。在均匀背景中,$S_i$ 的 PDF 可以通过令 $\beta_i=0$ 或 $m_i=0$ 得到。

2. CA-S-SUM 和 OS-S-SUM 的性能模型

1) S 类局部检测统计量在 SUM 融合条件下的基本模型

这里考虑的是图 13.1 所示的并行网络结构,各局部处理器分别将各自的局部检测统计量传送到融合中心,在融合中心按 SUM 融合形成全局检测统计量 G,然后进行式(13.31)定义的全局判决。用 CA-S-SUM 和 OS-S-SUM 分别表示基于 CA-S 和 OS-S 局部检测统计量和 SUM 融合的上述分布式 CFAR 检测方案。

在假设 $S_i(i=1,2,\cdots,L)$之间统计独立的条件下,全局检测统计量 G 大于 0 的概率为

$$\Pr[G>0]=\int\limits_{\sum_{i=1}^{L}y_i>0}\cdots\int\prod_{i=1}^{L}f_{S_i}(y_i)\mathrm{d}y_1\mathrm{d}y_2\cdots\mathrm{d}y_L \tag{13.37}$$

经过复杂的推导,可得 G 大于零的概率为

$$\Pr[G\geqslant 0]=\sum_{i=1}^{L}C_i\sum_{j=0}^{n_i-1}\left(\frac{Q_i}{\mu_i}\right)^j\frac{(-1)^j}{j!}\frac{\mathrm{d}^{(j)}\Phi_Z(u)}{\mathrm{d}u^{(j)}}\bigg|_{u=Q_i/\mu_i} \tag{13.38}$$

其中,$Q_i=T/(1+\lambda_i)$,$\Phi_Z(u)$是 Z 的特征函数。由此可见,求得 Z 的特征函数即可得 $\Pr[G>0]$。

2) CA-S-SUM 的性能模型

对于 CA-S,Z_i 的特征函数为

$$\Phi_{Z_i}(u)=\frac{1}{[1+u\mu_i(1+\beta_i)]^{m_i}(1+u\mu_i)^{N_i-m_i}} \tag{13.39}$$

由于 $Z_i(i=1,2,\cdots,L)$间是统计独立的,因此 Z 的特征函数是 $Z_i(i=1,2,\cdots,L)$的特征函数之积。于是由式(13.38)可得 CA-S-SUM 的全局检测统计量 G 大于 0 的概率为

$$\Pr[G\geqslant 0]=\sum_{i=1}^{L}C_i\sum_{j=0}^{n_i-1}\left(\frac{Q_i}{\mu_i}\right)^j\frac{(-1)^j}{j!}\frac{\mathrm{d}^{(j)}}{\mathrm{d}u^{(j)}}\left[\prod_{l=1}^{L}\frac{1}{[1+u\mu_l(1+\beta_l)]^{m_l}(1+u\mu_l)^{N_l-m_l}}\right]\bigg|_{u=Q_i/\mu_i} \tag{13.40}$$

表面上看式(13.40)与 $\mu_i(i=1,2,\cdots,L)$有关,因此不是 CFAR 的。实际上,可以证明 $\mu_i(i=1,2,\cdots,L)$之间的相对比例系数 $\rho_i(i=1,2,\cdots,L)$确定时,CA-S-SUM 的 $\Pr[G\geqslant 0]$与 $\mu_i(i=1,2,\cdots,L)$无关,这时是 CFAR 的。

3) OS-S-SUM 的性能模型

对于 OS-S,Z_i 的特征函数为

$$\Phi_{Z_i}(u)=H_i\left[N_i,k_i;m_i,\beta_i;-\frac{a_i}{\mu_i u+a_i}\right] \tag{13.41}$$

由于 $Z_i(i=1,2,\cdots,L)$间是统计独立的,因此 Z 的特征函数是 $Z_i(i=1,2,\cdots,L)$的特征函数之积。于是,由式(13.38)可得 OS-S-SUM 的全局检测统计量 G 大于 0 的概率为

$$\Pr[G\geqslant 0]=\sum_{i=1}^{L}C_i\sum_{j=0}^{n_i-1}\left(\frac{Q_i}{\mu_i}\right)^j\frac{(-1)^j}{j!}\frac{\mathrm{d}^{(j)}}{\mathrm{d}u^{(j)}}\left[\prod_{l=1}^{L}H_l\left[N_l,k_l;m_l,\beta_l;-\frac{a_l}{\mu_l u+a_l}\right]\right]\Bigg|_{u=Q_i/\mu_i} \tag{13.42}$$

当 $\mu_i(i=1,2,\cdots,L)$之间的相对比例系数确定时,OS-S-SUM 的 $\Pr[G\geqslant 0]$与 $\mu_i(i=1,2,\cdots,L)$无关,这时是 CFAR 的。

通过适当地设置式(13.40)和式(13.42)中各变量的值可以分别计算 CA-S-SUM 和 OS-S-SUM 在均匀背景和多目标环境中的检测概率及杂波边缘环境中的虚警概率。

3. 基于 S 类局部检测统计量的分布式 CFAR 检测的性能分析

在 ρ 值匹配且 $\rho=1$ 的情况下,COS 具有最好的性能,因为它的处理方式使局部观测信息损失小。OS-S-SUM 与 COS 极为接近。

当 $\rho\neq 1$ 时,即使 ρ 值匹配,由于各局部传感器背景噪声功率水平不相等,基于 S 类局部检测统计量的方案及集中式检测的性能都严重下降。此时最合理的处理方法是 R 类局部检测统计量,即在局部处理中完成 CFAR 处理。

当 ρ 值失配时,给出的几种方案受影响的严重程度依次降低的顺序为 COS、OS-S-SUM。COS 的性能下降到了不能接受的程度。然而基于 R 类局部检测统计量的方案不受 ρ 的影响。因此,在 ρ 未知或变化的环境中,无论是考虑 ρ 的影响,还是考虑数据压缩,基于 R 类局部检测统计量的方案在上述几种方案中都是首选。

13.4 分布式 MIMO 雷达 CFAR 检测

对于常规雷达目标检测来说,目标角闪烁通常是一个不利因素,然而分布式 MIMO 雷达的新颖之处就在于它能有效地利用目标角闪烁来改善雷达的目标检测性能[27]。其基本原理是,分布式 MIMO 雷达利用波形分集和分布式多视角观测,获得关于目标的多个观测通道回波数据,然后通过通道间局部检验统计量非相参积累的方式形成最终的 CFAR 检测器[26-29]。

13.4.1 目标回波经典线性模型及检测器设计

图 13.8 所示的 MIMO 分布孔径雷达系统包括 M 个发射子阵和 N 个接收子阵,阵元数分别为 M_m 和 $N_n(m=1,2,\cdots,M,n=1,2,\cdots,N)$,总的发射和接收阵元数分别为 $\widetilde{M}=\sum_{m=1}^{M}M_m$ 和 $\widetilde{N}=\sum_{n=1}^{N}N_n$。相参脉冲数为 L,采用脉内正交编码波形,编码长度为 p,并跟通常文献一样假设各发射子阵所辐射的 M 个

脉冲串波形是相互理想正交的。

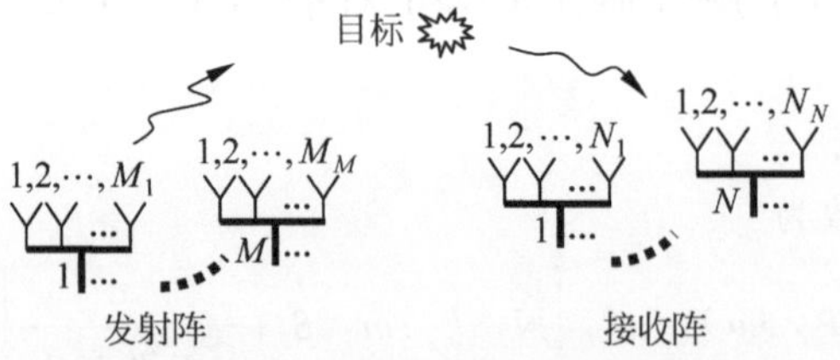

图 13.8 MIMO 分布孔径雷达系统示意图

利用脉冲波形之间的正交性，对此雷达系统各个接收子阵获取的 L 个脉冲回波进行匹配滤波之后，得到对应于 MN 个观测通道的 MN 个观测向量

$$\boldsymbol{x}_{mn}=\boldsymbol{s}_{mn}+\boldsymbol{z}_{mn}=\boldsymbol{c}_{mn}\cdot\boldsymbol{a}_n\otimes\boldsymbol{v}_{mn}+\boldsymbol{z}_{mn} \tag{13.43}$$

其中，$\boldsymbol{s}_{mn}$ 是第 mn 个观测通道的目标回波向量，$\boldsymbol{a}_n$ 是第 n 个接收子阵的导向向量；而 $\boldsymbol{c}_{mn}$ 服从零均值、方差为 σ_s^2 的复高斯分布，简记为 $\boldsymbol{c}_{mn}=\mathrm{CN}(0,\sigma_s^2)$，其中，$\sigma_s^2$ 表示各观测通道中的目标功率水平；$\boldsymbol{v}_{mn}=[1\quad\cdots\quad \mathrm{e}^{\mathrm{j}2\pi f_{\mathrm{d},mn}L}]^{\mathrm{T}}$ 表示目标在第 mn 个观测通道中的多普勒向量。$f_{\mathrm{d},mn}$ 通常是未知的，需要在 MN 个观测通道中执行联合优化搜索[30-31]。由于这里主要讨论检测性能分析问题，故为了简单起见，假设各通道中的 $f_{\mathrm{d},mn}$ 都已知。另外，基于噪声快拍独立同分布假设，第 mn 个通道中的噪声向量 $\boldsymbol{z}_{mn}$ 服从 $\mathrm{CN}(0,\boldsymbol{R}_{z,mn})$ 分布，$\boldsymbol{R}_{z,mn}=\boldsymbol{R}_0\otimes\boldsymbol{I}_{L\times L}$，$\boldsymbol{R}_0$ 是每个通道中噪声观测向量的空域协方差矩阵。

将所有通道的观测向量堆积成一个列向量，$\boldsymbol{x}=\left[\boldsymbol{x}_{11}^{\mathrm{T}}\quad\cdots\quad\boldsymbol{x}_{MN}^{\mathrm{T}}\right]^{\mathrm{T}}$，即可得如下观测模型

$$\boldsymbol{x}=\boldsymbol{s}+\boldsymbol{z}=\boldsymbol{H}\cdot\boldsymbol{c}+\boldsymbol{z} \tag{13.44}$$

其中，$\boldsymbol{s}=\boldsymbol{H}\cdot\boldsymbol{c}$ 表示目标回波的经典线性模型，$\boldsymbol{H}$ 称为通道模式矩阵，由 MN 个观测通道的空时导向向量组成

$$\begin{aligned}\boldsymbol{H}&=\mathrm{diag}(\boldsymbol{h}_{11},\cdots,\boldsymbol{h}_{M1},\cdots,\boldsymbol{h}_{M2},\cdots,\boldsymbol{h}_{1N},\cdots,\boldsymbol{h}_{MN})\\&=\mathrm{diag}(\boldsymbol{a}_1\otimes\boldsymbol{v}_{11},\cdots,\boldsymbol{a}_1\otimes\boldsymbol{v}_{M1},\cdots,\boldsymbol{a}_2\otimes\boldsymbol{v}_{M2},\cdots,\boldsymbol{a}_N\otimes\boldsymbol{v}_{1N},\cdots,\boldsymbol{a}_N\otimes\boldsymbol{v}_{MN})\end{aligned} \tag{13.45}$$

显然，$\boldsymbol{H}$ 的各列是相互正交的，故可称 $\boldsymbol{H}$ 为正交通道模式矩阵。未知的通道系数向量 $c\sim\mathrm{CN}(0,\sigma_s^2\boldsymbol{I}_{MN\times MN})$，而 $z\sim\mathrm{CN}(0,\mathrm{diag}(\boldsymbol{R}_{z,11},\cdots,\boldsymbol{R}_{z,MN}))$。

根据式(13.44)所示的观测模型可得如下二元假设检验，

$$\begin{cases}\mathrm{H}_0:\ x=z & \sim\mathrm{CN}(0,\sigma^2\boldsymbol{R}_{z0})\\ \mathrm{H}_1:\ x=\boldsymbol{H}\cdot\boldsymbol{c}+z & \sim\mathrm{CN}(\boldsymbol{H},\sigma^2\boldsymbol{R}_{z0})\end{cases} \tag{13.46}$$

其中，假设 $\boldsymbol{c}$ 的统计特性未知，故将其建模为确定性未知量；采用两步法，得到如下的 AMF 检测器

$$T_{\mathrm{AMF}}=\boldsymbol{x}^{\mathrm{H}}\hat{\boldsymbol{R}}_{z0}^{-1}\boldsymbol{H}(\boldsymbol{H}^{\mathrm{H}}\hat{\boldsymbol{R}}_{z0}^{-1}\boldsymbol{H})^{-1}\boldsymbol{H}^{\mathrm{H}}\hat{\boldsymbol{R}}_{z0}^{-1}\boldsymbol{x} \tag{13.47}$$

其中，$\hat{\boldsymbol{R}}_{z0}=\dfrac{1}{K}\sum_{k=1}^{K}\boldsymbol{x}_k\boldsymbol{x}_k^{\mathrm{H}}$，$\boldsymbol{x}_k$ 是参考距离单元中的观测向量，K 为参考距离单元数量。

13.4.2 MIMO 分布孔径雷达 AMF 检测器性能分析

1. AMF 检测器结构分析

注意到，在 MN 个观测通道都是相互统计独立的假设下，色噪声背景的空时协方差矩阵 $\hat{\boldsymbol{R}}_{z0}$ 和目

标模式矩阵 $\boldsymbol{H}$ 都是分块对角阵，且每一分块都对应着一个观测通道，因此，式(13.47)可变形为

$$T_{\mathrm{AMF}}=\sum_{m=1}^{M}\sum_{n=1}^{N}\frac{\left|\boldsymbol{h}_{mn}^{\mathrm{H}}\left((\hat{\sigma}_{c,mn}^{2}+\hat{\sigma}_{0,mn}^{2})\hat{\boldsymbol{R}}_{z0,mn}\right)^{-1}\boldsymbol{x}_{mn}\right|^{2}}{\boldsymbol{h}_{mn}^{\mathrm{H}}\left[(\hat{\sigma}_{c,mn}^{2}+\hat{\sigma}_{0,mn}^{2})\hat{\boldsymbol{R}}_{z0,mn}\right]^{-1}\boldsymbol{h}_{mn}}=\sum_{m=1}^{M}\sum_{n=1}^{N}l_{mn} \tag{13.48}$$

其中，$\boldsymbol{x}_{mn}$ 是第 mn 个通道中的观测向量，$\boldsymbol{h}_{mn}$ 如式(13.45)所示，表示第 mn 个通道中的空时导向向量。由式(13.48)可以看出 AMF 检测器的结构特点，即先在每个观测通道中分别独立地估计色噪声背景空时协方差矩阵 $\boldsymbol{R}_{z0,mn}$，然后对检验单元的观测数据和通道模式矩阵进行白化操作，并通过归一化匹配滤波处理形成该通道的局部检测统计量 l_{mn}，最后对这 MN 个观测通道输出的局部检测统计量进行同等权重的非相参求和。

式(13.48)中 AMF 检测器施给每个通道中的权重都是 1，这种做法是基于各个通道对于目标观测来说都是统计相同的。然而实际中，飞行器隐身技术、强杂波干扰抑制技术及多径效应都可能导致部分观测通道中的目标回波功率极小，这意味着这些观测通道失效，也意味着各个观测通道不是统计相同的。在这种情况下，式(13.48)所示的全部观测通道等权重地非相参积累结构形式不一定能获得最好的检测性能，而正确的做法是，对不同的通道施以不同的权重，目标回波功率强的通道施以较大的权重，目标回波功率弱的通道施以较小的权重，甚至直接删除那些目标回波功率极低的通道。关于通道选择的分布式 MIMO 雷达检测算法请参考文献[32]。

2. AMF 检测器极限检测性能分析

MIMO 分布孔径雷达 AMF 检测器极限检测性能分析是基于已知噪声空时协方差矩阵 $\boldsymbol{R}_{z0}$ 和通道系数相关矩阵 $\sigma_s^2\boldsymbol{R}_s$ 展开的。此时，AMF 检测器对应的观测模型变为 $\boldsymbol{x}=\boldsymbol{H}'\cdot\boldsymbol{\beta}+\boldsymbol{z}$，其中，$\boldsymbol{H}'=\boldsymbol{H}\boldsymbol{G}_s$，$\boldsymbol{R}_s=\boldsymbol{G}_s\boldsymbol{G}_s^{\mathrm{H}}$，$\boldsymbol{\beta}\sim\mathrm{CN}(0,\sigma_s^2\boldsymbol{I}_{MN\times MN})$。那么 AMF 检测器的输出信噪比表达式推导如下

$$\begin{aligned}\frac{\text{AMF 检测器输出信号功率}}{\text{AMF 检测器输出噪声功率}}&=\frac{\mathrm{E}\{\boldsymbol{s}^{\mathrm{H}}\boldsymbol{R}_{z0}^{-1}\boldsymbol{H}'(\boldsymbol{H}'^{\mathrm{H}}\boldsymbol{R}_{z0}^{-1}\boldsymbol{H}')^{-1}\boldsymbol{H}'^{\mathrm{H}}\boldsymbol{R}_{z0}^{-1}\boldsymbol{s}\}}{\mathrm{E}\{\boldsymbol{z}^{\mathrm{H}}\boldsymbol{R}_{z0}^{-1}\boldsymbol{H}'(H'^{\mathrm{H}}\boldsymbol{R}_{z0}^{-1}\boldsymbol{H}')^{-1}\boldsymbol{H}'^{\mathrm{H}}\boldsymbol{R}_{z0}^{-1}\boldsymbol{z}\}}\\&=\frac{\mathrm{tr}(\boldsymbol{R}_{z0}^{-1}\boldsymbol{H}'(\boldsymbol{H}'^{\mathrm{H}}\boldsymbol{R}_{z0}^{-1}\boldsymbol{H}')^{-1}\boldsymbol{H}'^{\mathrm{H}}\boldsymbol{R}_{z0}^{-1}\boldsymbol{H}'\cdot\mathrm{E}\{\boldsymbol{\beta}\boldsymbol{\beta}^{\mathrm{H}}\}\cdot\boldsymbol{H}'^{\mathrm{H}})}{\mathrm{tr}(\boldsymbol{R}_{z0}^{-1}\boldsymbol{H}'(\boldsymbol{H}'^{\mathrm{H}}\boldsymbol{R}_{z0}^{-1}\boldsymbol{H}')^{-1}\boldsymbol{H}'^{\mathrm{H}}\boldsymbol{R}_{z0}^{-1}\mathrm{E}\{\boldsymbol{z}\boldsymbol{z}^{\mathrm{H}}\})}\\&=\frac{\sigma_s^2\cdot\mathrm{tr}(\boldsymbol{R}_s\boldsymbol{H}^{\mathrm{H}}\boldsymbol{R}_{z0}^{-1}\boldsymbol{H})}{MN}=\frac{\displaystyle\sum_{m=1}^{M}\sum_{n=1}^{N}\frac{\sigma_s^2\cdot\boldsymbol{h}_{mn}^{\mathrm{H}}\boldsymbol{R}_{z0,mn}^{-1}\boldsymbol{h}_{mn}}{\sigma_{c,mn}^2+\sigma_{0,mn}^2}}{MN}\\&=\frac{\displaystyle\sum_{m=1}^{M}\sum_{n=1}^{N}\frac{\sigma_s^2\cdot\boldsymbol{h}_{mn}^{\mathrm{H}}R_{z0,0}^{-1}\boldsymbol{h}_{mn}}{\sigma_c^2+\sigma_0^2}}{MN}=\frac{\displaystyle\sum_{m=1}^{M}\sum_{n=1}^{N}\mathrm{SNR}_{mn}}{MN}\end{aligned} \tag{13.49}$$

上式中利用了 $\boldsymbol{R}_s$ 的对角线元素是 1 的特点，其中 SNR_{mn} 表示第 mn 个通道经过白化、匹配滤波处理之后的输出信噪比。式(13.49)表明，AMF 检测器的输出信噪比等于各观测通道白化处理后的剩余信噪比关于通道数取平均。这意味着 AMF 检测器的输出信噪比不是各白化观测通道输出的剩余信噪比之和，所以单纯增加观测通道并不一定能提高检测器的输出信噪比，进而改善检测性能。

3. AMF 检测器的解析性能表达式

这里采用文献[33]中的方法推导 $\boldsymbol{R}_{z0}$ 和 $\sigma_s^2\boldsymbol{R}_s$ 都未知条件下的 MIMO 分布孔径雷达 AMF 检测器的解析性能表达式，如式(13.50)和式(13.51)所示

$$p_{\text{fa}}(\eta)=1-\mathrm{E}_b\left\{\text{betacdf}\left(\frac{b\eta}{K+b\eta},MN,K-\widetilde{N}L+1\right)\right\} \tag{13.50}$$

$$p_{\text{d}}(\eta)=\mathrm{E}_\varepsilon\left\{1-\mathrm{E}_b\left\{\mathrm{e}^{-b\cdot\varepsilon}\sum_{k=0}^{\infty}\frac{(b\cdot\varepsilon)^k}{k!}\text{betacdf}\left(\frac{b\eta}{K+b\eta},MN+k,K-\widetilde{N}L+1\right)\right\}\right\} \tag{13.51}$$

其中，$p_{\text{fa}}(\eta)$表示虚警概率，$p_{\text{d}}(\eta)$表示检测概率，两者都是检测门限 η 的函数，随机变量 b 服从 $\beta_{K-\widetilde{N}L+MN+1,\widetilde{N}L-MN}$ 分布，$\mathrm{E}_b\{\cdot\}$表示关于随机变量 b 取均值；由式(13.50)可知，AMF 检测器的虚警概率只与尺寸参数 $M,N,K,\widetilde{N},L$ 有关，而与背景统计特性参数无关，因此，该检测器是 CFAR 的。$\varepsilon=\sum_{i=1}^{MN}\text{SNR}_i$ 表示各通道输出信噪比之和，其中各通道输出信噪比 SNR_i 都是服从指数分布的随机变量，而且相互之间可能还存在相关性，这正是与文献[33]的主要区别之处。若直接考虑 ε 的统计分布，则很难得出封闭的检测概率表达式。于是借鉴文献[33]对随机变量 b 的处理办法，采用样本均值代替数值积分来计算 $\mathrm{E}_\varepsilon\{\cdot\}$。$\varepsilon$ 的样本值通过式(13.52)获得

$$\varepsilon=\boldsymbol{c}^{\mathrm{H}}(\boldsymbol{H}^{\mathrm{H}}\hat{\boldsymbol{R}}_{z0}^{-1}\boldsymbol{H})\boldsymbol{c}=\sigma_s^2\cdot\tilde{\boldsymbol{c}}^{\mathrm{H}}(\boldsymbol{R}_s^{1/2})^{\mathrm{H}}(\boldsymbol{H}^{\mathrm{H}}\hat{\boldsymbol{R}}_{z0}^{-1}\boldsymbol{H})\boldsymbol{R}_s^{1/2}\tilde{\boldsymbol{c}} \tag{13.52}$$

其中，$\boldsymbol{R}_s$ 由仿真条件来设定，$\boldsymbol{R}_s^{1/2}$ 表示 $\boldsymbol{R}_s$ 的 Cholesky 分解，$\tilde{c}\sim\text{CN}(0,\sigma_s^2\boldsymbol{I}_{MN\times MN})$，$\sigma_s^2$ 是检测性能曲线的自变量之一，因此通过仿真 $\tilde{c}$ 来得到即可得到ε 的样本值。

13.4.3 仿真分析

在高斯色噪声及相关观测通道条件下，本节将通过计算机仿真验证关于 MIMO 分布孔径雷达 AMF 检测器性能的分析。仿真条件：发射子阵的数量 M 与接收子阵的数量 N 均取值为 3，$M_m=N_n=8$，$L=16$，目标具有严重的角闪烁现象，并考虑 6 种仿真场景，分别是"白色背景＋独立通道""白色背景＋强相关通道""白色背景＋弱相关通道""强相关背景＋独立通道""弱相关背景＋独立通道""弱相关背景＋弱相关通道"，其中通道相关性采用下式建模

$$r(i)=\rho^i,\quad i=0,1,\cdots,MN-1 \tag{13.53}$$

"强相关通道"指"通道相关矩阵的秩为 1，即 $\rho=1$"，"弱相关通道"指 $\rho=0.5$，而"独立通道"指 $\rho=0$。"背景"均是"高斯色噪声背景"，由"高斯杂波＋接收机热噪声"组成，而高斯杂波的相关性也按式(13.53)建模，"强相关背景"指"$\rho=1$，杂噪比 $\text{CNR}=10\lg(\sigma_c^2/\sigma_\theta^2 0)=60\text{dB}$，且除一个通道外，其余通道中的目标被杂波完全淹没"，而"弱相关背景"指"$\rho=0.5$，$\text{CNR}=20\text{dB}$，且只有少数观测通道中的目标被杂波淹没"，而"白色背景"指没有杂波。

图 13.9 分别给出了 MIMO 分布孔径雷达 AMF 检测器在 6 种不同场景中检测性能的 Monte-Carlo 仿真结果与解析计算结果，其中 $\text{SNR}=10\lg(\sigma_s^2/\sigma_\theta^2)$ 表示"各观测通道中白化和积累处理之前的信噪比"。

由图 13.9 可知，相关观测通道和色噪声背景都会使得检测性能相对于"白色背景＋独立通道"情况有不同程度的下降。相关观测通道和色噪声背景导致检测性能下降的原因有两个方面，一是它们造成各个虚拟的独立观测通道实际输出的目标功率水平并不相同；二是 AMF 检测器是对各个实际的观测通道输出进行等权重的非相参积累，因此在色噪声背景中，AMF 检测器没有考虑各个实际的独立观测通道输出目标功率水平的不同，而在相关观测通道中，AMF 检测器既不是对虚拟独立观测通道进行非相参积累(因为通道相关性未知，故无法得到虚拟独立观测通道)，也不像相控阵雷达那样对完全相关通道进行相参积累。

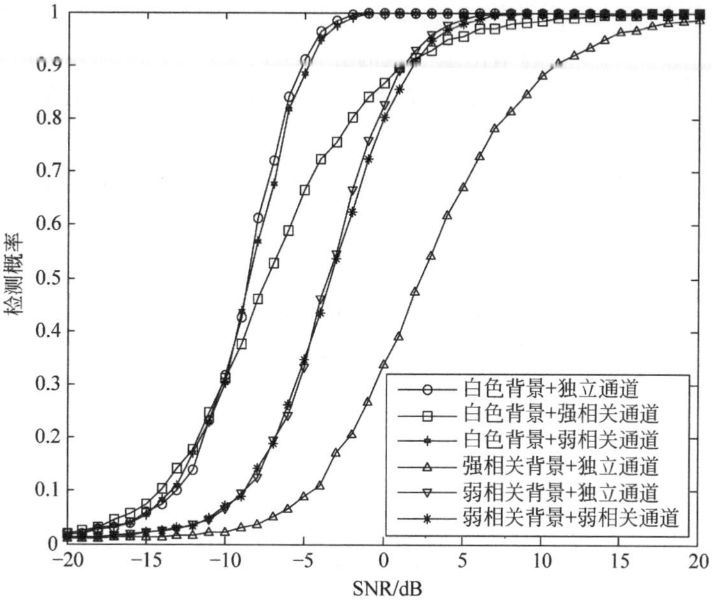

(a) Monte-Carlo仿真结果

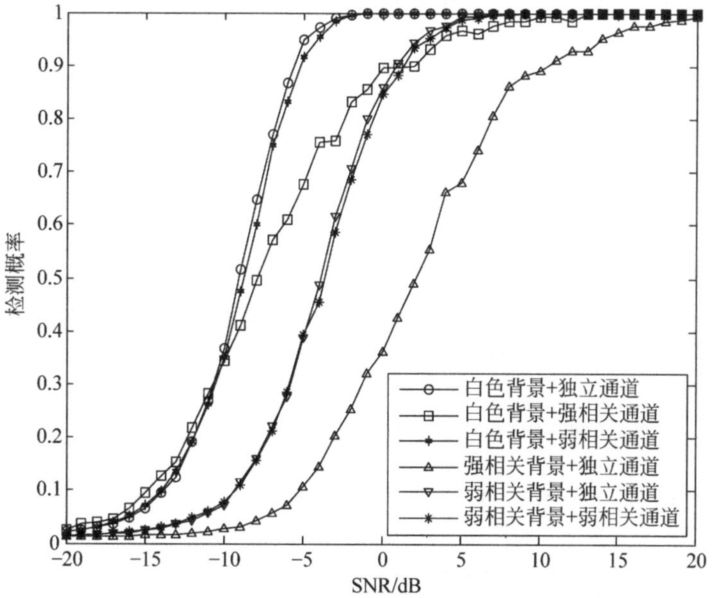

(b) 解析计算结果

图 13.9 AMF 检测器在六种场景中的性能比较

13.5 小结

本章讨论了多传感器分布式 CFAR 检测，具体包括基于二元局部判决的分布式 CA-CFAR 检测和分布式 OS-CFAR 检测、基于 R 类和 S 类局部检测统计量的分布式 CFAR 检测。

13.2 节研究了基于二元局部判决的分布式 CFAR 检测问题，介绍了分布式 CA-CFAR 和分布式 OS-CFAR 的基本原理，并基于“或”和“与”融合规则详细讨论了这两种分布 CFAR 检测局部标称化因子的优化方法。在分布式 CFAR 检测中,"或"规则优于"与"规则,在均匀背景条件下,CA-CFAR 优于 OS-CFAR,这与单传感器 CFAR 处理结论是一致的。

13.3 节介绍了基于局部检测统计量的分布式 CFAR 检测。基于 R 类局部检测统计量的分布式 CFAR 检测克服了集中式 CFAR 检测中局部背景噪声功率水平间相对关系的变化对 CFAR 性能的影响,并且在简单的求和融合时也具有接近最优的性能,而且数据形式非常简单,所以要求 CFAR 处理在局部处理中完成。

阵列雷达是现代雷达的主流形式,分布式阵列的信号检测也是多传感器分布式检测中有意义的研究内容。文献[34]研究了多传感器阵列信号的并行分布式 CFAR 检测。随着 MIMO 体制的兴起,分布式 MIMO 检测逐渐引起关注,其中也有 CFAR 处理问题。13.4 节以分布式 MIMO 雷达为研究对象,讨论了 AMF CFAR 检测器的设计及其性能,在各观测通道背景统计独立的条件下,AMF 检测器本质上就是各局部检测统计量非相参积累的结果。MIMO 发射正交波形从多个方向观测目标,有助于动目标检测,文献[35]研究了分布式 MIMO 雷达对动目标的检测。文献[36]将广义似然比检测拓展到分布式 MIMO 中。文献[37]在分布式 MIMO 的各空间分集通道中分别做 CFAR 检测,将通过检测的通道的检测统计量送到融合中心。实际上,若融合中心的全局检测统计量可以表示为局部检测统计量的函数,这些研究内容都可以归结基于局部检测统计量的分布式 CFAR 检测。

关于多传感器分布式 CFAR 检测方法还有其他方法。例如,文献[38]研究了利用模糊逻辑融合的分布式 CFAR 检测,文献[39]研究了基于遗传算法的并行分布式 CFAR 检测的最优化问题。文献[40]对分布式检测的研究成果进行了系统的梳理。

参考文献

[1] Varshney P K. Distributed detection and data fusion[M]. New York: Springer-Verlag, 1996.

[2] 何友, 王国宏, 关欣, 等. 信息融合理论及应用[M]. 北京: 电子工业出版社, 2010.

[3] 关键, 何友, 彭应宁. 多传感器分布式检测综述[J]. 系统工程与电子技术, 2000, 22(12): 11-15.

[4] Tenney R R, Sandell N R. Detection with distributed sensors[J]. IEEE Transactions on Aerospace and Electronic Systems, 1981, 17(4): 501-510.

[5] Reibman A R, Nolte L W. Design and performance comparison of distributed detection networks[J]. IEEE Trans. on AES, 1987, 23(6): 789-797.

[6] Tang Z B, Pattipati K R, Kleinman D L. Optimization of distributed detection networks: Part II-tree structures[J]. IEEE Trans. on SMC, 1993, 23(1): 211-221.

[7] Barkat M, Varshney P K. Decentralized CFAR signal detection[J]. IEEE Trans. on AES, 1989, 25(2): 141-149.

[8] Barkat M, Varshney P K. Adaptive cell-averaging CFAR detection in distributed sensor networks[J]. IEEE Trans. on AES, 1991, 27(3): 424-429.

[9] Elias-Fuste A R, Broquetas-Ibars A, Antequera J P, et al. CFAR data fusion center with inhomogeneous receivers[J]. IEEE Trans. on AES, 1992, 28(1): 276-285.

[10] Uner M K, Varshney P K. Distributed CFAR detection in homogeneous and nonhomogeneous backgrounds[J]. IEEE Trans. on AES, 1996, 32(1): 84-97.

[11] Gini F, Lombardini F, Verrazzani L. Decentralized CFAR detection with binary integration in Weibull Clutter[J]. IEEE Trans. on AES, 1997, 33(2): 396-407.

[12] 王明宇. 复杂环境下雷达 CFAR 检测与分布式雷达 CFAR 检测研究[D]. 西安: 西北工业大学, 2001.

[13] 梁元辉. 分布式 CFAR 融合检测算法研究[D]. 成都: 西南交通大学, 2012.

[14] Amirmehrabi H, Viswanathan R. A new distributed constant false alarm rate detector[J]. IEEE Trans. on AES,

1997，33(1)：85-97.
[15] 关键. 多传感器分布式恒虚警率检测(CFAR)算法研究[D]. 北京：清华大学，2000.
[16] Guan Jian，Peng Ying-Ning，He You. Three types of distributed CFAR detection based on local test statistic[J]. IEEE Trans. on AES，2002，38(1)：278-288.
[17] 关键，何友，彭应宁. 基于局部观测信噪比的新的分布式 CFAR 检测[J]. 清华大学学报，1999，39(1)：51-54.
[18] 关键，何友，彭应宁. 分布式 OS-CFAR 检测在多脉冲非相干积累条件下的性能分析[J]. 电子科学学刊，2000，22(5)：747-752.
[19] 魏玺章. 分布式 CFAR 检测理论研究[D]. 长沙：国防科技大学，2002.
[20] 严军. 多传感器检测系统融合与关联算法研究[D]. 北京：清华大学，2006.
[21] 严军，关键，彭应宁. 分布式检测系统的混合融合算法[J]. 清华大学学报(自然科学版)，2006，46(1)：46-49.
[22] 严军，关键，彭应宁. 分布式检测系统的幂求和融合准则[J]. 系统工程与电子技术，2006，28(1)：7-10.
[23] 夏畅雄，彭应宁，关键. 基于局部检测统计量的反馈多传感器分布式检测[J]. 清华大学学报(自然科学版)，2005，45(1)：33-36.
[24] 刘向阳，彭应宁，关键. 两种基于局部广义符号统计量的融合准则[J]. 清华大学学报(自然科学版)，2006，46(7)：1223-1226.
[25] 邵志强. 分布式雷达的信号融合检测方法研究[D]. 西安：西安电子科技大学，2019.
[26] Guan Jian，Huang Yong. Detection performance analysis for MIMO radar with distributed apertures in Gaussian colored noise[J]. Sci China Ser F-Inf，2009，52(9)：1689-1696.
[27] Fishler E，Haimovich A，Blum R，et al. MIMO radar：an idea whose time has come[C]. Philadelphia：Proceedings of the IEEE Radar Conference，2004.
[28] 陈沁根. 分布式 MIMO 雷达数据融合检测算法研究[D]. 南京：南京大学，2019.
[29] 廖羽宇. 统计 MIMO 雷达检测理论研究[D]. 成都：电子科技大学，2012.
[30] Li Jian，Petre S. MIMO Radar Signal Processing[M]. WILEY & SONS Publication，2009.
[31] Lehmann N，Haimovich A M，Blum R S，et al. MIMO-radar application to moving target detection in homogeneous clutter[C]. Waltham：In Proceedings of the 14th Annual Workshop on Adaptive Sensor Array and Multi-channel Processing，2006.
[32] 黄勇，黄涛，刘宁波，等. 基于均匀性判定规则的统计 MIMO 雷达多通道融合检测技术[J]. 海军航空工程学院学报. 2018，33(1)：101-105.
[33] Kraut S，Louis L S，McWhorter L T. Adaptive subspace detectors[J]. IEEE Transactions on Signal Processing，2001，49(1)：1-16.
[34] 严军，关键，彭应宁. 多传感器阵列信号的分布式恒虚警检测[J]. 清华大学学报(自然科学版)，2004，44(7)：950-953.
[35] He Q，Lehmann，Nikolaus H，et al. MIMO Radar Moving Target Detection in Homogeneous Clutter[J]. IEEE Transactions on Aerospace & Electronic Systems，2010，46(3)：1290-1301.
[36] Zhou S，Liu H. Signal fusion-based target detection algorithm for spatial diversity radar[J]. Iet Radar Sonar & Navigation，2011，5(3)：204-214.
[37] 胡勤振. 分布式 MIMO 雷达目标检测若干关键技术研究[D]. 西安：西安电子科技大学，2016.
[38] Hammoudi Z，Soltani F. Distributed CA-CFAR and OS-CFAR detection using fuzzy spaces and fuzzy fusion rules[J]. IEE Proceedings-Radar Sonar and Navigation，2004，151(3)：135-142.
[39] Yu Ze，Zhou Yinqing. Parallel Distributed CFAR Detection Optimization Based on Genetic Algorithm with Interval Encoding[J]. Chinese Journal of Aeronautics 2010，23：351-358.
[40] 刘向阳，许稼，彭应宁. 多传感器分布式信号检测理论与方法[M]. 北京：国防工业出版社，2017.

第 14 章 多维 CFAR 处理

CHAPTER 14

14.1 引言

雷达技术的发展遵循由简单到复杂、由初级到高级的一般规律，并呈现由低维度探测到高维度探测逐步演进的趋势[1]。20 世纪前半叶是雷达发展的初步阶段，这一时期的雷达技术以单站、单传感器构型和一维时域信号处理为主。从 20 世纪 60 年代脉冲多普勒雷达（1959 年）、相控阵雷达（1960 年）和合成孔径雷达（1966 年）的诞生，到固态有源相控阵（1980 年）、星载合成孔径雷达（1988 年）的出现，雷达技术相继取得重大突破，单站、多传感器简单构型（模拟相控阵）和二维信号处理（距离-多普勒、距离-方位）成为主要特征。从 20 世纪 90 年代至今，在电子信息技术和各领域应用需求的共同推动下，各种雷达新体制、新系统和新方法仍呈现飞速发展的态势，如数字阵列雷达、三维合成孔径雷达、综合脉冲孔径雷达、多输入多输出（Multiple Input Multiple Output，MIMO）雷达[2]等相继诞生，这一时期的雷达技术呈多站、多传感器构型和多维信号处理的特点。作为当前多维信号处理技术研究领域常用的典型雷达体制，MIMO 雷达是在综合脉冲孔径雷达（Synthetic Impulse and Aperture Radar，SIAR）的基础上，由无线通信领域中的 MIMO 技术演变而来的。根据收发天线的间距大小，可将 MIMO 雷达分为分布式 MIMO 雷达和集中式 MIMO 雷达（相参 MIMO 雷达、MIMO 阵列雷达），两者都基于波形分集（Waveform Diversity，WD）的思想。不同之处在于，前者通过从不同视角观测目标，可以克服目标的角闪烁效应；后者可以看作数字阵列雷达的广义形式，且相对于数字阵列雷达具有体制上的优势[3]。

雷达体制由一维向多维发展，本质上就是雷达系统能够提供的信息处理自由度越来越多，表现在信号处理领域就是“多维信号处理”，而本章要讨论的“多维 CFAR 处理”就是多维信号处理中以 CFAR 检测为目的的那一部分内容。多维信号处理能够进一步增强对微弱目标的探测能力，提升雷达系统的抗干扰、抗杂波性能，适应更加复杂的工作环境和满足不断提高的对高性能雷达的需求。

“多维”的概念很宽泛，为了方便后续讨论，这里从信息来源的角度界定本章中“多维”的范围，包括阵元维（数字阵列、多站布阵）、脉冲维（相参/非相参脉冲串）、波形维（脉冲压缩波形、脉间编码波形、频率分集波形等）、扫描帧维（雷达对目标的一次重访为一个扫描帧）及极化维（极化分集、多极化雷达、全极化雷达）等。

基于上述对“多维”范围的界定，本章内容安排如下。14.2 节讨论阵列雷达 CFAR 检测，这属于“阵元-脉冲”二维联合 CFAR 检测，具体包括复高斯背景中的秩 1 目标 CFAR 检测与子空间目标 CFAR 检测等内容。14.3 节利用 MIMO 雷达提供的脉间编码波形自由度，结合三种优化准则，讨论基于自适应空时编码设计的“阵元-波形”二维联合 CFAR 检测问题。14.4 节结合 MIMO 雷达讨论基于空-时-距自适应处理的自适应聚焦、检测一体化处理技术，这属于“阵元-脉冲-波形”三维联合 CFAR 检测。14.5 节从扫描帧间融合检测及极化 CFAR 等角度，简要讨论了其他几种多维 CFAR 检测技术。另外，从多雷达融合检测的角度来讲，第 13 章讨论的多传感器分布式 CFAR 处理也属于多维 CFAR 检测的范畴，因此本章不单独讨论多雷达层面的多维 CFAR 检测。

14.2 阵列雷达 CFAR 检测

14.2.1 信号模型与二元假设检验

1. 观测向量

假设一维线阵天线有 M 个阵元，每个阵元同时发射一个包含 N 个单载频脉冲的脉冲串，接收时对每个脉冲只做一次快拍，即每个距离分辨单元采样一个点，那么一个距离单元中，由 N 次快拍得到的这些观测构成 $MN\times 1$ 维的观测向量[4]，记为

$$\boldsymbol{x}=[\boldsymbol{x}_1^{\mathrm{T}}\quad \boldsymbol{x}_2^{\mathrm{T}}\quad \cdots\quad \boldsymbol{x}_N^{\mathrm{T}}]^{\mathrm{T}}=[\boldsymbol{x}_{11}\quad \cdots\quad \boldsymbol{x}_{1M}\quad \cdots\quad \boldsymbol{x}_{N1}\quad \cdots\quad \boldsymbol{x}_{NM}]^{\mathrm{T}} \tag{14.1}$$

其中，$\boldsymbol{x}_n=[x_{n1}\quad x_{n2}\quad \cdots\quad x_{nM}]^{\mathrm{T}}(n=1,2,\cdots,N)$是第 n 个脉冲的回波在 M 个阵元上做一次快拍构成的 M 维观测向量。

2. 目标模型

对于脉间不起伏的目标回波（对应于 Swerling 模型中的 0 型、Ⅰ型和Ⅲ型）来说，各个阵元和脉冲上获得的目标回波复幅度（含模值和初始相位）是相同的，但这 MN 个样本值之间存在由阵元间距和多普勒频率引起的相位差，且相邻两元素的相位差是固定的，因此，可将 $MN\times 1$ 维目标回波向量记为

$$\boldsymbol{x}_{\mathrm{s}}=b\boldsymbol{s} \tag{14.2}$$

其中，b 表示目标观测向量的复幅度，$\boldsymbol{s}$ 称为目标导向向量，其元素为

$$\boldsymbol{s}_{nm}=\exp\{-\mathrm{j}2\pi[(m-1)d\sin\theta/\lambda-(n-1)f_{\mathrm{d}}]\} \tag{14.3}$$

其中，θ 和 f_{d} 分别表示目标方向角和目标多普勒频率，在 CFAR 检测领域中，这两个目标参数通常是通过搜索得到的，因此可以认为是已知的，进而可假设目标导向向量 $\boldsymbol{s}$ 是已知的。式(14.2)所表示的目标信号模型称为秩 1 目标模型[5]。其他参数：d 表示阵元间距，λ 表示波长，m 表示第 m 个阵元，n 表示第 n 个脉冲。

对于脉间快起伏的目标回波（对应 Swerling 模型中的Ⅱ型和Ⅳ型）来说，一次快拍各个阵元上获得的目标回波复幅度（含模值和初始相位）是相同的，但不同快拍间的回波复幅度是统计独立的，因此 $MN\times 1$ 维目标回波向量 $\boldsymbol{x}_{\mathrm{s}}$ 可表示为

$$\begin{bmatrix} b_1 s_{11} & \cdots & b_N s_{1N} \\ \vdots & \ddots & \vdots \\ b_1 s_{M1} & \cdots & b_N s_{MN} \end{bmatrix} \xrightarrow{\text{矢量化}} \boldsymbol{x}_{\mathrm{s}}=\begin{bmatrix} b_1\boldsymbol{s}_1 \\ \vdots \\ b_N\boldsymbol{s}_N \end{bmatrix}=\begin{bmatrix} \boldsymbol{s}_1 & \cdots & 0 \\ \vdots & \ddots & \vdots \\ 0 & \cdots & \boldsymbol{s}_N \end{bmatrix}\cdot\begin{bmatrix} b_1 \\ \vdots \\ b_N \end{bmatrix}=\boldsymbol{S}\boldsymbol{b} \tag{14.4}$$

其中，$\boldsymbol{S}$ 表示目标所在的 N 维子空间，$\boldsymbol{b}$ 表示目标的 N 维复幅度向量。

对于更一般的情况，$MN\times 1$ 维目标回波向量 $\boldsymbol{x}_{\mathrm{s}}$ 可表示为

$$\boldsymbol{x}_{\mathrm{s}}=\boldsymbol{\Psi}\boldsymbol{\alpha} \tag{14.5}$$

式(14.5)所表示的目标信号模型称为子空间目标模型[6]，其中，$\boldsymbol{\alpha}$ 表示目标的 r 维复幅度向量，$\boldsymbol{\Psi}$ 表示目标所在的 r 维目标子空间；在目标检测领域，$\boldsymbol{\Psi}$ 可通过先验信息来构建，亦可通过搜索来获得，因此

在上述二元检测问题中,一般假设$\boldsymbol{\Psi}$是已知的。

值得注意的是,从上述建模过程来看,目标的复幅度要么是随机变量,要么是随机向量,但是在CFAR检测过程中,由于缺乏关于目标复幅度统计特性的先验信息(如概率密度函数),因此一般将目标的复幅度(向量)建模为确定性未知量,并采用极大似然估计对其进行估计。

3. 杂波模型

将$MN\times 1$维杂波观测向量记为$\boldsymbol{x}_{\mathrm{c}}$,用零均值复高斯模型[5,7]对其复幅度进行建模,则其概率密度函数的表达式为

$$f(\boldsymbol{x}_{\mathrm{c}})=\frac{1}{\pi^{MN}\left|\boldsymbol{R}_{\mathrm{c}}\right|}\exp\{-\boldsymbol{x}_{\mathrm{c}}^{\mathrm{H}}\boldsymbol{R}_{\mathrm{c}}^{-1}\boldsymbol{x}_{\mathrm{c}}\} \tag{14.6}$$

其中,$\boldsymbol{R}_{\mathrm{c}}$表示杂波协方差矩阵。

4. 检测模型

阵列雷达目标检测问题就是解决如下二元假设检验问题

$$\begin{cases}\boldsymbol{x}=\boldsymbol{x}_{\mathrm{s}}+\boldsymbol{x}_{\mathrm{c}}+\boldsymbol{x}_{\mathrm{n}}, & \mathrm{H}_1\\ \boldsymbol{x}=\boldsymbol{x}_{\mathrm{c}}+\boldsymbol{x}_{\mathrm{n}}, & \mathrm{H}_0\end{cases} \tag{14.7}$$

其中,$\boldsymbol{x}$表示检测单元中的观测向量,$\boldsymbol{x}_{\mathrm{s}}$、$\boldsymbol{x}_{\mathrm{c}}$、$\boldsymbol{x}_{\mathrm{n}}$分别表示目标观测向量、杂波观测向量和噪声观测向量,$\boldsymbol{x}_{\mathrm{cn}}=\boldsymbol{x}_{\mathrm{c}}+\boldsymbol{x}_{\mathrm{n}}$表示由杂波观测向量与噪声观测向量构成的背景观测向量,且$\boldsymbol{x}_{\mathrm{c}}$与$\boldsymbol{x}_{\mathrm{n}}$相互统计独立,$\boldsymbol{x}_{\mathrm{n}}$也服从零均值复高斯分布,协方差记为$\boldsymbol{R}_{\mathrm{n}}$;在本章后续内容中,$\boldsymbol{x}_{\mathrm{s}}$分别采用秩1目标模型和子空间目标模型建模,$\boldsymbol{x}_{\mathrm{cn}}$采用零均值复高斯模型建模,协方差记为$\boldsymbol{R}_{\mathrm{cn}}$。下面分别给出不同检测器对应的模型假设情况。

模型假设1:采用秩1目标模型和复高斯背景模型,假设检测单元的背景协方差矩阵与参考单元的背景协方差矩阵相同,且估计检测单元背景协方差矩阵时不考虑检测单元观测向量的影响;基于此假设设计的检测器称为AMF检测器[8]。

模型假设2:采用秩1目标模型和复高斯背景模型,假设检测单元的背景协方差矩阵与参考单元的背景协方差矩阵相同,且估计检测单元背景协方差矩阵时考虑检测单元观测向量的影响;基于此假设设计的检测器称为Kelly检测器[5]。

模型假设3:采用秩1目标模型和复高斯背景模型,假设检测单元的背景协方差矩阵与参考单元的背景协方差矩阵仅相差一个比例因子;基于此假设设计的检测器称为Kalson检测器[9]。

模型假设4:采用子空间目标模型和复高斯背景模型,假设检测单元的背景协方差矩阵与参考单元的背景协方差矩阵相同,且估计检测单元背景协方差矩阵时不考虑检测单元观测向量的影响;基于此假设设计的检测器称为MSD检测器[6]。

模型假设5:采用子空间目标模型和复高斯背景模型,假设检测单元的背景协方差矩阵与参考单元的背景协方差矩阵相同,且估计检测单元背景协方差矩阵时考虑检测单元观测向量的影响;基于此假设设计的检测器称为ASD检测器[10]。

模型假设6:采用子空间目标模型和复高斯背景模型,假设检测单元的背景协方差矩阵与参考单元的背景协方差矩阵仅相差一个比例因子;基于此假设设计的检测器称为COS^2检测器[10]。

上述6种情况中,一个共同的假设前提是,检测单元与参考单元的背景协方差矩阵结构都是相同的,最多只相差一个比例因子,这是自适应处理能够进行杂波抑制的物理基础。

关于上述6种阵列信号检测器及其在不同雷达体制以及杂波背景中的研究情况,有兴趣的读者可

参见文献[11-33]；另外，本章只讨论了复高斯背景，而对于复合高斯杂波条件下检测器的推导，有兴趣的读者可以查阅相关文献[34-37]，本书不再赘述。

14.2.2 秩1目标模型下的阵列雷达目标检测器

秩1目标模型条件下，二元假设检验问题可表示为

$$\begin{cases} \boldsymbol{x}=b\boldsymbol{s}+\boldsymbol{x}_{\mathrm{cn}}, & \mathrm{H}_1 \\ \boldsymbol{x}=\boldsymbol{x}_{\mathrm{cn}}, & \mathrm{H}_0 \end{cases} \tag{14.8}$$

其中，b 为未知的目标复幅度，$\boldsymbol{s}$ 为已知的目标导向向量，$\boldsymbol{x}_{\mathrm{cn}}\sim\mathrm{CN}(\boldsymbol{0},\boldsymbol{R}_{\mathrm{cn}})$ 表示 $\boldsymbol{x}_{\mathrm{cn}}$ 服从均值为零向量，协方差为 $\boldsymbol{R}_{\mathrm{cn}}$ 的复高斯分布。$\boldsymbol{x}$ 为检测单元数据，另有 K 个参考单元观测向量，记为，$\boldsymbol{x}_k$，$k=1,2,\cdots K$，仅包含背景观测向量；检测单元观测向量与参考单元观测向量相互统计独立。

1. AMF检测器

AMF检测器的推导是建立在14.2.1节中的模型假设1基础上的[8]。对于式(14.8)的检测问题，先假设 $\boldsymbol{R}_{\mathrm{cn}}$ 已知，于是在 H_0 假设下，检测单元中的观测向量的概率密度函数为

$$f_0(\boldsymbol{x})=\frac{1}{\pi^{MN}\left|\boldsymbol{R}_{\mathrm{cn}}\right|}\mathrm{e}^{-\boldsymbol{x}^{\mathrm{H}}\boldsymbol{R}_{\mathrm{cn}}^{-1}\boldsymbol{x}} \tag{14.9}$$

H_1 假设下的检测单元观测向量的概率密度函数为

$$f_1(\boldsymbol{x})=\frac{1}{\pi^{MN}\left|\boldsymbol{R}_{\mathrm{cn}}\right|}\mathrm{e}^{-(\boldsymbol{x}-b\boldsymbol{s})^{\mathrm{H}}\boldsymbol{R}_{\mathrm{cn}}^{-1}(\boldsymbol{x}-b\boldsymbol{s})} \tag{14.10}$$

对式(14.10)中的 b 求极大似然估计，可得

$$\hat{b}=\frac{\boldsymbol{s}^{\mathrm{H}}\boldsymbol{R}_{\mathrm{cn}}^{-1}\boldsymbol{x}}{\boldsymbol{s}^{\mathrm{H}}\boldsymbol{R}_{\mathrm{cn}}^{-1}\boldsymbol{s}} \tag{14.11}$$

将式(14.11)代入似然比检验 Λ，并化简可得

$$\Lambda=\frac{f_1(\cdot)}{f_0(\cdot)}\xrightarrow{\text{化简}}l=\frac{\left|\boldsymbol{s}^{\mathrm{H}}\boldsymbol{R}_{\mathrm{cn}}^{-1}\boldsymbol{x}\right|^2}{\boldsymbol{s}^{\mathrm{H}}\boldsymbol{R}_{\mathrm{cn}}^{-1}\boldsymbol{s}} \tag{14.12}$$

采用样本协方差矩阵 $\boldsymbol{M}$ 估计背景协方差矩阵 $\boldsymbol{R}_{\mathrm{cn}}$

$$\boldsymbol{M}=\frac{1}{K}\sum_{k=1}^{K}\boldsymbol{x}_k\boldsymbol{x}_k^{\mathrm{H}} \tag{14.13}$$

从而得到AMF检测器

$$T_{\mathrm{AMF}}=\frac{\left|\boldsymbol{s}^{\mathrm{H}}\boldsymbol{M}^{-1}\boldsymbol{x}\right|^2}{\boldsymbol{s}^{\mathrm{H}}\boldsymbol{M}^{-1}\boldsymbol{s}} \tag{14.14}$$

2. Kelly检测器

Kelly检测器的推导是建立在14.2.1节中的模型假设2基础上的[5]。不同于AMF检测器的推导过程，在Kelly检测器的推导过程中，由于需要考虑检测单元观测向量的影响，因此不能直接用样本协方差矩阵作为检测单元背景协方差矩阵的估计，而是需要分别对 H_1 和 H_0 条件下检测单元中的背景协方差矩阵进行估计。

对于式(14.8)的检测问题，在 H_0 假设下，参考单元与检测单元观测向量的联合概率密度函数为

$$f_0(\boldsymbol{x},\boldsymbol{x}_1,\cdots,\boldsymbol{x}_K)=\left[\frac{1}{\pi^{MN}\left|\boldsymbol{R}_{\mathrm{cn},0}\right|}\mathrm{e}^{-\mathrm{tr}(\boldsymbol{R}_{\mathrm{cn},0}^{-1}\boldsymbol{T}_0)}\right]^{K+1} \tag{14.15}$$

$$\boldsymbol{T}_0=\frac{1}{K+1}\left(\boldsymbol{x}\boldsymbol{x}^{\mathrm{H}}+\sum_{k=1}^{K}\boldsymbol{x}_k\boldsymbol{x}_k^{\mathrm{H}}\right) \tag{14.16}$$

其中,tr(·)表示对矩阵求迹。在 H_1 假设下,参考单元与检测单元观测向量的联合概率密度函数为

$$f_1(\boldsymbol{x},\boldsymbol{x}_1,\cdots,\boldsymbol{x}_K)=\left[\frac{1}{\pi^{MN}\left|\boldsymbol{R}_{\mathrm{cn},1}\right|}\mathrm{e}^{-\mathrm{tr}(\boldsymbol{R}^{-1}{}_{\mathrm{cn},1}\boldsymbol{T}_1)}\right]^{K+1} \tag{14.17}$$

$$\boldsymbol{T}_1=\frac{1}{K+1}\left[(\boldsymbol{x}-b\boldsymbol{s})(\boldsymbol{x}-b\boldsymbol{s})^{\mathrm{H}}+\sum_{k=1}^{K}\boldsymbol{x}_k\boldsymbol{x}_k^{\mathrm{H}}\right] \tag{14.18}$$

对式(14.15)和式(14.17)分别关于 $\boldsymbol{R}_{\mathrm{cn}}$ 求偏导可得,无目标时,$\boldsymbol{R}_{\mathrm{cn},0}$ 的极大似然估计为 $\hat{\boldsymbol{R}}_{\mathrm{cn},0}=\boldsymbol{T}_0$;有目标时,$\boldsymbol{R}_{\mathrm{cn},1}$ 的极大似然估计为 $\hat{\boldsymbol{R}}_{\mathrm{cn},1}=\boldsymbol{T}_1$。将 $\hat{\boldsymbol{R}}_{\mathrm{cn},1}$ 代入式(14.17),并求 b 的极大似然估计,可得 $\hat{b}=\dfrac{\boldsymbol{s}^{\mathrm{H}}\boldsymbol{M}^{-1}\boldsymbol{x}}{\boldsymbol{s}^{\mathrm{H}}\boldsymbol{M}^{-1}\boldsymbol{s}}$,其中 $\boldsymbol{M}$ 如式(14.13)所示。然后,将 $\hat{b}$、$\hat{\boldsymbol{R}}_{\mathrm{cn},0}$ 和 $\hat{\boldsymbol{R}}_{\mathrm{cn},1}$ 代入式(14.12)所示的似然比检验 Λ,化简后得到 Kelly 检测器

$$T_{\mathrm{Kelly}}=T_{\mathrm{GLRT}}=\frac{\left|\boldsymbol{s}^{\mathrm{H}}\boldsymbol{M}^{-1}\boldsymbol{x}\right|^2}{\boldsymbol{s}^{\mathrm{H}}\boldsymbol{M}^{-1}\boldsymbol{s}(K+\boldsymbol{x}^{\mathrm{H}}\boldsymbol{M}^{-1}\boldsymbol{x})} \tag{14.19}$$

3. Kalson 检测器

Kalson 检测器的推导是建立在 14.2.1 节中的模型假设 3 基础上的[9]。类似于 Kelly 检测器的推导过程,Kalson 检测器推导过程中也需要考虑检测单元观测向量对检测单元背景协方差矩阵估计的影响;同时,又因为检测单元的背景协方差矩阵与参考单元的背景协方差矩阵相差一个比例因子,因此需要考虑 H_0 假设与 H_1 假设下该比例因子的不同;于是,需要估计的参数包括 $\boldsymbol{R}_{\mathrm{cn},0}$、$\boldsymbol{R}_{\mathrm{cn},1}$、$\sigma_0^2$、$\sigma_1^2$、$b$。

有目标时的似然函数为

$$\begin{aligned}f(\boldsymbol{x}\mid\mathrm{H}_1)&=\frac{1}{\left[\pi^{MN}\sigma_1^{2MN/(K+1)}\left|\boldsymbol{R}_{\mathrm{cn},1}\right|\right]^{K+1}}\mathrm{e}^{-\frac{(\boldsymbol{x}-b\boldsymbol{s})^{\mathrm{H}}\boldsymbol{R}_{\mathrm{cn},1}^{-1}(\boldsymbol{x}-b\boldsymbol{s})}{\sigma_1^2}-\sum_{k=1}^{K}\boldsymbol{x}_k^{\mathrm{H}}\boldsymbol{R}_{\mathrm{cn},1}^{-1}\boldsymbol{x}_k}\\&=\left[\frac{1}{\pi^{MN}\sigma_1^{2MN/(K+1)}\left|\boldsymbol{R}_{\mathrm{cn},1}\right|}\mathrm{e}^{-\mathrm{tr}(\boldsymbol{R}_{\mathrm{cn},1}^{-1}\boldsymbol{T}_1)}\right]^{K+1}\end{aligned} \tag{14.20}$$

无目标时的似然函数为

$$\begin{aligned}f(\boldsymbol{x}\mid\mathrm{H}_0)&=\frac{1}{\left[\pi^{MN}\sigma_0^{2MN/(K+1)}\left|\boldsymbol{R}_{\mathrm{cn},0}\right|\right]^{K+1}}\mathrm{e}^{-\frac{\boldsymbol{x}^{\mathrm{H}}\boldsymbol{R}_{\mathrm{cn},0}^{-1}\boldsymbol{x}}{\sigma_0^2}-\sum_{k=1}^{K}\boldsymbol{x}_k^{\mathrm{H}}\boldsymbol{R}_{\mathrm{cn},0}^{-1}\boldsymbol{x}_k}\\&=\left[\frac{1}{\pi^{MN}\sigma_0^{2MN/(K+1)}\left|\boldsymbol{R}_{\mathrm{cn},0}\right|}\mathrm{e}^{-\mathrm{tr}(\boldsymbol{R}_{\mathrm{cn},0}^{-1}\boldsymbol{T}_0)}\right]^{K+1}\end{aligned} \tag{14.21}$$

其中,$\boldsymbol{T}_0=\dfrac{1}{K+1}\left(\dfrac{\boldsymbol{x}\boldsymbol{x}^{\mathrm{H}}}{\sigma_0^2}+\displaystyle\sum_{k=1}^{K}\boldsymbol{x}_k\boldsymbol{x}_k^{\mathrm{H}}\right)$, $\boldsymbol{T}_1=\dfrac{1}{K+1}\left[\dfrac{(\boldsymbol{x}-b\boldsymbol{s})(\boldsymbol{x}-b\boldsymbol{s})^{\mathrm{H}}}{\sigma_1^2}+\displaystyle\sum_{k=1}^{K}\boldsymbol{x}_k\boldsymbol{x}_k^{\mathrm{H}}\right]$。

对上面两个似然函数关于 $\boldsymbol{R}_{\mathrm{cn}}$ 分别求偏导可得,无目标时,$\boldsymbol{R}_{\mathrm{cn},0}$ 的极大似然估计为 $\hat{\boldsymbol{R}}_{\mathrm{cn},0}=\boldsymbol{T}_0$;有目标时,$\boldsymbol{R}_{\mathrm{cn},1}$ 的极大似然估计为 $\hat{\boldsymbol{R}}_{\mathrm{cn},1}=\boldsymbol{T}_1$。将 $\hat{\boldsymbol{R}}_{\mathrm{cn},0}=\boldsymbol{T}_0$,$\hat{\boldsymbol{R}}_{\mathrm{cn},1}=\boldsymbol{T}_1$ 代入似然比表达式中,并对 σ_0^2 和 σ_1^2 分别求极大似然估计得

$$\hat{\sigma}_1^2=\frac{K-N+1}{KN}(\boldsymbol{x}-b\boldsymbol{s})^{\mathrm{H}}\boldsymbol{M}^{-1}(\boldsymbol{x}-b\boldsymbol{s}),\quad \hat{\sigma}_0^2=\frac{K-N+1}{KN}\boldsymbol{x}^{\mathrm{H}}\boldsymbol{M}^{-1}\boldsymbol{x} \tag{14.22}$$

其中，$\boldsymbol{M}$ 如式(14.13)所示。将 $\hat{\boldsymbol{R}}_{\text{cn},0}$、$\hat{\boldsymbol{R}}_{\text{cn},1}$、$\hat{\sigma}_1^2$、$\hat{\sigma}_0^2$ 代入似然比表达式，并对 b 求极大似然估计得，$\hat{b}=\dfrac{\boldsymbol{s}^{\text{H}}\boldsymbol{M}^{-1}\boldsymbol{x}}{\boldsymbol{s}^{\text{H}}\boldsymbol{M}^{-1}\boldsymbol{s}}$。将各个未知参数的极大似然估计代入似然比表达式中，化简可得 Kalson 检测器

$$T_{\text{Kalson}}=\frac{|\boldsymbol{s}^{\text{H}}\boldsymbol{M}^{-1}\boldsymbol{x}|^2}{\boldsymbol{s}^{\text{H}}\boldsymbol{M}^{-1}\boldsymbol{s}\cdot\boldsymbol{x}^{\text{H}}\boldsymbol{M}^{-1}\boldsymbol{x}} \tag{14.23}$$

14.2.3 子空间目标模型下的阵列雷达目标检测器

子空间目标模型条件下，目标观测向量为 $\boldsymbol{x}_{\text{s}}=\boldsymbol{\Psi}\boldsymbol{\alpha}$，$\boldsymbol{\alpha}$ 表示未知的目标复幅度向量，$\boldsymbol{\Psi}$ 表示目标子空间。于是，二元假设检验问题可表示为

$$\begin{cases}\boldsymbol{x}=\boldsymbol{\Psi}\boldsymbol{\alpha}+\boldsymbol{x}_{\text{cn}}, & \text{H}_1\\ \boldsymbol{x}=\boldsymbol{x}_{\text{cn}}, & \text{H}_0\end{cases} \tag{14.24}$$

其中，$\boldsymbol{x}_{\text{cn}}\sim\text{CN}(\boldsymbol{0},\boldsymbol{R}_{\text{cn}})$ 表示 $\boldsymbol{x}_{\text{cn}}$ 服从均值为零向量，协方差为 $\boldsymbol{R}_{\text{cn}}$ 的复高斯分布。$\boldsymbol{x}$ 为检测单元数据，另有 K 个参考单元观测向量，记为 $\boldsymbol{x}_k$，$k=1,2,\cdots,K$，仅包含背景观测向量；检测单元观测向量与参考单元观测向量相互统计独立。

1. MSD 检测器

MSD 检测器的推导是建立在 14.2.1 节中的模型假设 4 基础上的[6]。推导过程中，先假设背景协方差矩阵已知，然后结合似然比检验和未知参数$\boldsymbol{\alpha}$ 的极大似然估计得到检验统计量，接着把基于参考单元数据获得的协方差矩阵的极大似然估计代入这个检验统计量中，从而得到 MSD 检测器。

$K+1$ 个数据向量 $\{\boldsymbol{x}_0,\boldsymbol{x}_1,\cdots,\boldsymbol{x}_K\}$ 是统计独立的，那么，H_1 条件下，此 $K+1$ 个数据向量的联合概率密度函数为

$$f_1=f_1(\boldsymbol{x})\prod_{k=1}^{K}f(\boldsymbol{x}_k)=\frac{1}{\pi^{MN}|\boldsymbol{R}_{\text{cn}}|}\text{e}^{-(\boldsymbol{x}-\boldsymbol{\Psi}\boldsymbol{\alpha})^{\text{H}}\boldsymbol{R}_{\text{cn}}^{-1}(\boldsymbol{x}-\boldsymbol{\Psi}\boldsymbol{\alpha})}\prod_{k=1}^{K}\frac{1}{\pi^{MN}|\boldsymbol{R}_{\text{cn}}|}\text{e}^{-\boldsymbol{x}_k^{\text{H}}\boldsymbol{R}_{\text{cn}}^{-1}\boldsymbol{x}_k} \tag{14.25}$$

H_0 条件下，此 $K+1$ 个数据向量的联合概率密度函数为

$$f_0=f_0(\boldsymbol{x})\prod_{k=1}^{K}f(\boldsymbol{x}_k)=\frac{1}{\pi^{MN}|\boldsymbol{R}_{\text{cn}}|}\text{e}^{-\boldsymbol{x}^{\text{H}}\boldsymbol{R}_{\text{cn}}^{-1}\boldsymbol{x}}\prod_{k=1}^{K}\frac{1}{\pi^{MN}|\boldsymbol{R}_{\text{cn}}|}\text{e}^{-\boldsymbol{x}_k^{\text{H}}\boldsymbol{R}_{\text{cn}}^{-1}\boldsymbol{x}_k} \tag{14.26}$$

计算似然比，并化简得到等价的似然比函数

$$l=\boldsymbol{x}^{\text{H}}\boldsymbol{R}_{\text{cn}}^{-1}\boldsymbol{x}-(\boldsymbol{x}-\boldsymbol{\Psi}\boldsymbol{\alpha})^{\text{H}}\boldsymbol{R}_{\text{cn}}^{-1}(\boldsymbol{x}-\boldsymbol{\Psi}\boldsymbol{\alpha}) \tag{14.27}$$

计算$\boldsymbol{\alpha}$ 的极大似然估计，

$$\frac{\partial l}{\partial\boldsymbol{\alpha}}=\frac{\partial[\boldsymbol{x}^{\text{H}}\boldsymbol{R}_{\text{cn}}^{-1}\boldsymbol{x}-(\boldsymbol{x}-\boldsymbol{\Psi}\boldsymbol{\alpha})^{\text{H}}\boldsymbol{R}_{\text{cn}}^{-1}(\boldsymbol{x}-\boldsymbol{\Psi}\boldsymbol{\alpha})]}{\partial\boldsymbol{\alpha}}=[\boldsymbol{\Psi}^{\text{H}}2\boldsymbol{R}_{\text{cn}}^{-1}(\boldsymbol{x}-\boldsymbol{\Psi}\boldsymbol{\alpha})]=\boldsymbol{0}_{r\times1} \tag{14.28}$$

从而得到$\boldsymbol{\alpha}$ 的极大似然估计为$\hat{\boldsymbol{\alpha}}=\dfrac{\boldsymbol{\Psi}^{\text{H}}\boldsymbol{R}_{\text{cn}}^{-1}\boldsymbol{x}}{\boldsymbol{\Psi}^{\text{H}}\boldsymbol{R}_{\text{cn}}^{-1}\boldsymbol{\Psi}}$。将$\hat{\boldsymbol{\alpha}}$ 代入 l 中，得

$$l=\boldsymbol{x}^{\text{H}}\boldsymbol{R}_{\text{cn}}^{-1}\boldsymbol{\Psi}(\boldsymbol{\Psi}^{\text{H}}\boldsymbol{R}_{\text{cn}}^{-1}\boldsymbol{\Psi})^{-1}\boldsymbol{\Psi}^{\text{H}}\boldsymbol{R}_{\text{cn}}^{-1}\boldsymbol{x} \tag{14.29}$$

在这里，采用式(14.13)所示的样本协方差矩阵 $\boldsymbol{M}$ 来估计检测单元背景的协方差矩阵$\boldsymbol{R}_{\text{cn}}$，于是，得到如下匹配子空间检测器

$$T_{\text{MSD}}=\boldsymbol{x}^{\text{H}}\boldsymbol{M}^{-1}\boldsymbol{\Psi}(\boldsymbol{\Psi}^{\text{H}}\boldsymbol{M}^{-1}\boldsymbol{\Psi})^{-1}\boldsymbol{\Psi}^{\text{H}}\boldsymbol{M}^{-1}\boldsymbol{x} \tag{14.30}$$

2. ASD 检测器

ASD 检测器的推导是建立在 14.2.1 节中的模型假设 5 基础上的[10]。不同于 MSD 检测器的推导

过程,在ASD检测器的推导过程中,由于需要考虑检测单元观测向量的影响,因此不能直接用样本协方差矩阵作为检测单元背景协方差矩阵的估计,而是需要分别对 H_1 和 H_0 条件下检测单元中的背景协方差矩阵进行估计。

H_1 和 H_0 条件下的概率密度函数可变形为

$$f_0=\frac{1}{(\pi^{MN})^{K+1}|\boldsymbol{R}_{\mathrm{cn},0}|^{K+1}}\mathrm{e}^{-\boldsymbol{x}^{\mathrm{H}}\boldsymbol{R}_{\mathrm{cn},0}^{-1}\boldsymbol{x}-\sum\limits_{k=1}^{K}\boldsymbol{x}_k^{\mathrm{H}}\boldsymbol{R}_{\mathrm{cn},0}^{-1}\boldsymbol{x}_k}=\left[\frac{1}{\pi^{MN}|\boldsymbol{R}_{\mathrm{cn},0}|}\mathrm{e}^{-\mathrm{tr}(\boldsymbol{R}_{\mathrm{cn},0}^{-1}\boldsymbol{T}_0)}\right]^{K+1} \tag{14.31}$$

其中,$\boldsymbol{T}_0=\frac{1}{K+1}(\boldsymbol{x}\boldsymbol{x}^{\mathrm{H}}+\sum\limits_{k=1}^{K}\boldsymbol{x}_k\boldsymbol{x}_k^{\mathrm{H}})$,那么

$$|\boldsymbol{T}_0|=\left|\frac{K}{K+1}\left(\frac{1}{K}\boldsymbol{x}\boldsymbol{x}^{\mathrm{H}}+\boldsymbol{M}\right)\right|=\left(\frac{K}{K+1}\right)^{MN}|\boldsymbol{M}|\left(1+\frac{1}{K}\boldsymbol{x}^{\mathrm{H}}\boldsymbol{M}^{-1}\boldsymbol{x}^{\mathrm{H}}\right) \tag{14.32}$$

$$\begin{aligned}f_1&=\frac{1}{(\pi^{MN})^{K+1}|\boldsymbol{R}_{\mathrm{cn},1}|^{K+1}}\mathrm{e}^{-(\boldsymbol{x}-\boldsymbol{\Psi}\boldsymbol{\alpha})^{\mathrm{H}}\boldsymbol{R}_{\mathrm{cn},1}^{-1}(\boldsymbol{x}-\boldsymbol{\Psi}\boldsymbol{\alpha})-\sum\limits_{k=1}^{K}\boldsymbol{x}_k^{\mathrm{H}}\boldsymbol{R}_{\mathrm{cn},1}^{-1}\boldsymbol{x}_k}\\&=\left[\frac{1}{\pi^{MN}|\boldsymbol{R}_{\mathrm{cn},1}|}\mathrm{e}^{-\mathrm{tr}(\boldsymbol{R}_{\mathrm{cn},1}^{-1}\boldsymbol{T}_1)}\right]^{K+1}\end{aligned} \tag{14.33}$$

其中,$\boldsymbol{T}_1=\frac{1}{K+1}[(\boldsymbol{x}-\boldsymbol{\Psi}\boldsymbol{\alpha})(\boldsymbol{x}-\boldsymbol{\Psi}\boldsymbol{\alpha})^{\mathrm{H}}+\sum\limits_{k=1}^{K}\boldsymbol{x}_k\boldsymbol{x}_k^{\mathrm{H}}]$,那么

$$\begin{aligned}|\boldsymbol{T}_1|&=\left|\frac{K}{K+1}\left[\frac{1}{K}(\boldsymbol{x}-\boldsymbol{\Psi}\boldsymbol{\alpha})(\boldsymbol{x}-\boldsymbol{\Psi}\boldsymbol{\alpha})^{\mathrm{H}}+\boldsymbol{M}\right]\right|\\&=\left(\frac{K}{K+1}\right)^{MN}|\boldsymbol{M}|\left[1+\frac{1}{K}(\boldsymbol{x}-\boldsymbol{\Psi}\boldsymbol{\alpha})^{\mathrm{H}}\boldsymbol{M}^{-1}(\boldsymbol{x}-\boldsymbol{\Psi}\boldsymbol{\alpha})\right]\end{aligned} \tag{14.34}$$

求 H_1 和 H_0 条件下的协方差 $\boldsymbol{R}_{\mathrm{cn},0}$ 和 $\boldsymbol{R}_{\mathrm{cn},1}$ 的极大似然估计。计算过程简单描述如下:$f=\frac{1}{a|\boldsymbol{R}|}\cdot\mathrm{e}^{-\mathrm{tr}(\boldsymbol{R}^{-1}\boldsymbol{T})}$,对 f 关于 $\boldsymbol{R}$ 求偏导,并令其为0得到 $\hat{\boldsymbol{R}}=\boldsymbol{T}$。因为在似然比中,$H_1$ 和 H_0 条件下的协方差估计是不一样的,且分别在分子和分母中,故 $\boldsymbol{R}_{\mathrm{cn},0}$ 和 $\boldsymbol{R}_{\mathrm{cn},1}$ 可以分开独立估计。因此可得 $\hat{\boldsymbol{R}}_{\mathrm{cn},0}=\boldsymbol{T}_0$、$\hat{\boldsymbol{R}}_{\mathrm{cn},1}=\boldsymbol{T}_1$。

将 $\hat{\boldsymbol{R}}_{\mathrm{cn},0}=\boldsymbol{T}_0$、$\hat{\boldsymbol{R}}_{\mathrm{cn},1}=\boldsymbol{T}_1$ 代入似然比表达式求解 $\boldsymbol{\alpha}$ 的极大似然估计,也可以代入 f_1 中进行求解,因为只有 f_1 中才有目标未知参数 $\boldsymbol{\alpha}$。得到 $\boldsymbol{\alpha}$ 的极大似然估计为 $\hat{\boldsymbol{\alpha}}=\frac{\boldsymbol{\Psi}^{\mathrm{H}}\boldsymbol{M}^{-1}\boldsymbol{x}}{\boldsymbol{\Psi}^{H}\boldsymbol{M}^{-1}\boldsymbol{\Psi}}$。最后将 $\boldsymbol{R}_{\mathrm{cn},0}$、$\boldsymbol{R}_{\mathrm{cn},1}$、$\boldsymbol{\alpha}$ 的极大似然估计代入似然比检验中,可化简得到ASD检测器

$$T_{\mathrm{ASD}}=\frac{\boldsymbol{x}^{\mathrm{H}}\boldsymbol{M}^{-1}\boldsymbol{\Psi}(\boldsymbol{\Psi}^{\mathrm{H}}\boldsymbol{M}^{-1}\boldsymbol{\Psi})^{-1}\boldsymbol{\Psi}^{\mathrm{H}}\boldsymbol{M}^{-1}\boldsymbol{x}}{K+\boldsymbol{x}^{\mathrm{H}}\boldsymbol{M}^{-1}\boldsymbol{x}} \tag{14.35}$$

3. COS^2 检测器

COS^2 检测器的推导是建立在14.2.1节中的模型假设6基础上的[10]。类似ASD检测器的推导过程,COS^2 检测器推导过程中也需要考虑检测单元观测向量对检测单元背景协方差矩阵估计的影响;同时,又因为检测单元的背景协方差矩阵与参考单元的背景协方差矩阵相差一个比例因子,因此需要考虑 H_0 假设与 H_1 假设下该比例因子的不同;于是,需要估计的参数包括 $\boldsymbol{R}_{\mathrm{cn},0}$、$\boldsymbol{R}_{\mathrm{cn},1}$、$\sigma_0^2$、$\sigma_1^2$、$\boldsymbol{\alpha}$。

$$f(\boldsymbol{x} \mid \mathrm{H}_1)=\frac{1}{[\pi^{MN}\sigma_1^{2MN/(K+1)}|\boldsymbol{R}_{\mathrm{cn},1}|]^{K+1}}\mathrm{e}^{-\frac{(\boldsymbol{x}-\boldsymbol{\Psi s})^{\mathrm{H}}\boldsymbol{R}_{\mathrm{cn},1}^{-1}(\boldsymbol{x}-\boldsymbol{\Psi s})}{\sigma_1^2}-\sum_{k=1}^{K}\boldsymbol{x}_k^{\mathrm{H}}\boldsymbol{R}_{\mathrm{cn},1}^{-1}\boldsymbol{x}_k}$$

$$=\left[\frac{1}{\pi^{MN}\sigma_1^{2MN/(K+1)}|\boldsymbol{R}_{\mathrm{cn},1}|}\mathrm{e}^{-\mathrm{tr}(\boldsymbol{R}_{\mathrm{cn},1}^{-1}\boldsymbol{T}_1)}\right]^{K+1} \tag{14.36}$$

$$f(\boldsymbol{x} \mid \mathrm{H}_0)=\frac{1}{[\pi^{MN}\sigma_0^{2MN/(K+1)}|\boldsymbol{R}_{\mathrm{cn},0}|]^{K+1}}\mathrm{e}^{-\frac{\boldsymbol{x}^{\mathrm{H}}\boldsymbol{R}_{\mathrm{cn},0}^{-1}\boldsymbol{x}}{\sigma_0^2}-\sum_{k=1}^{K}\boldsymbol{x}_k^{\mathrm{H}}\boldsymbol{R}_{\mathrm{cn},0}^{-1}\boldsymbol{x}_k}$$

$$=\left[\frac{1}{\pi^{MN}\sigma_0^{2MN/(K+1)}|\boldsymbol{R}_{\mathrm{cn},0}|}\mathrm{e}^{-\mathrm{tr}(\boldsymbol{R}_{\mathrm{cn},0}^{-1}\boldsymbol{T}_0)}\right]^{K+1} \tag{14.37}$$

式(14.36)和式(14.37)分别给出了有目标时和无目标时的似然函数。采用与 ASD 检测器相同的推导过程,可得 $\hat{\boldsymbol{R}}_{\mathrm{cn},0}=\boldsymbol{T}_0$、$\hat{\boldsymbol{R}}_{\mathrm{cn},1}=\boldsymbol{T}_1$。然后,计算 σ_0^2 和 σ_1^2 的极大似然估计。先将 $\hat{\boldsymbol{R}}_{\mathrm{cn},0}=\boldsymbol{T}_0$、$\hat{\boldsymbol{R}}_{\mathrm{cn},1}=\boldsymbol{T}_1$ 代入似然比表达式中,得到如下似然比表达式

$$l=\frac{\left[\dfrac{1}{\sigma_1^{2MN/(K+1)}|\boldsymbol{T}_1|}\right]^{K+1}}{\left[\dfrac{1}{\sigma_0^{2MN/(K+1)}|\boldsymbol{T}_0|}\right]^{K+1}}=\left\{\frac{\sigma_0^{2MN/(K+1)}\left(1+\dfrac{1}{K\sigma_0^2}\boldsymbol{x}^{\mathrm{H}}\boldsymbol{M}^{-1}\boldsymbol{x}\right)}{\sigma_1^{2MN/(K+1)}\left[1+\dfrac{1}{K\sigma_1^2}(\boldsymbol{x}-\boldsymbol{\Psi\alpha})^{\mathrm{H}}\boldsymbol{M}^{-1}(\boldsymbol{x}-\boldsymbol{\Psi\alpha})\right]}\right\}^{K+1} \tag{14.38}$$

跟前面一样,σ_0^2 只在式(14.38)的分子中,σ_1^2 只在式(14.38)的分母中,故可以分开独立进行估计。经观察,二者等价于对 $f=(\sigma^2)^{N/(K+1)}\left(1+\dfrac{1}{K\sigma^2}\boldsymbol{u}^{\mathrm{H}}\boldsymbol{M}^{-1}\boldsymbol{u}\right)$关于$\sigma^2$ 求极大似然估计。利用$\dfrac{\partial f}{\partial(\sigma^2)}=0$ 可得,$\sigma^2=\dfrac{K-MN+1}{KMN}\boldsymbol{u}^{\mathrm{H}}\boldsymbol{M}^{-1}\boldsymbol{u}$。于是

$$\hat{\sigma}_1^2=\frac{K-MN+1}{KMN}(\boldsymbol{x}-\boldsymbol{\Psi\alpha})^{\mathrm{H}}\boldsymbol{M}^{-1}(\boldsymbol{x}-\boldsymbol{\Psi\alpha}),\quad \hat{\sigma}_0^2=\frac{K-MN+1}{KMN}\boldsymbol{x}^{\mathrm{H}}\boldsymbol{M}^{-1}\boldsymbol{x} \tag{14.39}$$

将 $\hat{\boldsymbol{R}}_{\mathrm{cn},0}$、$\hat{\boldsymbol{R}}_{\mathrm{cn},1}$、$\hat{\sigma}_1^2$、$\hat{\sigma}_0^2$ 代入似然比,可解得$\boldsymbol{\alpha}$ 的极大似然估计$\hat{\boldsymbol{\alpha}}=\dfrac{\boldsymbol{\Psi}^{\mathrm{H}}\boldsymbol{M}^{-1}\boldsymbol{x}}{\boldsymbol{\Psi}^{\mathrm{H}}\boldsymbol{M}^{-1}\boldsymbol{\Psi}}$。最后将各个未知参数的极大似然估计代入似然比检验中,经化简可得到 COS^2 检测器

$$T_{\mathrm{COS}^2}=\frac{\boldsymbol{x}^{\mathrm{H}}\boldsymbol{M}^{-1}\boldsymbol{\Psi}(\boldsymbol{\Psi}^{\mathrm{H}}\boldsymbol{M}^{-1}\boldsymbol{\Psi})^{-1}\boldsymbol{\Psi}^{\mathrm{H}}\boldsymbol{M}^{-1}\boldsymbol{x}}{\boldsymbol{x}^{\mathrm{H}}\boldsymbol{M}^{-1}\boldsymbol{x}} \tag{14.40}$$

14.2.4 阵列雷达目标检测器的性质与性能

1. 6 种检测器的能量投影性质

用样本协方差矩阵 $\boldsymbol{M}$ 对检测单元观测向量 $\boldsymbol{x}$、目标导向向量 $\boldsymbol{s}$ 及目标子空间$\boldsymbol{\Psi}$ 分别进行白化操作,得到白化后的观测向量 $\tilde{\boldsymbol{x}}$、白化后的目标导向向量 $\tilde{\boldsymbol{s}}$ 和白化后的目标子空间$\tilde{\boldsymbol{\Psi}}$

$$\tilde{\boldsymbol{x}}=\boldsymbol{M}^{-1/2}\boldsymbol{x},\quad \tilde{\boldsymbol{s}}=\boldsymbol{M}^{-1/2}\boldsymbol{s},\quad \tilde{\boldsymbol{\Psi}}=\boldsymbol{M}^{-1/2}\tilde{\boldsymbol{\Psi}} \tag{14.41}$$

于是,上述 6 种检测器变形为[6,10]

$$T_{\mathrm{AMF}}=\tilde{\boldsymbol{x}}^{\mathrm{H}}\tilde{\boldsymbol{s}}(\tilde{\boldsymbol{s}}^{\mathrm{H}}\tilde{\boldsymbol{s}})^{-1}\tilde{\boldsymbol{s}}^{\mathrm{H}}\tilde{\boldsymbol{x}}=\|\boldsymbol{P}_{\hat{\boldsymbol{s}}}\tilde{\boldsymbol{x}}\|^2 \tag{14.42}$$

$$T_{\mathrm{Kelly}}=T_{\mathrm{GLRT}}=\frac{\tilde{\boldsymbol{x}}^{\mathrm{H}}\tilde{\boldsymbol{s}}(\tilde{\boldsymbol{s}}^{\mathrm{H}}\tilde{\boldsymbol{s}})^{-1}\tilde{\boldsymbol{s}}^{\mathrm{H}}\tilde{\boldsymbol{x}}}{K+\tilde{\boldsymbol{x}}^{\mathrm{H}}\tilde{\boldsymbol{x}}}=\frac{\|\boldsymbol{P}_{\hat{\boldsymbol{s}}}\tilde{\boldsymbol{x}}\|^2}{K+\|\tilde{\boldsymbol{x}}\|^2} \tag{14.43}$$

$$T_{\text{Kalson}}=\frac{\tilde{\boldsymbol{x}}^{\mathrm{H}}\tilde{\boldsymbol{s}}(\tilde{\boldsymbol{s}}^{\mathrm{H}}\tilde{\boldsymbol{s}})^{-1}\tilde{\boldsymbol{s}}^{\mathrm{H}}\tilde{\boldsymbol{x}}}{\tilde{\boldsymbol{x}}^{\mathrm{H}}\tilde{\boldsymbol{x}}}=\frac{\|\boldsymbol{P}_{\hat{s}}\tilde{\boldsymbol{x}}\|^{2}}{\|\tilde{\boldsymbol{x}}\|^{2}} \tag{14.44}$$

$$T_{\text{MSD}}=\tilde{\boldsymbol{x}}^{\mathrm{H}}\tilde{\boldsymbol{\Psi}}(\tilde{\boldsymbol{\Psi}}^{\mathrm{H}}\tilde{\boldsymbol{\Psi}})^{-1}\tilde{\boldsymbol{\Psi}}^{\mathrm{H}}\tilde{\boldsymbol{x}}=\|\boldsymbol{P}_{\tilde{\boldsymbol{\Psi}}}\tilde{\boldsymbol{x}}\|^{2} \tag{14.45}$$

$$T_{\text{ASD}}=\frac{\tilde{\boldsymbol{x}}^{\mathrm{H}}\tilde{\boldsymbol{\Psi}}(\tilde{\boldsymbol{\Psi}}^{\mathrm{H}}\tilde{\boldsymbol{\Psi}})^{-1}\tilde{\boldsymbol{\Psi}}^{\mathrm{H}}\tilde{\boldsymbol{x}}}{K+\tilde{\boldsymbol{x}}^{\mathrm{H}}\tilde{\boldsymbol{x}}}=\frac{\|\boldsymbol{P}_{\tilde{\boldsymbol{\Psi}}}\tilde{\boldsymbol{x}}\|^{2}}{K+\|\tilde{\boldsymbol{x}}\|^{2}} \tag{14.46}$$

$$T_{\text{COS}^2}=\frac{\tilde{\boldsymbol{x}}^{\mathrm{H}}\tilde{\boldsymbol{\Psi}}(\tilde{\boldsymbol{\Psi}}^{\mathrm{H}}\tilde{\boldsymbol{\Psi}})^{-1}\tilde{\boldsymbol{\Psi}}^{\mathrm{H}}\tilde{\boldsymbol{x}}}{\tilde{\boldsymbol{x}}^{\mathrm{H}}\tilde{\boldsymbol{x}}}=\frac{\|\boldsymbol{P}_{\tilde{\boldsymbol{\Psi}}}\tilde{\boldsymbol{x}}\|^{2}}{\|\tilde{\boldsymbol{x}}\|^{2}} \tag{14.47}$$

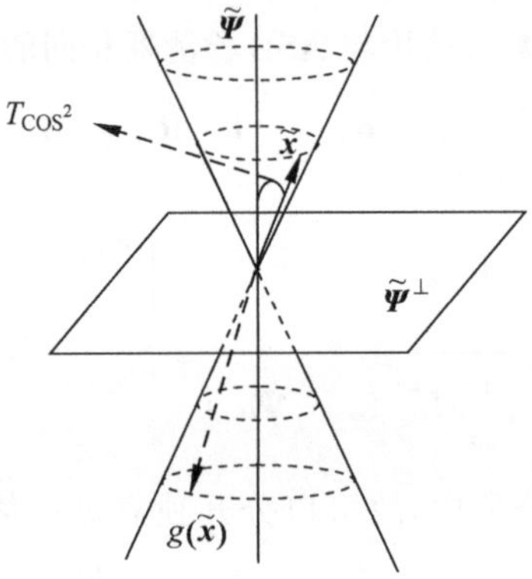

图 14.1 T_{COS^2} 的几何解释

其中，$\boldsymbol{P}_{\hat{s}}=\tilde{s}(\tilde{s}^{\mathrm{H}}\tilde{s})^{-1}\tilde{s}^{\mathrm{H}}$ 表示“往白化后的秩 1 目标空间投影”；$\|\cdot\|^{2}$ 表示向量所包含的能量，即向量中每个元素平方之和；$\boldsymbol{P}_{\tilde{\boldsymbol{\Psi}}}=\tilde{\boldsymbol{\Psi}}(\tilde{\boldsymbol{\Psi}}^{\mathrm{H}}\tilde{\boldsymbol{\Psi}})^{-1}\tilde{\boldsymbol{\Psi}}^{\mathrm{H}}$ 表示“往白化后的目标子空间投影”。上述 6 个式子表明[6,10]，AMF 检测器和 MSD 检测器本质上就是“白化后观测向量往白化后目标空间投影的能量”；而另外 4 种检测器本质上是“白化后观测向量往白化后目标空间投影的能量”与“白化后观测向量的能量”的简单函数。图 14.1 给出了式(14.47)描述的 T_{COS^2} 的几何解释，即“白化后观测向量 $\tilde{\boldsymbol{x}}$ 与白化后目标空间$\tilde{\boldsymbol{\Psi}}$之间的余弦平方”，其中 $g(\tilde{\boldsymbol{x}})$表示对 $\tilde{\boldsymbol{x}}$ 的旋转变换与尺度变换。显然，对位于锥形表面上的 $\tilde{\boldsymbol{x}}$ 进行旋转变换和尺度变换，都不会改变余弦平方的值，也就是说，T_{COS^2} 关于 $\tilde{\boldsymbol{x}}$ 的旋转变换和尺度变换是不变的。

2. 6 种检测器的 CFAR 性质

文献[10]指出，复高斯杂波背景下的 AMF 检测器、Kelly 检测器、Kalson 检测器、MSD 检测器、ASD 检测器和 COS^2 检测器都能分别转化为 5 个独立随机变量的不同组合，即所谓的统计等价分解，且在 H_0 假设下，这 5 个随机变量都只与检测问题涉及的参考单元数 K，阵元数 M 和脉冲数 N 有关，而与背景分布参数无关，因此表明这 6 种检测器都是 CFAR 检测器。下面以 MSD 检测器为例，给出其统计等价分解的推导过程，而其余检测器的统计等价分解的推导过程可参考文献[10]及本节的推导过程。

下面对 T_{MSD} 进行分解，分解过程中要注意样本值与真实值的区别。

(1) 第一步：预处理。

针对式(14.30)，令 $\boldsymbol{A}=K\boldsymbol{M}$，那么

$$T_{\text{MSD}}=K(\boldsymbol{x}^{\mathrm{H}}\boldsymbol{A}^{-1}\boldsymbol{\Psi}(\boldsymbol{\Psi}^{\mathrm{H}}\boldsymbol{A}^{-1}\boldsymbol{\Psi})^{-1}\boldsymbol{\Psi}^{\mathrm{H}}\boldsymbol{A}^{-1}\boldsymbol{x}) \tag{14.48}$$

(2) 第二步：白化。

注意：由于这里是要解析的分析统计性能，所以这里的白化不是用样本协方差矩阵 $\boldsymbol{M}$ 来实现的，而是用真实的背景协方差矩阵$\boldsymbol{R}_{\text{cn}}$ 来实现。这里为了简化书写，本节后续内容中将采用 $\boldsymbol{R}$ 来代替$\boldsymbol{R}_{\text{cn}}$。

对检测单元和参考单元中的观测向量和目标子空间都进行白化操作，其中目标子空间的秩为 p，$\tilde{\boldsymbol{x}}=\boldsymbol{R}^{-1/2}\boldsymbol{x}$，$\tilde{\boldsymbol{\Psi}}=\boldsymbol{R}^{-1/2}\boldsymbol{\Psi}$。

$$\boldsymbol{B}=\sum_{k=1}^{K}(\boldsymbol{R}^{-1/2}\boldsymbol{x}_k)(\boldsymbol{R}^{-1/2}\boldsymbol{x}_k)^{\mathrm{H}}=\boldsymbol{R}^{-1/2}\left(\sum_{k=1}^{K}\boldsymbol{x}_k\boldsymbol{x}_k^{\mathrm{H}}\right)(\boldsymbol{R}^{-1/2})^{\mathrm{H}}=\boldsymbol{R}^{-1/2}\boldsymbol{A}(\boldsymbol{R}^{-1/2})^{\mathrm{H}} \tag{14.49}$$

其中,$\boldsymbol{B}$ 表示 K 倍白化样本协方差矩阵,那么,T_{MSD} 变形为

$$T_{\mathrm{MSD}}=K[\tilde{\boldsymbol{x}}^{\mathrm{H}}\boldsymbol{B}^{-1}\tilde{\boldsymbol{\Psi}}(\tilde{\boldsymbol{\Psi}}^{\mathrm{H}}\boldsymbol{B}^{-1}\tilde{\boldsymbol{\Psi}})^{-1}\tilde{\boldsymbol{\Psi}}^{\mathrm{H}}\boldsymbol{B}^{-1}\tilde{\boldsymbol{x}}] \tag{14.50}$$

其中,$\boldsymbol{B}^{-1}=(\boldsymbol{R}^{1/2})^{\mathrm{H}}\boldsymbol{A}^{-1}\boldsymbol{R}^{1/2}$。根据文献[10]中的定理 1 可知,$\boldsymbol{B}$ 服从 $\mathrm{CW}[K,MN;\boldsymbol{I}]$ 分布,即参数为 $[K,MN;\boldsymbol{I}]$ 的复 Wishart 分布,$\boldsymbol{I}$ 表示单位阵。

(3) 第三步:酉变换。

酉变换将观测向量旋转到一个新的坐标系统中。构建如下酉变换

$$\boldsymbol{U}=\left[\tilde{\boldsymbol{\Psi}}(\tilde{\boldsymbol{\Psi}}^{\mathrm{H}}\tilde{\boldsymbol{\Psi}})^{-1/2},\frac{\boldsymbol{P}_{\tilde{\boldsymbol{\Psi}}}^{\perp}\tilde{\boldsymbol{x}}}{\sqrt{\tilde{\boldsymbol{x}}^{\mathrm{H}}\boldsymbol{P}_{\tilde{\boldsymbol{\Psi}}}^{\perp}\tilde{\boldsymbol{x}}}},\tilde{\boldsymbol{U}}\right] \tag{14.51}$$

其中,$\boldsymbol{U}$ 的各列是基,前 p 个基向量由目标子空间决定,第 $p+1$ 个基向量由白化观测向量 $\tilde{\boldsymbol{x}}$ 中不在目标子空间的分量来决定,剩余的 $MN-p-1$ 个基向量可通过与前 $p+1$ 个基向量正交来求得。

酉变换 $\boldsymbol{U}$ 的作用,一是将 $\tilde{\boldsymbol{x}}$ 中的目标能量集中到前 p 个元素中,二是将 $\tilde{\boldsymbol{x}}$ 中的所有能量集中到前 $p+1$ 个非零元素中,分别如式(14.52)和式(14.53)所示。

$$\boldsymbol{U}^{\mathrm{H}}\tilde{\boldsymbol{x}}=\boldsymbol{U}^{\mathrm{H}}(\tilde{\boldsymbol{\Psi}}\alpha+\tilde{\boldsymbol{x}}_{\mathrm{cn}})=\begin{bmatrix}(\tilde{\boldsymbol{\Psi}}^{\mathrm{H}}\tilde{\boldsymbol{\Psi}})^{-1/2}\tilde{\boldsymbol{\Psi}}^{\mathrm{H}}\\ \dfrac{\tilde{\boldsymbol{x}}^{\mathrm{H}}\boldsymbol{P}_{\tilde{\boldsymbol{\Psi}}}^{\perp}}{\sqrt{\tilde{\boldsymbol{x}}^{\mathrm{H}}\boldsymbol{P}_{\tilde{\boldsymbol{\Psi}}}^{\perp}\tilde{\boldsymbol{x}}}}\\ \tilde{\boldsymbol{U}}^{\mathrm{H}}\end{bmatrix}\tilde{\boldsymbol{\Psi}}\alpha+\boldsymbol{U}^{\mathrm{H}}\tilde{\boldsymbol{x}}_{\mathrm{cn}}=\begin{bmatrix}(\tilde{\boldsymbol{\Psi}}^{\mathrm{H}}\tilde{\boldsymbol{\Psi}})^{1/2}\boldsymbol{\alpha}\\ 0\\ \boldsymbol{0}\end{bmatrix}+\boldsymbol{U}^{\mathrm{H}}\tilde{\boldsymbol{x}}_{\mathrm{cn}} \tag{14.52}$$

$$\boldsymbol{U}^{\mathrm{H}}\tilde{\boldsymbol{x}}=\begin{bmatrix}(\tilde{\boldsymbol{\Psi}}^{\mathrm{H}}\tilde{\boldsymbol{\Psi}})^{-1/2}\tilde{\boldsymbol{\Psi}}^{\mathrm{H}}\\ \dfrac{\tilde{\boldsymbol{x}}^{\mathrm{H}}\boldsymbol{P}_{\tilde{\boldsymbol{\Psi}}}^{\perp}}{\sqrt{\tilde{\boldsymbol{x}}^{\mathrm{H}}\boldsymbol{P}_{\tilde{\boldsymbol{\Psi}}}^{\perp}\tilde{\boldsymbol{x}}}}\\ \tilde{\boldsymbol{U}}^{\mathrm{H}}\end{bmatrix}\tilde{\boldsymbol{x}}=\begin{bmatrix}(\tilde{\boldsymbol{\Psi}}^{\mathrm{H}}\tilde{\boldsymbol{\Psi}})^{-1/2}\tilde{\boldsymbol{\Psi}}^{\mathrm{H}}\tilde{\boldsymbol{x}}\\ \sqrt{\tilde{\boldsymbol{x}}^{\mathrm{H}}\boldsymbol{P}_{\tilde{\boldsymbol{\Psi}}}^{\perp}\tilde{\boldsymbol{x}}}\\ \tilde{\boldsymbol{U}}^{\mathrm{H}}\tilde{\boldsymbol{x}}\end{bmatrix}=\begin{bmatrix}(\tilde{\boldsymbol{\Psi}}^{\mathrm{H}}\tilde{\boldsymbol{\Psi}})^{1/2}\boldsymbol{\alpha}+(\tilde{\boldsymbol{\Psi}}^{\mathrm{H}}\tilde{\boldsymbol{\Psi}})^{-1/2}\tilde{\boldsymbol{\Psi}}^{\mathrm{H}}\tilde{\boldsymbol{x}}_{\mathrm{cn}}\\ \sqrt{\tilde{\boldsymbol{x}}^{\mathrm{H}}\boldsymbol{P}_{\tilde{\boldsymbol{\Psi}}}^{\perp}\tilde{\boldsymbol{x}}}\\ \boldsymbol{0}\end{bmatrix} \tag{14.53}$$

对观测数据做酉变换后,必然改变了观测数据中目标模型的模式矩阵,如下所示

$$\boldsymbol{U}^{\mathrm{H}}\tilde{\boldsymbol{\Psi}}=\begin{bmatrix}(\tilde{\boldsymbol{\Psi}}^{\mathrm{H}}\tilde{\boldsymbol{\Psi}})^{-1/2}\tilde{\boldsymbol{\Psi}}^{\mathrm{H}}\\ \dfrac{\tilde{\boldsymbol{x}}^{\mathrm{H}}\boldsymbol{P}_{\tilde{\boldsymbol{\Psi}}}^{\perp}}{\sqrt{\tilde{\boldsymbol{x}}^{\mathrm{H}}\boldsymbol{P}_{\tilde{\boldsymbol{\Psi}}}^{\perp}\tilde{\boldsymbol{x}}}}\\ \tilde{\boldsymbol{U}}^{\mathrm{H}}\end{bmatrix}\tilde{\boldsymbol{\Psi}}=\begin{bmatrix}(\tilde{\boldsymbol{\Psi}}^{\mathrm{H}}\tilde{\boldsymbol{\Psi}})^{1/2}\\ 0\\ \boldsymbol{0}\end{bmatrix}=\tilde{\tilde{\boldsymbol{\Psi}}} \tag{14.54}$$

$$\boldsymbol{U}^{\mathrm{H}}\tilde{\boldsymbol{x}}=\begin{bmatrix}(\tilde{\boldsymbol{\Psi}}^{\mathrm{H}}\tilde{\boldsymbol{\Psi}})^{-1/2}\tilde{\boldsymbol{\Psi}}^{\mathrm{H}}\tilde{\boldsymbol{x}}\\ \sqrt{\tilde{\boldsymbol{x}}^{\mathrm{H}}\boldsymbol{P}_{\tilde{\boldsymbol{\Psi}}}^{\perp}\tilde{\boldsymbol{x}}}\\ 0\end{bmatrix}=\sigma\begin{bmatrix}\tilde{\tilde{\boldsymbol{n}}}\\ g\\ 0\end{bmatrix}=\tilde{\tilde{\boldsymbol{x}}} \tag{14.55}$$

$$\begin{bmatrix}\tilde{\tilde{\boldsymbol{n}}}\\ g\\ 0\end{bmatrix}=\begin{bmatrix}\dfrac{(\tilde{\boldsymbol{\Psi}}^{\mathrm{H}}\tilde{\boldsymbol{\Psi}})^{-1/2}\tilde{\boldsymbol{\Psi}}^{\mathrm{H}}\tilde{\boldsymbol{x}}}{\sigma}\\ \dfrac{\sqrt{\tilde{\boldsymbol{x}}^{\mathrm{H}}P_{\tilde{\boldsymbol{\Psi}}}^{\perp}\tilde{\boldsymbol{x}}}}{\sigma}\\ 0\end{bmatrix}\tag{14.56}$$

因此,将酉变换后的新的观测数据及其模式矩阵代入检验统计量 T_{MSD} 中可得新的检验统计量,为了简化书写,仍记为 T_{MSD}

$$T_{\mathrm{MSD}}=K[\tilde{\tilde{\boldsymbol{x}}}^{\mathrm{H}}\boldsymbol{C}^{-1}\tilde{\tilde{\boldsymbol{\Psi}}}(\tilde{\tilde{\boldsymbol{\Psi}}}^{\mathrm{H}}\boldsymbol{C}^{-1}\tilde{\tilde{\boldsymbol{\Psi}}})^{-1}\tilde{\tilde{\boldsymbol{\Psi}}}^{\mathrm{H}}\boldsymbol{C}^{-1}\tilde{\tilde{\boldsymbol{x}}}]\tag{14.57}$$

其中,$\boldsymbol{C}=\boldsymbol{U}^{\mathrm{H}}\boldsymbol{B}\boldsymbol{U}$。应用文献[10]中的引理可知,$\boldsymbol{C}$ 服从参数为 $[K,MN;\boldsymbol{I}]$ 的复 Wishart 分布。根据式(14.56),由于 $\tilde{\boldsymbol{x}}$ 服从 $\mathrm{CN}[\tilde{\boldsymbol{\Psi}}\boldsymbol{\alpha},\boldsymbol{I}_p]$分布,故 $\tilde{\tilde{\boldsymbol{n}}}$ 服从 $\mathrm{CN}\left[\dfrac{(\tilde{\boldsymbol{\Psi}}^{\mathrm{H}}\tilde{\boldsymbol{\Psi}})^{1/2}}{\sigma}\boldsymbol{\alpha},\boldsymbol{I}_p\right]$分布,$\boldsymbol{I}_p$ 表示 $p\times p$ 维的单位阵;g^2 服从具有 $MN-p$ 自由度的卡方分布,$g^2\sim\chi^2_{MN-p}[0]$。

(4) 第四步:协方差矩阵分块。

由 $\tilde{\tilde{\boldsymbol{x}}}$、$\tilde{\tilde{\boldsymbol{\Psi}}}$ 的形式可知,$\boldsymbol{C}^{-1}$ 中只有左上$(p+1)\times(p+1)$维模块有用。故可对 $\boldsymbol{C}$ 进行如下分块,$\boldsymbol{C}=\begin{bmatrix}\boldsymbol{C}_{(p+1)(p+1)} & \boldsymbol{C}^{\mathrm{H}}_{(MN-p-1)(p+1)}\\ \boldsymbol{C}_{(MN-p-1)(p+1)} & \boldsymbol{C}_{(MN-p-1)(MN-p-1)}\end{bmatrix}$,由文献[10]中的定理 1 可知,$\boldsymbol{C}^{-1}$ 中左上$(p+1)\times(p+1)$维模块 $\boldsymbol{D}=\boldsymbol{C}_{(p+1)(p+1)}-\boldsymbol{C}^{\mathrm{H}}_{(MN-p-1)(p+1)}\boldsymbol{C}^{-1}_{(MN-p-1)(N-p-1)}\boldsymbol{C}_{(MN-p-1)(p+1)}$ 服从复 Wishart 分布,$\boldsymbol{D}\sim\mathrm{CW}[K-MN+p+1,p+1,\boldsymbol{I}_{p+1}]$。于是,$T_{\mathrm{MSD}}$ 变为

$$T_{\mathrm{MSD}}=K(\tilde{\tilde{\boldsymbol{x}}}^{\mathrm{H}}_{p+1}\boldsymbol{D}^{-1}\tilde{\tilde{\boldsymbol{\Psi}}}_{p+1}(\tilde{\tilde{\boldsymbol{\Psi}}}^{\mathrm{H}}_{p+1}\boldsymbol{D}^{-1}\tilde{\tilde{\boldsymbol{\Psi}}}_{p+1})^{-1}\tilde{\tilde{\boldsymbol{\Psi}}}^{\mathrm{H}}_{p+1}\boldsymbol{D}^{-1}\tilde{\tilde{\boldsymbol{x}}}_{p+1})\tag{14.58}$$

其中,$\tilde{\tilde{\boldsymbol{\Psi}}}_{p+1}=\begin{bmatrix}(\tilde{\boldsymbol{\Psi}}^{\mathrm{H}}\tilde{\boldsymbol{\Psi}})^{1/2}\\ 0\end{bmatrix}$,$\tilde{\tilde{\boldsymbol{x}}}_{p+1}=\begin{bmatrix}\sigma\tilde{\tilde{\boldsymbol{n}}}\\ \sigma g\end{bmatrix}$。

再次利用文献[10]中的定理 1 来考察 $\boldsymbol{D}^{-1}$,矩阵分块为 $\boldsymbol{D}=\begin{bmatrix}\boldsymbol{D}_{pp} & \boldsymbol{D}^{\mathrm{H}}_{1p}\\ \boldsymbol{D}_{1p} & \boldsymbol{D}_{11}\end{bmatrix}$,于是

$$\boldsymbol{D}^{-1}=\begin{bmatrix}\boldsymbol{E}^{-1} & -\boldsymbol{E}^{-1}\boldsymbol{D}^{\mathrm{H}}_{1p}\boldsymbol{D}^{-1}_{11}\\ -\boldsymbol{D}^{-1}_{11}\boldsymbol{D}_{1p}\boldsymbol{E}^{-1} & \boldsymbol{D}^{-1}_{11}\boldsymbol{D}_{1p}\boldsymbol{E}^{-1}\boldsymbol{D}^{\mathrm{H}}_{1p}\boldsymbol{D}^{-1}_{11}+\boldsymbol{D}^{-1}_{11}\end{bmatrix}\tag{14.59}$$

其中,$\boldsymbol{E}=\boldsymbol{D}_{pp}-\boldsymbol{D}^{\mathrm{H}}_{1p}\boldsymbol{D}^{-1}_{11}\boldsymbol{D}_{1p}\sim\mathrm{CW}[K-MN+p,p,\boldsymbol{I}_p]$。

将式(14.59)代入式(14.58)中,化简得

$$T_{\mathrm{MSD}}=K(\sigma\tilde{\tilde{\boldsymbol{n}}}-\sigma\boldsymbol{D}^{\mathrm{H}}_{1p}\boldsymbol{D}^{-1}_{11}g)^{\mathrm{H}}\boldsymbol{E}^{-1}(\sigma\tilde{\tilde{\boldsymbol{n}}}-\sigma\boldsymbol{D}^{\mathrm{H}}_{1p}\boldsymbol{D}^{-1}_{11}g)\tag{14.60}$$

(5) 第五步:将 T_{MSD} 中 $\boldsymbol{D}$ 的各个分块独立化

显然 $\boldsymbol{D}$ 的各分块矩阵$\boldsymbol{D}_{pp}$、$\boldsymbol{D}_{1p}$、$\boldsymbol{D}_{11}$ 相互之间不是独立的,这能从 $\boldsymbol{D}$ 的 Wishart 分布中看出来。再次应用文献[10]中的定理 1,可以获得相互统计独立的新分块矩阵 $\boldsymbol{E}$、h_2^2、$\boldsymbol{h}_3$,

$$\begin{aligned}&\boldsymbol{E}=\boldsymbol{D}_{pp}-\boldsymbol{D}^{\mathrm{H}}_{1p}\boldsymbol{D}^{-1}_{11}\boldsymbol{D}_{1p}\sim\mathrm{CW}[K-MN+p,p,\boldsymbol{I}_p]\\ &h_2^2=\boldsymbol{D}_{11}\sim\chi^2_{K-MN+p+1}[0]\quad(\text{标量})\\ &\boldsymbol{h}_3=-\frac{\boldsymbol{D}_{p1}}{\sqrt{\boldsymbol{D}_{11}}}\sim\mathrm{CN}[\boldsymbol{0},\boldsymbol{I}_p]\quad(\text{矢量})\end{aligned}\tag{14.61}$$

将这些变量代入式(14.60)中,可得

$$T_{\mathrm{MSD}}=K\left(\sigma\tilde{\tilde{\boldsymbol{n}}}+\sigma g\frac{\boldsymbol{h}_3}{h_2}\right)^{\mathrm{H}}\boldsymbol{E}^{-1}\left(\sigma\tilde{\tilde{\boldsymbol{n}}}+\sigma g\frac{\boldsymbol{h}_3}{h_2}\right) \tag{14.62}$$

(6) 第六步:再次酉变换

若 $p=1$,则 $\boldsymbol{E}$、$\boldsymbol{h}_3$ 和 $\tilde{\tilde{\boldsymbol{n}}}$ 都将是标量,那么 T_{MSD} 就已经是完全简化了的。

若 $p>1$,则 $\boldsymbol{E}$ 是矩阵,需简化。令$\boldsymbol{\lambda}=\sigma\tilde{\tilde{\boldsymbol{n}}}+\sigma g\dfrac{\boldsymbol{h}_3}{h_2}$,构造酉变换,$\boldsymbol{V}=\begin{bmatrix}\dfrac{\boldsymbol{\lambda}}{\sqrt{\boldsymbol{\lambda}^{\mathrm{H}}\boldsymbol{\lambda}}} & \tilde{\boldsymbol{V}}\end{bmatrix}$。其中,$\tilde{\boldsymbol{V}}^{\mathrm{H}}\boldsymbol{\lambda}=\boldsymbol{0}$。该酉变换旋转的是$\boldsymbol{\lambda}$,$\boldsymbol{V}^{\mathrm{H}}\boldsymbol{\lambda}=\begin{bmatrix}\sqrt{\boldsymbol{\lambda}^{\mathrm{H}}\boldsymbol{\lambda}}\\ \boldsymbol{0}\end{bmatrix}$,则$\boldsymbol{\lambda}=\boldsymbol{V}\begin{bmatrix}\sqrt{\boldsymbol{\lambda}^{\mathrm{H}}\boldsymbol{\lambda}}\\ \boldsymbol{0}\end{bmatrix}$。得到

$$T_{\mathrm{MSD}}=K\left(\boldsymbol{V}\begin{bmatrix}\sqrt{\boldsymbol{\lambda}^{\mathrm{H}}\boldsymbol{\lambda}}\\ \boldsymbol{0}\end{bmatrix}\right)^{\mathrm{H}}\boldsymbol{E}^{-1}\left(\boldsymbol{V}\begin{bmatrix}\sqrt{\boldsymbol{\lambda}^{\mathrm{H}}\boldsymbol{\lambda}}\\ \boldsymbol{0}\end{bmatrix}\right)=K\left\{\begin{bmatrix}\sqrt{\boldsymbol{\lambda}^{\mathrm{H}}\boldsymbol{\lambda}} & \boldsymbol{0}\end{bmatrix}\boldsymbol{G}^{-1}\begin{bmatrix}\sqrt{\boldsymbol{\lambda}^{\mathrm{H}}\boldsymbol{\lambda}}\\ \boldsymbol{0}\end{bmatrix}\right\} \tag{14.63}$$

其中,$\boldsymbol{G}=\boldsymbol{V}^{\mathrm{H}}\boldsymbol{E}\boldsymbol{V}$。$\boldsymbol{G}$ 是 $\boldsymbol{E}$ 经过酉变换后的样本协方差矩阵,其统计特性与 $\boldsymbol{E}$ 是一样的,故 $\boldsymbol{G}\sim\mathrm{CW}[K-MN+p,p,\boldsymbol{I}_p]$。很显然,$\boldsymbol{G}^{-1}$ 中起作用的只是其前 1×1 模块,因此还可进行矩阵分块。

(7) 第七步:再次矩阵分块

$$\boldsymbol{G}=\begin{bmatrix}\boldsymbol{G}_{11} & \boldsymbol{G}_{1(p-1)}\\ \boldsymbol{G}_{(p-1)1} & \boldsymbol{G}_{(p-1)(p-1)}\end{bmatrix},\quad \boldsymbol{G}^{-1}=\begin{bmatrix}(\boldsymbol{G}_{11}-\boldsymbol{G}_{(p-1)1}^{\mathrm{H}}\boldsymbol{G}_{(p-1)(p-1)}^{-1}\boldsymbol{G}_{(p-1)1})^{-1} & *\\ * & *\end{bmatrix}=\begin{bmatrix}h_1^{-2} & *\\ * & *\end{bmatrix}$$

其中,$h_1^2=\boldsymbol{G}_{11}-\boldsymbol{G}_{(p-1)1}^{\mathrm{H}}\boldsymbol{G}_{(p-1)(p-1)}^{-1}\boldsymbol{G}_{(p-1)1}$。根据文献[10]中的定理1可知,$h_1^2\sim\chi_{K-MN+1,1}^2[0]$。那么,$T_{\mathrm{MSD}}$ 变形为

$$T_{\mathrm{MSD}}=K\left(\begin{bmatrix}\sqrt{\boldsymbol{\lambda}^{\mathrm{H}}\boldsymbol{\lambda}} & \boldsymbol{0}\end{bmatrix}\begin{bmatrix}h_1^{-2} & *\\ * & *\end{bmatrix}\begin{bmatrix}\sqrt{\boldsymbol{\lambda}^{\mathrm{H}}\boldsymbol{\lambda}}\\ \boldsymbol{0}\end{bmatrix}\right)=K\frac{\boldsymbol{\lambda}^{\mathrm{H}}\boldsymbol{\lambda}}{h_1^2} \tag{14.64}$$

将$\boldsymbol{\lambda}=\sigma\tilde{\tilde{\boldsymbol{n}}}+\sigma g\dfrac{\boldsymbol{h}_3}{h_2}$代入此式,即可得 MSD 检测器的统计等价分解表达式

$$\begin{cases}T_{\mathrm{MSD}}=\boldsymbol{x}^{\mathrm{H}}\boldsymbol{M}^{-1}\boldsymbol{\Psi}(\boldsymbol{\Psi}^{\mathrm{H}}\boldsymbol{M}^{-1}\boldsymbol{\Psi})^{-1}\boldsymbol{\Psi}^{\mathrm{H}}\boldsymbol{M}^{-1}\boldsymbol{x}\\ T_{\mathrm{MSD}}=K\dfrac{\sigma^2}{h_1^2}\left\|\tilde{\tilde{\boldsymbol{n}}}+g\dfrac{\boldsymbol{h}_3}{h_2}\right\|^2,\quad \sigma=1\\ \tilde{\tilde{\boldsymbol{n}}}\sim\mathrm{CN}\left[\dfrac{(\tilde{\boldsymbol{\Psi}}^{\mathrm{H}}\tilde{\boldsymbol{\Psi}})^{1/2}}{\sigma}\boldsymbol{\alpha},\boldsymbol{I}_p\right]\Leftrightarrow\mathrm{CN}[\text{白化后的幅度信噪比矢量},\boldsymbol{I}_p]\\ g^2\sim\chi_{MN-p}^2[0]\\ h_1^2\sim\chi_{K-MN+1,1}^2[0]\\ h_2^2\sim\chi_{K-MN+p+1}^2[0]\\ \boldsymbol{h}_3\sim\mathrm{CN}[\boldsymbol{0},\boldsymbol{I}_p]\end{cases} \tag{14.65}$$

其中,正态随机矢量 $\tilde{\tilde{\boldsymbol{n}}}$、$\boldsymbol{h}_3$,瑞利随机变量 g、h_1、h_2 五者之间都是统计独立的。

对于其他检测器(如 Kelly 检测器、Kalson 检测器、ASD 检测器、COS^2 检测器)来说,MSD 检测器仅是其分子部分,因此还需要推导分母部分的统计等价分解表达式。幸运的是,这些检测器的分子分母都是标量,故分解过程完全可以分子分母分开独立进行。但应当注意的是,"分开独立进行"并不意味着

"分子分母能经历不同的变换",而恰恰相反,"分子分母必须经历相同的变换"。对此,本文不再赘述,感兴趣的读者可参考文献[10]。表 14.1 给出了 6 种检测器解析分解式,其中还给出了 $K\to\infty$时,各个检测器的统计等价分解的极限形式。

表 14.1　6 种检测器的统计等价分解

检　测　器		检测器及其统计等价分解	检测器极限形式
秩 1 目标检测器	AMF 检测器	$T_{\mathrm{AMF}}=\dfrac{\lvert \boldsymbol{s}^{\mathrm{H}}\boldsymbol{M}^{-1}\boldsymbol{x}\rvert^{2}}{\boldsymbol{s}^{\mathrm{H}}\boldsymbol{M}^{-1}\boldsymbol{s}}, T_{\mathrm{AMF}}=K\dfrac{\sigma^{2}}{h_1^{2}}\left\lVert \tilde{\tilde{n}}+g\dfrac{h_3}{h_2}\right\rVert^{2}, \sigma=1$	$T_{\mathrm{AMF}}\to\lVert\tilde{\tilde{n}}\rVert^{2}$
	Kelly 检测器	$T_{\mathrm{Kelly}}=\dfrac{\lvert \boldsymbol{s}^{\mathrm{H}}\boldsymbol{M}^{-1}\boldsymbol{x}\rvert^{2}}{\boldsymbol{s}^{\mathrm{H}}\boldsymbol{M}^{-1}\boldsymbol{s}(K+\boldsymbol{x}^{\mathrm{H}}\boldsymbol{M}^{-1}\boldsymbol{x})}, T_{\mathrm{Kelly}}=\dfrac{\mathcal{K}^{2}}{1+\mathcal{K}^{2}}, \mathcal{K}^{2}=\dfrac{\left\lVert\tilde{\tilde{n}}+g\dfrac{h_3}{h_2}\right\rVert^{2}}{h_1^{2}\left(\dfrac{1}{\sigma^{2}}+\dfrac{g^{2}}{h_2^{2}}\right)}, \sigma=1$	$K\cdot T_{\mathrm{Kelly}}\to\lVert\tilde{\tilde{n}}\rVert^{2}$
	Kalson 检测器	$T_{\mathrm{Kalson}}=\dfrac{\lvert \boldsymbol{s}^{\mathrm{H}}\boldsymbol{M}^{-1}\boldsymbol{x}\rvert^{2}}{\boldsymbol{s}^{\mathrm{H}}\boldsymbol{M}^{-1}\boldsymbol{s}\cdot\boldsymbol{x}^{\mathrm{H}}\boldsymbol{M}^{-1}\boldsymbol{x}}, T_{\mathrm{Kalson}}=\dfrac{\boldsymbol{F}}{1+\boldsymbol{F}}, \boldsymbol{F}=\dfrac{\left\lVert\tilde{\tilde{n}}\dfrac{h_2}{g}+h_3\right\rVert^{2}}{h_1^{2}}$	$T_{\mathrm{Kalson}}\to\dfrac{\left\lVert\dfrac{\tilde{\tilde{n}}}{g}\right\rVert^{2}}{1+\left\lVert\dfrac{\tilde{\tilde{n}}}{g}\right\rVert^{2}}$
	变量说明	$\tilde{\tilde{n}}\sim\mathrm{CN}\left[\dfrac{(\boldsymbol{s}^{\mathrm{H}}\boldsymbol{s})^{1/2}}{\sigma}b,1\right]$ ⇔CN[白化后的幅度信噪比矢量,1] $g^{2}\sim\chi^{2}_{MN-1}[0]$ $h_1^{2}\sim\chi^{2}_{K-MN+1,1}[0]$ $h_2^{2}\sim\chi^{2}_{K-MN+2}[0]$ $h_3\sim\mathrm{CN}[0,1]$ 正态随机变量 $\tilde{\tilde{n}}$、h_3,瑞利随机变量 g、h_1、h_2 五者之间都是统计独立的	当 $K\to\infty$ 时,$h_1^{2}\to K\to\infty$,$h_2^{2}\to K\to\infty$,g 和 h_3 不受影响
子空间目标检测器	MSD 检测器	$T_{\mathrm{MSD}}=\boldsymbol{x}^{\mathrm{H}}\boldsymbol{M}^{-1}\boldsymbol{\Psi}(\boldsymbol{\Psi}^{\mathrm{H}}\boldsymbol{M}^{-1}\boldsymbol{\Psi})^{-1}\boldsymbol{\Psi}^{\mathrm{H}}\boldsymbol{M}^{-1}\boldsymbol{x}, T_{\mathrm{MSD}}=K\dfrac{\sigma^{2}}{h_1^{2}}\left\lVert\tilde{\tilde{\boldsymbol{n}}}+g\dfrac{\boldsymbol{h}_3}{h_2}\right\rVert^{2}, \sigma=1$	$T_{\mathrm{MSD}}\to\lVert\tilde{\tilde{\boldsymbol{n}}}\rVert^{2}$
	ASD 检测器	$T_{\mathrm{ASD}}=\dfrac{\boldsymbol{x}^{\mathrm{H}}\boldsymbol{M}^{-1}\boldsymbol{\Psi}(\boldsymbol{\Psi}^{\mathrm{H}}\boldsymbol{M}^{-1}\boldsymbol{\Psi})^{-1}\boldsymbol{\Psi}^{\mathrm{H}}\boldsymbol{M}^{-1}\boldsymbol{x}}{K+\boldsymbol{x}^{\mathrm{H}}\boldsymbol{M}^{-1}\boldsymbol{x}},$ $T_{\mathrm{ASD}}=\dfrac{\mathcal{K}^{2}}{1+\mathcal{K}^{2}}, \mathcal{K}^{2}=\dfrac{\left\lVert\tilde{\tilde{\boldsymbol{n}}}+g\dfrac{\boldsymbol{h}_3}{h_2}\right\rVert^{2}}{h_1^{2}\left(\dfrac{1}{\sigma^{2}}+\dfrac{g^{2}}{h_2^{2}}\right)}, \sigma=1$	$K\cdot T_{\mathrm{ASD}}\to\lVert\tilde{\tilde{\boldsymbol{n}}}\rVert^{2}$

续表

检测器		检测器及其统计等价分解	检测器极限形式
子空间目标检测器	COS^2 检测器	$T_{\cos^2}=\dfrac{\boldsymbol{x}^{\mathrm{H}}\boldsymbol{M}^{-1}\boldsymbol{\Psi}(\boldsymbol{\Psi}^{\mathrm{H}}\boldsymbol{M}^{-1}\boldsymbol{\Psi})^{-1}\boldsymbol{\Psi}^{\mathrm{H}}\boldsymbol{M}^{-1}\boldsymbol{x}}{\boldsymbol{x}^{\mathrm{H}}\boldsymbol{M}^{-1}\boldsymbol{x}}$, $T_{\cos^2}=\dfrac{\boldsymbol{F}}{1+\boldsymbol{F}}$, $\boldsymbol{F}=\dfrac{\left\Vert \tilde{\tilde{\boldsymbol{n}}}\dfrac{h_2}{g}+\boldsymbol{h}_3\right\Vert^2}{h_1^2}$	$T_{\cos^2}\to\dfrac{\left\Vert\dfrac{\tilde{\tilde{\boldsymbol{n}}}}{g}\right\Vert^2}{1+\left\Vert\dfrac{\tilde{\tilde{\boldsymbol{n}}}}{g}\right\Vert^2}$
	变量说明	$\tilde{\tilde{\boldsymbol{n}}}\sim\mathrm{CN}\left[\dfrac{(\tilde{\boldsymbol{\Psi}}^{\mathrm{H}}\tilde{\boldsymbol{\Psi}})^{1/2}}{\sigma}\boldsymbol{\alpha},\boldsymbol{I}_p\right]\Leftrightarrow$CN[白化后的幅度信噪比矢量,$\boldsymbol{I}_p$] $g^2\sim\chi^2_{MN-p}[0]$ $h_1^2\sim\chi^2_{K-MN+1,1}[0]$ $h_2^2\sim\chi^2_{K-MN+p+1}[0]$ $\boldsymbol{h}_3\sim\mathrm{CN}[\boldsymbol{0},\boldsymbol{I}_p]$ 正态随机矢量 $\tilde{\tilde{\boldsymbol{n}}}$、$\boldsymbol{h}_3$,瑞利随机变量 g、h_1、h_2 五者之间都是统计独立的	当 $K\to\infty$ 时,$h_1^2\to K\to\infty$,$h_2^2\to K\to\infty$,g 和 $\boldsymbol{h}_3$ 不受影响

3. 6 种检测器的检测概率与虚警概率表达式

这里仍以 MSD 检测器为例,在其等价分解的基础上,推导其检验统计量的概率密度函数,进而推导其检测概率和虚警概率表达式。其他 5 种检测器的相关推导都是类似的[10],本书以表格形式给出了 AMF、Kelly、Kalson 检测器的检测概率和虚警概率表达式,如表 14.2 所示。

表 14.2 $p=1$ 时的 AMF、Kelly、Kalson 检测器的检测概率与虚警概率表达式

	AMF 检测器	Kelly 检测器	Kalson 检测器
虚警概率	$P_{\mathrm{fa}}=\int_0^1\left(1+\dfrac{b}{K}\eta\right)^{MN-K-1}f(b)\mathrm{d}b$	$P_{\mathrm{fa}}=(1+\eta)^{MN-K-1}$	$P_{\mathrm{fa}}=\int_0^1[1+(1-b)\eta]^{MN-K-1}f(b)\mathrm{d}b$
门限	数值求解	$\eta=P_{\mathrm{fa}}^{1/(MN-K-1)}-1$	数值求解
检测概率	$P_{\mathrm{d}}=\int_0^1\ell\left(b;\dfrac{b}{K}\eta\right)f(b)\mathrm{d}b$	$P_{\mathrm{d}}=\int_0^1\ell(b;\eta)f(b)\mathrm{d}b$	$P_{\mathrm{d}}=\int_0^1\ell[b;(1-b)\eta]f(b)\mathrm{d}b$
门限函数	$u=\dfrac{b}{K}\eta$	$u=\eta$	$u=(1-b)\eta$

$$\ell(b;u)=\mathrm{e}^{-b\cdot\mathrm{SNR}}\left\{(1+u)^{MN-K-1}+\sum_{k=1}^{\infty}\frac{(b\cdot\mathrm{SNR})^k}{k!}\right.$$

$$\left.\left[1-\frac{1}{B(1+k,K-MN+1)}\int_0^u\frac{t^k}{(1+t)^{1+(K-MN+1)+k}}\mathrm{d}t\right]\right\}$$

其中,b 服从复 $\beta_{K-MN+2,MN-1}$ 分布,概率密度函数如下[38]

$$\mathrm{CB}_{v_1,v_2}(x)=\begin{cases}\dfrac{1}{B(v_1,v_2)}x^{v_1-1}(1-x)^{v_2-1}, & 0<x<1\\ 0, & \text{其他}\end{cases},\quad B(v_1,v_2)\text{表示参数为}(v_1,v_2)\text{的 beta 函数}$$

1) MSD 检测器的概率密度函数

第一步:分离出向量 $\tilde{\tilde{\boldsymbol{n}}}$ 的均值

用 $\bar{\tilde{\boldsymbol{n}}}$ 表示 $\tilde{\boldsymbol{n}}$ 的 p 维均值向量，$\bar{\tilde{\boldsymbol{n}}}=\dfrac{(\tilde{\boldsymbol{\Psi}}^{\mathrm{H}}\tilde{\boldsymbol{\Psi}})^{1/2}}{\sigma}\boldsymbol{\alpha}$，那么

$$T_{\mathrm{MSD}}=K\frac{\sigma^2}{h_1^2}\left\|\tilde{\boldsymbol{n}}+g\frac{\boldsymbol{h}_3}{h_2}\right\|^2\xlongequal{\text{分布上等于}}K\frac{\sigma^2}{h_1^2}\left\|\boldsymbol{n}_0+\bar{\tilde{\boldsymbol{n}}}+g\frac{\boldsymbol{h}_3}{h_2}\right\|^2 \tag{14.66}$$

其中，$\boldsymbol{n}_0$ 和 $\boldsymbol{h}_3$ 都服从分布 $\mathrm{CN}[\boldsymbol{0},\boldsymbol{I}_p]$。$\boldsymbol{n}_0+g\dfrac{\boldsymbol{h}_3}{h_2}$ 所服从的分布与 $\boldsymbol{m}\sqrt{1+\dfrac{g^2}{h_2^2}}$ 所服从的分布是一样的，其中 $\boldsymbol{m}$ 服从分布 $\mathrm{CN}[\boldsymbol{0},\boldsymbol{I}_p]$。因此，有

$$T_{\mathrm{MSD}}\xlongequal{\text{分布上等于}}K\frac{\sigma^2\left(1+\dfrac{g^2}{h_2^2}\right)}{h_1^2}\left\|\boldsymbol{m}+\frac{1}{\sqrt{1+\dfrac{g^2}{h_2^2}}}\bar{\tilde{\boldsymbol{n}}}\right\|^2 \tag{14.67}$$

第二步：得出条件复 F 分布(CCF 分布)

令 $b=\dfrac{1}{1+\dfrac{g^2}{h_2^2}}$，那么 b 服从 $\beta_{K-MN+p+1,MN-p}$ 分布，自由度参数分别是 $K-MN+p+1$ 和 $MN-p$，那么，$T_{\mathrm{MSD}}\xlongequal{\text{分布上等于}}\dfrac{\|\xi\|^2}{h_1^2}\cdot\dfrac{K\sigma^2}{b}$。因此，在以 b 为条件的情况下，有

$$\tilde{T}_{\mathrm{MSD}}=\frac{K-MN+1}{p}\frac{b}{K\sigma^2}T_{\mathrm{MSD}}\xlongequal{\text{分布上等于}}\frac{\dfrac{\|\xi\|^2}{p}}{\dfrac{h_1^2}{K-MN+1}} \tag{14.68}$$

$\tilde{T}_{\mathrm{MSD}}$ 表示复非中心 F 分布(记为 CF)的随机变量，非中心参数为 $b\cdot\mathrm{SNR}$，复自由度为 p 和 $K-MN+1$，其中 SNR 表示信噪比，ξ 在以 b 为条件下服从复随机正态分布 $\mathrm{CN}[\bar{\tilde{\boldsymbol{n}}}\cdot\sqrt{b},\boldsymbol{I}_p]$。

2) 检测概率表达式

由于 $\{\tilde{T}_{\mathrm{MSD}}|b\}\sim\mathrm{CF}_{p,K-MN+1}[b\cdot\mathrm{SNR}]$，条件检测概率表达式为

$$\begin{aligned}P_{\mathrm{d}}\mid b&=\Pr\{T_{\mathrm{MSD}}\mid b>\eta\}=\Pr\left\{\tilde{T}_{\mathrm{MSD}}>\frac{b\cdot(K-MN+1)}{K\sigma^2\cdot p}\eta\right\}\\&=\int_{\eta_b}^{\infty}\mathrm{CF}_{p,K-MN+1}[b\cdot\mathrm{SNR}]\,\mathrm{d}(\tilde{T}_{\mathrm{MSD}})\end{aligned} \tag{14.69}$$

其中，门限 $\eta_b=\dfrac{b}{K\sigma^2}\dfrac{K-MN+1}{p}\eta$，概率密度函数 $\mathrm{CF}_{p,K-MN+1}[b\cdot\mathrm{SNR}](x)$ 为

$$\mathrm{CF}_{p,K-MN+1}[b\cdot\mathrm{SNR}](x)=\begin{cases}\mathrm{e}^{-b\cdot\mathrm{SNR}}\displaystyle\sum_{k=0}^{\infty}\frac{(b\cdot\mathrm{SNR})^k}{k!}\frac{\left(\dfrac{p}{K-MN+1}\right)^{p+k}}{B(p+k,K-MN+1)}\frac{x^{p+k-1}}{\left(1+\dfrac{p}{K-MN+1}x\right)^{p+(K-MN+1)+k}}, & x>0\\0, & x<0\end{cases} \tag{14.70}$$

对 η_b 作如下代换

$$\frac{b}{K\sigma^2}\frac{K-MN+1}{p}\eta < x < \infty \Rightarrow \eta < \frac{p}{(K-MN+1)\dfrac{b}{K\sigma^2}}x < \infty$$

$$\text{令 } t = \frac{p}{(K-MN+1)\dfrac{b}{K\sigma^2}}x \Rightarrow \mathrm{d}x = \frac{(K-MN+1)\dfrac{b}{K\sigma^2}}{p}\mathrm{d}t \tag{14.71}$$

然后将式(14.69)～式(14.71)代入检测概率表达式 $P_{\mathrm{d}} = \int_0^1 (P_{\mathrm{d}} \mid b) f(b)\mathrm{d}b$ 中，其中 $f(b)$ 为 b 所服从的 $\beta_{K-MN+p+1,MN-p}$ 分布概率密度函数，再结合 MSD 检测器中的 $\sigma^2=1$，可求得检测概率表达式如下

$$P_{\mathrm{d}} = \int_0^1 \left[\int_\eta^\infty \mathrm{e}^{-b\cdot\mathrm{SNR}} \sum_{k=0}^{\infty} \frac{(b\cdot \mathrm{SNR})^k}{k!} \frac{\dfrac{b}{K}}{B(p+k, K-MN+1)} \frac{\left(\dfrac{b}{K}t\right)^{p+k-1}}{\left(1+\dfrac{b}{K}t\right)^{p+(K-MN+1)+k}}\mathrm{d}t\right]\cdot f(b)\mathrm{d}b \tag{14.72}$$

3）虚警概率表达式

在式(14.72)中，令 SNR＝0，即可得到虚警概率表达式

$$P_{\mathrm{fa}} = \int_0^1 \left[\int_\eta^\infty \frac{\dfrac{b}{K}}{B(p, K-MN+1)} \frac{\left(\dfrac{b}{K}t\right)^{p-1}}{\left(1+\dfrac{b}{K}t\right)^{p+(K-MN+1)}}\mathrm{d}t\right]\cdot f(b)\mathrm{d}b \tag{14.73}$$

当 $p=1$ 时，对应的是 AMF 检测器的检测概率与虚警概率，如下所示

$$P_{\mathrm{fa}} = \int_0^1 \ell_0(b)\cdot f(b)\mathrm{d}b = \mathrm{E}[\ell_0(b)] = \frac{\sum_{i=1}^{M}\ell_0(b_i)}{M} \tag{14.74}$$

$$P_{\mathrm{d}} = \int_0^1 \ell_1(b)\cdot f(b)\mathrm{d}b = \mathrm{E}[\ell_1(b)] = \frac{\sum_{i=1}^{M}\ell_1(b_i)}{M} \tag{14.75}$$

其中，

$$\ell_0(b) = \left(1+\frac{b}{K}\eta\right)^{MN-K-1} \tag{14.76}$$

$$\ell_1(b) = \mathrm{e}^{-b\cdot\mathrm{SNR}}\cdot\left\{\left(1+\frac{b}{K}\eta\right)^{MN-K-1} + \sum_{k=1}^{\infty}\frac{(b\cdot\mathrm{SNR})^k}{k!}\left[1-\frac{\displaystyle\int_0^{\frac{b}{K}\eta}\frac{u^k}{(1+u)^{1+(K-MN+1)+k}}\mathrm{d}u}{B(1+k, K-MN+1)}\right]\right\} \tag{14.77}$$

4. 6 种检测器的检测性能分析

这里利用前面给出的检测性能解析表达式、统计等价分解及其极限形式，来对比分析这 6 种检测器的检测性能。

图 14.2 利用这 6 种检测器的统计等价分解的极限形式，对比分析了不同信号子空间维数及不同信号能量分配情况下这 6 种检测器的检测性能。图 14.2 中，“2 倍 SNR 集中于 1 维”和“1 倍 SNR 集中于 1 维”针对的是信号子空间维数 $p=1$ 的 AMF、Kelly 和 Kalson 检测器，而仿真中所用 SNR 的数值分别是横坐标所示 SNR 的 2 倍和 1 倍；“2 倍 SNR 均分于 2 维”“2 倍 SNR 均分于 3 维”和“2 倍 SNR 均分于 10 维”针对的是信号子空间维数 $p>1$ 的 MSD、ASD 和 COS^2 检测器，而仿真中每维信号子空间所用 SNR 的数值分别是图 14.2 中横坐标所示 SNR 的 2 倍的 1/2、1/3 和 1/10。

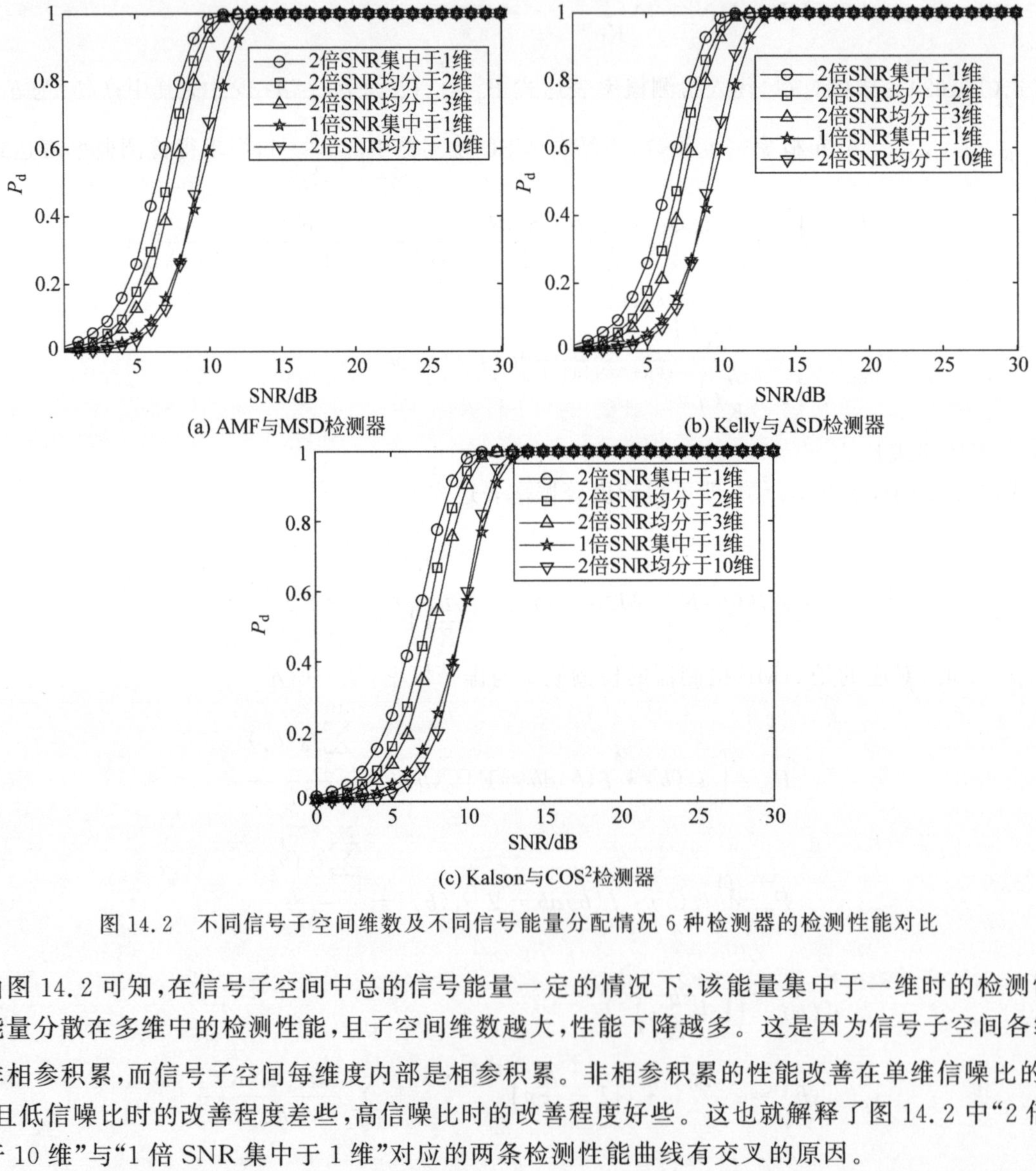

图 14.2　不同信号子空间维数及不同信号能量分配情况 6 种检测器的检测性能对比

由图 14.2 可知，在信号子空间中总的信号能量一定的情况下，该能量集中于一维时的检测性能优于该能量分散在多维中的检测性能，且子空间维数越大，性能下降越多。这是因为信号子空间各维度之间是非相参积累，而信号子空间每维度内部是相参积累。非相参积累的性能改善在单维信噪比的$\sqrt{p}\sim p$ 倍，且低信噪比时的改善程度差些，高信噪比时的改善程度好些。这也就解释了图 14.2 中“2 倍 SNR 均分于 10 维”与“1 倍 SNR 集中于 1 维”对应的两条检测性能曲线有交叉的原因。

由图 14.2 的结果还可得出，对检测性能来说，$p=1$ 与 $p>1$ 的区别仅在于信噪比的不同，其他检测性能规律对于两者来说都是类似的，因此后续只分析 $p=1$ 的情况。

图 14.3 给出了信号子空间维数 $p=1$ 时，AMF、Kelly 和 Kalson 3 种检测器在 $K\to\infty$的条件下，其检测性能随观测向量维数 MN 的变化情况，此时的检测性能指的就是极限检测性能，其中观测向量维数 MN 值是 $MN=M\times N$，检测性能曲线是通过统计等价分解的极限形式仿真得出的。由图 14.3 可知，在 $K\to\infty$

的条件下，AMF 和 Kelly 检测器的极限检测性能不随 MN 值的变化而变化，而 Kalson 检测器的极限检测性能与 MN 值有关，MN 值越大，其极限检测性能越好。这一点从检测器的统计等价分解的极限形式上也可以看出。AMF 和 Kelly 检测器的极限形式只与 $\|\tilde{\tilde{n}}\|^2$ 有关，即只与 SNR 和子空间维数 p(此处 $p=1$)有关；而 Kalson 检测器除了与 $\|\tilde{\tilde{n}}\|^2$ 有关外，还与 g 有关，而 g 与 MN 有关。

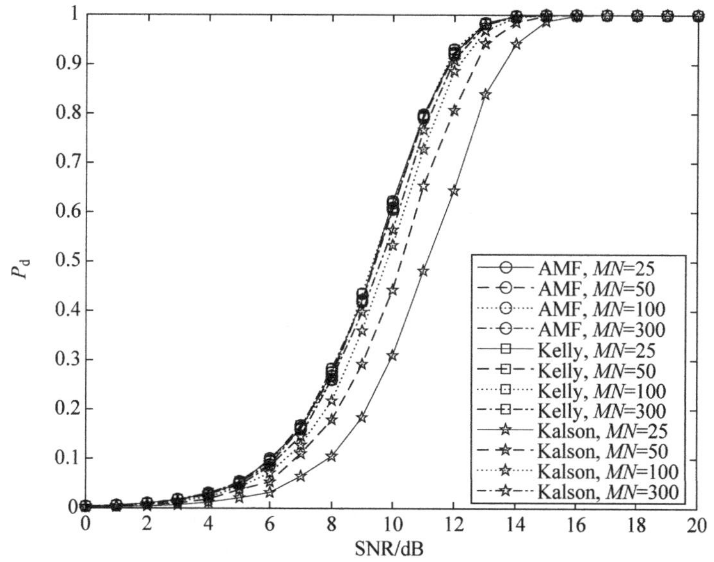

图 14.3　$p=1$ 时的 AMF、Kelly、Kalson 的极限检测性能曲线

对于 AMF、Kelly 和 Kalson 三种检测器，图 14.4(a)、图 14.4(b)和图 14.4(c)分别给出了三种检测器的极限检测性能曲线、基于检测性能解析表达式的检测性能曲线和基于统计等价分解的检测性能曲线。由图 14.4 可知以下特点。

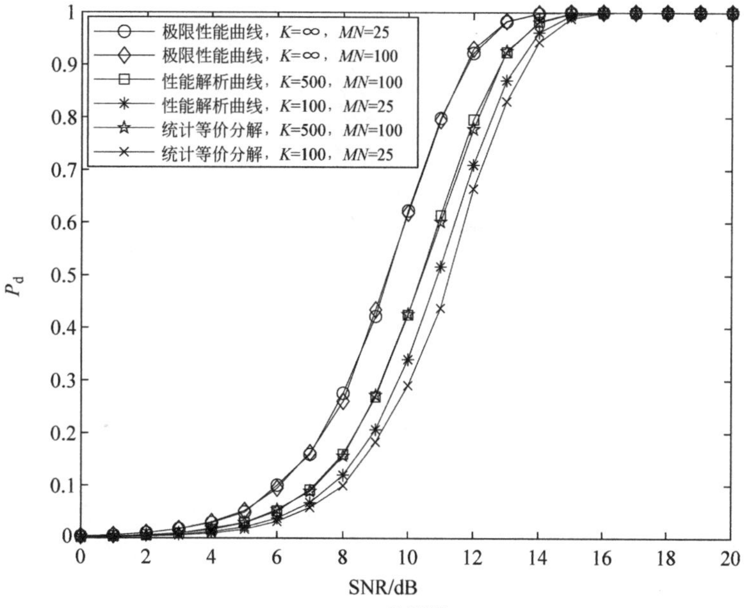

(a) AMF检测器

图 14.4　三种检测器的三类检测性能曲线对比

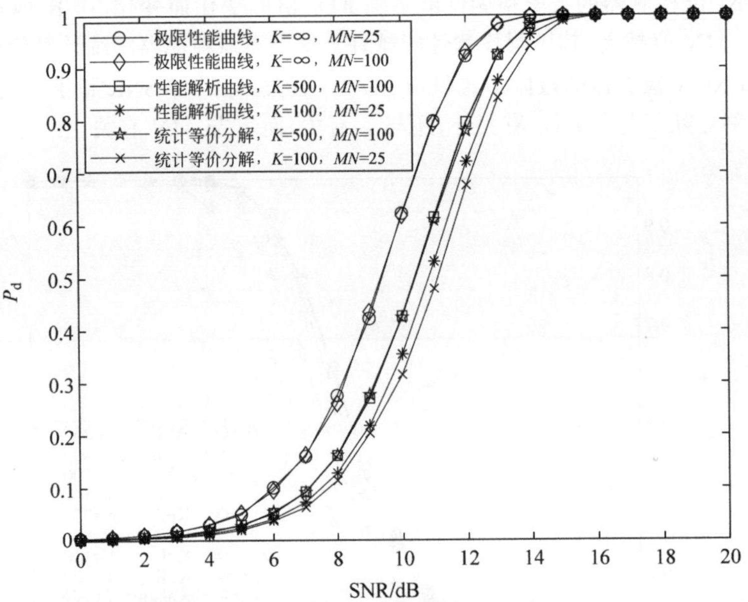

(b) Kelly检测器

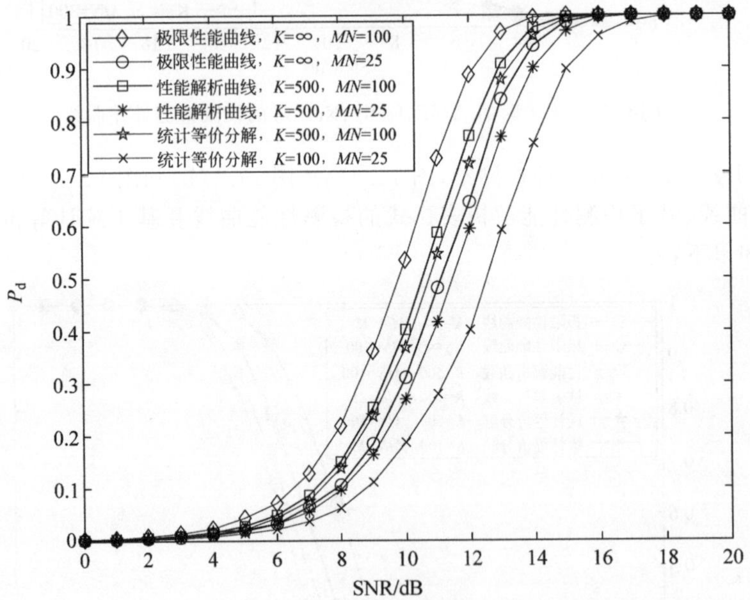

(c) Kalson检测器

图 14.4 （续）

(1) 对每种检测器来说，极限检测性能曲线是其检测性能上限，其中 Kalson 检测器的极限检测性能曲线还与 MN 值有关。

(2) 在相同的 K 与 MN 参数组条件下，基于检测性能解析表达式得到的检测性能曲线优于基于统计等价分解得到的检测性能曲线，这是因为解析性能表达式中不涉及参数估计问题，因此没有参数估计带来的性能损失，这一点在 Kalson 检测器上表现得更加明显。

(3) 对每种检测器来说，K 值越大，其基于统计等价分解得到的检测性能曲线越接近于基于检测性能解析表达式得到的检测性能曲线。

从检测性能曲线来看，这三个检测器中，Kalson 检测器的检测性能较差，而 AMF 与 Kelly 检测器的性能相差不大，原因在于 Kalson 检测器考虑了检测单元与参考单元背景功率水平不一致的问题，并对此具备了自适应能力，因而消耗了一定的自由度，导致检测性能损失，而 AMF 和 Kelly 检测器没有考虑这一问题，直接认为两者背景功率水平一致。另外，AMF 与 Kelly 检测器的性能差异主要体现在导向向量偏离目标的情况下，而当导向向量与目标真实导向向量没有偏离时，二者性能的确相差不大。关于 AMF 与 Kelly 检测器的更多性能对比分析请参考文献[39]。

14.3 基于自适应空时编码设计的二维联合 CFAR 检测

雷达技术的发展使得雷达阵列能够在不同的阵元脉冲上辐射不同的波形，从而为充分利用发射端的空时自由度提供了物理基础。根据观测场景自适应地调整发射波形能显著地改善雷达性能，例如，调整发射波形使之匹配目标特征能够增强目标的可检测性、增强对杂波和干扰的抑制能力及对多径现象的鲁棒性[40]。Friedlander 提出了自适应波形设计的目标子空间框架[41]，并将研究结果推广到 MIMO 阵列雷达的目标检测问题中，讨论了基于最大化信号干扰噪声比的最优波形设计方法及几种次优方法[40,42-43]。由于 Friedlander 主要研究脉冲内的波形设计，因此目标的多普勒特征及杂波复幅度的空时相关性都没有反映到波形设计中。De Maio 和 Lops[44]基于互信息提出了"分布式 MIMO 雷达"检测器的空时编码设计准则，其中考虑了目标多普勒特征，但假设杂波的统计特性(协方差矩阵)与发射波形无关[42,45]。另外，其他学者还从最大化目标处的发射功率和匹配发射波束方向图[46-47]、目标参数估计与跟踪[48]及目标识别与分类[49-50]等角度出发对 MIMO 雷达的波形设计问题展开了深入研究。

本节将在 Friedlander 和 De Maio 等研究的基础上，从改善 MIMO 阵列雷达检测性能的角度出发，讨论脉间空时编码的设计问题。与前人研究相比，本节研究的基于自适应空时编码设计的 MIMO 阵列雷达检测技术[51]有以下 3 个特点。

(1) 考虑了目标和杂波的空时特征(即方位-多普勒特征)及杂波复幅度的空时相关性对编码设计的影响。

(2) 本节的"自适应"有两重含义：一是发射的空时编码依赖于被观测场景的信息，二是被观测场景的信息是通过自适应估计得到的。

(3) 推导了多目标(更一般地说是多秩目标)条件下空时编码的设计公式，文献[40]对此也进行了研究，但其给出的设计过程都是针对秩 1 目标的。

14.3.1 信号模型及 MSD 检测器

假设 MIMO 阵列雷达系统由 M 个发射阵元和 N 个接收阵元组成，间距分别为 d_t 和 $d_r=\lambda/2$，其中 $\lambda/2\leqslant d_t\leqslant N\lambda/2$，$\lambda$ 是波长。发射阵与接收阵同基地共线布置，每个发射阵元发射 L 个编码脉冲形成的相参脉冲串，各个脉冲上的码字为 $c_{m,l}$，$m=1,2,\cdots,M$，$l=1,2,\cdots,L$。由 $c_{m,l}$ 形成的空时编码矩阵记为$\boldsymbol{C}=[\boldsymbol{c}_1,\boldsymbol{c}_2,\cdots,\boldsymbol{c}_L]$，其中，$\boldsymbol{c}_l=[c_{1,l},c_{2,l},\cdots,c_{M,l}]^{\mathrm{T}}$。每个距离环上有 Q 个方位单元。设当前距离环中第 $q(q=1,2,\cdots,Q)$个方位单元散射体在第 l 个脉冲回波中的复幅度为 $\rho_{q,l}$，那么第 $i(i=1,2,\cdots,N)$个接收阵元接收到的来自此散射体的回波信号为

$$r_{q,i}(t)=b_{q,i}\sum_{m=1}^{M}\sum_{l=1}^{L}a_{q,m}d_{q,l}\rho_{q,l}c_{m,l}s\left[t-\tau-(l-1)T_p\right] \tag{14.78}$$

其中,T_p 是脉冲重复周期,$s(t)$是脉冲波形,假设其具有单位能量,脉宽为 τ_p。$a_{q,m}$ 和 $b_{q,i}$ 分别是该散射体对应的发射导向向量 $\boldsymbol{a}_q$ 和接收导向向量 $\boldsymbol{b}_q$ 的第 m 和第 i 个元素;$d_{q,l}$ 是该散射体的多普勒在第 l 个脉冲上引起的相移因子;τ 是当前距离单元对应的延迟时间。将 N 个阵元接收到的来自该散射体的回波匹配 $s(t)$的滤波器,然后采样得到 $NL\times1$ 维向量

$$\boldsymbol{r}_q=\boldsymbol{A}_q\boldsymbol{\Lambda}\boldsymbol{\rho}_q \tag{14.79}$$

其中,$\boldsymbol{\Lambda}=\mathrm{diag}(\boldsymbol{c}_1,\boldsymbol{c}_2,\cdots,\boldsymbol{c}_L)$,$\boldsymbol{A}_q=\mathrm{diag}(d_{q,1},d_{q,2},\cdots,d_{q,L})\otimes(\boldsymbol{b}_q\boldsymbol{a}_q^{\mathrm{T}})$,$\boldsymbol{\rho}_q=[\rho_{q,1},\rho_{q,2},\cdots,\rho_{q,L}]^{\mathrm{T}}$ 表示该散射体在各个脉冲上的回波复幅度形成的向量。下面根据目标和杂波的不同模型给出式(14.79)的具体形式,其中,对于当前距离单元中第 $s(s=1,2,\cdots,S)$个目标来说,$NL\times1$ 维观测向量 $\boldsymbol{r}_q$ 用 $\boldsymbol{r}_{\mathrm{t},s}$ 代替,总的目标回波表示为 $\boldsymbol{r}_{\mathrm{t}}$;对于当前距离单元中第 $q(q=1,2,\cdots,Q)$个方位杂波块来说,则用 $\boldsymbol{r}_{\mathrm{c},q}$ 表示其回波,总的杂波回波用 $\boldsymbol{r}_{\mathrm{c}}$ 代替;在这些符号中,下标 t 表示该符号表征的是目标的属性;类似地,下标 c 和下标 n 分别意味着该符号表征的是杂波和噪声属性。

1. 目标回波模型

假设当前距离单元中有 S 个目标,且每个目标的回波复幅度在一个相参处理间隔中是不变的,因此,第 $s(s=1,2,\cdots,S)$个目标的回波复幅度向量可表示为 $\rho_{\mathrm{t},s}=[\rho_{\mathrm{t},s,1},\rho_{\mathrm{t},s,2},\cdots,\rho_{\mathrm{t},s,L}]^{\mathrm{T}}=\rho_{\mathrm{t},s}\cdot\mathbf{1}_{L\times1}$,其中,$\rho_{\mathrm{t},s}$ 表征了第 s 个目标的回波复幅度,$\mathbf{1}_{L\times1}$ 是 $L\times1$ 维的全 1 向量,这意味着目标可能是 Swerling Ⅰ型的,也可能是非起伏目标。那么根据式(14.79)可推导出 S 个目标的回波向量为

$$\boldsymbol{r}_{\mathrm{t}}=\sum_{s=1}^{S}\boldsymbol{r}_{\mathrm{t},s}=\sum_{s=1}^{S}\boldsymbol{A}_{\mathrm{t},s}\boldsymbol{\Lambda}\boldsymbol{\rho}_{\mathrm{t},s}=\sum_{s=1}^{S}\boldsymbol{A}_{\mathrm{t},s}\boldsymbol{c}\boldsymbol{\rho}_{\mathrm{t},s}=[\boldsymbol{A}_{\mathrm{t},1}\boldsymbol{c},\boldsymbol{A}_{\mathrm{t},2}\boldsymbol{c},\cdots,\boldsymbol{A}_{\mathrm{t},S}\boldsymbol{c}]\,[\rho_{\mathrm{t},1},\rho_{\mathrm{t},2},\cdots,\rho_{\mathrm{t},S}]^{\mathrm{T}}=\boldsymbol{B}_{\mathrm{t}}\boldsymbol{\rho}_{\mathrm{t}} \tag{14.80}$$

其中,$\boldsymbol{B}_{\mathrm{t}}=[\boldsymbol{A}_{\mathrm{t},1}\boldsymbol{c},\boldsymbol{A}_{\mathrm{t},2}\boldsymbol{c},\cdots,\boldsymbol{A}_{\mathrm{t},S}\boldsymbol{c}]$ 表示目标的模式矩阵,$\boldsymbol{c}=\mathrm{vec}(\boldsymbol{C})$ 称为空时编码波形,$\boldsymbol{\rho}_{\mathrm{t}}=[\rho_{\mathrm{t},1},\rho_{\mathrm{t},2},\cdots,\rho_{\mathrm{t},S}]^{\mathrm{T}}$ 是由 S 个目标的回波复幅度形成的 $S\times1$ 维目标回波复幅度向量,且假设$\boldsymbol{\rho}_{\mathrm{t}}\sim\mathrm{CN}(\boldsymbol{0},\sigma_{\mathrm{t}}^2\boldsymbol{I}_{S\times S})$,$\sigma_{\mathrm{t}}^2$ 是每个目标的功率水平。

2. 杂波回波模型

总的杂波向量 $\boldsymbol{r}_{\mathrm{c}}$ 是距离环中 Q 个方位单元中的杂波散射体回波的向量和,由式(14.79)可得

$$\boldsymbol{r}_{\mathrm{c}}=\sum_{q=1}^{Q}\boldsymbol{r}_{\mathrm{c},q}=\sum_{q=1}^{Q}\boldsymbol{A}_{\mathrm{c},q}\boldsymbol{\Lambda}\boldsymbol{\rho}_{\mathrm{c},q}=[\boldsymbol{A}_{\mathrm{c},1}\boldsymbol{\Lambda},\boldsymbol{A}_{\mathrm{c},2}\boldsymbol{\Lambda},\cdots,\boldsymbol{A}_{\mathrm{c},Q}\boldsymbol{\Lambda}]\,[\boldsymbol{\rho}_{\mathrm{c},1}^{\mathrm{T}},\boldsymbol{\rho}_{\mathrm{c},2}^{\mathrm{T}},\cdots,\boldsymbol{\rho}_{\mathrm{c},Q}^{\mathrm{T}}]^{\mathrm{T}}=\boldsymbol{B}_{\mathrm{c}}\boldsymbol{\rho}_{\mathrm{c}} \tag{14.81}$$

其中,$\boldsymbol{B}_{\mathrm{c}}=[\boldsymbol{A}_{\mathrm{c},1}\boldsymbol{\Lambda},\boldsymbol{A}_{\mathrm{c},2}\boldsymbol{\Lambda},\cdots,\boldsymbol{A}_{\mathrm{c},Q}\boldsymbol{\Lambda}]$表示杂波的模式矩阵,$\boldsymbol{\rho}_{\mathrm{c}}=[\boldsymbol{\rho}_{\mathrm{c},1}^{\mathrm{T}},\boldsymbol{\rho}_{\mathrm{c},2}^{\mathrm{T}},\cdots,\boldsymbol{\rho}_{\mathrm{c},Q}^{\mathrm{T}}]^{\mathrm{T}}$ 表示杂波复幅度向量。根据 Q 个杂波点形成的复幅度向量$\boldsymbol{\rho}_{\mathrm{c}}$ 在方位维和脉冲维的不同相关性质,杂波协方差矩阵 $\boldsymbol{R}_{\mathrm{c}}$ 的形式可分为以下 3 种情况。

(1) $\boldsymbol{\rho}_{\mathrm{c}}$ 在方位维和脉冲维均存在部分相关性,用空时相关阵 $\boldsymbol{M}_{\mathrm{c}}=\mathrm{E}\{\boldsymbol{\rho}_{\mathrm{c}}\boldsymbol{\rho}_{\mathrm{c}}^{\mathrm{H}}\}$ 来描述。此时杂波协方差矩阵为$\boldsymbol{R}_{\mathrm{c}}=\boldsymbol{B}_{\mathrm{c}}\boldsymbol{M}_{\mathrm{c}}\boldsymbol{B}_{\mathrm{c}}^{\mathrm{H}}$。$\mathrm{E}\{\cdot\}$表示求期望。

(2) $\boldsymbol{\rho}_{\mathrm{c}}$ 在方位维有部分相关性,在脉冲维完全相关,它是情况(1)的特例,此时 $\boldsymbol{\rho}_{\mathrm{c},q}=[\rho_{\mathrm{c},q,1},\rho_{\mathrm{c},q,2},\cdots,\rho_{\mathrm{c},q,L}]^{\mathrm{T}}=\rho_{\mathrm{c},q}\cdot\mathbf{1}_{L\times1}$,$\boldsymbol{\rho}_{\mathrm{c}}=\boldsymbol{\rho}_{\mathrm{cs}}\otimes\mathbf{1}_{L\times1}$,$\boldsymbol{\rho}_{\mathrm{cs}}=[\rho_{\mathrm{c},1},\rho_{\mathrm{c},2},\cdots,\rho_{\mathrm{c},Q}]^{\mathrm{T}}$,因此

$$\boldsymbol{r}_{\mathrm{c}}=\boldsymbol{B}_{\mathrm{c}}\boldsymbol{\rho}_{\mathrm{c}}=[\boldsymbol{A}_{\mathrm{c},1}\boldsymbol{\Lambda},\boldsymbol{A}_{\mathrm{c},2}\boldsymbol{\Lambda},\cdots,\boldsymbol{A}_{\mathrm{c},Q}\boldsymbol{\Lambda}](\boldsymbol{\rho}_{\mathrm{cs}}\otimes\mathbf{1}_{L\times1})=[\boldsymbol{A}_{\mathrm{c},1}\boldsymbol{c},\boldsymbol{A}_{\mathrm{c},2}\boldsymbol{c},\cdots,\boldsymbol{A}_{\mathrm{c},Q}\boldsymbol{c}]\boldsymbol{\rho}_{\mathrm{cs}}=\boldsymbol{B}_{\mathrm{cs}}\boldsymbol{\rho}_{\mathrm{cs}}$$

其中,$\boldsymbol{B}_{\mathrm{cs}}=[\boldsymbol{A}_{\mathrm{c},1}\boldsymbol{c},\boldsymbol{A}_{\mathrm{c},2}\boldsymbol{c},\cdots,\boldsymbol{A}_{\mathrm{c},Q}\boldsymbol{c}]$,杂波协方差矩阵为$\boldsymbol{R}_{\mathrm{c}}=\boldsymbol{B}_{\mathrm{cs}}\boldsymbol{M}_{\mathrm{cs}}\boldsymbol{B}_{\mathrm{cs}}^{\mathrm{H}}$,其中,$\boldsymbol{M}_{\mathrm{cs}}=\mathrm{E}\{\boldsymbol{\rho}_{\mathrm{cs}}\boldsymbol{\rho}_{\mathrm{cs}}^{\mathrm{H}}\}$ 称为杂波复幅度方位维相关矩阵。

(3) $\boldsymbol{\rho}_c$ 在方位维相互独立，在脉冲维完全相关，它是情况(2)的特例。此时 $\boldsymbol{M}_{cs}=\sigma_c^2\boldsymbol{I}_{Q\times Q}$，$\boldsymbol{R}_c=\sigma_c^2\boldsymbol{B}_{cs}\boldsymbol{B}_{cs}^{\mathrm{H}}$，其中，$\sigma_c^2$ 是各个杂波散射体的功率水平。文献[40]中对杂波空域相关性质的假设类似于这种情况。

3. MSD 检测器

上述3种情况对应的研究方法都是一样的，因此不失一般性，本节只讨论情况(2)。根据以上信号模型，可以给出如下二元假设检验

$$\begin{cases}\mathrm{H}_0: \boldsymbol{r}=\boldsymbol{r}_c+\boldsymbol{r}_n=\boldsymbol{B}_{cs}\boldsymbol{\rho}_{cs}+\boldsymbol{r}_n\\ \mathrm{H}_1: \boldsymbol{r}=\boldsymbol{r}_t+\boldsymbol{r}_c+\boldsymbol{r}_n=\boldsymbol{B}_t\boldsymbol{\rho}_t+\boldsymbol{B}_{cs}\boldsymbol{\rho}_{cs}+\boldsymbol{r}_n\end{cases} \tag{14.82}$$

其中，噪声向量 $\boldsymbol{r}_n\sim\mathrm{CN}(\boldsymbol{0},\boldsymbol{R}_n)$，$\boldsymbol{R}_n=\sigma_n^2\boldsymbol{I}_{NL\times NL}$ 中的噪声功率水平 σ_n^2 和 $\boldsymbol{R}_c$ 中的 $\boldsymbol{M}_{cs}$ 都是未知的，模式矩阵 $\boldsymbol{B}_t$ 和 $\boldsymbol{B}_{cs}$ 都是已知的。与式(14.82)相对应的 MSD 检测器如下

$$T_{\mathrm{MSD}}=\boldsymbol{r}^{\mathrm{H}}\boldsymbol{R}_{cn}^{-1}\boldsymbol{B}_t(\boldsymbol{B}_t^{\mathrm{H}}\boldsymbol{R}_{cn}^{-1}\boldsymbol{B}_t)^{-1}\boldsymbol{B}_t^{\mathrm{H}}\boldsymbol{R}_{cn}^{-1}\boldsymbol{r}=\|\boldsymbol{W}_t^{\mathrm{H}}\boldsymbol{r}\|^2 \tag{14.83}$$

其中，$\boldsymbol{R}_{cn}=\boldsymbol{R}_c+\boldsymbol{R}_n=\boldsymbol{B}_{cs}\boldsymbol{M}_{cs}\boldsymbol{B}_{cs}^{\mathrm{H}}+\sigma_n^2\boldsymbol{I}_{NL\times NL}$，$NL\times S$ 维权矩阵 $\boldsymbol{W}_t=\boldsymbol{R}_{cn}^{-1}\boldsymbol{B}_t(\boldsymbol{B}_t^{\mathrm{H}}\boldsymbol{R}_{cn}^{-1}\boldsymbol{B}_t)^{-1/2}$。由文献[44]可知，式(14.83)描述的 MSD 检测器是 CFAR 的。

式(14.83)的 MSD 检测器中包含了发射的空时编码波形 $\boldsymbol{c}$。由14.2节的分析可知，MSD 检测器的检测性能由其输出 SCNR ε 决定

$$\varepsilon=\frac{p_t}{p_{cn}}=\frac{\mathrm{E}\{\|\boldsymbol{W}_t^{\mathrm{H}}\boldsymbol{r}_t\|^2\}}{\mathrm{E}\{\|\boldsymbol{W}_t^{\mathrm{H}}(\boldsymbol{r}_c+\boldsymbol{r}_n)\|^2\}}=\frac{\sigma_t^2\mathrm{tr}(\boldsymbol{B}_t^{\mathrm{H}}\boldsymbol{R}_{cn}^{-1}\boldsymbol{B}_t)}{S} \tag{14.84}$$

其中，p_t 表示输出目标功率，p_{cn} 表示输出杂噪功率，$\mathrm{tr}(\cdot)$ 表示求矩阵的迹。

当其他参数都一定时，最大化 MSD 检测器的检测性能等价于最大化其输出 SCNR ε。因此，就最大化检测概率而言，最大化 MIMO 阵列雷达 MSD 检测器的输出 SCNR 是进行空时编码波形设计的最优准则，简称为最大化 SCNR 准则。为了方便，本节将初始探测时所用的正交空时编码波形记为 $\boldsymbol{c}_0$，其中，$\boldsymbol{c}_0=\mathrm{vec}(\boldsymbol{C}_0)$，$\boldsymbol{C}_0\boldsymbol{C}_0^{\mathrm{H}}=\boldsymbol{I}_{M\times M}$。

14.3.2 自适应空时编码设计

1. 最大化 SCNR 准则

最大化 SCNR 准则是最优准则，简记为 Max SCNR 准则。由式(14.84)可知，最大化 MSD 检测器的输出 SCNR ε 等价于最大化

$$\mathrm{tr}(\boldsymbol{B}_t^{\mathrm{H}}\boldsymbol{R}_{cn}^{-1}\boldsymbol{B}_t)=\sum_{s=1}^{S}\boldsymbol{c}^{\mathrm{H}}\boldsymbol{A}_{t,s}^{\mathrm{H}}\boldsymbol{R}_{cn}^{-1}\boldsymbol{A}_{t,s}\boldsymbol{c}=\boldsymbol{c}^{\mathrm{H}}\Big(\sum_{s=1}^{S}\boldsymbol{A}_{t,s}^{\mathrm{H}}\boldsymbol{R}_{cn}^{-1}\boldsymbol{A}_{t,s}\Big)\boldsymbol{c} \tag{14.85}$$

其中，$\boldsymbol{A}_{t,s}(s=1,2,\cdots,S)$ 是已知的；而 $\boldsymbol{R}_{cn}$ 是未知的，也就是 $\boldsymbol{M}_{cs}$ 和 σ_n^2 是未知的。式(14.85)是文献[40]中结论在“多目标和未知 $\boldsymbol{R}_{cn}$”条件下的推广，仍可采用该文献中的迭代优化算法求取最优的空时编码波形，但其中 $\boldsymbol{R}_{cn}$ 的计算分为两步，一是反复计算 $\boldsymbol{B}_{cs}$，因为 $\boldsymbol{R}_{cn}$ 中只有 $\boldsymbol{B}_{cs}$ 与 $\boldsymbol{c}$ 有关，二是事先估计 $\boldsymbol{M}_{cs}$ 和 σ_n^2。$\boldsymbol{M}_{cs}$ 和 σ_n^2 的估计方法将在后面讨论。执行最优准则的迭代优化过程描述如下：

(1) 初始探测：利用正交空时编码波形 $\boldsymbol{c}_0$ 获得 $\boldsymbol{B}_{cs}$，并用 $\boldsymbol{c}_0$ 进行初始探测，利用参考距离单元中的回波估计 $\boldsymbol{M}_{cs}$ 和 σ_n^2；并将这些估计值及 $\boldsymbol{c}_0$ 和已知的 $\boldsymbol{A}_{t,s}$ 代入式(14.85)中，得到的结果存为 $f(0)$。

(2) 第 k 次迭代：计算 $\boldsymbol{G}=\sum_{s=1}^{S}\boldsymbol{A}_{t,s}^{\mathrm{H}}\boldsymbol{R}_{cn}^{-1}\boldsymbol{A}_{t,s}$，并求出 $\boldsymbol{G}$ 的最大特征值对应的特征向量 $\boldsymbol{g}$，于是，第 k 次迭代得到的空时编码波形为 $\boldsymbol{c}_k=\boldsymbol{g}/\|\boldsymbol{g}\|$；利用 $\boldsymbol{c}_k$ 获得 $\boldsymbol{B}_{cs}$，进而利用初始探测时估计得到的 $\boldsymbol{M}_{cs}$

和σ_n^2求得$\boldsymbol{R}_{cn}$；将$\boldsymbol{c}_k$、$\boldsymbol{R}_{cn}$及已知的$\boldsymbol{A}_{t,s}$代入式(14.85)中，得到的结果存为$f(k)$。

(3) 令迭代次数k增1，重新执行步骤(2)，直到$f(k)$的值趋向稳定为止。

将上述迭代优化过程得到的空时编码波形记为$\text{opt}_\boldsymbol{c}_1$。需要指出的是，当杂噪比(Clutter-to-Noise-Ratio，CNR)很小或等于0时就会出现一种特殊情况，即$\boldsymbol{R}_{cn}=\sigma_n^2\boldsymbol{I}_{NL\times NL}$，此时式(14.85)简化为

$$\text{tr}(\boldsymbol{B}_t^H\boldsymbol{R}_{cn}^{-1}\boldsymbol{B}_t)=\boldsymbol{c}^H\left(\sum_{s=1}^{S}\boldsymbol{A}_{t,s}^H\boldsymbol{R}_{cn}^{-1}\boldsymbol{A}_{t,s}\right)\boldsymbol{c}=N\sigma_n^{-2}\cdot\text{tr}\left(\sum_{s=1}^{S}\boldsymbol{a}_{t,s}^*\boldsymbol{a}_{t,s}^T\boldsymbol{C}\boldsymbol{C}^H\right) \tag{14.86}$$

因此，最大化式(14.85)就等价于最大化

$$\text{tr}(\boldsymbol{B}_t^H\boldsymbol{R}_{cn}^{-1}\boldsymbol{B}_t)=\text{tr}\left(\sum_{s=1}^{S}\boldsymbol{a}_{t,s}^*\boldsymbol{a}_{t,s}^T\boldsymbol{C}\boldsymbol{C}^H\right) \tag{14.87}$$

其中，$\boldsymbol{a}_{t,s}$是已知的第s个目标的发射导向向量。式(14.87)表明，对于此特殊情况，基于最大化SCNR准则的空时编码设计问题可以归结为Stoica P等人研究的基于最大化目标处的发射功率和发射波束方向图匹配准则的波形设计问题[50]，特别地，当$S=1$时，$\text{opt}\_\boldsymbol{c}_1=\boldsymbol{1}_{L\times 1}\otimes\boldsymbol{a}_t$。

从上面的分析可知，最优准则需要已知目标的方位-多普勒信息，并要求自适应估计$\boldsymbol{M}_{cs}$和σ_n^2。但是这些信息在某些条件下可能得不到，因此有必要研究需要较少先验知识的次优设计准则。借鉴文献[40]的研究思路，讨论最大化目标功率准则和最小化杂波功率准则这两种次优的设计准则，并将正交空时编码波形作为进行性能比较的参考基点。本节不考虑该文献中的最大化信杂比准则，因为在自适应估计$\boldsymbol{M}_{cs}$和σ_n^2的前提下，该次优准则所需的先验信息量与最优准则相同，但性能不如最优准则，因此不作研究。

2. 最大化目标功率准则

最大化目标功率准则是一种次优准则，简记为Max SP准则。Max SP准则指的是最大化检测器的输出目标功率p_t，它不需要背景的信息。

$$\begin{aligned}p_t&=\text{E}\{\|\boldsymbol{W}_t^H\boldsymbol{r}_t\|^2\}=\sigma_t^2\cdot\boldsymbol{w}_t^H\left(\boldsymbol{I}_{S\times S}\otimes\sum_{s=1}^{S}\boldsymbol{A}_{t,s}\boldsymbol{c}\boldsymbol{c}^H\boldsymbol{A}_{t,s}^H\right)\boldsymbol{w}_t=\sigma_t^2\cdot\boldsymbol{w}_t^H\boldsymbol{Q}_{st}\boldsymbol{w}_t\\&=\sigma_t^2\boldsymbol{c}^H\left(\sum_{s=1}^{S}\boldsymbol{A}_{t,s}^H\boldsymbol{W}_t\boldsymbol{W}_t^H\boldsymbol{A}_{t,s}\right)\boldsymbol{c}=\sigma_t^2\cdot\boldsymbol{c}^H\boldsymbol{Q}_{wt}\boldsymbol{c}\end{aligned} \tag{14.88}$$

其中，$\boldsymbol{w}_t=\text{vec}(\boldsymbol{W}_t)$。根据式(14.88)，最大化目标功率准则就是寻找$\boldsymbol{w}_t$和$\boldsymbol{c}$，使得$\boldsymbol{w}_t^H\boldsymbol{Q}_{st}\boldsymbol{w}_t$和$\boldsymbol{c}^H\boldsymbol{Q}_{wt}\boldsymbol{c}$最大化。式(14.88)也是文献[40]中结论在多目标条件下的推广，而设计过程中所采用的迭代优化算法与该文献类似，基本过程如下：

(1) 初始探测：利用正交空时编码波形$\boldsymbol{c}_0$和已知的$\boldsymbol{A}_{t,s}$，计算$\sum_{s=1}^{S}\boldsymbol{A}_{t,s}\boldsymbol{c}_0\boldsymbol{c}_0^H\boldsymbol{A}_{t,s}^H$，并求解其前$S$个最大特征值对应的特征向量$\boldsymbol{g}_1,\boldsymbol{g}_2,\cdots,\boldsymbol{g}_S$，因此$\boldsymbol{W}_t=[\boldsymbol{g}_1,\boldsymbol{g}_2,\cdots,\boldsymbol{g}_S]$，将$\boldsymbol{w}_t=\text{vec}(\boldsymbol{W}_t)$代入$\boldsymbol{w}_t^H\boldsymbol{Q}_{st}\boldsymbol{w}_t$中，得到的结果存为$f(0)$。

(2) 第k次迭代：利用已得到的$\boldsymbol{W}_t$，计算$\boldsymbol{Q}_{wt}=\sum_{s=1}^{S}\boldsymbol{A}_{t,s}^H\boldsymbol{W}_t\boldsymbol{W}_t^H\boldsymbol{A}_{t,s}$，并求出$\boldsymbol{Q}_{wt}$的最大特征值对应的特征向量$\boldsymbol{g}$，于是，第$k$次迭代得到的空时编码波形为$\boldsymbol{c}_k=\boldsymbol{g}/\|\boldsymbol{g}\|$；计算$\sum_{s=1}^{S}\boldsymbol{A}_{t,s}\boldsymbol{c}_k\boldsymbol{c}_k^H\boldsymbol{A}_{t,s}^H$，并求解其前$S$个最大特征值对应的特征向量$\boldsymbol{g}_1,\boldsymbol{g}_2,\cdots,\boldsymbol{g}_S$，于是，$\boldsymbol{W}_t=[\boldsymbol{g}_1,\boldsymbol{g}_2,\cdots,\boldsymbol{g}_S]$，将$\boldsymbol{w}_t=\text{vec}(\boldsymbol{W}_t)$代入$\boldsymbol{w}_t^H\boldsymbol{Q}_{st}\boldsymbol{w}_t$中，得到的结果存为$f(k)$。

(3) 令迭代次数 k 增 1,重新执行步骤(2),直到 $f(k)$的值趋向稳定为止。

将上述迭代优化过程得到的空时编码波形记为 opt_$\boldsymbol{c}_2$。显然,对于特殊情况 $\boldsymbol{R}_{\text{cn}}=\sigma_{\text{n}}^2\boldsymbol{I}_{NL\times NL}$,由式(14.88)中 p_{t} 的表达式可以直接推导出式(14.87),这表明,对于此特殊情况,最大化目标功率准则等价于最优准则。

3. 最小化杂波功率准则

最小化杂波功率准则也是一种次优准则,简记为 Min CP 准则。Min CP 准则指的是最小化检测器的输出杂波功率 p_{c},它不需要目标的信息。

$$\begin{aligned}p_{\text{c}}&=\text{E}\{|\boldsymbol{v}_{\text{t}}^{\text{H}}\boldsymbol{r}_{\text{c}}|^2\}=\boldsymbol{v}_{\text{t}}^{\text{H}}(\boldsymbol{B}_{cs}\boldsymbol{M}_{cs}\boldsymbol{B}_{cs}^{\text{H}})\boldsymbol{v}_{\text{t}}=\boldsymbol{v}_{\text{t}}^{\text{H}}\boldsymbol{Q}_{sc}\boldsymbol{v}_{\text{t}}\\&=\boldsymbol{c}^{\text{H}}\left(\sum_{i=1}^{Q}\sum_{j=1}^{Q}\boldsymbol{M}_{cs}(i,j)\boldsymbol{D}_{i,j}\right)\boldsymbol{c}=\boldsymbol{c}^{\text{H}}\boldsymbol{Q}_{wc}\boldsymbol{c}\end{aligned}\tag{14.89}$$

其中,$\boldsymbol{D}=[\boldsymbol{A}_{\text{c},1},\boldsymbol{A}_{\text{c},2},\cdots,\boldsymbol{A}_{\text{c},Q}]^{\text{H}}\boldsymbol{v}_{\text{t}}\boldsymbol{v}_{\text{t}}^{\text{H}}[\boldsymbol{A}_{\text{c},1},\boldsymbol{A}_{\text{c},2},\cdots,\boldsymbol{A}_{\text{c},Q}]$,$\boldsymbol{D}_{i,j}$ 是$\boldsymbol{D}$ 中第(i,j)个 $ML\times ML$ 维子矩阵。迭代优化方法的过程与文献[40]类似,是该方法在"未知 $\boldsymbol{R}_{\text{cn}}$"条件下的推广,基本过程如下:

(1) 初始探测:利用正交空时编码波形 $\boldsymbol{c}_0$ 获得 $\boldsymbol{B}_{cs}$,并用 $\boldsymbol{c}_0$ 进行初始探测,利用参考距离单元中的回波估计 $\boldsymbol{M}_{cs}$;利用这些值计算$\boldsymbol{Q}_{sc}=\boldsymbol{B}_{cs}\boldsymbol{M}_{cs}\boldsymbol{B}_{cs}^{\text{H}}$,并对 $\boldsymbol{Q}_{sc}$ 进行特征分解,获得其最小特征值对应的特征向量 $\boldsymbol{g}$,于是,$\boldsymbol{v}_{\text{t}}=\boldsymbol{g}/\|\boldsymbol{g}\|$;计算$\boldsymbol{v}_{\text{t}}^{\text{H}}\boldsymbol{Q}_{sc}\boldsymbol{v}_{\text{t}}$,并将得到的结果存为 $f(0)$。

(2) 第 k 次迭代:利用得到的$\boldsymbol{v}_{\text{t}}$ 计算 $\boldsymbol{D}$,从而得到 $\boldsymbol{Q}_{wc}$;对 $\boldsymbol{Q}_{wc}$ 进行特征分解,获得其最小特征值对应的特征向量 $\boldsymbol{h}$,于是,第 k 次迭代得到的空时编码波形 $\boldsymbol{c}_k=\boldsymbol{h}/\|\boldsymbol{h}\|$;再利用 $\boldsymbol{c}_k$ 获得 $\boldsymbol{B}_{cs}$,然后用 $\boldsymbol{B}_{cs}$ 和$\boldsymbol{M}_{cs}$ 计算 $\boldsymbol{Q}_{sc}=\boldsymbol{B}_{cs}\boldsymbol{M}_{cs}\boldsymbol{B}_{cs}^{\text{H}}$,并对 $\boldsymbol{Q}_{sc}$ 进行特征分解,获得其最小特征值对应的特征向量 $\boldsymbol{g}$,于是,$\boldsymbol{v}_{\text{t}}=\boldsymbol{g}/\|\boldsymbol{g}\|$;计算$\boldsymbol{v}_{\text{t}}^{\text{H}}\boldsymbol{Q}_{sc}\boldsymbol{v}_{\text{t}}$,并将得到的结果存为 $f(k)$。

(3) 令迭代次数 k 增 1,重新执行步骤(2),直到 $f(k)$的值趋向稳定时为止。

将上述迭代优化过程得到的空时编码波形记为 opt_$\boldsymbol{c}_3$。最小化杂波功率准则不需要目标信息,这意味着它将"平等地"对待杂波子空间以外的所有目标。当 $\boldsymbol{R}_{\text{cn}}=\sigma_{\text{n}}^2\boldsymbol{I}_{NL\times NL}$ 时,观测背景是白色的,这意味着该准则将"平等地"对待整个空间中的所有散射体,这与正交空时编码设计的思想是相同的,因此得到的 opt_$\boldsymbol{c}_3$ 就应当是正交空时编码波形 $\boldsymbol{c}_0$,而实际上 $\boldsymbol{c}_0$ 就是白色背景且没有目标信息条件下的最优空时编码波形。然而问题是,基于式(14.89)的迭代优化方法并不适用于这种特殊情况,因为此时 $\boldsymbol{M}_{cs}$ 的估计近似为零矩阵,导致 p_c 几乎为 0。解决此问题的方法是,在估计 $\boldsymbol{M}_{cs}$ 的过程中判断其是否近似为零矩阵,若是,则直接令 opt_$\boldsymbol{c}_3=\boldsymbol{c}_0$。

4. 杂波复幅度方位维相关矩阵和噪声功率水平的估计

杂波复幅度方位维相关矩阵 $\boldsymbol{M}_{cs}$ 和噪声功率水平 σ_{n}^2 的估计过程如下。

(1)发射具有正交空时编码波形 $\boldsymbol{c}_0$ 的相参脉冲串照射观测区域,利用接收到的无目标的 K 个参考距离单元观测向量 $\boldsymbol{r}_k(k=1,2,\cdots,K)$估计 $\hat{\boldsymbol{R}}_{\text{cn}}$,$\hat{\boldsymbol{R}}_{\text{cn}}=\dfrac{1}{K}\sum\limits_{k=1}^{K}\boldsymbol{r}_k\boldsymbol{r}_k^{\text{H}}$。

(2) 采用 $\hat{\boldsymbol{R}}_{\text{cn}}$ 的特征分解或谱分析方法估计噪声子空间的功率水平 $\hat{\sigma}_{\text{n}}^2$。

(3) 采用最小二乘法求解:$\hat{\boldsymbol{M}}_{cs}=(\boldsymbol{B}_{cs}^{\text{H}}\boldsymbol{B}_{cs})^{-1}\boldsymbol{B}_{cs}^{\text{H}}\hat{\boldsymbol{R}}_{cn}\boldsymbol{B}_{cs}(\boldsymbol{B}_{cs}^{\text{H}}\boldsymbol{B}_{cs})^{-1}-\hat{\sigma}_{\text{n}}^2(\boldsymbol{B}_{cs}^{\text{H}}\boldsymbol{B}_{cs})^{-1}$。

上述步骤中,关键是正确地估计 $\hat{\sigma}_{\text{n}}^2$。但与以往特征分解或谱估计方法的要求不同,这里并不要求严格地区分杂波子空间与噪声子空间,而只需区分白色区域与有色区域,因此低 CNR 条件下估计出来的 $\boldsymbol{M}_{cs}$ 和σ_{n}^2 也能很好地满足后续空时编码波形设计要求。

14.3.3 仿真与分析

下面通过仿真比较分析 4 种空时编码波形 $\boldsymbol{c}_0$、opt_$\boldsymbol{c}_1$、opt_$\boldsymbol{c}_2$ 和 opt_$\boldsymbol{c}_3$ 在 MSD 检测器输出空时谱的能量聚集性、检测性能以及输出 SCNR 方面的差异，其中输出空时谱的能量聚集性的含义是“空时编码波形将能量聚集于期望探测目标所在方位-多普勒单元中的能力，包括方位-多普勒分辨力和聚集能量大小两方面”。仿真参数如下：$M=5,N=5,L=16,\lambda=0.03\text{m},T_p=20\text{kHz}$，载机速度 $v_p=150\text{m/s}$；当 $d_t=d_r$ 时，$Q=N+(L-1)=20$，当 $d_t=Nd_r$ 时，$Q=N+N\cdot(M-1)+(L-1)=40$；杂波复幅度的方位维相关特性用高斯谱描述；$S=2$，且两个目标的速度分别为 $v_1=35\text{m/s}$ 和 $v_2=-50\text{m/s}$。

1. MSD 检测器输出的空时谱

图 14.5 给出了 $d_t=Nd_r$、CNR=80dB 时 4 种空时编码波形 $\boldsymbol{c}_0$、opt_$\boldsymbol{c}_1$、opt_$\boldsymbol{c}_2$ 和 opt_$\boldsymbol{c}_3$ 对应的 MSD 检测器输出的空时谱。其中，两目标的多普勒-方位坐标分别是(0.37,0.52)和(−0.42,−0.52)，每个目标的输入 SCNR≈−60dB，MSD 检测器中的模式矩阵 $\boldsymbol{B}_t$ 匹配于上述两个目标。图 14.6 给出了 $d_t=d_r$ 条件下的相应结果，而其余条件均与图 14.5 一样。

根据图 14.5 和图 14.6 的仿真结果，可对 4 种空时编码波形的性能作如下总结。

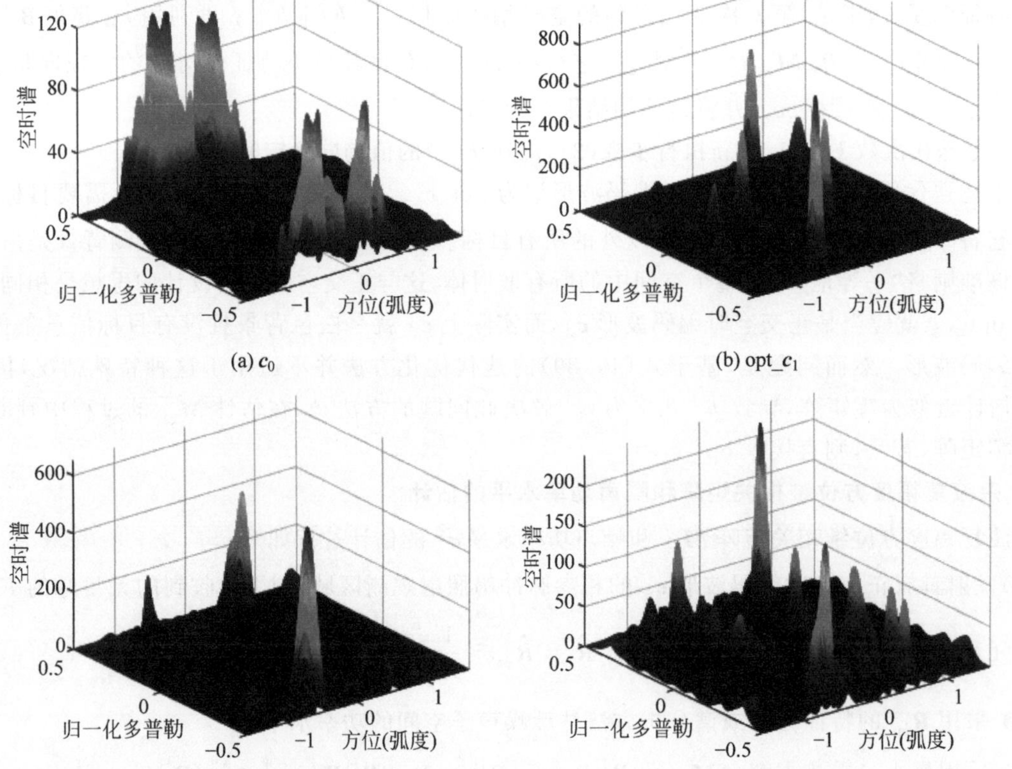

(a) c_0　(b) opt_c_1　(c) opt_c_2　(d) opt_c_3

图 14.5　$d_t=Nd_r$ 时不同空时编码波形对应的 MSD 检测器输出的空时谱

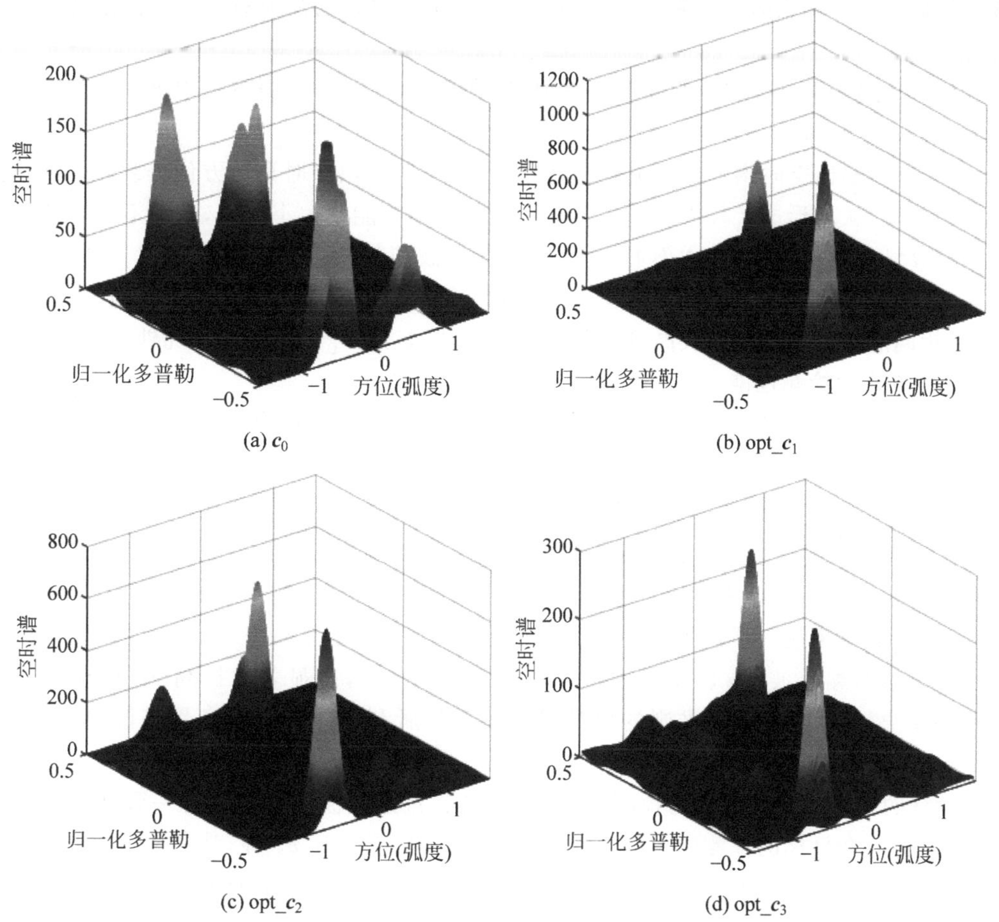

图 14.6 $d_t=d_r$ 时不同空时编码波形对应的 MSD 检测器输出的空时谱

(1) 在图 14.5 和图 14.6 中，当 CNR＝80dB，且经过杂波抑制后，输出空时谱的能量聚集性按照 opt_$\boldsymbol{c}_1$、opt_$\boldsymbol{c}_2$、opt_$\boldsymbol{c}_3$、$\boldsymbol{c}_0$ 的顺序依次递减，且不同目标处的能量聚集性是不一样的。出现上述现象的原因描述如下。

① $\boldsymbol{c}_0$ 是非自适应的，不能根据观测场景的信息调整自身各元素之间的相位关系。在有色背景中，杂波抑制过程破坏了 $\boldsymbol{c}_0$ 的空域正交性，导致方位分辨力下降；而各发射脉冲串之间的正交性降低了多普勒分辨力。同时目标信息未知，使得 $\boldsymbol{c}_0$ 不能有效地将发射能量聚集于目标，导致能量被分散，因此图 14.5 与图 14.6 中(a)的空时谱值都偏小。

② opt_$\boldsymbol{c}_1$、opt_$\boldsymbol{c}_2$ 和 opt_$\boldsymbol{c}_3$ 对于观测场景都具有不同程度的自适应能力，因此都具有较好的方位-多普勒分辨能力。但在聚集能量的大小方面，opt_$\boldsymbol{c}_1$ 在设计过程中能根据杂波信息自适应地将能量聚集于受杂波抑制过程影响较小的目标，因而聚集的能量最高；opt_$\boldsymbol{c}_2$ 在设计过程中没有考虑杂波背景，因此不能控制杂波抑制过程对目标能量的影响，其聚集在目标处的能量会因杂波抑制而受到损失；对于 opt_$\boldsymbol{c}_3$ 来说，由于其没有已知目标信息，因此在能量聚集能力方面是三者中最差的，能量的扩散程度远高于前两者。

③ 多目标条件下得到的空时编码波形对不同目标的作用效果并不相同，这与目标所在的方位-多普勒位置有关，也与所采用的优化准则有关。本节的优化准则都是从改善检测性能的角度提出来的，并

不关心各个目标上形成的空时谱是否均等。由此也会造成所设计的波形对不同的目标具有不同的检测性能。

(2) 图 14.5 与图 14.6 在仿真条件上的差别是，前者中 $d_t=Nd_r$，而后者中 $d_t=d_r$，对比两者的仿真结果可以发现，这个条件上的差异带来了以下性能上的差异。

① $d_t=Nd_r$ 时输出空时谱的方位分辨力优于 $d_t=d_r$ 时输出空时谱的方位分辨力。这是由于前者发射阵元间距较宽，从而形成了较大的虚拟孔径。但就聚集的能量大小而言，前者不如后者。

② $d_t=Nd_r$ 时的杂波抑制能力不如 $d_t=d_r$ 时的杂波抑制能力。原因在于，$d_t=Nd_r$ 时的杂波子空间维数明显大于 $d_t=d_r$ 时的杂波子空间维数，而两种情况中可供利用的空时编码波形的自由度是一样的，因此以相同的自由度去应对较高维数的杂波子空间，显然 $d_t=Nd_r$ 的情况处于劣势。

2. MSD 检测器的检测性能

图 14.7 给出了四种空时编码波形 $\boldsymbol{c}_0$、opt_$\boldsymbol{c}_1$、opt_$\boldsymbol{c}_2$ 和 opt_$\boldsymbol{c}_3$ 对应的 MSD 检测器的检测性能曲线。其中，对两个目标以及两种发射阵元间距都分别进行了考虑。除信噪比变化外，其余条件与 14.3.3 节中的(1)相同。图 14.7 中“波形 0”代表 $\boldsymbol{c}_0$，“波形 1”代表 opt_$\boldsymbol{c}_1$，“波形 2”代表 opt_$\boldsymbol{c}_2$，“波形 3”代表 opt_$\boldsymbol{c}_3$；图 14.7(a)对应于 $d_t=Nd_r$，而图 14.7(b)对应于 $d_t=d_r$；横坐标 SNR(dB)表示杂波抑制之前的目标功率与噪声功率之比。根据图 14.7 的仿真结果可总结出以下 3 点。

(1) 对于每个目标来说，不同空时编码波形的检测能力按 opt_$\boldsymbol{c}_1$、opt_$\boldsymbol{c}_2$、opt_$\boldsymbol{c}_3$、$\boldsymbol{c}_0$ 的顺序依次减弱，而且相对于正交空时编码波形，通过自适应空时编码波形设计，能显著地提高 MSD 检测器的检测性能。这个结果是显然的，无论是从准则的设计思路还是从所利用的先验信息量来说，Max SCNR 准则显然是最好的，而没有任何先验信息的正交空时编码波形显然具有最差的检测性能。

(2) 所设计的波形对两个目标的检测能力并不相同，且这些波形也并不是一致地对某个目标的检测能力强，而对另一个目标的检测能力弱，例如图 14.7(a)中，除 opt_$\boldsymbol{c}_3$ 外，其余 3 种波形对目标 2 的检测能力强于对目标 1。产生这些现象的原因正如前文所总结的，既与目标所在的方位-多普勒位置有关，也与所采用的优化准则有关。极端的情况是，如果某个目标刚好淹没在杂波谱中，Max SCNR 准则会倾向于将所有发射能量聚集于其他目标，而 Max SP 准则和 Min CP 准则不会如此处理。导致的结果是，后两个准则分配给该目标的能量将会在后续杂波抑制过程中被抑制掉。

(3) $d_t=d_r$ 时的检测性能优于 $d_t=Nd_r$ 时的检测性能，原因在于 $d_t=Nd_r$ 时，所设计波形的杂波抑制效果不如前者，这在 14.3.3 节的(1)中已有讨论。

结合 14.3.1 节和 14.3.2 节，可以对发射阵元间距的选择作以下总结：当需要较高的方位分辨力时，应当选择较大的发射阵元间距；而当需要较好的检测性能时，选择半个波长的发射阵元间距较好；但是选择半个波长的发射阵元间距容易造成目标谱被杂波谱淹没而丢失目标，因为此时的观测空间维数相对于 $d_t=Nd_r$ 时的观测空间维数低很多，相应的无杂波子空间的维数也远低于后者。因此综合考虑，在保证较好的检测性能的条件下，以选择较大的发射阵元间距为宜。

3. MSD 检测器输出的 SCNR

14.3.1 节指出，最大化 MSD 检测器的检测性能等价于最大化其输出 SCNR ε。根据式(14.84)，不同的空时编码波形正好对应着不同的 ε，因此，衡量 opt_$\boldsymbol{c}_1$、opt_$\boldsymbol{c}_2$ 和 opt_$\boldsymbol{c}_3$ 条件下 MSD 检测器的检测性能相当于衡量 $\boldsymbol{c}_0$ 条件下检测性能的改善程度，可以等价地通过比较它们之间对输出 SCNR 的改善程度来描述。

从 14.3.2 节的分析中可以看出，CNR 是影响自适应空时编码波形设计的重要因素。当 CNR 很低以至无须考虑杂波的存在时，Max SCNR 准则与 Max SP 准则在最大化检测器的输出 SCNR 上是等价

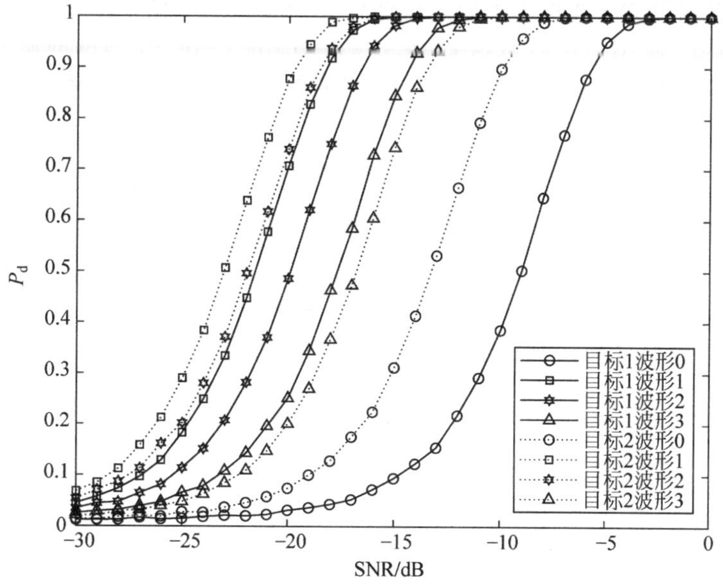

(a) $d_t=Nd_r$时的结果

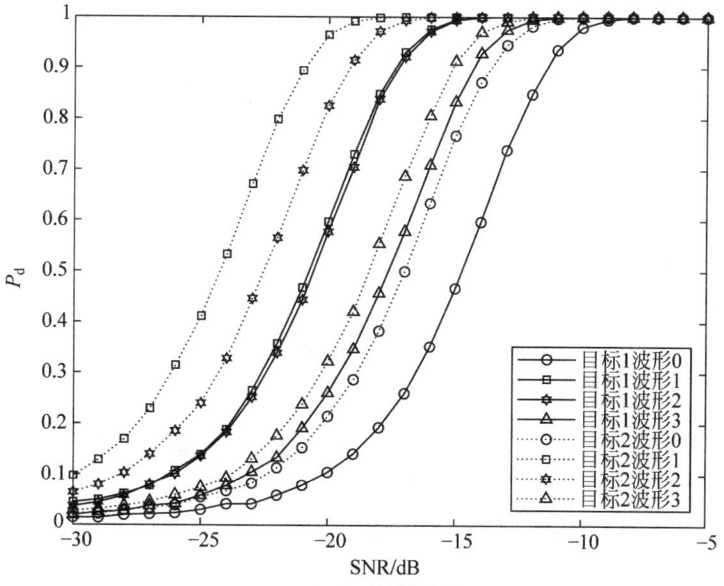

(b) $d_t=d_r$时的结果

图 14.7　四种空时编码波形条件下 MSD 检测器的检测性能比较

的，而 Min CP 准则设计出的波形接近正交空时编码波形。随着 CNR 的逐渐增大，$\boldsymbol{c}_0$、opt_$\boldsymbol{c}_1$、opt_$\boldsymbol{c}_2$ 和 opt_$\boldsymbol{c}_3$ 对应的输出 SCNR 都会逐渐下降，但本节关心的是它们之间的相对变化，具体情况如图 14.8 所示。

图 14.8 中低 CNR 部分的仿真结果验证了 14.3.2 节中相应分析的正确性；随着 CNR 的增加，$\boldsymbol{c}_0$、opt_$\boldsymbol{c}_1$、opt_$\boldsymbol{c}_2$ 和 opt_$\boldsymbol{c}_3$ 对应的输出 SCNR 都在逐渐下降，但下降的程度各不相同，如图 14.8(a)所示。图 14.8(b)中，以 $\boldsymbol{c}_0$ 对应的输出 SCNR 作为参考，随着 CNR 的增加，opt_$\boldsymbol{c}_1$ 对输出 SCNR 的改善程度逐渐增加并趋向平稳；而 opt_$\boldsymbol{c}_2$ 的改善程度随着 CNR 的增加而增加到一定程度后转而有下降的趋

势，这是由于其对应的输出 SCNR 的下降速度超过 $\boldsymbol{c}_0$ 对应的下降速度，而 $\boldsymbol{c}_0$ 对应的下降速度的减小来源于此时它对应的输出 SCNR 已经较低，没有多少下降的空间；opt_$\boldsymbol{c}_3$ 对应的输出 SCNR 的变化趋势比较缓慢，其对应的改善程度的增加主要是由于 $\boldsymbol{c}_0$ 对应的输出 SCNR 的下降引起的。

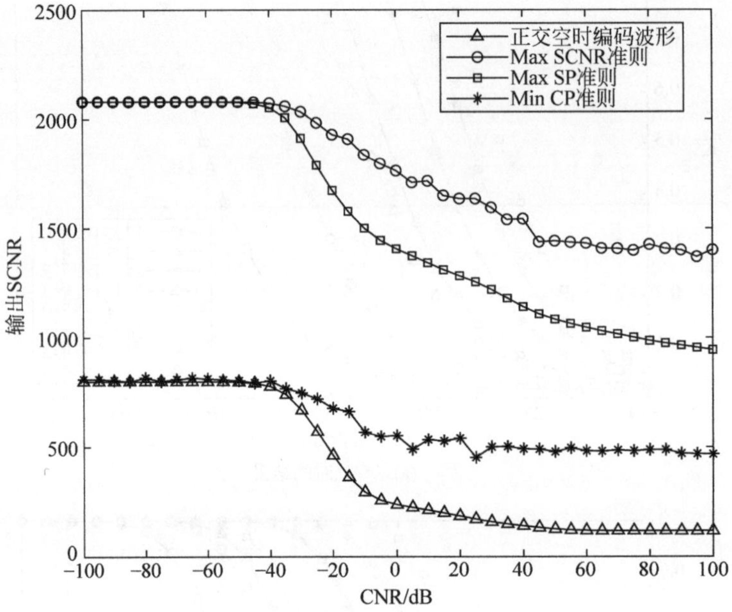

(a) 输出SCNR与CNR的关系曲线

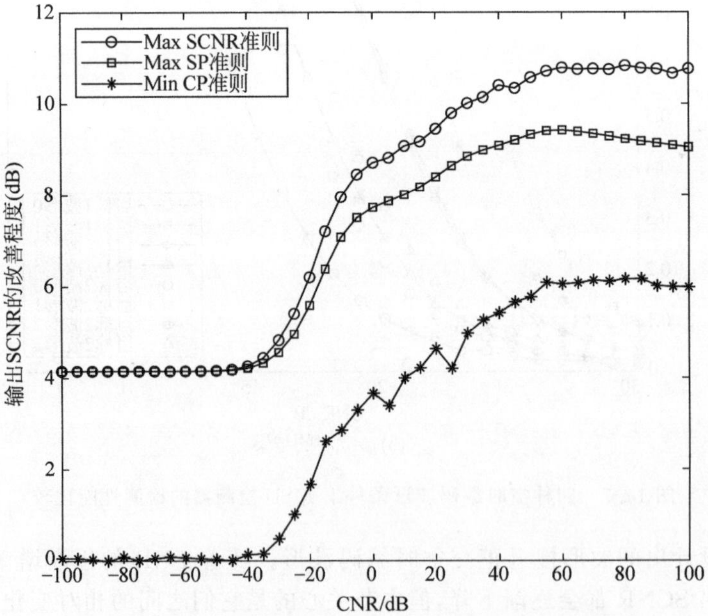

(b) 三种设计准则对应的输出SCNR相对于c_0的改善程度

图 14.8　四种空时编码波形对应的 MSD 检测器输出 SCNR 之间的比较

本节从改善 MIMO 阵列雷达检测性能的角度出发，基于初步探测获取的反馈信息，研究了脉冲间自适应空时编码的设计问题，考虑了多目标环境、杂波复幅度的空时相关性及其自适应估计等较为复杂的情况；推导了最大化 SCNR 准则，最大化目标功率准则和最小化杂波功率准则下自适应空时编码波形的设计公式，阐述了这 3 种准则之间以及它们与正交空时编码设计和发射方向图匹配设计准则之间的联系与区别；描述了空时编码波形设计过程中所需的杂波复幅度空时相关矩阵和噪声功率水平的自适应估计方法；最后通过数值仿真比较分析了 3 种自适应空时编码波形 opt_$\boldsymbol{c}_1$、opt_$\boldsymbol{c}_2$ 和 opt_$\boldsymbol{c}_3$ 在 MSD 检测器输出空时谱的能量聚集性、检测性能及输出 SCNR 方面相对于正交空时编码波形 $\boldsymbol{c}_0$ 的改善程度。

研究结果表明，相对于非自适应的正交空时编码设计，3 种具有不同自适应能力的设计准则得出的空时编码波形都能在空时谱的能量聚集能力、检测器的输出 SCNR 及检测性能方面获得非常明显的改善。

14.4 基于空-时-距三维联合的自适应检测

常规阵列雷达信号处理过程一般都会采用如图 14.9 所示的级联处理方案[52]。而 MIMO 阵列雷达常用的信号处理方案中，一般将发射波形分离作为第一步，因为这样可以得到虚拟孔径，进而采用如图 14.9 所示的级联处理方案。级联处理方案的优势在于处理流程简单、技术成熟；但是其缺点也是较为明显的，在每级处理中只利用了部分自由度，没有联合利用系统提供的全部自由度，从而造成系统性能损失。

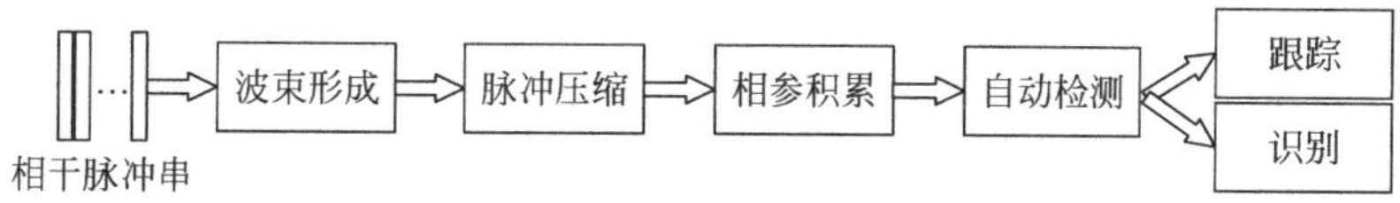

图 14.9 传统相参雷达信号处理流程

对于 MIMO 阵列雷达来说，实现发射波形有效分离的前提是设计一组完全正交的波形和对应的一组性能优异的滤波器。理想的正交波形需要满足互相关水平和自相关旁瓣趋零。但是在有限编码长度和良好多普勒容忍性的约束条件下，很难得到这样一组性能优异的波形，因此基于匹配滤波的方法难以做到波形的有效分离，进而对后续的波束形成、相参积累和目标检测等流程造成负面影响。

众所周知，联合利用空域和时域抑制地面杂波的 STAP 技术是一种优秀的空时二维自适应处理技术，而 MIMO 阵列与 STAP 的结合进一步提高了空时自由度，因而有利于获得更佳的空时二维杂波抑制性能，但是其前提是发射波形被有效分离，否则 MIMO STAP 将面临性能下降的问题。此外，STAP 在通过邻近单元的数据估计当前检测单元的背景协方差矩阵时，需要满足 RMB 准则，即至少需要不少于 2 倍自由度的统计独立同分布数据，且不含目标。显然，实际中很难保证这一点，当邻近单元存在干扰目标时，基于 RMB 准则的方法面临性能下降的问题。

针对基于匹配滤波的方法难以做到波形有效分离的问题，基于 APC(Adaptive Pulse Compression，自适应脉冲压缩)的波形分离技术通过自适应地估计每个单元和波形的滤波器权重，可以有效提高波形分离的效果。

针对常规阵列雷达干扰目标环境下基于 RMB 准则的 STAP 方法性能下降的问题，在匹配滤波或者自适应脉冲压缩处理之后采用 IAA(Iterative Adaptive Approach，迭代自适应方法)实现角度-多普勒聚焦是一个有效的解决方案。但是对于 MIMO 阵列雷达，尤其是发射波形非理想正交或存在邻近干扰时，这种级联的处理方式效果不好。

IAA 与 APC 在本质上是相同的，因此 APC 同样可以实现空-距[53]和时-距[54]的自适应处理，并且得到优于级联处理的性能增益。由于脉冲压缩、波束形成和多普勒滤波在本质上是相似的，都是基于相关原理实现对期望信号的相参积累，因此借鉴空时自适应处理和自适应脉冲压缩的思想将三者联合起来，联合利用"阵元-脉冲-波形"三维资源，实现"空-时-距"三维联合自适应处理(Space-Time-Range Adaptive Processing，STRAP)，充分利用整个系统的自由度，从而获得最优的积累增益，同时还可以得到以下益处：

(1) 解决匹配滤波无法有效分离发射波形的问题；

(2) 解决级联处理在多目标环境下性能下降的问题；

(3) 降低对大量统计独立同分布数据的要求；

(4) 更高的自由度带来更低的旁瓣和更好的杂波抑制效果；

(5) 不受相干源对消的影响。

本节基于自适应脉冲压缩框架和最小方差无畸变响应原则，进行空-时-距三维联合自适应处理以获得 MIMO 雷达高分辨率的角度-多普勒-距离像，在计算协方差矩阵时，分在脉压(匹配滤波)前考虑邻近距离单元的干扰目标与在脉压(匹配滤波)后考虑邻近距离单元的干扰目标两种处理方式，得到了两种空-时-距三维联合自适应处理方法；然后在准确估计协方差矩阵的基础上，进一步提出自适应聚焦、检测一体化处理方法。相关研究还可参考文献[55-61]。

14.4.1 MIMO 雷达信号模型

考虑一个 MIMO 雷达均匀线阵，包含 N_{T} 个发射阵元，阵元间距为 d_{T} 以及 N_{R} 个接收阵元，阵元间距为 d_{R}。窄带发射波形 $\boldsymbol{S}=[\boldsymbol{s}_1,\boldsymbol{s}_2,\cdots,\boldsymbol{s}_{N_{\mathrm{T}}}]^{\mathrm{T}}$，其中第 n 行是长度为 N_{W} 的离散波形。

假设脉内多普勒可以忽略并且多个目标可以同时存在于一个角度-距离单元内，则角度、多普勒频率和距离单元分别为 θ、f_{d} 和 l 的第 m 个脉冲回波经过下变频和模数转换后可写为

$$\boldsymbol{X}(m,l,\theta,f_{\mathrm{d}})=\alpha(l,\theta,f_{\mathrm{d}})\mathrm{e}^{\mathrm{j}2\pi(m-1)f_{\mathrm{d}}T_{\mathrm{r}}}\boldsymbol{b}(\theta)\boldsymbol{a}^{\mathrm{T}}(\theta)\boldsymbol{S} \tag{14.90}$$

其中，$\alpha(l,\theta,f_{\mathrm{d}})$ 表示目标散射系数，$\boldsymbol{a}(\theta)$ 为发射导向向量，$\boldsymbol{b}(\theta)$ 为接收导向向量，T_{r} 为脉冲重复间隔(PRI)。将 N_{P} 个相参脉冲串堆成向量的形式可得

$$\boldsymbol{x}_{\mathrm{t}}(l,\theta,f_{\mathrm{d}})=\alpha(l,\theta,f_{\mathrm{d}})\boldsymbol{V}(\theta,f_{\mathrm{d}}) \tag{14.91}$$

其中

$$\boldsymbol{V}(\theta,f_{\mathrm{d}})=\boldsymbol{S}^{\mathrm{T}}\boldsymbol{a}(\theta)\otimes\boldsymbol{b}(\theta)\otimes\boldsymbol{u}(f_{\mathrm{d}}) \tag{14.92}$$

$\otimes$代表 Kronecker 积，$\boldsymbol{u}(f_{\mathrm{d}})$ 为对应多普勒频率 f_{d} 的时域导向向量。当第 l 个距离单元附近有多个干扰目标时，邻近干扰可建模为

$$\boldsymbol{x}_{\mathrm{j}}(l+p,\theta,f_{\mathrm{d}})=\alpha(l+p,\theta,f_{\mathrm{d}})\boldsymbol{V}(p,\theta,f_{\mathrm{d}}) \tag{14.93}$$

其中，$\boldsymbol{V}(p,\theta,f_{\mathrm{d}})=[\boldsymbol{J}^{\mathrm{T}}(p)\boldsymbol{S}^{\mathrm{T}}\boldsymbol{a}(\theta)]\otimes\boldsymbol{b}(\theta)\otimes\boldsymbol{u}(f_{\mathrm{d}})$，$\boldsymbol{J}(p)$ 表示移位矩阵，定义如下

$$\boldsymbol{J}_{i,j}(p)=\begin{cases}1, & i-j+p=0\\ 0, & i-j+p\neq 0\end{cases} \tag{14.94}$$

综合考虑当前处理单元附近 $2P$ 个距离单元、N_{D} 个多普勒单元和 N_{S} 个角度单元散射体的干扰，其中 P 最大取$(N_{\mathrm{W}}-1)$即可，因此降低了对大量统计独立同分布数据的需求。总的干扰模型可写为

$$\boldsymbol{x}_{\mathrm{j}}(l)=\sum_{\theta}\sum_{f_{\mathrm{d}}}\sum_{p=-P,p\neq 0}^{P}\boldsymbol{x}_{\mathrm{j}}(l+p,\theta,f_{\mathrm{d}}) \tag{14.95}$$

总的“目标＋干扰＋噪声”信号模型为

$$\boldsymbol{y}(l)=\boldsymbol{x}_{\mathrm{t}}(l,\theta,f_{\mathrm{d}})+\boldsymbol{x}_{\mathrm{j}}(l)+\boldsymbol{n} \tag{14.96}$$

其中，$\boldsymbol{n}$ 为服从圆周对称的复高斯随机向量，均值为 $\mathbf{0}$，协方差均值为 $\sigma^2\boldsymbol{I}$。

对于机载 MIMO 雷达，必须考虑地杂波的影响。由于平台的运动，机载 MIMO 雷达的杂波在空域、时域和距离上都有分布。STAP 技术充分利用了杂波方位 θ_{c} 与多普勒 $f_{\mathrm{d,c}}$ 的内在关系，即

$$f_{\mathrm{d,c}}=\beta\theta_{\mathrm{c}} \tag{14.97}$$

其中，β 表示依赖平台运动、PRI 和 d_{R} 的固定常数。第 l 个距离环的杂波可建模为

$$\boldsymbol{x}_{\mathrm{c}}(l)=\sum_{\theta_{\mathrm{c}}}\sum_{p=-P}^{P}\alpha(l+p,\theta_{\mathrm{c}},\beta\theta_{\mathrm{c}})\boldsymbol{V}(p,\theta_{\mathrm{c}},\beta\theta_{\mathrm{c}}) \tag{14.98}$$

匹配滤波器组常被用来分离发射波形，为了保证滤波后的噪声仍是白噪声，则匹配滤波器可采用如下形式

$$\boldsymbol{S}_{\mathrm{MF}}=\boldsymbol{S}^{\mathrm{H}}(\boldsymbol{S}\boldsymbol{S}^{\mathrm{H}})^{-1/2} \tag{14.99}$$

其中，$(\cdot)^{-1/2}$ 表示 Hermitian 矩阵的平方根。这里假设波形协方差矩阵是满秩的。总的“目标＋干扰＋杂波＋噪声”模型可写为

$$\boldsymbol{y}(l)=\boldsymbol{x}_{\mathrm{t}}(l,\theta,f_{\mathrm{d}})+\boldsymbol{x}_{\mathrm{j}}(l)+\boldsymbol{x}_{\mathrm{c}}(l)+\boldsymbol{n} \tag{14.100}$$

14.4.2 匹配滤波后的空-时-距自适应处理

这里讨论的是，匹配滤波之后邻近距离单元干扰目标的影响，以及利用基于 MVDR 准则的迭代自适应处理抑制干扰目标的距离旁瓣。经过匹配滤波（脉冲压缩）之后的 N_{R} 个接收阵元与 N_{P} 个发射阵元的信号可写为

$$\hat{\boldsymbol{y}}(l)=(\boldsymbol{S}_{\mathrm{MF}}^{\mathrm{T}}\otimes\boldsymbol{I}_{N_{\mathrm{R}}}\otimes\boldsymbol{I}_{N_{\mathrm{P}}})\boldsymbol{y}(l)=\sum_{\theta}\sum_{f_{\mathrm{d}}}\sum_{p=-P}^{P}\alpha(l+p,\theta,f_{\mathrm{d}})\hat{\boldsymbol{V}}(p,\theta,f_{\mathrm{d}})+\hat{\boldsymbol{n}} \tag{14.101}$$

$$\hat{\boldsymbol{V}}(p,\theta,f_{\mathrm{d}})=[\boldsymbol{S}_{\mathrm{MF}}^{\mathrm{T}}\boldsymbol{J}^{\mathrm{T}}(p)\boldsymbol{S}^{\mathrm{T}}\boldsymbol{a}(\theta)]\otimes\boldsymbol{b}(\theta)\otimes\boldsymbol{u}(f_{\mathrm{d}}) \tag{14.102}$$

$\hat{\boldsymbol{n}}=(\boldsymbol{S}_{\mathrm{MF}}^{\mathrm{T}}\otimes\boldsymbol{I}_{N_{\mathrm{R}}}\otimes\boldsymbol{I}_{N_{\mathrm{P}}})\boldsymbol{n}$ 为滤波后的噪声项。在式(14.101)基础上可得匹配滤波的输出为

$$\hat{\alpha}_{\mathrm{MF}}(l,\theta,f_{\mathrm{d}})=\frac{\hat{\boldsymbol{V}}^{\mathrm{H}}(\theta,f_{\mathrm{d}})\hat{\boldsymbol{y}}(l)}{\hat{\boldsymbol{V}}^{\mathrm{H}}(\theta,f_{\mathrm{d}})\hat{\boldsymbol{V}}(\theta,f_{\mathrm{d}})} \tag{14.103}$$

其中，$\hat{\boldsymbol{V}}(\theta,f_{\mathrm{d}})\triangleq\hat{\boldsymbol{V}}(0,\theta,f_{\mathrm{d}})$。式(14.103)的匹配滤波在白噪声背景下可以获得最大输出信噪比，但是匹配滤波的分辨率不高且难以抑制干扰目标。基于 MVDR 准则的匹配滤波后的空-时-距自适应处理，通过构造如下代价函数来推导最优滤波器系数

$$J(l,\theta,f_{\mathrm{d}})=\mathrm{E}[|\alpha(l,\theta,f_{\mathrm{d}})-\hat{\boldsymbol{w}}^{\mathrm{H}}(l,\theta,f_{\mathrm{d}})\hat{\boldsymbol{y}}(l)|^2] \tag{14.104}$$

在单位增益约束 $\hat{\boldsymbol{w}}^{\mathrm{H}}(l,\theta,f_{\mathrm{d}})\hat{\boldsymbol{V}}(\theta,f_{\mathrm{d}})=1$ 下，利用拉格朗日乘子法最小化可得

$$\hat{\boldsymbol{w}}(l,\theta,f_{\mathrm{d}})=\{\mathrm{E}[\hat{\boldsymbol{y}}(l)\hat{\boldsymbol{y}}^{\mathrm{H}}(l)]\}^{-1}\left\{\mathrm{E}[\alpha^{*}(l,\theta,f_{\mathrm{d}})\hat{\boldsymbol{y}}(l)]-\frac{\lambda}{2}\hat{\boldsymbol{V}}(\theta,f_{\mathrm{d}})\right\} \tag{14.105}$$

其中，$(\cdot)^{*}$ 表示共轭运算，λ 表示拉格朗日乘子。假设每个角度-多普勒-距离单元散射系数互不相关，且与噪声不相关，可得

$$\hat{\boldsymbol{w}}(l,\theta,f_{\mathrm{d}})=\frac{\hat{\boldsymbol{R}}^{-1}(l)\hat{\boldsymbol{V}}(\theta,f_{\mathrm{d}})}{\hat{\boldsymbol{V}}^{\mathrm{H}}(\theta,f_{\mathrm{d}})\hat{\boldsymbol{R}}^{-1}(l)\hat{\boldsymbol{V}}(\theta,f_{\mathrm{d}})} \tag{14.106}$$

其中，$\hat{\boldsymbol{R}}(l)=\sum_{\theta}\sum_{f}\sum_{p=-P}^{P}\hat{\boldsymbol{y}}(l+p,\theta,f_{\mathrm{d}})\hat{\boldsymbol{y}}^{\mathrm{H}}(l+p,\theta,f_{\mathrm{d}})+\hat{\boldsymbol{R}}_{\mathrm{NSE}}$ 表示第 l 个距离环经过匹配滤波之后的协方差矩阵，$\hat{\boldsymbol{R}}_{\mathrm{NSE}}$ 为噪声协方差矩阵。关于噪声的先验信息可通过 $\hat{\boldsymbol{R}}_{\mathrm{NSE}}$ 在式(14.106)中使用。式(14.106)的形式与常用的MVDR波束形成器是相似的，不同之处在于这里的方法是迭代实施的，因此可简写为MF-RMVDR。

14.4.3 空-时-距自适应处理

受上述匹配滤波后的空-时-距自适应处理所启发，这里考虑更直接的空-时-距自适应处理，即匹配滤波前的空-时-距自适应处理，简写为STRAP。不同于文献[62]中IAA方法需要迭代全部距离单元的是，STRAP方法仅使用检测单元附近 $2P$ 个距离单元。

与14.4.2节的推导过程类似，直接对式(14.100)应用匹配滤波可得

$$\breve{\alpha}_{\mathrm{MF}}(l,\theta,f_{\mathrm{d}})=\frac{\boldsymbol{V}^{\mathrm{H}}(\theta,f_{\mathrm{d}})\boldsymbol{y}(l)}{\boldsymbol{V}^{\mathrm{H}}(\theta,f_{\mathrm{d}})\boldsymbol{V}(\theta,f_{\mathrm{d}})} \tag{14.107}$$

采用MVDR准则和拉格朗日因子法推导可得

$$\breve{\boldsymbol{w}}(l,\theta,f_{\mathrm{d}})=\frac{\boldsymbol{R}^{-1}(l)\boldsymbol{V}(\theta,f_{\mathrm{d}})}{\boldsymbol{V}^{\mathrm{H}}(\theta,f_{\mathrm{d}})\boldsymbol{R}^{-1}(l)\boldsymbol{V}(\theta,f_{\mathrm{d}})} \tag{14.108}$$

其中，$\boldsymbol{R}(l)=\sum_{\theta}\sum_{f}\sum_{p=-P}^{P}\boldsymbol{y}(l+p,\theta,f_{\mathrm{d}})\boldsymbol{y}^{\mathrm{H}}(l+p,\theta,f_{\mathrm{d}})+\boldsymbol{R}_{\mathrm{NSE}}$ 为表示第 l 个距离环总的协方差矩阵。根据 $\hat{\boldsymbol{R}}(l)$ 和 $\boldsymbol{R}(l)$ 的形式可知，其为每个角度-多普勒-距离单元协方差矩阵之和(或平均值)，而传统的基于RMB准则的估计方法则是直接通过接收信号估计协方差矩阵，其需要大量统计独立同分布数据抑制不同单元之间的互相关量，当参考单元不是统计独立同分布时，协方差矩阵的估计也就不满足无偏性了。

14.4.4 算法实施与矩阵快速更新

1. 迭代处理

由于协方差矩阵 $\hat{\boldsymbol{R}}(l)$ 和 $\boldsymbol{R}(l)$ 的估计与周围单元的散射系数有关，因此算法需要采用迭代处理，即采用前一次的结果估计当前的协方差矩阵。为了加速程序的运行速度，可以将导向向量 $\boldsymbol{V}(p,\theta,f_{\mathrm{d}})$ 和 $\hat{\boldsymbol{V}}(p,\theta,f_{\mathrm{d}})$ 预先存储，其中的角度和多普勒参数空间通过网格划分为离散形式。对于匹配滤波前的空-时-距自适应处理，首先根据式(14.101)将接收信号按照波形分离以获得虚拟孔径。然后，通过式(14.103)和(14.107)得到散射系数的初始估计值，进而得到协方差矩阵的估计值和滤波器权重。新的散射系数估计值分别通过 $\hat{\alpha}(l,\theta,f_{\mathrm{d}})=\hat{\boldsymbol{w}}^{\mathrm{H}}(l,\theta,f_{\mathrm{d}})\hat{\boldsymbol{y}}(l)$ 和 $\breve{\alpha}(l,\theta,f_{\mathrm{d}})=\breve{\boldsymbol{w}}^{\mathrm{H}}(l,\theta,f_{\mathrm{d}})\boldsymbol{y}(l)$ 得到。重复上述过程，直到相邻两次迭代的估计值之差或期望的迭代次数达到要求。一般来讲，可以在5次迭代实现收敛。两种方法的具体实施步骤如图14.10所示。

2. 快速矩阵更新方法

由于 $\hat{\boldsymbol{R}}(l)$ 和 $\boldsymbol{R}(l)$ 的维度分别为 $N_{\mathrm{T}}N_{\mathrm{R}}N_{\mathrm{P}}$ 和 $N_{\mathrm{W}}N_{\mathrm{R}}N_{\mathrm{P}}$，如果每次估计某一距离单元的三维散射系数时都重新计算协方差矩阵，那么计算代价将非常大，所以具体实现上需要优化算法。各个距离单元的回波在时间上的叠加如图14.11所示，因此相邻两个距离单元的协方差会有重叠，利用矩阵求逆公式可得到一种协方差矩阵快速更新策略。

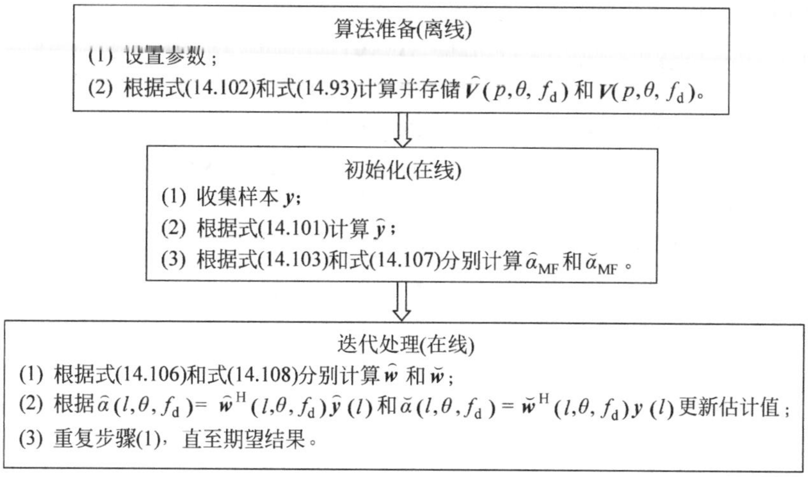

图 14.10 方法实施步骤图

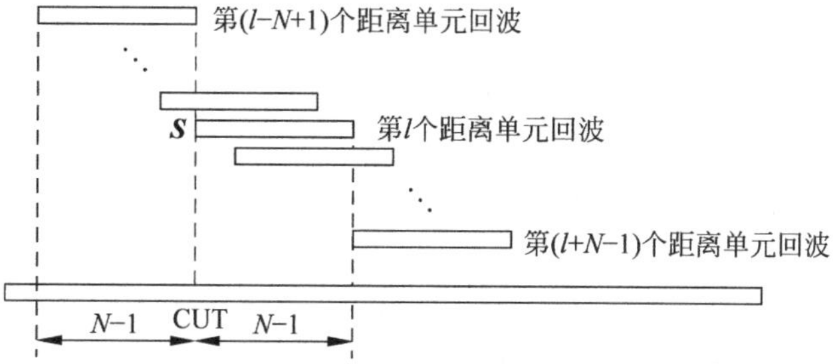

图 14.11 不同距离单元回波叠加示意图

协方差矩阵$\boldsymbol{R}(l-1)$可写为

$$\boldsymbol{R}(l-1)=\begin{bmatrix}\boldsymbol{B} & \boldsymbol{A}^{H}\\ \boldsymbol{A} & \boldsymbol{C}\end{bmatrix} \tag{14.109}$$

其中，$\boldsymbol{A}$是$(N_PN_RN_W-N_RN_P)\times N_RN_P$维复矩阵，$\boldsymbol{B}$是$N_RN_P\times N_RN_P$维复矩阵，$\boldsymbol{C}$是$(N_PN_RN_W-N_RN_P)\times(N_PN_RN_W-N_RN_P)$维复矩阵，且有

$$\begin{bmatrix}\boldsymbol{B}\\ \boldsymbol{A}\end{bmatrix}=\sum_{\theta}\sum_{f_d}\sum_{p=-P}^{P}|\breve{\alpha}(l+p-1,\theta,f_d)|^2\tilde{\boldsymbol{s}}_{-p}(\theta)\boldsymbol{J}^{T}(p)\boldsymbol{S}^{T}\boldsymbol{a}(\theta)\otimes$$
$$[\boldsymbol{b}(\theta)\boldsymbol{b}^{H}(\theta)]\otimes[\boldsymbol{u}(f_d)\boldsymbol{u}^{H}(f_d)] \tag{14.110}$$

其中，$\tilde{s}_p$表示$\boldsymbol{S}^{T}\boldsymbol{a}(\theta)$的第$p$个元素。则下一距离单元的协方差矩阵可写为

$$\boldsymbol{R}(l)=\begin{bmatrix}\boldsymbol{C} & \boldsymbol{D}\\ \boldsymbol{D}^{H} & \boldsymbol{E}\end{bmatrix} \tag{14.111}$$

其中，$\boldsymbol{D}$是$(N_PN_RN_W-N_RN_P)\times N_RN_P$维复矩阵，$\boldsymbol{E}$是$N_RN_P\times N_RN_P$维复矩阵，且

$$\begin{bmatrix}\boldsymbol{D}\\ \boldsymbol{E}\end{bmatrix}=\sum_{\theta}\sum_{f_d}\sum_{p=0}^{P}|\breve{\alpha}(l+p,\theta,f_d)|^2\tilde{\boldsymbol{s}}_{P-p-1}(\theta)\boldsymbol{J}^{T}(p)\boldsymbol{S}^{T}\boldsymbol{a}(\theta)\otimes$$
$$[\boldsymbol{b}(\theta)\boldsymbol{b}^{H}(\theta)]\otimes[\boldsymbol{u}(f_d)\boldsymbol{u}^{H}(f_d)] \tag{14.112}$$

两者之间的关系如下

$$\boldsymbol{R}(l)=\boldsymbol{P}\boldsymbol{R}(l-1)\boldsymbol{P}^{\mathrm{T}}+\begin{bmatrix}\boldsymbol{D}-\boldsymbol{A} & \boldsymbol{0}\\ \boldsymbol{E}-\boldsymbol{B} & \boldsymbol{I}\end{bmatrix}\begin{bmatrix}\boldsymbol{0} & \boldsymbol{I}\\ \boldsymbol{D}^{\mathrm{H}}-\boldsymbol{A}^{\mathrm{H}} & \boldsymbol{0}\end{bmatrix}=\boldsymbol{Q}+\boldsymbol{U}\boldsymbol{V} \tag{14.113}$$

其中，$\boldsymbol{P}$ 表示置换矩阵，满足 $\boldsymbol{P}^{-1}=\boldsymbol{P}^{\mathrm{T}}$。根据矩阵求逆公式可得

$$\boldsymbol{R}^{-1}(l)=\boldsymbol{Q}^{-1}-\boldsymbol{Q}^{-1}\boldsymbol{U}(\boldsymbol{I}+\boldsymbol{V}\boldsymbol{Q}^{-1}\boldsymbol{U})^{-1}\boldsymbol{V}\boldsymbol{Q}^{-1} \tag{14.114}$$

其中，$\boldsymbol{Q}^{-1}=\boldsymbol{P}\boldsymbol{R}^{-1}(l-1)\boldsymbol{P}^{\mathrm{T}}$。

对于匹配滤波后的空-时-距自适应处理，注意到

$$\hat{\boldsymbol{V}}(\theta,f_{\mathrm{d}})=(\boldsymbol{S}_{\mathrm{MF}}^{\mathrm{T}}\otimes\boldsymbol{I}_{N_{\mathrm{R}}}\otimes\boldsymbol{I}_{N_{\mathrm{P}}})\boldsymbol{V}(p,\theta,f_{\mathrm{d}}) \tag{14.115}$$

则协方差矩阵 $\hat{\boldsymbol{R}}(l)$ 可写为

$$\hat{\boldsymbol{R}}(l)=(\boldsymbol{S}_{\mathrm{MF}}^{\mathrm{T}}\otimes\boldsymbol{I}_{N_{\mathrm{R}}}\otimes\boldsymbol{I}_{N_{\mathrm{P}}})\boldsymbol{R}(l)(\boldsymbol{S}_{\mathrm{MF}}^{\mathrm{T}}\otimes\boldsymbol{I}_{N_{\mathrm{R}}}\otimes\boldsymbol{I}_{N_{\mathrm{P}}})^{\mathrm{H}} \tag{14.116}$$

于是，$\bar{\boldsymbol{R}}^{-1}(l)$ 可通过 $\boldsymbol{R}^{-1}(l)$ 与波形的伪逆矩阵计算得到。

14.4.5 自适应聚焦和检测一体化处理

传统的雷达信号处理流程如图 14.9 所示，一般需要通过接收波束形成、脉冲压缩和相参积累（杂波抑制）获得一个空间谱，然后在这个空间谱上做自动目标检测。事实上，前面三个流程的基本原理都是基于相关理论实现对期望信号的积累（聚焦），并抑制掉干扰信号。另外，根据文献[63]，自适应滤波加单元平均恒虚警检测器等价于自适应匹配滤波检测器。因此，可以进行自适应聚焦和检测一体化处理，以空-时-距自适应处理为例，其检测问题可表述为

$$\begin{cases}\mathrm{H}_0:\ \boldsymbol{y}(l)=\boldsymbol{x}_{\mathrm{j}}(l)+\boldsymbol{x}_{\mathrm{c}}(l)+\boldsymbol{n}\\ \mathrm{H}_1:\ \boldsymbol{y}(l)=\boldsymbol{x}_{\mathrm{t}}(l,\theta,f_{\mathrm{d}})+\boldsymbol{x}_{\mathrm{j}}(l)+\boldsymbol{x}_{\mathrm{c}}(l)+\boldsymbol{n}\end{cases} \tag{14.117}$$

在迭代估计得到当前距离单元的协方差矩阵 $\boldsymbol{R}(l)$ 后，可直接在原始回波信号 $\boldsymbol{y}(l)$ 上做检测。由文献[64]可知，当观测数据的统计属性确定后，即其分布满足尺度变换不变性，常见的自适应检测器具有 CFAR 属性。以自适应匹配滤波检测器为例，其统计量可写为

$$\frac{|\boldsymbol{R}^{-1}(l)\boldsymbol{V}(\theta,f_{\mathrm{d}})\boldsymbol{y}(l)|^2}{\boldsymbol{V}^{\mathrm{H}}(\theta,f_{\mathrm{d}})\boldsymbol{R}^{-1}(l)\boldsymbol{V}(\theta,f_{\mathrm{d}})}\underset{\mathrm{H}_0}{\overset{\mathrm{H}_1}{\gtrless}}\gamma \tag{14.118}$$

其中，γ 为检测阈值。

14.4.6 仿真与分析

这里将通过数值仿真分析比较上述两种方法（MF-RMVDR 与 STRAP）与传统级联处理的性能，分别从聚焦和检测两个方面评估联合处理相对于级联处理的优势，其中，前者考察各种处理方案对空间谱的聚焦能力，后者考察各方案的检测性能。

1. 聚焦结果分析

聚焦性能分析分别考虑了两种场景，即对空监视场景和对地动目标指示场景。所考虑的 MIMO 雷达由 $N_{\mathrm{T}}=4$ 个间距 $d_{\mathrm{T}}=2\lambda$ 的发射天线和 $N_{\mathrm{R}}=4$ 个间距 $d_{\mathrm{R}}=0.5\lambda$ 的接收天线构成，其中 λ 表示波长。一个 CPI 的脉冲数 $N_{\mathrm{P}}=8$，脉冲重复周期 $T_{\mathrm{r}}=0.0005\mathrm{s}$。发射波形采用文献[65]设计的长度 $N_{\mathrm{W}}=40$ 的多相编码信号，有着很好的相关特性。观测场景的每个距离环被划分为 $N_{\mathrm{S}}\times N_{\mathrm{D}}=81\times81$ 个角度和多普勒单元，并分布有如表 14.3 所示的 12 个目标，其中噪声功率设为恒定的 $-20\mathrm{dB}$。被用作对比分析的三种级联处理方法分别为级联 MF-MF、级联 MF-IAA 和

级联 APC-IAA。

表 14.3 目标描述

角度-多普勒-距离单元(θ,f_d,l)	SNR/dB	角度-多普勒-距离单元(θ,f_d,l)	SNR/dB
(0,0.2,20)	20	(0.1,0,18)	25
(0.2,0,20)	20	(−0.1,0,18)	25
(0,−0.2,20)	20	(−0.3,0.2,24)	30
(−0.2,0,20)	20	(0.3,0.2,24)	30
(−0.1 0,22)	25	(0.3,0.2,16)	30
(0.1,0,22)	25	(−0.3,0.2,16)	30

(1) 对空监视场景

首先考虑的是对空监视场景，假设只存在接收机噪声。级联 MF-IAA 方法、级联 APC-IAA 方法、MF-RMVDR 和 STRAP 方法的迭代次数分别设置为 10、10、5 和 5 次。第 20 个距离单元的角度-多普勒聚焦结果如图 14.12 所示。由图 14.12(a)可知，在第 20 个距离单元有 4 个目标分布于角度-多普勒空间中，但是采用匹配滤波进行波形分离的图 14.12(b)和图 14.12(c)都存在着严重的邻近单元旁瓣干扰问题。图 14.12(d)的匹配滤波后空-时-距自适应处理可在一定程度上抑制距离旁瓣，但是稍逊于级联 APC-IAA 处理，这是由于自适应脉冲压缩在噪声环境下可以有效地分离波形并抑制距离旁瓣[66]。图 14.12(f)的空-时-距联合自适应处理将所有旁瓣压到噪声水平，且主瓣很窄，这充分说明了联合处理的优势。

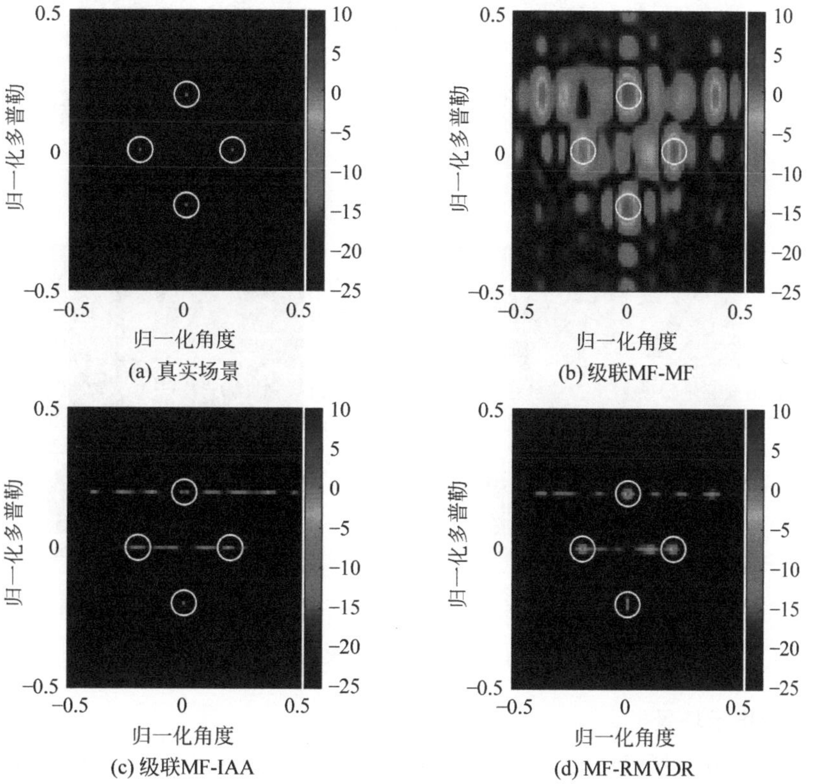

图 14.12 角度-多普勒聚焦结果

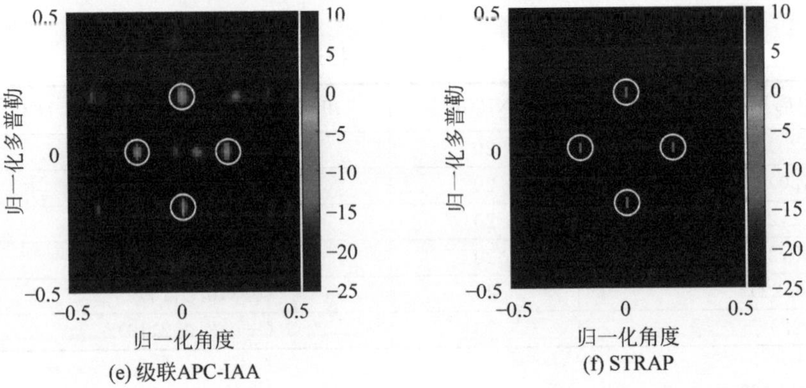

图 14.12 （续）

图 14.13 显示了几种方法在第 41 个多普勒单元处的角度-距离聚焦结果。由表 14.3 可知，第 41 个多普勒单元有 6 个目标，分别位于 18、20 和 22 距离单元。与预期相符的是空-时-距自适应处理有着最高的分辨力和最低的旁瓣。

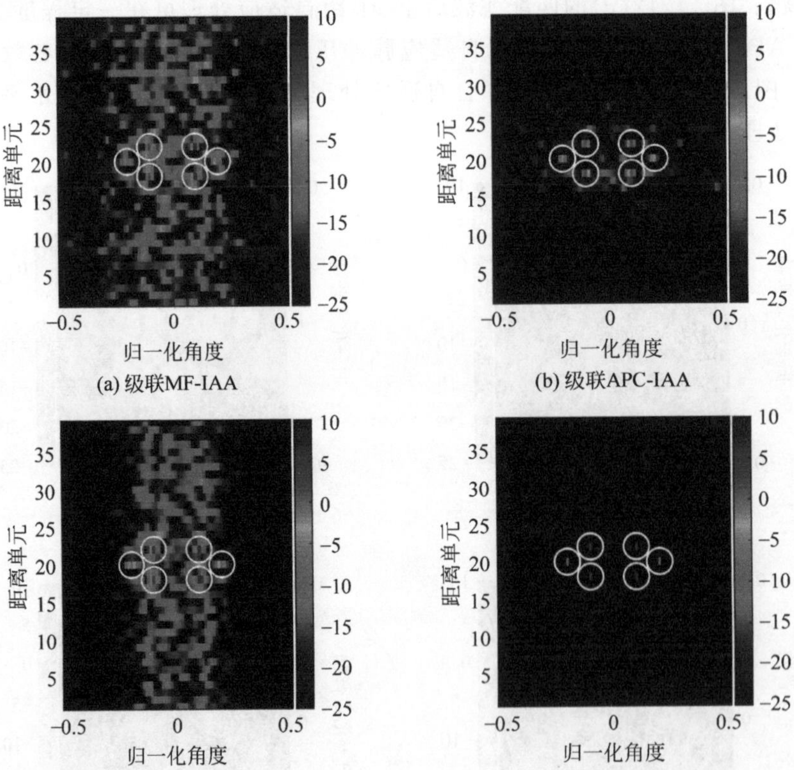

图 14.13 角度-距离聚焦结果

（2）地面动目标指示场景

测试地杂波对不同方法聚焦性能的影响。将杂噪比（Clutter-to-Noise Ratio，CNR）设置为 30dB，每个距离环划分为 1000 个杂波块。对于侧视和小擦地角机载雷达，杂波的多普勒频率与方位角的正弦线性相关。当平台的折叠系数 $\beta=1$ 时，如图 14.14(a)所示的对角线被称为杂波脊。图 14.14(b)、(c)和(f)表明传统的级联处理方案在强杂波下几乎难以分辨目标，杂波脊展宽严重。与对空监视场景不同的是，图 14.14(d)表明匹配滤波后的空-时-距自适应处理优于级联 APC-IAA，可以较好地抑制杂波，零多普勒附近的两个目标可以显示出来，但是邻近距离单元的大目标干扰没有被很好地抑制。图 14.14(f)表明，直接空-时-距自适应处理有着较为显著的优势，可以很好地抑制距离旁瓣和杂波。

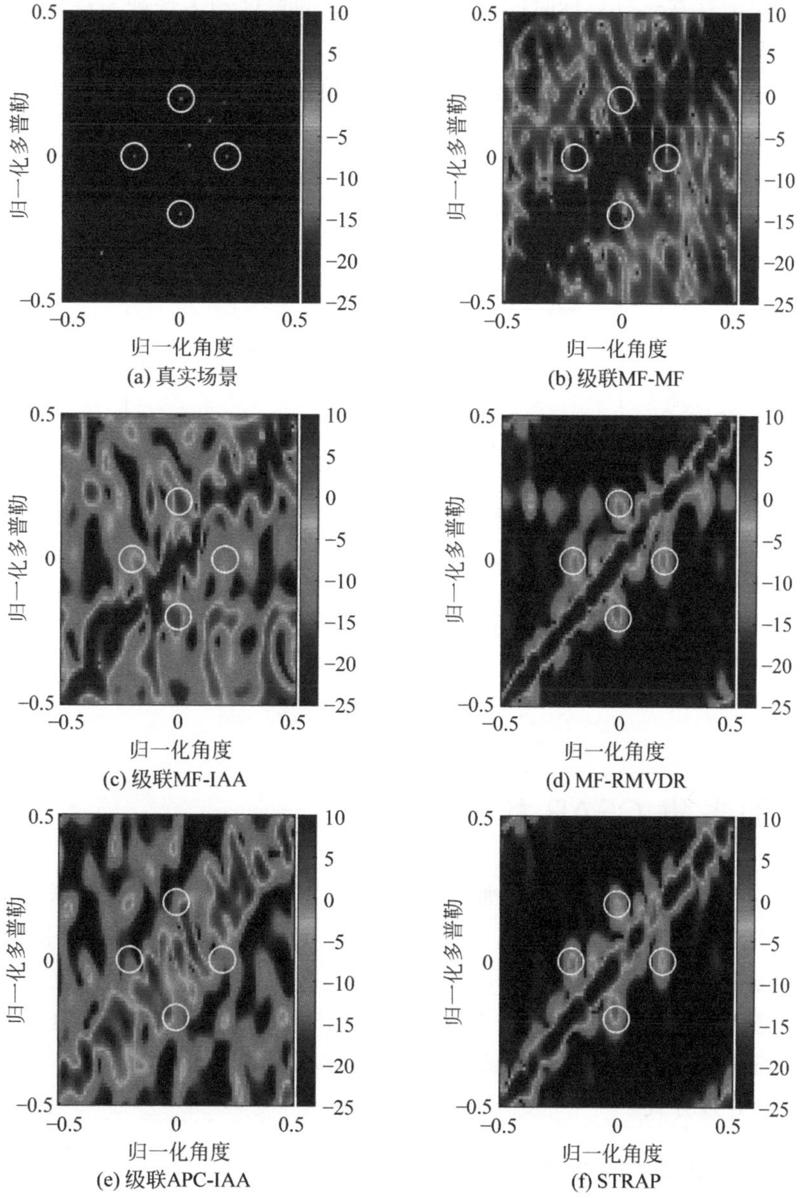

图 14.14 角度-多普勒聚焦结果

2. 检测结果分析

现在通过仿真比较级联处理与联合处理的检测性能，作为对比的级联方案在通过 MF-IAA 和 APC-IAA 得到空间谱后采用单元平均恒虚警检测器。考虑地面动目标指示场景，雷达参数设置为

$N_T = N_R = 3, N_P = 4, N_W = 10$,发射波形采用伪随机噪声序列,其余设置与前面相同。目标归一化角度和多普勒分别取 0 和 0.25,当 $P_{fa} = 10^{-1}$ 时,各种方案的检测性能曲线如图 14.15 中的点实线所示。图 14.15 中的信杂比(Signal-to-Clutter Ration,SCR)定义为脉压之前的信杂比,每个信杂比下的检测概率通过 1000 次仿真统计得到。从图 14.15 中可以看出,两种空-时-距自适应处理方案的检测性能差别不大,并且显著优于级联处理。在所设定的仿真参数下,相对于联合处理,级联处理的信杂比损失大于 10dB。考虑邻近单元存在干扰目标时的情况,干扰目标的角度和多普勒与待检测目标相同,位置相差 5 个距离单元,功率为 0dB,检测结果如图 14.15 虚线所示。此时,由于干扰目标及其旁瓣的影响,传统的级联处理方案性能进一步下降,而空-时-距联合处理检测器与单目标时的性能相当,这表明自适应聚焦、检测一体的方案对多目标检测的稳健性很好。

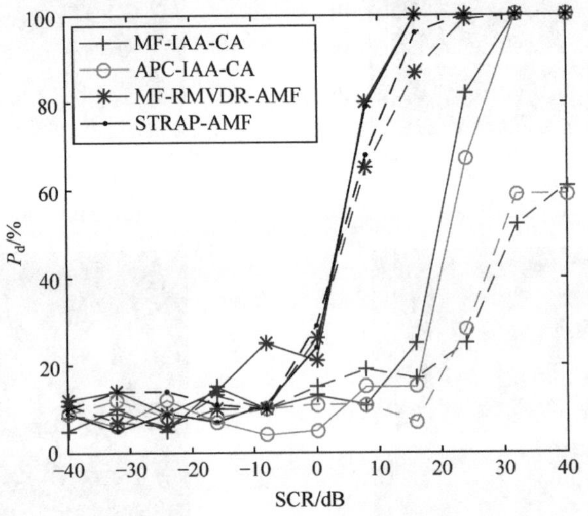

图 14.15 检测性能对比

14.5 其他多维 CFAR 检测

本章前面分别讨论了“阵元-脉冲”二维联合 CFAR 检测、“阵元-波形”二维联合 CFAR 检测及“阵元-脉冲-波形”三维联合 CFAR 检测。从信号来源的角度来看,上述讨论主要利用了阵元维、脉冲维和波形维的信息。引言中指出,除上述信息来源外,多雷达、多次扫描及各极化通道也都能为雷达 CFAR 检测提供丰富的信息来源。在常规的阵元、脉冲、波形信息利用的基础上,再联合利用多雷达[67-72]、多次扫描及各极化通道的信息进行 CFAR 检测,也属于多维 CFAR 检测的范畴,尽管后续几种信息的利用通常是基于非相参处理方式。

14.5.1 扫描间融合 CFAR 检测

长时间积累处理是检测低信杂比目标的常用方法,其中相参积累能够获得较大的积累增益,但这种做法会受到许多实际因素的制约,例如目标的速度可能会使得目标在长时间相参积累过程中出现距离走动,而目标的散射特性也可能会使得长时间积累过程中得到的目标回波脉冲之间并不满足相参性。因此,短时间相参积累、长时间(扫描间)非相参积累的处理方式更容易适应实际需求。扫描间融合

CFAR 检测的基本原理是，在单次扫描形成局部检验统计量的基础上，利用检测前跟踪（Track-Before-Detect，TBD）的搜索策略[73-79]，实现同一目标多次扫描对应的局部检验统计量的融合检测；对于 MIMO 阵列雷达来说，扫描间融合 CFAR 检测实现了在单次扫描融合“阵元-脉冲”等信息形成局部检验统计量的基础上，再联合多次扫描信息构成“阵元-脉冲-扫描间”至少三维 CFAR 检测的目的。

1. 扫描间融合检测问题的 GLRT

MIMO 阵列雷达系统由 M 个发射阵元和 N 个接收阵元组成，一个相参处理间隔包含 L 个相参脉冲，采用脉内正交编码，从而可以得到 MN 个阵元组成的虚拟接收天线；对同一个分辨单元的回波连续进行两次相参积累处理的时间间隔就是雷达天线的扫描周期，因此对同一个分辨单元的回波进行相参积累处理的次数就是雷达天线的扫描次数，此数量设为 Q。

在对各个正交脉冲波形进行相应的匹配滤波之后，将第 $q(q=1,2,\cdots,Q)$ 次扫描中获得的所有观测排列成 $v=MNL$ 维长向量 $\boldsymbol{x}_q$，则可形成如下观测模型

$$\boldsymbol{x}_q=\boldsymbol{s}_q+\boldsymbol{z}_q=\sum_{s=1}^{S}\boldsymbol{h}_{q,s}\boldsymbol{c}_{q,s}+\boldsymbol{z}_q=\boldsymbol{H}_q\boldsymbol{c}_q+\boldsymbol{z}_q \tag{14.119}$$

其中，S 为目标数量；$\boldsymbol{c}_q=[\boldsymbol{c}_{q,1}^{\mathrm{T}},\boldsymbol{c}_{q,2}^{\mathrm{T}},\cdots,\boldsymbol{c}_{q,S}^{\mathrm{T}}]^{\mathrm{T}}$ 是第 q 次扫描时 S 个目标回波的未知复幅度向量，$\boldsymbol{c}_{q,s}$ 是第 q 次扫描时第 $s(s=1,2,\cdots,S)$ 个目标的 $MN\times 1$ 维复幅度向量，且假设 $\boldsymbol{c}_{q,s}\sim\mathrm{CN}(0,\sigma_{q,s}^2\boldsymbol{R}_{q,s})$，$\sigma_{q,s}^2$ 和 $\boldsymbol{R}_{q,s}$ 未知；$\boldsymbol{H}_q=[\boldsymbol{h}_{q,1},\boldsymbol{h}_{q,2},\cdots,\boldsymbol{h}_{q,S}]$ 为第 q 次扫描时 S 个目标形成的空时模式矩阵；$\boldsymbol{h}_{q,s}$ 为第 q 次扫描第 s 个目标的 $MNL\times 1$ 维空时导向向量。“杂波＋热噪声”观测向量 $\boldsymbol{z}_q$ 是零均值、协方差矩阵为 $\boldsymbol{H}_{z,q}$ 的复高斯随机向量，且假设各个观测通道中，检验单元与参考单元的背景是相同的。那么第 q 次扫描时式(14.119)对应的二元假设检验为

$$\begin{cases}\mathrm{H}_0: \boldsymbol{x}_q=\boldsymbol{z}_q & \sim\mathrm{CN}(\boldsymbol{0},\boldsymbol{R}_{z,q})\\ \mathrm{H}_1: \boldsymbol{x}_q=\boldsymbol{H}_q\boldsymbol{c}_q+\boldsymbol{z}_q & \sim\mathrm{CN}(\boldsymbol{H}_q\boldsymbol{c}_q,\boldsymbol{R}_{z,q})\end{cases} \tag{14.120}$$

其中，$\boldsymbol{c}_q$ 为确定性未知量。由 14.2 节的分析可得 MSD 检测器，其检验统计量如下

$$T_{q,\mathrm{MSD}}=\boldsymbol{x}_q^{\mathrm{H}}\hat{\boldsymbol{R}}_{z,q}^{-1}\boldsymbol{H}_q(\boldsymbol{H}_q^{\mathrm{H}}\hat{\boldsymbol{R}}_{z,q}^{-1}\boldsymbol{H}_q)^{-1}\boldsymbol{H}_q^{\mathrm{H}}\hat{\boldsymbol{R}}_{z,q}^{-1}\boldsymbol{x}_q \tag{14.121}$$

其中，$\hat{\boldsymbol{R}}_{z,q}$ 是 $\boldsymbol{R}_{z,q}$ 的极大似然估计。

对于 TBD 问题来说，需要沿着目标运动航迹对式(14.121)的输出值进行非相参积累，于是得到 MIMO 阵列雷达多目标 TBD 问题的 GLRT 如下

$$\underset{\boldsymbol{P}_q\in\phi(\boldsymbol{P}_{q-1})}{\mathrm{Max}}\left(\sum_{q=1}^{Q}T_{q,\mathrm{MSD}}\right)\underset{\mathrm{H}_0}{\overset{\mathrm{H}_1}{\gtrless}}\gamma \tag{14.122}$$

其中，由于各次扫描之间的时间间隔较长（如秒级），因此可以认为 Q 次扫描得到的 Q 个检测统计量 $T_{q,\mathrm{AMF}}$ 是相互统计独立的；$\boldsymbol{P}_q\in\phi(\boldsymbol{P}_{q-1})$ 表示两次连续扫描之间目标位置信息的传递关系，$\boldsymbol{P}_q$ 是第 q 次扫描时 S 个目标位置信息的集合。

由式(14.121)和式(14.122)可知，统计 MIMO 雷达多目标 TBD 问题的 GLRT 可简化为如下两个步骤。

(1) 基于第 q 次扫描时第 i 个距离-方位分辨单元中的观测向量 $\boldsymbol{x}_{q,i}$，计算检测统计量 $T_{q,\mathrm{AMF}}$，并将输出值记为 $T_{q,i}$，为了便于描述，本章将 $T_{q,i}$ 称为“检验统计量值”。由此形成了对应于第 q 次扫描的数据空间，称为“原始数据帧”，记为 D_q。D_q 中记录的关于第 $i(i=1,2,\cdots,\Pi)$ 个观测数据的信息有：

扫描次序 q、距离 $r_{q,i}$、方位 $\theta_{q,i}$ 及检验统计量值 $T_{q,i}$，其中，Π 表示数据空间中分辨单元的数量。需要指出的是，后续仿真过程中为了便于显示，将极坐标的距离-方位数据转化为直角坐标数据($x_{q,i}=r_{q,i}\cos\theta_{q,i}$，$y_{q,i}=r_{q,i}\sin\theta_{q,i}$)。

(2)根据位置信息的传递关系 $\boldsymbol{P}_q\in\phi(\boldsymbol{P}_{q-1})$，在 Q 次扫描获取的 Q 帧原始数据空间中进行搜索，并积累 $T_{q,i}$，在这里将 Q 帧原始数据形成的空间称为原始搜索网络。然后取积累值最大者进行门限判决。

2. 扫描间搜索策略

式(14.122)中 GLRT 的最优解需要执行两维位置信息(距离与方位)的 S 维联合搜索。这里介绍两种搜索技术，一是基于动态规划的多维联合搜索技术，这是一种最优搜索策略，其基本原理是通过数据点组合操作，将观测空间中多目标航迹的联合搜索过程转化为扩展观测空间中单个"目标组合"航迹的搜索过程，同时结合虚假航迹筛选技术与不连续航迹剔除技术，有效降低虚警和漏警，但其缺陷在于搜索量很大。另一种搜索技术是基于 PHT-STC 的多次单维搜索技术，这是一种次优搜索策略，其基本原理是将观测空间中多目标航迹的联合搜索过程转化为 Hough 参数空间里逐个参数单元中的单目标航迹搜索过程，即将多维航迹联合搜索转化为多个一维航迹搜索，克服了多目标航迹搜索过程中相互干扰的问题；同时，通过求解 Hough 空间不同参数单元上的第二门限，克服了直接在 Hough 参数空间设置统一门限及采用二值积累等做法所带来的各目标之间相互影响的缺陷。这里需要指出的是，"最优"与"次优"仅仅是就搜索策略而言的，而不是指检测性能上的"最优"与"次优"，检测性能除了与搜索策略有关外，还与搜索策略对信噪比的要求有关。

14.5.2 极化 CFAR 检测

极化雷达可以测量雷达反射目标的散射矩阵，并且因此获得目标远场的完整信息。这个极化信息可以用在极化信号处理中。极化信号处理可以处理多通道信号，而不是像许多目前常用的雷达那样只处理一个通道的信息。

进行极化信号处理时，重要的是不同反射目标的散射矩阵的差别，因为这些差别使得极化信号处理方法相对于常用的单通道处理方法在性能上有所提高。文献[80]提供了一类地杂波，雨杂波和一个小喷气飞机在不同视角上的测量散射矩阵数据。基于这些测量结果的差别，Wanielik 和 Stack[80] 提出了利用包含在散射矩阵中的所有信息的极化(Polarimetric，P)-CFAR 检测器，它是 Wanielik 在文献[81]中提出的 CFAR 检测器的一个扩展。这些测量数据的差别说明了不同反射目标的极化差别，并且也说明了飞机反射体的视角依赖性和使用非极化雷达时出现的信息损失。这些数据说明了与常用的 CFAR 检测器相比，利用全部散射矩阵信息的 P-CFAR 检测器有可能使性能得到很大的提高。

与使用一系列标量幅度进行判决的传统 CFAR 检测器不同，P-CFAR 基于一个向量序列$\{\boldsymbol{s}(i),i=1,2,\cdots,2N-1\}$进行判决，其中，$2N$ 是参考单元数。如果极化雷达是双基的，向量 $\boldsymbol{s}(i)=[s_{11}(i),s_{21}(i),s_{12}(i),s_{22}(i)]^{\mathrm{T}}$ 包含散射矩阵的全部元素 $s_{jk}(i)$。对于单基地雷达情况，$s_{12}(i)=s_{21}(i)$，向量 $\boldsymbol{s}(i)$只包含这些元素中的 3 个。对于接收极化的雷达，向量 $\boldsymbol{s}(i)$只包含散射矩阵的两个元素 $s_{11}(i)$和 $s_{21}(i)$。当进行 CFAR 处理时，需要对比检测单元向量 $\boldsymbol{s}(N+1)$和由雷达扫描图的一小部分中选取的参考向量，判断 $\boldsymbol{s}(N+1)$是否包含与参考向量相似的向量信息，或者是否是一个异物。如果有

$$[\boldsymbol{s}(N+1)]^{\mathrm{T}}\cdot\boldsymbol{J}^{-1}\cdot[\boldsymbol{s}(N+1)]>S \tag{14.123}$$

则可能是一个目标。其中，$\boldsymbol{J}$ 是参考向量的协方差矩阵，S 是在保证给定的虚警概率条件下选择的阈值。因为协方差矩阵是未知的，需要由参考单元估计。这个矩阵通常由式(14.124)估计

$$J_1=\frac{1}{N}\sum_{i=1}^{N}\{s(i)s^{\mathrm{T}}(i)+[s(N+1+i)][N+1+i]^{\mathrm{T}}\} \tag{14.124}$$

当数据样本来自于协方差矩阵为J的过程，并且数据中不包含其他干扰时，J_1是J的最好的估计。在多目标和杂波边缘环境中，文献[82]介绍了两种J的估计方法。相对于式(14.124)的估计方法，这两种估计方法在非均匀杂波环境中获得增强的CFAR性能。一种方法是对每个参考样本$s(i)$赋予一定的权值。对表现为干扰目标的向量赋予较其他向量低的权值。这种方法被称为加权采样法。第二种方法用聚类算法[83]来处理不同类中的数据，假设一些类中包含干扰，而另一些包含目标。

文献[84]的分析表明，即使在常用的CFAR检测器失利的情况下，文献[81]中介绍的利用极化信息判决目标存在的CFAR检测器也显示出了较高的性能。并且预料文献[82]中提出的增强的CFAR方案在多目标和非均匀干扰环境中会有较好的表现。在单目标和均匀杂波的正常环境中，相对于文献[81]中描述的极化CFAR检测器，它们应该只有一个很小的损失。

文献[84]和[85]研究了高斯和非高斯背景中相参雷达的自适应极化目标检测问题，推导了这两种背景环境下的自适应极化相参检测方案，并将已有的GLRT极化检测器推广到多通道情况中。针对高斯背景，文献[84]推导了极化检测器的检测性能解析表达式，由此表明其具有CFAR性质，但是当背景统计特性偏离高斯分布时，该极化检测器的性能明显下降；为此，针对非高斯背景，文献[85]提出了一种不依赖于纹理分量的GLRT检测器，分析结果表明，该检测器在具有未知参数的复合高斯杂波背景中具有CFAR性质。相关的研究还可参考文献[86-91]。

14.6 小结

本章首先从信息来源的角度对“多维”进行了界定，然后重点从阵列雷达CFAR检测的角度讨论了“阵元-脉冲”二维联合CFAR检测，从基于自适应空时编码设计的MIMO阵列雷达目标检测的角度讨论了“阵元-波形”二维联合CFAR检测，以及从空时距三维联合自适应处理的角度讨论了“阵元-脉冲-波形”三维联合CFAR检测。上述多维联合处理主要是多维联合相参处理，主要是利用了阵元维、脉冲维和波形维的信息。在此基础上，讨论了利用多次扫描以及各极化通道信息的多维CFAR检测，与前面不同的是，对这几维信息的利用通常基于非相参处理方式。

参考文献

[1] 杨建宇. 雷达技术发展规律和宏观趋势分析[J]. 雷达学报，2012.1，1(1)：19-27.

[2] Bliss D W，Forsythe K W. Multiple-input multiple output(MIMO) radar and imaging: degrees of freedom and resolution[C]. The Thirty-Seventh Asilomar Conference on Signals，Systems & Computers，2003.

[3] 赵永波，刘宏伟. MIMO雷达技术综述[J]. 数据采集与处理，2018,33(3)：389-399.

[4] Reed I S，Mallett J D，Brennan L E. Rapid convergence rate in adaptive arrays[J]. IEEE Transactions on Aerospace and Electronic Systems，1974，10(6)：853-863.

[5] Kelly E J. An adaptive detection algorithm[J]. IEEE Transactions on Aerospace and Electronic Systems，1986，22(1)：115-133.

[6] Scharf L L，Friedlander B. Matched subspace detectors[J]. IEEE Transactions on Signal Processing，1994，42(8)：2146-2157.

[7] Kay S M. Fundamentals of statistical signal processing volume I&II: Estimation theory & detection theory[M]. Prentice-Hall，Inc.，1993.

[8] Robey F C, Fuhrmann D R, Kelly E J, et al. A CFAR adaptive matched filter detector[J]. IEEE Transactions on Aerospace and Electronic Systems, 1992, 28(1): 208-216.

[9] Kalson S Z. Adaptive array CFAR detection[J]. IEEE Transactions on Aerospace and Electronic Systems, 1995, 31(2): 534-542.

[10] Kraut S, Scharf L L, McWhorte L T. Adaptive subspace detectors[J]. IEEE Transactions on Signal Processing, 2001, 49(1): 1-16.

[11] Burgess K A, Barry D V V. Subspace-based adaptive generalized likelihood ratio detection[J]. IEEE Transactions on Signal Processing, 1996, 44(4): 912-927.

[12] Raghavan R S, Pulsone N, McLaughlin D J. Performance of the GLRT for adaptive vector subspace detection[J]. IEEE Transactions on Aerospace and Electronic Systems, 1996, 32(4): 1473-1487.

[13] Gini F, Farina A. Matched subspace CFAR detection of hovering helicopters[J]. IEEE Transactions on Aerospace and Electronic Systems, 1999, 35(4): 1293-1305.

[14] Kraut S, Scharf L L. The CFAR adaptive subspace detector is a scale-invariant GLRT[J]. IEEE Transactions on Signal Processing, 1999, 47(9): 2538-2541.

[15] Liu J, Li H B, Himed B. Persymmetric adaptive target detection with distributed MIMO radar[J]. IEEE Transactions on Aerospace and Electronic Systems, 2015, 51(1): 372-382.

[16] Liu J, Han J W, Zhang Z J, et al. Target detection exploiting covariance matrix structures in MIMO radar[J]. Signal Processing, 2019, 154: 174-181.

[17] Guan J, Huang Y, He Y. A CFAR detector for MIMO array radar with adaptive pulse compression-capon filter[J]. Science China Information Sciences, 2011, 41(10): 1268-1282.

[18] 关键，黄勇，何友. MIMO阵列雷达检测器分析[J]. 电子学报，2010，38(9)：2107-2111.

[19] 黄勇，关键. MIMO阵列雷达Two-Step检测器及其性能分析[J]. 信号处理，2009，25(8A)：461-464.

[20] 关键，黄勇. 一种简便的MIMO阵列雷达CFAR检测器[J]. 信号处理，2010，26(3)：467-472.

[21] Guan Jian, Zhang Yan fei, Huang Yong. Adaptive subspace detection of range-distributed target in compound-Gaussian clutter[J]. Digital Signal Processing, 2009, 19(1): 66-78.

[22] 董云龙，黄勇，关键. 非均匀杂波中统计MIMO雷达的相参检测[J]. 中国电子科学研究院学报，2011，6(2)：200-203.

[23] 董云龙，黄勇，关键. MIMO雷达目标子空间建模与检测性能分析[J]. 火力指挥与控制，2012，37(1)：25-28.

[24] 赵翔. MIMO雷达中的空时处理与检测研究[D]. 成都：电子科技大学，2018.

[25] 韩金旺. MIMO雷达自适应目标检测算法研究[D]. 西安：西安电子科技大学，2019.

[26] 高永婵. 复杂场景下多通道阵列自适应目标检测算法研究[D]. 西安：西安电子科技大学，2015.

[27] 王泽玉. 雷达目标自适应检测算法研究[D]. 西安：西安电子科技大学，2018.

[28] 邓萍. 基于多秩子空间的宽带雷达目标检测方法研究[D]. 西安：西安电子科技大学，2020.

[29] 丁昊，薛永华，黄勇，等. 均匀和部分均匀杂波中子空间目标的斜对称自适应检测方法[J]. 雷达学报，2015，4(4)：418-430.

[30] 刘维建，简涛，杨海峰，等. 适用于子空间信号失配的参数可调多通道自适应检测器[J]. 电子与信息学报，2016，38(2)：3011-3017.

[31] Liu Wei jian, Xie Wen chong, Liu Jun, et al. Adaptive double subspace signal detection in Gaussian background-part I: homogeneous environments[J]. IEEE Transactions on Signal Processing, 2014, 62(9): 2345-2357.

[32] Liu Wei jian, Xie Wen chong, Liu Jun, et al. Adaptive double subspace signal detection in Gaussian background-part II: partially homogeneous environments[J]. IEEE Transactions on Signal Processing, 2014, 62(9): 2358-2369.

[33] 王作珍. 自适应子空间信号检测理论和技术研究[D]. 成都：电子科技大学，2020.

[34] Gini F, Farina A, Greco M V. Detection of multidimensional gaussian random signals in compound-Gaussian clutter plus thermal noise[J]. Proceedings of ICSP'98, 1998: 1650-1653.

[35] Gini F, Farina A. Vector subspace detection in compound-gaussian clutter part I: survey and new results[J]. IEEE Transactions on Aerospace and Electronic Systems, 2002,38(4): 1295-1311.

[36] Gini F, Farina A, Montnari M. Vector subspace detection in compound-gaussian clutter part II: performance analysis[J]. IEEE Transactions on Aerospace and Electronic Systems, 2002, 38(4): 1312-1323.

[37] Desai M N, Mangoubi R S. Robust gaussian and non-gaussian matched subspace detection[J]. IEEE Transactions on Signal Processing, 2003, 51(12): 3115-3127.

[38] 刘维建，谢文冲，王永良. 多通道复信号检测中的几种常用复统计分布[J]. 电子学报，2013，41(6)：1238-1241.

[39] Kalson S Z. An adaptive array detector with mismatched signal rejection[J]. IEEE Transactions on Aerospace and Electronic Systems, 1992, 28(1): 195-207.

[40] Friedlander B. Waveform design for MIMO radars[J]. IEEE Transactions on Aerospace and Electronic Systems, 2007, 43(3): 1227-1238.

[41] Friedlander B. A subspace framework for adaptive radar waveform design[C]. Pacific Grove: Proceedings of the 39th Asilomar Conference on Signal, Systems and Computers, 2005.

[42] Li Jian, Petre S. MIMO Radar Signal Processing[M]. New York: WILEY & SONS Publication, 2009.

[43] Friedlander B. Waveform design for MIMO radar with space-time constraints[C]. Pacific Grove: Proceedings of the Asilomar Conference on Signal, Systems and Computers, 2007.

[44] De Maio A, Lops M. Design principles of MIMO radar detectors[J]. IEEE Transactions on Aerospace and Electronic Systems, 2007, 43(3): 886-898.

[45] Liu Zheng, Zhao Rui li, Liu Yun fu, et al. Performance analysis on MIMO radar waveform based on mutual information and minimum mean-square error estimation[C]. Guilin: Proceedings of the IET, International Radar Conference, 2009.

[46] Hu L B, Liu H W, Zhou S H, et al. Convex optimization applied to transmit beampattern synthesis and signal waveform design for MIMO radar[C]. Guilin: Proceedings of the IET, International Radar Conference, 2009.

[47] Petre S, Li Jian, Xie Yao. On probing signal design for MIMO radar[J]. IEEE Transactions on Signal Processing, 2007, 55(8): 4151-4161.

[48] Leshem A, Naparstek O, Nehorai A. Information theoretic adaptive radar waveform design for multiple extended targets[J]. IEEE Journal of Selected Topics in Signal Processing, 2007, 1(1): 42-55.

[49] Yang Yang, Blum R. MIMO radar waveform design based on mutual information and minimum mean-square error estimation[J]. IEEE Transactions on Aerospace and Electronic Systems, 2007, 43(1): 330-343.

[50] Naghibi T, Behnia F. Convex optimization and MIMO radar waveform design in the presence of clutter[C]. Monastir: Proceedings of the 2008 International Conference on Signals, Circuits and Systems, 2008.

[51] 黄勇，关键，董云龙. MIMO阵列雷达检测中的自适应空时编码设计[J]. 电子与信息学报，2010，32(8)：1831-1836.

[52] Richards M A. Fundamentals of radar signal processing[M]. New York: McGraw-Hill Education and Publishing House of Electronics Industry, 2014.

[53] Higgins T, Blunt S D, Shackelford A K. Space-range adaptive processing for waveform-diverse radar imaging[C]. 2010 IEEE Radar Conference, 2010.

[54] Higgins T, Blunt S D, Shackelford A K. Time-range adaptive processing for pulse agile radar[C]. 2010 International Waveform Diversity and Design Conference, 2010.

[55] 陈宝欣. MIMO雷达多维联合处理方法研究[D]. 烟台：海军航空大学，2019.

[56] Chen Bao xin, Chen Xiao long, Huang Yong, et al. Transmit beampattern synthesis for FDA radar[J]. IEEE Antennas & Wireless Propagation Letters, 2018, 17(01): 98-101.

[57] Chen Bao xin, Huang Yong, Chen Xiao long, et al. Multiple-Frequency CW radar and the array structure for uncoupled angle-range indication[J]. IEEE Antennas & Wireless Propagation Letters, 2018, 17(12): 2203-2207.

[58] 陈宝欣，关键，董云龙，等. 多频连续波雷达与角度-距离联合估计方法[J]. 电子学报，2020，48(2)：375-383.

[59] 陈宝欣，黄勇，陈小龙，等. 基于迭代超分辨的单快拍DOA估计方法[J]. 信号处理，2019，35(5)：775-780.

[60] Chen Bao xin，Huang Yong，Chen Xiao long，et al. Space-time-range adaptive processing for MIMO radar imaging [C]. Brisbane：International Conference on Radar，2018.

[61] Chen Xiao long，Chen Bao xin，Guan Jian，et al. Space-range-doppler focus-based low-observable moving target detection using frequency diverse array MIMO radar[J]. IEEE Access，2018，6：43892-43904.

[62] Roberts W，Stoica P，Li Jian，et al. Iterative adaptive approaches to MIMO radar imaging[J]. IEEE Journal of Selected Topics in Signal Processing，2010，4(1)：5-20.

[63] Robey F C，Fuhrmann D R，Kelly E J，et al. A CFAR adaptive matched filter detector[J]. IEEE Transactions on Aerospace and Electronic Systems，1992，28(1)：208-216.

[64] Guan Jian，Peng Ying ning，He You. Proof of CFAR by the use of the invariant test[J]. IEEE Transactions on Aerospace and Electronic Systems，2000，36(1)：336-339.

[65] Deng Hai. Polyphase code design for orthogonal netted radar systems[J]. IEEE Transactions on Signal Processing，2004，52(11)：3126-3135.

[66] Li Ning，Tang Jun，Peng Ying ning. Adaptive pulse compression of MIMO radar based on GSC[J]. Electronics Letters，2008，44(20)：1217-2118.

[67] Fishler E，Haimovich A，Blum R，et al. MIMO radar：an idea whose time has come[C]. Philadelphia：Proceedings of the IEEE Radar Conference，2004.

[68] Guan Jian，Huang Yong. Detection performance analysis for MIMO radar with distributed apertures in Gaussian colored noise[J]. Sci China Ser F-Inf Sci，2009，52(9)：1689-1696.

[69] 陈沁根. 分布式MIMO雷达数据融合检测算法研究[D]. 南京：南京大学，2019.

[70] 廖羽宇. 统计MIMO雷达检测理论研究[D]. 成都：电子科技大学，2012.

[71] Lehmann N，Haimovich A M，Blum R S，et al. MIMO-radar application to moving target detection in homogeneous clutter[C]. Waltham：In Proceedings of the 14th Annual Workshop on Adaptive Sensor Array and Multi-channel Processing，2006.

[72] 黄勇，黄涛，刘宁波，等. 基于均匀性判定规则的统计MIMO雷达多通道融合检测技术[J]. 海军航空工程学院学报，2018，33(1)：101-105.

[73] 曲长文，黄勇，苏峰. 基于动态规划的多目标检测前跟踪算法[J]. 电子学报，2006，34(12)：2138-2141.

[74] 关键，黄勇. MIMO雷达多目标检测前跟踪算法研究[J]. 电子学报，2010，38(06)：1449-1453.

[75] 曲长文，黄勇，苏峰，等. 基于坐标变换与随机Hough变换的抛物线运动目标检测算法[J]. 电子与信息学报，2005，27(10)：1573-1575.

[76] Huang Yong，Jiang Guo feng，Qiu Kai lan，et al. Radar track-before-detect algorithm of multitarget based on the dynamic programming[C]. Proceedings of 2006 CIE international conference on radar，2006.

[77] Huang Yong，Guan Jian. A track-before-detect algorithm for statistical mimo radar multitarget detection[C]. Washington DC：In Proceedings of 2010 IEEE Radar Conference，Crystal Gateway Marriott，2010.

[78] 秦文利，郑娜娥，顾帅楠. MIMO雷达弱目标检测前跟踪算法[J]. 太赫兹科学与电子信息学报，2017，15(4)：595-600.

[79] 战立晓，汤子跃，朱振波. 雷达微弱目标检测前跟踪算法综述[J]. 现代雷达，2013，35(4)：45-53.

[80] Wanielik G，Stock D J R. Measured scattering-matrix-data and a polarimetric CFAR detector which works on this data：Proceedings of IEEE International Radar Conference，1990[C]. Germany：IEEE，Radar1990，1990.

[81] Wanielik G，Stock D J R. Use of radar polarimetric information in CFAR and classification algorithms：Proceedings of IEEE International Radar Conference[C]. Paris：IEEE，Radar1989，1989.

[82] Wanielik G，Stock D J R. A new clutter rejection method using a robust polarimetric CFAR detector[J]. IEEE Aerospace and Electronic Systems Magazine，1990，5(6)：7-10.

[83] Seber G A F. Multivariate observations[M]. New York：John Wiley，1984.

[84] Pastina D，Lombardo P，Bucciarelli T. Adaptive polarimetric target detection with coherent radar，I. detection

against Gaussian background[J]. IEEE Trans. on AES, 2001, 37(4): 1194-1206.

[85] Lombardo P, Pastina D, Bucciarelli T. Adaptive polarimetric target detection with coherent radar, II. detection against non-Gaussian background[J]. IEEE Trans. on AES, 2001, 37(4): 1207-1220.

[86] Liu L D, Wu S J, Sun X W. CFAR polarimetric adaptive subspace detector[J]. Acta Electronica Sinica, 2005, 33(9): 1553-1556.

[87] Lombardo P, Pastina D, Corsale E. Polarimetric coherent adaptive detection against compound-Gaussian clutter[C]. Waltham: Proceedings of the 1999 IEEE Radar Conference-Radar into the Next Millennium, 1999.

[88] 关键，刘宁波，张建，等. 海杂波的多重分形关联特性与微弱目标检测[J]. 电子与信息学报，2010，32(1)：54-61.

[89] 吴迪军，徐振海，张亮，等. 极化空时自适应匹配滤波检测器[J]. 电子学报，2013，41(4)：744-750.

[90] 吴迪军. 机载雷达极化空时自适应处理技术研究[D]. 长沙：国防科学技术大学，2012.

[91] 雷世文. 不同环境下目标的极化匹配检测理论和技术研究[D]. 成都：电子科技大学，2015.

第 15 章 基于特征的 CFAR 处理

CHAPTER 15

15.1 引言

分形理论的研究对象是自然界中不光滑和不规则的几何体,它深刻揭示了实际系统与随机信号中广泛存在的自相似性和标度不变性。分形理论的提出拓宽了人们的视野,并已在自然科学的多个领域获得广泛应用,尤其是在雷达信号处理[1-2]、图像处理[3-4]、语音信号处理[5]等领域具有重要的应用价值和广阔的发展前景。自 1993 年 T. Lo 首次将单一分形维数应用于海杂波中目标检测以来[1],分形理论在雷达目标检测方向的应用取得了较好的发展,并取得了一系列有意义的成果。分形理论在雷达目标检测中的应用经历了由浅入深、由简单到复杂的发展历程。雷达目标的分形检测方法主要关心的是时间序列结构的变化,而不是完全依赖于信号幅度的强弱,从而在一定程度上可以摆脱 SCR 的束缚,但在 SCR 很低时分形检测方法同样是无效的,而且随着分形理论应用的复杂化,所得到的检测算法越来越复杂,难以具有实时性。另外,由于分形参数估计对数据量要求高,在时域中直接采用分形特征检测海面运动目标难以实现。参考相参雷达中的动目标指示(MTI)处理方法,若可以在频域中引入分形分析方法,则一方面时域回波变换到频域后运动目标回波的能量得到了有效积累,提升了 SCR,另一方面可以降低分形特征参数估计过程中对数据量的需求。目前可以找到的与此研究方向较为接近的是文献[6]和[7],其中提到,如果一个信号是分形的,那么它的功率谱密度与频率呈幂律关系,这种过程称为 $1/f^{\alpha}$ 噪声,但并未进一步研究。也有相关文献研究 FBM 的谱特性,但其主要目的是通过功率谱估计谱指数得到时域序列的单一 Hurst 指数[8-9],而不是将分形分析方法引入频域中。

本章首先将以分数布朗运动(FBM)为例,系统阐述 FBM 在时域具有自相似(分形)特性的前提下,FBM 的频谱也将具有自相似(分形)特性,即傅里叶变换具有保持原序列自相似性的特性,然后将分形分析方法引入对雷达实测海杂波数据频谱的分析中,分析频谱的分形特性及相应分形特性的影响因素,同时还将进一步研究目标回波对海杂波频谱分形特性的影响,为设计海杂波中的目标检测算法奠定基础。在分形特征处理的基础上,进一步结合深度学习的处理方法,探索脉压与检测一体化处理方法,提升目标检测性能。

15.2 海杂波时域分形特征与 CFAR 检测

海表面作为一个永不停息随机起伏的散射面,其形态受天气因素影响(如海表面的风、雨等)很大,往往天气条件越恶劣则越容易导致高的海情,此时,通过雷达获得的海杂波往往起伏十分尖锐,这一现象在低分辨率雷达中经常能遇见，在高分辨率雷达中这一现象则更为明显。近年来，很多研究人员对吉赫兹频率范围的低擦地角海杂波进行了大量研究[6-23]。研究表明，高海情和低擦地角使得海面回波具有一个十分重要的特征，即海杂波中包含着一些幅度很强且与目标十分相像的回波簇，这样的回波簇称为海尖峰[24]。从统计学角度来看，海尖峰在整个海杂波背景中应属于小概率事件（“罕见”事件），根据高斯模型可知，海尖峰主要出现于高斯分布的尾部[25]，因此若海杂波序列中存在较多幅度较强的海尖峰，则需要采用“厚拖尾”分布才能达到较好的拟合效果。无论在高分辨率雷达还是低分辨率雷达中，海尖峰的存在都可能严重影响雷达对海面微弱目标的检测与跟踪性能，因此对低擦地角和高海情条件下的海尖峰开展研究将有助于深入理解海尖峰特性，并对现代雷达信号处理与目标检测具有重要意义。

在过去的几十年中，研究人员主要从实验测量分析和数学模型分析两个方面对海尖峰进行了详细研究。在实验测量分析方面，研究人员通过人工造浪同时实地测量的方式，研究了海面涌浪、碎浪、飞沫等不同类型海浪的电磁散射效应，试图将海面回波信号中出现的尖峰与物理海面的某些急速起伏现象联系起来，结果发现海面碎浪可以产生很强的尖峰回波，海尖峰与海面碎浪现象匹配程度最高，可以达到80%[12,26-29]；在数学模型分析方面，已有文献对海尖峰的空间与时间相关性[30-31]、多普勒特性[21,32]和统计特性[33-34]进行了相关研究，研究结果表明，海杂波包含两种组成成分，一种是漫散射背景，其通常具有低的后向散射功率、HH/VV极化率和多普勒速度；另一种是海尖峰，其通常具有高的后向散射功率、HH/VV极化率和多普勒速度。漫散射背景可以采用倾斜调制的Bragg散射进行很好的建模，而海尖峰却与起伏剧烈的碎浪密切相关[13]，建模较困难。2003年，Fred Posner等提出采用两个量来描述海杂波，即平均尖峰持续时间和平均尖峰间隔[24]，而这两个描述量则是基于3个海尖峰判定参数的，即尖峰幅度门限、最小尖峰宽度（最小尖峰持续时间）和最小尖峰间隔时间。海杂波序列中的采样点簇需同时满足上述3个条件才算是一个海尖峰。在以上判定条件基础之上，M. Greco等分析了IPIX雷达海杂波数据，发现平均尖峰持续时间和平均尖峰间隔是服从指数分布的[35]，但从最终指数分布拟合效果来看，指数分布对所占比例较大的短持续时间海尖峰的拟合效果并不好。

实际上，由于强海尖峰的存在（一般而言，其持续时间也相对较长），海杂波序列中存在的海尖峰的尖峰持续时间和尖峰间隔的分布类型通常具有厚拖尾。对于这种厚拖尾现象，本节采取不同的角度重新考虑这一问题，即将每个海尖峰的持续时间（或尖峰间隔）看作随机分布于正半轴实直线上的点，则由这一系列的点可以构成一个可数集合，然后采用代数分形模型——Paretian泊松模型[36-37]进行建模，研究海尖峰持续时间与海尖峰发生频数之间的代数自相似关系并估计代数分形参数。

15.2.1 海尖峰判定

1. 海尖峰判定方法

对于来自某一距离单元的海杂波时间序列，其采样点簇需满足如下3个条件才能判定为海尖峰[24,35]，即尖峰幅度门限、最小尖峰宽度（最小尖峰持续时间）、最小尖峰间隔时间。尖峰幅度门限，是指采样点的幅度必须超过一定的门限才可能属于某一个海尖峰；最小尖峰宽度是指采样点幅度连续保持在尖峰幅度门限之上的时间必须大于或等于规定的最小尖峰宽度；最小尖峰间隔时间是指如果高于尖峰幅度门限的连续采样点之后出现采样点的幅度低于尖峰幅度门限，那么采样点幅度低于门限的时间不能超过规定的最小尖峰间隔时间，如果超过了这一时间，那么这一簇采样点就被认为是由两个或多个海尖峰组成的，并且每个海尖峰都须满足这3个条件。

2. 实测海杂波数据

这里采用X波段雷达数据(X-26#)进行验证与分析。数据26#来源于“Osborn Head Database”，是由加拿大McMaster大学利用X波段的IPIX雷达开展对海探测实验采集得到的，数据采集时雷达天线工作于驻留模式，对某一方位海面长时间照射，观察目标为一漂浮于海面上包裹着金属网的塑料球体，数据采集过程中天气晴朗，海面风向为东北风，风速约20km/h，更详细的情况请见文献[38]。在使用过程中主要涉及26#数据的HH和VV两种极化方式，每种极化方式包含14个距离单元的回波数据，其中目标单元位于第6～8距离单元，其余为纯海杂波单元(共计11个距离单元)，每个距离单元包含2^{17}个采样点，采样频率为1kHz，则每个距离单元时间序列的持续时间约为131s，X-26#数据的距离-时间-幅度三维图和单个距离单元的时域波形图分别如图15.1和图15.2所示。由于本节主要研究海杂波背景中的海尖峰判定及海尖峰特性，因此，本节只采用11个纯海杂波单元的数据进行后续分析。

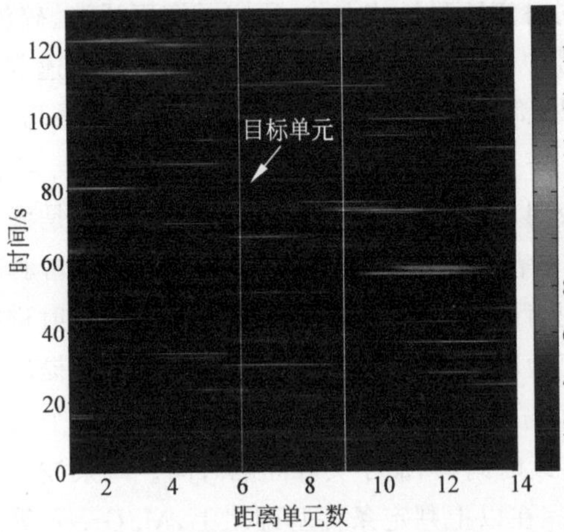

(a) HH极化

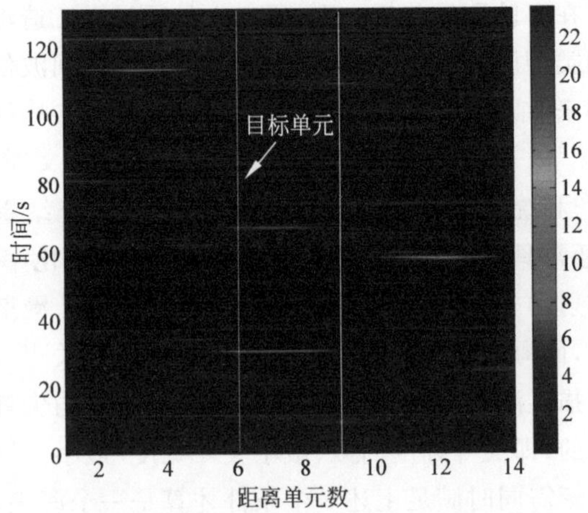

(b) VV极化

图15.1 X-26#数据距离-时间-幅度三维图

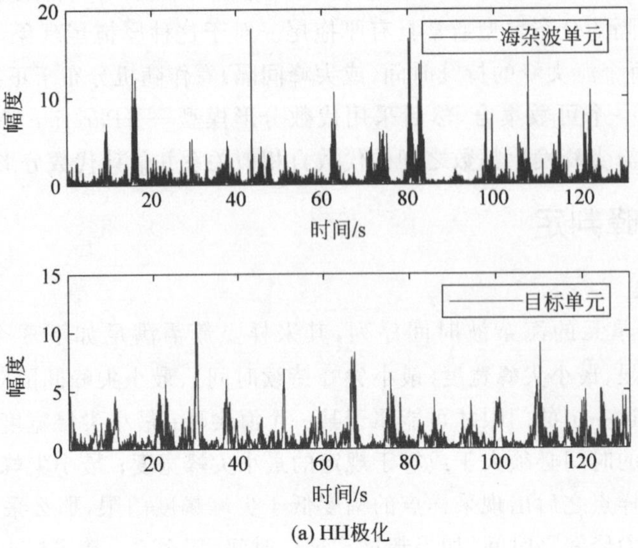

(a) HH极化

图15.2 X-26#海杂波和目标单元时域波形图

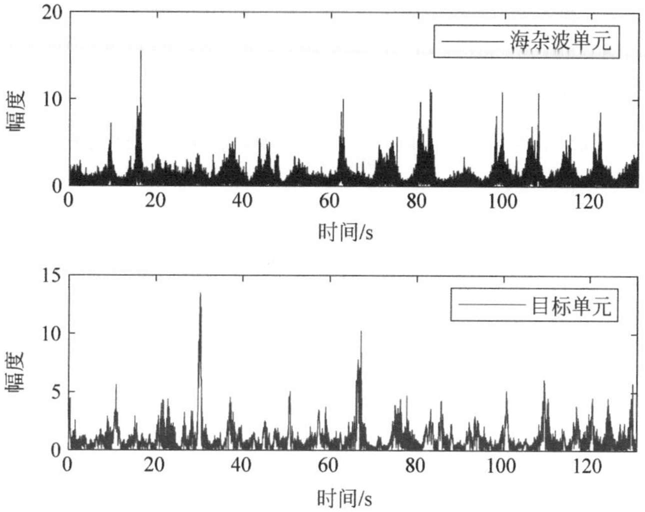

(b) VV极化

图 15.2 (续)

3. 实测数据处理结果

为指导后续合理设置海尖峰判定的三个参数，尤其是最小尖峰宽度、最小尖峰间隔时间两个参数，图 15.3 给出了 X-26＃海杂波数据的时间相关函数 $R_t(k)$ 曲线。时间相关函数 $R_t(k)$ 采用如下公式计算[39]

$$R_t(k)=\frac{\sum_{i=0}^{N-1} z_i z_{i+k}^*}{\sum_{i=0}^{N-1} |z_i|^2} \tag{15.1}$$

其中，z_i 代表海杂波序列的第 i 个采样点；“$*$”代表取共轭；N 是海杂波序列的总长度。由图 15.3 可以看到，单个距离单元的海杂波时间序列的强相关时间不超过 0.05s，弱相关时间持续时间较长，可达到几秒钟的时间。因此，根据图 15.3 所示结果，当两个采样点之间的间隔时间超过 0.1s 时，可以近似认为二者是不相关的，在雷达海杂波的实际物理应用背景下，当两个海尖峰之间的间隔超过 0.1s 时，可以近似认为二者之间是相互独立的。

根据前文的分析结果并结合文献[35]的相关结论，下面给出海尖峰三个判定参数的经验值，然后根据这三个判定参数从海杂波序列中提取海尖峰。尖峰幅度门限 T_s 可取为海杂波平均功率的 M_A 倍的平方根，公式表示如下：

$$T_s=\sqrt{\frac{M_A}{N_{sc}}\sum_{i=1}^{N_{sc}} |z_i|^2} \tag{15.2}$$

其中，N_{sc} 表示海杂波采样点的总数目；z_i 代表海杂波序列的第 i 个采样点；M_A 的经验取值区间为[3,6]，本节在计算过程中取 $M_A=5$。最小尖峰宽度设为 0.1s，最小尖峰间隔设为 0.5s。实际上，这三个参数并不是固定不变的，其数值往往随着观测条件以及海情的变化而有所变化。

图 15.4 和图 15.5 分别给出了 X-26＃数据多个海杂波单元和单个海杂波单元中判定出的海尖峰情况。图 15.4 中非海尖峰采样点的幅度被设为 0，海尖峰采样点的幅度保持原值不变，因此

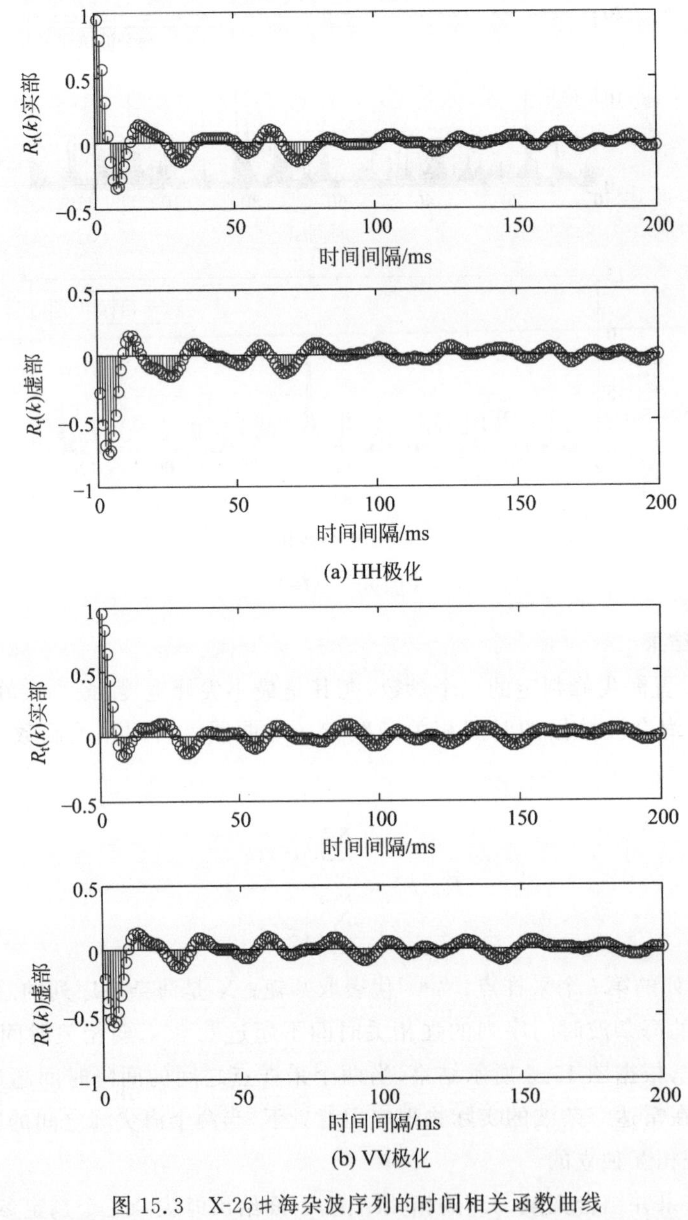

图 15.3　X-26#海杂波序列的时间相关函数曲线

图 15.4 中亮度较高的区域表示海尖峰,其他的暗色背景区域表示非海尖峰的海杂波;图 15.5 中则用深蓝色标记海尖峰(灰度图形中则显示为黑色),非海尖峰的海杂波则用浅灰色进行标记。由图 15.4 和图 15.5 可以看到,海尖峰在各个海杂波单元中均存在,且在某些区域强度较大,与目标回波类似。图 15.4 中 HH 极化和 VV 极化条件下海尖峰的数目分别为 205 和 193,而在图 15.5 所示的单距离单元条件下,HH 极化和 VV 极化条件下海尖峰的数目分别为 21 和 17。可见,HH 极化条件下海尖峰的数目要多于 VV 极化条件下的海尖峰数目,这说明 HH 极化条件下海杂波起伏程度相对较剧烈;此外,对比图 15.5(a)和(b)还可以发现,HH 极化条件下的海尖峰幅度从总体上要高于 VV 极化条件下的海尖峰幅度。

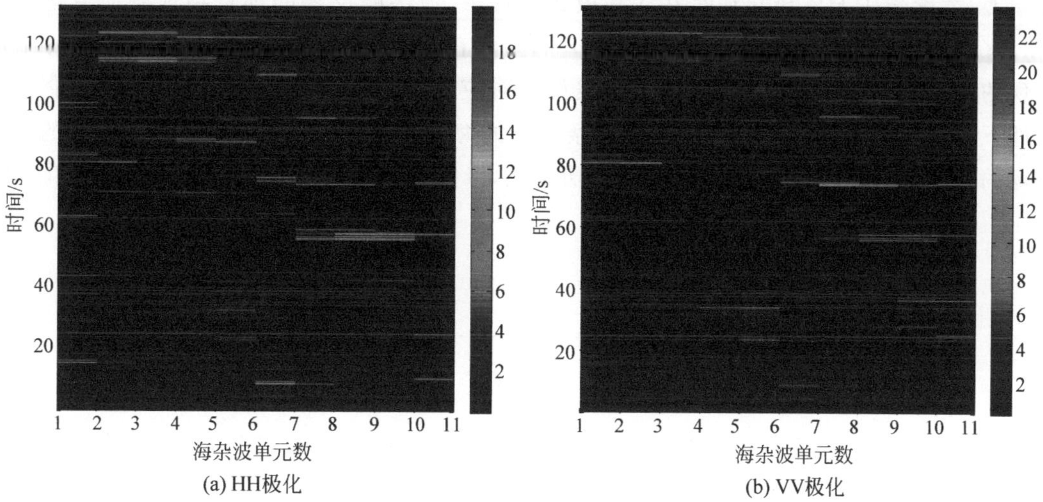

图 15.4　X-26＃海杂波背景中的海尖峰情况(多距离单元)

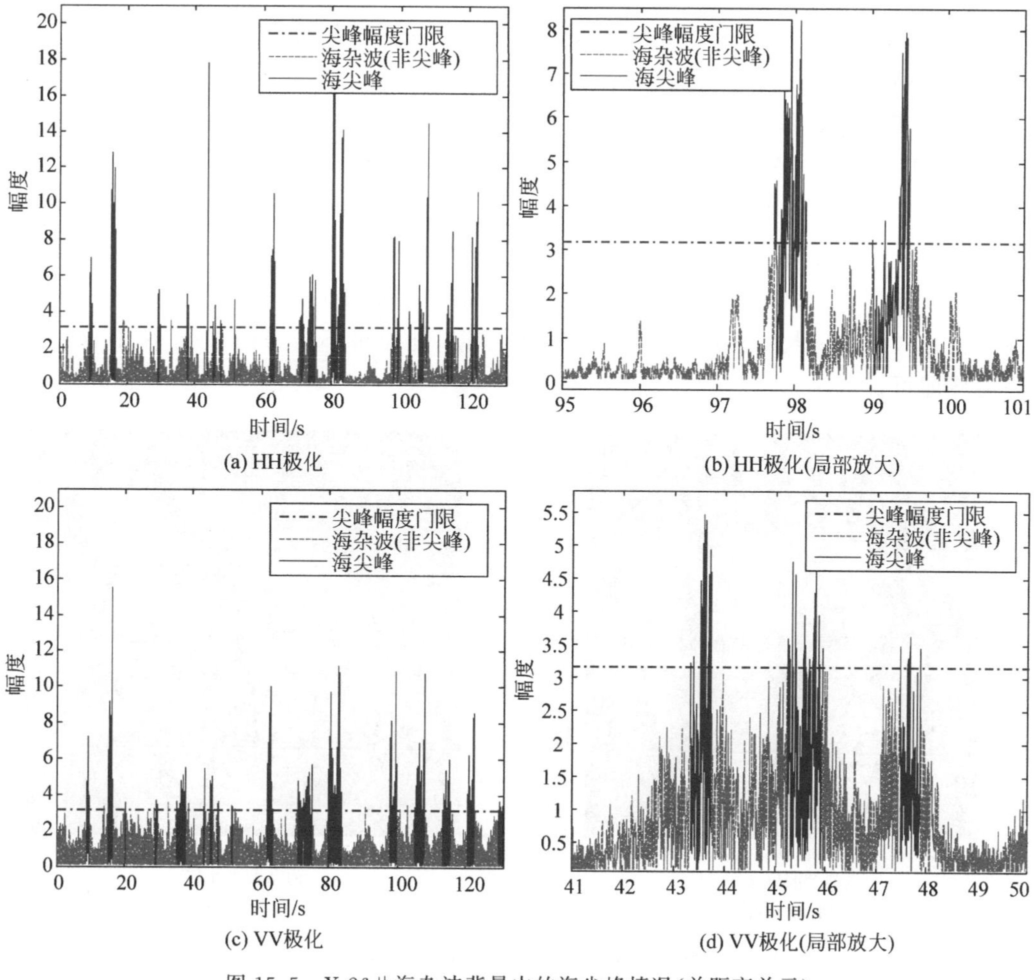

图 15.5　X-26＃海杂波背景中的海尖峰情况(单距离单元)

实际上,无论是在 HH 极化还是在 VV 极化条件下,海尖峰反映的都是物理海面急速起伏的碎浪,既然海尖峰反映的是同一个物理现象,那么海尖峰在 HH 极化条件下和 VV 极化条件下应该是对应出现的。此外,在 HH 极化和 VV 极化条件下对应出现的海尖峰中,根据前面的结论可知,HH 极化条件下的海尖峰幅度一般高于 VV 极化条件下的海尖峰幅度,因此这里引入极化率的概念[35],即海尖峰的 HH 极化成分与 VV 极化成分的功率比,以符号 Λ 表示,则极化率 $\Lambda \geqslant 1$。极化率计算公式如下:

$$\Lambda = \frac{\sum_{i=1}^{M_s} |z_i^{HH}|^2}{\sum_{i=1}^{M_s} |z_i^{VV}|^2} \tag{15.3}$$

其中,z_i^{HH} 代表 HH 极化条件下一个海尖峰的第 i 个采样点;z_i^{VV} 代表 VV 极化条件下同一个海尖峰的第 i 个采样点;M_s 表示这一个海尖峰所包含的采样点的数目。这里需说明的是,根据海尖峰的三个判定参数在 HH 极化和 VV 极化条件下分别判断得到的海尖峰不一定能一一对应起来,对于在一种极化条件下存在而在另一种极化条件下找不到对应项的海尖峰不必考虑极化率的问题,直接从判定的海尖峰中剔除。对于在 HH 极化和 VV 极化条件下对应存在的海尖峰,M_s 则表示二者共同拥有的那一部分采样点的数目。

在图 15.4 和图 15.5 基础上,图 15.6 和图 15.7 分别给出了考虑极化率后判断出的海尖峰情况,即双极化(HH & VV)条件下的海尖峰情况。在此条件下,图 15.6 中的海尖峰数目为 127,HH 极化与 VV 极化条件下海尖峰数目相同,图 15.7 中所示的单个距离单元中的海尖峰数目在 HH 极化和 VV 条件下均为 14。可见,在考虑极化率的条件下,一些仅在某一种极化条件下出现的海尖峰被剔除掉了,海尖峰数目均有所降低。这里需说明的是,在实际雷达工作中,一般仅工作在某一种极化方式下,此时无法获得对应的双极化数据,此时可不必考虑极化率,根据三个判定条件提取海尖峰即可。

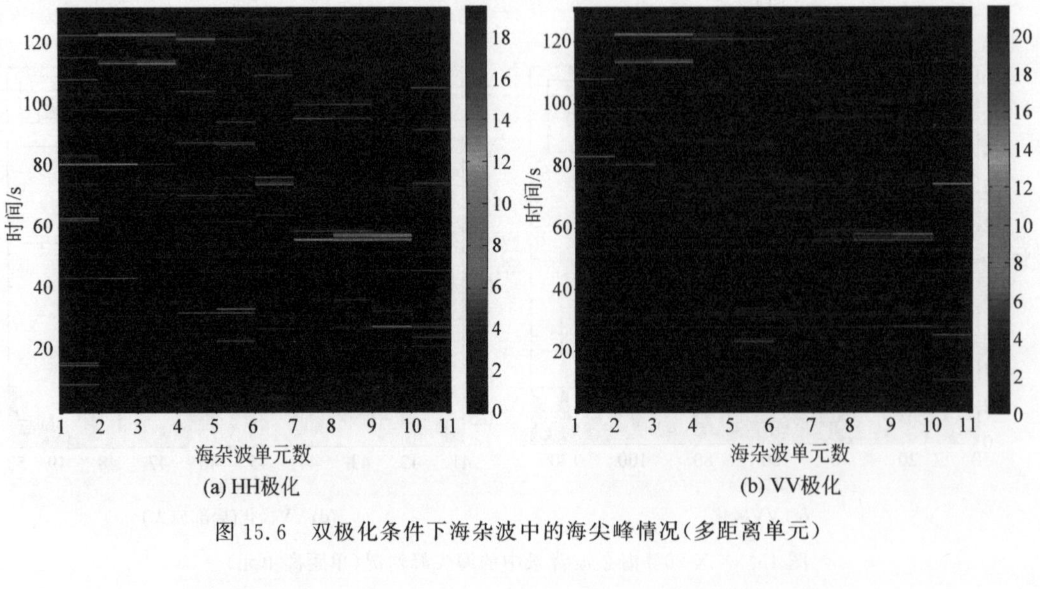

图 15.6 双极化条件下海杂波中的海尖峰情况(多距离单元)

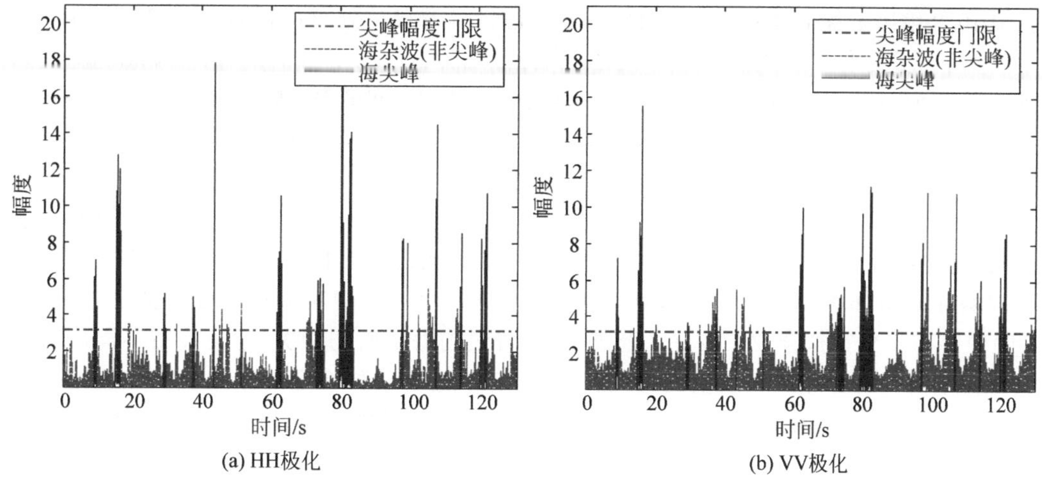

图 15.7 双极化条件下海杂波中的海尖峰情况(单距离单元)

15.2.2 海尖峰描述参数及统计特性

本节主要介绍两个常用的海尖峰统计特征描述参数,即平均尖峰持续时间和平均尖峰间隔[24,35]。图 15.8 给出了参数 M_A 取不同值条件下海尖峰的平均持续时间和平均尖峰间隔曲线,可见,当参数 M_A 的取值在区间[3,6]时,海尖峰的平均持续时间相对较平稳,而海尖峰的平均尖峰间隔则随着参数 M_A 的增大而呈现近似线性增加的趋势。此外,由图 15.8 还可发现,HH 极化条件下的平均尖峰持续时间要短于 VV 极化条件下的平均尖峰持续时间,但 VV 极化下的平均尖峰持续时间随 M_A 的增大下降较快;HH 极化条件下的平均尖峰间隔整体上要稍大于 VV 极化条件下的平均尖峰间隔,二者随 M_A 增大的增长率基本相同。之所以出现这种现象,究其原因,可能是由 HH 极化条件下海杂波的起伏剧烈程度要高于 VV 极化条件下引起的,海杂波起伏越"陡峭",越容易出现持续时间短的海尖峰。

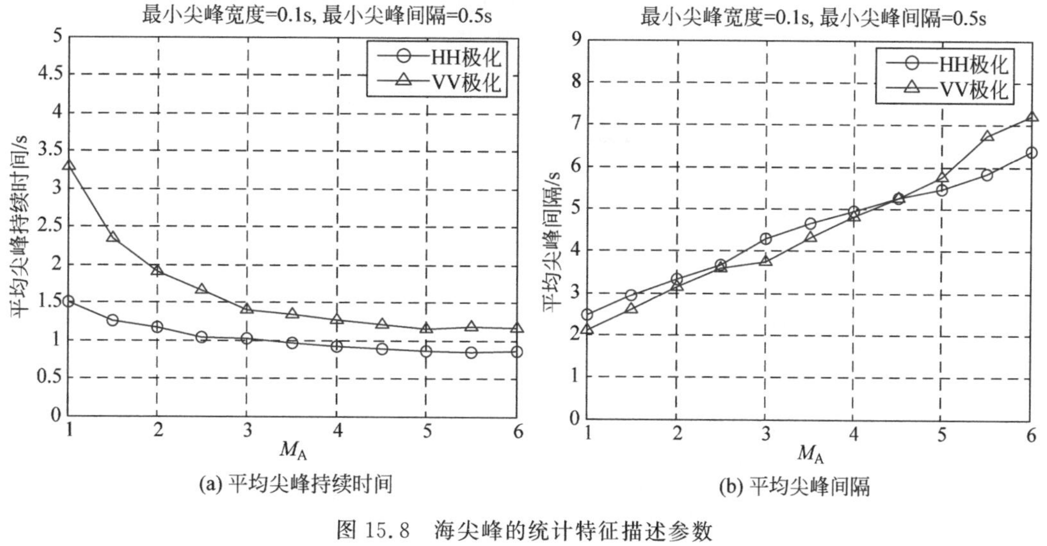

图 15.8 海尖峰的统计特征描述参数

根据图 15.8 分析结果,下面的分析过程主要针对统计特征相对较稳定的尖峰持续时间进行,实际上,对尖峰间隔时间进行分析采用的分析方法相同,且也可以得到类似的结果,因此,下面对尖峰间隔时间不再详述。图 15.9 给出了各个海尖峰持续时间的直方图统计及统计分布拟合的结果，其中，在对数正态分布和韦布尔分布拟合过程中参数估计方法采用最大似然估计方法。由图 15.9(a)～图 15.9(c)可以发现，大部分海尖峰的持续时间集中在 0～2s 的范围内，超过 2s 持续时间的海尖峰数目较少，但正是这一部分海尖峰导致了海尖峰分布的"拖尾"现象发生。为定量化说明对数正态分布和韦布尔分布对海尖峰持续时间概率密度函数的拟合效果，表 15.1 给出了图 15.9 中所示分布拟合的χ^2 检验结果，其中，置信度水平 $\alpha=0.05$。由表 15.1 可知，无论是韦布尔分布还是对数正态分布拟合，检验统计量χ^2 的值均落于拒绝域的范围内，这说明在置信度水平 $\alpha=0.05$ 条件下，不能认为海尖峰持续时间的概率密度函数是韦布尔分布或者对数正态分布的，相对而言，在图 15.9(a)和图 15.9(c)中对数正态分布拟合效果较好，而在图 15.9(b)中韦布尔分布拟合效果较好。此外，海尖峰还与海情密切相关，这意味着不同的海情也可能引起海尖峰持续时间的分布类型发生变化，这些因素对于基于某种特定分布的检测方法而言是至关重要的，一旦分布类型发生变化，与检测方法预先设定不符，检测性能可能急剧下降或失去 CFAR 能力。

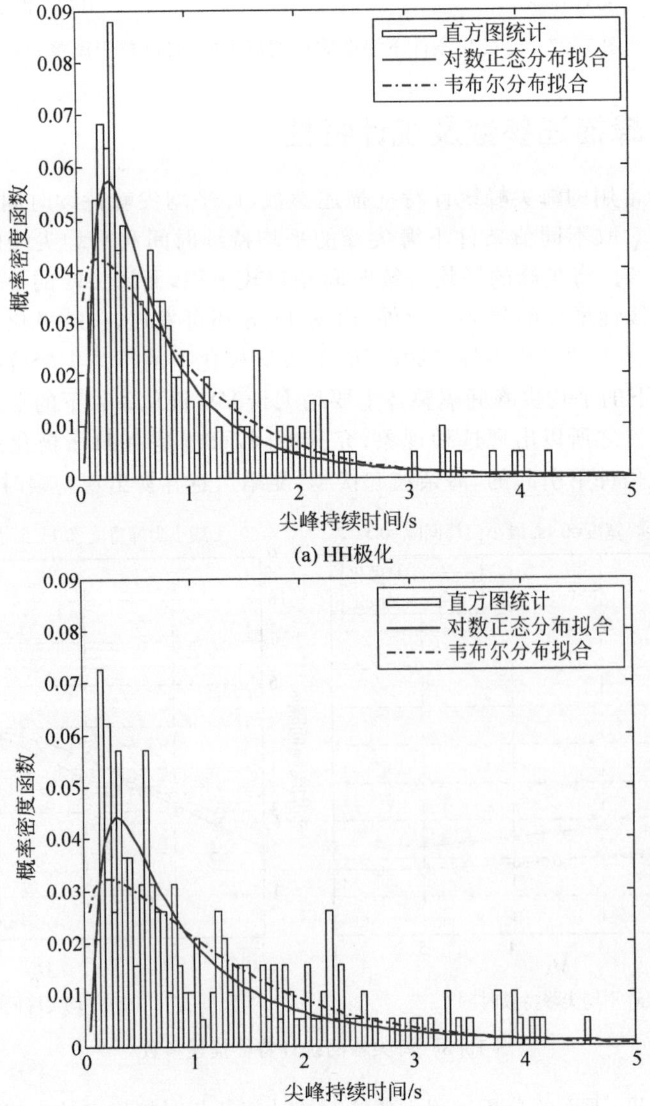

图 15.9　海尖峰持续时间直方图统计及分布拟合

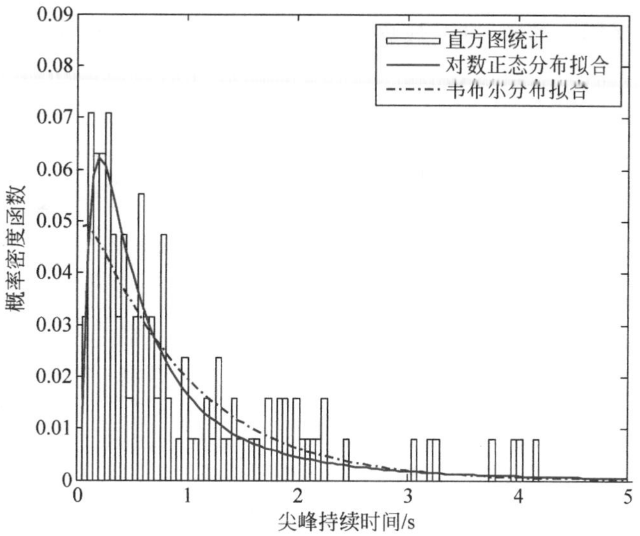

(c) 双极化(HH & VV)

图 15.9 (续)

表 15.1 图 15.9 所示分布拟合的χ^2 检验结果(置信度水平 $\alpha=0.05$)

	HH 极化	VV 极化	双极化
检验统计量 χ^2(对数正态分布拟合)	18.3675	18.1256	19.2453
检验统计量 χ^2(韦布尔分布拟合)	23.6222	15.0773	20.4779
拒绝域	$\chi^2>\chi^2_{0.95}(7)$,即$\chi^2>14.067$		

15.2.3 海尖峰的 Paretian 泊松模型

对于统计分布拟合效果较差且难以准确判断分布类型的问题,本节采用 Paretian 泊松过程建模海尖峰的持续时间。首先,简单回顾从海尖峰背景中提取海尖峰的过程:①海尖峰的幅度必需超过设定的尖峰幅度门限;②每个尖峰的持续时间必须超过 0.1s;③相邻两个海尖峰之间的间隔必须超过 0.5s。此外,根据海尖峰判定部分中得到的结论,即两个相邻的海尖峰可以近似认为是独立的,则所有的海尖峰可以形成一个内部各要素间相互独立的序列,且各个要素可以采用相同的分布类型进行建模,比如可以认为海尖峰的持续时间是定义于实数轴正半轴上泊松分布。在以上前提条件下,即可采用 Paretian 泊松过程建模海尖峰序列,每个海尖峰采用持续时间 t 作为其特征量,而不是传统建模中经常采用的幅度特征,此时,可以得到独立同分布(IID)序列 $\omega=\{t_1,t_2,\cdots,t_N\}$。对于固定的采样率而言,海尖峰的持续时间与每个海尖峰所包含的采样点数是一一对应的,因此,以每个海尖峰包含的采样点数为特征同样可以构成一个 IID 序列,在不至于混淆的情况下,仍表示为 $\omega=\{t_1,t_2,\cdots,t_N\}$,且序列 ω 中的随机变量均服从参数为 $\lambda(t_r)$的泊松分布。根据 Paretian 泊松过程,$\lambda(t_r)$服从幂律关系,转化为对数方式显示如下

$$\log_2\lambda(t_r)=-(1+\alpha)\log_2 t_r+\log_2(c\alpha) \quad (t>0) \tag{15.4}$$

即泊松参数 $\lambda(t_r)$与海尖峰包含采样点数 t_r 在双对数坐标下满足线性关系,且 $\lambda(t_r)=\Pr\{t>t_r\}$,实际上,$t_r$ 就是通常所说的尺度。在实际系统中,式(15.4)所示的线性关系通常不能在整个尺度区间内成立,而是仅在某一个子区间内成立,这样的尺度区间定义为代数尺度不变区间(Algebraic Scale Invariant Interval,ASII)。

图 15.10 给出了双对数坐标下泊松参数 $\lambda(t_r)$与尺度 t_r 之间的关系曲线,同时给出了最小二乘(LS)直线拟合结果。由图 15.10 可以看到,泊松参数 $\lambda(t_r)$与尺度 t_r 之间的线性均在某一个尺度区间

内成立，即代数尺度不变区间。因此，可以判断序列 ω 是代数分形的，同时也验证了采用 Paretian 泊松过程建模海尖峰持续时间序列是可行的。

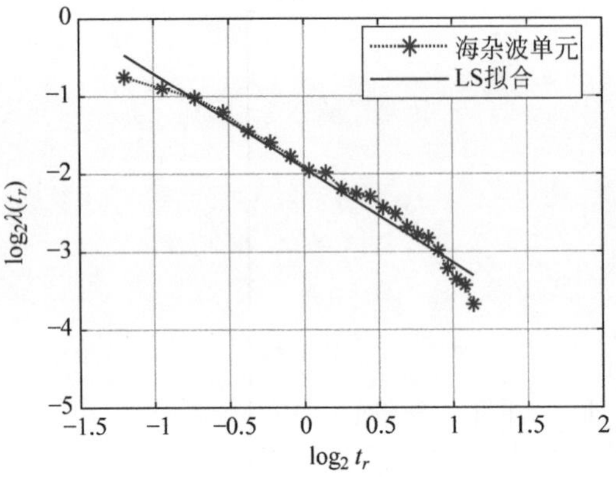

(a) HH极化

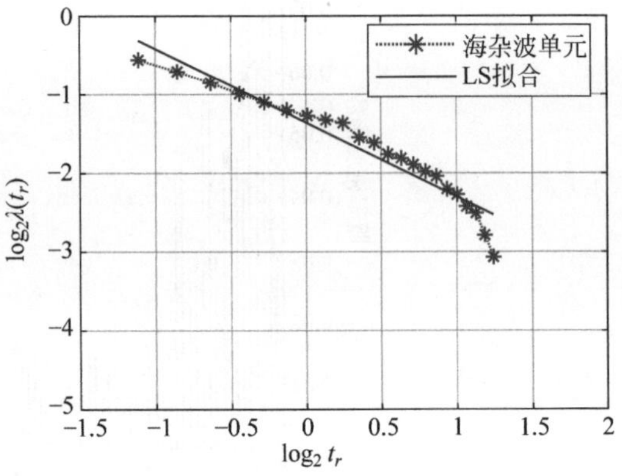

(b) VV极化

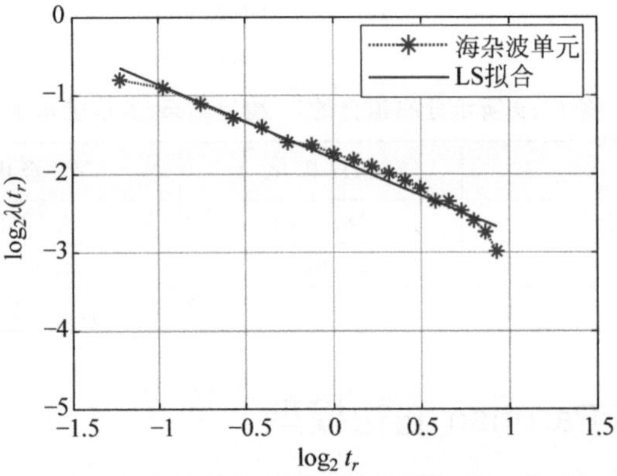

(c) 双极化(HH & VV)

图 15.10　双对数坐标下泊松参数 $\lambda(t_r)$ 曲线及 LS 直线拟合效果

15.2.4 目标检测及性能分析

1. 海杂波与目标代数分形特性差异

本节主要研究由海杂波单元与目标单元提取的尖峰的代数分形特性差异，并寻求合适的代数分形参数设计目标检测方法。图 15.11 分别给出了不同极化条件下海杂波单元和目标单元尖峰的代数分形特性曲线，可见，由目标单元得到的代数分形特性曲线在某个尺度区间内仍然存在近似线性特性，通过直观观察可以发现，海杂波单元与目标单元的代数分形特性曲线有轻微的区别，为量化二者之间的区别，这里对图 15.11 中所有的直线拟合结果进行一元线性回归分析，并计算 Paretian 指数 α（与拟合直线斜率相对应）、Paretian 幅度 c（与拟合直线截距相对应）、残差平方和 Q 及相应的代数尺度不变区间 ASII，分析结果如表 15.2 所示。由表 15.2 观察可知，在所有参数中 Paretian 指数 α 和残差平方和 Q 在海杂波中存在目标时变大，且这种差异在各个极化条件下均存在。海杂波单元与目标单元之间 Paretian 指数的差异表明，目标的存在将会改变海杂波的起伏结构，换言之，当海面存在目标时，散射回波中依然包含尖峰，但是这些尖峰的代数分形特性与纯海杂波单元中海尖峰的代数分形特性是不同的；海杂波单元与目标单元之间残差平方和的差异表明，目标的存在使得对数泊松参数 $\log_2\lambda(t_r)$ 偏离拟合直线，也即目标的存在会使得海尖峰与 Paretian 泊松模型失配。

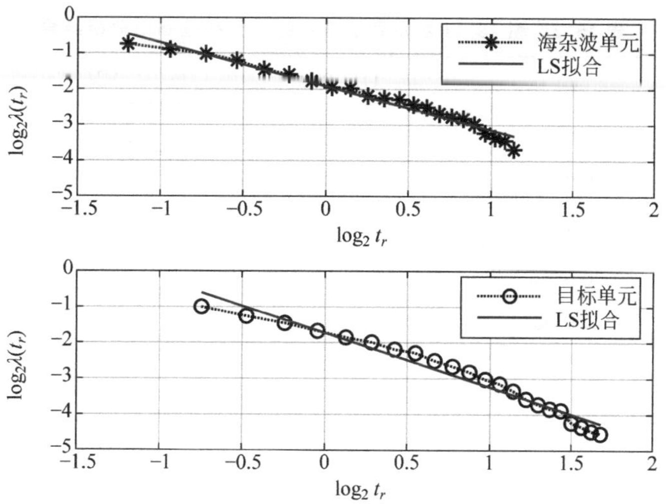

(a) HH极化

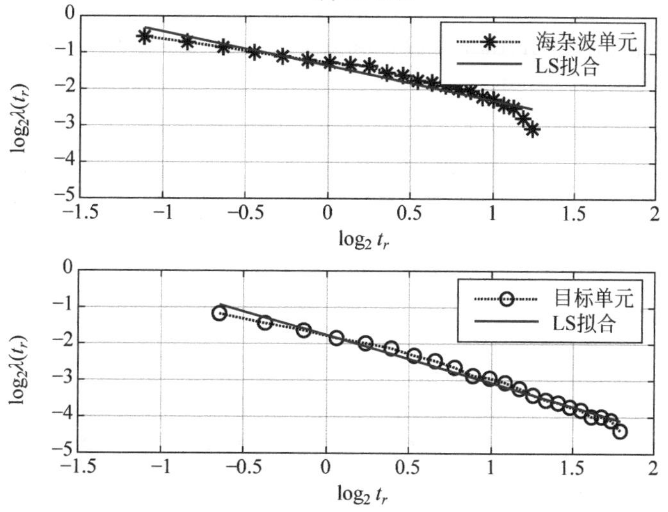

(b) VV极化

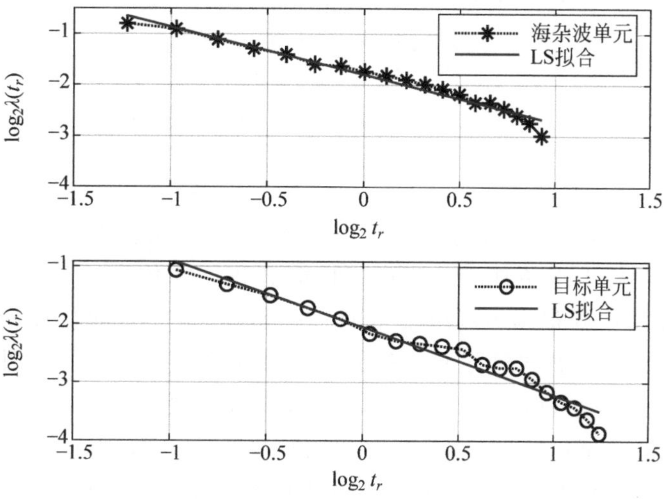

(c) 双极化(HH & VV)

图 15.11 海杂波单元与目标单元尖峰的代数分形特性曲线

表 15.2 图 15.11 所示直线拟合的一元线性回归分析结果

		HH 极化	VV 极化	双极化(HH & VV)
Paretian 指数 α	海杂波单元	0.21	−0.22	−0.06
	目标单元	0.30	−0.19	0.11
Paretian 幅度 c	海杂波单元	1.22	−1.70	−4.80
	目标单元	0.87	−1.9	2.61
残差平方和 Q	海杂波单元	210.95	136.32	215.56
	目标单元	236.78	159.53	234.24
ASII	海杂波单元	[0.43,2.20]	[0.28,2.27]	[0.42,1.90]
	目标单元	[0.41,2.08]	[0.27,2.25]	[0.45,2.02]

2. 目标检测方法

基于上述分析,下面将采用 Paretian 指数 α 和残差平方和 Q 进行目标检测方法设计,并在实测数据基础上进行目标检测性能分析。图 15.12 给出了基于 Paretian 泊松模型的海杂波中目标检测方法流程图,其具体工作流程如下:当雷达接收到海面回波信号时,首先根据海尖峰判定方法从回波信号中提取尖峰,然后采用 Paretian 泊松过程建模尖峰序列,在此基础上便可采用一元线性回归分析方法估计 Paretian 指数 α 和残差平方和 Q,最后将估计得到的参数与给定虚警概率 P_{fa} 条件下的门限进行比较,若其超过门限则判断为目标单元,否则判断为海杂波单元。这里需说明的是,Paretian 指数和残差平方和可以分别用于目标检测,将二者同时置于图 15.12 中是为便于画图及说明,检测过程中二者可不必同时参与。

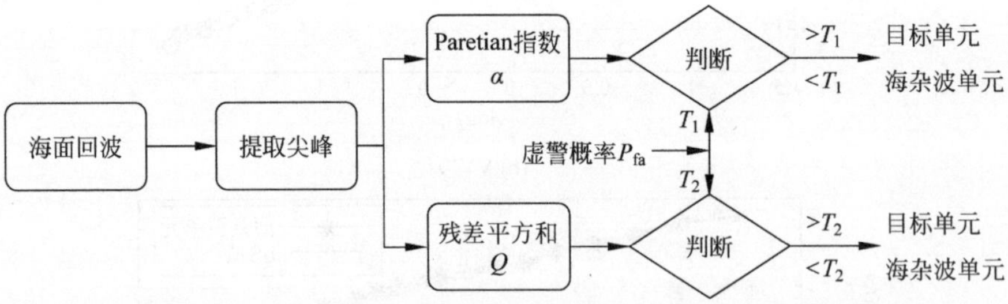

图 15.12 基于 Paretian 泊松模型的海杂波中目标检测方法流程图

3. 检测性能分析

下面分析图 15.12 所示目标检测方法的检测性能,由于无法获得一系列指定 SCR 条件下的实测数据,因此这里采用蒙特卡洛仿真的方法进行性能分析。在仿真过程中,海杂波直接采用 X 波段实测雷达数据,目标回波则根据 Swerling Ⅱ 目标模型仿真产生,即目标回波幅度服从瑞利分布且在脉冲间起伏,然后将海杂波与目标回波叠加构成目标单元回波信号。图 15.13 分别给出了利用 Paretian 指数 α 和残差平方和 Q 的检测概率曲线,并且为便于对比,图 15.13 还给出了双参数 CFAR 检测器[40]的检测概率曲线,其中非相参积累脉冲数 N_{nci} 分别为 1024、2048 和 4096。双参数 CFAR 检测器根据独立同分布的噪声或杂波在非相参积累后近似服从高斯分布的性质,用不同参考单元估计积累后噪声或杂波的均值和标准差,然后将待检测单元减去均值,再与标准差估计值和阈值因子的乘积相比较，从而实现CFAR检测。由于IPIX雷达数据每个距离单元的时间序列长度为2^{17}，因此在计算检测概率过程中，时间序列被分成互不交叠的16段，每段包含8192个采样点，最终的检测概率是由这16段数据分别计算得到的检测概率平均后的结果，并且图15.13中所有检测概率曲线对应的虚警概率均为$P_{fa}=10^{-3}$。另外，这里需说明的是，在图15.13(c)中所示的双极化条件下，由于双参数CFAR恒虚警检测器不能同时运用HH极化和VV极化数据进行检测，因此图15.13(c)中给出的双参数CFAR检测器性能曲线是在仅采用VV极化海杂波数据条件下得到的。

由图15.13(a)和图15.13(b)可以看到，当非相参积累脉冲个数N_{nci}小于2048时，图15.12所示目标检测方法的检测性能明显优于双参数CFAR检测器，即令$N_{nci}=2048$，当SCR<−1dB时，图15.12所示目标检测方法的检测性能也优于双参数CFAR检测器。对于图15.13(c)所示的双极化情况，由于

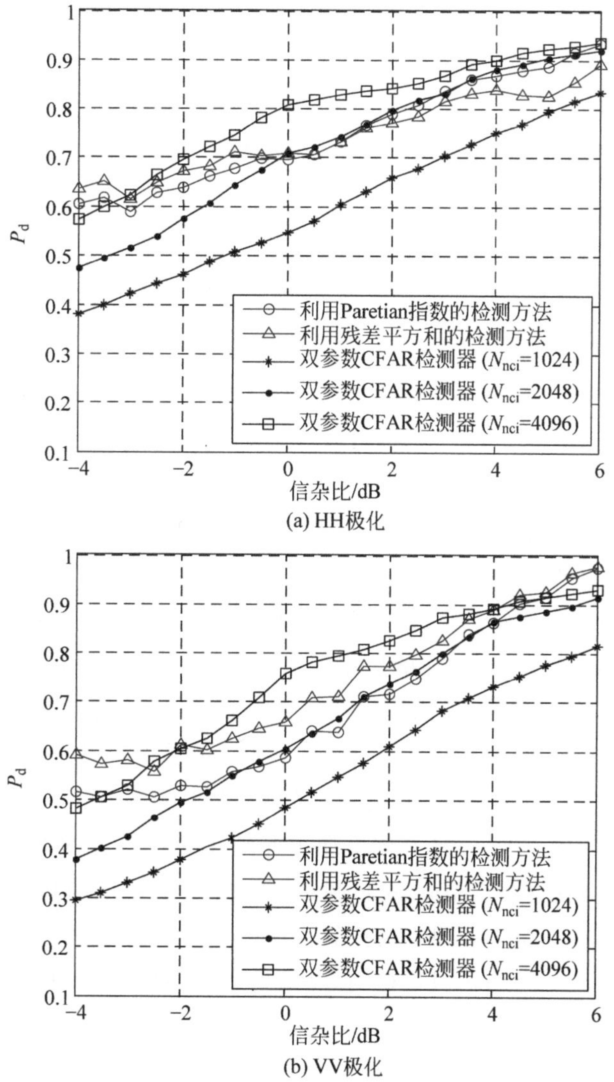

图15.13 不同SCR条件下图15.12所示检测方法的性能曲线($P_{fa}=10^{-3}$)

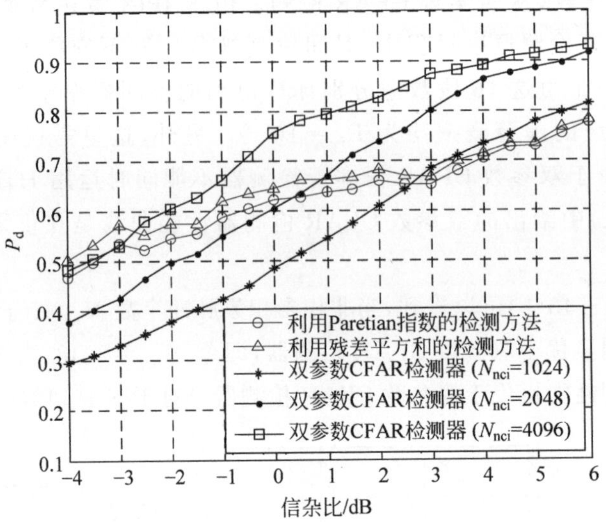

(c) 双极化(HH&VV)

图 15.13 （续）

考虑了极化率因素，提取到的海尖峰数目减少，这一方面可能使得海杂波单元与目标单元的区分度降低，另一方面使得进行直线拟合可用数据也减少，从而拟合误差变大，最终使得双极化条件下的检测概率低于单一极化条件下的检测概率，但在 SCR＜－1dB 时，图 15.12 所示目标检测方法的检测性能仍优于 N_{nci}＝2048 时双参数 CFAR 检测器的检测性能。另外，分别比较图 15.13(a)～图 15.13(c)中利用 Paretian 指数和利用残差平方和的目标检测方法可以发现，当 SCR＜0dB 时，利用残差平方和的目标检测方法的检测性能稍优于利用 Paretian 指数的目标检测方法，且这一优越性在 VV 极化条件下表现相对较明显；在 SCR＝0dB 左右，图 15.12 所示目标检测方法的最佳检测性能在 HH 极化和 VV 极化条件下分别可达到 70％和 65％。

进一步观察图 15.13 中利用 Paretian 指数和利用残差平方和的目标检测方法的检测概率曲线可以发现，检测概率曲线的总体趋势是随着 SCR 增加，检测概率也逐步上升，然而，比较两个相邻近 SCR 条件下的检测概率可发现，较大的 SCR 对应的检测概率不一定大。这是因为，SCR 可以影响尖峰提取的第一步，即确定尖峰幅度门限，当两个 SCR 数值之间差异较大时，与其相对应的检测概率则主要受 SCR 的影响，从而大的 SCR 对应大的检测概率，即检测概率曲线的总体趋势是上升的；然而，当两个 SCR 数值之间差异较小(基本处于同一水平)时，起决定作用的因素不再是 SCR，而是尖峰的代数分形结构，在这种情况下，相对较小的 SCR 对应的检测概率便可能超过相对较大的 SCR 对应的检测概率，可见，图 15.12 所示的目标检测方法并不严格受 SCR 制约，其在一定程度上可以摆脱 SCR 的束缚。此外，由于尖峰是海面回波中的高幅度部分，仅研究尖峰的特性可以在一定程度上降低噪声和低幅度海杂波的影响。最后需指出的是，图 15.12 所示目标检测方法需要大量的采样点来提取尖峰并估计参数以降低统计误差，因此其实时性较差，综合考虑该检测方法的处理过程，其较适合于检测在高分辨率雷达中经常遇到的扩展目标情况。

15.3　海杂波频域分形特征与CFAR检测

15.3.1　分数布朗运动在频域中的分形特性

分数布朗运动(FBM)是由经典的布朗运动推广而来的，因此本节首先给出布朗运动的定义[41]。布朗运动 $B(t)$是满足下列条件的随机过程：

(1) $B(0)=0$(过程从原点开始)，且 $B(t)$为 t 的连续函数；

(2) 对任意的 $t\geqslant 0$ 和 $h>0$，增量 $B(t+h)-B(t)$服从均值为 0，方差为 h 的正态分布；

(3) 若 $0\leqslant t_1\leqslant t_2\leqslant\cdots\leqslant t_n$，则增量 $B(t_2)-B(t_1),B(t_3)-B(t_2),\cdots,B(t_n)-B(t_{n-1})$相互独立。

由(1)和(2)可得，对于每个 t，$B(t)$自身也服从均值为 0，方差为 t 的正态分布，且 $B(t)$的增量是平稳的，因此 $B(t)$的概率密度函数可表示为

$$P_{B(t)}(x)=\frac{1}{\sqrt{2\pi t}}\exp\left(-\frac{x^2}{2t}\right) \tag{15.5}$$

将式(15.5)作如下尺度变换

$$t\rightarrow\kappa t,\quad x\rightarrow\kappa^{1/2}x \tag{15.6}$$

则有如下标度规律

$$\begin{aligned}P_{B(\kappa t)}(\kappa^{1/2}x)&=\frac{1}{\sqrt{2\pi\kappa t}}\exp\left[-\frac{(\kappa^{1/2}x)^2}{2\kappa t}\right]\\&=\frac{1}{\sqrt{\kappa}}\cdot\frac{1}{\sqrt{2\pi t}}\exp\left\{-\frac{x^2}{2t}\right\}\\&=\kappa^{-1/2}P_{B(t)}(x)\end{aligned} \tag{15.7}$$

且在式(15.6)所示的标度变换下，总分布概率保持不变，即

$$\int_{-\infty}^{\infty}P_{B(t)}(x)\mathrm{d}x=\int_{-\infty}^{\infty}P_{B(\kappa t)}(\kappa^{1/2}x)\mathrm{d}(\kappa^{1/2}x)=1 \tag{15.8}$$

式(15.8)表明，$B(t)$与 $\kappa^{-1/2}B(\kappa t)$具有相同的概率分布，即在统计意义下二者是自相似的，可采用如下形式表达

$$B(t)\overset{\text{s.t.a}}{=}\kappa^{-1/2}B(\kappa t) \tag{15.9}$$

其中，$\overset{\text{s.t.a}}{=}$表示在统计意义下相等。

分数布朗运动 $B_H(t)$首先由 Mandelbrot 从布朗运动推广而来[42-43]，所使用的公式如下所示

$$B_H(t)-B_H(0)=\frac{1}{\Gamma(H+1/2)}\int_{-\infty}^{t}\varphi(t-s)\mathrm{d}B(s) \tag{15.10}$$

其中

$$\varphi(t-s)=\begin{cases}(t-s)^{H-1/2} & 0\leqslant s\leqslant t\\(t-s)^{H-1/2}-(-s)^{H-1/2} & -\infty\leqslant s<0\end{cases} \tag{15.11}$$

$\Gamma(\cdot)$为伽马函数，由第二类欧拉积分确定，即 $\Gamma(z)=\int_0^{\infty}\mathrm{e}^{-t}t^{z-1}\mathrm{d}t\,(z>0)$，其中，e 为自然底数，$B(t)$为一维布朗运动曲线，$H$ 为单一 Hurst 指数。

FBM 的定义[41]与布朗运动定义的前两个条件相似,即

(1) $B_H(0)=0$,且 $B_H(t)$为 t 的连续函数;

(2) 对任意的 $t\geqslant 0$ 和 $h>0$,增量 $B_H(t+h)-B_H(t)$服从均值为 0,方差为 h^{2H} 的正态分布。

此定义也蕴含增量 $B_H(t+h)-B_H(t)$是平稳的,但与布朗运动不同的是,FBM($H\neq 1/2$ 时)不具有独立的增量。当 $H=1/2$ 时,式(15.10)所示的 FBM 即简化为布朗运动,此时有

$$\varphi(t-s)=\begin{cases}1, & 0\leqslant s\leqslant t\\ 0, & -\infty\leqslant s<0\end{cases} \tag{15.12}$$

使

$$B_H(t)-B_H(0)=\frac{1}{\Gamma(1)}\left[\int_{-\infty}^{0}0\cdot \mathrm{d}B(s)+\int_{0}^{t}1\cdot \mathrm{d}B(s)\right]=\int_{0}^{t}\mathrm{d}B(s) \tag{15.13}$$

可见,布朗运动实际上是 FBM 的一个特殊形式。对于 FBM 曲线,若时间尺度从 t 变到 κt,则在布朗运动是统计自相似的前提下,式(15.10)变为

$$\begin{aligned}B_H(\kappa t)-B_H(0)&=\frac{1}{\Gamma(H+1/2)}\int_{-\infty}^{\kappa t}\varphi(\kappa t-\kappa s)\mathrm{d}B(\kappa s)\overset{\text{s.t.a}}{=}\frac{1}{\Gamma(H+1/2)}\int_{-\infty}^{t}\kappa^{H-1/2}\varphi(t-s)\kappa^{1/2}\mathrm{d}B(s)\\&=\kappa^{H}\cdot\frac{1}{\Gamma(H+1/2)}\int_{-\infty}^{t}\varphi(t-s)\mathrm{d}B(s)\\&=\kappa^{H}[B_H(t)-B_H(0)]\end{aligned} \tag{15.14}$$

为不失一般性,可假设 $B_H(0)=0$,则 $B_H(t)\overset{\text{s.t.a}}{=}\kappa^{-H}B_H(\kappa t)$,即 FBM 也是统计自相似的。在 $H=1/2$ 时,有 $B_H(t)\overset{\text{s.t.a}}{=}\kappa^{-1/2}B_H(\kappa t)$,与布朗运动相符。

为进一步研究 FBM 在频域的自相似性,首先需对 FBM 的位移曲线进行傅里叶变换,把原来在时域内以时间 t 为变量的函数 $B_H(t)$变换为频域内以频率 f 为变量的函数 $F_B(f)$,即

$$F_B(f)=\int_{0}^{\infty}B_H(t)\mathrm{e}^{-\mathrm{j}2\pi f t}\mathrm{d}t \tag{15.15}$$

其中,$B_H(t)$定义在区间$(0,T)$内。由于频谱振幅的平方正比于功率,单位时间内的功率谱密度 $S_B(f)$可定义如下[44]

$$S_B(f)=\frac{|F_B(f)|^2}{T} \tag{15.16}$$

在此基础上,可以研究尺度变换条件下频谱与功率谱的自相似特性,采用尺度变换 $t'=\kappa t$,则由式(15.14)可得 $B_H(t')\overset{\text{s.t.a}}{=}\kappa^{H}B_H(t)$,代入式(15.15)中可得

$$F_B(f)\overset{\text{s.t.a}}{=}\int_{0}^{\kappa T}\frac{B_H(t')}{\kappa^{H}}\mathrm{e}^{-\mathrm{j}2\pi f t'/\kappa}\mathrm{d}(t'/\kappa)=\frac{1}{\kappa^{H+1}}\int_{0}^{\kappa T}B_H(t')\mathrm{e}^{-\mathrm{j}2\pi\frac{f}{\kappa}t'}\mathrm{d}t'=\frac{1}{\kappa^{H+1}}F_B\left(\frac{f}{\kappa}\right) \tag{15.17}$$

将式(15.17)代入式(15.16)中,整理可得

$$S_B(f)\overset{\text{s.t.a}}{=}\frac{1}{\kappa^{2H+1}}\cdot\frac{\left|F_B\left(\frac{f}{\kappa}\right)\right|^2}{\kappa T}=\frac{1}{\kappa^{2H+1}}S_B\left(\frac{f}{\kappa}\right) \tag{15.18}$$

由式(15.17)和式(15.18)可知,频率标度变为原来的 $1/\kappa$ 后,频谱变为原来的 κ^{H+1} 倍,而功率谱密度变为原来的 κ^{2H+1} 倍[44]。这说明 FBM 的频谱、功率谱密度都是频率的幂函数,均具有自相似性。为与时域中的无标度区间区分,在频域中自相似性成立的区间称为频率无标度区间。

综上所述，FBM的频谱或功率谱密度函数是分形的，这为在频域中应用分形理论奠定了基础。以往的研究中通过功率谱进行分形分析很大程度上是为时域中的分形参数估计服务的，而下面的分析将立足于频域，采用分形理论直接对海杂波的频谱(或功率谱)序列进行分析与参数估计。

15.3.2　海杂波频谱的单一分形特性

本节将采用X波段与S波段实测海杂波数据进行分析与验证，其中X波段海杂波数据来自Osborn Head Database的26＃数据(X-26＃)，本节主要使用在HH极化与VV极化两种情况下的数据，其SCR为0～6dB，更具体的情况可参考文献[38]；另外一组海杂波数据(S-1＃)是某S波段雷达对海照射采集得到的，数据采集时天线工作在驻留模式且处于VV极化模式下，观察目标为一远离雷达缓慢运动的小渔船(渔船运动方向与雷达天线指向有约30°夹角，由于雷达天线不动，因此渔船在雷达视野内出现一段时间后消失)，SCR为0～3dB。图15.14给出了三组海杂波数据的时域归一化幅度波形图。

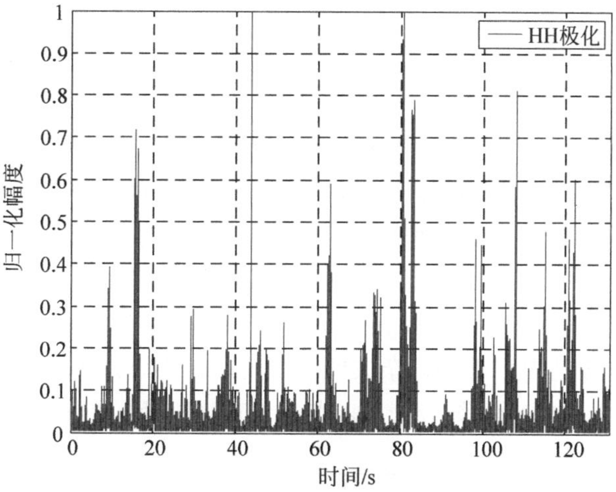

(a) X波段HH极化

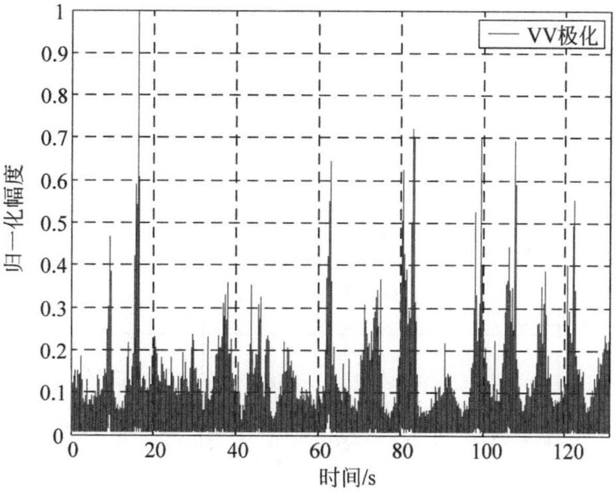

(b) X波段VV极化

图15.14　实测海杂波数据归一化时域波形图

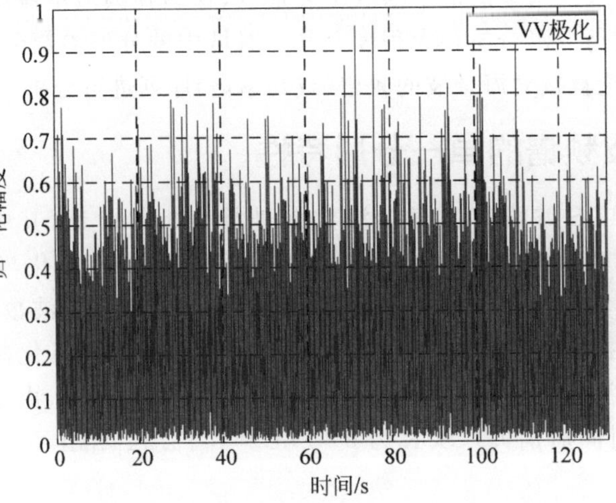

(c) S波段VV极化

图 15.14 （续）

根据时域海杂波幅度统计特性分析结论，再结合 FBM 的定义可知，采用 FBM 建模海杂波具有一定的合理性，且这一合理性已有不少文献进行了研究，本节对此不再赘述，而将在前人研究的基础上分析 X 波段与 S 波段海杂波频谱的分形特性。图 15.15 给出了三组数据杂波单元与目标单元的频谱，由图 15.15 可以看到，X 波段与 S 波段海杂波的多普勒谱均具有一定偏移，大约在 −100～100Hz 的范围内，这主要是因为海面是一个永不停息运动着的散射体，当海浪(Bragg 波)向雷达运动时，海杂波多普勒谱的中心频率偏向正值方向；当海浪(Bragg 波)远离雷达运动时，海杂波多普勒谱的中心频率则偏向负值方向。此外，由图 15.15 还可以看到海杂波的多普勒谱在中心频率附近具有一定的展宽，这主要是由大尺度波浪的振动效应造成的。由图 15.15(a)与图 15.5(b)所示的目标单元频谱可以发现，目标能

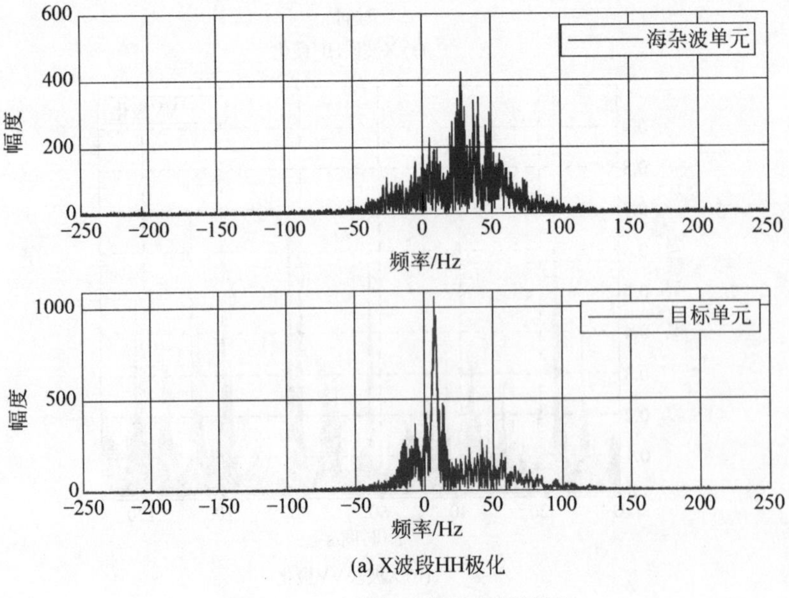

(a) X波段HH极化

图 15.15 实测雷达数据的频谱图

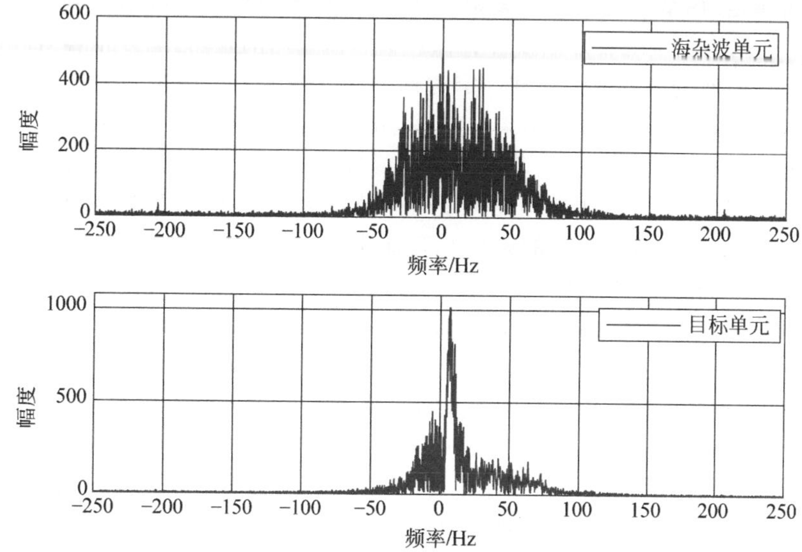

(b) X波段VV极化

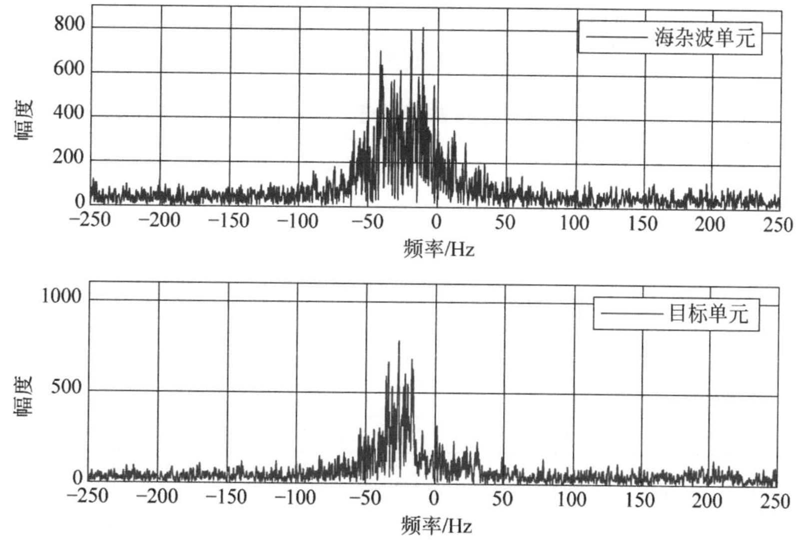

(c) S波段VV极化

图 15.15　(续)

量得到了较好的积累，且目标的多普勒频率靠近零频，但不绝对为零，这主要是因为目标随海面的运动而具有微弱的速度；而由图 15.15(c)所示的目标单元频谱可以发现 S 波段雷达观测目标的多普勒中心频率为负值，这与小渔船远离雷达的运动状态相符，但能量积累效果不如 X 波段条件下明显，原因如下：频谱计算是针对由某一距离单元所得到的回波时间序列进行的，图 15.15 中计算频谱时所采用的时间序列较长，若目标静止(如 X 波段数据)，则在某一距离单元目标回波一直存在，时间序列越长积累效果越好；若目标运动(如 S 波段数据)，则对某一距离单元而言，由其得到的长时间序列可能只在一段时间包含目标回波，而在其他时刻目标已不在此距离单元，从而回波只包含海杂波(此处不考虑热噪声等)，这种条件下，将回波序列变换到频域就可能引起目标能量积累效果不明显。另外，S 波段雷达所观察的目标较小，SCR 较低，这就进一步使目标的多普勒频率混叠于杂波频谱中，难以区分。

下面采用"随机游走"模型[6]建模海杂波频谱，则在双对数坐标下由频谱得到的配分函数 $F(m)$ 与频率尺度 m 在某一区间内应呈现较好的线性关系。这里需指出的是，自然界中客观存在的物体并不是严格分形的，分形体所特有的自相似性只在一定的标度区间内成立。图 15.16 给出了三组实测数据杂波单元与目标单元频谱的自相似性分析结果(这里采用的时间序列长度为 2^{16} 点，FFT 点数为 2^{16})。可以发现，海杂波单元的频谱在一定的尺度范围内近似线性均成立(存在无标度区间，X 波段雷达数据无标度区间范围为 $2^5\sim2^{10}$，S 波段雷达数据无标度区间范围为 $2^7\sim2^{10}$)，另外还可观察到海杂波单元与目标单元曲线在无标度区间内具有一定的差异。下面对这一差异进行量化分析，表 15.3 列出了三组实测数据的海杂波单元与目标单元曲线的一元线性回归分析结果，其中回归显著性检验的显著水平 α 设定为 0.001，r 检验过程中拒绝域临界点分别为 $r_{n,\alpha}=r_{96,0.001}=0.3270$(X 波段雷达数据)，$r_{n,\alpha}=r_{84,0.001}=0.3482$(S 波段雷达数据)。由表 15.3 可以看到，一元线性回归均具有极高的显著性水平，且相关系数 R 均非常接近于 1，这说明在无标度区间内采用直线对实测数据进行拟合具有很好的效果。另外，比较表 15.3 中所列出的几个参数可以发现，频域单一 Hurst 指数与残差平方和 Q 在出现目标时均会增大，且这一变化在三组数据中都比较稳定(但有无目标时参数的变化量有所差异)，这为后续在频域中进行目标分形检测算法研究奠定了基础。之所以会出现上述变化，是因为目标的出现使海杂波序列不规则程度有所降低，从而导致其在一定程度上偏离 FBM 模型，而 FBM 模型的刻画参数，即单一 Hurst 指数反映了所研究分形体的不规则程度，单一 Hurst 指数越小表明不规则程度越大，所以目标出现时单一 Hurst 指数变大，同时残差平方和 Q 也有所增大。

表 15.3 雷达实测数据频谱一元线性回归分析

雷达实测数据类型		频域单一 Hurst 指数(拟合直线斜率)	拟合直线截距 b	残差平方和 Q	相关系数 R	回归显著性 r 检验($\alpha=0.001$)
X 波段(HH 极化)	杂波单元	0.0617	6.6679	0.0480	0.9541	极其显著
	目标单元	0.1423	6.8672	0.0486	0.9907	极其显著
X 波段(VV 极化)	杂波单元	0.0274	7.1928	0.0306	0.9836	极其显著
	目标单元	0.1449	7.0041	0.0503	0.9908	极其显著
S 波段(VV 极化)	杂波单元	0.2919	5.8021	0.0139	0.9982	极其显著
	目标单元	0.3155	5.5451	0.0323	0.9965	极其显著

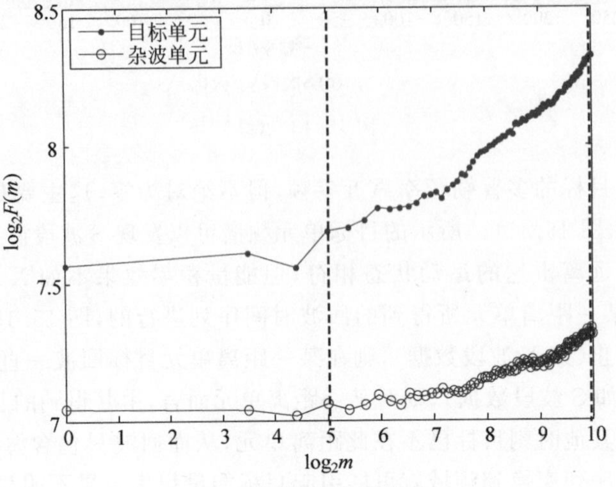

(a) X波段HH极化

图 15.16 实测雷达数据频谱的自相似性分析结果

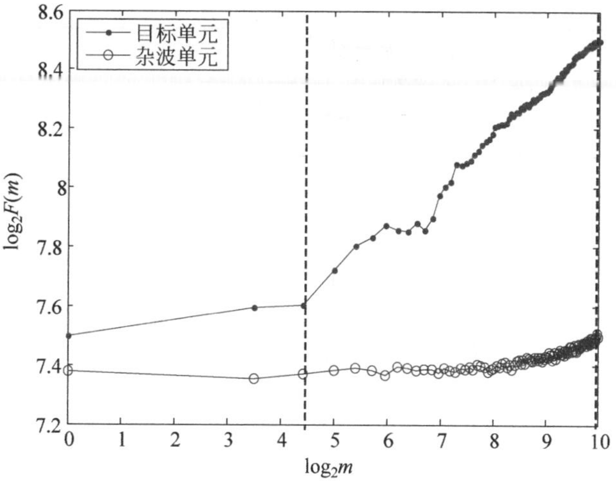

(b) X波段VV极化

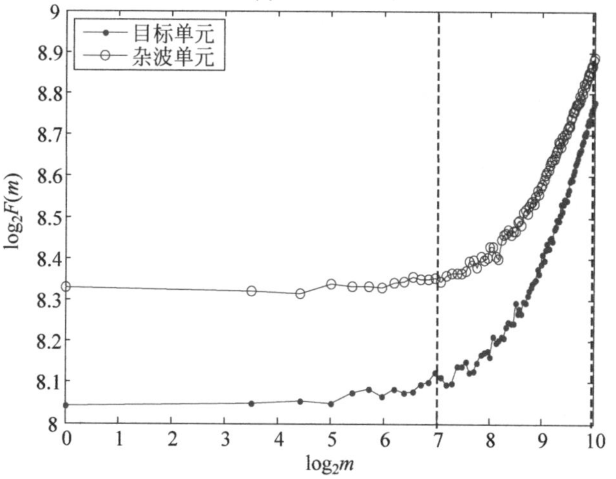

(c) S波段VV极化

图 15.16 （续）

15.3.3 海杂波频谱单一分形参数的影响因素

由于在进行频谱单一分形特性分析之前需对时域海杂波数据进行快速傅里叶变换(FFT)，而 FFT 所采用的时间序列长度 L_t 及 FFT 点数 L_f 都将直接影响到单一分形特性分析的结果，因此本节将分析序列长度 L_t 及 FFT 点数 L_f 对频谱单一分形特性的影响。图 15.17～图 15.19 给出了三组海杂波数据在序列长度分别为 1024 点和 8192 点时采用不同的 FFT 点数进行单一分形特性分析的曲线，可以看出，无论是 X 波段还是 S 波段数据，对于同一组数据取同一序列长度计算得到的结果，均存在尺度区间使得近似线性成立同时目标单元的斜率大于杂波单元的斜率，并且随着 FFT 点数的增加，这一尺度区间逐步向大尺度移动，基本在 2^{16} 点 FFT 时取得最佳效果。对比图 15.17 与图 15.18 可以发现，图 15.18 所示的 X 波段 VV 极化下的杂波与目标的区分效果要优于 X 波段 HH 极化，可见，在频域中目标对 VV 极化海杂波频谱的影响要大于 HH 极化。这里需说明的是，比较区分效果首先应比较无标度区间的范围，若无标度区间范围较大，则再比较直线拟合得到的斜率或线性程度等特征；若无标度区间较小导致只有较少的点参与拟合运算，这样得到的结果没有意义，从而认为区分效果较差。

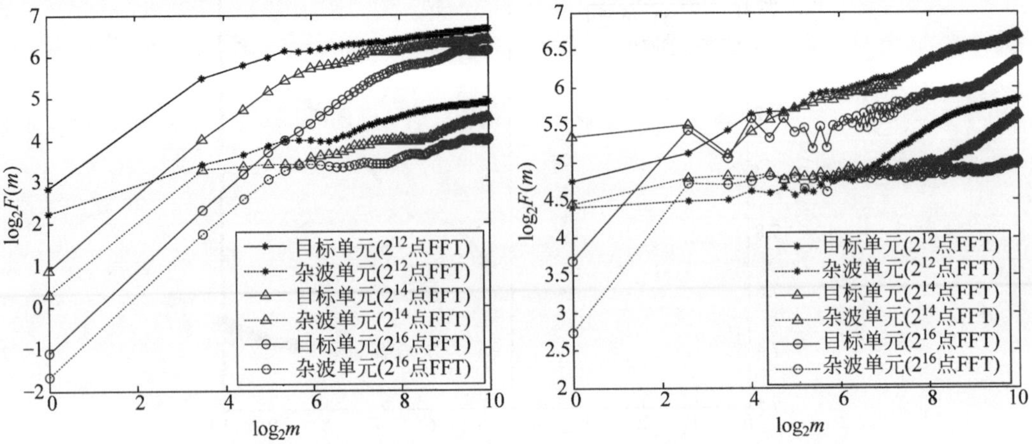

(a) 时域序列长度为1024点　　(b) 时域序列长度为8192点

图 15.17　不同 FFT 点数对频谱单一分形特性的影响(X 波段 HH 极化)

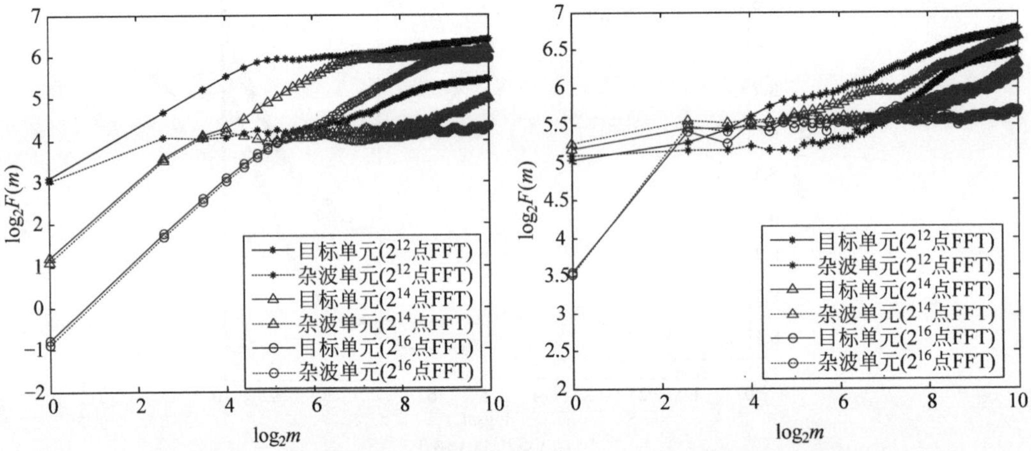

(a) 时域序列长度为1024点　　(b) 时域序列长度为8192点

图 15.18　不同 FFT 点数对频谱单一分形特性的影响(X 波段 VV 极化)

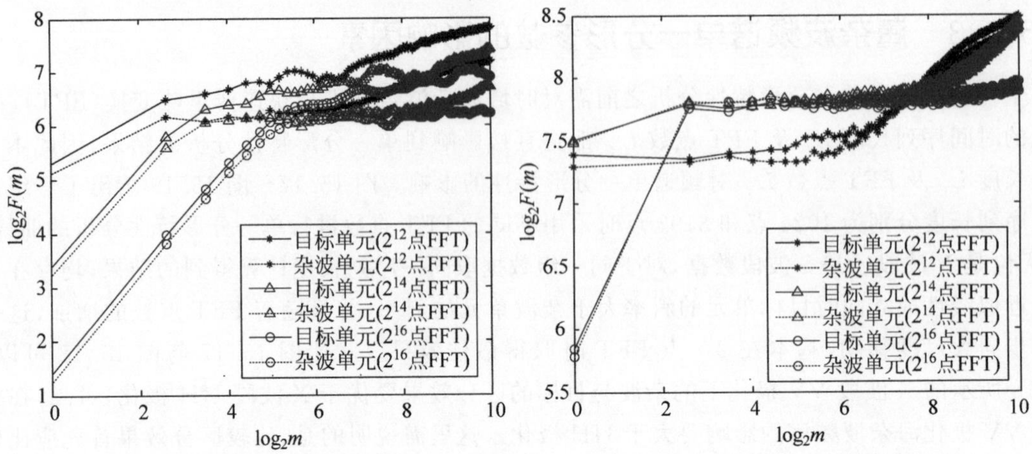

(a) 时域序列长度为1024点　　(b) 时域序列长度为8192点

图 15.19　不同 FFT 点数对频谱单一分形特性的影响(S 波段 VV 极化)

再分别对比图 15.17～图 15.19 中的(a)与(b)可以发现，在 X 波段数据下后者杂波与目标的区分效果要优于前者，而在 S 波段数据下前者的区分效果却优于后者，这是因为 X 波段雷达探测目标为一固定目标，FFT 所采用的时间序列越长，目标能量积累效果越好，在频谱中则表现为对原海杂波频谱的影响程度越大，因此时间序列长度越长，区分效果越好；而 S 波段雷达探测目标为一运动目标，若在截取的时间序列内目标没有移出当前距离单元，则序列越长目标能量积累效果越好，反之，若目标移出了当前距离单元，则截取的时间序列越长，目标能量无法持续积累，而杂波能量却相对上升，此时截取的时间序列越长，区分效果反而越差。因此，对运动目标而言，需要考虑积累时间选取问题，单纯地增加时间序列长度，效果可能适得其反。另外，比较 X 波段数据(图 15.17 和图 15.18)与 S 波段数据(图 15.19)可以发现，在相同参数设置条件下，X 波段数据杂波与目标的区分程度比 S 波段数据更明显，这主要是由 SCR 不同引起的，X 波段数据 SCR 相对较高，经过 FFT 积累后 SCR 进一步升高，从而 X 波段数据目标与海杂波频谱的单一分形特征差异较大。对于图 15.17～图 15.19 还需要指出的一点是，在尺度 m 较小时，杂波单元与目标单元的频谱均表现出较好的线性且斜率基本相同，这与傅里叶变换本身所具有的尺度不变特性以及“随机游走”模型的计算方式有关。“随机游走”模型是每次间隔 m 个采样点对频谱序列进行重采样，然后平方并取均值，这样在尺度较小时，抽取得到的杂波与噪声能量相对较高，目标能量相对较低，从而在此段区间杂波与目标单元的单一 Hurst 指数主要由杂波与噪声决定，而且采用的 FFT 点数越大，此类标度区间范围越大。

图 15.20 给出了在同样 FFT 点数下当采取的时间序列长度不同时单一分形特性分析的结果，参考上述分析中得到的结论，这里均采用 2^{16} 点的 FFT，而时间序列长度分别取为 2^{10}、2^{13} 和 2^{16} 点。由图 15.20 可以看到，X 波段雷达采用的时域序列长度越长，杂波单元频谱的配分函数整体上越接近于线性(单一分形特性越明显)，同时杂波单元与目标单元的区分效果越好；S 波段雷达采用的时间序列越长，杂波单元与目标单元的区分效果反而越差，这仍然是因为 X 波段雷达探测的目标是静止的，而 S 波段雷达探测的目标是运动的，当截取的时间序列较长时运动目标已经移出了所处理的距离单元。

综合图 15.19 与图 15.20 所示结果可得到如下结论：对于静止目标，截取的时间序列长度越长，杂波与目标区分效果越好，但考虑到运算量以及实时性，一般取 2^{10}～2^{13} 点即可；对于运动目标，截取时间序列时需考虑目标不能运动出一个距离单元，并非越长越好。另外，在选取 FFT 点数时以 2^{14}～2^{16} 点为宜。

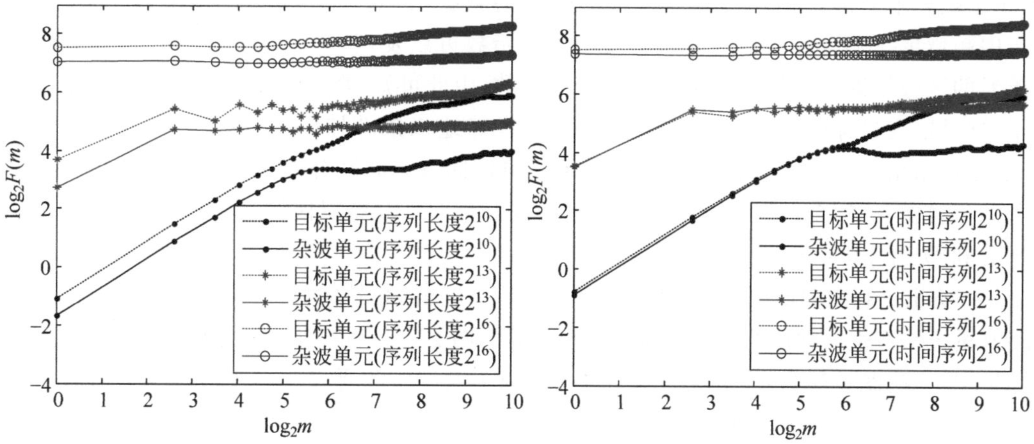

(a) X波段HH极化　　(b) X波段VV极化

图 15.20　不同时间序列长度对频谱单一分形特性的影响(采用 2^{16} 点 FFT)

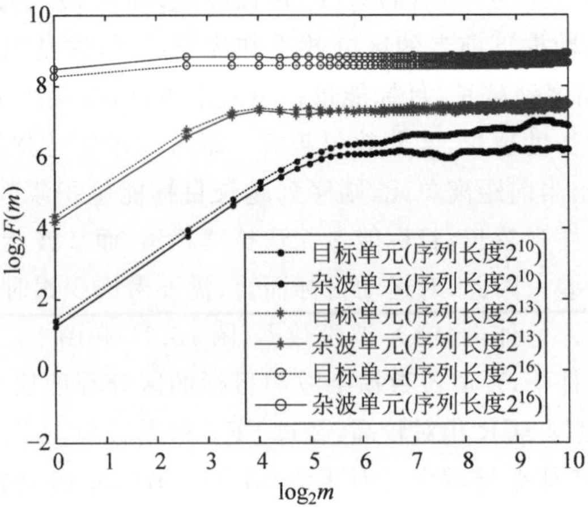

(c) S波段VV极化

图 15.20 (续)

15.3.4 目标检测与性能分析

由海杂波单一分形特性分析可知,频域单一 Hurst 指数 H_f 与残差平方和 Q 具有在频域中区分海杂波与目标回波的能力,但二者对海杂波与目标的区分能力高低有所差异,为定量比较二者对海杂波与目标回波的区分能力,下面根据表 15.3 分别计算目标出现前后频域单一 Hurst 指数 H_f 与残差平方和 Q 的相对变化率:对于频域单一 Hurst 指数,目标出现前后的相对变化率分别为 130.63%(X 波段 HH 极化数据)、428.83%(X 波段 VV 极化数据)和 81.21%(S 波段 VV 极化数据);对于残差平方和,目标出现前后的相对变化率分别为 1.25%(X 波段 HH 极化数据)、64.38%(X 波段 VV 极化数据)和 15.97%(S 波段 VV 极化数据)。显然,相对变化率越大,海杂波与目标回波的频域分形参数间的差异越稳定,同时频域单一分形参数对海杂波与目标回波的区分能力越强,因此下面将选用频域单一 Hurst 指数 H_f 设计相应的目标检测方法。

图 15.21 给出了三组海杂波数据的频域单一 Hurst 指数,作为比较,图 15.21 中还同时给出了三组海杂波数据的时域单一 Hurst 指数。这里需注明的是,在计算图 15.21 中时域和频域单一 Hurst 指数过程中,每个距离单元的数据均被划分成互不交叠的数据段,每个数据段包含 1024 个采样点。比较图 15.21 中频域单一 Hurst 指数与时域单一 Hurst 指数可以发现,频域单一 Hurst 指数对海杂波单元与目标单元的区分效果明显优于时域单一 Hurst 指数,由此可知,在频域中目标回波对海杂波的影响能力提高了,这是因为目标回波经过 FFT 后能量得到很好地积累,同时海杂波能量积累效果不明显。此外,图 15.22 还给出了残差平方和 Q 对海杂波单元与目标单元的区分效果,可见,其对海杂波与目标回波的区分能力较弱,且很多情况下海杂波与目标回波混叠在一起难以区分,这与前文中利用相对变化率对残差平方和 Q 的分析结论相一致。

图 15.23 给出了利用频域单一分形特征,即频域单一 Hurst 指数的目标检测方法流程图,其中,门限 T 采用 CFAR 的方法产生。常用的 CFAR 检测器可以分为两类,即参量 CFAR 检测器和非参量 CFAR 检测器,参量 CFAR 检测器对背景杂波的概率分布类型通常具有特定的要求,由于频域单一 Hurst 指数的概率分布类型未知,这限制了参量 CFAR 检测器的应用,并可能由于分布类型不匹配而带来严重的性能损失,而非参量 CFAR 检测器没有这一束缚,其仅要求检测单元和参考单元具有相同的分布类型及由脉冲串各个脉冲得到的秩之间是相互独立的,考虑到具体的应用背景,图 15.23 中采用的是广义符号检测器。

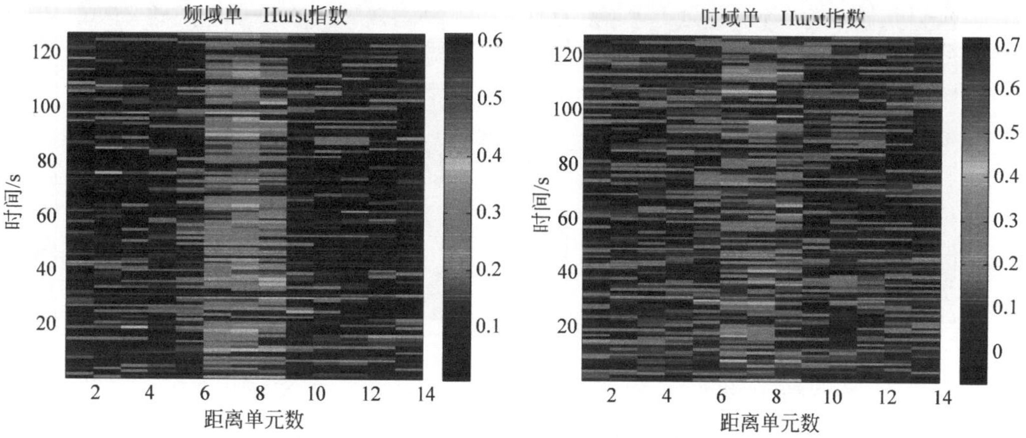

(a) X波段HH极化

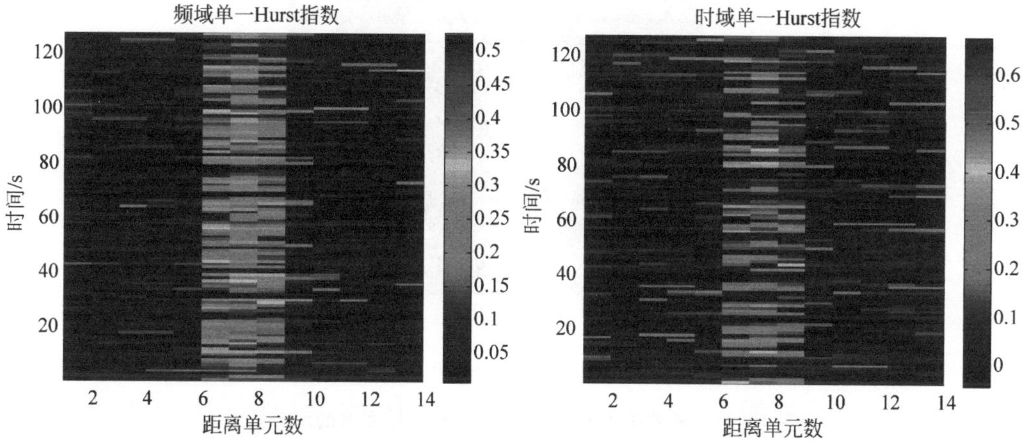

(b) X波段VV极化

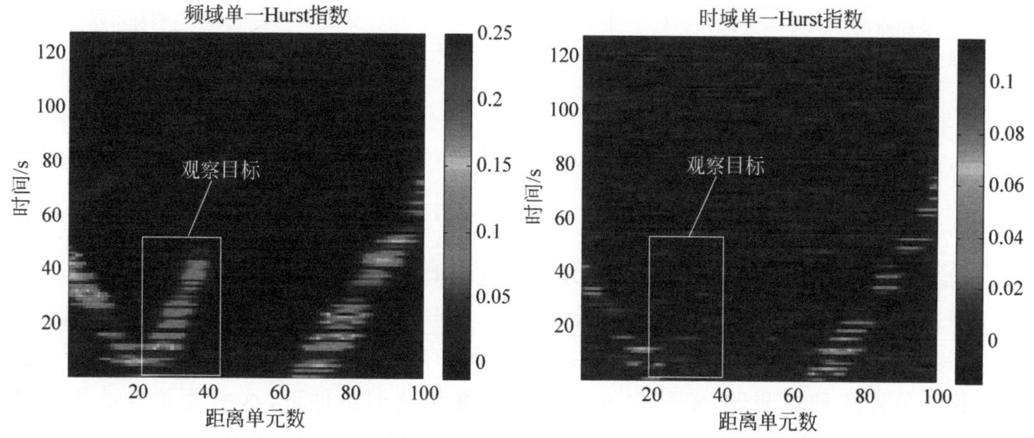

(c) S波段VV极化

图 15.21　实测雷达数据的频域与时域单一 Hurst 指数

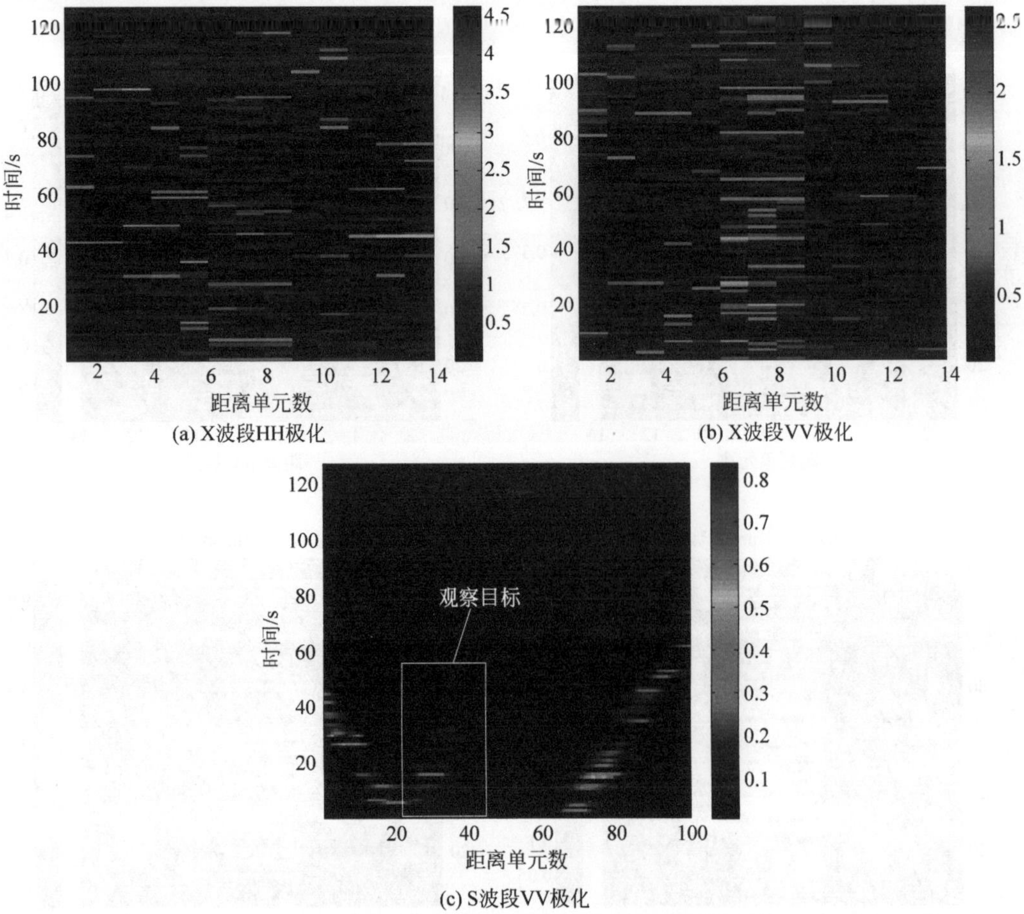

图 15.22　残差平方和 Q 对海杂波与目标回波的区分效果

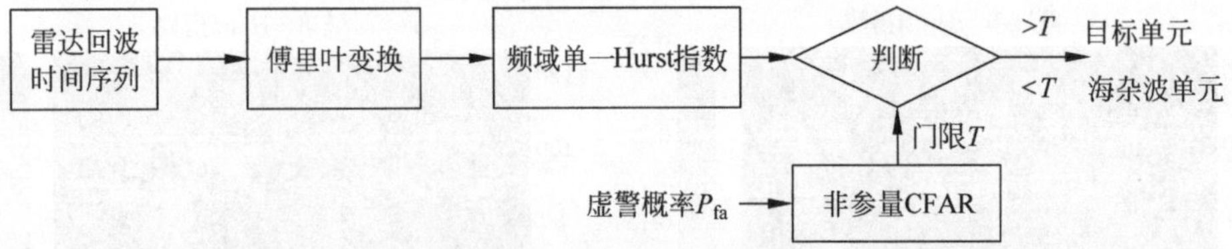

图 15.23　利用频域单一分形特征的目标 CFAR 检测流程图

表 15.4 给出了图 15.23 中所示目标检测方法的检测概率，其中，虚警概率 P_{fa} 设为 10^{-4}。作为对比，表 15.4 中还给出了一种经典的频域 CFAR 检测器，即频域单元平均(CA)CFAR 检测器和一种利用时域单一 Hurst 指数的 CFAR 检测器[2]。由表 15.4 可以明显看到，图 15.23 所示目标检测方法的检测性能明显优于另外两种检测器，尤其在 S 波段 VV 极化数据条件下，这一优越性更为明显。比较图 15.23 所示目标检测方法和频域 CA-CFAR 检测器与时域分形 CFAR 检测器可知，由于图 15.23 所示目标检测方法和频域 CA-CFAR 检测器均进行了傅里叶变换，此二者的检测概率均高于时域分形 CFAR 检测器的检测概率，可见，通过相参积累，即傅里叶变换提升 SCR 可以有效提升目标检测方法的检测性能；此外，比较图 15.23 所示目标检测方法与频域 CA-CFAR 检测器可发现，二者第一步均进行

了傅里叶变换，但图 15.23 所示目标检测方法具有更高的检测概率，这是因为，频谱幅度的统计分布类型与 CA-CFAR 检测器所要求的瑞利背景分布类型不同，从而检测性能下降，实际上该检测器也在一定程度上失去了 CFAR 能力。

表 15.4　所提检测方法在三组实测数据条件下的检测概率($P_{fa}=10^{-4}$)

检测方法	X 波段 HH 极化	X 波段 VV 极化	S 波段 VV 极化
图 15.23 所示检测方法	83.47%	95.31%	76.19%
频域 CA-CFAR 检测器	67.63%	78.44%	51.26%
时域分形 CFAR 检测器	54.17%	62.83%	<10%
处理 10 个距离单元的平均耗时	0.667s	0.678s	0.742s

另外，在进行 FFT 之前可以进行补零，这使得估计频域单一 Hurst 指数最少仅需要 1000 左右采样点即可，而若要得到较稳定的时域单一 Hurst 指数则至少需要 2000 左右采样点，可见，在频域进行分形处理可以在一定程度上降低对数据量的需求，这对于检测方法在实际中运用是十分有意义的。下面以区域搜索警戒雷达为例说明图 15.23 所示目标检测方法在实际中如何工作。由于估计频域单一 Hurst 指数需要的采样点数较多，雷达不能通过一次扫描就完全获得，因此该检测方法需要在雷达开机一小段时间(一般在几秒到十几秒范围内)并获得足够的采样点数量后才开始工作，计算各个距离单元的频域单一 Hurst 指数用于后续的 CFAR 检测。随着雷达的不断扫描，在每个距离单元均不断获得新的回波数据，方便起见，这里假设每次扫描每个距离单元可以获得的新回波采样点数为 n，此时，每个距离单元原有的长回波序列采用如下方式进行更新，即抛掉原序列前 n 个采样点，将新获得的 n 个采样点追加于原序列末尾，这样就形成一个新的长回波序列，然后便可估计新的频域单一 Hurst 指数用于 CFAR 检测，雷达每扫描一次这一过程便循环一次。毫无疑问，估计分形参数所需要的数据量越小，则序列采样点的更新速度越快，本节所提检测方法对目标出现也越敏感。为说明本节所提检测方法的实时处理速度，表 15.4 还给出了本节所提检测方法处理 10 个距离单元的平均耗时情况，其中，测试平台的硬软件情况为：CPU 为英特尔 Core I5 2.8G，RAM 为 4G DDR Ⅲ 1333。在实际雷达系统中，由于需处理的数据量非常大，就目前耗时情况而言仍难以实时运用，因此此方面仍需要更为深入的研究。

15.4　海杂波时/频域多特征与目标检测

由于海面目标的复杂性和多样性，通常无法获取所有种类目标回波，因此通常将纯杂波回波看作正常观测，含目标回波看作异常观测，从回波中提取在两种观测模式下具有明显差异的特征，用于海杂波与目标的区分[45-46]。同时，由于不同特征是从不同侧面对回波特性进行描述的，因此将多特征构造成特征向量，联合运用以增强海杂波与目标的差异度。由多特征构成的多维特征空间，需在其中找到一个可以包含几乎所有训练样本特征的最小判决区域，即构造出一个检测器：当回波特征向量落在判决区域外时，判断该观测为异常，即有目标存在。多维特征与分类器的联合运用为海杂波中目标检测问题提供了新的解决途径。

15.4.1　特征提取与分析

假设雷达在一个波位发射长度为 N 的相干脉冲串，通过 I/Q 通道接收，可以得到每个距离单元的复回波数据。含目标回波包含目标回波、海杂波和噪声；如果不包含目标，则雷达回波只包含海杂波和噪声。待估计单元周围的纯杂波单元为参考单元，用来估计海杂波的特性。雷达目标检测问题因此可以表述成如下二元假设问题

$$\begin{cases} \mathrm{H}_0: \begin{cases} x(n)=c(n), & n=1,2,\cdots,N \\ x_p(n)=c_p(n), & p=1,2,\cdots,P \end{cases} \\ \mathrm{H}_1: \begin{cases} x(n)=s(n)+c(n) & n=1,2,\cdots,N \\ x_p(n)=c_p(n), & p=1,2,\cdots,P \end{cases} \end{cases} \tag{15.19}$$

假设 H_0 表示待估计单元内不包含目标；假设 H_1 表示待估计单元内包含目标，当信噪比较高时可以忽略噪声的影响。$x(n)$、$s(n)$和 $c(n)$分别为雷达回波、目标回波和海杂波。

$x_p(n)=c_p(n)(p=1,2,\cdots,P)$是待估计单元周围参考单元，假设海面是局部均匀的，参考单元内的海杂波特性与待估计单元杂波特性基本一致，即可以用参考单元估计杂波特性。NT_r 为雷达在一个波位的驻留时间，其中 T_r 是脉冲重复周期。

1. 相对平均幅度 RAA

长度为 N 的信号，平均幅度定义如下

$$\bar{A}(\boldsymbol{x})=\frac{1}{N}\sum_{n=1}^{N}|x(n)| \tag{15.20}$$

其中 $\boldsymbol{x}$ 代表回波时间序列。区分回波强度的大小是传统雷达检测目标的主要根据，RAA 通过计算待估计单元周围的回波强度，计算相对平均幅度。$\bar{A}(\boldsymbol{x})$和 $\bar{A}(\boldsymbol{x}_p)$分别代表待估计单元平均回波幅度和周围参考单元的平均回波幅度。相对平均幅度计算如下

$$\xi_1(\boldsymbol{x})\equiv\frac{\bar{A}(\boldsymbol{x})}{1/P\sum_{p=1}^{P}\bar{A}(\boldsymbol{x}_p)} \tag{15.21}$$

对于距离非平稳的海杂波，式(15.34)中的分母可以使该特征具有一定的恒虚警特性。下面利用 IPIX 雷达数据＃311 的 HV 极化对 RAA 的分类能力进行分析，数据被分为长度为 1024 的数据段，观测时间约为 1s。一个典型的纯杂波单元和目标所在单元的回波序列幅度图如图 15.24(a)和图 15.24(b)所示。通过计算该组数据所有 10 个纯杂波单元共 20320 个样本的 RAA，所得直方图如图 15.24(c)所示。由于海尖峰的存在，直方图具有很强的拖尾；2032 个目标所在单元回波序列 RAA 的直方图如图 15.24(d)所示。由于目标在海面上随波浪上下起伏，目标的 RCS 有着剧烈的起伏，因此目标的 RAA 较分散，从 1.2 到 18。可以看出，目标单元的 RAA 和纯杂波单元的 RAA 直方图分布有一定差别，可以作为用于目标检测的一个特征，但由于两个直方图分布有较多地方重叠，仅 RAA 无法使检测器获得较好的检测性能。

2. 相对多普勒峰高 RPH

雷达照射的大面积海面具有大量不同速度的散射结构，这是导致海杂波具有较宽多普勒带宽的原因。海杂波带宽与很多因素有关，如雷达工作频率、海况和极化等，海面漂浮目标由于径向速度较小，其多普勒峰通常淹没在主杂波区内。对于 X 波段 IPIX 雷达来说，其主要海杂波带宽在 100～150Hz。由于人造目标在惯性的作用下，在较短时间内目标速度没有改变或仅有较小范围改变，因此目标回波的能量比海杂波更集中，分布在有限几个多普勒单元内。图 15.25 中分别给出了两种有无目标存在时的回波多普勒谱。

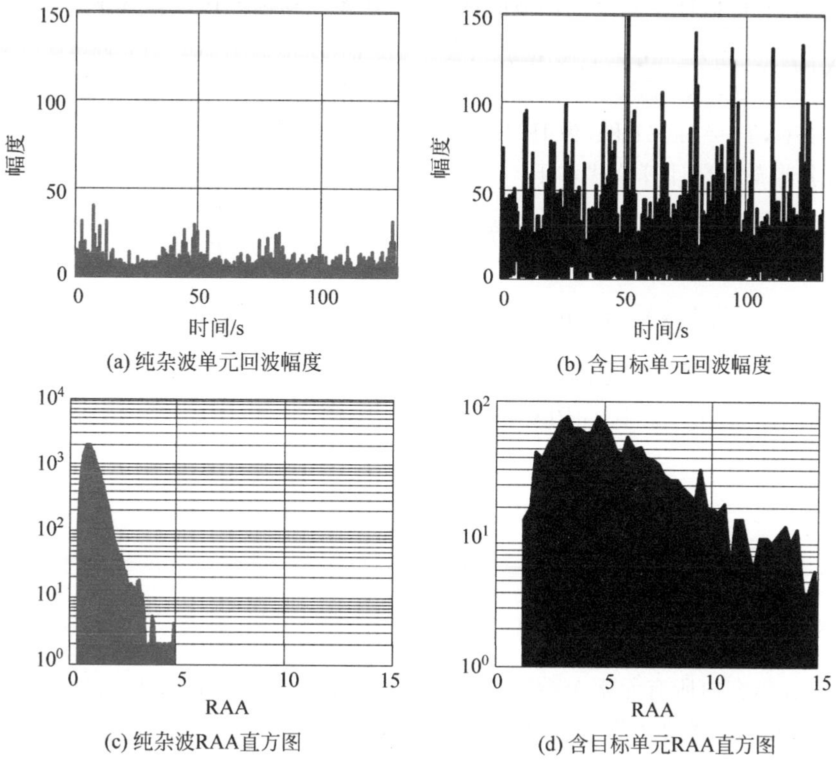

图 15.24　回波幅度图和相对平均幅度的直方图

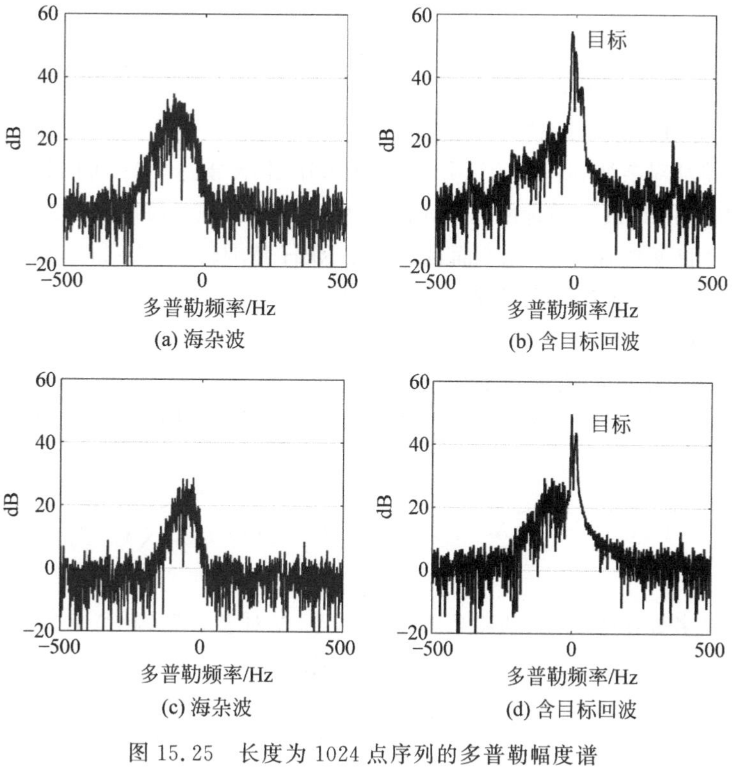

图 15.25　长度为 1024 点序列的多普勒幅度谱

$x(n)$为长度为 N 的待估计单元雷达回波时间序列,其多普勒幅度谱计算如下

$$X(f_d)=\frac{1}{\sqrt{N}}\left|\sum_{n=1}^{N}x(n)\exp(-2\pi f_d nT_r)\right|,\quad -\frac{1}{2T_r}\leqslant f_d\leqslant\frac{1}{2T_r} \tag{15.22}$$

其中,f_d 为多普勒频率,$T_r=0.001$ 秒,为 IPIX 雷达的脉冲重复间隔。

多普勒偏移和多普勒峰高计算公式如下

$$\begin{aligned}\text{Peak}(\boldsymbol{x})&=\max_{f_d}\left\{X(f_d),-\frac{1}{2T_r}\leqslant f_d\leqslant\frac{1}{2T_r}\right\}\\ f_d^{\max}(\boldsymbol{x})&=\operatorname*{argmax}_{f_d}\left\{X(f_d),-\frac{1}{2T_r}\leqslant f_d\leqslant\frac{1}{2T_r}\right\}\end{aligned} \tag{15.23}$$

令 δ_1 为可供参考的多普勒单元范围,δ_2 为目标最大可能占据的多普勒带宽。δ_1 和 δ_2 随着雷达特性和目标运动特性的不同而不同,对于 IPIX 雷达,按经验选取 $\delta_1=50\text{Hz}$,$\delta_2=5\text{Hz}$。多普勒幅度谱的相对峰高定义如下

$$\text{RPH}(\boldsymbol{x})\equiv\frac{\text{peak}(\boldsymbol{x})}{\frac{1}{\#\Delta}\sum_{f_d\in f_d^{\max}(\boldsymbol{x})+\Delta}X(f_d)} \tag{15.24}$$

其中,Δ 代表所有的多普勒参考单元组成的集合,$\Delta=\left[-\frac{1}{2}\delta_1,-\frac{1}{2}\delta_2\right]\cup\left[\frac{1}{2}\delta_2,\frac{1}{2}\delta_1\right]$,$\cup$代表两个集合的并集,$\#\Delta$ 代表 Δ 集合内所有元素个数。通常含目标回波的 RPH 较大而纯杂波回波 RPH 较小。由于海杂波的能量分布在较宽的多普勒区间,绝对的多普勒峰高很难区分杂波和目标。

利用待估计单元的 RPH 和周围参考单元的 RPH,构造第二个特征如下

$$\xi_2(\boldsymbol{x})=\frac{\text{RPH}(\boldsymbol{x})}{1/P\sum_{p=1}^{P}\text{RPH}(\boldsymbol{x}_p)} \tag{15.25}$$

其中,$\boldsymbol{x}_p$ 代表纯杂波参考单元的回波,第二个特征可以表述为在二维距离多普勒平面的相对多普勒峰高,与多普勒单元 CFAR 的比率检测统计量类似[47]。

下面验证第二个特征对杂波和目标的区分能力,仍采用 IPIX 雷达#311 的 HV 极化数据,观测时间约为 1 秒。10 个纯杂波单元 RPH 和目标所在单元的 RPH 的直方图分别如图 15.26(a)和图 15.26(b)所示。可以看到,混杂波的 RPH 集中在 0.6~2.2 而目标单元 RPH 范围较大,从 0.6~19.6。两个直方图之间的重叠意味着单一利用 RPH 无法对目标完成有效的检测。

3. 相对多普勒谱熵 RVE

纯杂波单元的多普勒幅度谱能量分布在较宽的带宽(100~150Hz)内,而目标所在单元的多普勒谱能量分布在较窄的范围内。熵可以描述系统或者数据的混乱程度,同样利用多普勒谱向量熵可用来判断目标是否存在[48],定义为

$$\text{VE}(\boldsymbol{x})\equiv-\sum_{f_d}\hat{X}(f_d)\lg\hat{X}(f_d),\quad \hat{X}(f_d)=\frac{X(f_d)}{\sum_{f_d}X(f_d)} \tag{15.26}$$

其中,$\hat{X}(f_d)$为归一化多普勒幅度谱,利用参考距离单元,相对多普勒熵 RVE 定义为

$$\xi_3(\boldsymbol{x})\equiv\frac{\text{VE}(\boldsymbol{x})}{1/P\sum_{p=1}^{P}\text{VE}(\boldsymbol{x}_p)} \tag{15.27}$$

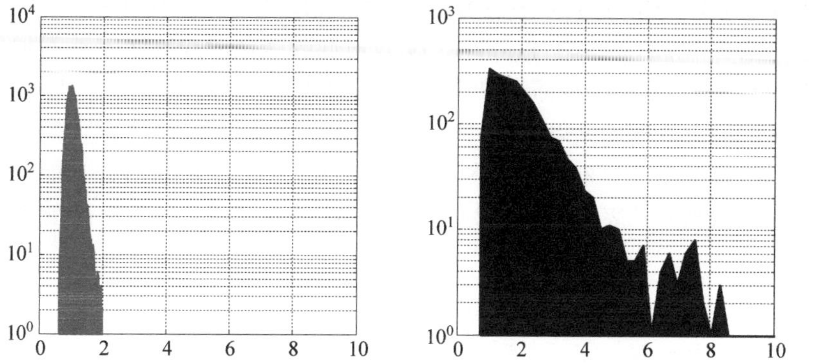

图 15.26 相对多普勒峰高直方图

类似地，利用前面的数据，纯杂波单元和目标所在单元 RVE 的直方图如图 15.27(a)和图 15.27(b)所示。可以看到，纯杂波单元与目标所在单元相比具有较大的 RVE，说明海杂波的多普勒谱更混乱，因此该特征具有一定的目标检测能力。

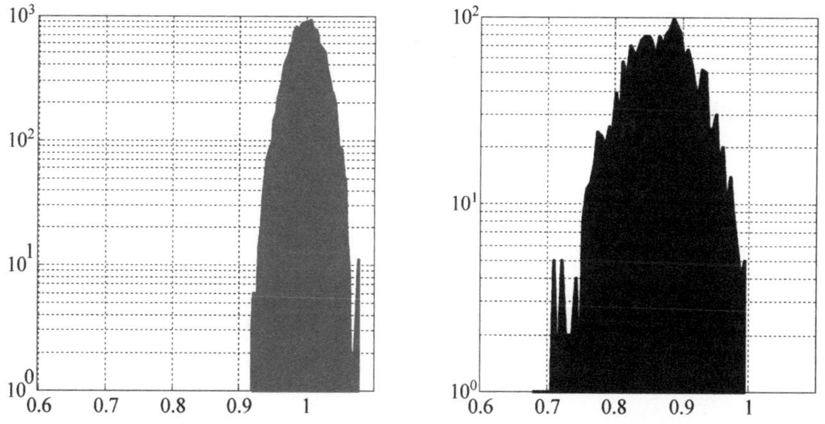

图 15.27 相对多普勒熵的直方图

在对所有 10 组数据进行验证后，揭露出来几个现象。第一，单一特征的检测能力对于每组数据来说不尽相同；第二，这三种特征区分目标与杂波有很强的互补性；第三，无论哪一个特征区分能力与 SCR 有很大关系，当 SCR 降低时，区分能力都会降低。

4. 杂波和目标在特征空间的可分性分析

上述分析表明，三个特征都具有区分目标与杂波的能力，但是单独一种特征无法很好地完成目标检测。下面将回波数据映射到由上述三特征构成的三维特征空间，分析在特征空间中目标与杂波的可分性。对 IPIX 雷达 #311 HV 极化数据[38]的纯杂波单元及目标所在单元回波进行特征提取，将三个特征表示为三维特征向量。图 15.28 给出了观测时间为 0.512s、1.024s、2.048s 及 4.096s 时纯杂波与目标回波特征在三维空间中的分布情况。可以明显看出，随着观测时间的增加，纯杂波和目标回波两类数据的特征在三维空间的分离程度逐渐提高。但是，三种特征有着不同的变化趋势。随着观测时间的增加，两种模式的 RAA 和 RVE 区分能力逐步增强。原因是长时间的观测平滑了海杂波的海尖峰分量，增强了目标的回波，改善了 RAA 检测目标的能力；长时间观测同时提高了多普勒谱的分辨率，有助于利用 RVE 来区分目标。但是，由于目标是漂浮在水面上的球体，随着波浪上下起伏，径向速度在较短的时间内可认为基本无变化，目标能量集中在一个或个别几个多普勒单元内，然而随着观测时间的增加，径向速度呈现出差异，目标能量分布于多个多普勒单元，从而 RPH 特征的分类能力有了一定的下降。经多组 IPIX 雷达实测数据验证表明，当 SCR 大于 5dB 时，纯杂波回波和目标存在时回波在三维特征空间能够有效分离。

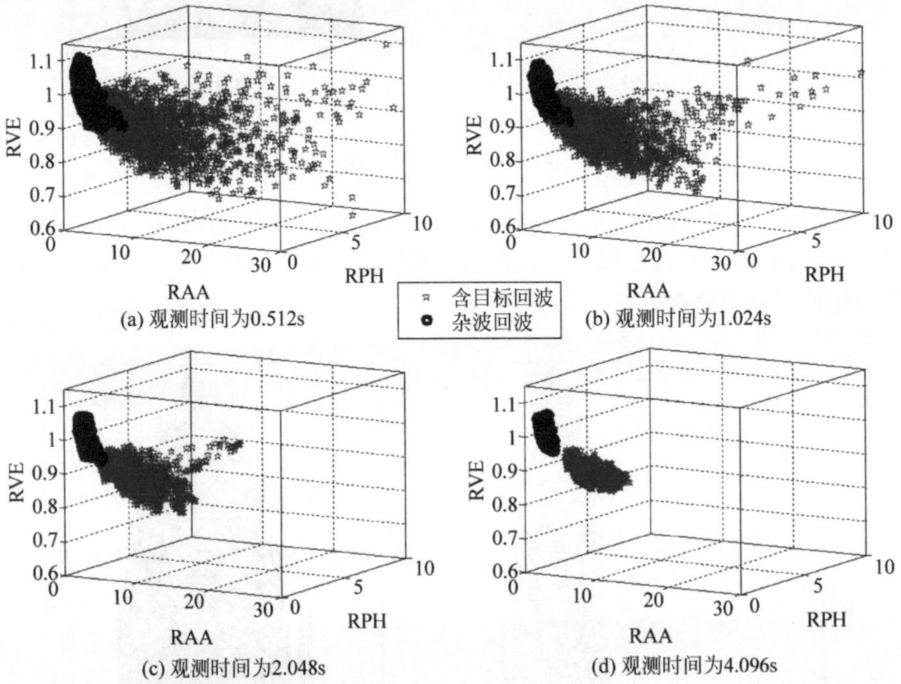

图 15.28 纯杂波与目标特征在三维空间中的分布

15.4.2 三维特征检测器

当雷达开机时接收到大量回波数据，从纯杂波数据中可提取大量特征向量，这些特征向量在三维特征空间中服从某种未知的条件概率密度函数 $p(\boldsymbol{\xi}\mid H_0)$，$\boldsymbol{\xi}=[\xi_1(\boldsymbol{x}),\xi_2(\boldsymbol{x}),\xi_3(\boldsymbol{x})]^{\mathrm{T}}\in\mathbf{R}^3$。这些特征向量构成了纯杂波的训练样本 $\Lambda_{H_0}=\{\boldsymbol{\xi}_i\in\mathbf{R}^3:i=1,2,\cdots,I\}$。由于海面目标的复杂性，目标回波可能来自雷达照射区域不同类型的漂浮物，获取各种类型目标的回波及特征向量难度很大。实际中没有训练样本用来学习 H_1 假设下的概率密度函数。因此，可以利用纯杂波特征训练样本训练得到检测判决区域，当待检测回波的特征落在判决区域内时判断没有目标存在；反之，当落到判决区域外时，即认为有目标存在。

该问题与异常检测中的单分类问题相类似[45]，海杂波特征向量被看作正常观测，目标特征向量被看作异常观测。在实际中，由于纯杂波的数据较容易获得，因此单分类问题在于如何确定训练样本 Λ_{H_0} 的概率密度函数 $p(\boldsymbol{\xi}\mid H_0)$ 在给定分位点条件下的最小支撑区域，进而作为检测器的检测判决区域[49]，该分位点由检测器的虚警概率确定。在三维特征空间中，当两种条件概率密度已知时，给定虚警概率条

件下的检测判决区域根据奈曼-皮尔逊准则可写为

$$\begin{aligned}&\max_{\Omega}\left\{P_{\mathrm{d}}=1-\iiint_{\Omega}p(\xi \mid H_1)\,\mathrm{d}\xi\right\}\\&\text{s.t.}\iiint_{\Omega}p(\xi \mid H_0)\,\mathrm{d}\xi=1-P_{\mathrm{fa}}\end{aligned} \tag{15.28}$$

其中，P_{d} 为检测概率，P_{fa} 为虚警概率，Ω 为检测判决区域。假设在检测判决区域 Ω 包含在一个更大的均匀分布区域 Θ 内，则检测概率 P_{d} 等于 $1-|\Omega|/|\Theta|$，其中 $|\Omega|$ 表示 Ω 的体积，$|\Theta|$ 表示 Θ 的体积，参考文献[49]，并将判决区域限制为有界凸集，式(15.28)可由如下优化问题描述

$$\begin{aligned}&\min_{\Omega\in \mathrm{C}}\{|\Omega|\}\\&\text{s.t.}\quad \{i:\xi_i\in\Omega\}=[I\times(1-P_{\mathrm{fa}})]\end{aligned} \tag{15.29}$$

其中，C 表示三维特征空间中的所有有界凸集，$\#$A 表示集合 A 的元素个数，I 表示训练样本总数。解决这个优化问题其实是找到包含 $I\times(1-P_{\mathrm{fa}})$ 个训练样本的最小凸区域，这样就将目标检测问题转化为异常检测中的单分类问题。通常训练样本数需满足 $I\times P_{\mathrm{fa}}\geqslant 10$，例如，若需训练满足虚警概率 $P_{\mathrm{fa}}=0.001$ 的凸包，则至少需要 $I=10000$ 个样本。这里在寻求最优解时采用一种快速凸包学习算法[50]。

图 15.29 给出了一种基于三维特征的目标检测器[50]，其分为训练和检测两个步骤。在训练时，对纯杂波样本进行采集，提取纯杂波特征向量，训练得到满足虚警概率的检测判决区域；在检测时，对待检测(CUT)单元样本进行采集，提取特征向量，进行决策。

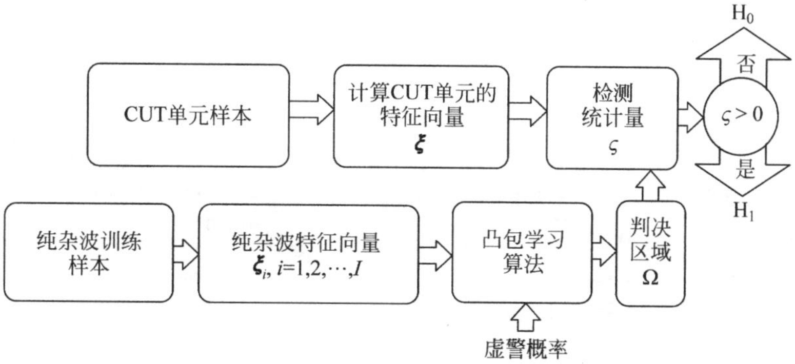

图 15.29 基于三维特征的检测器框图

在进行训练步骤时，需要注意如下几点：首先，由于海杂波随着海况变化而变化，当海况改变时，检测判决区域需要重新训练，因此需要在线进行凸包学习；其次，在实际中，很难知道待检测单元周围是否是纯杂波单元，为了避免其他目标或者异常单元对检测单元的影响，特征均值可以用序列统计量代替；再次，由于雷达照射方向角的不同，特征特性也会有一定变化，但是在较小范围内可认为是不变的，因此多个临近方位角可以共用同一个检测判决区域；最后，在获取纯杂波训练样本时会有少量异常样本，如未知目标或异常散射体，这些异常样本需要从训练样本中剔除，因此在训练之前需要进行预处理，根据 Mahalanobis 距离[49]，当异常点距离样本中心较远时需要将其剔除。

15.4.3 检测性能分析

这里采用 1993 年 IPIX 雷达采集[38]的 10 组驻留数据对三维特征检测器的检测性能进行验证。多组数据包含 10 个纯杂波距离单元，每个距离单元回波数据长度为 2^{17} 个采样点。将数据分为小段数据，

数据长度 N 分别为 512、1024、2048 和 4096，由于 IPIX 雷达脉冲重复频率为 1000Hz，则对应的观测时间分别为 0.512s、1.024s、2.048s 和 4.096s。由于数据量限制，训练样本数据无法满足虚警概率为 0.001 时的实验条件，因此这里采用重叠的方法增加样本数目。每段长为 2^{17} 的时间序列 $x(n)$ 被分为长度为 N 的向量 $\boldsymbol{c}_i(n)$

$$\boldsymbol{c}_i(n)=x[64(i-1)+1:64(i-1)+N],\quad i=1,2,\cdots,\frac{2^{17}}{N} \tag{15.30}$$

从 10 个纯杂波单元中可得到超过 20000 个纯杂波训练样本，利用这些样本可以提取出纯杂波特征向量。利用快速凸包学习算法[50]，可以得到满足虚警概率为 0.001 的检测判决区域。同样，目标所在单元的时间序列也被分成同样长度 N 的向量，得到 2000 个待检测样本，检测概率可通过检测出的目标样本数与目标总样本数的比值得到。

将基于三维特征的检测器与基于分形的检测器进行性能比较[51-52]。对于基于分形的检测器，10 组实测数据可以共用一个检测器。由于基于三维特征的检测器利用多普勒信息，其对海况条件敏感，为公平比较，这里将 10 组数据检测结果进行平均。由于海杂波时间序列在时间尺度为 0.001～4s 范围内表现出分形特性[51]，因此在时间尺度为 0.016～2.048s 范围内计算 Hurst 指数。基于联合分形的检测器包含两个分形特征，通过去趋势波动分析(DFA)计算 Hurst 指数及在尺度为 0.256s 时的截距进行目标检测[52]，检测判决区域利用二维凸包学习算法得到。

图 15.30 是三个检测器在观测时间为 4.096s、四种极化方式数据下的检测性能曲线。可以看到，基于三维特征的检测器有更好的检测性能，当虚警概率较高时基于联合分形的检测器性能略优于基于三维特征的检测器。基于分形的检测器与基于联合分形的检测器在虚警为 0.01 时，检测曲线会有交叉，相比基于联合分形的检测器，基于分形的检测器在虚警较低时有较好的检测性能。

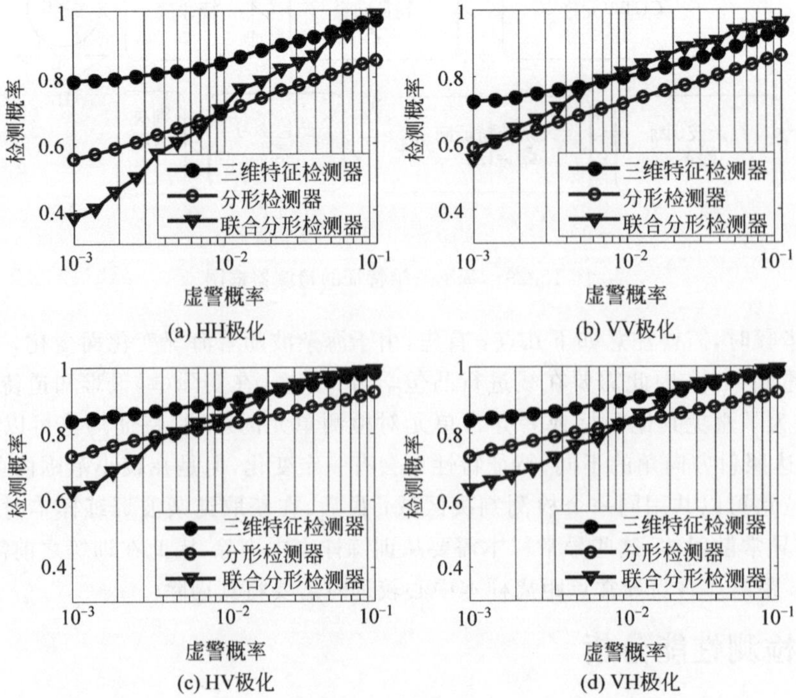

图 15.30 基于三维特征的检测器与两种基于分形特征的检测器性能曲线

自适应海杂波抑制和目标回波匹配累积是检测的有效途径，自适应归一化匹配滤波器（ANMF）检测器[18-19,51]在对长拖尾非高斯背景下运动目标检测中有很好的检测效果，因此将基于三维特征的检测器与ANMF检测器进行比较。由于ANMF检测器需要较多的参考单元，这里利用IPIX雷达1998年采集的数据#202225作为验证数据。该数据包含28个连续的距离单元，每个距离单元数据长度为60000个采样点，距离分辨率为30m，在第24距离单元有一个漂浮的小船，为目标所在单元，第23和第25距离单元为受目标影响的距离单元，其余25个距离单元为纯杂波单元，该数据在HH、VV、HV和VH极化下的信杂比分别为5.17、4.66、26.26和27.47。图15.31(a)为该数据HV极化下回波幅度图，图15.31(b)为目标所在单元的时频谱。

由于小船随着海浪起伏，目标的多普勒频率在很短时间内可以认为不变，但当观测时间增加到512ms时，目标能量在多普勒域扩展到多个多普勒单元内，传统ANMF检测器无法直接使用，为了比较两种检测器在较长观测时间下的性能，这里采用了组合ANMF检测器[50]，该检测器组合使用多个短时间条件下的ANMF检测统计量，当目标多普勒频移有变化时，可以有效地判决是否有目标的存在。对于HH、VV、HV和VH四种极化条件下的实测数据，组合ANMF检测器的检测概率分别为0.41、0.39、0.61和0.65，基于三维特征的检测器检测概率为0.59、0.58、0.81和0.89。图15.31(c)和图15.31(d)分别为组合ANMF检测器和三维特征检测器在HH极化下的检测结果图。黑色的方块代表检测出有目标，目标所在单元为24，因此除去24距离单元外，其余距离单元检测出的目标即为虚警。可以看到，两个检测器均有3个虚警点，但基于三维特征的检测器在24距离单元检测出的目标的次数更多，黑色方块更加连续，表现出更好的检测性能。

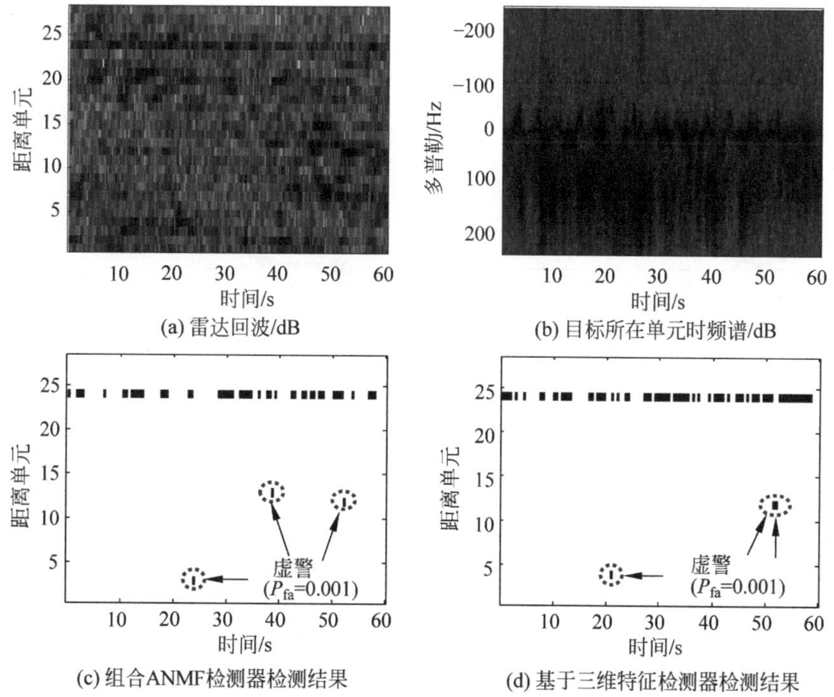

图15.31　基于三维特征的检测器与组合ANMF检测器的检测性能比较

为进一步比较不同信杂比下两种检测器的性能，在纯杂波数据中增加仿真目标回波$s(n)$，如下所示。

$$s(n)=a[1+0.8\cos(2\pi T_r n/4+\varphi_a)]\exp[80\mathrm{i}\sin(2\pi T_r n/3+\varphi_p)],\quad n=1,2,\cdots,2^{17} \tag{15.31}$$

其中,a 为一个正值,用来调整回波信杂比,雷达脉冲重复周期为 $T_r=0.001\mathrm{s}$,φ_a 和 φ_p 为在$[-\pi,\pi]$间随机的初始相位。模拟信号 $s(n)$分为两部分,前半部分模拟随时间起伏的目标回波幅度,起伏周期为4s,起伏范围设为 19dB(参考实测数据);后半部分模拟随时间变化的目标多普勒频移,假设为周期为3s 的正弦函数,幅度为 80/3。虚警概率为 0.001 时,信杂比从 5dB 变化到 28dB,分别对四种极化海杂波数据下两种检测器的性能进行比较。从图 15.32 中可以看到,三维特征检测器有更好的检测性能,其原因为目标的多普勒谱位于海杂波的主杂波区域,因此组合 ANMF 的杂波抑制能力有所下降。除此之外还发现,组合 ANMF 在 HH 和 VV 极化下的检测概率高于 HV 和 VH 极化,这是由交叉极化下的杂噪比低于同极化造成的。

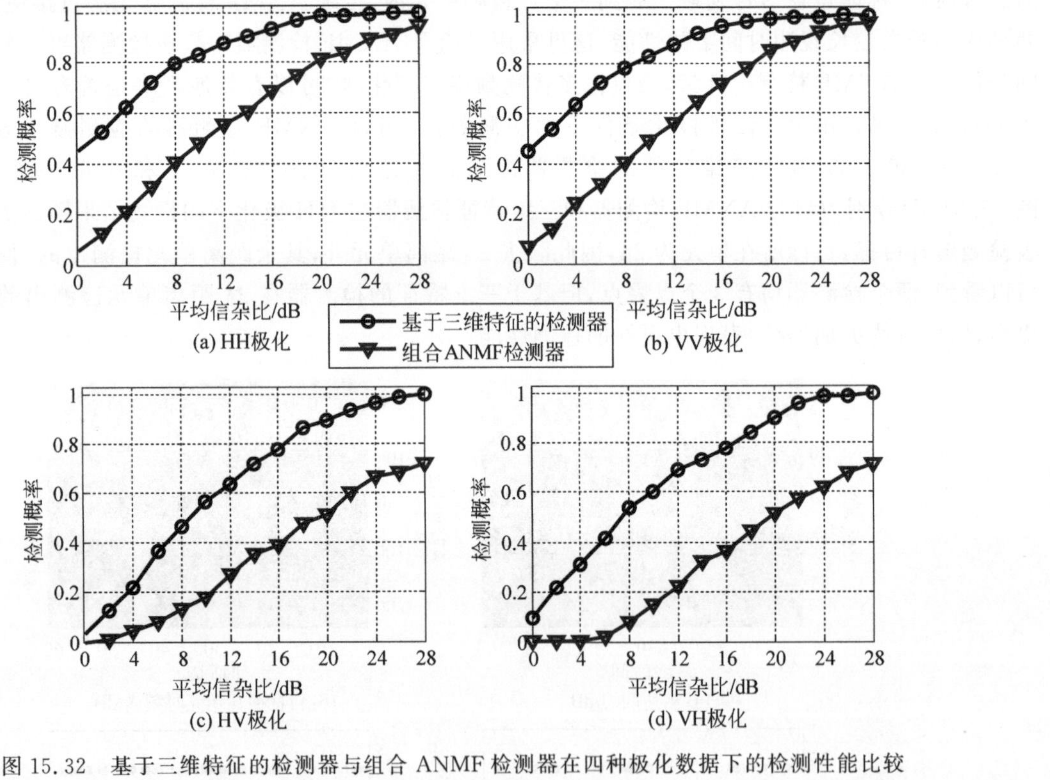

图 15.32 基于三维特征的检测器与组合 ANMF 检测器在四种极化数据下的检测性能比较

15.5 基于深度学习的目标检测

雷达目标自动检测是将待检测单元的脉冲压缩输出与自适应于背景噪声、干扰等的阈值进行比较,从而获得具有恒定虚警概率的检测器。在实际中,干扰目标的位置、数量和强度分布是随机的,杂波边缘的位置是变化的,因此很难获得一个可以兼顾所有情况的 CFAR 检测器。各种 CFAR 检测器还未考虑大目标旁瓣对小目标的遮挡问题,由于大目标旁瓣的遮蔽效应,小目标很难超过检测门限。此外,目标多普勒的影响又会进一步降低脉冲压缩的输出信(杂)噪比,从而影响后续的检测过程。

本节综合考虑脉压旁瓣、多目标及其多普勒的影响,基于深度循环神经网络构建脉压、检测一体化处理框架,并将其推广到多脉冲处理,采用端到端的监督训练方式进行网络训练,在仿真和实测数据集上对目标检测性能进行分析。

15.5.1 基于深度循环神经网络的脉压、检测一体化

1. 基于深度循环神经网络的检测器模型构建

在众多种类的神经网络中，循环神经网络被认为是非常适用于处理序列数据的神经网络，在语音识别[53]、机器翻译[54]等领域取得巨大成功，并在雷达领域开始崭露头角[55-56]。显而易见的是，雷达的一个脉冲的快时间数据也是一个时间序列。因此，深度循环神经网络也适用于雷达目标检测。

循环神经网络通过在每个时间步上共享权重达到降低网络参数的目的，并通过状态向量实现记忆的功能。一个经典的循环神经网络单元如图 15.33 左半部分所示，其中 $\boldsymbol{y}$ 为输入向量，$\boldsymbol{U}$ 为输入层到隐藏层的权重矩阵，$\boldsymbol{W}$ 是隐藏层上一状态到下一状态的循环权重，$\boldsymbol{V}$ 为隐藏层到输出层的权重矩阵。循环神经网络的输出不仅取决于当前时刻的输入，还取决于过去时刻的状态。RNN 的展开形式如图 15.33 右半部分所示，用公式可表示为

$$\boldsymbol{s}_t = f(\boldsymbol{U}\boldsymbol{y}_t + \boldsymbol{W}\boldsymbol{s}_{t-1}) \tag{15.32}$$

$$\boldsymbol{o}_t = g(\boldsymbol{V}\boldsymbol{s}_t) \tag{15.33}$$

其中，f 和 g 分别表示隐藏层和输出层的元素级激活函数。对于基于神经网络的雷达目标检测器而言，前人的研究多是采用脉压后信号作为全连接网络(多层感知机)的输入，未考虑目标多普勒、波形旁瓣、失配滤波等造成的检测性能损失。得益于深度学习强大的近似能力，本节直接采用脉压前的数据 $\boldsymbol{y}$ 作为网络输入。对待检测单元影响最大的是其前后各 $N-1$ 个距离单元的数据，这是因为从脉压的角度来说旁瓣区域即这 $2N-2$ 个距离单元，而双向 RNN 结合了从序列起点开始移动和从序列末尾开始移动的两个 RNN 结构，对这种具有前后文依赖关系的场景提供了学习能力。Graves 等[53]认为采用深度的 RNN 结构可以得到更好的映射。因此，本节采用堆叠多层双向 RNN 单元的方式增强网络的近似能力。对于网络的输出，有两种方式可以选择：每个时间步对应每个距离单元的检测结果，即多输入多输出模式或者所有时间步对应当前检测单元的检测结果，即多输入单输出模式。假定各个距离单元之间统计独立，则上述多输出模式可看作多个单输出模式的独立组合。从便于评估模型的角度考虑，本节采用单输出的模式，即网络的最终输出对应一个待检测单元的检测结果。将每层前后向 RNN 单元的最终状态汇集到一起，作为最终输出层的输入，如图 15.34 所示。这种直接映射的方式也称残差连接，可以使网络层数堆叠的很深而不发生退化。

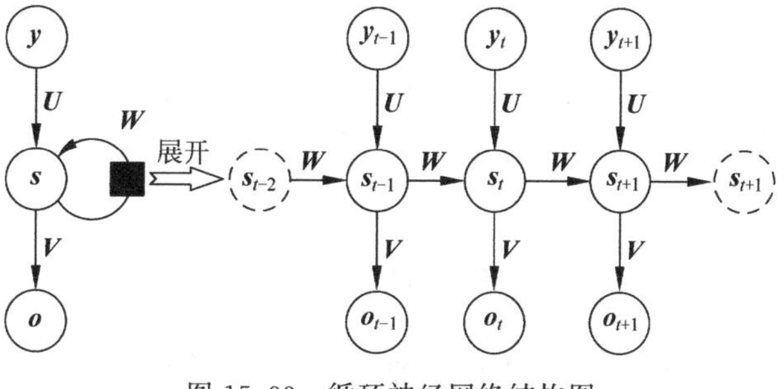

图 15.33 循环神经网络结构图

原始的 RNN 结构在处理较长序列数据时容易发生梯度消失或爆炸，难以学习长时间的特征。为此，长短期记忆网络(Long Short-Term Memory，LSTM)和门控循环单元(Gated Recurrent Unit，GRU)等结构被引入以解决长时间的记忆问题。然而，LSTM 和 GRU 结构采用的 tanh 和 sigmoid 激

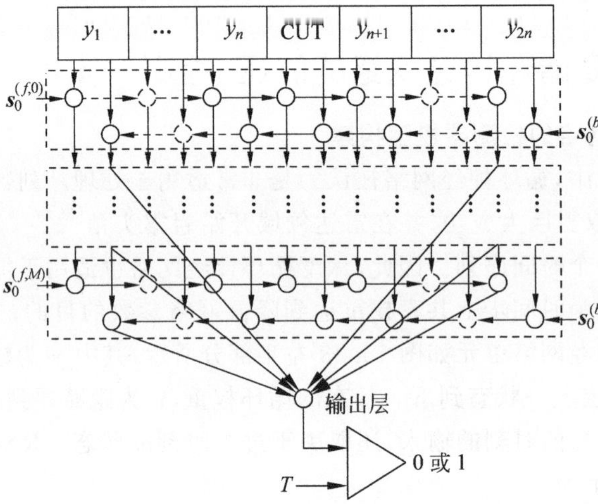

图 15.34 基于深度 RNN 的检测器框架

活函数会导致不同层之间的梯度衰退,并且 LSTM 和 GRU 结构较为复杂,计算量大。文献[57]借鉴了 LSTM 和 GRU 采用的门结构,提出了一种独立循环神经网络(Independent RNN,IndRNN)结构,通过将循环层改为点乘并采用 ReLU 激活函数,缓解了长时间和深层次带来的梯度消失和爆炸问题。三种激活函数的表达式如下

$$\text{sigmoid}(x)=\frac{1}{1+\mathrm{e}^{-x}} \tag{15.34}$$

$$\tanh(x)=\frac{\mathrm{e}^{x}-\mathrm{e}^{-x}}{\mathrm{e}^{x}+\mathrm{e}^{-x}} \tag{15.35}$$

$$\text{ReLU}(x)=\max(0,x) \tag{15.36}$$

本节分别采用原始 RNN、GRU 和 IndRNN 单元构建不同的深度神经网络检测器以选择最合适的网络结构。

2. 网络训练策略

神经网络常用的损失函数有均方根误差和交叉熵等,一般前者适用于回归问题,后者适用于分类问题。María-Pilar Jarabo-Amores 等[58-59]证明了使用 MSE 或交叉熵作为损失函数的神经网络检测器对所有可能的 P_{fa} 是最优的,因此本节采用交叉熵作为损失函数。定义神经网络的输出为 $F(\boldsymbol{y},\boldsymbol{\Theta})$,其中$\boldsymbol{\Theta}$ 表示网络的权重向量。$F(\boldsymbol{y},\boldsymbol{\Theta})$将输入 $\boldsymbol{y}$ 映射为两种假设或两种类别,即 H_0 和 H_1。假设 Ψ_i 表示对应于假设 H_i 的所有可能的输入向量的集合,则对于期望的输出$\{0,1\}$,交叉熵损失函数定义如下

$$J(\boldsymbol{\Theta})=-\frac{1}{N_{\text{B}}}\left\{\sum_{y\in\Psi_1}\lg F(\boldsymbol{y},\boldsymbol{\Theta})+\sum_{y\in\Psi_0}\lg\left[1-F(\boldsymbol{y},\boldsymbol{\Theta})\right]\right\} \tag{15.37}$$

其中,N_{B} 为每一训练批次的样本数。采用梯度下降算法作为网络的优化算法,其基本原理如下

$$\boldsymbol{\Theta}^{(i+1)}=\boldsymbol{\Theta}^{(i)}-\eta(i)\cdot\nabla_{\boldsymbol{\Theta}}J \tag{15.38}$$

其中,$i=0,1$ 表示迭代次数,$\eta(i)$为学习率,$\nabla_{\boldsymbol{\Theta}}J$ 是关于$\boldsymbol{\Theta}$ 的梯度。式(15.38)的学习率是一个较难调的超参数,一些自适应的梯度下降算法如 Adagrad、Adadelta 和 Adam(Adaptive Moment Estimation)等对 η 的变化较为稳健。由于 Adam 算法在同等的数据量的情况下占用内存较少,且对超参数相对稳

健，这里选择 Adam 算法作为优化算法。神经网络的权重通过反向传播逐层更新。

训练时采用交叉验证的方式确定模型中的超参数：①将数据集划分为训练集和验证集；②根据经验或者其他方式设定一组超参数；③在训练的过程中定期在验证集上计算模型当前的损失或者检测概率；④如果当前超参数的模型在验证集上都不理想，则改变超参数，重复上述过程；⑤选择在验证集上性能最好的模型用于后续评估测试。

3. 深度神经网络检测器的检验输出

一个训练好的神经网络检测器，其输出可近似为最大后验概率，即

$$F(\boldsymbol{y},\boldsymbol{\Theta})\approx\frac{P(\mathrm{H}_1)\Lambda}{P(\mathrm{H}_1)\Lambda+P(\mathrm{H}_0)}=P(\mathrm{H}_1\mid\boldsymbol{y})\tag{15.39}$$

其中，Λ 表示似然比，因此可以通过设定一个拒绝阈值得到最终的检测判决。根据文献[59]，此时的神经网络检测器可以称为最佳 NP 检测器或最佳贝叶斯检测器。需要注意的是，神经网络对最优检测器的近似性能依赖于网络的结构、隐含层的层数、训练算法等超参数因素，如果设置不当，网络输出将与实际的概率密度函数相差较大。

4. 多脉冲扩展

上述方案主要针对单脉冲场景，而对于多脉冲场景，有两种方案可以选择：神经网络检测之后的二值积累方案和直接神经网络多维联合检测，如图 15.35 所示。

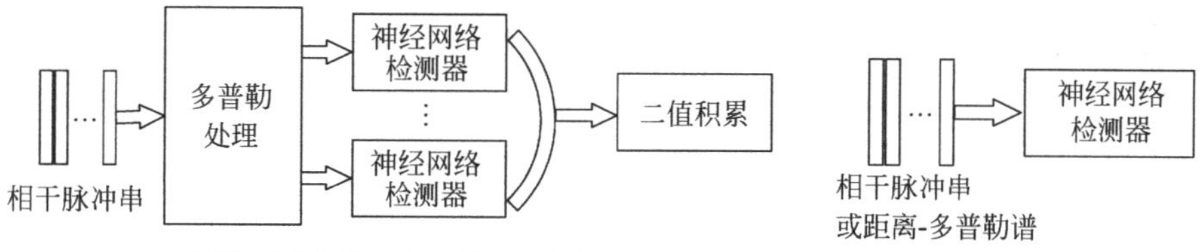

(a) 神经网络检测之后的二值积累方案　(b) 直接神经网络多维检测

图 15.35　神经网络检测之后的二值积累方案

对于前者，相当于对多普勒处理后的信号重复了多次深度神经网络(Neural Network，NN)检测后，再进行一个二值积累判决。而后者是将多个相干脉冲作为输入，相当于通过神经网络实现了脉压、相干积累和检测过程的一体化。由于输入维度和变量增多，需要更多的隐含层和更高效的网络结构。此外，如果需要指示动目标，可以将神经网络检测器的输出维度增加至 K 个(多普勒单元数)，则经过归一化指数函数后的第 i 个输出可写为

$$F_i(\boldsymbol{y},\boldsymbol{\Theta})=\frac{\mathrm{e}^{z_i}}{\sum_{k=1}^{K}\mathrm{e}^{z_k}}\approx P(\mathrm{H}_i\mid\boldsymbol{y})\tag{15.40}$$

其中，z_i 表示神经网络的第 i 个输出，H_i 表示第 i 个多普勒单元有目标。

5. 复散射系数估计

由式(15.40)可知，神经网络检测器的输出是最大后验概率，然而在很多时候，我们需要得到目标的复增益估计值。假设经过神经网络检测器后检测得到 M 个目标的距离单元索引分别为 $l_1,l_2,\cdots,l_M$。则包含这 M 个目标回波的 L 个距离单元 $\tilde{\boldsymbol{y}}=[y(0)\quad y(1)\quad\cdots\quad y(L-1)]^{\mathrm{T}}$ 可表示为

$$\tilde{\boldsymbol{y}}=\boldsymbol{S}\tilde{\boldsymbol{x}}+\tilde{\boldsymbol{v}}\tag{15.41}$$

其中，$\tilde{\boldsymbol{r}}=[r(l_1) \quad r(l_0) \quad \cdots \quad r(l_M)]^{\mathrm{T}}$ 表示目标检测单元的复增益，$\tilde{\boldsymbol{v}}=[v(0) \quad v(1) \quad \cdots \quad v(L-1)]^{\mathrm{T}}$ 为噪声矢量，$\boldsymbol{S}$ 是 $L\times M$ 的发射波形与距离像脉冲响应的卷积矩阵，其第 i 列可写为

$$\boldsymbol{S}_i=[\underbrace{0 \quad \cdots \quad 0}_{l_i} \quad s_0 \quad \cdots \quad s_{N-1} \quad 0 \quad \cdots \quad 0]^{\mathrm{T}} \tag{15.42}$$

采用最小二乘估计目标复增益可得

$$\tilde{\boldsymbol{x}}_{\text{NN-LS}}=(\boldsymbol{S}^{\mathrm{H}}\boldsymbol{S})^{-1}\boldsymbol{S}^{\mathrm{H}}\tilde{\boldsymbol{y}} \tag{15.43}$$

其中，$\tilde{\boldsymbol{x}}_{\text{NN-LS}}$ 表示神经网络检测结果通过最小二乘估计得到的目标复增益，从均方误差的角度来看，式(15.43)在加性白噪声下是最优的。如果得到了目标的多普勒信息，则可以将式(15.43)中的发射波形替换为对应的多普勒频移版本。

15.5.2 仿真与分析

本节首先介绍了训练集和测试集的构建，包括雷达、目标和噪声模型中使用的参数，然后测试了不同超参数和网络结构对模型性能的影响，最后详细评估了基于深度学习的雷达脉压、检测一体化方案相对于传统脉压、检测级联处理的性能，并在一个实测的数据集上做了验证。深度神经网络模型的构建采用强大的 TensorFlow 框架，实验环境为一台英伟达(NVIDIA)DGX-Station 工作站，主要配置为一颗 Intel 至强 E5-2698v4 20 核处理器，内存 256GB，4 块 NVIDIA Telsa V100 GPU，每块显卡 32GB 显存，共 128GB。

1. 仿真数据集构建

根据表 15.5 所示参数构建训练神经网络需要的数据集。神经网络检测器的输入为 59 个距离单元的数据，最多有 7 个目标随机分布其中。目标归一化多普勒频率在[−0.24rad，0.24rad]之间均匀分布，对应最大速度约 2.2 马赫的目标被一个脉宽 3μs 的 X 波段雷达脉冲照射所产生的多普勒频移。目标个数从 0～7 变化分别合成 50000 个数据，即总共 400000 个数据用于训练神经网络检测器，并用同样的方式生成 1000 个数据作为验证集。

表 15.5 系统参数

频段	10GHz	目标类型	Swerling 0
带宽	10MHz	目标归一化多普勒范围	[−0.24，0.24]
脉宽	3μs	最多目标数	7
采样频率	10MHz	距离单元数	59
噪声功率	0dB	检测单元	30
信噪比范围	[−15dB，45dB]		

2. 网络结构对模型训练的影响

不同网络结构和参数对神经网络检测器的性能影响很大，过于复杂的网络结构会造成过拟合或者网络退化，而过于简单的网络将影响近似性能。本节比较了不同循环神经网络如原始 RNN、LSTM、GRU 和 IndRNN，以及不同的网络层数对模型训练的影响。一般来讲，作为 LSTM 的简化版本，GRU 有着与之相当的性能，并且计算量较小，因此本节只对比了基本 RNN、GRU 和 IndRNN 之间的性能差异。初始学习率设为 0.0005，批大小为 1000，最大迭代次数为 100000 次，则不同结构和不同参数的模型训练损失如图 15.36 所示，其中 R/8/64 表示网络采用 8 层基本 RNN 单元，每层的隐含层维度为 64，

其余类推。由图 15.36 可知，网络参数最多的 G/8/128 训练损失最小。图 15.37 给出了训练损失最小的前两个网络模型的验证损失。由图 15.37 可知，两个模型的验证损失都收敛，未产生严重的过拟合，G/8/128 要优于 G/4/128。

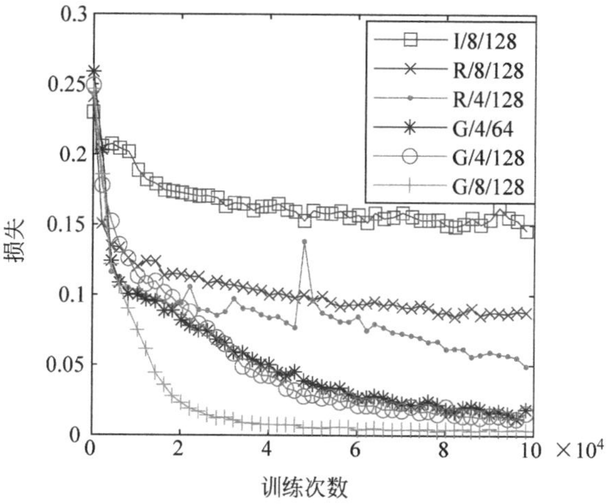

图 15.36　不同网络模型的训练损失

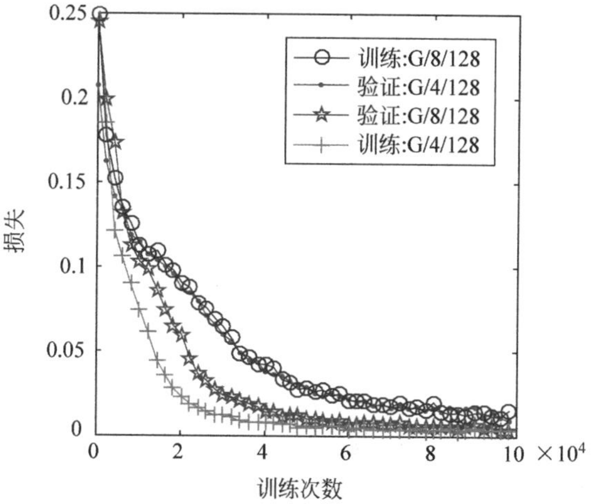

图 15.37　两种最优模型的验证损失

3. 检测性能分析

衡量一个检测器的好坏时，主要考虑三种典型情况下的性能：一是参考单元内的杂波强度是均匀的；二是参考单元内有多目标的情况；三是参考单元内杂波强度突然发生变化的杂波边缘情况。最后一种情况主要发生在陆海交界等地方，可以通过电子地图屏蔽处理，故暂不考虑这种情况。本节考虑以下三种情况时，基于深度学习的脉压、检测一体化方案与传统脉压检测级联处理的性能差异：单目标环境、参考单元内有一干扰大目标的环境、随机干扰目标场景。深度神经网络检测器采用训练和验证时收敛较好的 G/4/128 和 G/8/128 结构。作为对比的传统方法采用脉压级联恒虚警检测器的方案，其中脉压采用加汉明窗的匹配滤波。对于不同环境，CA-CFAR 和 OS-CFAR 分别被用于对比，并且其单侧的参考单元和保护单元数分别为 27 和 2。由于神经网络的输出可以看作后验概率，如果采用根据虚警概率确定门限的仿真方法可能会发生检测阈值等于 1 的情况，因此神经网络检测器的评估应该先设定阈值再计算不同阈值下的检测概率和虚警概率。基于此，本节通过仿真得到了不同信噪比下的受试者工作特征(Receiver Operating Characteristic，ROC)曲线作为不同方法的评估依据，其中信噪比定义为脉压后的信号与噪声之比，变化范围为 0～57dB，步长 3dB。

1) 单目标环境

首先考虑参考单元内没有干扰目标，即背景是均匀的瑞利分布，各种方案的 ROC 曲线如图 15.38 所示。从图 15.38(a)中可以看出，CA-CFAR 的检测性能最好，这是由于 CA-CFAR 可以看作在均匀瑞利背景下非起伏目标的最优检测器。OS-CFAR 在单目标环境下相对于 CA-CFAR 略有损失。而基于深度学习的两个检测器与 OS-CFAR 性能相当，其中 G/4/128 略优于 OS-CFAR，而 G/8/128 略逊于 OS-CFAR。同时注意到，图 15.38(a)的 ROC 曲线相比于关于未知均匀分布初相和非起伏幅度的单脉冲信号的检测曲线有一定的性能损失，这是由于本节在仿真过程中考虑了脉压旁瓣，而在传统检测器研究中一般只考虑脉压后的点目标，不关心旁瓣的影响。

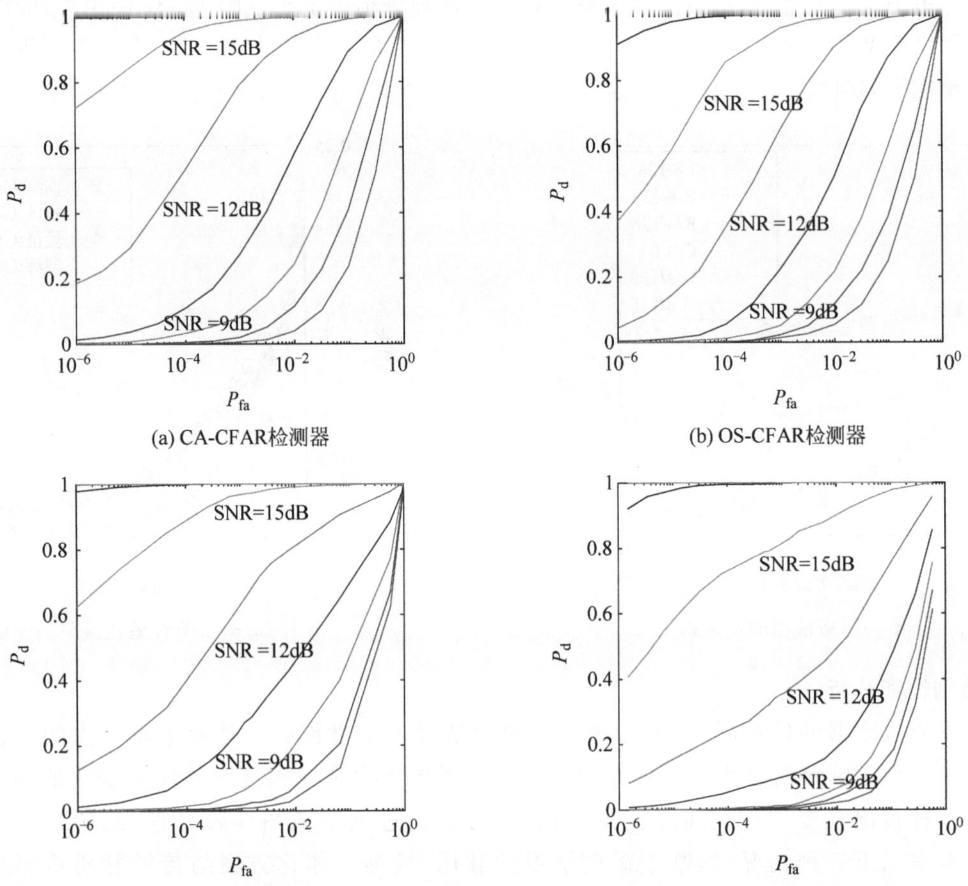

(a) CA-CFAR检测器　(b) OS-CFAR检测器　(c) G/4/128检测器　(d) G/8/128检测器

图 15.38　目标时不同检测器的 ROC 曲线

2）参考单元内有一干扰大目标

当参考单元内存在干扰目标，尤其是干扰目标较大时，其距离旁瓣会影响待检测单元，此时，干扰目标的位置会对检测性能造成影响。因此，分别考察干扰目标位置固定和随机变化时各种检测器的性能。

考虑参考单元内有一个干扰大目标，信噪比在 55～60dB 范围内均匀变化，且干扰目标的多普勒随机变化。由于 CA-CFAR 在多目标环境下性能严重下降，因此未在此列出。图 15.39(a)、图 15.39(c)、图 15.39(e)为干扰大目标位置固定时的 ROC 曲线，图 15.39(b)、图 15.39(d)、图 15.39(f)为干扰目标位置在参考单元内随机变化时的 ROC 曲线。从图 15.39 中可以看出，由于干扰目标旁瓣的原因，干扰目标位置变化会降低各个检测器的检测性能。当干扰目标位置固定时，基于深度学习的检测器性能明显优于 OS-CFAR 检测器。由于 G/8/128 结构网络层数更多，在这种情况下要显著优于层数较浅的 G/4/128 结构。当干扰目标位置随机变化时，基于深度学习的检测器性能也优于 OS-CFAR。

3）随机干扰目标场景

考虑参考单元内有 6 个干扰目标，信噪比在 0～60dB 范围内随机变化，且目标位置和多普勒随机变化。从图 15.40 所示的 ROC 曲线可以看出，受干扰目标位置、旁瓣及多普勒频移等多因素影响，OS-

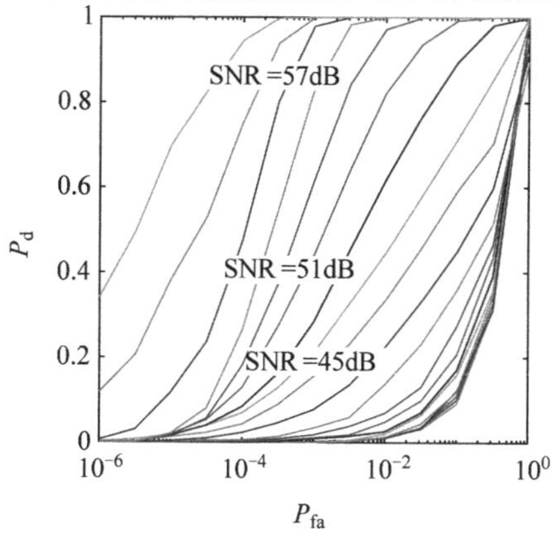

(a) OS-CFAR检测器（干扰目标固定）

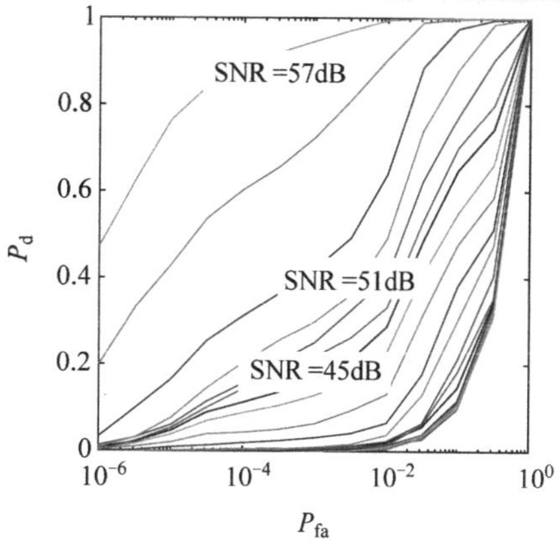

(b) OS-CFAR检测器（干扰目标变化）

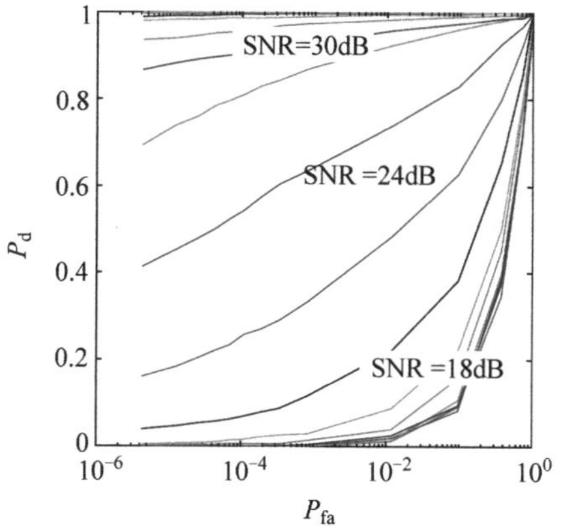

(c) G/4/128检测器（干扰目标固定）

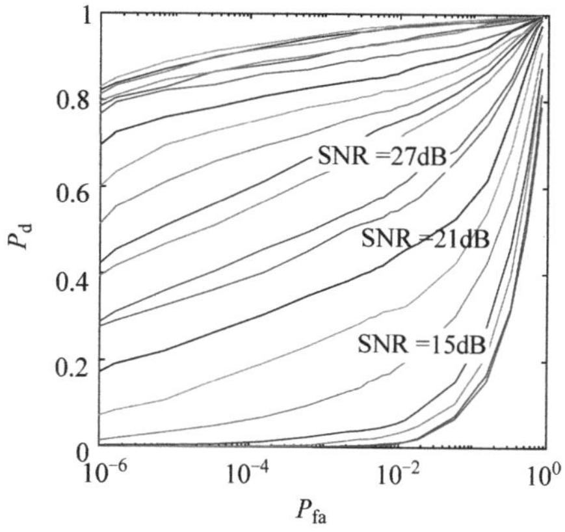

(d) G/4/128检测器（干扰目标变化）

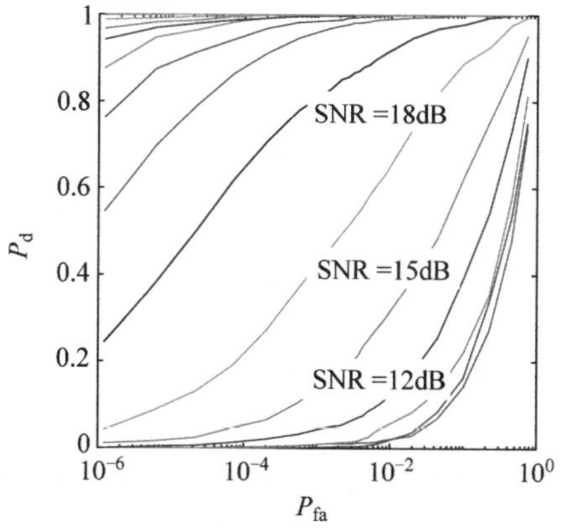

(e) G/8/128检测器（干扰目标固定）

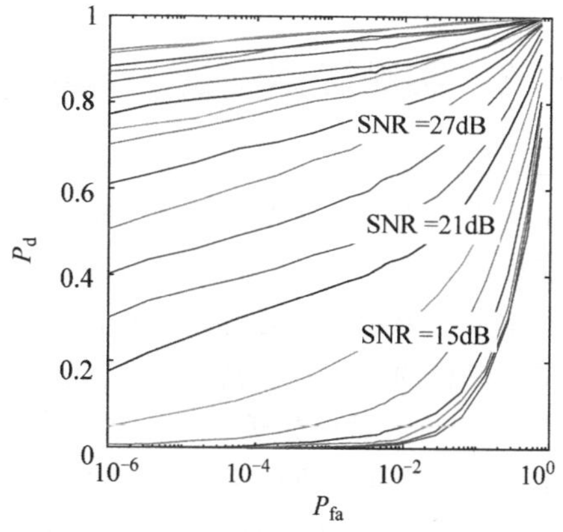

(f) G/8/128检测器（干扰目标变化）

图 15.39　存在大目标干扰时不同检测器的 ROC 曲线

CFAR 检测器的性能严重下降，而两种结构的深度学习检测器性能较好，且相对于位置随机的一个大目标时性能仅略微下降。

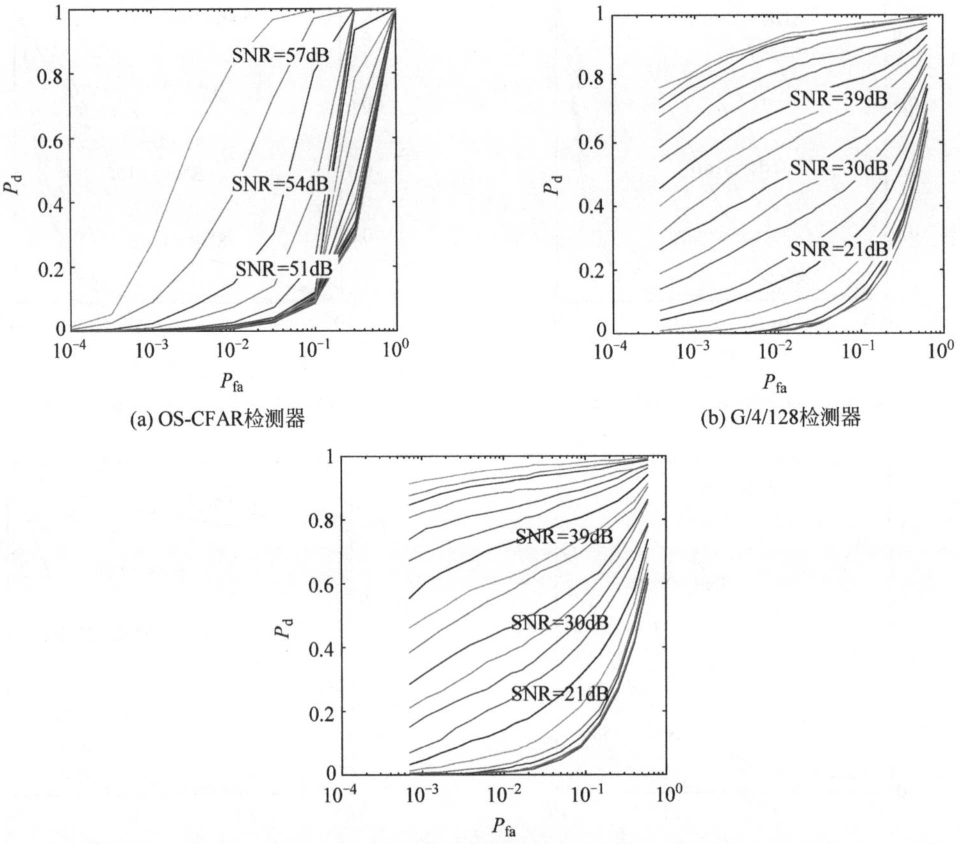

(a) OS-CFAR检测器

(b) G/4/128检测器

(c) G/8/128检测器

图 15.40　不同检测器在随机目标场景的 ROC 曲线

15.5.3　实测数据验证

本节通过仿真比较了深度学习检测器相对于传统处理方法在性能上的提升，验证了所提方案的有效性。为了更进一步验证深度神经网络检测器的实际可行性，本节通过合作目标试验，获得了大量的实测数据。在此基础上，对其中部分数据，通过数据预处理和分析，结合目标的运动轨迹，实现了对数据的标注。由于本节主要考虑的是脉压与检测的一体化处理，因此本节的实测数据是通过一个 X 波段的脉冲多普勒雷达获得的，而常规的脉冲多普勒雷达可以看作是 MIMO 的特例。

1. 试验介绍与数据预处理

试验雷达被架设于高度约 80m 的海边，雷达的极化方式为 HH，对海发射。合作目标为一艘渔船和一架无人机，渔船和无人机在指定区域内往复运动，雷达采用跟踪模式持续照射目标。此外，实验区域内还会有一些非合作渔船目标出没。实验期间海况较低，海面波浪较小，波峰呈镜面状且未破碎，有效波高在 0.2m 左右，海况等级约为 2 级。

一组试验数据的脉压和多普勒处理结果如图 15.41 所示，其中脉压结果采用加汉明窗的匹配滤波，多普勒处理采用 FFT 计算。大目标渔船位于 2200 号距离单元，小目标无人机位于 2116 号距离单元。

由图 15.41(b)可知,受杂波和大目标旁瓣影响,小目标几乎完全被遮挡。

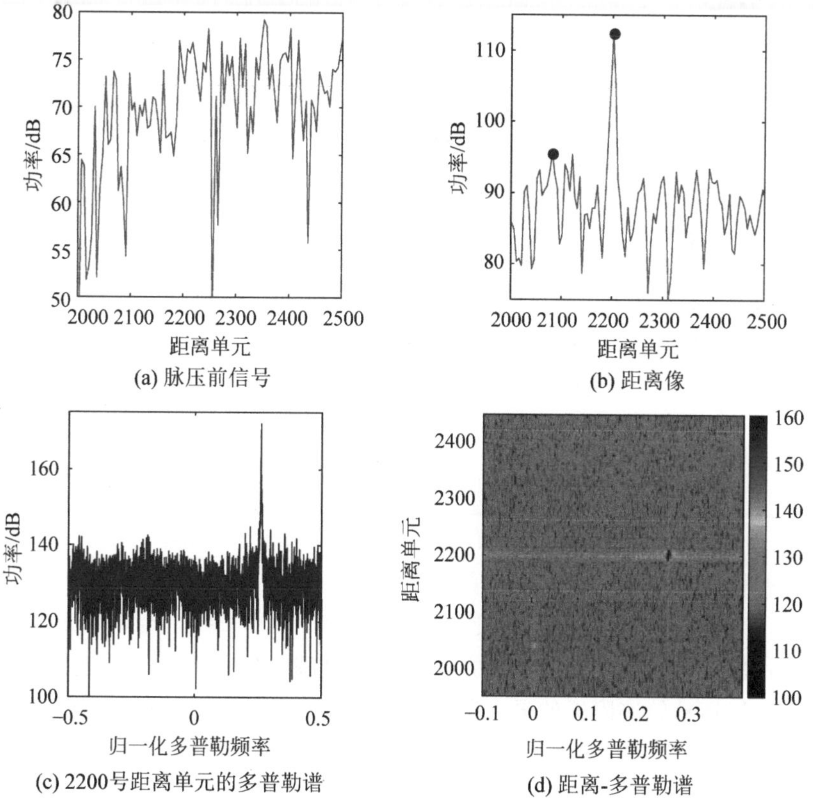

(a) 脉压前信号　(b) 距离像

(c) 2200号距离单元的多普勒谱　(d) 距离-多普勒谱

图 15.41　实测数据

由于背景分布类型会影响检测器的检测性能,因此在验证本节方法之前,首先对海杂波的幅度分布特性进行分析。幅度分布建模的理论模型包括瑞利、韦布尔、对数正态和 K 分布,其中,除对数正态分布外,其余均属于复合高斯分布族,分析结果如图 15.42 所示,估计出的模型参数及拟合误差如表 15.6 所示。从总体建模结果来看,对数正态分布与累积分布函数(CDF)曲线有较大偏离,表明该模型不适用于该情况下的海杂波建模,而对于其余三种模型,建模曲线较为接近,且与经验 CDF 的吻合度较好。

表 15.6　幅度分布参数估计值

模　　型	参　　数	参数估计值	拟 合 误 差
瑞利	参数 a	0.2021	0.0458
韦布尔	形状参数 c	1.9815	0.0410
	尺度参数 b	0.2852	
对数正态	形状参数 σ	0.6459	0.7071
	尺度参数 μ	−1.5454	
K	形状参数 γ	67.7259	0.0298
	尺度参数 c	57.5970	

从模型参数来看,韦布尔分布模型的形状参数接近 2,且 K 分布的形状参数明显大于 10,在该情况下这两种模型均在理论上近似退化为瑞利分布。尽管 K 分布的拟合误差相对更小,但均可假定海

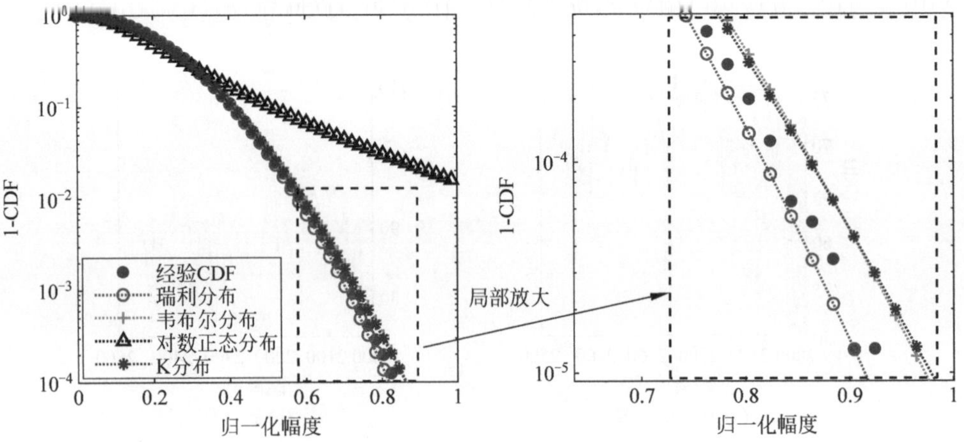

图 15.42 幅度分布概率密度函数拟合曲线

杂波建模为瑞利分布类型,与仿真场景一致。采用更多数据进行分析,均可得到类似结果,这里不再赘述。

为了对数据进行标注,本节采用距离-多普勒谱结合目标运动轨迹的方法确定目标所在的距离单元。在实测数据中,一般已不满足点目标假设,即使目标的尺寸小于一个距离单元,由于采样的原因,其脉压后的主瓣仍会占据多个距离单元。传统方法在检测过程之后会有一个点迹凝聚的过程。因此,本节在数据标注时将目标运动轨迹附近的最大散射点的位置作为目标真实位置。具体做法如下:首先在目标粗位置(实验时由 GPS 信息获取)附近的距离-多普勒谱上确定目标的潜在距离单元;然后通过一段时间内的多个潜在距离单元和对应的时间拟合出目标的运动轨迹;滑窗处理多个脉冲的距离-多普勒谱,并搜索最大值的位置,如果最大值位置在上述运动轨迹附近 2 个距离单元内,则以最大值位置作为目标真实位置,否则通过在运动轨迹上插值作为目标真实位置。由此得到 20000 条标注好的数据用于训练。

2. 实测数据结果与分析

实测数据采用网络结构为 G/8/64 的模型,用于测试的实验数据来自两个合作目标相向行驶的一次实验,同样按照前述标注方法获得 10000 条数据。由于目标散射截面积起伏等原因,小目标脉压后的信杂比最低时小于 0dB,平均信杂比约为 9dB。分别采用 CA-CFAR 检测、OS-CFAR 检测器和 G/8/64 的深度神经网络检测器进行处理,其中 CFAR 的单侧参考单元和保护单元分别取 16 和 1,并在检测之前采用加汉明窗的匹配滤波做脉压处理,实验结果如所示图 15.43 所示。

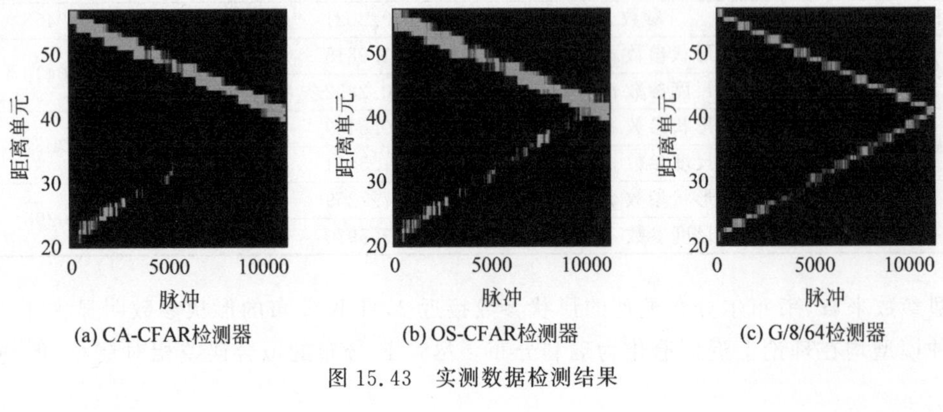

图 15.43 实测数据检测结果

从图15.43(a)中可以看出,CA-CFAR检测器在两目标相距约16个距离单元时性能开始降低,这是由于此时参考单元中开始有干扰大目标存在,而图15.43(b)中的OS-CFAR则在靠近大目标处仍能较好地检测到小目标,但是仍有较多漏警。图15.43(c)中的深度学习检测器呈现出最优的检测效果,即使靠近大目标附近仍能很好地检测到小目标。对图15.43中第20～40距离单元处的小目标检测结果进行统计,三种检测器对应的检测概率和虚警概率如表15.7所示。在相同量级的虚警概率下,G/8/64检测器对小目标的检测概率最大。

表15.7 小目标的检测概率和虚警概率

检 测 器	P_d	P_{fa}
CA-CFAR检测器	0.3101	0.0015
OS-CFAR检测器	0.3948	0.0060
G/8/64检测器	0.6686	0.0023

15.6 小结

本章主要讨论了几种利用特征差异的海杂波中目标CFAR检测方法,具体包括利用海尖峰分形特征的目标检测方法、利用频域分形特征的目标CFAR检测方法和基于深度学习的脉压、检测一体化处理技术,并采用实测数据验证了检测方法的性能。

15.2节主要研究了海杂波的时域分形特性,采用Paretian泊松模型建模海杂波,形成相应的特征并构造目标检测方法。实测数据验证表明,所提出的利用海尖峰分形特征的目标检测方法性能优于双参数CFAR检测方法,但该方法需要大量的雷达回波采样点提取尖峰并估计参数以降低统计误差,因此其实时性有待进一步提升。

基于分数布朗运动,15.3节首先说明了傅里叶变换保持原序列分形特性的性质,在已有的关于时域海杂波单一分形建模研究基础之上,研究海杂波频谱的单一分形特性,得到相应的分形特征,并对分形参数的影响因素进行分析;利用对海杂波和目标回波有明显区别的频域分形特征设计海杂波中的目标检测方法,并基于实测数据进行性能分析。得益于傅里叶变换有效提升信杂比的优点,所提出的目标检测方法比时域中经典的分形目标检测方法具有更好的性能。

15.4节构建了利用多特征进行目标检测的框架,利用从雷达回波中提取的三种特征与凸包分类方法相结合检测海面漂浮小目标,并利用实测数据进行了性能分析,实验结果表明,在较短观测时间内,基于三维特征的检测器相比较两种基于分形的单一特征检测器和组合ANMF检测器有更好的检测性能。

传统基于模型驱动的级联处理方法对模型假设依赖严重,如果数据与假设模型不符,则处理性能将下降,为此,15.5节讨论了基于深度学习的数据驱动方法,构建了基于深度循环神经网络的脉压、检测一体化处理框架,并分别通过仿真和实测数据集详细评估了深度学习与传统级联处理的性能。结果显示,基于深度循环神经网络的脉压、检测一体化方案可以取得和传统处理方法相当的性能,在复杂环境下可以超越传统处理方法的性能,有效降低了对各种模型假设的要求。

参考文献

[1] Lo T，Leung H，Haykin S. Fractal Characterisation of Sea-Scattered Signals and Detection of Sea-Surface Targets[J]. IEE Proceedings-F. 1993，140(4)：243-250.

[2] Shen X L，Song Z Y，Zhu Y F，et al. Fractal detector design and application in maritime target detection[J]. Journal of Systems Engineering and Electronics，2017，28(01)：27-35.

[3] 郭睿，臧博，张双喜，等. 高分辨 SAR 复杂场景中的人造目标检测[J]. 电子与信息学报，2010，32(12)：3018-3021.

[4] 陈小龙，刘宁波，宋杰，等. 海杂波 FRFT 域分形特征判别及动目标检测方法[J]. 电子与信息学报，2011，33(4)：823-830.

[5] 包永强，赵力，邹采荣. 噪声环境下语音分形特征的提取和分析[J]. 电子与信息学报，2007，29(3)：585-588.

[6] Hu J，Tung W，Gao J B. Detection of Low Observable Targets within Sea Clutter by Structure Function based Multifractal Analysis[J]. IEEE Trans. A&P，2006，54(1)：136-143.

[7] Gao J B，Cao Y，Lee J M. Principal component analysis of $1/f$ noise[J]. Phys. Lett. A，2003，3(14)：392-400.

[8] Chang Y C，Chang S. A Fast Estimation Algorithm on the Hurst Parameter of Discrete-Time Fractional Brownian Motion[J]. IEEE Trans. Signal Processing，2002，50(3)：554-559.

[9] Kim T S，Kim S. Singularity spectra of fractional Brownian motions as a multi-fractal[J]. Chaos，Solitons & Fractals，2004，19：613-619.

[10] McLaughlin D，Allan J，N，Twarog E. M. High resolution polarimetric radar scattering measurements of low grazing angle sea clutter[J]. IEEE Journal of Oceanography Engineering，1995，20(3)：166-178.

[11] West J C，Sturm J M，Ja S J. Low-grazing scattering from breaking water waves using an impedance boundary MM/GTD approach[J]. IEEE Trans. Antennas Propagat.，1998，46(1)：93-100.

[12] Liu Y，Frasier S J，McIntosh R E. Measurement and classification of low-grazing-angle radar sea spikes[J]. IEEE Trans. Antennas Propagat.，1998，46(1)：27-40.

[13] Melief H W，Greidanus H，Genderen P. Analysis of sea spikes in radar sea clutter data[J]. IEEE Trans. Geoscience and Remote Sensing，2006，44(4)：985-993.

[14] 丁昊，董云龙，刘宁波，等. 海杂波特性认知研究进展与展望[J]. 雷达学报，2016，5(5)：499-516.

[15] 刘宁波，董云龙，王国庆，等. X 波段雷达对海探测试验与数据获取[J]. 雷达学报，2019，8(5)：656-667.

[16] 丁昊，刘宁波，董云龙，等. 雷达海杂波测量试验回顾与展望[J]. 雷达学报，2019，8(3)：281-302.

[17] 徐雅楠，刘宁波，丁昊，等. 利用 CNN 的海上目标探测背景分类方法[J]. 电子学报，2019，47(12)：2505-2514.

[18] 许述文，白晓惠，郭子薰，等. 海杂波背景下雷达目标特征检测方法的现状与展望[J]. 雷达学报，2020，9(4)：684-714.

[19] 郭子薰，水鹏朗，白晓惠，等. 海杂波中基于可控虚警 K 近邻的海面小目标检测[J]. 雷达学报，2020，9(4)：654-663.

[20] Song C，Xiuwen L. Statistical analysis of X-band sea clutter at low grazing angles[C]. Bangkok：2020 International Conference on Big Data & Artificial Intelligence & Software Engineering(ICBASE)，2020.

[21] Rosenberg L，Duk V，Ng B W H. Detection in Sea Clutter Using Sparse Signal Separation[J]. IEEE Transactions on Aerospace and Electronic Systems，2020，56(6)：4384-4394.

[22] Guo Z X，Shui P L. Anomaly Based Sea-Surface Small Target Detection Using K-Nearest Neighbor Classification[J]. IEEE Transactions on Aerospace and Electronic Systems，2020，56(6)：4947-4964.

[23] 刘宁波，丁昊，黄勇，等. X 波段雷达对海探测试验与数据获取年度进展[J]. 雷达学报，2021，10(1)：173-182.

[24] Posner F，Gerlach K. Sea spike demographics at high range resolutions and very low grazing angles[C]. Huntsville：IEEE Radar Conference，2003.

[25] Greco M，Stinco P，Gini F. Statistical analysis of sea clutter spikes[C]. Rome：Proc. of the 6th European Radar

Conference, 2009.
[26] Ding H, Liu N, Huang Y, et al. Covariance Matrix Estimation in Compound-Gaussian Sea Clutter with Discrete Spikes [C]. Chongqing: IEEE International Conference on Signal, Information and Data Processing (ICSIDP), 2019.
[27] Smith M J, Poulter E M, McGregor J A. Doppler radar measurements of wave groups and breaking waves[J]. Journal of Geophysical Research, 1996, 101(C6): 14269-14282.
[28] Sun H Q, Wan X Y. Research on the sea spike suppression based on range domain characteristic of relatively high resolution radar[C]. Journal of Physics: Conference Series, 2020.
[29] Fuchs J, Regas D, Waseda T, et al. Correlation of hydrodynamic features with LGA radar backscatter from breaking waves[J]. IEEE Trans. Geoscience and Remote Sensing. 1999, 37(5): 2442-2460.
[30] Fred L P. Experimental observations at very low grazing angles of high range resolution microwave backscatter from the sea[R]. Naval Research Laboratory Report, NRL/MR/53 10-98-8326, 1998.
[31] Fred L. P. Spiky sea clutter at high range resolutions and very low grazing angles[J]. IEEE Trans. Aerospace and Electronic Systems, 2002, 38(1): 58-73.
[32] Trizna D B, Hansen J P, Hwang P. Laboratory studies of radar sea spikes at low grazing angles[J]. Journal of Geophysical Research, 1991, 96(C7): 12529-12537.
[33] Wit J M, Schouten M W. Discriminating sea spikes in incoherent radar measurements of sea clutter[R]. TNO Report, TNO-DV 2008 A067, 2008.
[34] Greco M, Gini F, Rangaswamy M. Statistical analysis of measured polarimetric clutter data at different range resolutions[J]. IEE Proc. Radar Sonar Navigation, 2006, 153(6): 473-481.
[35] Greco M, Stinco P, Gini F. Identification and analysis of sea radar clutter spikes[J]. IET Radar Sonar Navigation, 2010, 4(2): 239-250.
[36] Eliazar I, Klafter J. Paretian Poisson processes[J]. Journal of Statistical Physics, 2008, 131(3): 487-504.
[37] Eliazar I, Klafter J. Correlation cascades of Lévy-driven random processes[J]. Physica A, 2007, 376(2): 1-26.
[38] Drosopoulos A. Description of the OHGR database[R]. Ottawa: Defence Research Establishment, Tech. Note No. 94-14, 1994.
[39] 欧阳文. 基于海杂波模型的目标检测方法研究[D]. 烟台：海军航空工程学院，2005.
[40] 孟华东，王希勤，王秀坛，等. 与初始噪声分布无关的恒虚警处理器[J]. 清华大学学报(自然科学版)，2001，41(7)：51-53.
[41] Falconer K. 分形几何：数学基础及其应用[M]. 北京：人民邮电出版社，2007.
[42] Mandelbrot B B. The Fractal Geometry of Nature[M]. San Francisco: WH Freeman, 1982.
[43] Decreusefond L, Üstünel A S. Stochastic analysis of the fractional Brownian motion[R]. American Mathematical Society 1991 subject classifications, Primary 60H07; Secondary 60G18, 2007: 1-39.
[44] 孙霞，吴自勤，黄畇. 分形原理及其应用[M]. 合肥：中国科学技术大学出版社，2006.
[45] Chandola V, Banerjee A, and Kumar V. Anomaly detection: a Survey[J]. ABM Computing Surveys, 2009, 41(3): 15.1-15.58.
[46] Steinwart I, Scovel C, Hush D. A Classification Framework for Anomaly Detection[J]. Journal of Machine Learning Research, 2005, 34: 211-232.
[47] Hinz J O, Hvlters M, Zolzer U, et al, Presegmentativn-based adaptive CFAR detection for HFSWR[C]. Atlanta: 2012 IEEE Radar Conference, 2012.
[48] Wang X, Liu J, Liu H. Small Target Detection in Sea Clutter Based on Doppler Spectrum Features[C]. Shanghai: 2006 CIE International Conference on Radar, 2006.
[49] Hartigan J A. Estimation of a convex density contour in two dimensions[J]. Journal of the American Statistical Association, 1987, 82(397): 267-270.
[50] 李东宸. 海杂波中小目标的特征检测方法[D]. 西安：西安电子科技大学，2016.

[51] Hu J, Gay J B, Posner F L, et al. Target Detection Within Sea Clutter: A Comparative Study by Fractal Scaling Analyses[J]. Fractals, 2006, 14: 187-204.

[52] Xu X K. Low Observable Targets Detection by Joint Fractal Properties of Sea Cutter: An Experimental Study of IPIX OHGR Datasets[J]. IEEE Transactions on Antennas and Propagation, 2010, 58(4): 1425-1429.

[53] Graves A, Mohamed A, Hinton G. Speech Recognition with Deep Recurrent Neural Networks[C]. Vancouver: ICASSP, 2013.

[54] Pascanu R, Gulcehre C, Cho K, et al. How to Construct Deep Recurrent Neural Networks[C]. Computer Science, 2014.

[55] Chu Y, Fei J, Hou S. Adaptive Global Sliding-Mode Control for Dynamic Systems Using Double Hidden Layer Recurrent Neural Network Structure[J]. IEEE Transactions on Neural Networks and Learning Systems, 2020, 31(4): 1297-1309.

[56] Moalla M, Frigui H, Karem A, et al. Application of Convolutional and Recurrent Neural Networks for Buried Threat Detection Using Ground Penetrating Radar Data[J]. IEEE Transactions on Geoscience and Remote Sensing, 2020, 58(10): 7022-7034.

[57] Li S, Li W, Cook C, et al. Independently Recurrent Neural Network(IndRNN): Building a Longer and Deeper RNN[C]. Salt Lake City: Computer Vision and Pattern Recognition, 2018.

[58] Jarabo-Amores P, Rosa-Zurera M, Gil-Pita R, et al. Sufficient Condition for an Adaptive System to Approximate the Neyman-Pearson Detector[C]. Bordeaux: IEEE/SP 13th Workshop on Statistical Signal Processing, 2005.

[59] Jarabo-Amores M, Rosa-Zurera M, Gil-Pita R, et al. Study of Two Error Functions to Approximate the Neyman-Pearson Detector Using Supervised Learning Machines[J]. IEEE Transactions on Signal Processing, 2009, 57(11): 4175-4181.

第16章 回顾、建议与展望

CHAPTER 16

本书的目的是系统和深入地讨论雷达目标检测和CFAR处理问题，通过大量地收集文献和系统整理，本书形成了CFAR处理理论体系，为开展该领域的工作提供了丰富的数学模型、分析结论和参考资料。

16.1 回顾

16.1.1 形成CFAR处理理论体系

在出版本书第一版时，就本书作者当时所掌握的情况来看，无论是在国内还是在国际上，尚无专著对CFAR处理方法进行系统和全面的分析与讨论。本书试图建立一个完整的体系，将CFAR处理理论划分成若干个研究领域。

(1) 从体制上看，分为低分辨(点目标)条件下的和高分辨(一维距离扩展目标和二维成像目标)条件下的CFAR处理方法，以及低维(单传感器)和高维(阵列、多传感器分布式)的CFAR处理方法。

(2) 从背景杂波统计特性方面，分为高斯(指数分布)和非高斯(韦布尔分布、对数正态分布、α稳定分布、复合高斯分布)杂波中的CFAR处理方法。

(3) 按照信号处理方式，分为由空域(利用空间上相邻参考单元的样本)到时域(杂波图)、由时域到变换域(频域、小波域、FRFT域、HHT域、稀疏域等)、由参量处理到非参量处理、由单一的幅度特征到分形等多种特征的CFAR处理方法。

本书第3～15章的内容就是按照这种顺序逐步递进安排的。为了本书的完整性，第2章系统介绍了雷达目标检测的经典理论，即已知噪声或杂波平均功率水平的固定门限检测，以此作为进行CFAR处理研究的基础。本书对具有代表性的CFAR处理方法进行了比较分析。性能分析统一在均匀背景、杂波边缘和多目标环境三种典型背景情况下。在均匀背景和多目标环境中，以检测概率、平均判决阈值ADT和CFAR损失作为重要的性能指标；在杂波边缘环境中，以对虚警概率的控制能力作为重要的依据。总之，对于一种CFAR处理方法来说，在保持一定的虚警概率的同时，提高检测概率是其主要目标。

本书适用于从事雷达系统研究与设计的专业人员，尤其是作为重要的参考资料，将有助于从事CFAR研究的学者和从事雷达教学的教师对CFAR处理形成清晰、全面、系统和正确的认识。

16.1.2 提出GOS类CFAR检测器并建立统一模型

本书作者提出了具有自动筛选技术的GOS类CFAR检测器，并对它们的性能进行了分析和比较。

它们是 CA、SO、GO、OS 和 TM 的扩展，即更广义的 CA、SO、GO、OS 和 TM。在 R_1、R_2 和 k_1、k_2 设置为某些特定值时，这些新的 GOS 类 CFAR 检测器就变成了上述五种检测器之一，并且在 R_1、R_2 和 k_1、k_2 变化时，GOS 类检测器可以由上述检测器的一种过渡到另一种。因此，GOS 类检测器兼有这五种检测器的特点。并且可供选择的 k_1、k_2 参数使之设计更具灵活性。因而这类 CFAR 检测器具有重要的实际应用价值。

OS 是 TM 的特例，TM 涵盖了 OS 和 CMLD 等情形，因此上述的检测器中 OS 处理过程若用 TM 代替，上述的检测器就成为基于 TM 的 GOS 类 CFAR 检测器的特例。因此，基于 TM 的 GOS-CFAR 的处理模型和检测性能数学模型是这类检测器的统一模型，本书作者完成了这项建立统一模型的工作，提出了 MTM、TMGO、TMSO-CFAR 检测器，将其统称为 TCGS-CFAR 检测器。

16.1.3 延伸自适应 CFAR 检测

第 6 章之前讨论的 CFAR 检测器在形成检测阈值时也是自适应的，而第 6 章单独归纳出一类自适应 CFAR 检测方法，是因为它们在形成检测阈值的过程中不需要任何关于干扰(如杂波边缘和干扰目标)的先验信息，可以自动适应于干扰的变化，因而在实际工作环境中，自适应 CFAR 检测方法具有很强的优越性。CCA、E-CFAR 及一系列基于逐个参考样本多步删除规则的自适应 CFAR 检测方案都是针对多目标干扰环境设计的自适应 CFAR 检测器，HCE-CFAR 是针对杂波边缘环境设计的，而 VI-CFAR 则基于一定的判据综合利用 CA、GO 和 SO-CFAR。上述自适应 CFAR 检测器都基于包络大小来筛选参考样本，本书作者提出的 ESECA-CFAR 检测器从实测数据中目标回波包络形状的特点出发，综合利用包络大小与形状信息来设计参考样本筛选规则，因此对复杂非均匀环境具有更强的自适应能力，同时也为自适应 CFAR 检测器的设计延伸了一个方向，即利用回波特征信息来设计参考样本筛选规则。

16.1.4 发展多传感器分布式 CFAR 检测

增加传感器数量是提高探测性能的重要途径，在雷达探测领域中的应用就是多传感器分布式检测。经典的多传感器分布式检测有两级处理过程，第一级是二元量化处理，即局部判决，第二级是对局部判决结果融合。在这种情况下，CFAR 处理应该在哪一级实现？如何实现？这些都是多传感器分布式检测中 CFAR 处理需要研究的问题。本书作者阐述了在局部处理中完成 CFAR 处理的思路，提出基于局部检测统计量的分布式 CFAR 检测，并设计了 R 类局部检测统计量等几种具体形式，很好地解决了多传感器分布式 CFAR 处理问题。

16.1.5 将 CFAR 处理由时域和频域拓展到多种变换域

对于目标信号和杂波噪声在时域或频域难以分开的情况，变换到另一个域处理是现代信号处理的一个主要途径。现代信号处理理论提供了很多变换工具，如短时傅里叶变换、小波变换等时频分析工具，还有 FRFT、HHT 等。变换域处理将雷达信号变换至不同变换域，通过基函数分解、滤波、相参积累、经验模态分解等方法，在改善信杂噪比的同时，增强目标与背景噪声或杂波之间的特征差异。本书详细介绍了频域、小波域、FRFT 域、Hilbert-Huang 变换域和稀疏域的雷达目标检测方法，使变换域不局限于频域，拓展了变换域处理的内涵，这是一个很有发展前景的方向。本书补充了 FRFT、HHT 和稀疏时频分布等变换域 CFAR 处理，并通过雷达海上目标实测数据进行了验证，为提升雷达低可观测目标检测能力提供了新的方法。

16.1.6　将CFAR处理的信息源维度由一维扩展到多维并形成多维CFAR检测

雷达体制由一维向多维发展，使得雷达系统能够提供给CFAR检测的信息处理自由度越来越多，进而形成多维CFAR检测。本书详细阐述了复高斯背景中的"阵元-脉冲"二维联合CFAR检测，提出了基于自适应空时编码设计的"阵元-波形"二维联合CFAR检测技术、基于空-时-距自适应处理的"阵元-脉冲-波形"三维联合CFAR检测技术及基于TBD思想的扫描帧间融合检测技术。这些CFAR检测技术联合利用雷达体制提供的脉冲维、阵元维、波形维及扫描帧维等多维度信息，通过扩大观测空间维度提高杂波抑制能力及联合自适应处理提高各维度旁瓣抑制能力，来改善目标信杂噪比、提升目标检测性能。多维CFAR处理将会朝着自适应聚焦、检测跟踪识别一体化处理方向发展，是一个很有潜力的研究领域。

16.1.7　将幅度特征拓展到分形等多种特征

以往的CFAR检测多以幅度作为检测统计量。实际上检测就是区分目标与背景杂波，检测器设计即寻找使目标与背景杂波差异最大化的特征，幅度是特征之一。随着数学和信号处理理论的发展，很多新的特征使目标与背景杂波差异更为明显。尤其对于杂波中的微弱目标检测，目标的幅度特征已经不占优势，需要寻找其他新的特征。第15章介绍了利用分形等特征和深度学习方法进行分类的雷达目标检测方法。

16.2　问题与建议

16.2.1　性能分析与评价方法

在分析多参数的CFAR检测器时，由多个参数组合会产生很多种可能。例如，OS-OS类CFAR检测器具有两个可变参数 k_1 和 k_2，对于较大的 R_1 和 R_2 可产生上百种不同参数的CFAR检测器。逐一地比较它们之间的性能是不可能的，而且也是不全面的。所以，应将分析方式更加定量化。例如，能否利用 P_d、P_{fa}、T 或ADT对 k_1 和 k_2 的偏导和混合偏导，以及它们之间变化规律的三维图形来精确地分析 k_1、k_2 对性能的影响。

在选择CFAR检测器参数时还存在一个难题，即很难权衡多种情况下的性能。例如，具有较好检测性能的参数可能使检测器在杂波边缘环境中的性能不太理想，或者有良好抗边缘杂波性能的参数却对检测性能极为不利。因此，需要研究性能综合评价的准则，如一个综合性能指数 E，将使综合性能指数 E 产生极值的参数作为最佳参数。

16.2.2　加强对目标特性的研究

CFAR处理是针对剩余杂波的处理技术，主要是控制剩余杂波产生的虚警，似乎与目标的特性无关，并且多数CFAR处理方法也主要从杂波特性的角度进行研究。然而从检测的角度考虑，不仅要看虚警控制能力，还要看检测能力，即检测概率。因此，需要考虑目标特性对CFAR处理方法性能的影响，尤其在积累过程中主要利用的就是杂波与目标的特性差异，这就需要研究和掌握目标特性[1-2]。

16.2.3 拓展 CFAR 研究思路

经典的 CFAR 处理方法主要是在时域或频域上对信号幅度样本进行处理。现代雷达技术的发展使雷达信号维度在时间、空间、频率、波形、视角上得到了很大的扩展,而且现代信号处理理论的发展也为雷达目标检测提供了更多的变化域处理手段,如小波域、FRFT 域、稀疏表示域等。此外、分形、混沌等非线性数学理论也为信号检测提供了新的检测统计量,且相比于经典的基于幅度的检测统计量能更好地区分目标与杂波。实际上,CFAR 阈值就是高维信号空间中的目标与杂波子空间的分割曲面。所以,建议在多维、多域、多特征的框架下寻找解决 CFAR 处理问题的思路。

16.2.4 注重新体制雷达中的 CFAR 处理研究

随着 CFAR 技术的广泛应用,雷达目标 CFAR 处理已拓展至超宽带成像、低频(UHF)、无源被动、分布式多基地、MIMO、多波段多极化、天基等新体制雷达应用领域,其中涉及的部分信号检测模型与传统模型差异较大,背景杂波统计特性亦出现新变化,新体制雷达面临具体应用背景时,可能涉及非高斯杂波精细建模、干扰抑制、非参量检测、杂波样本等方面的新问题。建议结合新体制雷达的具体信号形式、检测模型和应用场景,开展 CFAR 处理方法研究。

16.3 研究方向展望

几十年来,尽管雷达目标检测 CFAR 处理理论已有了很大的发展,但随着探测技术发展,仍然有很多新的问题有待进一步研究与探索。下面仅阐述其中的一些主要研究方向。

16.3.1 多维信号 CFAR 处理

拓展观测维度是探测手段不断发展的一个重要体现,因此多维信号处理也就成为信号处理领域的一个主要发展方向。经典的 CFAR 检测主要在时域和频域上进行处理,维度局限于时间维,如脉冲雷达的快时间维与慢时间维。相控阵雷达使雷达信号处理扩展到空间维,而数字阵列雷达的逐步实用化[3-4],进一步推动了阵列信号处理理论与实践的深入发展,将数字波束形成、空时自适应处理与目标检测结合起来,形成一个 CFAR 检测统计量,同时完成目标空时能量的积累、杂波与干扰抑制及 CFAR 检测,已成为阵列信号 CFAR 处理的一个特点。

现代雷达技术的不断进步使得雷达信号处理又有了更多的维度。譬如时间维还有帧间和长时间的多时相,此外还有空间(距离、方位、俯仰)、频率(多波段、频率分集、步进频)、极化(单极化、全极化)、通道(阵元)、波形(MIMO)、视角(空间分集),以及涡旋电磁波的轨道角动量等。高维信号蕴含目标和杂波的高维特性,更有助于区分目标与杂波,但是高维信号需要有效的表示手段才能展现高维特性,如向量、矩阵及张量[5-8]、图数据[9]等。因此,多维 CFAR 处理应当重点关注多维信号表征及如何从表征域中提取目标与背景的差异特征,这应该是一个很有意义的研究方向。

16.3.2 背景杂波辨识与智能处理

雷达的工作环境一般是变化的,背景杂波的特性也在时间和空间上变化。CFAR 处理的基本方法就是由参考单元样本筛选出与检测单元背景杂波特性一致的,然后由此估计背景杂波功率水平,进而形成检测阈值。因而,背景杂波的辨识是非常基本且重要的步骤。

背景杂波的辨识应在多维、多域、多特征的框架下进行。先进的数学理论(如信息几何[10-15]和流行的机器学习方法)具有更强的学习能力,记忆、选择和抽象能力,以及对数据或环境条件不确定性的容忍能力。

雷达技术发展至今对人的依赖还是很重的,有经验的雷达操作员的视觉观察处理在目标检测和虚警控制方面常常胜于众多的CFAR处理方法。人的视觉观察机理非常复杂,在图像信息提取方面非常有效,是算法至今难以替代的。将视觉处理技术的一些方法应用到CFAR处理中也是非常值得研究的。

16.3.3 信号处理新方法应用与多特征CFAR处理

现代信号处理手段日益丰富,如信息几何[11]、神经网络[16]、图信号处理[17]、高阶统计量[17]、非整数阶矩估计[18]、模糊信息处理[19]及FRFT等变换域处理。在不同信号处理方法中,雷达回波中的目标和杂波可能体现出不同特征。如何有效提取出显著的差异性特征,并将多种特征综合表达,以及应用于CFAR处理是雷达目标检测中的关键工作。

16.3.4 其他领域的CFAR处理

CFAR处理不仅在雷达目标检测中是重要处理环节,在很多信号检测中都有应用。作为探测手段,除了电磁波,声波也是一种有效的探测手段,可以在液体、固体、气体中传播。例如,在水下,雷达失去了威力,而声信号却能在水下信息感知中发挥作用,还有地声和空气声在探测中也有广泛应用。声信号检测中也常常用到CFAR处理。

在水声信号检测中,CFAR处理也是重要的研究内容[20]。类似对雷达背景杂波的统计特性描述,声呐信号处理中也要对背景混响的统计特性建模。早期也用高斯分布,随着声呐分辨率的提高,混响统计特性偏离高斯分布,可用复合高斯模型,或用非高斯拟合分布描述,如Pearson分布、Pareto分布,能够高分辨率描述声呐混响的统计分布,拟合效果优于Lognormal、Weibull和K分布。

文献[21]和[22]在声呐信号检测CFAR处理方面做了系列研究工作,在Pareto分布和Pearson分布混响背景下分析了OSSO和OSGO-CFAR检测器性能;在ParetoⅡ型分布混响条件下,文献[23]提出了基于几何平均(GM)CFAR的距离扩展目标检测方法;在抗多目标干扰方面,文献[24]提出了UMCASO-CFAR检测器,在前后滑窗分别用无偏最小方差估计(UMVE)和单元平均(CA)形成局部估计,然后选两者中最小作为混响背景功率水平估计;文献[25]提出了基于自动删除单元平均(ACCA)CFAR的距离扩展目标检测方法。潜艇减振降噪、敷设消声瓦等低辐射技术的发展使低频声呐技术备受关注。但是低频噪声平稳性差、干扰多。针对低频噪声背景中主动声呐脉冲信号的检测问题,文献[26]研究了频域CFAR检测方法。Verma在文献[27]中将VI-CFAR用于声呐目标检测,并与CA-CFAR进行了比较分析。文献[28]提出用VI-CFAR评估背景特性,然后自适应地在CA、GO、OS、ACMLD-CFAR中选择检测方法,分析表明,该方法在混响边缘和离散干扰等非均匀背景下性能良好。在非瑞利分布混响处理方面,文献[29]将混响数据进行转换处理,将非瑞利分布转换为近似瑞利分布,然后进行目标检测。

除水声信号处理外,在空气声信号处理方面(如车辆检测、智能地雷、语音识别、脚步声识别),以及在固体声信号处理中(如管道检测等),还有在电网信号检测[30]中都有CFAR处理的应用。

参考文献

[1] 关键. 雷达海上目标特性综述[J]. 雷达学报, 2020, 9(4): 674-683.

[2] 黄培康，殷红成，许小剑. 雷达目标特性[M]. 北京：电子工业出版社，2005.

[3] 张光义，赵玉洁. 相控阵雷达技术[M]. 北京：电子工业出版社，2006.

[4] 张明友. 数字阵列雷达和软件化雷达[M]. 北京：电子工业出版社，2008.

[5] 黄克智，陆明万. 张量分析[M]. 北京：清华大学出版社，2003.

[6] Kolda T G, Bader B W. Tensor Decompositions and Applications[J]. SIAM Review, 2009, 51(3): 455-500.

[7] 张星. 组合空间和光谱特性的高光谱图像异常检测与目标识别方法研究[D]. 长沙：国防科学技术大学，2016.

[8] 李翠平. 张量匹配子空间检测及应用[D]. 西安：西安电子科技大学，2019.

[9] Su N Y, Chen X L, Guan J, et al. Maritime target detection based on radar graph data and graph convolutional network[J]. IEEE Geoscience & Remote Sensing Letters, 2022, 19: 1-5. DOI: 10.1109/LGRS.2021.3133473.

[10] 黎湘，程永强，王宏强，等. 信息几何理论与应用研究进展[J]. 中国科学：信息科学，2013,43(6)：707-732.

[11] 赵兴刚，王首勇. 雷达目标检测的信息几何方法[J]. 信号处理，2015，31(6)：631-637.

[12] HUA X Q, CHENG Y Q, WANG H Q, et al. Matrix CFAR detectors based on symmetrized Kullback-Leibler and total Kullback-Leibler divergences[J]. Digital Signal Processing, 2017, 69: 106-116.

[13] 赵兴刚，王首勇. 基于 K-L 散度和散度均值的改进矩阵 CFAR 检测器[J]. 中国科学：信息科学，2017,47(2)：247-259.

[14] Aub R, De Maio A, Pallotta L. A geometric approach to covariance matrix estimation and its applications to radar problems[J]. IEEE Transactions on Signal Processing, 2018, 66(4): 907-922.

[15] 赵文静，金明录，刘文龙. 基于谱范数的矩阵 CFAR 检测器[J]. 电子学报，2019，47(9)：1951-1956.

[16] Chen X L, Su N Y, Huang Y, et al. False-alarm-controllable radar detection for marine target based on multi features fusion via CNNs[J]. IEEE Sensors Journal, 2021, 21(7): 9099-9111.

[17] Zebiri K, Mezache A. Radar CFAR detection for multiple-targets situations for Weibull and log-normal distributed clutter[J]. Signal Image and Video Process., 2021: 1-8. https://doi.org/10.1007/s11760-021-01905-6.

[18] Gouri A, Mezache A, Oudira H. Radar CFAR detection in Weibull clutter based on zlog(z) estimator[J]. Remote Sensing Letters, 2020, 11(6): 581-589.

[19] Xu Y, Yan S, Ma X, et al. Fuzzy soft decision CFAR detector for the K distribution data[J]. IEEE Transactions on AES, 2015, 51(4): 3001-3013.

[20] 曲超. 主动声呐恒虚警检测方法研究[D]. 北京：中国科学院声学所，2008.

[21] 郝程鹏，蔡龙. Pearson 分布混响背景下均值类恒虚警检测器的性能分析[J]. 弹箭与制导学报，2009，29(2)：227-230.

[22] 魏嘉，徐达，闫晟，等. OSSO 和 OSGO 恒虚警检测器在 Pareto 分布混响背景下的性能分析[J]. 信号处理，2019，35(9)：1599-1606.

[23] 罗海力，徐达，陈模江，等. Pareto Ⅱ 型分布混响中距离扩展目标 CFAR 检测[J]. 水下无人系统学报，2020，28(1)：18-23.

[24] 殷超然，闫林杰，郝程鹏，等. 均匀混响背景下抗多目标干扰恒虚警检测器设计[J]. 水下无人系统学报，2019，27(4)：434-441.

[25] 孙梦茹，郝程鹏，刘明刚. 具有提升抗干扰能力的距离扩展目标模糊 CFAR 检测方法[J]. 信号处理，2019，35(9)：1580-1589.

[26] 梁增，马启明，杜栓平. 低频脉冲信号的频域恒虚警检测[J]. 声学技术，2016，35(1)：68-72.

[27] Verma A K. Variability index constant false alarm rate(VI-CFAR) for sonar target detection[C]. Anna University Chennai India: IEEE-International Conference on Signal processing, 2008.

[28] 卢术平，胡鹏，丁烽. 复杂非均匀背景下的鲁棒声呐恒虚警检测算法[J]. 声学技术，2020，39(6)：744-751.

[29] 许彦伟，张宝华，张春华，等. 非瑞利海洋混响抑制技术研究[J]. 声学学报，2012，37(5)：489-494.

[30] Hua G, Liao H, Wang Q Y, et al. Detection of electric network frequency in audio recordings-from theory to practical detectors[J]. IEEE Transactions on Information Forensics and Security, 2021, 16: 236-248.

英文缩略语

AC(approach cell) 逼近单元
ACF(autocorrelation function) 自相关函数
ACGO(adaptive censored greatest-of) 自适应删除最大
ACMLD(automatic censored mean level detector) 自动删除均值检测器
ACCA(automatic censored cell averaging) 自动删除平均
AD(analog digital) 模数(转换)
Adam(adaptive moment estimation) 自适应矩估计
ADT(average decision threshold) 平均判决阈值
AE(asymptotic efficiency) 渐近效率
A-GMSD(adaptive-GMSD) 自适应广义匹配子空间检测器
AL(adaptive length) 自适应长度
AMF(adaptive matched filter) 自适应匹配滤波器
ANMF(adaptive normalized matched filter) 自适应归一化匹配滤波器
ANN(artificial neural networks) 人工神经网络技术
AOS(adaptive order statistic) 自适应有序统计量
APC(adaptive pulse compression) 自适应脉冲压缩
ARE(asymptotic relative efficiency) 渐近相对效率
ASD(adaptive subspace detector) 自适应子空间检测器
ASII(algebraic scale-invariant interval) 代数尺度不变区间
ASN(average sample number) 平均采样数
ATD(average threshold deviation) 平均阈值偏差
BLIE(best linear invariant estimate) 最优线性不变估计
BLUE(best linear unbiased estimate) 最优线性无偏估计
BLUGO(best linear unbiased with greatest of selection) 最优线性无偏最大选择
CA(cell averaging) 单元平均
CAL(cell averaging logarithm) 对数单元平均
CCA(censored cell-averaging) 删除单元平均
CCDF(complementary CDF) 概率分布函数的补
CDF(cumulative distribution function) 概率分布函数
CF(clutter feature) 杂波属性
CFAR(constant false alarm rate) 恒虚警率
CGO(censored greatest of) 删除最大
CLT(central limit theory) 中心极限定理

CM(censored method) 删除技术
CMLD(censored mean level detector) 删除均值检测器
CNR(clutter to noise ratio) 杂波与噪声(的平均功率)之比
CPI(coherent processing interval) 相参处理间隔
DFT(discrete Fourier transform) 离散傅里叶变换
DWT(discrete wavelet transform) 离散小波变换
E(excision) 删除
ET(estimation test) 估计检验
EVD(eigenvalue decomposition) 特征值分解
FAR(false alarm rate) 虚警率
FBM(fractional Brownian motion) 分数布朗运动
FFT(fast Fourier transform) 快速傅里叶变换
FSS(fixed sample size) 固定样本数
GASD(generalized adaptive subspace detector) 广义自适应子空间检测器
GCMLD(generalized censored mean-level detector) 广义删除均值检测器
GLRT(generalized likelihood ratio test) 广义似然比检验
GMGD(generalized multivariate Gamma distribution) 广义多变量 Gamma 分布
GMSD(generalized matched subspace detector) 广义匹配子空间检测器
GO(greatest of) 最大
GOOSE(greatest of order statistics estimate) 最大有序统计量估计
GOS(generalized order statistics) 广义有序统计量
GOSCA(generalized order statistics cell averaging) 广义有序统计量单元平均
GOSGO(generalized order statistics with greatest of) 广义有序统计量的最大
GOSSO(generalized order statistics with smallest of) 广义有序统计量的最小
GRU(gated recurrent unit) 门控循环单元
GST(generalized sign test) 广义符号检验
GTL-CMLD(generalized two-level censored mean level detector) 广义两级删除均值检测器
HCE(heterogeneous clutter estimate) 非均匀杂波估计
HERM(helicopter rotors modulation) 直升机主旋翼调制
HRR(high range resolution) 高距离分辨率
IAA(iterative adaptive approach) 迭代自适应方法
IID(independent and identically distribute) 独立同分布
IndRNN(independent RNN) 独立循环神经网络
INR(interference to noise ratio) 干扰与噪声(的平均功率)之比
LMAP(logarithm maximum a posteriori) 对数最大后验
LMS(least mean square) 最小二乘
LOG(logarithm) 对数
LSTM(long short-term memory) 长短期记忆
LTS(local test statistic) 局部检测统计量

MAX(maximum) 最大
Max SP(maximum signal power) 最大化目标功率
MBLU(modified best linear unbiased) 修正的最佳线性无偏
MDS(multiple dominant scattering) 多主散射点
MEMO(median morphological filter) 中值形态滤波
MGF(moment generation function) 矩母函数
MGST(modified generalized sign test) 修正的广义符号检验
MIMO(multiple input multiple output) 多输入多输出
ML(mean level) 均值
MLD(mean level detector) 均值检测器
MLH(maximum likelihood) 最大似然
MMSE(minimum mean square error) 最小均方差
MNLT(memoryless non linear transform) 无记忆非线性变换
MOSCA(mean of order statistics and cell averaging) 有序统计量和单元平均的平均
MOSCM(mean of order statistics and censored mean) 有序统计量和删除均值的平均
MOSTM(mean of order statistics and trimmed mean) 有序统计量和削减均值的平均
MQBW(modified quasi-best weighted) 修正的准最佳加权
MRS(modified rank squared) 修正的秩方
MSD(matched subspace detector) 匹配子空间检测器
MSE(mean square error) 均方误差
MSLR(main side lobe ratio) 主(杂波)与旁瓣(杂波)之比
MTD(moving target detection) 动目标检测
MTI(moving target indicator) 动目标指示
MVDR(minimum variance distortionless response) 最小方差无失真响应
MX-MLD(maximum mean level detector) 最大均值检测器
MX-OSD(maximum orders statistics detector) 最大有序统计量检测器
NCCS(noncentral Chi-square) 非中心 Chi 方
NMF(normalized matched filter) 归一化匹配滤波器
NN(neural network) 神经网络
NP(Neyman-Pearson) 奈曼-皮尔逊
NSCM(normalized sample covariance matrix) 归一化采样协方差矩阵
ODV(ordered data variability) 有序数据可变性
OEP(odd-even processing) 奇偶处理
OLF(optimum linear filter) 最优线性滤波
OS(order statistics) 有序统计量
OSCA(order statistics and cell averaging) 有序统计量和单元平均
OSCAGO(order statistics and cell averaging greatest of) 有序统计量和单元平均的最大
OSCASO(order statistics and cell averaging smallest of) 有序统计量和单元平均的最小
OSCMGO(order statistics and censored mean greatest of) 有序统计量和删除均值的最大

OSCMSO(order statistics and censored mean smallest of) 有序统计量和删除均值的最小
OSD(order statistics detector) 有序统计量检测器
OSGO(order statistics with greatest of) 有序统计量的最大
OSSO(order statistics with smallest of) 有序统计量的最小
OSTA(order statistics with threshold adjustable) 阈值可调有序统计量
OSTMGO(order statistics and trimmed mean with greatest of) 有序统计量和削减均值的最大
OSTMSO(order statistics and trimmed mean with smallest of) 有序统计量和削减均值的最小
OSTWO(order statistics with two parameters) 两参数的有序统计量
P(polarimetric) 极化
PD(pulse Doppler) 脉冲多普勒
PDF(probability density function) 概率密度函数
PFLOM(phased fractional lower-order moment) 分数低阶矩
PG(processing gain) 处理增益
PHT-STC(polar Hough transform successive target cancellation) 极坐标 Hough 变换-逐目标消除
PRF(pulse repetition frequency) 脉冲重复频率
PRI(pulse repetition interval) 脉冲重复间隔
PSD(power spectrum density) 功率谱密度
PT(permutation test) 置换检验
QBW(quasi-best weighted) 准最佳加权
RAA(relative average amplitude) 相对平均幅度
RCS(radar cross section) 雷达横截面积
RF(radio frequency) 射频
RNN(recurrent neural network) 循环神经网络
ROC(receiver operating characteristic) 接收机工作特性曲线
RPH(relative Doppler peak height) 相对多普勒峰高
RS(rank squared) 秩方
RVE(relative vector-entropy) 相对多普勒谱熵
S(switching) 转换
SαS(symmetric alpha stable) 对称 alpha 稳定
SCM(sample covariance matrix) 采样协方差矩阵
SCNR(signal to clutter and noise ratio) 信号与杂波加噪声(的平均功率)之比
SCR(signal to clutter ratio) 信号与杂波(的平均功率)之比
SDADT(standard deviation of the average decision threshold) 平均判决门限标准差
SE(structuring element) 结构元素
SIAR(synthetic impulse and aperture radar) 综合脉冲孔径雷达
SIR(signal to interfering ratio) 信号与干扰目标(的平均功率)之比
SIRP(spherically invariant random processes) 球不变随机过程
SLC(side lobe canceller) 旁瓣对消器
SNR(signal to noise ratio) 信号与噪声(的平均功率)之比

SO(smallest of) 最小
STAP(spatial temporal adaptive processing) 空时自适应处理
STRAP(space-time-range adaptive processing) "空-时-距"自适应处理
TM(trimmed mean) 剔除平均
ULA(uniformly linear array) 均匀线阵
UMV(uniformly minimum variance) 一致最小方差
UMP(uniformly most powerful) 一致最优势
VI(variability index) 变化指数
VI(variable index) 变量索引
VTM(variably trimmed mean) 可变削减平均
WCA(weighted cell averaging) 加权单元平均
WD(waveform diversity) 波形分集
WFFT(weighted fast Fourier transform) 加权快速傅里叶变换
ZP(zero padding) 填零处理